ASTRONOMY AND ASTROPHYSICS LIBRARY

Springer-Verlag Berlin Heidelberg GmbH

Gerhard Börner

The Early Universe

Facts and Fiction

Fourth Edition
With 244 Figures Including 40 Color Figures,
19 Tables, and 80 Exercises

Springer

Professor Gerhard Börner
Max-Planck-Institut für Astrophysik
Karl-Schwarzschild-Strasse 1
85740 Garching
Germany

Cover picture:
Vantage point. A look into the deep sky (inspired by C.D. Friedrich, and the PLANCK-team)

Previous editions appeared in the series Texts and Monographs in Physics.

Library of Congress Cataloging-in-Publication Data applied for.
A catalog record for this book is available from the Library of Congress.
Bibliographic information published by Die Deutsche Bibliothek
Die Deutsche Bibliothek lists this publication in the Deutsche Nationalbibliogra e;
detailed bibliographic data is available in the Internet at http://dnb.ddb.de

Fourth Edition 2003
Second Corrected Printing 2004

ISSN 0941-7834
ISBN 978-3-642-07915-3 ISBN 978-3-662-05267-9 (eBook)
DOI 10.1007/978-3-662-05267-9

springeronline.com

Originally published by Springer-Verlag Berlin Heidelberg New York in 2003
Softcover reprint of the hardcover 4th edition 2003

Typesetting by the author
Data conversion by LE-TEX, Leipzig
Cover design: *design & production* GmbH, Heidelberg

Printed on acid-free paper 55/3141/tr - 5 4 3 2 1 0

“Truth is stranger than fiction,
but it is because fiction is obliged
to stick to possibilities: truth isn't.”

(Mark Twain “Pudd'nhead Wilson's new calendar”)

To Mara

Preface to the Fourth Edition

Fourteen years is a long time, and especially in the field of cosmology new observational results and new theoretical ideas seem to appear at a steadily increasing rate. It is a challenge to try to review the current status, to give a reasonably fair account of new developments, and not to increase the size of the book out of all proportion. So this fourth edition is practically a new book, with many chapters and sections newly written, not just updated.

I have kept the original layout of the book with three parts concerned with (I) the standard model, (II) some basic implications of quantum field theory, and (III) questions of structure formation. I have given special emphasis to the new observations of the anisotropies of the cosmic microwave background, and attempted to explain their importance for cosmology as well as for structure formation models. There have also been improved measurements in almost every cosmologically relevant field, from the Hubble constant to element abundances and galaxy distribution statistics. Quite surprisingly, the standard cosmological models can still accommodate all these new observations.

Galaxy formation and structure formation have become the central research field in cosmology. This is treated in Part III, which has been completely rewritten. Details of galaxy formation have been left out, and instead only the formation of structure by dark matter is dealt with in detail. Since this has also become the focus of my own research, it seems appropriate here to express my thanks to my two young Chinese colleagues, Houjun Mo (Munich) and Yipeng Jing (Shanghai), for our many years of intense and enjoyable collaboration.

The connection between cosmology and fundamental theories has become ever more intricate. We now have to explain the mysterious dark energy, the spectrum of the CMB anisotropies, and the dark matter constituents – all these force us to consider field theories in the very early universe. I have made the decision to stick to the more basic effective field theories, like GUTs, and omit ideas connected with superstring, and quantum gravity. Deplorable as this is, it helps to keep the book to a manageable size. My main reasons for leaving out these fields are that they still have only a weak link with experiments, and that it seems almost impossible to distinguish short-lived flares of excitement from approaches that will have some weight in the future, let alone to sort out fact from fiction.

The earlier editions were conceived as a monograph, but nevertheless often used as a course textbook. This fourth edition is organized so as to make it more useful as a textbook, i.e. it contains a problem section at the end of each chapter, and almost every chapter begins with an easy-to-read introduction which presents the flow of the argument, before the more technical treatment starts.

I would welcome readers', suggestions, criticisms, and error notifications (e-mail: grb@mpa-garching.mpg.de).

Munich, November 2002 *Gerhard Börner*

Preface to the First Edition

Connections between particle physics and cosmology have received much attention in recent years. There is no better place to study the fundamental interaction of elementary particles than the early universe with its extremely high thermal energies. Unfortunately the observers appear on the stage several billion years too late, and can find only leftovers from the brilliant beginning. For the theorists, however, the early universe is a wonderful playground. The impact of elementary particle theories on cosmology is unavoidable, and for any sketch of a fundamental theory, cosmological consequences follow. Thus many speculative scenarios have been suggested and partly explored.

This book is an attempt to describe the present status of this rapidly changing field. On the one hand I imagine an interested astronomer (or graduate student) who wants to understand in some detail the implications of particle physics. On the other, I see an interested particle physicist who wishes to get acquainted with the basic facts of cosmology. Thus I have tried to avoid the expert-oriented style of a review, and to give explanations for the non-expert in each specific topic. The references should help the reader to find easy access to the literature. The bibliography reflects the fact that it is impossible to attempt any kind of completeness in this field.

The classical big-bang picture of cosmology has to be included in such an account, especially since the last decade has brought powerful new equipment and exciting observational results. Part II of the book contains an introduction to some of the basic concepts of gauge theories, especially the electroweak standard model and GUTs. The description is semiclassical – appropriate for a perturbation approach. Cosmological aspects such as relic particles, baryon-asymmetry, and inflation are discussed in some detail. Finally, in Part III, the problem of galaxy formation is presented with emphasis on the effects of non-baryonic dark matter.

Quantum gravity, Kaluza-Klein and Superstring theories have been omitted. Spacetime is always viewed as a classical background model on which quantum fields act.

A modest aim was to sort out speculations from well-established results, which in cosmology is harder than elsewhere. I hope that some of the enthusiasm and excitement present in this field can also be found in this book.

Munich, April 1988 *Gerhard Börner*

Acknowledgements

Many of my colleagues helped me with discussions and drew my attention to new references.

Special thanks are due to M. Bartelmann (Munich), Y.P. Jing (Shanghai), and H.J. Mo (Munich), who helped with managing the electronic files, supplied pictures, and advice. Mrs. R. Mayr-Ihbe helped with producing many of the pictures.

I am very grateful to Mrs. M. Depner, R. Jurgeleit, and G. Kratschmann, who took on the tedious job of transferring the printed manuscript into a tex-file.

My brother, Klaus Börner, designed the cartoons.

Contents

Part I. The Standard Big-Bang Model

Part II. Particle Physics and Cosmology

Part I

The Standard Big-Bang Model

1. The Cosmological Models

"The simple is the seal of the true."
"Beauty is the splendour of truth."

S. Chandrasekhar (1984) on Einstein's theory of General Relativity

"Innocent light-minded men, who think that astronomy can be learnt by looking at the stars without knowledge of mathematics will, in the next life, be birds."

Plato, Timaeos

Before the observational evidence is presented in detail it would be convenient to establish a few theoretical concepts. The basic motivation is to describe the expansion of the universe in as simple terms as possible. In this restricted sense the classical term "cosmological model" usually means a geometrical description of the space-time structure and the smoothed-out matter and radiation content of an expanding universe. Clearly the theory then has the task of explaining the origin of the structures that are observed. How simple can the universe have been originally in order to still allow for the evolution of galaxies, stars, and ourselves? This question will be discussed in Part III.

A "cosmological model" must take into account the fundamental fact that the motion of celestial bodies is dominated by the gravitational force. Einstein's theory of General Relativity (GR) is taken as the fundamental theory of gravity throughout this book. The Newtonian theory of gravity is a limiting case of GR appropriate for a description of systems with weak gravitational fields and small relative velocities $v \ll c$ (c = velocity of light; $c = 1$, unless it appears explicitly in a formula).

There is not really much choice – so far GR has passed all experimental tests with flying colours, while competitors, such as the Jordan–Brans–Dicke theory, for example, have been reduced to insignificance by recent observations (for an extensive review on the experimental tests of GR, see [Will 1981]; compare also [Hellings et al. 1983]).

There is a continuing discussion of possible corrections to Newton's law of gravitational attraction. The claim that experiments require small repulsive corrections on a typical scale of $\approx 100\,\mathrm{m}$ [Fishbach et al. 1986] has been abandoned. But attempts to modify the $1/r^2$ law on galactic and larger scales cannot easily be refuted. These changes have been postulated to eliminate the evidence for dark matter which is derived from an application of Newton's

Before you go full steam ahead, it is wise to check whether you are on the right track

law to the dynamics of galaxies and clusters ([Milgrom 1983; Saunders 1986]). We will have to wait for new, precise measurements before this modification can be laid to rest. (Earlier experiments already indicate that such correction terms are very small [Braginsky 1974; Dicke 1964].)

Since there are many excellent textbooks describing various cosmological models in detail, we shall analyse only those few points in the following which contribute to making this book self-contained. (Extensive discussions can be found in a number of textbooks [Weinberg 1972; Heidmann 1980, Misner et al. 1973; Hawking and Ellis 1973; Landau and Lifshitz 1979; Peebles 1993; Peacock 1999].)

1.1 Friedmann–Lemaître Space-Time

The Friedmann–Lemaître (FL) space-times are a very simple family of solutions of GR. They rely on a smoothed-out matter content, and therefore can be a good approximation to our universe with its various distinct objects only in some average sense.

In the context of GR, a space-time is a 4-dimensional differentiable manifold M with a metric $g_{\mu\nu}$ which has the signature $(+---)$. The metric tensor determines distances, time intervals, and the causal order of events. Einstein's field equations relate $g_{\mu\nu}$ to the energy-momentum tensor $T_{\mu\nu}$:

$$R_{\mu\nu} - \frac{1}{2} R g_{\mu\nu} - \Lambda g_{\mu\nu} = 8\pi G T_{\mu\nu}. \tag{1.1}$$

$R_{\mu\nu}$ is the contracted curvature tensor, $R = R^{\mu}_{\mu}$ its trace. Λ is a hypothetical constant, called the "cosmological constant". It was introduced originally by Einstein to obtain a static cosmological model as a solution of (1.1).

Cosmological observations indicate that the length $|\Lambda|^{-1/2}$ is at least of the order of 1 Gpc – 1 parsec (pc) = 3.2 light-years – so the Λ-term is negligible, except perhaps on the largest extragalactic scales.

In matter-free space $T_{\mu\nu}=0$. For an electromagnetic field $F_{\mu\nu}$,

$$T_{\mu\nu} = \frac{1}{4\pi}\left(F_{\mu\lambda}F_{\nu\kappa}g^{\lambda\kappa} - \frac{1}{4}g_{\mu\nu}F_{\sigma\tau}F_{\kappa\lambda}g^{\sigma\kappa}g^{\lambda\tau}\right),$$

while for a perfect fluid or a radiation field

$$T_{\mu\nu} = (\varrho + p)V_\mu V_\nu - pg_{\mu\nu}$$

with density ϱ, pressure p, and 4-velocity V_μ, or

$$T_{\mu\nu}(x) = \int p_\mu p_\nu f(x,p) d\pi,$$

where $f(x,p)$ is the one-particle distribution function of an ideal gas. (The kinetic theory within GR has been developed by several authors; see e.g. [Ehlers 1971].)

Most known exact solutions of (1.1) describe spaces of high symmetry. The FL spaces are such highly symmetric space-times. They have exact spherical symmetry about every point. This implies [Hawking and Ellis 1973] that the space-time is spatially homogeneous and admits a six-parameter group of isometries whose orbits are space-like 3-surfaces of constant curvature. Minkowski space and de Sitter space are examples of FL space-times which have an even higher symmetry. One can choose coordinates so that the line element reads

$$ds^2 = dt^2 - R^2(t)d\sigma^2, \tag{1.2}$$

where $d\sigma^2$ is the line element of a Riemannian 3-space of constant curvature, independent of time. Its line element can be written as

$$d\sigma^2 = d\chi^2 + f^2(\chi)(d\theta^2 + \sin^2\theta d\varphi^2), \tag{1.3}$$

where $f(\chi)$ depends on the sign of the curvature. By a suitable rescaling of the "expansion" factor $R(t)$ the curvature can be normalized to $K = +1, -1$, or 0.

The function:

$$f(\chi) = \begin{cases} \sin\chi & \text{for} \quad K = +1, \\ \chi & \text{for} \quad K = 0, \\ \sinh\chi & \text{for} \quad K = -1 \end{cases} \tag{1.4}$$

determines how the area of a sphere $\chi = \text{const.}$ changes with radius χ. The coordinate χ runs from 0 to ∞, if $K = 0$ or $K = -1$. If $K = +1$ it varies from 0 to π. When $K = -1$ the 3-spaces are said to have a hyperbolic or

Lobachevskian geometry, and when $K = 0$ a planar or Euclidean geometry. For $K = +1$ the geometry is spherical. We shall see below that the sign of the curvature determines the ultimate fate of the universe model: For $\Lambda = 0$, models with $K = 0$ or $K = -1$ will expand forever, while models with $K = +1$ will eventually recollapse.

It should be pointed out here that the curvature does not determine whether the universe is infinite or finite in extent. The statement that if the universe is finite its geometry must be locally spherical and that if the geometry is locally hyperbolic the universe must be infinite holds only for simply connected spaces. There are many examples where the global topology of locally homogeneous spaces can be quite involved [Thurston and Weekes 1984; Wolf 1967; Heidmann 1980]. An instructive example of a compact space of constant negative curvature is the 3-manifold discovered by H. Seifert and C. Weber in 1932 [Thurston and Weekes 1984], which can be visualized as a dodecahedron where opposite faces are identified in a mathematically consistent way. It corresponds to a finite universe that expands forever.

All locally spherical ($K = +1$) constant curvature spaces are compact.

The metric (1.3) requires via (1.1) that the energy-momentum tensor is of the perfect fluid type

$$T_{\mu\nu} = \varrho V_\mu V_\nu - p(g_{\mu\nu} - V_\mu V_\nu). \tag{1.5}$$

The density and pressure are functions of t only, and the flow lines with tangent V_ν are the curves $(\chi, \theta, \varphi) =$ constant. The coordinates chosen in (1.3) are "comoving". For these coordinates $V_\nu = (1, 0, 0, 0)$.

The expansion factor $R(t)$ changes with time as the spatial distance of any two particles of a perfect fluid, i.e., of 2 "galaxies" without peculiar motions. Such comoving galaxies are considered as the representative particles with a perfect fluid motion. It must be tested observationally, of course, whether a gas of galaxy particles can be approximated satisfactorily by a perfect fluid. For two points with comoving coordinates (χ, θ, φ) and $(\chi_0, \theta, \varphi)$ the spatial distance at a time t is

$$d = R(t)(\chi - \chi_0);$$

the corresponding kinematic relative speed is thus $V_{\text{kin}} = (\dot{R}/R)d$.

Light propagates along the null geodesics of space-times. This can be studied most easily by changing to a new time coordinate η:

$$dt = R(t)d\eta. \tag{1.6}$$

Then

$$ds^2 = R^2(\eta)[d\eta^2 - d\chi^2 - f^2(\chi)d\Omega^2].$$

"conformal coordinates"; $d\Omega^2 \equiv d\theta^2 + \sin^2\theta d\varphi^2$.

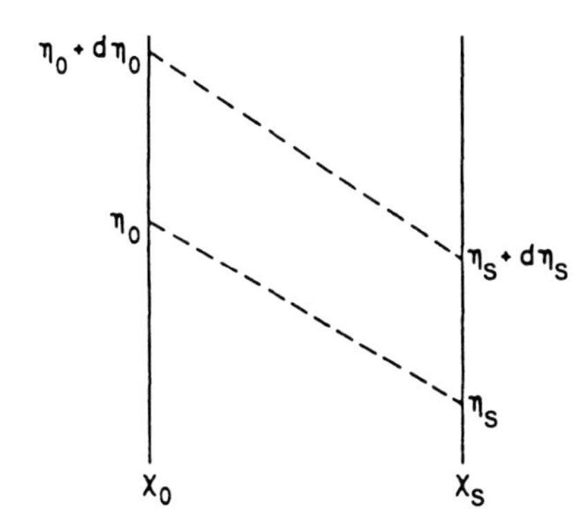

Fig. 1.1. Signals emitted at time η_S and $\eta_S + d\eta_S$ at location χ_S are received at times η_0 and $\eta_0 + d\eta_0$ at $\chi_0; d\eta_0 = d\eta_S$

"Radial" propagation (no loss of generality) of signals (i.e. $\theta = \varphi = \text{const.}$) gives $d\eta = \pm d\chi$ along the light-cone; for rays towards an observer at $\chi = 0$,

$$\chi = \eta_0 - \eta, \tag{1.7}$$

where η_0 denotes the present epoch. Two signals emitted at times η_S and $\eta_S + d\eta_S$ by a source following the mean motion at χ_S will be received at times η_0 and $\eta_0 + d\eta_0$ at $\chi = 0$. Because of (1.7) $d\eta_S = d\eta_0$ (Fig. 1.1). Expressed in the proper time t this condition reads

$$\frac{dt_0}{R(t_0)} = \frac{dt_S}{R(t_S)};$$

or, expressed in terms of a frequency $\omega \alpha (dt)^{-1}$,

$$\omega_0 R(t_0) = \omega_S R(t_S).$$

The emitted frequency ω_S differs from the observed one ω_0:

$$\frac{\omega_S}{\omega_0} = \frac{R_0}{R(t_S)}.$$

This corresponds to a shift of spectral lines

$$z \equiv \frac{\lambda_0 - \lambda_S}{\lambda_S} = \frac{\omega_S}{\omega_0} - 1 = \frac{R_0}{R(t_S)} - 1.$$

We know from Slipher's and Hubble's observations that this shift is a redshift:

$$1 + z = \frac{R_0}{R(t_S)} > 1. \tag{1.8}$$

This redshift of the spectral lines arises from the relative motion of source and observer, and may therefore be called a Doppler shift. In general space-times the frequency shift cannot be separated into a part due to gravitational effects and a part due to motion.

For nearby galaxies we can see that (1.8) corresponds to the elementary, non-relativistic Doppler formula. Expanding $R(t) = R_0 + (t - t_0)\dot{R}_0$, we find

$$z = \frac{R_0}{R_0 - \Delta t \dot{R}_0} - 1 \simeq \Delta t \frac{\dot{R}_0}{R_0} = d\frac{\dot{R}_0}{R_0} \equiv V_0 .$$

Hence we have

$$V_0 = z = dH_0 \tag{1.9}$$

where $H_0 \equiv \dot{R}_0/R_0$, the "Hubble constant". Equation (1.9) is "Hubble's Law", the linear relation between redshift and distance. It leads to the conclusion that the universe is expanding.

There is a universal time coordinate in the FL models. This cosmic time t corresponds to the proper time of observers following the mean motion V_μ. When we use these models to accommodate cosmological observations, we assume that they approximate in some sense to the real universe. In other words, we use as a working hypothesis: At any time and at any point in the universe there exists a mean motion such that all cosmological properties are isotropic with respect to the reference frame following that mean motion (cf. [Ehlers 1976]). The presently known measurements do not contradict this hypothesis; the isotropy of the 3 K background in fact gives strong support to it. To what extent this hypothesis is actually tested by observations will be discussed in Chaps. 2 to 4. Inserting (1.2)–(1.4) into the field equations (1.1) leads to 2 equations:

$$3\left(\frac{\dot{R}^2}{R^2} + \frac{K}{R^2}\right) = 8\pi G\rho + \Lambda,$$

$$-2\frac{\ddot{R}}{R} - \frac{\dot{R}^2}{R^2} - \frac{K}{R^2} = 8\pi G p - \Lambda.$$

The Einstein equation also implies a continuity equation:

$$\dot{\rho} = -3\frac{\dot{R}}{R}(\rho + p).$$

A little rearrangement leads to

$$\ddot{R} = -\frac{4\pi}{3}(\varrho + 3p)GR + \frac{1}{3}\Lambda R, \tag{1.10}$$

$$\dot{R}^2 = \frac{8\pi}{3}G\varrho R^2 + \frac{1}{3}\Lambda R^2 - K, \tag{1.11}$$

$$(\rho R^3)^{\cdot} + p(R^3)^{\cdot} = 0; \tag{1.12}$$

$\dot{\varrho}, \dot{R}$ stand for $d\varrho/dt, dR/dt$. Whenever $\dot{R} \neq 0$ any two of (1.10)–(1.12) imply the third. These are the Friedmann equations [Friedmann 1922; Lemaître

1927], and they determine the dynamics of the cosmological model if an equation of state, e.g. $p = f(\varrho, t)$ or $p = f(\varrho)$, is given.

At the present epoch $t = t_0$ the pressure is very small: $p/\varrho c^2 \leq 10^{-4}$. Let us take $p = 0$; then from (1.12),

$$\varrho R^3 = \text{const.} \equiv \frac{3M}{4\pi} \tag{1.13}$$

where M is a constant mass. Equation (1.11) can then be written as

$$\dot{R}^2 - \frac{2MG}{R} - \frac{1}{3}\Lambda R^2 = -K \equiv \frac{E}{M} \tag{1.14}$$

with $E \equiv -KM$. This is an energy-conservation equation for a comoving volume of matter; the constant E is the sum of the kinetic and potential energy, the term ΛR^2 is a kind of oscillator energy (for $\Lambda > 0$). Equation (1.13) expresses the conservation of mass.

At the present epoch $t = t_0$ ($p_0 = 0$), (1.10) and (1.11) are given by

$$\begin{aligned} \ddot{R}_0 &= -\frac{4\pi}{3}\varrho_0 G R_0 + \frac{1}{3}\Lambda R_0, \\ \dot{R}_0^2 &= \frac{8\pi G}{3}\varrho_0 R_0^2 + \frac{1}{3}\Lambda R_0^2 - K, \end{aligned}$$

and can be rewritten in terms of the Hubble constant $H_0 \equiv \dot{R}_0/R_0$ and the "deceleration parameter" $q_0 \equiv -(\ddot{R}R/\dot{R}^2)|_0$ as

$$\begin{aligned} \frac{1}{3}\Lambda &= \frac{4\pi}{3}G\varrho_0 - q_0 H_0^2, \\ \frac{K}{R_0^2} &= H_0^2(2q_0 - 1) + \Lambda. \end{aligned} \tag{1.15}$$

It is common to define a density parameter

$$\Omega \equiv \varrho_0/\varrho_c \equiv \frac{8\pi G}{3} H_0^{-2} \varrho_0. \tag{1.16}$$

This parametrization brings out the fact that there is essentially only one scale in these models, namely the Hubble constant H_0. All other quantities can be given as dimensionless ratios. The term

$$\varrho_c \equiv 3H_0^2(8\pi G)^{-1}$$

plays the role of a critical density.

Substituting for Λ,

$$\frac{K}{R_0^2} = \frac{H_0^2}{2}(3\Omega_0 - 2q_0 - 2), \tag{1.17}$$

and setting

$$\Omega_\Lambda \equiv \frac{\Lambda}{3H_0^2},$$

we have

$$\Omega_\Lambda = \Omega_0/2 - q_0.$$

Then

$$K/R_0^2 H_0^2 = (\Omega_0 + \Omega_\Lambda - 1).$$

$$K = \begin{cases} +1 & \text{for} \quad \Omega_0 + \Omega_\Lambda - 1 > 1 \\ 0 & \text{for} \quad \Omega_0 + \Omega_\Lambda - 1 = 1 \\ -1 & \text{for} \quad \Omega_0 + \Omega_\Lambda - 1 < 1. \end{cases} \tag{1.18}$$

Since q_0 and Ω_0 are in principle measurable quantities, it is possible to determine whether our universe is a hyperbolic or spherical FL space-time. We can of course define these quantities for all cosmic epochs. Then the evolution of the universe is described by functions of time

$$\begin{aligned} H(t) &\equiv \dot{R}/R, \\ q(t) &= \frac{-\ddot{R}R}{\dot{R}^2}, \\ \Omega(t) &= \frac{8\pi G}{3} H^{-2} \rho, \\ \Omega_\Lambda(t) &= \Lambda/3H^{-2}. \end{aligned}$$

The FL equations then take the form

$$\begin{aligned} q &= \Omega/2 - \Omega_\Lambda(t), \\ K/R^2 &= H^2(\Omega + \Omega_\Lambda(t) - 1). \end{aligned} \tag{1.19}$$

Inserting (1.17) and the relation $\rho = \rho_0(R/R_0)^{-3}$ (which hold for $p = 0$) into (1.11), one finds

$$(\dot{R}/R)^2 = H_0^2 \left\{ 1 - \Omega_0 - \Omega_\Lambda + \frac{\Omega_0 R_0}{R} + \Omega_\Lambda R^2/R_0^2 \right\}.$$

Substituting $x = R/R_0$,

$$(t - t_M)H_0 = \int_{x_M}^{x} \left\{ 1 - \Omega_0 - \Omega_\Lambda + \Omega_0/y + \Omega_\Lambda y^2 \right\}^{-1/2} dy, \tag{1.20}$$

where t_M is a time in the past for which the approximation $p = 0$ still holds. Since $t_M \ll t_0$ and $R_M \ll R_0$, there is the following approximate relation for t_0, the age of the model universe:

$$H_0 t_0 = \int_0^1 \left\{1 - \Omega_0 - \Omega_\Lambda + \Omega_0/y + \Omega_\Lambda y^2\right\}^{-1/2} dy. \tag{1.21}$$

For the simplest model, the Einstein–de Sitter model, we have $\Omega_\Lambda = 0$ and $\Omega_0 = 1$, thus $H_0 t_0 = 2/3$. The age t_0 is just $2H_0^{-1}/3$, two-thirds the Hubble time. We also see from (1.21) that for $\Omega_\Lambda = 0$, $t_0 H_0$ increases for decreasing Ω_0. The maximal value is reached by letting $\Omega_0 \to 0$. Then $H_0 t_0$ is equal to one. For $\Omega_\Lambda > 0$, and adjusted appropriately, any value of $H_0 t_0$ can be reached (see Sect. 1.4)

The approximation $p \ll \varrho$ is reasonable now and for most of the past. But the universe contains the 3 K background radiation field, and because of its equation of state $p = \varrho/3$, the continuity equation (1.12) leads to an increase of the radiation energy density ϱ_γ as $\rho_\gamma R^4 = \text{const}$. Thus at a sufficiently early epoch, radiation will be the dominant form of energy. This leads inevitably to a hot origin of the universe. The crucial assumption is the interpretation of the 3 K microwave signals as a cosmic background radiation. Proponents of a cold origin (e.g. [Wickramasinghe et al. 1975]) have the difficult task of explaining the 3 K radiation as non-cosmic, due to the emission of specific sources. Up to now no convincing or appealing model has been suggested. It seems almost impossible to generate the 3 K background from discrete sources (see Chap. 2 for a detailed discussion).

1.2 The Initial Singularity

Without solving the Friedmann equation it is seen from (1.10) that the present expansion $\dot{R}_0 > 0$, together with the requirement $\varrho + 3p - (\Lambda/4\pi G) \geq 0$, i.e. $\ddot{R} \leq 0$, leads to a concave graph of $R(t)$ vs. t. If the condition $\varrho + 3p - (\Lambda/4\pi G) \geq 0$ holds for all times, then $\ddot{R} \leq 0$ always, and R(t) must necessarily be zero at some finite time in the past (Fig. 1.2). Moreover, the time since then is bounded by H_0^{-1}, the "Hubble time" ($t_0 \leq H_0^{-1}$).

Let us call this instant $t = 0$. Thus $R(0) = 0$, and our present epoch t_0 is the cosmic time elapsed since the initial singularity when $t = 0$. This "point" is a true singularity of the space-time: the energy density $\varrho \to \infty$ for $R \to 0$. One can construct an invariant quantity from the components of the space-time curvature tensor (even if $K = 0$), which will also diverge as $R \to 0$ [Hawking and Ellis 1973]. The "point" $R = 0$ does not belong to the space-time, and it is not clear at present how close to $R = 0$ we should accept the validity of the classical theory of GR. Most physicists would agree that the state of matter in a classical model of the universe can still be described

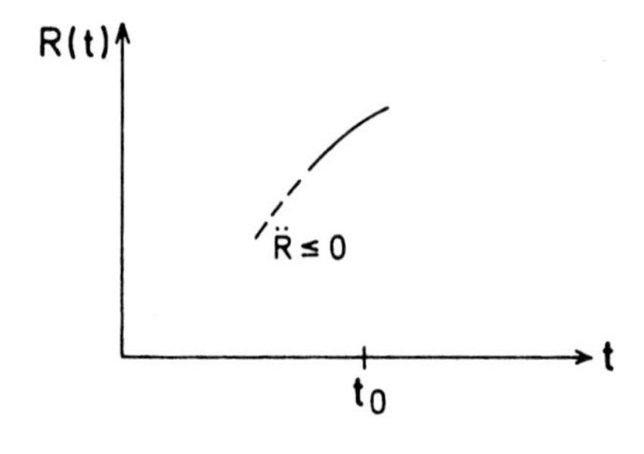

Fig. 1.2. $R(t)$ is a concave function of t. Therefore $R(t) \to 0$ for a finite t, if $\ddot{R} \leq 0$

reasonably accurately up to temperatures and densities close to the limits of terrestrial laboratory experiments. This would correspond to $t \approx 10^{-4}$ s. Such an initial singularity is a remarkable feature of some of the FL cosmologies. It means that the universe we know came into being some finite time ago.

The condition $\varrho + 3p - (\Lambda/4\pi G) \geq 0$ is sufficient, but not necessary, for the existence of a singularity. Observations do not unambiguously give $\varrho + 3p - (\Lambda/4\pi G) \geq 0$ now (compare the discussion in Chap. 2), but if this inequality was satisfied at some past epoch, this implies a singularity, provided $\varrho + 3p$ was never smaller before that epoch. The inequality $\varrho > 0$ is certainly satisfied by the kinds of matter known to us. The inequality $\varrho + 3p > 0$ is certainly satified up to nuclear matter density. We shall see in Sect. 1.4 that the singularity can be avoided only by a special choice of the cosmological constant Λ, and then $\ddot{R}$ is always positive. This would make it difficult to explain the background radiation and would make it virtually impossible to explain the cosmic abundance of helium as a pre-galactic phenomenon.

The notion that the occurence of the singularity might depend on the exact spherical symmetry and spatial homogeneity is unfounded. Several interesting theorems have been proved ([Hawking and Penrose 1970]; see also [Hawking and Ellis 1973]), which give a range of general conditions sufficient for the occurence of a singularity. These "singularity theorems" are among the outstanding achievements in the theory of GR. They remain slightly unsatisfactory, however, since they are quite indirect – as is very common in mathematics. They are not derived from a straightforward collapse calculation, but from the demonstration that an assumed space-time with certain features, e.g. "completeness", cannot be fitted together properly. Consider as an example the following theorem [Hawking and Ellis 1973, Sect. 8.2 Th. 2].

Theorem 1. *The following conditions on a time-oriented space-time (M, g) are incompatible:*

1. $R_{ab}K^aK^b \geq 0$ *for all time-like vectors K (this can be translated into conditions on the stress–energy tensor when ϱ and p_a exist: $\varrho + p_a \geq 0$; $\varrho + \Sigma p_a \geq 0$, provided $\Lambda \leq 0$).*
2. *On every time-like or null geodesic there exists a point at which the tangent vector K satisfies $K_{[\alpha}R_{\beta]\gamma\varepsilon[\delta}K_{\lambda]} \neq 0$. ($A_{[\alpha\beta]}$ designates the antisym-*

metrized quantity $\frac{1}{2}(A_{\alpha\beta} - A_{\beta\alpha})$.*) This condition is technically necessary, but not of fundamental importance. It ensures that the curvature tensor does not satisfy a very specialized equation along any geodesic.*

3. *The space-time is causal: there are no closed time-like curves.*
4. *The space-time M contains a closed trapped 2-surface (that is, a 2-surface both of whose families of future-pointing, normal null geodesics are converging). This is one possible formulation of a "black-hole condition".*
5. *The space-time is complete for time-like and null geodesics.*

Several slightly different forms of such theorems can be found in the standard textbooks [Hawking and Ellis 1973; Wald 1984]. In general the theorems show that the causal Cauchy development of a 3-surface is incomplete if there is an energy condition and some equivalent to the existence of a closed, trapped 2-surface. There exists at least one geodesic which cannot be continued to arbitarily large positive or negative values of its affine parameter.

How are these incompletenes theorems connected with singularities? Can singularities be avoided by extensions of Cauchy developments which may be acausal in certain space-time regions?

The question of causality violation in relation to singularities has been elucidated recently [Tipler 1977]. The boundary of a Cauchy development must be achronal (no two points are connected by a time-like curve). Too much focusing of its null geodesic generators by matter or tidal forces leads to a singularitiy. Thus in many cases a failure of Cauchy predictability is accompanied by singularities – one exception is causality violation starting within a spatially bounded region.

A universe whose curvature satisfies condition (1) of the theorem quoted above, has singularities if there exists a compact space-like slice whose normals are everywhere converging [Ellis and Hawking 1973, Sect. 8.2, Th 2]. As an application of this theorem it can be shown [Tipler 1976] that the "bouncing back" of a universe model before it runs into a singularity – i.e. the change from a contraction to an expansion phase – cannot even occur via causality violation in a high-density phase, if compact space-like slices existed in successive low-density periods and condition (1) above was satisfied.

In those cases where the incompleteness of the space-time cannot be avoided by extensions of the Cauchy development, one has to consider boundary points from inside the given space-time (a review of the problems incurred is given in [Geroch 1968; Ellis and Schmidt 1987]). At present the "affine b-boundary" $\partial_b M$ [Schmidt 1971] provides a reasonable construction of boundary points (Fig. 1.3). Regular boundary points are points in $\partial_b M$ which are regular points of an extension of M. All other points are singular.

Quasi-regular singularities represent topological obstructions to extensions, like the vertex of a cone. The curvature tensor components have finite limits in a frame parallel propagated along any curve terminating at the singularity. All other singular points in $\partial_b M$ have at least one curve along which the components of the curvature tensor have no limit in a parallel

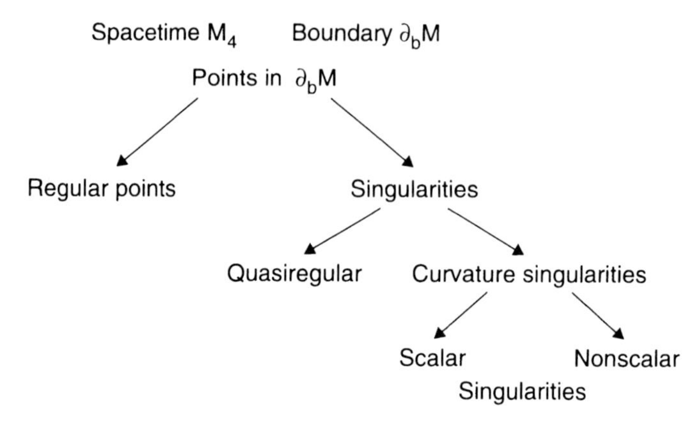

Fig. 1.3. A scheme for the classification of boundary points

propagated frame. All types of singularities occur in solutions of Einstein's equations [Hawking and Ellis 1973; Ellis and Schmidt 1977].

Very little is known about the collection of all boundary points of a specific space-time. Not even for the Schwarzschild and FL space-times have the total b-boundaries been constructed. It is clear, however, that many space-times have curvature singularities, and the theorem quoted above shows that their occurence does not depend on any symmetry of a space-time. The inescapable conclusion of an initial singularity is that the description of the universe by an FL model leads to the existence of an early epoch of high density, when all the structures we see now – galaxies, stars, etc. – did not yet exist.

What was there before the initial singularity? This question cannot be answered within the context of classical GR, because the concepts of this theory cease to be applicable at the singularity.

The causal structure of the FL space-times can conveniently be studied by using the form of the line element

$$ds^2 = R^2(\eta)(d\eta^2 - d\chi^2 - f^2(\chi)d\Omega^2)$$

for $K = +1, f(\chi) = \sin\chi$ and this is conformal to the Einstein static space (Fig. 1.4).

This space-time can be visualized – by suppressing θ and φ, each point represents a 2-sphere, such as the cylinder $x^2 + y^2 = 1$ in 3-dimensional Minkowski space. The conformal structure of the FL space-times then corresponds to a part of the Einstein static model, determined by the values taken by η. The mapping of such a finite part of the Einstein cylinder onto the plane gives a conformal picture of the space-time-preserving angles, i.e. the light-cone structure. These so-called "Penrose diagrams" are a very instructive, pictorial way to describe the conformal structure of a space-time.

For $K = +1$, $p = \Lambda = 0$, one has $0 < \eta < \pi$, i.e. the initial singularity ($\eta = 0$) is space-like, as well as the singularity in the future ($\eta = \pi$). The

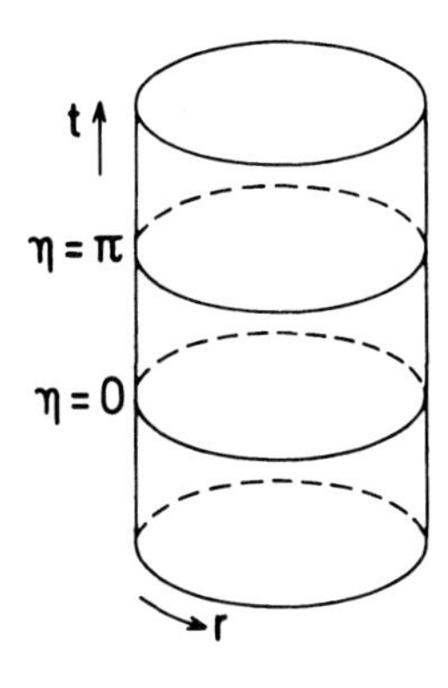

Fig. 1.4. The Einstein cylinder; $\eta = 0$ and $\eta = \pi$ are the boundaries of a $K = +1$ FL space-time

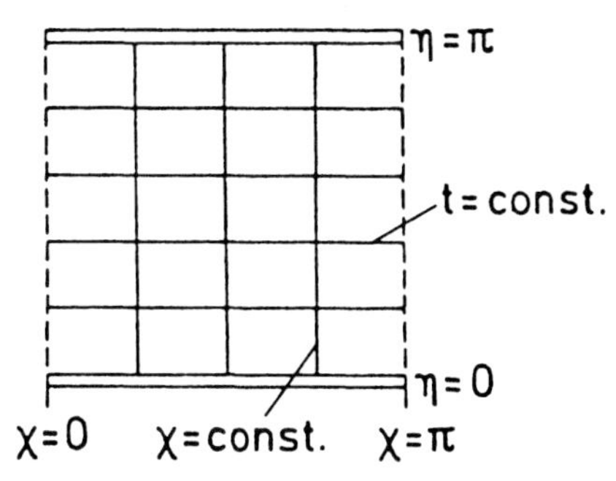

Fig. 1.5. Penrose diagram of a $K = +1$ FL space-time

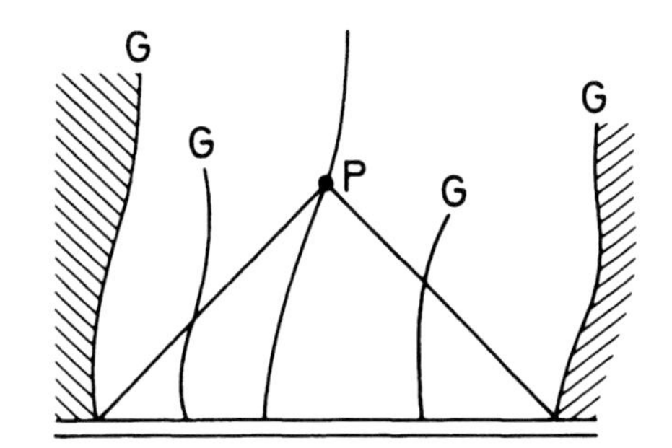

Fig. 1.6. The past light-cone of an observer at P covers only a fraction of the objects (G) in the universe

Penrose diagram for this space-time is shown in Fig. 1.5. There are particle horizons in this space-time (Fig. 1.6). An observer can only see objects inside his past light-cone. The particle horizon separates all the objects that are yet invisible to the observer at P. Since the future singularity $\eta = \pi$ is also space-like, there are events that an observer will never see (future event horizon, Fig. 1.7).

For $K = 0, K = -1$, and $p = \Lambda = 0$ the Penrose diagram is shown in Fig. 1.8 (all figures from [Hawking and Ellis 1973]). Future infinity is a null surface, and there are no future event horizons for the comoving observers in these spaces.

Light signals emitted at $t(\eta_0 - \chi)$ and received at time $t_0(\eta_0)$ at the point $\chi = 0$ come from points located on the sphere with area

$$O_L = 4\pi(R(\eta_0 - \chi))^2 \begin{cases} \sin^2\chi & K = +1, \\ \chi^2 & K = 0, \\ \sinh^2\chi & K = -1. \end{cases} \tag{1.22}$$

O_L increases from 0 at $\chi = 0$, reaches a maximum, and decreases to disappear at $\chi = \eta$ (where $R(\eta_0 - \chi) = 0$). Figure 1.9 shows this reconverging past light-cone.

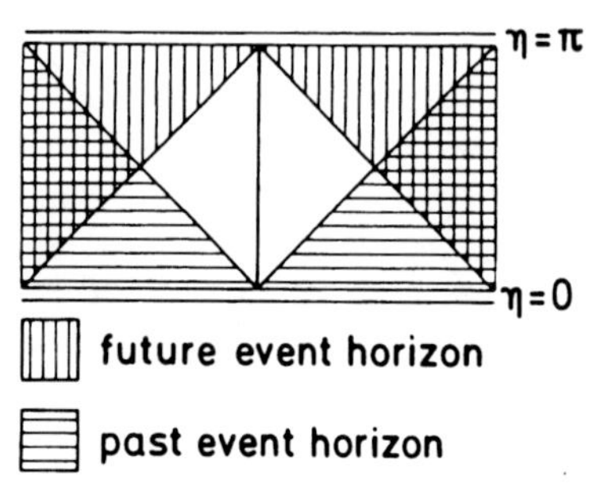

Fig. 1.7. Future and past event horizons in a $K = +1$ model

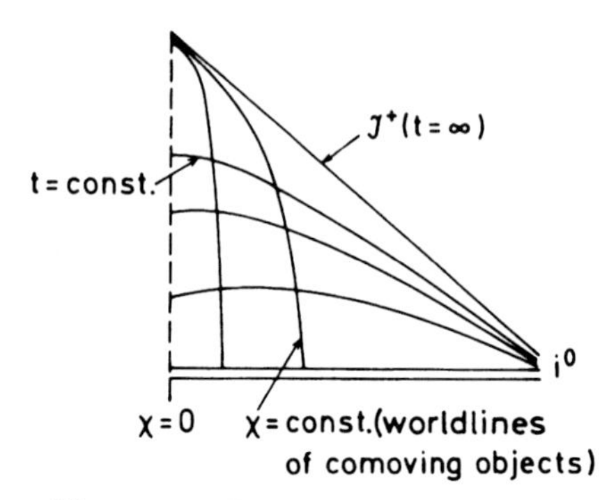

Fig. 1.8. Penrose diagram of $K = 0, -1$ FL models. Future infinity is a null surface

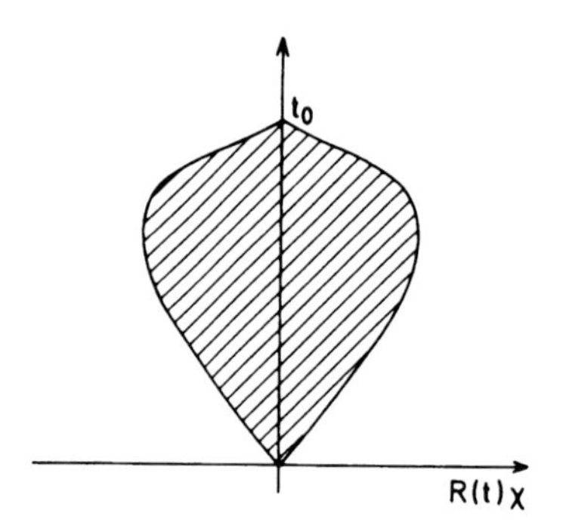

Fig. 1.9. The past light-cone at t_0 reconverges for $t \to 0$

1.3 Light Propagation in an FL Model

Astronomical observations of distant sources have to be connected with the propagation of light in an FL model. Three different pieces of knowledge help to interpret the incoming electromagnetic signals received by an observer:

i) The redshift z can be measured as the redshift of spectral lines.
ii) Liouville's equation (e.g. [Ehlers 1971]) allows a simple relation for the specific intensity I_ω ((erg/cm^2)s^{-1} (unit solid angle) $^{-1}$ (unit frequency interval)$^{-1}$) to be proved:

$$I_\omega/\omega^3 = \text{const.}$$

independent of the observer and constant along a ray.
When we compare intensities at the source $I_{\omega,S}$ and at the observer $I_{\omega,0}$, this results in:

$$I_{\omega,0} = \left(\frac{\omega_0}{\omega_S}\right)^3 I_{\omega,S} = \frac{I_{\omega,S}}{(1+z)^3} \,. \tag{1.23}$$

iii) The flux from a point source is observed by means of a bundle of null geodesics with a small solid angle $d\Omega_S$ at the source and cross-sectional

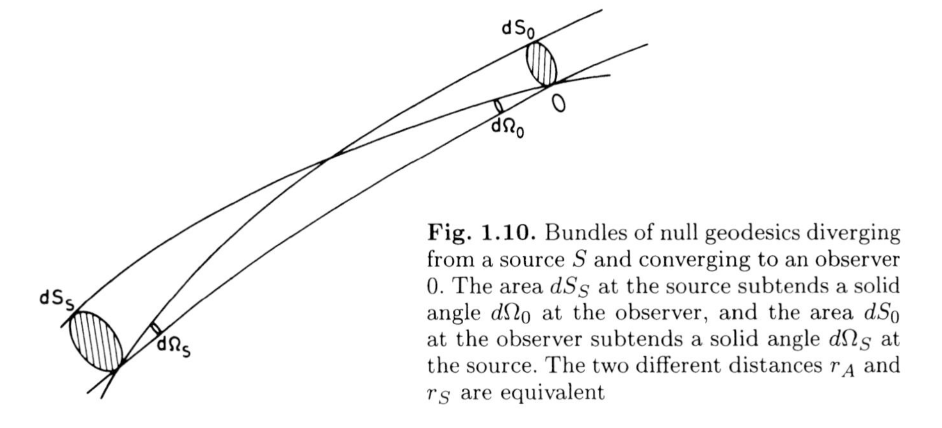

Fig. 1.10. Bundles of null geodesics diverging from a source S and converging to an observer 0. The area dS_S at the source subtends a solid angle $d\Omega_0$ at the observer, and the area dS_0 at the observer subtends a solid angle $d\Omega_S$ at the source. The two different distances r_A and r_S are equivalent

area dS_0 at the observer. Therefore a luminosity distance for a point source can be defined [Ellis 1971] by:

$$dS_0 = r_S^2 d\Omega_S \tag{1.24}$$

(cf. Fig. 1.10).

The solid angle $d\Omega_S$ cannot be measured, however, and therefore r_S is not a measurable quantity either.

Another distance can be defined, which is in principle observable, by considering an extended object of cross-sectional area dS_S, subtending the solid angle $d\Omega_0$ at the observer, Then by

$$dS_S = r_A^2 \, d\Omega_0 \tag{1.25}$$

a "distance from apparent size" r_A can be measured, in principle, if the solid angle subtended by some object is measured, and if the intrinsic cross-sectional area can be found from astrophysical considerations (Fig. 1.10). When there are no anisotropies, $r_A\delta = D$, where D is the linear extent and δ the angular diameter of an object (Fig. 1.11). In an FL cosmology $r_A = R(t)f(\chi)$.

These two different distance definitions between a given galaxy and the observer, r_S and r_A are essentially equivalent [Etherington 1933; Penrose 1966; Ellis 1971, p. 153].

$$r_S^2 = r_A^2(1+z)^2 \qquad \text{“Reciprocity theorem”}.$$

This result is a consequence of the geodesic deviation equation. It is true for a general space-time, not just for FL models. For $z = 0$, equal surface elements dS_0 and dS_S subtend equal solid angles $d\Omega_S$ and $d\Omega_0$, irrespective of the curvature of space-time. The factor $(1+z)^2$ is the special relativistic correction to solid-angle measurements.

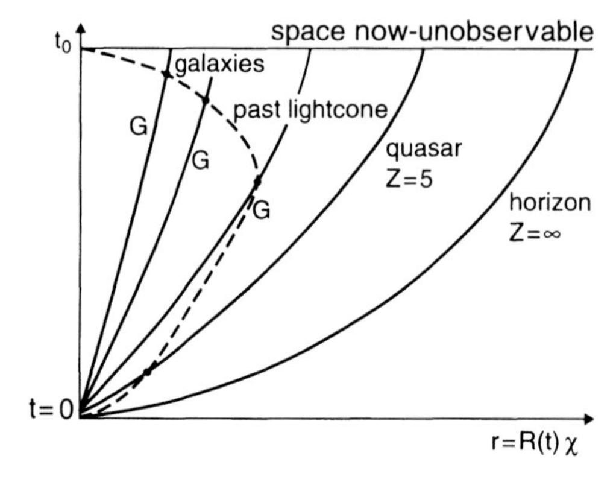

Fig. 1.11. The observational situation in an FL model. Galaxies are observed as they cross our past light-cone. Space at the present time is not observable

The so-called "corrected luminosity distance r_C" is defined from the flux $L/4\pi$ through a unit sphere, centred on the source, in the locally Euclidean space near the source [Kristian and Sachs 1966] (L is the luminosity of the source) by:

$$r_C^2 = F^{-1}(1+z)^{-2}\frac{L}{4\pi},$$

where F is the flux at the observer.

The reciprocity theorem states that

$$r_C^2 = r_A^2(1+z)^2 = r_S^2.$$

Since the corrected luminosity distance r_C can, in principle, be measured for any source with known intrinsic luminosity – by measuring the flux from the source – this relation can, in principle, be used to determine r_A. In practice, r_C and r_A cannot be measured independently.

It has been pointed out [Ellis 1971] that confusion has been caused in the literature by the fact that r_c, r_s and r_d – defined by $r_d^2 \equiv L/4\pi F$ – have all been called "luminosity distance". There has been a certain amount of discussion as to the correct redshift factors in various formulae. As suggested by [Ellis 1971], it seems reasonable to call r_d a "luminosity distance" and $r_C = r_S$ a "corrected luminosity distance" which is, however, not measurable over cosmic distances. These different distance concepts all reduce to the one, usual, special-relativistic distance concept in the case of

1. slowly moving nearby sources, and
2. a static situation in a flat space-time.

Even within the homogeneous FL models, the matter along our past light-cone is sufficient to refocus the light rays. This "gravitational lens" effect can be seen most simply for the Einstein–de Sitter model with $K = 0, d\Omega_\Lambda = 0$ and $\Omega_0 = 1$. Here $R(t) = R_0(t/t_0)^{2/3}$ and $H_0 = 2/(3t_0)$, for the matter-dominated era.

Along null geodesics $ds^2 = 0$, i.e. $dt = \pm R(t)dr$. The null geodesic from $(t = 0, r = 0)$ to (t_0, r_M) gives the coordinate value r_M of the particle horizon:

$$r_M = \int_0^{t_0} \frac{dt}{R(t)} dt = 3t_0/R_0,$$

and the physical size of the particle horizon at t_0 is

$$d_M = R_0 r_M = 3t_0 = 2/H_0.$$

The shape of the past light-cone through $(r = 0, t_0)$ is given by $d(t)$

$$d(t) = R(t)r = d_M \left[(t/t_0)^{2/3} - (t/t_0) \right].$$

Since $(1 + z) \equiv (t/t_0)^{2/3}$ we find

$$d(t) = d_M \left[(1 + z)^{-1} - (1 + z)^{-3/2} \right].$$

$d(t)$ has a maximum at $z = 1.25$.

This means that the angular diameter of any given object decreases to a minimum, and then starts increasing again as the object is moved further down the past light-cone of the observer (Fig. 1.12). A gravitational lens effect of this kind which leads to an anomalously large source solid angle, when the source is past the maximum of $d(t)$, can also lead to an anomalously large source brightness. Some or all of the distances defined above will be double-valued under such circumstances. It should be noted that the true past light-cone will have a much more complicated, chaotic shape due to the gravitational lens effect of clumped mass concentrations which produce multiple images of a distant source.

So far we have always considered the bundle of null geodesics emanating from a point. In practice one usually observes extended sources, and instead of measuring the flux from the source the direct measurements tell us the flux per unit solid angle from the source $dF/d\Omega$ – i.e. the intensity of radiation from the source. Considering a source of area dA, the intensity is

$$I \equiv \frac{dF}{d\Omega_0} = \frac{L}{4\pi dA}(1 + z)^{-4};$$

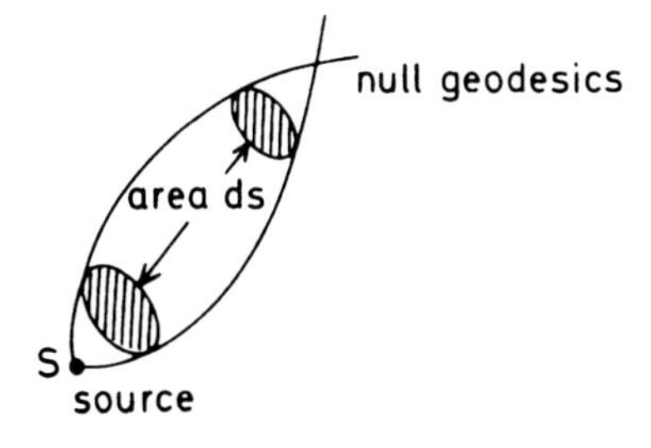

Fig. 1.12. The refocussing of light rays from a source S. The cross-sectional areas of the bundle at two different locations are the same. As a result, the source will appear to be anomalously bright with anomalously large angular diameter at these points

where the factor $I_S \equiv L/4\pi dA$ is just determined by the source characteristics. It is the surface brightness of the source. For a given source the intensity is independent of the distance, and depends only on the redshift z!

To find the flux F from an extended source, the procedure is to measure I and then to integrate over the image to obtain F. Therefore in general the solid angle subtended by the source has to be estimated before the flux F can be deduced from direct mesurements (except in the case of point-like quasars or very distant galaxies). As [Ellis 1971] points out, we may lose useful information if we consider only the flux F rather than the intensity and solid-angle information that is combined to give F.

It is interesting to note that the expression for I is involved in Olbers's paradox ("the darkness of the night sky"). This shows that we can explain the dark night sky either by assuming a very low surface brightness for the sources along almost all null geodesics from us (perhaps because of the source evolution that must occur in view of the finite source lifetime), or that the redshift increases indefinitely along almost all null geodesics.

Let us use r_A to relate luminosity and observed flux:

$$\frac{dF}{d\Omega_0} = \int I_{\omega,0} d\omega_0 = \int \frac{I_{\omega,s}}{(1+z)^3} d\omega_0 = (1+z)^{-4} \frac{L}{dS_S 4\pi};$$

where L is the total luminosity of the source. Then

$$dF = \frac{L}{(1+z)^4} r_A^{-2} \frac{1}{4\pi}. \tag{1.26}$$

Now, if $\Lambda = 0, \chi$ can be expressed terms of z, q_0, H_0 (e.g. [Mattig 1958]), and r_A:

$$\begin{aligned} r_A &= R(t) f(\chi) && (1.27) \\ &= H_0^{-1} q_0^{-2} (1+z)^{-2} \left\{ z q_0 + (q_0 - 1)[(1 + 2q_0 z)^{1/2} - 1] \right\} && \text{for} \quad q_0 \pm 0, \end{aligned}$$

$$r_A = \frac{1}{2} H_0^{-1} \left[1 - \frac{1}{(1+z)^2} \right] \quad \text{for} \quad q_0 = 0.$$

For small z this gives the linear relation

$$(1+z)^2 r_A \equiv d = z H_0^{-1}.$$

1.4 Explicit Solutions

Let us finish discussion of the FL models with a short representation of explicit solutions. We choose the measurable quantities q_0, H_0, ρ_0 – or equivalently Ω_0 to parametrize the cosmological models. First of all, consider the case $\Lambda = 0$; then $\Omega_0 = 2q_0$ and q_0 can be used to characterize the solutions.

At present the universe contains matter and radiation without appreciable interaction with equations of state $p = 0$ (for matter), and $p = \rho c^2/3$ (for radiation). Only the zero-pressure approximation will be treated. This covers the largest part of the cosmic evolution. The early phase with $p = \rho/3$ has a slightly different behaviour, and this calculation is left as an exercise.

For $p = 0, \Lambda = 0$ the integration of the Friedmann equations proceeds in the following steps. Rewrite the equation

$$\dot{R}^2 + K = \frac{8\pi G}{3}\rho R^2 \tag{1.28}$$

in terms of a new time variable η, where $d\eta \equiv dt/R$. This "conformal" time η has already been used in (1.6),

$$\left(\frac{\dot{R}}{R}\right)^2 = (dR/d\eta)^2 R^4 = \frac{8\pi G}{3}\rho_0(R_0/R)^3 - \frac{K}{R^2}; \tag{1.29}$$

or

$$R^{-1/2}dR/d\eta \equiv 2\frac{d}{d\eta}R^{1/2} = \left(\frac{8\pi G}{3}\rho_0 R_0^3 - KR\right)^{1/2}.$$

We introduce the abbreviation

$$\frac{8\pi G}{3}\rho_0 R_0^3 \equiv R_u$$

for the "Schwarzschild radius" $R_u = 2GM$ of the mass $M = \frac{4\pi}{3}\rho_0 R_0^3$. Integration of (1.29) leads to:

$$\begin{aligned}\eta/2 &= \int_0^{R^{1/2}} dR^{1/2}(R_u - KR)^{-1/2} \\ &= \begin{cases} \sin^{-1}(R^{1/2}R_u^{-1/2}) & \text{for} \quad K = +1, \\ R^{1/2}R_u^{-1/2} & \text{for} \quad K = 0, \\ \sinh^{-1}(R^{1/2}R_u)^{-1/2} & \text{for} \quad K = -1. \end{cases}\end{aligned}$$

Inverting the formulae, and replacing R_u by

$$R_u = \begin{cases} 2q_0 H_0^{-1}\left|2q_0 - 1\right|^{-3/2} & \text{for} \quad K = \pm 1, \\ H_0^2 R_0^3 & \text{for} \quad K = 0, \end{cases}$$

results in

$$R = \begin{cases} q_0 H_0^{-1}(2q_0 - 1)^{-3/2}(1 - \cos\eta) & \text{for} \quad K = +1, \\ H_0^2 R_0^3/4\eta^2 & \text{for} \quad K = 0, \\ q_0 H_0^{-1}(1 - 2q_0)^{-3/2}(\cosh\eta - 1) & \text{for} \quad K = -1. \end{cases}$$

Integrating $dt = Rd\eta$, we obtain:

$$t = \begin{cases} q_0 H_0^{-1}(2q_0 - 1)^{-3/2}(\eta - \sin\eta) & \text{for} \quad K = +1, \\ H_0^2 R_0^3/12\eta^3 & \text{for} \quad K = 0, \\ q_0 H_0^{-1}(1 - 2q_0)^{-3/2}(\sinh\eta - \eta) & \text{for} \quad K = -1. \end{cases} \tag{1.30}$$

For $K = 0$ we find the especially simple relation:

$$R(t)/R_0 = (3H_0 t/2)^{2/3}. \tag{1.31}$$

For $t = t_0 : H_0 t_0 = 2/3$
For the $K = +1$ spherical model

$$\begin{aligned} R/R_0 &= q_0(2q_0 - 1)^{-1}(1 - \cos\eta), \\ H_0 t &= q_0(2q_0 - 1)^{-3/2}(\eta - \sin\eta), \end{aligned}$$

where $0 \le \eta \le 2\pi$.

This describes an increase of $R(t)$ until a maximal expansion stage is reached ($\eta = \pi$), and from then on $R(t)$ decreases, the universe contracts. Since we are now measuring expansion, the universe is still in the ascending branch of the $R(t)$ graph in Fig. 1.13.

For the $K = -1$ hyperbolic model we simply replace $\cos\eta$ by $\cosh\eta$, $\sin\eta$ by $\sinh\eta$ and change signs appropriately:

$$\begin{aligned} R/R_0 &= q_0(1 - 2q_0)^{-1}(\cosh\eta - 1), \\ H_0 t &= q_0(1 - 2q_0)^{-3/2}(\sinh\eta - \eta); \end{aligned}$$

here $0 < \eta < \infty$.

This describes the ever-increasing expansion factor $R(t)$ in Fig. 1.13. Finally the solutions with $\Lambda \neq 0$ will be discussed briefly. The early phases – when $R \to 0$ – will be radiation-dominated, and the contribution of Λ will be negligible. Therefore only a matter-dominated phase will be considered. In general these solutions can only be obtained by numerical integration. A few general features can be pointed out in a qualitative discussion.

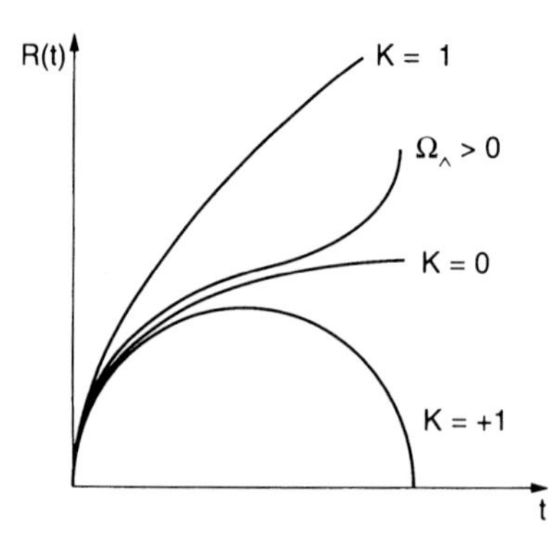

Fig. 1.13. The expansion factor $R(t)$ for $K = +1, -1, 0$ FL models

It is convenient to introduce dimensionless quantities by setting

$$x \equiv R/R_0; \tau = tH_0. \tag{1.32}$$

Then the Friedmann equation reads

$$\dot{x}^2 = \Omega_0/x + \Omega_\Lambda x^2 - \Omega_0 - \Omega_\Lambda + 1. \tag{1.33}$$

Here the relation $K/R_0^2 H_0^2 = \Omega_0 + \Omega_\Lambda - 1$ has been used. We can discuss (1.33) like the one-dimensional Newtonian motion of a test particle in a potential:

$$V(x) = -\Omega_0/x - \Omega_\Lambda x^2$$

with energy $1 - \Omega_0 - \Omega_\Lambda$. For $\Omega_\Lambda < 0$ the potential $V(x)$ is monotonic, and all possible solutions evolve from an initial singularity into a final singularity (Fig. 1.14).

For $\Omega_\Lambda > 0$ the potential $V(x) < 0$ for all values of x, i.e. for the energy $(\Omega_\Lambda + \Omega_0 - 1) \leq 0$ the solutions correspond to expansion from a singularity (Fig. 1.15). $V(x)$ has a maximum at $x_m = (\Omega_0/2\Omega_\Lambda)^{1/3}$, and for $\Omega_\Lambda + \Omega_0 > 1$ there is a critical value $\Omega_{\Lambda c}$ such that $dx/d\tau = 0$ at x_m.

$\Omega_{\Lambda c}$ can be found as the solution of the equation

$$(\Omega_{\Lambda c} + \Omega_0 - 1)^3 = \frac{27}{4}\Omega_0^2 \Omega_{\Lambda c}. \tag{1.34}$$

For $\Omega_\Lambda < \Omega_{\Lambda c}$ and $\Omega_{\Lambda c} + \Omega_0 - 1 > 0$ we still find expansion forever, starting from an initial singularity.

When $\Omega_\Lambda = \Omega_{\Lambda c}$ and $K = +1$ (i.e. $\Omega_{\Lambda c} + \Omega_0 - 1 > 0$) there is a static solution, the "Einstein universe", with a constant radius:

$$x_E = (\Omega_0/2\Omega_{\Lambda c})^{1/3}.$$

In Fig. 1.16 we can see, how this radius defines a limiting case in the parameter space of the $\Omega_\Lambda > 0$ models. A small change in Ω_Λ leads away

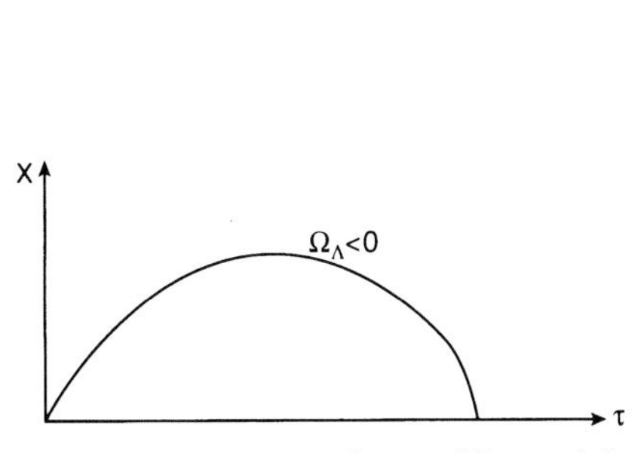

Fig. 1.14. x vs. τ for an FL model with $\Omega_\Lambda < 0$

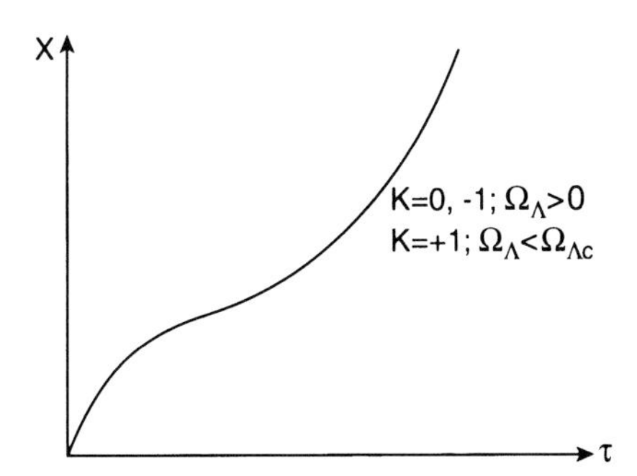

Fig. 1.15. x vs. τ for an FL model with $K = 0, -1; \Omega_\Lambda > 0$, or with $K = +1; \Omega_\Lambda < \Omega_{\Lambda c}$

from the static universe to an expanding solution. There is a solution starting from $x = 0$, and "creeping" towards x_E, while another type of solution starts expanding from the finite radius x_E (the "Eddington–Lemaître universe"). If Ω_Λ is only slightly smaller than $\Omega_{\Lambda c}$ an intermediate case between the Einstein and Eddington–Lemaître models arises: the cosmological model starts at $x = 0$, and then stays for a long time (depending on how close Ω_Λ is to $\Omega_{\Lambda c}$) near x_E, with $\dot{x} \approx 0$, before it expands as in the Eddington–Lemaître case (Fig. 1.17). In Fig. 1.18 $x(t)$ is displayed for different values Ω_Λ and for $\Omega_0 = 0.2$. It is clear from those figures that we can reach any desired value for the procuct $H_0 t_0$ by appropriatly choosing Ω_Λ close to $\Omega_{\Lambda c}$. Since measurements of H_0 and t_0 may force us to introduce a cosmological constant to achieve consistency within the simple cosmological models we should consider Ω_Λ as another fundamental cosmological quantity to be determined by observations.

For $\Omega_\Lambda > \Omega_{\Lambda c}$ there is the possibility of a model contracting from infinity to a minimal radius, and expanding again. This model, as well as

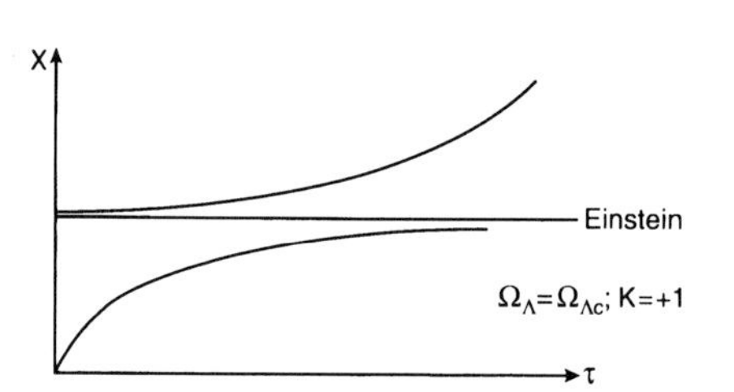

Fig. 1.16. x vs. τ for the static Einstein, and the Eddington-Lemaître model

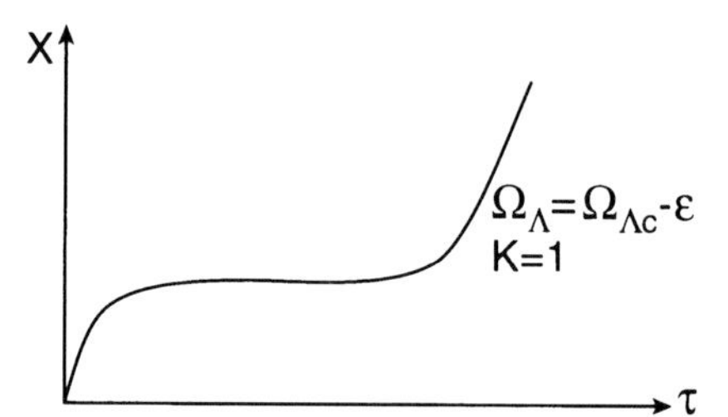

Fig. 1.17. For $K = +1, \Omega_\Lambda = \Omega_{\Lambda c} - \epsilon$ the FL model shows a long epoch of almost constant x

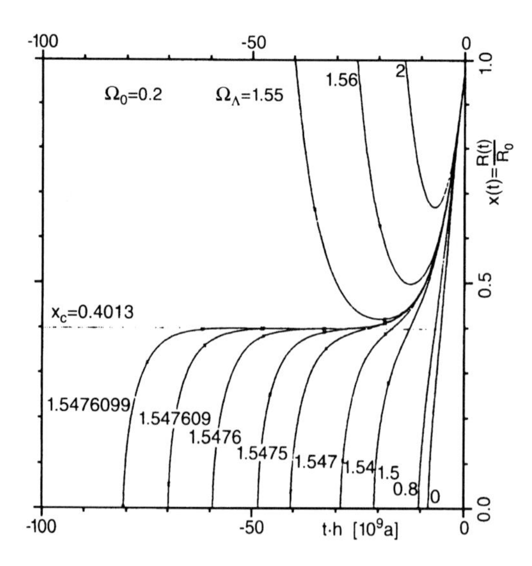

Fig. 1.18. Various FL models with Ω_Λ less and greater than $\Omega_{\Lambda c}$. x is plotted against time in units of the Hubble time

the Eddington–Lemaître model, avoids the initial singularity. Both models, however, seem to be in conflict with existing observational results: it is quite instructive to follow this argument. The epoch of maximal contraction, at $x_m = R_m/R_0$, corresponds to a maximal redshift $1 + z_m \equiv x_M^{-1}$. It is straightforward to show for the Eddington–Lemaître model that the following relation holds:

$$\Omega_0 = \frac{2x_M^3}{(1 - 3x_M^2 + 2x_M^3)}. \tag{1.35}$$

For all models without a big bang, one can show [Börner and Ehlers 1988] that the density parameter Ω_0 must satisfy:

$$\Omega_0 \leq \frac{2}{z_m^2(z_m + 3)}.$$

(This follows from $\dot{x}_m = 0$, and $\Omega_\Lambda \geq \Omega_{\Lambda c}$.) Since there are quasars with $z \approx 5$, we find $\Omega_0 \leq 0.01$ – in contradiction to the measurements. It is interesting that the existence of objects with large redshifts is sufficient to reach that conclusion. An appeal to the cosmic background radiation is not necessary, nor an appeal to observational limits on Ω_Λ.

Exercises

1.1

a) Show that for $\dot{R} \neq 0$ any two Friedmann equations imply the third.
b) Discuss the solutions of the Friedmann equations for an equation of state $p = K\varrho^\gamma$ (K, γ constants).

1.2 Consider an anisotropic, homogeneous universe model [Kramer et al. 1980]:

$$ds^2 = dt^2 - X^2(t)dx^2 - Y^2(t)dy^2 - Z^2(t)dz^2.$$

With $S^3 \equiv XYZ$ one finds for a "dust" equation of state ($p = 0$) that

$$\dot{X}\dot{Y}Z + \dot{X}Y\dot{Z} + X\dot{Y}\dot{Z} = 8\pi G\varrho S^3$$

and

$$(\dot{X}YZ)^{\cdot} = 4\pi G\varrho S^3;$$

similarly for $X\dot{Y}Z$ and $XY\dot{Z}$.

The continuity equation is

$$\varrho S^3 = \text{const.}$$

Integrate these equations and discuss the solutions, esp. for $t \to 0$.

1.3 Derive (1.35). Show that for FL models without a big bang the density parameter Ω_m must satisfy $\Omega_0 \leq \frac{2}{Z_m^2(Z_m+3)}$.

1.4 Derive (1.27) for r_A. Consider $\Omega_\Lambda \neq 0$, and try to derive a formula for the luminosity distance $d_L = r_A(1+Z)^2$.

(Hint: Show that [Carroll et al. 1992]

$$d_L = \frac{c(1+z)}{H_0\sqrt{|\,\Omega_k\,|}} \operatorname{sinn}\left\{\sqrt{|\,\Omega_k\,|}\int_0^z \frac{dz'}{\sqrt{P(z')}}\right\};$$

where $\Omega_k = 1 - \Omega_m - \Omega_\Lambda, \operatorname{sinn} x = \begin{cases}\sin x, \Omega_k > 0\\ \sinh x, \Omega_k < 0\end{cases};$

$P(z) \equiv (1+z)^2(1+\Omega_m z) - z(2+z)\Omega_\Lambda$.)

Also calculate d_L for $\Omega_k = 0$. Give an expansion in terms of z, to estimate the first correction terms to the linear Hubble law $d_L H_0 = cz$.

1.5 De Sitter space is an FL model with $\Lambda \neq 0$ and $\varrho = 0$. Show that the solutions of the FL equations for this case have $R(t)$ exponentially growing. Since de Sitter space does not contain matter, not only "comoving", but other coordinate systems are useful:

Embedding de Sitter space in a 5-dimensional Minkowski space-time with

$$\eta_{AB} = \operatorname{diag}(1,-1,-1,-1,-1), (A,B = 0,\dots,5),$$

leads to a description as the 4-dimensional hypersurface

$$\zeta^A \eta_{AB} \zeta^B = -S^2.$$

Draw this surface suppressing 2-space coordinates.

Various coordinate systems can be introduced:

a) Conformal coordinates x^μ are obtained by a stereographic projection of the de Sitter hypersurface from the point $(0,0,0,0,S)$ onto the hyperplane tangent at $(0,0,0,0,-S)$.
Show that

$$\zeta^\mu = \phi(x^2)x^\mu; \zeta^4 = -S\phi(x^2)\left(1+\frac{x^2}{4S^2}\right).$$

$$\left(x^2 = x^\mu \eta_{\mu\nu} x^\nu; \mu, \nu = 0, \dots, 3; \phi^{-1}(x^2) = \left(1 - \frac{x^2}{4S^2}\right)\right).$$

What are the 3-spaces of constant x^0? Compute the line element ds^2 on the hypersurface.

b) $(\lambda, \boldsymbol{y})$ coordinates:

$$\begin{aligned}
\zeta^0 &= \frac{S}{2}(\lambda - \lambda^{-1}) - \frac{1}{2S} y^2 \lambda^{-1}, \\
\zeta^{1,2,3} &= \lambda^{-1} y^{1,2,3}, \\
\zeta^4 &= -\frac{S}{2}(\lambda + \lambda^{-1}) + \frac{1}{2S} y^2 \lambda^{-1} \\
&(\lambda \neq 0; -\infty < y^i; \lambda < \infty).
\end{aligned}$$

Compute ds^2 in these coordinates. What are the 3-spaces of constant time in this coordinate system? Show that for $S \to \infty$, 4-dimensional Minkowski-space is recovered.

c) Static coordinates (r, t, θ, φ):

$$\begin{aligned}
\zeta^0 &= (S^2 - r^2)^{1/2} \sinh t/s; \\
\zeta^1 &= (S^2 - r^2)^{1/2} \cosh t/s; \\
\zeta^2 &= r \sin\theta \cos\varphi; \\
\zeta^3 &= r \sin\theta \sin\varphi; \\
\zeta^4 &= r \cos\theta;
\end{aligned}$$

where $(0 \leq r \leq S)$.
Compute ds^2 and compare this to the Schwarzschild metric. Which quantity corresponds to the Schwarzschild radius? Show that this is a static metric. Why is de Sitter space itself not static?

1.6 Compute the angle θ subtended by a fixed physical length D in $K = 0, +1, -1$ FL models! Show the following:

a) at a given redshift z the angular size is smallest for $K = -1$, largest for $K = +1$ models;
b) the dependence of θ on z for large z;
c) the angular size of a comoving length is constant for large z;
d) for $K = 0$ there is a minimal θ at $z = 5/4$, i.e. for larger z the apparent angular size of objects increases again.

2. Facts – Observations of Cosmological Significance

"There are probably few features of theoretical cosmology that could not be completely upset and rendered useless by new observational discoveries."

H. Bondi

"I hate reality, but it is still the only place where I can get a decent steak."

Woody Allen

In this chapter various observations relevant to cosmology will be introduced. We shall see how they fit into a "standard model" – one of the homogeneous, isotropic Friedmann–Lemaître models presented in the last chapter – and to what extent they determine the parameters. In Chap. 1 we pointed out that the cosmological model and the present epoch (t_0) can be determined by the measurement of two dimensionless numbers: $q_0, \Omega_0 \equiv \varrho_0/\varrho_C$, and a scale constant H_0. $\Lambda = 0$ is equivalent to $2q_0 = \Omega_0$.

A rough orientation may be helpful

It would seem sensible to begin with a broad survey of astronomical terminology to get acquainted with the various objects and structures to be discussed. The cheapest way to start with cosmological observations is to step out of your front or back door at night, and look at the sky. Why is it dark? (If there were a constant density of unchanging stars the night sky should be brilliant like the surface of the Sun – Olbers's paradox). Then a meteorite might fall into your yard, and you could analyse it, measuring its constituents, and especially its content of radioactive isotopes. In this way you could determine its age. The high art of cosmochronology does just that – it measures ages and events in the history of rocks, meteorites, and lunar minerals (see Sect. 2.1). With a little bit of theory, an educated guess can be made at the age of the elements. Again, when you look up at the sky you see the brilliant band of the Milky Way, consisting of billions of stars that are all radiating energy; actually, with the naked eye you can see only about one thousand of our nearest neighbours.

A measure of the flux received from a star is the "apparent brightness". Astronomers use an old, revered scale for the energy flux, which has been in use since the time of the ancient Greek astronomer Hipparchos.

There are many systems of magnitudes but all give the magnitude m by a relationship

$$m = m_0 + 2.5 \log F_0/F,$$

where m_0 is an arbitrary reference point for the system used. F is the flux received, and F_0 the standard corresponding to m_0. An increase of 5 in magnitude corresponds to a decrease in flux by a factor of 0.01.

Originally the Pole Star ($m_\mathrm{v} = 2^m.12$) was used as a prominent reference point. Because this star unfortunately shows small fluctuations, a sequence of stars – exactly measured and tested for constancy of brightness – was set down in 1922, the "International Polar Sequence". Early photometric measurements were done visually. Then photographic plates imitating the spectral sensitivity of the eye were developed. Consequently one characterized the stars by visual (m_v) or photographic (m_p) magnitudes. Nowadays a combination of plates and photoelectric devices allows one to choose maximum detector sensitivity at any wavelength interval between the ultraviolet (UV) and the infrared (IR) parts of the spectrum. What, then, is the physical meaning of the different magnitude scales? (See [Unsöld 1991] for a precise explanation.)

The specific flux S_λ from a star at a wavelength λ is received at the Earth by an instrument with a response function E_λ, which describes the reaction of a linear device to an incident unit flux. (E_λ is in general appreciably large only in a small wavelength interval.) The actual measurement is proportional to the integral

$$\int_0^\infty S_\lambda E_\lambda d\lambda$$

and the apparent brightness m of the star (up to a normalization constant) is given by

$$m = -2.5\log \int_0^\infty S_\lambda E_\lambda d\lambda + \text{const.}$$

(The extinction in the Earth's atmosphere has to be considered in addition – this is easy since in general E_λ is concentrated strongly around a specific λ.) Visual magnitudes m_v are defined accordingly by a function E_λ centred on $\lambda \sim 5400\text{Å}$ ("the isophote wavelength"). For photographic magnitudes m_p, the centre is at $\lambda \sim 4200\text{Å}$. Photoelectric measurements often refer to a magnitude sharply centred on one wavelength $m(\lambda)$, i.e. $E(\lambda) \sim \delta(\lambda)$. A standard often in use is the UBV system ([Johnson and Morgan 1953]; U = ultraviolet, B = blue, V= visual). The corresponding sensitivity functions E_λ are centred on $\lambda_U \sim 3650\text{Å}$. $\lambda_B \sim 4400\text{Å}, \lambda_V \sim 5480\text{Å}$, for average star colours. The UBV magnitudes are related to each other such that U = B = V for stars of spectral type A0. Often the "colour indices" U-B or B-V are used to characterize the stars.

Let us list some examples of visual magnitudes:

Sun	$m = -26^m.86$
Full moon:	$m = -12^m.55$
Venus:	$m = -4^m.5$
Sirius:	$-1^m.8$
Pole Star:	$+2^m$
Limit for the naked eye:	$+5^m$; see [Unsöld 1991].

An absolute magnitude scale $M, M = M_0 + 2.5 \log L_0/L$, corresponding to a logarithmic scale of the intrinsic luminosity, is also used. The absolute magnitude M is the apparent brightness of the object at a distance of 10 parsecs. Thus if d is the distance in parsecs, then

$$M = m - 5\log[d/10]$$

or, equivalently,

$$d = 10^{(1+0.2(m-M))}.$$

M_v designates the power emitted in visible light, M_bol (the bolometric magnitude) is the total luminosity, which is really a theoretical concept involving corrections to measured values. Taking the Sun as an example, one has

$$M_\text{v} = 4^m \cdot 71,$$
$$M_\text{p} = 5^m.16,$$
$$M_\text{bol} = 4^m.62.$$

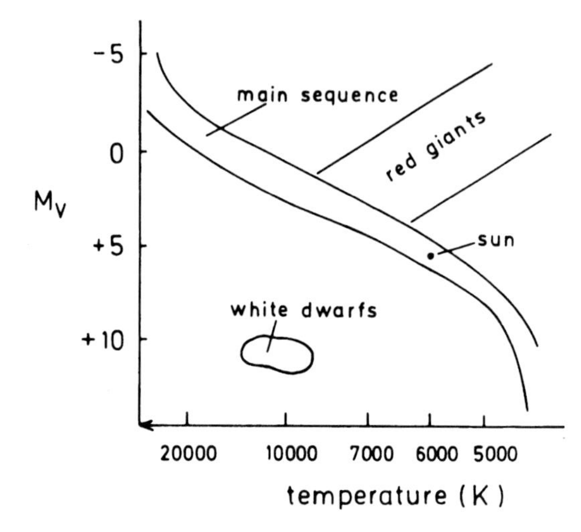

Fig. 2.1. Schematic HR diagram. M_V (corresponding to luminosity) is plotted versus temperature which is equivalent to a B-V colour index

The number $m - M$ is often used as a measure of the distance in place of d; it is called the "distance modulus".

The colour of a star is a clue to its surface temperature. The B-V colour index, for example, measures the ratio of luminosities in the blue and in the red part of the spectrum. Red stars, i.e. stars of low surface temperature, have a positive colour index. Blue stars with a high surface temperature have a negative colour index.

If the stars are now plotted on a diagram of M_v versus B-V index, as E.J. Hertzsprung and H.N. Russell did at the beginning of the last century, then such "HR" diagram clearly shows that M_v and B-V are related: most of the stars are on a line running diagonally from the upper left (blue, high M_v) to the lower right (red, low M_v). This line is called the main sequence (Fig. 2.1).

Not all stars, however, are located on the main sequence. Stars of high luminosity with a positive (B-V) index populate the upper right of the HR diagram to the right of the main sequence; they are the red giants. Stars of low luminosity and high temperature make up the "white dwarf" population in the lower left-hand corner.

The theory of stellar evolution gives convincing explanations for the distribution found in the HR diagram: stars on the main sequence derive their energy from the fusion of hydrogen into helium in their central cores. Our Sun is one of these peaceful main-sequence stars. The outer layers of the star are unchanged by this energy production in the interior until about 12% of the hydrogen has been fused into helium. Then the burning in the helium-enriched core stops, and the fusion process shifts to a thin spherical shell surrounding the core. The star leaves the main sequence and becomes a red giant. (There is apparently no simple physical argument for this behaviour, we have to take it as an outcome of the numerical computations that follow the evolution of a star through various equilibrium stages.)

A rough estimate of the time a star stays on the main sequence can be easily given: fusion of four protons to ^{4}He liberates 0.7% of the rest-mass energy, which, in equilibrium, is radiated away. Thus the characteristic time $t_{m.s.}$ to use up 12% of the mass M, for a luminosity L (energy emitted per unit time), is

$$t_{m.s.} = 0.12Mc^2 \times 0.007/L = 1.1 \times 10^{10}(M/M_\odot)(L_\odot/L)\,\text{years};$$

$M_\odot = 2 \times 10^{33}\,\text{g}, L_\odot = 4 \times 10^{33}\,\text{erg/s}$ are the mass and luminosity of the Sun. Supplementing this by a mass–luminosity relation derived from stellar evolution theory,

$$L/L_\odot \simeq (M/M_\odot)^3,$$

gives

$$t_{m.s.} \simeq 10^{10}(M/M_\odot)^{-2}\,\text{y}.$$

While the Sun will stay on the main sequence for about 10 billion years (half of this time has already passed!), more massive stars will branch off much earlier ($t_{m.s.} = 10^8$y for $M = 10M_\odot$), and stars of small mass will stay on longer ($t_{m.s.} = 4 \times 10^{10}$y for $M = 0.5M_\odot$). Thus the luminous stars with a negative (B-V) index (i.e. a strong blue component in the spectrum) are young stars, while the less luminous, reddish stars on the main sequence have a smaller mass and constitute an old population of stars.

Distances are determined within our stellar system by using the parallax caused by the Earth's orbit around the Sun and star streaming methods for nearby objects: the High Precision Parallax Collecting Satellite (Hipparcos) has measured the parallax of more than 10^5 single stars with a precision of ≈ 1.3 marcsec [ESA 1997]. The parallax of Cepheid variables can also be determined. These are pulsating stars whose pulsation period and colour give an exact measure of their absolute luminosity. The Hipparcos measurements give a direct calibration of the Cepheids' period–colour luminosity relation [Feast and Catchpole 1997; Oudmaijer et al. 1998]. Therefore $m - M$ can be obtained from measuring the apparent brightness m, the period and the colour.

The measurements reveal that the stars are arranged in a large disc-shaped system of radius ~ 15 kpc and thickness ~ 1 kpc, with a bulge at the centre. The young stars form a "Population I" of objects lying in the disc, and showing a normal abundance of elements. The old stars are distributed mainly in a halo surrounding the disc, and they form a "Population II" of somewhat metal-deficient objects. (Metal is here taken to include all elements heavier than He.) Prominent examples are the stars in the globular clusters that surround the disc in a roughly spherical halo (cf. Fig. 2.2). They probably contain the oldest stars in the Galaxy, and their age must be representative of the age of the Galaxy (see Sect. 2.1). Observations of extremely metal-deficient stars (see [Bond 1980]) seem to indicate the existence of an even older (?) Population III component of the Galaxy.

Fig. 2.2. HST picture of the globular cluster M15 (courtesy of STScI)

Radio observations have shown that our Galaxy rotates around its centre (once every 200 million years), and that the disc divides into spiral arms. By high-resolution imaging observations in the near-infrared the proper motions of 39 stars within 0.3 pc of the compact radio source Sagittarius A* at the galactic centre have been determined [Eckart and Genzel 1996, 1997]. There is strong evidence from these data that a black hole of about $2.5 \times 10^6 M_\odot$ exists at the centre of our Milky Way.

Gas and dust between the stars are the raw material out of which new stars are continuously born. Large molecular clouds have been identified as the places where star formation is still happening (Fig. 2.3). The interplay between nuclear physics, radiation pressure, and gravity determines the equlibrium configurations of stars – fusion reactors made from interstellar gas and dust – to have masses from about $0.1 M_\odot$ to $60 M_\odot$. After a long time on the main sequence the stars pass through a red-giant phase rather rapidly, and then end their thermonuclear activity. The final outcome depends on the mass of the core of the red giant. If it is below the "Chandrasekhar limit"

Fig. 2.3. HST picture of a star-forming region in the Eagle nebula (courtesy of STScI). You can see columns of cool hydrogen gas and dust illuminated by the hot UV radiation from new-born stars (*top*)

(the stability boundary for white dwarfs) $M \leq M_C = 1.4 M_\odot$, then eventually the core will contract to a white dwarf (a dense $\sim 10^6\,\mathrm{g/cm^3}$, small $\sim 10^9\,\mathrm{cm}$ object), while the outer layers of the red giant are expelled. For larger core masses, the core collapse proceeds to even higher densities and smaller radii (a neutron star or black hole is formed; in a certain mass range the core may also be completely disrupted) and triggers a supernova explosion of the star. The outer layers of the core and envelope are expelled in a gigantic eruption, and the exploding star appears exceedingly bright for some time (Figs. 2.4 to 2.6). The remnants of this explosion may again show some activity as pulsars (rotating neutron stars) or as X-ray sources if they are in a binary system and can accrete matter from a companion star.

Away from the disc of our Galaxy, many more stellar systems can be found. They appear in all kinds of shapes (see Figs. 2.7 to 2.12): systems with spiral arms like our own, and of similar size (e.g. M31), the Andromeda galaxy, at a distance of 750 kpc). There are elliptical galaxies which do not have spiral arms, but which appear as almost circular or elliptical discs. Elliptical galaxies can be very massive systems ($\sim 10^{13} M_\odot$), whereas the dwarf spheroidal galaxies, of similar appearance, are very small ($\sim 10^6 M_\odot$).

Fig. 2.4. The Large Magellanic Cloud (LMC) – an irregular galaxy at a distance of 55 kpc (ESO)

Fig. 2.5. The LMC region before February 23, 1987 (ESO)

Fig. 2.6. The same region as in Fig. 2.5, on February 24, 1987, a day after the supernova appeared

Fig. 2.7. The Andromeda galaxy (M31) at 2.1 million light-years is the nearest big spiral to the Milky Way. Our Galaxy is thought to look much like Andromeda (credit to J. Ware)

Fig. 2.8. The spiral galaxy NGC 4622 appears to rotate clockwise – contrary to expectations (STScI)

The galaxies can be arranged in a sequence according to their shape, the "Hubble sequence", but the sequence must not be interpreted in terms of a morphological evolution along this sequence.

The HST, the Keck telescope in Hawaii, and other large ground-based telescopes have increased the ability to study directly the evolution of galaxies at high redshift. An excess of faint blue galaxies exists at redshifts $z > 1$. This interpretation is consistent with galaxy formation models based on hierarchical clustering. The morphology of the excess faint blue galaxies has been revealed by HST, especially in the Hubble Deep Field images [Abraham 1996]: a high percentage are late-type spiral, irregular, and peculiar (possibly merging) galaxies.

For large galaxies at $z < 1$, there is no clear evidence of evolution, or of identification of young and old galaxies in the same way as there is for young and old stars.

Most of the galaxies (except some very close ones) show a redshift in their spectral lines, indicative of the general expansion of the universe (Fig. 2.13).

Fig. 2.9. The spiral galaxy M81. Emission from gaseous oxygen is blue, main bulk of starlight is yellow (48 inch Schmidt plate (Mt. Palomar) by H. Arp; image processing by Jean Lorre (JPL))

The precise measurement of the expansion rate (the determination of the Hubble constant) is plagued by various dificulties (Sect. 2.2). A distance scale of cosmic dimensions is very hard to establish, because even with the Hubble Space Telescope (HST) the most reliable indicators, the Cepheid variables, can only be used out to a distance modulus of $m - M = 32$. The Cepheid variables are still the best standard candles [Feast 1984]. The velocity fields seems to show local irregularities out to that value and only beyond $m - M = 35$ can a smooth Hubble flow be expected (i.e. motion which precisely follows the cosmic expansion). The HST has extended the Cepheid scale out to the Virgo cluster galaxies, but unfortunately no such convenient

Fig. 2.10. The Sombrero galaxy (NGC 4594) shows a dark layer of dust around its equator. This galaxy is type SA (VLT; P. Barthel et al.)

method exists for the measurement of very large distances. Instead, much less sharp, statistical correlations between the characteristics of some distant galaxy and its probable luminosity have to be employed; and there is, of course, an increase in uncertainty with distance.

The centres of galaxies are often places of violent activity, where huge eruptions take place, rapid flux variations are observed (Fig. 2.14), and generally a large amount of energy is radiated, from radio waves to X-rays. Some of these active galactic nuclei are so far away that the surrounding normal galaxy can no longer be seen, and they appear quasi-stellar (Fig. 2.15). These quasars exhibit redshifts in their emission lines of up to $z \simeq 5$, but normal galaxies are also seen at high redshifts $z \geq 6$. The intrinsic properties of quasars are not well known, however, and therefore these objects cannot be used as measuring devices for very large distances. The number counts of these objects give an indication of cosmological evolution (Sect. 2.7). The radio sources, objects of intense radio emission (Fig. 2.16) – which probably also can be detected at very large distance – can be used in a similar way. As the sources above a certain flux limit are counted, and as one proceeds from the bright sources to the fainter ones, one sees at first more sources than are to be expected from a uniform distribution in flat space, then, at the faint end, less sources than would be predicted. Apparently, bright radio sources

Fig. 2.11. The Cartwheel galaxy has a rim of newly formed, very bright stars (diameter 100 000 light-years). It has acquired its beautiful shape from the collision of two galaxies (AAO; credit to S. Lee and D.F. Malin)

and quasars were more frequent in the past than now, but they did not turn on earlier than at about $z \sim 5$.

Absorption lines along the line of sight to background quasars probe the gas clouds and/or protogalaxies at high redshift. These $Ly\alpha$ (Lyman alpha) absorbers may well be galaxies in formation. Estimates of the total amount of baryons present in these $Ly\alpha$ absorbers at a redshift $z \sim 3$ are consistent with the amount of baryons seen in galaxies today.

The galaxies are not distributed homogeneously, but they appear grouped together in various types of structures. These range from groups, like our local group (whose principal members are the Galaxy and the Andromeda galaxy) containing just a few members, to rich clusters (Fig. 2.17) with several thousand members (e.g. the Virgo Cluster of which our Local Group is a part). These clusters extend over distances of a few Mpc, but there is evidence that they are parts of even larger systems, the superclusters, extending over scales of ~ 25 Mpc. There seem to be filament-like chains of galaxies linking different superclusters into a network on scales of 100 Mpc or more. A cell-like structure may be a good picture, with most of the matter concentrated on the cell walls and very low-density regions ("voids") in between. Thus at scales of ~ 100 Mpc the matter distribution seems to be quite inhomogeneous,

Fig. 2.12. The Hubble Deep Field is one of mankind's most distant optical views of the universe. The dimmest galaxies in this picture are as faint as 30th magnitude. This 1 arcmin by 1 arcmin field in Ursa Major contains about 2500 galaxies. The HST was pointed at this single spot for 10 days (courtesy of R. Williams, HDF team (STScI))

Fig. 2.13. A few hundred of the galaxies in the Hubble Deep Field have their redshifts measured. The largest value obtained up to now is $z = 4.02$, obviously a very distant object. The redshifts indicate that all these galaxies are receding from our position (courtesy of Mark Dickinson (STSci) and Judy Cohen (Caltech, Keck telescope))

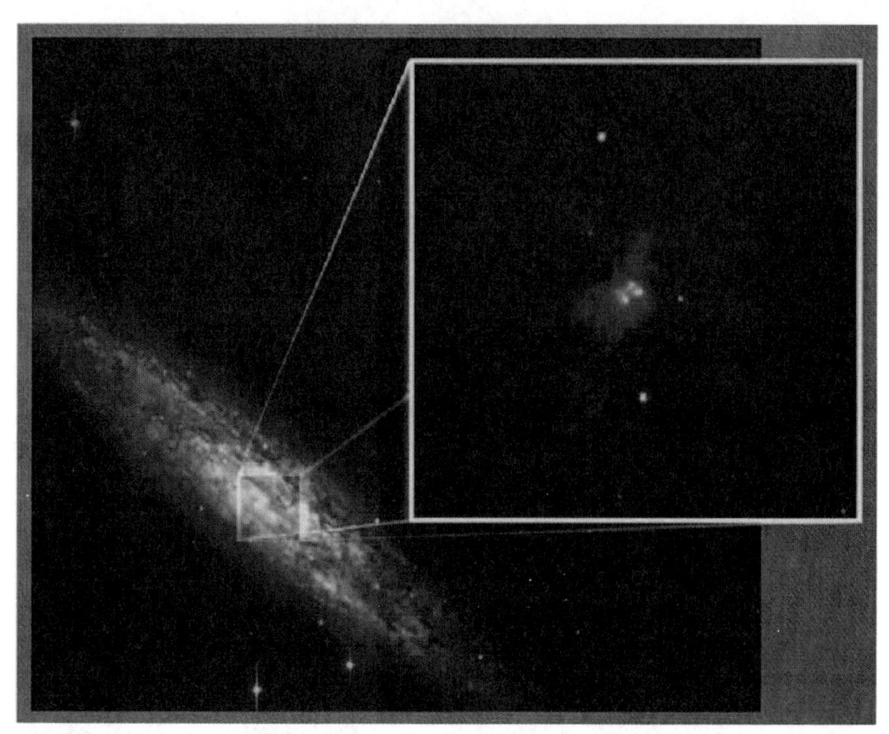

Fig. 2.14. The galaxy NGC 253 at a distance of 8 million light years shows prodigious star formation. In the core of the optical image (ESO) the observations of the Chandra X-ray observatory reveal the presence of hot gas, and at least four powerful X-ray sources within 3000 light-years of the centre (credit to K. Weaver (GSFC) et al.; ESO, NASA)

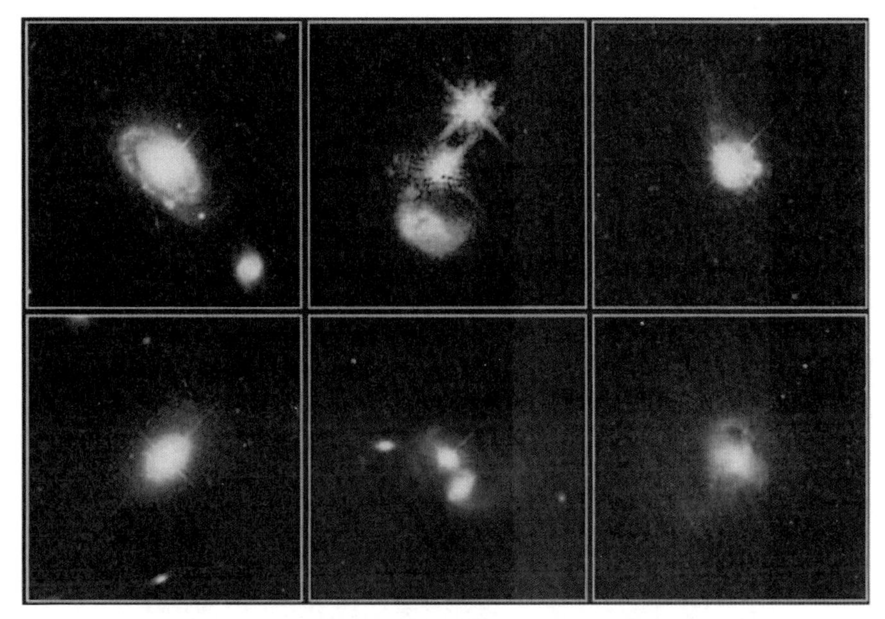

Fig. 2.15. This gallery of quasars (quasi-stellar objects) shows these bright objects (star-like with diffraction spikes) embedded in the matter of colliding and merging, disrupted galaxies. Quasars are probably powered by a large central black hole which accretes the infalling gas, dust, and stars (credit to J. Bahcall, M. Disney, and NASA)

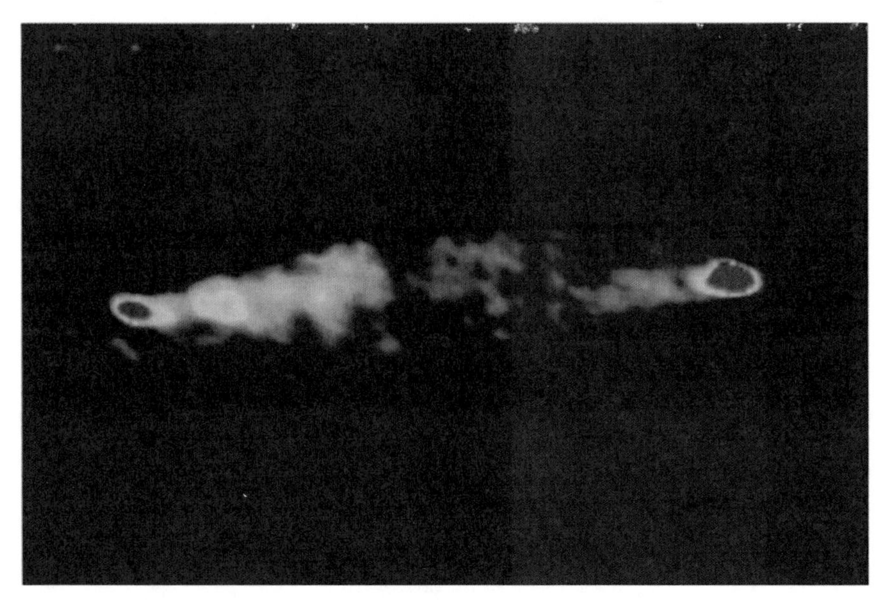

Fig. 2.16. An observation of the radiogalaxy NGC 0634-20 at 21 cm taken at the VLA array of radiotelescopes is shown here with coloured flux contours (*red to blue*: high to low flux) (Courtesy of MPI f. Radioastronomie, Bonn)

Fig. 2.17. The Coma cluster of galaxies is shown here. Each tiny speck of light in this photograph is one of the several thousand galaxies in this cluster. At a distance of 60 Mpc it has a diameter of a few Mpc (Kitt Peak National Obs., credit to O. Lopez-Cruz, I.K. Shelton)

and this may cast some doubt on the use of a smooth, homogeneous density distribution in the cosmological model. There are indications, however, from the isotropy of the background radiation (Sects. 2.5 and 2.6) as well as from radio source counts (Sect. 2.7) that at large distances the density contrasts become smaller and smaller, and blend – to a good approximation – into a homogeneous background. These different structures will be examined in some detail later.

2.1 Age Determinations

2.1.1 The Age of the Solar System

Ages of meteorites, lunar rocks, and minerals on Earth are obtained through measurements of isotropic ratios of radioactively decaying nuclei. The "age" measured is the date of an event when a system of parent and daughter elements becomes isolated – and afterwards follows, undisturbed, the exponential laws of radioactive decay. The α-decay pairs

$$^{238}\mathrm{U}\text{–}^{206}\mathrm{Pb}$$
$$^{235}\mathrm{U}\text{–}^{207}\mathrm{Pb}$$
$$^{232}\mathrm{Th}\text{–}^{208}\mathrm{Pb}$$

with half-lives of several billion years (see Table 2.1) (notation: $10^9\,\mathrm{y} \equiv 1\,\mathrm{Gigayear} \equiv 1\,\mathrm{Gyr}$) are often used in the measurement of the ages of rocks. Their decay times are comparable to the time intervals of interest, and so the isotopic ratios will change noticeably. In general the isotopic abundance by mass of a daughter element ${}^iD$ is the sum of a part ${}^iD_0$, initially present, and a part ${}^iD_r$ produced (at the point investigated – "in situ"), by the radioactive decay of a parent isotope ${}^KP$:

$$ {}^iD = {}^iD_r + {}^iD_0 = {}^iD_0 + {}^KP(e^{\lambda t} - 1). \tag{2.1}$$

A large number of people work diligently to erect the streamlined edifice of modern cosmology

Table 2.1. Half-lives of radio nuclei

Nuclides	Isotopic abundance (%)	Decay	Product	Half-life (y)
^{87}Rb	27.83	β^-	^{87}Sr	4.99×10^{10} (4.80×10^{10})
^{232}Th	100	α	^{208}Pb	1.39×10^{10} (1.41×10^{10})
^{238}U	99.28	α	^{206}Pb	4.50×10^{9} (4.47×10^{9})
^{235}U	0.72	α	^{207}Pb	7.13×10^{8} (7.04×10^{8})
^{147}Sm	15.0	α	^{143}Nd	1.06×10^{11}
^{187}Re	62.6	β^-	^{187}Os	$(3.5 - 5) \times 10^{10}$
^{129}I	–	β^-	^{129}Xe	1.7×10^{7} (1.57×10^{7})

Lists of determinations of radioactive decay rates can be found e.g. in [Lederer and Shirley 1978].

As it is easier to measure abundance ratios, one usually tries to find a stable isotope ${}^jD$ of the same chemical element D, and then refers all quantities to this – hopefully – unchanging isotope:

$$\frac{{}^iD}{{}^jD} - \left(\frac{{}^iD}{{}^jD}\right)_0 = \frac{{}^KP}{{}^jD}(e^{\lambda t} - 1). \tag{2.2}$$

In a plot of ${}^iD/{}^jD$ vs. ${}^KP/{}^jD$, this is a straight line with slope $(\tan\alpha) = (e^{\lambda t} - 1)$, if ${}^KP/{}^jD$ is variable and ${}^iD/{}^jD_0$ is constant.

This is the case for pieces of a meteorite or of a rock, if ${}^iD$ and ${}^jD$ are different isotopes of the same chemical element. Initial isotopic ratios should be equal in different probes ("separates") of one rock, since chemical fractionation cannot change these ratios. Thus one can expect $({}^iD/{}^jD_0)$ to be constant. Different separates can, however, have a different chemical history, and therefore the initial abundance ratios of chemical elements may vary throughout the separates, hence also ${}^KP/{}^jD$, if ${}^KP$ and ${}^jD$ are isotopes of different chemical elements. Then at least two different points ${}^KP/{}^jD$ vs. ${}^iD/{}^jD$ are needed to determine the slope $(e^{\lambda t} - 1)$, and the age t. The accuracy of the age determination depends, of course, on the magnitude of the variations of P/D through the rock.

The isochrone of ^{87}Rb/^{86}Sr vs. ^{87}Sr/^{86}Sr is shown for the Nakhla meteorite in Fig. 2.18.

In Table 2.1 a few of the most commonly used decay chains are listed.

Figure 2.19 shows, as another illustration, an age determination of the Allende meteorite, derived from considering the uranium decay chains. The known ration $^{238}\mathrm{U}/^{235}\mathrm{U} = 137.9$ has been used to eliminate uranium abundances in favour of the Pb isotopic ratios ^{207}Pb/^{204}Pb and ^{206}Pb/^{204}Pb.

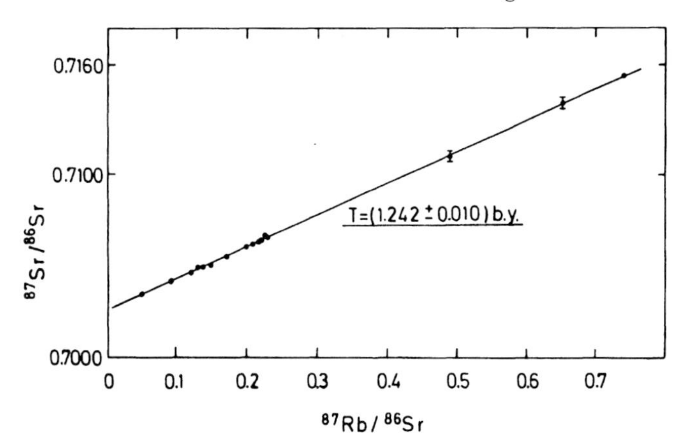

Fig. 2.18. Rubidium–strontium isochrone of the Nakhla meteorite minerals (after [Gale et al. 1975])

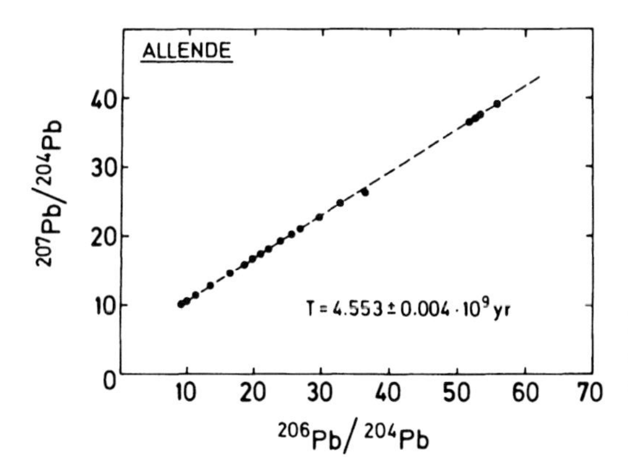

Fig. 2.19. Pb-isotopic ratios are used to determine the age of the Allende meteorite. (after [Tatsumoto et al. 1976]; cf. also [Kirsten 1978])

These measurements have become an art form in recent years – the art of cosmochronology – and have provided several solid facts on the evolution of our Solar System [Kirsten 1978]. The oldest lunar rocks were formed $4.5 - 4.6 \times 10^9$ years ago, while meteorites have ages of 4.57×10^9 years (Figs. 2.18 and 2.19). The Earth with its large thermal reservoir is still thermally active, i.e. processes of fractionation and chemical differentiation are still going on. It is thus very probable that the ages of rocks on the Earth do not correspond to the age of the planet. Some of the oldest rocks measured are Amitsoq-type gneisses found in Greenland. Rb-Sr ages are $(3.7 \pm 0.1) \times 10^9$ years. This establishes the existence of a granitic crust on Earth about 3.7×10^9 years ago. Since the ages of the meteorites and of the moon probably correspond to the age of the Solar System, one may conclude that for about 800 to 900 million years after its birth, the Earth existed without a solid crust.

The age of the Solar System is known with high accuracy $t_{as} = (4570 \pm 30) \times 10^6$ years, and there is even the possibility of determining small time intervals ($\delta \sim 10^6$ y) around this reference point t_{as}. This possibility arises from the existence of short-lived, but now absent, radioisotopes, whose decay products are found to be over-abundant when compared with the average isotopic abundances of these decay products.

The β^- decay of ^{129}I into ^{129}Xe has a half-life of 1.57×10^7 years. A deviation of up to a factor of 2 in the abundance ratio of $^{129}\mathrm{Xe}/^{127}\mathrm{I}$ (compared with the average solar values of 10^{-4}) has been measured in various meteorites. Since the nucleosynthesis of the elements more or less ended before the Solar System was formed, one may draw the conclusion that the time interval of $\Delta t \simeq 5 \times 10^6$ y deduced from the $^{129}\mathrm{Xe}/^{127}\mathrm{I}$ ratio is characteristic of the condensation time in the solar nebula. (The noble gas ^{129}Xe would not be mixed into the condensing meteorite, so all ^{129}Xe found must be a decay product.) The retention of short lived ^{129}I in meteorites means that these must have condensed in a time interval shorter than the half-life of this isotope. Thus cosmochemistry tell us that the Solar System condensed out of the solar nebula some 4.57×10^9 years ago, and in the extremely short time-span of a few million years.

Another puzzling fact has been the enrichment of ^{26}Mg in certain inclusions of the Allende meteorite (an excess of 1.3%). If this additional ^{26}Mg is a decay product of ^{26}Al, which has a half-life of only 0.716×10^6 y, then the conclusion is suggestive that the abundance ratio in Allende of $^{26}\mathrm{Al}/^{27}\mathrm{Al} \approx 0.6 \times 10^{-4}$ has been produced by a supernova explosion occurring in the vicinity of the Sun about 10^6 y before the condensation. One may speculate that perhaps this supernova explosion was the trigger for the formation process. Recent satellite observations [Mahoney et al. 1982, 1984], however, interpret a γ-line as a transition in the daughter nucleus ^{26}Mg after ^{26}Al decay. These observations indicate that there is always a large anount of ^{26}Al in interstellar space (due to supernovae, novae, and stellar winds). Thus the supernova trigger is not needed to explain the over-abundance in the Allende meteorite.

2.1.2 The Age of the Elements

These meteorite and rock measurements also give the abundance of the elements at the time when the Solar System was formed. The comparison of such isotopic abundances with abundances derived from models of nucleosynthesis gives ages for the elements. The basic idea is that the heavy elements formed in the hot interiors of massive stars ($\geq 20 M_\odot$), which evolve rapidly and end in a supernova explosion. The basic mechanisms were outlined by G. Burbidge, E. Burbidge, W. Fowler, and F. Hoyle in 1957 [Burbidge et al. 1957]. Within this framework the heavy nuclei with mass numbers above Pb and Bi are only produced in the so-called r-process (rapid neutron capture). This is a process of neutron capture on a short timescale. An astrophysical environment is needed with a high neutron density of $\simeq 10^{20}\,\mathrm{cm}^{-3}$. Then the rapid synthesis occours of neutron-rich nuclei with about 20 to 30 more neutrons than would correspond to β-stability. The subsequent decay back to the stability line completes the formation of the r-process abundance, with characteristic peaks at neutron magic numbers, where the largest half-lives are encountered[1]. Most of the neutron-rich nuclei formed are unknown in the laboratory, and one has to rely on theoretical models to describe their properties. One of the possible sites for this process was thought to be the high entropy environment (hot bubble) in a type II supernova explosion. Recent detailed investigations have shaken the original optimistic outlook [Takahashi and Janka, 1997; Hofman et al. 1997]. At present a convincing astrophysical site for the r-process is not known.

In the early eighties [Krumlinde et al. 1981; Thielemann et al. 1983; Thielemann and Metzinger 1983; Cowan et al. 1991] the production ratios of the actinide chronometer pairs $^{232}\mathrm{Th}/^{238}\mathrm{U}$, $^{235}\mathrm{U}/^{238}\mathrm{U}$, $^{244}\mathrm{Pu}/^{238}\mathrm{U}$ were recalculated.

For the first time detailed β-strength functions (describing the population of the various levels in a β-decay product), and the effect of β-delayed fission (fission of a nucleus after β-decay) have been incorporated. It turns out that β-delayed fission efficiently terminates the r-process at proton numbers $Z \approx 92$ (Fig. 2.20 gives a diagram of r-process paths). Thus with the present knowledge of β-strength functions and fission barrier heights no superheavy elements can be formed in this process.

The long-lived Th, U, and Pb isotopes are built up by α- and β-decay chains

$$\begin{aligned}
{}^{232}\mathrm{Th}(t_{1/2} &= 1.405 \times 10^{10}\,\mathrm{y}) \\
{}^{238}\mathrm{U}(t_{1/2} &= 4.468 \times 10^{9}\,\mathrm{y}) \\
{}^{235}\mathrm{U}(t_{1/2} &= 7.038 \times 10^{8}\,\mathrm{y}) \\
{}^{244}\mathrm{Pu}(t_{1/2} &= 8.26 \quad \times 10^{7}\,\mathrm{y}).
\end{aligned}$$

The production ratios in these recent computations [Krumlinde et al. 1981; Thielemann 1984] are compared (Table 2.2) to the early estimates of [Fowler and Hoyle 1960].

These nucleosynthesis calculations are certainly a big improvement. In particular, the low abundance of $^{244}\mathrm{Pu}$ was not obtained in most of the previous investigations. Theoretical extrapolations are involved, however, into a regime away from that known experimentally. Thus these numbers are theory-dependent, and will possibly be revised with improved input data.

---

[1] For reviews see [Hillebrandt 1978; Truran et al. 1978; Thielemann et al. 1979; Cowan et al. 1983; Cowan et al. 1991].

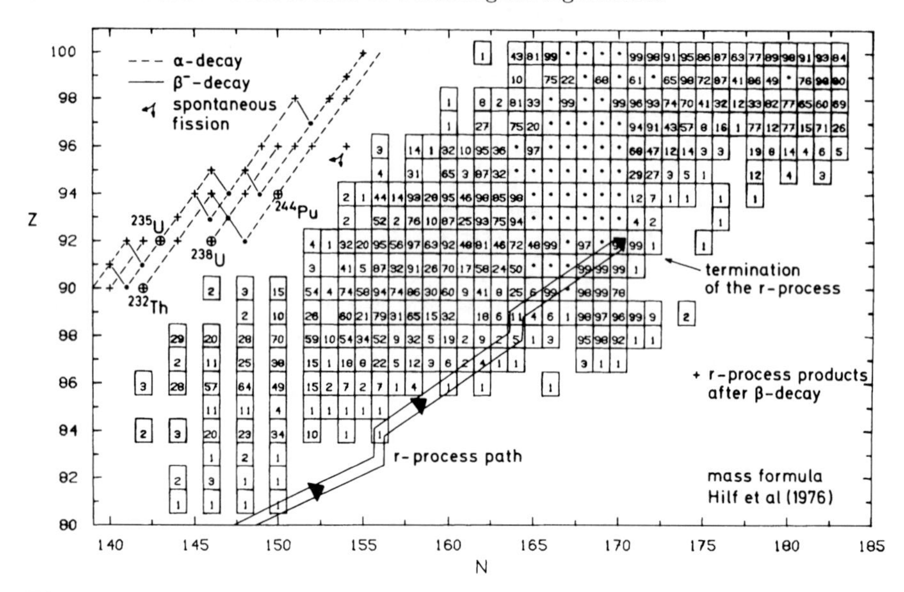

Fig. 2.20. The determination of the r-process by β-delayed fission at $Z = 92, N = 170$. β-decays of nuclei in the r-process path give rise to high-Z nuclei (marked "+"). Th, U, and Pu isotopes are built up by α- and β-decay chains. The percentage of fission in the daughter nucleus following the β-decay is given by the numbers in the boxes. *Dots* denote 100% fission. These (N, Z) positions act as a sink for the r-process [Thielemann et al. 1983]

Table 2.2. Production ratios of actinide pairs

^{232}Th/^{238}U	^{235}U/^{238}U	^{244}Pu/^{238}U	
1.65	1.65	(0.96)	(1960)
1.50	1.10	0.40	(1981)
1.39	1.24	0.12	(1983)

The production ratios together with the abundance ratios of these elements at the time of the formation of the Solar System can give us information about the duration of (r-process) nucleosynthesis in the Galaxy, and thus also about the age of the Galaxy. But to correlate these numbers a model for galactic evolution is necessary. Calculations of the production of these chronometric pairs of nuclides are fraught with uncertainties. In particular, as mentioned already, the lack of a convincing astrophysically occurring r-process scenario, as well as the fact that beyond ^{209}Bi only the chronometric nuclides themselves occur naturally, makes it almost impossible to estimate yield ratios with high accuracy [Takahashi 1998]. The determination of the age of the chemical elements is still very unreliable.

The evolution of the abundances of various nuclei is an interplay between their r-process production, their ejection into the interstellar medium by stellar winds, supernova and nova explosions, and their radioactive decay. The time dependence of the abundance $N_A(t)$ of a nucleus A can be expressed as the Laplace transform of a normalized production rate $P_A\psi(t)$ [Tinsley 1980; Thielemann 1984]

$$N_A(t) = \int_0^t \exp\{-(\lambda_A + \omega)(t - t')\}\, P_A\psi(t')dt', \tag{2.3}$$

where $\lambda_A \equiv \ln 2/t_{1/2}$ denotes the decay rate of nucleus A.

Dependence on the model adopted lies in the form assumed for the function $\psi(t)$ (the star formation rate), as well as in the value of ω, which gives the rate at which gas is used up in star formation. P_A is the r-process production rate. The history of nucleosynthesis with time may be schematically pictured as in Fig. 2.21. Some limits on such a wild and spiky behaviour of $\psi(t)$ can be set by observational constraints, but the models considered are actually very simple – perhaps too simple to be realistic.

A certain smoothing out of various irregularities can be achieved by considering the abundance ratios of pairs of nuclei A and B:

$$\frac{N_A(t)}{N_B(t)} = \frac{P_A}{P_B}\frac{f(\lambda_A;\omega,\psi,t)}{f(\lambda_B;\omega,\psi,t)}, \tag{2.4}$$

where

$$f(\lambda_A;\omega,\psi,t) \equiv \int_0^t \exp\{-(\lambda_A + \omega)(t - t')\}P_A\psi(t')dt'.$$

The function $\psi(t)$ exp (ωt) is an effective nucleosynthesis rate, which is often approximated by an exponential $\sim$ exp $(-\lambda_R t)$. The parameter λ_R can be limited by considering various models of galactic evolution [Thielemann and

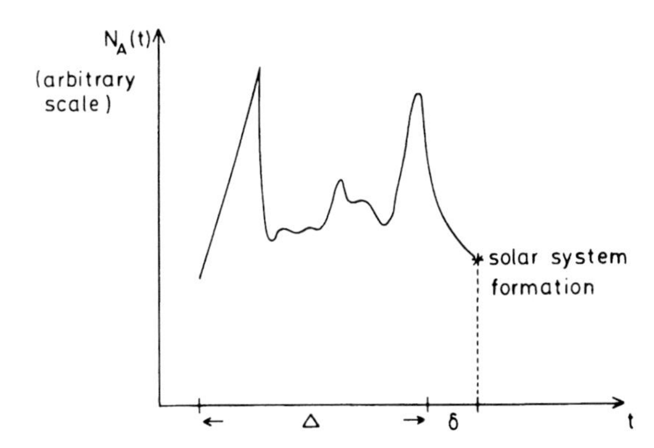

Fig. 2.21. Schematic picture of the evolution of nucleosynthesis with time

Truran 1985, Cowan et al. 1991]. It turns out that $-2 \times 10^{-10}\,\mathrm{y}^{-1} \leq \lambda_R \leq 0$ is a permitted range of λ_R in certain models. This value leads to a constant or slightly increasing rate of nucleosynthesis. The original "exponential model" (introduced by [Fowler 1972]), where continuous nucleosynthesis over a time Δ has been assumed with $\psi \sim \exp(-\lambda_R t)$, has been revised and refined several times [Thielemann et al. 1983; Thielemann and Truran 1985; Fowler and Meisl 1985; Takahashi 1998].

Calculated abundance ratios at the time of formation of the Solar System can then be fitted to the meteoritic abundances (see Table 2.3). In several models the actinide chronometers (cf. Table 2.2) are used, together with the pair $^{129}\mathrm{I}/^{127}\mathrm{I}$ (half-life 1.57×10^7 y), for which $P_A/P_B = 1.26$ is known from the empirical decomposition of solar abundances in r- and s-components [Käppeler et al. 1982]. The ratio of $^{129}\mathrm{I}/^{127}\mathrm{I}$ from meteorites is quite uncertain: (0.8–2.3) $\times 10^{-4}$.

It can be seen that new determinations have substantially reduced the $^{232}\mathrm{Th}/^{238}\mathrm{U}$ ratio to 2.32, or even 2.22, compared with earlier values of 2.50.

As a fifth pair the long-lived β-decay pair $^{187}\mathrm{Re} \rightarrow {}^{187}\mathrm{Os}$ (half-life $\sim$ 42 Gyr (Gyr $= 10^9$ years)) has been, and is considered for cosmo-chronometry [Thielemann and Truran 1985; Takahashi 1998]. This pair has the advantage that it is somewhat independent of r-process models, because the post-r-process cascades produce $^{187}\mathrm{Re}$ but not $^{187}\mathrm{Os}$ directly. $^{187}\mathrm{Os}$ is produced in the s-process, as well as by the β-decay of the unstable isotope $^{187}\mathrm{Re}$. The use of this pair as a chronometer requires an estimate of the s-process component in the observed $^{187}\mathrm{Os}$ abundance. The component from the decay of $^{187}\mathrm{Re}$ is the residual. The s-process can be modelled much better than the r-process, because it is non-explosive in nature, and its astrophysical sites are known. The calculation of the amount of $^{187}\mathrm{Os}$ created in the s-process has to take into account the first excited state ($^{187}\mathrm{Os}^*$) that will be populated in hot stellar interiors, where the temperatures are above the energy difference of 9.75 keV. But also in stellar interiors the decay of $^{187}\mathrm{Re}$ could be enormously enhanced, mainly because of bound-state β-decay to $^{187}\mathrm{Os}^*$. The β-decay half-life of fully ionized $^{187}\mathrm{Re}$ has recently been measured and found to be (32.9 ± 2.0) years [Bosch et al. 1996], more than one billion times shorter than the decay time of the neutral $^{187}\mathrm{Re}$. This surprising result establishes, on the other hand, a reliable basis for a calculation of the effects of stellar evolution

Table 2.3. Meteorite abundance ratios

$^{235}\mathrm{U}/^{238}\mathrm{U}$	$^{232}\mathrm{Th}/^{238}\mathrm{U}$	$^{244}\mathrm{Pu}/^{238}\mathrm{U}$	
0.317	2.48	0.005	[Symbalisty and Schramm 1981]
0.317	2.32	0.005	[Anders and Ebihara 1982]
0.315	2.22	0.005	[Cameron 1982]
		0.0068 ± 0.0019	[Hudson et al. 1984]

on the net cosmic abundance of ^{187}Re. A part of the cosmic ^{187}Re embedded in a new-born star could decay rapidly via ^{187}Os* into ^{187}Os, and would therefore exist as ^{187}Os in the ejecta, but also the transformation of ^{187}Os into ^{187}Re by electron capture could occur. Such "astration" effects can now be carefully estimated. [Takahashi 1998] describes a chemical evolution model [Yokoi et al. 1983] adapted to these new calculations. He derives an estimate of $(15 \pm 4) \times 10^9$ years for the age of the Galaxy, and looks optimistically towards further improvements.

This agrees quite well with estimates from other models which lead to a range of values for the age of the Galaxy:

$$11.8 \times 10^9\,\mathrm{y} \leq t_G \leq 19.8 \times 10^9\,\mathrm{y}. \tag{2.5}$$

Deplorable as the large uncertainties may be, the cosmologically important information that the mater around us has a finite age of between 12 and 20 billion years can be accepted.

2.1.3 The Age of Globular Clusters

The age of the chemical elements relies on precise measurements of radionuclides in the solar system, and on rather uncertain models of the chemical evolution of the Galaxy. Another, hopefully more reliable approach to setting limits on the age of the universe is to measure the age of the oldest stars. For white dwarfs, cooling ages can be estimated, from the evolutionary ages of globular cluster stars. Assuming an average mass of $0.6 M_\odot$ for the faintest white dwarfs, a minimum age of 8 or 12 billion years is obtained, depending on whether the white dwarf has a carbon or an oxygen core. The age of the oldest globular cluster is probably higher than the white dwarf cooling ages.

Even with a small telescope various star cluster can be seen in the Milky Way. Besides the less concentrated open star clusters, there are many highly concentrated, spherical assemblies of stars, the globular clusters. The catalogues of Messier (1784, M) and of Dreyer (1890, NGC) contain many such stellar associations. At the present time about 650 open clusters (among them the Hyades and the Pleiades) – each containing from 20 to about 1000 stars – and about 130 globular clusters – each containing 10^5 to 10^7 stars – are known in our Galaxy.

In neighbouring galaxies (M31, LMC) globular clusters have also been catalogued. In some cases a large number seems to be present, e.g. 15 000 for M87 [van den Bergh 1983]. Even in the spectra of very distant galaxies there is evidence for the existence of globular clusters. A typical diameter is ≈ 30 pc, so their star density is 1000 times larger than the density of stars in the vicinity of the Sun. The globular clusters in our Galaxy are distributed in a spherical halo with a typical radius of 15 kpc. These systems contain the oldest stars in our galaxy – they condensed in an early collapse phase of the pregalactic nebula.

In Fig. 2.22 the HR diagram for the globular cluster M15 is shown [after Salaris et al. 1997; Durrel and Harris 1993]. Clearly, the distribution of stars does not follow the main sequence all the way to the upper left of the HR diagram, but it branches off towards the right. Above a certain temperature the star clusters no longer contain main-sequence stars. Instead, the stars of higher luminosity populate a branch to the right of the HR main sequence, the red-giant branch. The assumption that all stars in a cluster were born at the same time is probably justified, and therefore the luminosity (or mass) of the stars sitting exactly at the turn-off point on the main sequence (MSTO) can determine the age of the cluster via the formulae given in the introduction (on p. 33). This sounds easy enough. The difficulties in precisely determining the relevant parameters are enormous. To determine observationally the luminosity of the turn-off position requires a knowledge of the distance. This can be bypassed by fitting sequences of stellar evolution models to the main sequence of the cluster. There are several problems with this procedure: there are uncertainties connected with the physics of the stars (e.g. the amount of convection present), and in addition the theoretical zero-age main sequence depends, in its position in the HR diagram, on the original metal and He content of the stars. Thus a star with the mass of the Sun, but with a lower content of metals, would spend less than 10 billion years on the main sequence. The main sequence turn-off is difficult to determine, distance estimates are uncertain, and the observational determination requires a precise measurement of the true colour of the faint, globular-cluster main

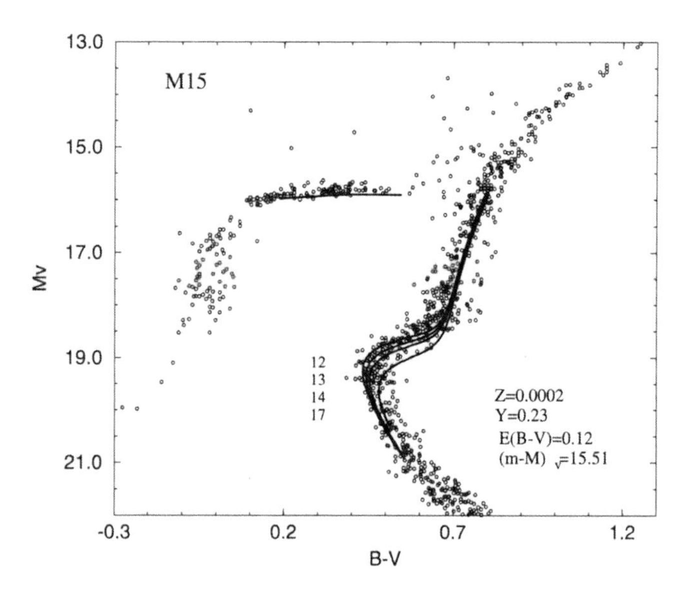

Fig. 2.22. Isochrones for different ages (given in Gyr in the figure) are fitted to the colour–magnitude diagram of the globular cluster M15. Composition, reddening, and distance modulus are displayed in the lower right corner

sequence stars, i.e. it also requires a knowledge of the reddening effect of the gas and dust between us and the object.

Since all the globular clusters are very distant, their main sequence is very faint. The best cases have apparent magnitudes of around 20^m to 19^m. Therefore it was only in 1952 that the first main sequences of globular clusters were measured, at Mt. Wilson and Palomar Observatories, by a group of young astronomers (among them A. Sandage and H. Arp) under the supervision of W. Baade. Even today only a few globular clusters have had their main sequence photometrically determined. But the new telescopes, especially the Hubble Space Telescope, are quickly improving the situation.

The range in ages is not very large, as the physics and the chemical composition are varied (not more than 15%) (see [Chaboyer 1995]) but the distance uncertainties introduce the biggest errors. A variation of 0.1 magnitudes in the distance modulus results in an age variation of $\approx$ 1.2 to 1.7 billion years (e.g. [Salaris et al. 1997]).

There are two approaches to an empirical distance determination: a fit to nearby main sequence stars, or the use of RR Lyrae stars as distance indicators. Only the second method seems capable of high enough accuracy. RR Lyrae stars burn helium in their cores, and they become unstable against pulsations in a certain temperature range. These helium burning stars populate a strip of almost constant luminosity to the left of the red giants in a magnitude–colour diagram of a globular cluster (see the example of M68 in Fig. 2.23). Their absolute magnitude can be expressed as an empirical relation of the form

$$M_v(RR) = \alpha[\mathrm{Fe/H}] + b, \tag{2.6}$$

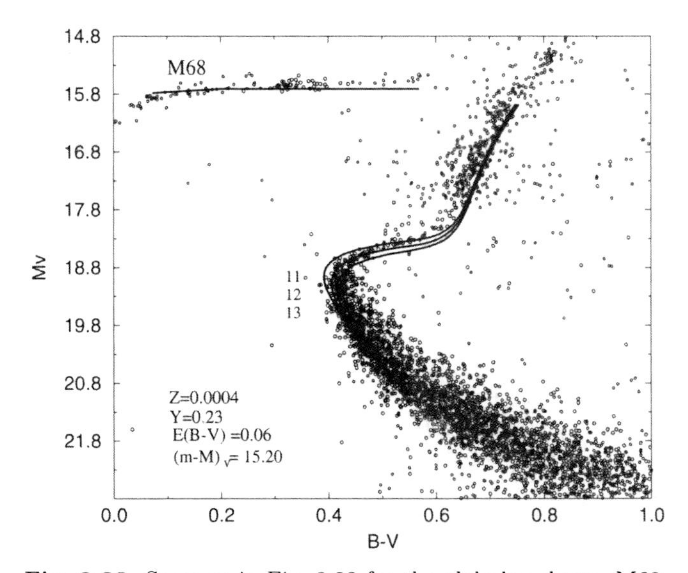

Fig. 2.23. Same as in Fig. 2.22 for the globular cluster M68

where [Fe/H] is the logarithm of the ratio of $N(\mathrm{Fe})/N(\mathrm{H})$ in solar units:

$$[\mathrm{Fe/H}] \equiv \ln\left[N(\mathrm{Fe})/N(\mathrm{H})\right] - \ln\left[N(\mathrm{Fe})/N(\mathrm{H})\right]_{\odot} \tag{2.7}$$

One such relation is [Clementini et al. 1995]

$$M_{\mathrm{v}}(RR) = (0.19 \pm 0.03)[\mathrm{Fe/H}] + (0.96 \pm 0.04). \tag{2.8}$$

The coefficients range – in different investigations – from 0.08 to 0.30 for α, and from 0.73 to 1.06 for b. As a consequence the ages obtained for globular clusters can vary considerably [Chaboyer et al. 1996]: for M68 different relations (2.6) lead to ages between 12.9 Gyr and 17.3 Gyr. Ignoring the uncertainty in α and b, as is often done, clearly limits the range of ages that can be obtained by this method, but the attentive reader will realize that the true uncertainty cannot be reduced so simply.

A relation with fixed coefficient has often been adopted ([Chaboyer et al. 1996] use $\alpha = 0.20$ and $b = 0.98$). Then M_{v} compared to the apparent luminosity of the cluster RR Lyrae star gives a distance modulus.

First results of trigonometric parallaxes of Galactic Cepheid stars directly obtained from observations made by the Hipparcos astrometry satellite [ESA 1997; Feast and Catchpole 1997] are of major importance for the calibration of the extragalactic distance scale. But they also influence the RR Lyrae measurements.

The zero point of the Cepheid period–luminosity relation was recalibrated, and a distance modulus for the Large Magellanic Cloud was derived, and applied to RR Lyrae observations in LMC globular clusters.

After correction for metallicity effects, a distance modulus of 18.70 ± 0.10 was derived for the LMC [Feast and Catchpole 1997]. This is almost 0.20 mag higher than the value of 18.50 commonly used up to now. The derivation has come under close scrutiny [Oudmaijer et al. 1998], because of a bias present in the Hipparcos star catalogue. The claim is that we more or less return to the old value of $(m - M)_{\mathrm{LMC}} = 18.50$ if the bias is properly accounted for (see below). For the RR Lyrae variable in the LMC this means $M_{\mathrm{v}}(RR) = 0.25 \pm 0.10$ at a metallicity $[\mathrm{Fe/H}] = -1.9$ for the new distance modulus and $M_{\mathrm{v}}(RR) = 0.45 \pm 0.10$ for the old value.

The age for the oldest globular cluster (mean $[\mathrm{Fe/H} = -1.9]$) has been derived as 14.56 Gyr [Chaboyer et al. 1996] based on $M_{\mathrm{v}}(RR) = 0.6$ mag at that metallicity. The new calibration increases the absolute magnitude of RR Lyrae stars, consequently gives a higher estimate of the MSTO luminosity of globular clusters, and therefore a reduction of the age. The revised ages for the oldest globular clusters are $\simeq 12$ Gyr.

This supports nicely the – presently still controversial – result that the age for globular clusters must be reduced substantially. This argument depends chiefly on two improvements: new improved opacities and equations of state were used for the stellar models, and the magnitude difference $\Delta(V)$ between the horizontal branch (HB) helium stars and the MSTO was computed consistently from the same theoretical stellar models as a function of age. The quantity $\Delta(V)$ is definitely a function of age, because the magnitude of the MSTO changes with age, while the magnitude of the HB stars is constant for the "zero age" helium stars, i.e. at the onset of the helium burning [Salaris et al. 1997]. The data in Figs. 2.22 and 2.23 also lend support to this idea of a constant luminosity for the HB helium stars in a globular cluster. The advantage of this method is that it relies just on the comparison of the observed $\Delta(V)$ with the observational quantity – it is independent of the distance. In fact, the computed HB luminosity can be compared to empirical fits to give a further test of models.

In Fig. 2.23 the colour–magnitude diagram of M68 is shown together with theoretical fits to the main sequence, red giant branch, and HB from the same stellar evolution models. An age of (12 ± 2) Gyr for M68 is derived using the $\Delta(V)$ method, considerably lower than previous estimates, and in agreement with the new Cepheid/RR Lyrae zero-point calibration [Salaris et al. 1997]. A slightly higher original helium content ($Y = 0.24$) and an increase of the metallicity by a factor of 2 ($z = 0.0004$) reduce the age by about 1 Gyr. (In astrophysical contexts "metallicity" is the total mass fraction of all elements except H and He. The solar metallicity, $Z_{\odot}$, is about 2%.)

In principle it is possible to determine the absolute age of M15 by the same method, but the errors are much larger: As discussed in [Salaris et al. 1997] the exact M_{v} of the MSTO lies between 19.2 and 19.4 which corresponds to the typical observational error of 0.001 mag for the (B-V) value of the turn-off. This uncertainty of 0.2 mag in the MSTO luminosity translates into an uncertainty age of approximately 3 Gyr around a mean value of 14 Gyr.

Therefore [Salaris et al. 1997] propose a method to derive the relative age of M15 with respect to M68 which goes back to [Vandenberg et al. 1990]. It relies on the possibility of deriving with high accuracy the relative ages of clusters with approximatly the same metallicity by considering the difference in (B-V) between the MSTO (where the almost vertical distribution of data points defines the colour very well; compare Figs. 2.22 and 2.23) and the base of the red giant branch (RGB). It is possible to conclude from a comparison between M68 and M15 that both clusters have the same age (within ~ 1 Gyr).

The cluster M92 which had been considerd the oldest so far with age estimates of up to 17 Gyr also seems to be of the same age. Although there is still some discussion among workers in the field it seems safe to conclude that new physics and better observational data lead to a substantial reduction in the age estimates for the globular clusters. Whether the range of values of 12 ± 2 Gyr is the last word remains to be seen – maybe a somewhat larger uncertainty would be appropriate – but in the following we shall use this estimate:

$$t_{g.c.} = (12 \pm 2) \times 10^9\,\text{y}. \tag{2.9}$$

Estimates of the age of the Galactic Disc by means of the luminosity function of the white dwarfs in the solar neighbourhood [Hernanz et al. 1994] give similar values: 10–12 Gyr. The interesting implication would be that the Galactic Disc began to form without a substantial time delay with respect to the Halo.

It is an open question whether another 1 Gyr should be added to $t_{g.c.}$ to allow enough time for the formation of the galaxy. This seems reasonable, but in some models of structure formation globular clusters can form very quickly, within 10^8 years. So, let us take (2.9) also as an estimate for the age of the universe.

2.2 The Hubble Constant H_0 – How Big is the Universe?

> "The search will continue. Not until the empirical resources are exhausted need we pass on to the dreamy realms of speculation."
>
> *E. Hubble, "The Realm of the Nebulae"*

Observations of the redshifted spectral lines of distant galaxies by Slipher, Wirtz, and Hubble, have led to the view of an expanding universe [Hubble 1929]. (An important theoretical argument was the demonstration [Weyl 1923] that the redshift is a coordinate-independent quantity.) The determination of the rate of this expansion has been a major undertaking of many astronomers over the past 60 years. The Hubble constant H_0 determines the size of the observable universe – at least in the context of the FL models. Of course, it is everybody's ambition to measure such a fundamental quantity as precisely as possible. The large value of $H_0 \sim 500\,\mathrm{km\,s^{-1}\,Mpc^{-1}}$ originally obtained by Hubble was even then in conflict with age determinations, and led to a critical attitude the big-bang models. Subsequent improved measurements, and especially the new lay-out of the distance ladder by Baade's recalibration of the Cepheid variables in 1952 (which increased cosmic distances roughly by a factor of 2), have led to values between $H_0 = 50$ and $H_0 = 100\,\mathrm{km\,s^{-1}\,Mpc^{-1}}$. There is still a lively debate about whether the true value of H_0 is small or large.

Weak characters liable to accept a mean value with appropriate uncertainties have no place in this battle between astronomers.

The range of values obtained reflects the difficulties in establishing reliable distances on cosmic scales. In principle, the situation is quite clear: if a galaxy follows the universal expansion perfectly, then in linear approximation (of a formula such as (1.8) for r_A)

$$cz = dH_0,$$

where z is the redshift, $1 + z = R_0/R$, and d is the distance.

The true value of the Hubble constant is still the subject of lively debate

The problem with a precise determination of H_0 is that for any single object it is almost certain that it does not follow exactly the overall expansion. All galaxies show peculiar motions of a few $100\,\mathrm{km\,s^{-1}}$, so

$$cz = dH_0 + v_p. \tag{2.10}$$

These peculiar motions occur because the galaxies are organized in large-scale structures which influence the velocity of their members. For instance, there seems to be a systematic deviation from the Hubble flow for a system containing our Local Group and the Virgo cluster at least. The infall of our galaxy towards the centre of the Virgo cluster makes it difficult to estimate the cosmic velocity of the centre of the Virgo cluster.

The peculiar velocities are comparable to dH_0 for nearby galaxies, and at distances where v_p becomes negligible the distance determination becomes increasingly uncertain.

This is the reason why for many years there was a split among observers in favour of a large distance scale ($H_0 \simeq 50\,\mathrm{km\,s^{-1}\,Mpc^{-1}}$) and those advocating a short distance scale ($H_0 \simeq 100\,\mathrm{km\,s^{-1}\,Mpc^{-1}}$). It traces back to the historic discrepancy between [Sandage and Tammann 1974, 1976, 1982a,b, 1984a] ($H_0 = 50$) and [de Vaucouleurs 1982] ($H_0 = 100$), both presenting their results usually with small error bars (less than 10%).

Let us look at a few salient points of the observational precedure. The problems of defining a scale over distances where the Cepheid variables are no longer visible, and where some statistical property (brightest object in a cluster, brightest red star in a spiral galaxy, etc.) has to be used to give a handle on the intrinsic luminosity, will be discussed below. The obvious observational difficulties of the "K" effect and the "aperture" effect are treated at present by all serious groups in a standard way:

- The finite bandwidth of receivers leads one to observe completely different parts of the spectrum in a strongly redshifted source, as compared with a source with a small redshift. The correction of this effect (the so-called "K" correction) can be done if the intrinsic spectral shape at the shorter wavelengths is known, either through theoretical estimates or through UV observations.
- Faint, extended objects (large m) almost merge into the background, and it is difficult to estimate where their edges are (the "aperture" effect). Corrections can be found by observing at different sensitivities.

These difficulties far outweigh the approximations usually made with respect to the cosmological model: in general, a Newtonian picture is used to accommodate the astronomical observations, where the distance d_L to a galaxy is given just by the ratio of its luminosity L and the flux F received: $d_L^2 = r_d^2 = L/4\pi F$. As has been shown in Chap. 1, this implies that $d_L = (1+z)^2 r_A$. In addition the "Newtonian model" implies a projection onto the surface $t_0 = \text{const.}$ – the world is measured as it is now, whereas the sources are really seen as they cross our past light-cone. Only a mixture of history and spatial distribution can be determined. Within an FL model it is trivial to apply the correction $d_L \to d_L(1+z)^{-2}$, but we should nevertheless keep in mind that "distance" is a theoretical concept, not a directly observable quantity, as are angles, fluxes and redshifts.

It is perhaps useful to keep referring to a simple picture, and therefore let us use the model with $K = 0, \Omega_0 = 1, p = 0$, and $\Lambda = 0$. In this Einstein–de Sitter model the Hubble constant is related to the present epoch by $H_0 = \frac{2}{3} t_0^{-1}$.

The part of the universe that is observable has a coordinate radius $r_H = \int_0^{t_0} dt/R(t)$ (along the light-cone $ds^2 = 0$; for radial directions then $dt = \pm R(t) dr$) given by

$$r_H = \frac{3t_0}{R_0}. \tag{2.11}$$

Referred to the present epoch t_0, the metric distance is

$$d_H = R_0 r_H = 3t_0 = 2H_0^{-1}$$

(actually $d_H = 2c/H_0$, but $c = 1$ as usual).

This is the radius of the particle horizon at the present time. The area distance r_A of a source at a redshift z is

$$r_A \equiv d_e = \frac{d_H}{(1+z)}\left(1 - \frac{1}{\sqrt{1+z}}\right), \qquad (2.12)$$

where d_e is the metric distance at the time of emission. At the present epoch this distance corresponds to

$$d_0 = d_H\left(1 - \frac{1}{\sqrt{1+z}}\right). \qquad (2.13)$$

The radius of the observable region, the "Hubble sphere" $d_H = 2/H_0$ is 12 000 Mpc for $H_0 = 50$, and 6000 Mpc for $H_0 = 100$. In what follows we want to refer the measurements to this sphere, and to indicate what fraction is covered by the different surveys. A few examples are sketched in Fig. 2.24.

The relation between flux, luminosity, r_A, H_0, q_0, z (cf. (1.24) and (1.25)) is translated into a relation for the logarithmic "distance modulus" $m - M$, expanded for $z \ll 1$, and using $H_0 = h_0 \cdot 100\,\mathrm{km\,s^{-1}\,Mpc^{-1}}$, written as

$$m - M = 48.65 - 5\log h_o + 5\log z + 1.086(1 - q_0)z. \qquad (2.14)$$

By looking at some objects with the same M but different m, the straight line of m vs. log z can be determined, and H_0 can be found, provided

$$\mid 1 - q_0 \mid \Delta \ll \Delta(m - 5\log cz).$$

In Fig. 2.25 the famous diagram plotting the 82 brightest galaxies in clusters [Sandage 1972] is reproduced – with a line of constant slope 0.2, which shows that these objects seem to follow nicely a linear Hubble law. The diagram covers a range of redshifts from 0.003, the very local region, to 0.3, about 12% of the Hubble radius. The value of the Hubble constant obtained from this analysis is $H_0 \sim 55\,\mathrm{km\,s^{-1}\,Mpc^{-1}}$.

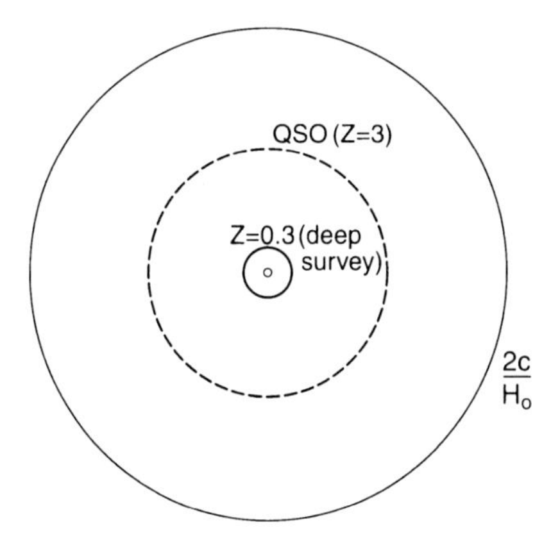

Fig. 2.24. Schematic representation of the depth of various astronomical observations, using (2.13). The epoch of recombination at $z = 10^3$ covers 97% of $2c/H_0$. A QSO (short for quasistellar object ≡ quasar) with $z = 3$ has d_0/d_H=0.5. Deep surveys $z \sim 0.3$ cover 12% of the Hubble radius. Radio source counts give model estimates of $z \leq 10$, with a typical z at 1 to 3

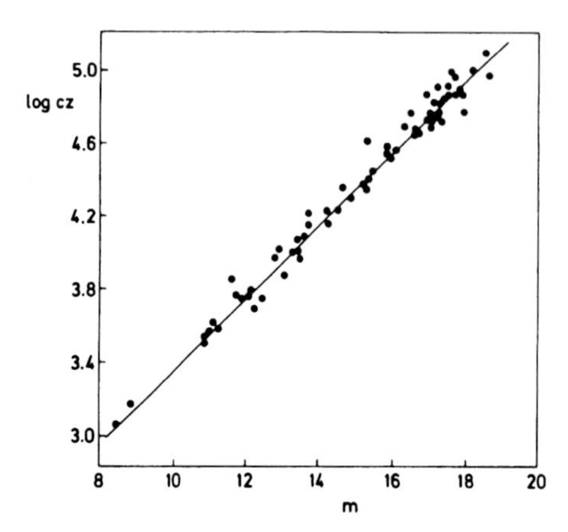

Fig. 2.25. Correlation of apparent magnitude of first ranked cluster galaxies with redshift cz of their parent clusters [Sandage 1972] ($c = 3 \times 10^5\,\mathrm{km\,s^{-1}}$)

The fundamental early work has been criticized mainly for the fact that it rested very precariously on distance indicators which cannot be used reliably as standard candles; e.g. [de Vaucouleurs 1981]. This lively debate has gone on a long time, and various points have been clarified in the course of the last 20 years (for an excellent review see [Rowan-Robinson 1985]).

2.2.1 Cepheids in Virgo

The situation was expected to improve dramatically with observations of the HST extending the Cepheid scale beyond 20 Mpc, i.e. out to the distance of the Virgo cluster. This programme has now produced its first results, and the distances to various galaxies in the Virgo cluster and in the Local Supercluster have been obtained.

The spiral galaxy M100 was observed especially thoroughly, with more than 50 Cepheids being analysed [Ferrarese et al. 1996]. With observations in the V-band (5500 Å) and the I-band (8000 Å) using the HST, a correction can be made for the effect of interstellar dust extinction (both internal obscuration in M100, and foreground effects due to dust in the Milky Way). A distance to M100 is thus obtained [Freedman et al. 1994a,1994b]:

$$d = 15.8 \pm 1.5\,\mathrm{Mpc}.$$

It is customary among astronomers to give distances in terms of the distance modulus. Then for M100:

$$(m - M) = 31.16 \pm 0.15,$$

the relation to the distance in Mpc being

$$d = 10^{-5+0.2(m-M)}\mathrm{Mpc}.$$

A few of these distance determinations are shown in Table 2.4.

Table 2.4. Cepheid Distances to galaxies in the Local Supercluster

Galaxy	Association	(m-M)	Authors
M100	Virgo	31.16 ± 0.15	[Freedman et al. 1994a]
NGC4571	Virgo	30.87 ± 0.15	[Pierce et al. 1994]
NGC4496	Virgo	31.13 ± 0.10	[Saha et al. 1997]
NGC4536	Virgo	31.10 ± 0.13	[Saha et al. 1997]
NGC4639	Virgo	32.00 ± 0.23	[Sandage et al. 1996]

The errors given here reflect the uncertainties in the Cepheid zero point, and in the photometric calibration. The calibration of the Cepheid "period–luminosity–colour" relation has been achieved now by direct trigonometric parallax measurements of Galactic Cepheids by the satellite Hipparcos [ESA 1997; Feast and Catchpole 1997]. The true absolute magnitude of the Cepheids follows the relation

$$\langle M_{\rm v} \rangle = -2.81 \log P - 1.43$$

as derived by [Feast and Catchpole 1997]. After correction for metallicity effects a distance modulus of 18.70 ± 0.10 for the Large Magellanic Cloud was obtained. This is 0.2 mag higher than the value of 18.50 commonly used. There is, however, a bias that must be considered when the Hipparcos results are applied. The cause of this so-called Lutz–Kelker bias [Lutz and Kelker 1973] is easy to see: a parallax π with a measurement error σ_π gives a certain distance d with an interval of uncertainty around it. Stars at a smaller distance and stars located further away can both be estimated to lie within the error at distance d. Since there are more stars outside than inside the distance range, more stars further away will be assumed to be at distance d. This effect causes a systematic bias such that measured parallaxes will on average result in a distance underestimate. In the case of the Hipparcos Cepheids [Oudmaijer et al. 1998; Groenewegen and de Jong 1998] the attempt to correct for this bias leads to a distance modulus of 18.56 ± 0.08 for the LMC, consistent with previous determinations. This LMC calibration is then used to determine Cepheid distances in the Virgo cluster.

While the distances are quite accurately known, it is uncertain whether these spiral galaxies give a correct estimate of the centre of mass of the Virgo cluster. Their distribution is extended and complex, and it does not correspond well to the X-ray emission maps of the Virgo cluster. There is the danger that these spirals are outlying members of the cluster, and the distance determination may carry a large systematic error. A cautious estimate seems to be [Mould et al. 1995]

$$d_{\rm Virgo} = (17.1 \pm 3.4)\,{\rm Mpc}.$$

This is a rather large uncertainty, and the value of H_0 measured by this approach is therefore also not as well determined as astronomers might wish. The Virgo cluster is close to our Local Group of galaxies which are attracted towards it. The infall pattern is not well known, and thus the true cosmic velocity of Virgo is also ill determined. This problem can be side-stepped in an elegant way by making use of distance determinations such us the Tully–Fisher or the fundamental plane method (see below).

Both methods can be used to obtain quite accurate relative distance values between Coma and Virgo. This leads to [Mould et al. 1995; Freedman et al. 1994a]

$$(m - M)_{\text{Coma}} - (m - M)_{\text{Virgo}} = 3.71 \pm 0.05.$$

Using $(m - M)_{\text{Virgo}} = 31.04 \pm 0.2$, the distance to Coma is determined as

$$(m - M)_{\text{Coma}} = 34.75 \pm 0.21$$

or

$$d_{\text{Coma}} = (89 \pm 9)\,\text{Mpc}.$$

At such distances the peculiar velocities are small compared to the Hubble expansion velocities, and thus the redshift of the Coma cluster corresponds to a value

$$v_{\text{Coma}} = 7146 \pm 86\,\text{km}\,\text{s}^{-1}.$$

The Hubble constant $H_0 = v_{\text{Coma}}/d_{\text{Coma}}$ is then

$$H_0 = (80 \pm 8)\,\text{km}\,\text{s}^{-1}\,\text{Mpc}^{-1}.$$

Taking into account an estimate of the systematic errors, a value

$$H_0 = (80 \pm 6 \pm 16)\,\text{km}\,\text{s}^{-1}\,\text{Mpc}^{-1}$$

is given [Mould et al. 1995].

Assuming a recession velocity of $1179 \pm 17\,\text{km}\,\text{s}^{-1}$ for Virgo [Jerjen and Tammann 1993] the value for the Hubble constant turns out to be

$$H_0 = (66 \pm 14)\,\text{km}\,\text{s}^{-1}\text{Mpc}^{-1}.$$

The mean values look quite different, but these H_0 measurements are really within each other's error bars. That is real progress! The variations in these estimates are due to systematic uncertainties:

- reddening corrections;
- zero point of the Cepheid PL relation;
- position of the centre of mass of Virgo;
- adopted recession velocity of Virgo.

The cosmic velocity of the Virgo cluster derived from these measurements

$$v_{\mathrm{Virgo}} = \frac{v_{\mathrm{Coma}}}{d_{\mathrm{Coma}}} \simeq (1400 \pm 400)\,\mathrm{km\,s^{-1}},$$

covers the values obtained in other astronomical observations. The mean value of $1400\,\mathrm{km\,s^{-1}}$ is rather high, however, compared to values of 1000 to $1200\,\mathrm{km\,s^{-1}}$ usually quoted. The mean value of $H_0 = 80\,\mathrm{km\,s^{-1}\,Mpc^{-1}}$ should therefore also be regarded with some caution. At this stage we must admit that the distance ladder out to the Virgo cluster does not seem a promising way to obtain a value of H_0 with errors less then 10%.

2.2.2 Type Ia Supernovae as Standard Candles

Supernovae of type Ia (SNIa) are very bright, and can be observed out to very great distances. (In Fig. 2.26 the light curves of 10 high-redshift supernovae are shown. For a review see [Leibundgut 2001].) Thus, if they are good standard candles the Virgo distance may be jumped over, and the true cosmic velocity field may be measured directly. SNIa are generally thought to be thermonuclear explosions of white dwarfs consisting mainly of carbon and oxygen. In all theoretical scenarios the element ^{56}Ni is synthesized in the explosion [Arnett et al. 1985; Niemeyer et al. 1996]. The optical luminosity of the supernova should – in theory – result from the thermalization of γ-rays and positrons, which are produced by the decay of ^{56}Ni to ^{56}Co to ^{56}Fe.

It is quite clear now that SNIa are not standard candles in the sense that they all have the same maximum absolute magnitude. The observations show that there is a substantial range (by a factor of 10) in the luminosities at maximum. Van den Bergh [Van den Bergh 1996] gives a range for the blue magnitude at maximum

$$-17.3 \leq M_B(\mathrm{max}) \leq -19.61$$

from nine SNIa which have a distance determined by the Cepheid method (Table 2.5).

A spread in the peak luminosity is not unexpected, because at present several theoretical scenarios are under discussion all of which seem to allow for a certain intrinsic variation of properties.

The model of the complete explosion of a white dwarf which collapses when it exceeds the Chandrasekhar mass limit is no longer considered adequate. Too many heavy elements would be produced in a supernova explosion of this type. More detailed models are now being worked out, where a nuclear burning front is ignited which moves through the white dwarf and eventually leads to a late or delayed detonation. White dwarfs, which are pushed beyond the Chandrasekhar limit by accretion, and then become unstable, merging white dwarfs, or white dwarfs below the Chandrasekhar limit, where a burning front is triggered by the detonation of Helium at the surface

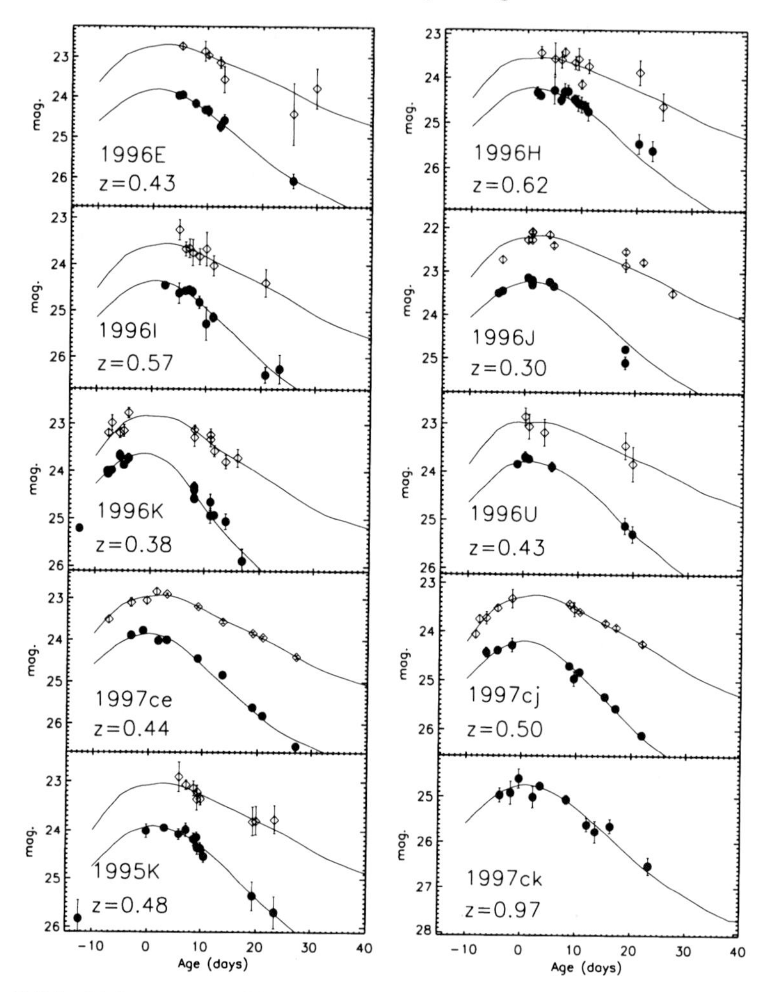

Fig. 2.26. Light curves of high-redshift SNIa. B (*filled symbols*) and V (*open symbols*) photometry is shown with B increased by 1 mag for ease of view. The *lines* are empirical model fits to the data, making use of the strong correlation between maximum luminosity and decline of the light curve ("MLCS" model) [from Riess et al. 1998]

– all these are possible theoretical scenarios [Höflich et al. 1996; Niemeyer et al. 1996; Hillebrandt and Niemeyer 2000]. The amount of ^{56}Ni produced in these models depends on the flame speed in the deflagration models, on the location of the burning front and of the delayed detonation, and on the mass of the white dwarf in sub-Chandrasekhar mass models. A substantial intrinsic variation of the amount of ^{56}Ni produced seems quite natural.

In addition, the calculation of the maximum luminosity makes use of intricate radiative transfer calculations in the expanding supernova shell,

Table 2.5. Cepheid distances to SNIa [van den Bergh 1996]

SN	Galaxy	Type	$(m - M)$	$M_B(\max)$	Δm_{15}
1885	M31	Sb	24.3 ± 0.1	-17.33	2.1
1937C	IC4182	Ir	28.36 ± 0.11	-19.53 ± 0.16	0.95
1960F	NGC 4496	SBc	31.1 ± 0.15	-20.40 ± 0.79	1.06
1972E	NGC5253	Pec	28.10 ± 0.12	-19.55 ± 0.21	0.94
1981B	NGC4536	Sc	31.10 ± 0.13	-19.61 ± 0.24	1.1
Virgo Cluster (8 SN in E and SO gal. before 1991)			31.02 ± 0.2	-18.76 ± 0.24	$\langle 1.1 \rangle$
1990N	NGC 4639	Sb	32.00 ± 0.23	-19.43 ± 0.23	1.01
1991T	NGC 4527	Sb	31.05 ± 0.15	-19.99 ± 0.21	0.94
1991bg	NGC 4374	E1	31.02 ± 0.2	-16.27 ± 0.2	1.95
1994D	NGC 4526	SO	31.02 ± 0.2	-19.43 ± 0.22	1.26

Δm_{15} is the decrease in magnitude 15 days after the peak luminosity was reached

where again parameters may change drastically with different empirical or theoretical input [Höflich et al. 1996].

But there is also a very helpful observational fact: the shape of the SNIa light curves seems to be correlated to $M_B(\max)$. The fast decaying ones are usually fainter, and the slow ones are in most cases more luminous at maximum. This seems to be a strong correlation, and one may use it to correct the standard candle assumption of a unique maximum luminosity. It has been shown [Riess et al. 1995] that such an empirical correction can reduce the scatter in estimated distances: for an independent sample out to $cz = 30\,000\,\mathrm{km\,s^{-1}}$ the scatter was reduced from 0.5 mag to 0.15 mag (see Fig. 2.27). This so-called "MLCS" method has been used in the same way for a large sample of near $(0.01 < z < 0.13)$ and far $(0.16 < z < 0.67)$ supernovae (see Fig. 2.28).

The dispersion of $M_B(\max)$ values in Table 2.5 is quite large. Even if the peculiar object SN 1991bg and SN 1960F (with uncertain $M_B(\max)$) are excluded, there still remains a range from $M_B(\max) = -19.61$ for SN 1981B to $M_B(\max) = -17.3$ for SN 1885.

The light curve shape method (MLCS) uses 1981B, 1972E, and 1990N for the calibration of SNIa distances [Riess et al. 1998]. It would, of course, be desirable to have more calibrators.

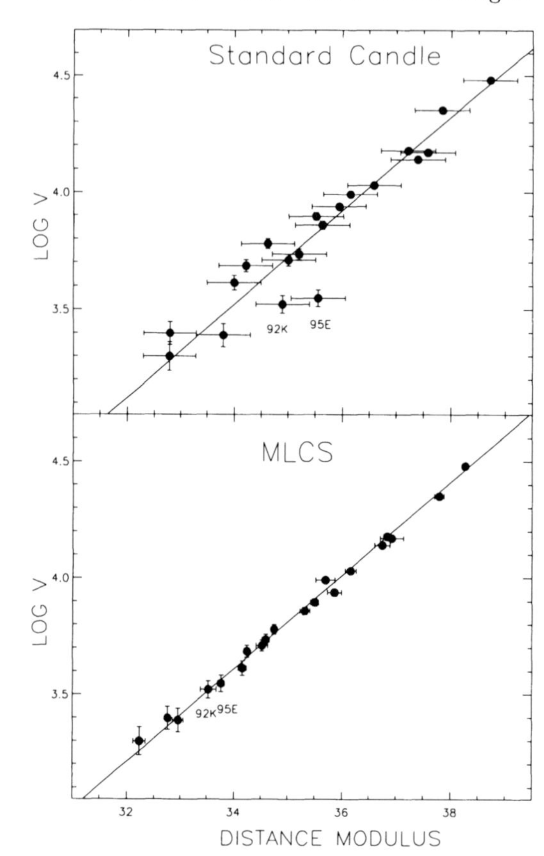

Fig. 2.27. The scatter around the Hubble relation is reduced when the empirical relation between the maximum luminosity and the light curve shape is applied [Riess et al. 1995]

Isn't it remarkable that the many violent physical processes occuring during the SN explosion, from the complicated radiation transfer to the rapidly expanding gas shell, conspire in such way that a tight and simple relation between luminosity and light curve decline results?

The value of H_0 obtained from presently available SNIa data is

$$H_0 = (65 \pm 7)\,\mathrm{km\,s^{-1}\,Mpc^{-1}} \tag{2.15}$$

[Riess et al. 1998]. The error is an estimate of the systematic uncertainties introduced by the SNIa calibration. Within the errors this value for H_0 is consistent with H_0 determined from the Cepheid distances to the Virgo cluster. It also agrees with other supernova determinations [Perlmutter et al. 1998].

Another approach which does not use MLCS, but instead eliminates some unsuitable SNIa, arrives at a value

$$(57 \pm 7)\,\mathrm{km\,s^{-1}\,Mpc^{-1}},$$

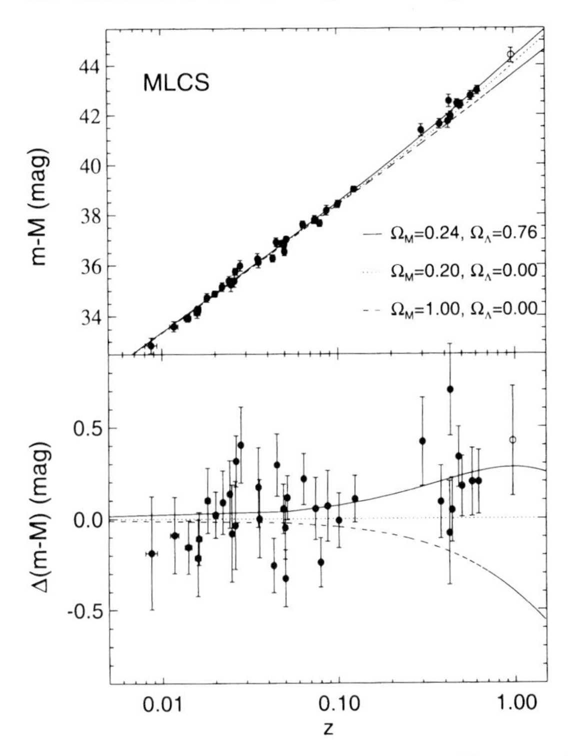

Fig. 2.28. The Hubble diagram for SNIa with the MLCS correction applied. 27 SNIa are in the nearby Hubble flow $0.01 < z < 0.13$, 10 are at high z ($0.16 \leq z \leq 0.62$). Supernova 1997ck at $z = 0.97$ is shown as an open circle. It does not have a spectroscopic classification and a colour measurement. Three cosmologies are indicated by *dashed* ($\Omega_m = 1, \Omega_\Lambda = 0$), *dotted* ($\Omega_m = 0.2, \Omega_\Lambda = 0$) and *solid* ($\Omega_m = 0.24, \Omega_\Lambda = 0.76$) *lines*. The Hubble constant measured from these data is $H_0 = (65 \pm 7)\,\mathrm{km\,s^{-1}\,Mpc^{-1}}$ [from Riess et al. 1998]

also in agreement with (2.15) [e.g. Tammann et al. 1997]. The difference from (2.15) seems to come mainly from giving much weight to SN1885 as a calibrator.

The tight Hubble relation followed by the SNIa, and the observation of such supernovae at high redshift makes it possible to place constraints on Ω_m and Ω_Λ, or equivalently q_0 and Ω_m.

Since big observation programmes are under way there will be more and more data, and our understanding of the physics of these supernovae will grow rapidly.

2.2.3 Some Statistical Aspects

A: The Malmquist Bias

The discussion of the preceding sections should have opened our eyes to some of the problems involved in a precision measurement of the Hubble constant. I think that the concept of "standard candles" in cosmology especially deserves closer scrutiny. Since this has always been an important issue in the history of H_0 let me recapitulate a few points.

Galaxies (and also SNIa) are distributed over a range of luminosities L (or magnitudes M), and the number of objects dN with luminosity between L and $L + dL$ is given by a distribution function $\Phi(L, z)$ (or $\Phi(M, z)$), the "luminosity function" $dN = \Phi(L, z)dL$.

In an analysis of the Hubble relation the data consist of a set of apparent magnitudes m_i and redshifts z_i ($i = 1, 2, \ldots, N$ for N objects). The cosmological model gives a relation

$$m = M + f(z). \tag{2.16}$$

For the linear Hubble expansion, for instance,

$$f(z) = 5 \log cz - 5 \log H_0 + 25 \tag{2.17}$$

(with c in $\mathrm{km\,s^{-1}}$ and H_0 in units of $\mathrm{km\,s^{-1}\,Mpc^{-1}}$). The expected value of m for a given, fixed value of z is then [Segal and Nicoll 1980]

$$E(m|z) = \frac{\int_{-\infty}^{m_c} m\Phi(m - f(z))dm}{\int_{-\infty}^{m_c} \Phi(m - f(z))dm}, \tag{2.18}$$

where the luminosity function $\Phi(M)$ and $m = M + f(z)$, have been used (evolutionary effects which would require $\Phi(M, z)$ are neglected here). The upper bound of the integration is m_c because generally astronomical samples such as catalogues of galaxies are flux-limited, i.e. only objects with $m < m_c$ are included. The truncation of m-values makes the evaluation of (2.18) somewhat tricky. This is the famous "Malmquist bias" of which the saying goes that in any sample of three astronomers two are convinced that the third one does not understand it [Malmquist 1920].

In Fig. 2.29 the effect of the trunction $m < m_c$ is demonstrated schematically. A sample of (m, z) data with points scattered uniformly within the ellipse gives a straight line fit to $E(m - z)$, if the average value of m for given z is determined by linear regression. The truncation $m < m_c$ leads to a significantly changed linear fit.

The astronomical explanation is simple. As distance increases, more and more of the faint objects drop below the flux limit. This leads to a deviation of the mean value of m (for fixed z) from the true value for the real, complete sample towards lower m. Thus, distances to the brighter objects are underestimated systematically.

In general there is no easy way to compensate for the Malmquist bias. Basically it is necessary to extend the (m, z) data set into the truncation region beyond m_c. A very mild assumption seems to be to take the same luminosity function $\Phi(M)$ everywhere.

For a long time it remained unresolved whether the distribution $\Phi(M)$, as well as the shape of the function f, can be determined from a set of observations (m_i, z_i). Meanwhile the possibility has been demonstrated [Woodroofe 1985], but the estimators are of bad quality, as are the astronomical data sets – statistically speaking.

The big advantage of standard candles can also be clearly seen from Fig. 2.29. The scattering diagram for good standard candles degenerates into a very narrow ellipse close to a straight line, and the truncation does not severely affect the fit.

The SNIa are clearly good candidates for standard candles, but they definitely show intrinsic variations in their peak luminosity. The semi-empirical correlation between the peak luminosity and the shape of the light curve reduces the scatter about the regression line of the linear Hubble relation, but rational tests, maybe even based on algorithmic prescriptions, of the standard candle hypotheses would be highly desirable.

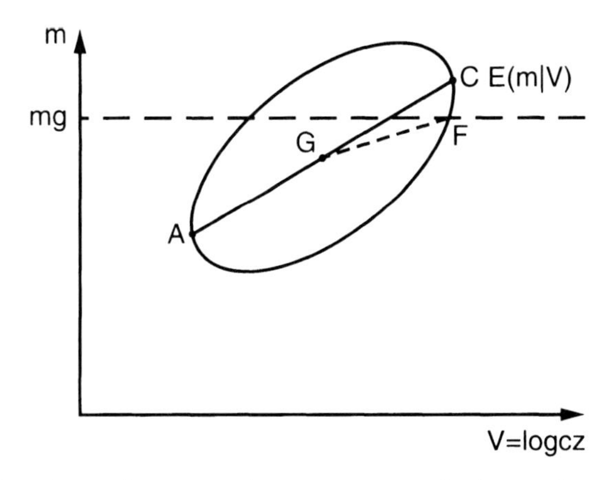

Fig. 2.29. The effect of a truncation in $m < m_g$ is depicted schematically for a distribution of m, z values. If the distribution is truncated at $m = m_g$, then the linear regression for $E(V/m)$ is no longer given by the line AC, but branches off at G to cut the contour $P(m, V) = \text{const.}$ (here schematically drawn as an ellipse) at point F. If we use linear regression nevertheless, we end up with the line AF which gives the wrong estimate of $E(V/m)$

B: A Redshift–Distance Square Law?

Historically the search for standard candles has led to the presentation of several unsuitable candidates, such as the brightest galaxy in a cluster, the upper limit on the red giant luminosity in a galaxy, or the brightest HII region in a galaxy. Very often it appeared that the intrinsic variation in luminosity of the proposed standard candles had been severely underestimated, or that unexpected correlations, such as the fact that brighter galaxies have brighter HII regions, made the distance indicator worthless.

A sarcastic person might here point out the remarkable fact that despite this long history of jumping to premature conclusions, of correcting or abandoning previously favoured distance indicators, the value of the Hubble constant remained always the same within narrow bounds (de Vaucoulers $H_0 = 100$, Sandage and Tammann $H_0 = 50$). (An account of H_0 debates, past and present, can be found in [Bonnell et al. 1996].)

An amusing climax was reached when some "heretics" [Segal and Nicoll 1980, 1996] challenged the establishment drastically by claiming that instead of the linear Hubble law $cz = H_0 d$, a relation $z \propto d^2$ fitted the astronomical data sets even better. Motivated by his – surrely somewhat exotic – "chronometric cosmology", which has $f(z) \sim \frac{5}{2}\log(\frac{z}{1+z})$, Segal modified the linear Hubble expansion law by introducing a parameter p:

$$f(z) = \frac{5}{p}\log cz - 5\log H_0 + 25. \tag{2.19}$$

Then a simultaneous estimate of the luminosity function Φ and p via (2.18) was carried out. The result, that p appeared to be consistently closer to $p = 2$ than $p = 1$, was surprising and annoying to the believers of the "Standard Model". The main criticism of the establishment has been that Segal and Co. do not correctly take into account the Malmquist bias.

Ullmann has investigated this controversy in detail [Ullmann 1997]. First of all we must remark that the estimate of $E(m|z)$ using (2.18) is better in general than the use of (2.16) with the assumption $M = M_0 =$ const. because this case is contained in (2.18) when Φ is a delta function.

Segal uses (2.18), and this is the optimal approach unless Φ is known. He uses a non-parametric method to evaluate (2.18), and his method corresponds to the algorithm constructed by Lynden-Bell for truncated distributions [Lynden-Bell 1971].

Let us briefly describe the basic features of this algorithm, when $f(z)$ is considered to be a known function, and Φ is to be determined. By a transformation of variables one can switch from the (m_i, z_i) data to corresponding (M_k, u_k) data with $u_k = m_k - M_k f(z_k)$. The number density of the objects in (M, u) space is then

$$N(M,u) = \sum_{k=1}^{n} \delta(M - M_k, u - u_k) = \Phi(M)D(u)H(m_c - (M+u)). \tag{2.20}$$

Here H is the Heaviside function

$$H(x) = \begin{cases} 1 & x > 0, \\ 0 & x < 0 \end{cases}$$

which describes the truncation; $D(u)$ is the number density at distance u; the sum runs over all objects $k = 1, \dots, n$.

The ansatz

$$\Phi(M) = \sum_1^n \psi_k \delta(M - M_k),$$

$$D(u) = \sum_1^n d_k \delta(u - u_k)$$

is inserted into (2.20). After integration over M or u as appropriate, equations for the constant coefficients ψ_k and d_k result:

$$1 = \psi_i \sum_{j=1}^{[i]} d_j,$$

$$1 = d_j \sum_{i=1}^{[j]} \psi_i \tag{2.21}$$

[i]: $\max[i]/M_i + u_i \leq m_c$ for $i, m, j = 1, \dots, n$.

From the solutions of (2.21) $\Phi(M)$ and $D(u)$ can be reconstructed, and thus using (2.20) the distribution in the whole region. Segal and his co-workers have developed an algorithm based on a similar ansatz. They use it a little differently, namely to determine Φ and to decide between $p = 1$ and $p = 2$ (taking (2.19) as the form of $f(z)$). The essential part of the algorithm is the construction of a histogram of the truncated sample ($m < m_c$). The luminosity function $\Phi(M)$ is approximated recursively by step-functions with fixed width and variable height. Simulations of model data sets with a definite Hubble relation imprinted have demonstrated conclusively that Segal's algorithm does indeed correct for the truncation $m < m_c$, i.e. the Malmquist effect [Ullmann 1997].

Other truncation and selection effects, as well as the presence of intrinsic and measurement errors in (2.16), are not corrected by this method. Generally the result – as shown by simulations [Ullmann 1997] – is that the estimates for both the variance of the luminosity function and the Hubble exponent are too large. This seems to be the main cause of Segal's controversial result.

There is a solid moral from this story: many astronomical samples are not suitable for an analysis of the Hubble relation according to (2.18). Astronomers have the ability to derive the Hubble law and the Hubble constant even from such data sets, because they know the form of $f(z)$ and the properties of unobservable galaxies.

2.2.4 Other Methods to Determine H_0

In view of the agreement reached (within $\pm 20\%$) between the method using Cepheids in Virgo and the method using SNIa as standard candles there is no urgent need to discuss other approaches to measure H_0. Most of these additional high-weight techniques need further refinement and development. In the following I'll mention just a few without a detailed discussion.

A: Direct Estimate from Gravitational Lens Systems

The prime example is the "double quasar" 0957 + 61. It was discovered in 1979 [Walsh et al. 1979] as a pair of quasars, 6 arcseconds apart, with identical redshifts and spectra. The initial guess that these were two images of one quasar produced by the gravitational lens effect of an intervening mass has been supported by subsequent observations. The lens is a cluster with a large central elliptical galaxy. (The "double quasar" is actually a triple system, but the third image is very faint.) Long before this object was discovered it had been pointed out [Refsdal 1964] that the time variability of an object mapped by a gravitational lens allows an estimate of the distances involved, and hence of H_0. The essential point is simply that the time delay (due to the different light paths) between a variation in one image, and the same variation in the other, gives the actual distance directly.

Such events have been found [first by Florentin-Nielsen 1984] – involving a patient watch on the quasar for many years. The time delay is estimated at

$$\Delta T = 417 \pm 3\,\mathrm{days}$$

and H_0 is derived as

$$H_0 = 66 \pm 15\,\mathrm{km\,s^{-1}\,Mpc^{-1}}$$

[Kundić et al. 1997].

The largest uncertainty in H_0 arises from the model for the mass distribution of the gravitational lens. The time delay along the different light paths depends very sensitively on details of the mass profile (see [Schneider et al. 1992] for in-depth explanations of gravitational lens physics).

A second gravitational lens system has also been investigated with a similar strong dependence of the time delay, and of H_0, on the mass model. For the quadruple gravitational lens PG 1115 + 080 different time delays between different images have been observed, namely $\sim 10d$ and $25d$. The value for the Hubble constant is:

$$H_0 = (51 \pm 14)\,\mathrm{km\,s^{-1}\,Mpc^{-1}}$$

[Keeton and Kochanek 1997].

B: H_0 from Tully–Fisher Distances

For spiral galaxies an empirical relation has been found between the rotational velocity and the luminosity of the galaxy [Öpik 1922; Tully and Fisher 1977]. The "Tully–Fisher" (TF) relation is the dependence of the luminosity in a certain frequency band on the line-width of the 21 cm line. This radio signal is emitted from the neutral hydrogen atoms in the galaxy, when the proton and electron spin directions change with respect to one another. The line-width Δv_{21} is a measure of the velocity dispersion of HI. There are "blue" and "infrared" TF relations, and for the H-band, e.g.

$$L_H \propto \Delta^4 v_{21}. \tag{2.22}$$

This correlation has been tested empirically quite extensively. The scatter of the relation turns out to be surprisingly small. Thus, once calibrated by Cepheids, the TF relation might be a good distance indicator. As an example for the use of the TF relation I quote $H_0 = (84 \pm 8)\mathrm{km\,s^{-1}\,Mpc^{-1}}$ [Lo et al. 1994].

C: Planetary Nebulae

The luminosity function of planetary nebulae has a characteristic shape with a pronounced cut-off. This has been used as a distance indicator, giving

$$H_0 = (86 \pm 18)\,\mathrm{km\,s^{-1}\,Mpc^{-1}}$$

for Virgo galaxies [Méndez et al. 1993], and

$$H_0 = (75 \pm 8)\,\mathrm{km\,s^{-1}\,Mpc^{-1}}$$

for galaxies in the Fornax cluster [McMillan et al. 1993].

D: Surface Brightness Fluctuations

The surface brightness fluctuations in elliptical galaxies lead to a mean value of

$$H_0 = (80 \pm 12)\,\mathrm{km\,s^{-1}\,Mpc^{-1}}$$

[Jacoby et al. 1992].

E: Sunyaev–Zel'dovich Effect

The Sunyaev–Zel'dovich (SZ) effect makes use of the interaction of clusters of galaxies and the 3 K background. Values of H_0 tend to be on the low side between 50 and 60 $\mathrm{km\,s^{-1}\,Mpc^{-1}}$, but this may be biased by observational selection effects [Birkinshaw and Hughes 1994]. The SZ effect will be discussed in Sect. 2.5.

2.2.5 Evidence for a Local Anisotropy of the Hubble Flow

The bright galaxies within $\sim 10\,\mathrm{Mpc}$ are concentrated toward the Virgo cluster, forming a flattened disc-like system that is called the Local Supercluster [de Vaucouleurs 1958]. It seems that the gravitational potential within this system accelerates our Local Group of galaxies towards the centre of the Virgo cluster.

Strong evidence for a peculiar motion of our system is provided by the observations of the background radiation. The measurements of a dipole anisotropy of the 3 K cosmic microwave background (CMB) have now firmly established a motion of our galaxy relative to CMB with a velocity $v_M = 627 \pm 22\,\mathrm{km\,s^{-1}}$ in the direction $(l, b) = (276° \pm 3°, 30° \pm 3°)$ in galactic coordinates (see references and further discussions in Sect. 2.5). As, apart from this dipole anisotropy, the CMB appears highly isotropic, it can be used to define perfectly the preferred frame of comoving coordinates in an FL model. The velocity v_M therefore defines our deviation from the overall Hubble flow. It is of great interest – for H_0 measurements for instance – whether the density contrast of the Local Supercluster can completely account for this peculiar motion.

The velocity vector points $\sim 45°$ away from the direction of the core of the Virgo cluster. Using determinations of our motion relative to the Local Group [Yahil et al. 1977], one finds that the peculiar velocity of the Local Group can be split up into the components (in $\mathrm{km\,s^{-1}}$).

$$(v_V, v_S, v_Z) = (410 \pm 25, 297 \pm 25, -308 \pm 25)$$

where v_V is the component towards Virgo, v_S transverse to this direction but in the plane of the Local Supercluster, v_Z is transverse to v_V and v_S.

Measurements of v_V are uncertain. The situation has been reviewed in detail [Davis and Peebles 1983a] with the conclusion that the Virgocentric component of this peculiar motion can be just the infall velocity towards Virgo. The widely different values obtained by various groups (ranging from $v_V = 40 \pm 160$ to 520 ± 75), allow estimates of

$$v_V = 200 \;\text{ to }\; 400\,\mathrm{km\,s^{-1}}.$$

Some interesting constraints can nevertheless be derived from a simple spherical model (Fig. 2.30). Within such a model the overdensity δ of the Virgo complex (calculated by averaging over the sphere whose surface is at the Local Group) can be related to v_V (within linear perturbation theory):

$$v_V = \frac{1}{3}\delta H_0 r \Omega_0^{0.6}. \tag{2.23}$$

This equation can be derived as follows [Peebles 1980]: in a Newtonian approximation, a spherical mass distribution (of radius r with mass $M + \delta M$), gives rise to a peculiar gravitational field

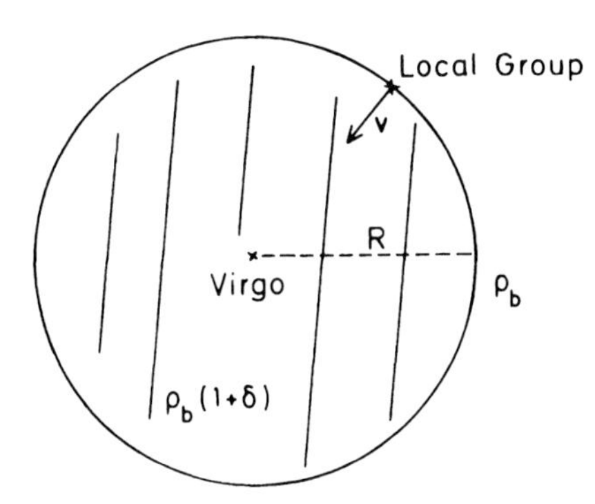

Fig. 2.30. A simple spherical model for the Virgocentric flow

$$g(r) = G\frac{\delta M}{r^2} = \frac{1}{2}H_0^2\frac{\delta M}{M}r\Omega_0$$

(with $GM = \frac{1}{2}H_0^2\Omega_0 r^3$ the background mass within r). The peculiar velocity field is given by

$$v = \frac{2f(\Omega_0)}{3H_0\Omega_0}g(r)$$

[Peebles 1980, paragraph 14], where $f(\Omega_0)$ is determined from solutions of the linear perturbation analysis ($f \equiv (R(t)/\delta_+)(d\delta_+/dR)$, cf. Sect. 11.2). A useful analytic approximation is $f(\Omega) = \Omega^{0.6}$ [Peebles 1980]. Then with $\delta_+ \equiv \delta M/M$ wc arrive at equation (2.23).

Thus a determination of Ω_0 might be possible from the infall, if δ can be guessed from galaxy counts, and v_V, as well as H_0 and r, is measured. Taking $H_0 r = 1170$ km s^{-1}, and values of $v_V = 200 \pm 50, \delta = 2.8 \pm 0.5$ [Davis and Peebles 1983] one obtains

$$\Omega_0 = 0.08 + \begin{matrix} 0.07 \\ -0.04 \end{matrix} \tag{2.24}$$

for the density parameter. A simple spherical model does not seem quite adequate, however, in view of the fact that the peculiar motion with respect to the rest-frame of the CMB has other components besides v_V. Why not decompose [Sandage and Tammann 1984b] the velocity vector into a Virgocentric component $v_V = (200 \pm 50)\,\mathrm{km\,s^{-1}}$ – corresponding to a local measurement of the infall towards Virgo – and into a second component pointing in the direction of the Hydra–Centaurus supercluster (Fig. 2.31) with a magnitude of

$$V_{\mathrm{HC}} = 460 \pm 70\,\mathrm{km\,s^{-1}}?$$

The Hydra–Centaurus supercluster [Materne and Hopp 1984] is a structure at $v_0 \simeq 3400\,\mathrm{km\,s^{-1}}$, with a projected extent of $40 \times 10\,\mathrm{Mpc}$. The Virgo complex might be an outlying component of this system. The search for the

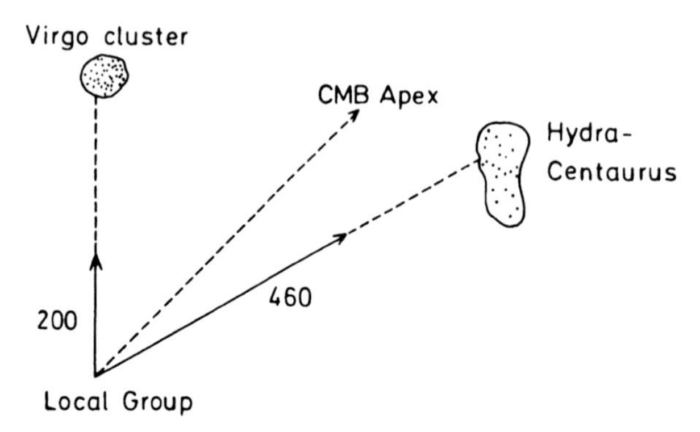

Fig. 2.31. Decomposition of the velocity vector for motion against the CMB. A component of $200\,\mathrm{km\,s^{-1}}$ is directed towards Virgo, and a component of $460\,\mathrm{km\,s^{-1}}$ points in the direction of Hydra–Centaurus

Local Group's Hydra–Centaurus velocity in the redshifts of field galaxies with $2500\,\mathrm{km\,s^{-1}} < v_0 < 5000\,\mathrm{km\,s^{-1}}$ might give indications of such a motion.

Further modifications of a smooth Hubble flow have been revealed – it seems – by spectroscopic and photometric measurements of elliptical galaxies [Dressler et al. 1987]. Data for 423 galaxies have been collected and analysed with a maximum redshift $\sim 6000\,\mathrm{km\,s^{-1}}$. Thus this survey covers about 2% of the Hubble sphere. The distances to these galaxies were inferred from an empirical relation between the central velocity dispersion in an elliptical galaxy, and its luminosity [Faber and Jackson 1976]. The virial theorem implies a dependence of the central velocity dispersion on the scale size; combined with an angular diameter defined by the surface brightness, this gives a measure of the distance, the "$D_n - \sigma$" method. The uncertainties in this distance estimate are probably large, but if one accepts the values within a $\pm 25\%$ error [Dressler et al. 1987], then one finds that the deviation of the velocity from a pure Hubble flow $v_{\mathrm{obs}} - v_H$ varies systematically over the celestial sphere. Several groups have investigated the redshift vs. distance relation according to these semi-empirical relations (Tully–Fisher, $D_n - \sigma$). It seems to be well established that a large-scale bulk motion exists involving a scale of $\sim$ 50 Mpc around us, and peculiar velocities $\sim 500\,\mathrm{km\,s^{-1}}$ [Dekel 1994].

If the CMB anisotropy is due to a motion of the Local Group of $\sim 600\,\mathrm{km\,s^{-1}}$, then this analysis shows that mass concentrations beyond $v \sim 5000\,\mathrm{km\,s^{-1}}$ must be held responsible. The suggestion is advanced that a shear motion is observed, caused by mass inhomogeneities just beyond $v \sim 5000\,\mathrm{km\,s^{-1}}$ [Dressler et al. 1987].

2.2.6 Side Remarks

In the following we shall use $H_0 = (65 \pm 7)$ kms^{-1} Mpc^{-1} as a canonical value, and introduce, whenever appropriate, a parameter $h = H_0/100$ into the formulae to allow for deviations from the canonical value.

H_0^{-1} is the present expansion time-scale. For the $K = 0$, $\Omega_\Lambda = 0$, $p = 0$ FL cosmologies, the age of the universe is

$$t_0 = \frac{2}{3h} \times 10^{10} \qquad \text{years},$$

i.e.

$$0.95 \times 10^{10}\,\text{y} < t_0 < 1.1 \times 10^{10}\,\text{y} \quad \text{for} \quad 0.6 < h < 0.7. \tag{2.25}$$

One can already point to a slight discrepancy between this range of t_0 and the values given in Sect. 2.1. I would like to stress, however, the much greater importance of the remarkable agreement between these different ages. (A universe with $\Lambda \neq 0$ could give $H_0^{-1} < t_0$.)

Distances have often been mentioned above. It should be emphasized that there are several "distances", which are derived concepts, each of them having an invariant meaning within GR. To illustrate this, let me define several distances in relativistic cosmology. (The following presentation is due to J. Ehlers (personal communication).)

An Einstein–de Sitter "dust" model (appropriate in the matter-dominated era) has

$$(K = 0, \Omega_\Lambda = 0, p = 0) : \frac{R(t)}{R_0} = \left(\frac{t}{t_0}\right)^{2/3} ; \; H_0 = \frac{2}{3t_0},$$

$$\varrho_0 = \varrho_c = \frac{3}{8\pi G} H_0^2 .$$

If t_e is the time of emission, z the redshift of the emitted photon, and r_0, r_e the metric distances of the source from "us" at t_0 and t_e respectively, then

$$t_e = t_0 (1+z)^{-3/2} ; \qquad r_0 = (1+z) r_e$$

$$r_0 = r_H (1 - (1+z)^{-1/2}); \qquad r_e = r_H ((1+z)^{-1} - (1+z)^{-3/2}),$$

where $r_H = 3ct_0 = 2c/H_0$ is the radius of our particle horizon at t_0("now"). The past null-cone of $(t_0, r = 0)$ is given by $r_e = r_H[(t_e/t_0)^{2/3} - (t_e/t_0)]$. The light–travel time of signal is $\Delta t = t_0 - t_e = (\frac{2}{3} H_0^{-1})(1 - (1+z)^{-3/2})$ (see Fig. 2.32). In addition we can define a Doppler velocity (special relativity):

$$\frac{v_r}{c} = z \frac{1 + (z/2)}{1 + z[1 + (z/2)]}$$

(assuming radial motion) and a conventional distance

$$d \equiv \frac{cz}{H_0} = \frac{1}{2} r_H z .$$

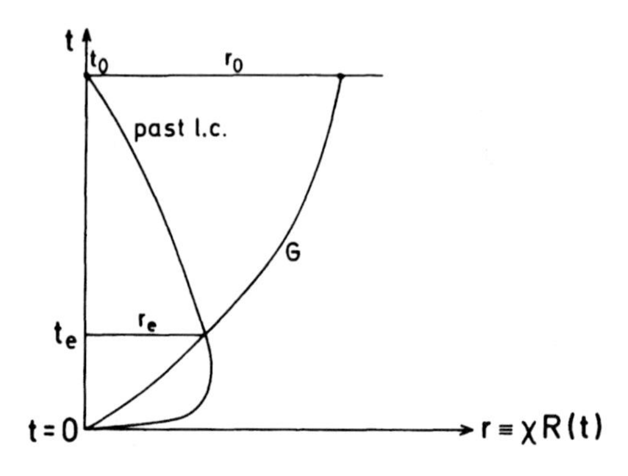

Fig. 2.32. Schematic diagram of our past light–cone, and world-lines of cosmic objects, $dr_e/dt = 0$ at $z = 1.25$ for $\Omega_0 = 1$. $t_e/t_0 = 0.3$ at this redshift

With $H_0 = (50$ or $100)$ the horizon radius is 12 000 Mpc or 6000 Mpc. Let us take a look at one of the most distant cosmic objects, a quasar with the redshift $z = 4.01$ [Warren et al. 1987]. The Doppler velocity is $v_r/c = 0.92$; the apparent kinematic velocity $\overline{v}_{kin} \equiv (r_0 - r_e)/\Delta t$ is beyond the velocity of light: $\overline{v}_{\text{kin}} = 1.46c$. The "Hubble distance" $d = 2r_H = 3.6r_0 = 18r_e$ is already twice as large as the horizon scale. The area distance of the object $r_0 = 0.55r_H$ is at 55% of the Hubble sphere; whereas at the time of emission the area distance $r_e = 0.11r_H$. The cosmic epoch at the time of emission $t_e = 0.09t_0$, and $\Delta t = 0.91t_0$. This quasar is beyond the maximum extent of our past light–cone in the $\Omega = 1$ model, which is at $z = 1.25, t_e = 0.3t_o$. In Fig. 2.32 (and in Fig. 1.11) the observational situation is sketched. Cosmic objects are observed as they cross the past light-cone of the space-time point $(r = 0, t_0) \equiv$ "us". Their redshift is determined by the time of horizon-crossing.

The present space distribution (the $t_0 =$ const. hypersurface) is unobservable to us. We can only observe a kind of historical cross-section. For nearby objects, with $z \ll 1$, a "Newtonian" picture, where the sources are considered to lie in the $t_0 =$ const. space, can be a good approximation.

2.2.7 Conclusion

There has been definite progress in the measurement of H_0. The Cepheid method for the distance ladder out to Virgo, and the SNIa method have led to results which overlap within the respective errors. The value of $H_0 = (65 \pm 7)\,\text{km}\,\text{s}^{-1}\,\text{Mpc}^{-1}$ given by a SNIa group [Garnavich et al. 1998] overlaps with all the other measurements. In the following we shall take this as a reference value. The ± 7 is an estimate of systematic errors, and this may change with future measurements or new insights.

The remarkable fact stands out that H_0^{-1}, the expansion time, is of the same magnitude as the age of the oldest stars $t_0 = (12 \pm 2)\,\text{Gyr}$. The fact that these two time-scales – one derived from the expansion of the system of galaxies, the other from a combination of nuclear physics, stellar evolution theory, and observation of stars – are not widely different, but agree more or less, reveals a deep truth about our universe: the time t_0, or H_0^{-1}, is a characteristic evolutionary time-scale of our universe, and it is reasonable to identify this time-scale with the "age of the universe" in FL cosmological models. If we take the measured values of H_0 and t_0, we find for the dimensionless number $t_0 H_0$ an interval of

$$0.6 \leq H_0 t_0 \leq 1.0. \tag{2.26}$$

As we have seen in Chap. 1, there is a relation between $H_0 t_0$ and Ω_m, Ω_Λ in FL models:

$$t_0 H_0 = \int_0^1 \frac{a^{1/2}}{(P(a))^{1/2}} da \tag{2.27}$$

where $P(a) = \Omega_\Lambda a^3 + \Omega_m + (1 - \Omega_m - \Omega_\Lambda)a$. In Fig. 2.33 lines of constant $H_0 t_0$ are plotted in their dependence on Ω_m and Ω_Λ. From these diagrams we can conclude that there is a range of cosmological models still compatible with the observations. We shall see in the following how this range shrinks by measurement of Ω_m and Ω_Λ.

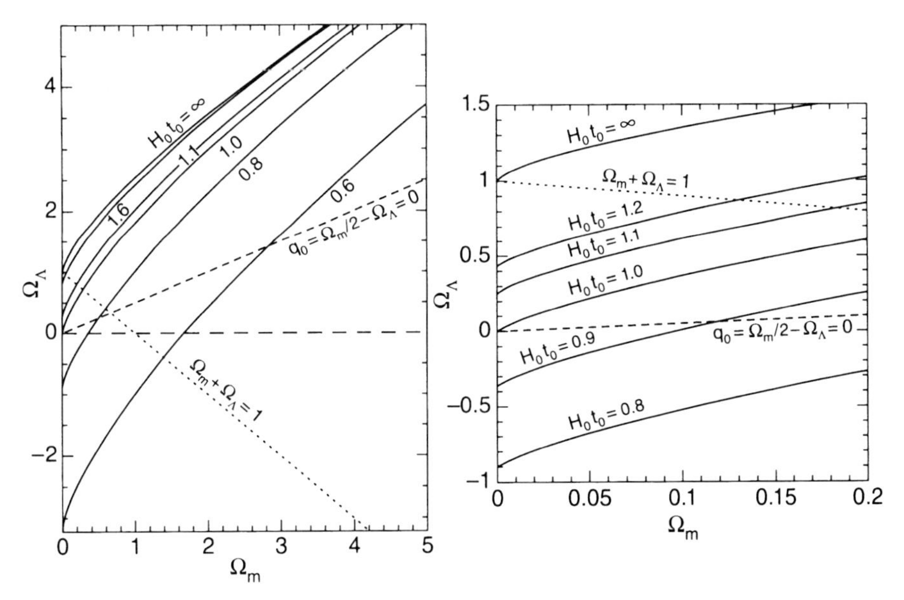

Fig. 2.33. Contours of constant $H_0 t_0$ are shown in an $(\Omega_m, \Omega_\Lambda)$ diagram. It is amazing that there is a concordance between the values of $H_0 t_0, \Omega_m, \Omega_\Lambda$ obtained in different ways within the simple FL models. A discordant set of parameters would immediately disprove these models, but they come out alright!

2.3 The Mean Density

The mean density of the universe is a fundamental quantity in cosmology. Recently it has been determined with high accuracy by the "Boomerang and Maxima" experiments on the CMB anisotropies (cf. Sect. 2.5 and Chap. 11):

$$\begin{aligned} \Omega_{\rm tot} &= 1.0 \\ \Omega_\Lambda &= 0.7 \\ \Omega_m &= 0.3 \\ H_0 &= 68 \end{aligned}$$

Another set of fit parameters is

$$\begin{aligned} \Omega_{\rm tot} &= 1.02 \pm 0.06 \\ \Omega_{\rm CDM} h^2 &= 0.13 \pm 0.05 \\ \Omega_B h^2 &= 0.022^{+0.004}_{-0.003} \\ \Omega_\Lambda + \Omega_B + \Omega_{\rm CDM} &= \Omega_{\rm tot} . \end{aligned}$$

In this section I want to discuss the basic steps and methods used by astronomers to derive a value for the mean density of matter. We can compare the results to the CMB values, and we shall see that not only the luminous matter must be taken into account, but also "dark matter" and "dark energy" which may even be the dominant component.

The distribution of the luminous matter (stars and galaxies) appears highly structured. It does not lend itself easily to an interpretation in terms of a smooth average density. The clusters, superclusters, and voids seen in the redshift catalogues, such us the CfA survey [Geller and Huchra 1989] or the Las Campanas Redshift Survey (LCRS) [Shectman et al. 1996] do not really give the impression of a uniform matter distribution. There seems to be only evidence for structure, not for homogeneity, from observations of luminous matter. The LCRS data, however, give some indication that the pattern of structure repeats itself over larger scales. There is no evidence for larger und larger structural elements as one looks at larger scales.

The two-point correlation functions of the northern and southern hemisphere of this data set agree quite precisely (see Chap. 10 for further discussion). This lends some support to the assumption that on large scales the density contrasts gradually merge into a uniform background. Thus the data do not contradict the assumption that the mean density, found by averaging over scales of several 100 Mpc, can be used as the homogeneous density ρ_0 of an FL universe.

This hypothesis is somewhat supported by the apparently uniform distribution of bright radio sources. The distribution of the 5000 radio sources (flux $S \geq 2$ janski (Jy)) of the 4C catalogue appears to be random [Peebles 1980; Peebles and Seldner 1978].

The best evidence for a uniform matter distribution is derived from the background radiation. The diffuse X-ray background is in large part the flux from active galaxies and clusters. The isotropy of the background sets a limit on fluctuations of $\delta\varrho/\varrho \sim 0.01$ to 0.1 on scales of 100 to 10^3 Mpc (a few degrees to 20°) [Fabian 1981] (see Sect. 2.6). Finally the 3 K background (see Sect. 2.5) is isotropic to $\delta T/T \sim 10^{-5}$ on angular scales from minutes to 180°. The matter distribution is not measured directly by this value, since matter and radiation are only weakly coupled at present. But the large-scale motions of the matter distribution must have been isotropic to an accuracy better than 1 in 10^3 (since $T \propto (1+z)^{-1/2}$).

All this evidence does not establish a mean density ϱ_0, but makes it plausible as a reasonable concept. The observations show that matter distribution and motion are very accurately isotropic around us. This leaves open, of course, the possibility that the universe is actually inhomogeneous, and is isotropic only around our world-line. It seems to be a useful working hypothesis, however, that our position is typical, and not unique (see, however, Chap. 4).

The observations involved in determining a mean mass density ρ_m are discussed in many textbooks (see e.g. [Peebles 1980, 1993]). The general procedure requires first a count of luminous objects to such an accuracy that a "mean luminosity density" $\langle L \rangle$ can be determined. Then the density parameter Ω_m is

$$\Omega_m = \langle L \rangle \times M/L/\rho_c,$$

$$\rho_c = \frac{3H_0^2}{8\pi G}.$$

The quantity M/L is the mass-to-light ratio of various kinds of objects, and the basic assumption is that the relation between light and mass in the universe is the same as in the objects measured. ρ_c is the critical density defined earlier. Ω_m is actually Ω_m at $t = t_0$, the present epoch, but for brevity we write Ω_m for $\Omega_m(t_0)$. We write $\Omega_m(t)$ when we consider the explicit time dependence in later chapters.

The Luminosity Function

Let $\phi(L)dL$ denote the number of galaxies per unit volume with luminosities in the interval between L and $L+dL$. The optical galaxy luminosity function has traditionally been fitted by a special analytic approximation [Schechter 1976]

$$\phi(L)dL = \phi^*(L/L^*)^\alpha \exp(-L/L^*)d(L/L^*). \tag{2.28}$$

For galaxies more luminous than a characteristic luminosity $L^*, \phi(L)$ drops exponentially with luminosity; for galaxies fainter than $L^*, \phi(L)$ approaches a power law with slope α. The quantity ϕ^* represents an overall normalization of the luminosity function.

In terms of absolute magnitudes instead of luminosities, (2.28) reads [Efstathiou and Silk 1983]

$$\phi(M)dM = 0.4 \ln 10\, \phi^*[10^{0.4(M^*-M)}]^{\alpha+1} \times \exp[-10^{0.4(M^*-M)}]dM. \quad (2.29)$$

Local galaxy surveys have found values of α that lie in the range $-1.2 \leq \alpha \leq -0.7$, giving a "flat" faint-end slope when the LF is plotted as a function of absolute magnitude.

Let me quote three representative determinations of the parameters α, M^*, and ϕ^* to give an idea of the uncertainty of the LF measurements:

[Efstathiou et al. 1988]	$M^* = -19.68 \pm 0.10 + 5 \log h$	(b_J magnitude)
	$\alpha = -1.07 \pm 0.05$	
	$\phi^* = (1.50 \pm 0.34) \times 10^{-2} \times h^3 \mathrm{Mpc}^{-3}$	
[Loveday et al. 1992]	$M^* = -19.50 \pm 0.13 + 5 \log h$	(b_J magnitude)
	$\alpha = -0.97 \pm 0.15$	
	$\phi^* = (1.40 \pm 0.17) \times 10^{-2} \times h^3 \mathrm{Mpc}^{-3}$	
[Lin et al. 1996]	$M^* = -20.29 \pm 0.02 + 5 \log h$	
	$\alpha = -0.70 \pm 0.05$	
	$\phi^* = (1.9 \pm 0.1) h^3 \mathrm{Mpc}^{-3}$	

Far larger uncertainties are connected with the fact that the faint galaxies, much fainter than M^*, are difficult to measure and easy to miss. There is growing evidence [Lin et al. 1996, Zucca et al. 1997] from recent surveys that for $M > -17$ the LF significantly exceeds the Schechter fit for galaxies brighter than $M = -17$.

A recent analysis confirms a rising faint end of the LF [Loveday 1998] which can be fitted by a power law with $\alpha = -2.76$ for absolute magnitudes below -14.

These preliminary results indicate that corrections must be expected for the galaxy LF. They also have important consequences with respect to possible evolutionary effects in the galaxy population.

When we use the Schechter fit to the LF, we find for the average luminosity density

$$\langle L \rangle = \int_0^\infty \phi(L) L dL = \phi^* L^* \Gamma(\alpha + 2); \quad (2.30)$$

where $\Gamma(x)$ is the gamma function. In the blue band of the spectrum, then,

$$\langle L_B \rangle = (1.9 \pm 0.06) L_\odot \quad \mathrm{hMpc}^{-3} \quad (2.31)$$

[Efstathiou et al. 1988].

The uncertainty in this value amounts to a factor of 2, when we consider the measurements by different groups, as we have listed them above. It can be even larger, because systematic effects like galaxy evolution, obscuration and other photometric difficulties, the contribution of dwarf galaxies and low surface brightness galaxies, are difficult to estimate at present. Measuring M/L in solar units, we can use (2.31) to arrive at

$$\Omega_m = (1450)^{-1} h^{-1} (M/L). \tag{2.32}$$

To find Ω_m, we have to estimate the M/L ratios of representative cosmic objects.

2.3.1 M/L Ratios for Galaxies

An individual galaxy can be described to a rough approximation by global physical quantities: its luminosity L, optical radius R, and internal velocity v. These can be related to three observable structural parameters: the effective radius r_e, the central velocity dispersion σ_0, and the mean effective surface brightness SB_e. There are empirical relations between these quantities, such as the Tully–Fisher relation for spirals [Tully and Fisher 1977] and the "fundamental plane" for ellipticals [Faber and Jackson 1976; Djorgovski and Davis 1987; Bender et al. 1992]. These can be used to obtain M/L ratios, as we shall discuss in some detail in the following.

The galaxies can also be arranged in a so-called Hubble sequence, giving a classification according to their morphology, differences in their degree of flattening, and in the amount to which their stellar components partake of a common orbital motion. The sequence leads from elliptical galaxies (circular to ellipsoidal discs in the projection on the sky, where the stars have essentially random orbital motions) to spiral (flat, disc-like objects with strongly organized circular rotation) and irregular galaxies.

For the Hubble types of galaxies strong correlations with their environment seem to exist: the field outside of clusters is dominated by spiral and irregular galaxies which make up more than 80% of the population, whereas in clusters E (ellipticals) and SO galaxies (circular disc without spiral arms) are the main components. Such a correlation [Dressler 1980] indicates some physical connection between environment and morphological type. The cosmologically important properties of galaxies have been compiled in various reviews and textbooks. [e.g. Kormendy 1982; Oort 1984; Jones 1976; Efstathiou and Silk 1983; Faber and Gallagher 1979; Toomre 1981; Combes et al. 1995].

a) Elliptical Galaxies

i) The intensity distribution accross the image of an elliptical galaxy is well approximated by an $r^{1/4}$ law [de Vaucouleurs 1959]

$$I(r) = I_e \exp[-7.67((r/r_e)^{1/4} - 1)]. \tag{2.33}$$

Here I_e is the intensity at an effective radius r_e (usually given in arcsec), defined such that within r_e half the total brightness is found. For nearby galaxies r_e is larger than 1 and the image observed is not affected by turbulence in the Earth's atmosphere.

ii) The observations have clearly shown that there exist two classes of ellipticals: less luminous E galaxies and also the central bulges of spiral galaxies seem to rotate moderately. Their rotation is responsible for the flattening of their shape. More luminous E galaxies, however, are rotators which are flattened by an anisotropic velocity distribution [Davies et al. 1993].

Detailed morphological investigations have shown [Bender 1988] that the isophotes (contours of constant brightness) of E galaxies are in general not perfect ellipses. Anisotropic galaxies have box-shaped – "boxy"– isophotes, while rotationally flattened ellipticals exhibit lemon-shaped – "discy" – isophotes due to the superposition of a bulge and a stellar disc.

Many bright ellipticals have kinematically decoupled cores where the stars rotate in the opposite sense to the main body [Bender 1990]. Most of these turn out to be rapidly spinning thick stellar discs which probably were produced by the merging of two massive star-dominated objects. These cores have masses in the range of $10^9 - 10^{10} M_\odot$ and radii of a few to several hundred parsecs.

The brightest Es apparently contain a large amount of hot gas ($T \simeq 10^7$ K) which emits thermal X-ray radiation. The same objects often also exhibit radio emission indicating an active galactic nucleus (AGN). These results suggest a division of elliptical galaxies into two classes [Bender et al. 1989]:

- Bright ($M_V < -22.0$ mag), anisotropic, boxy Es often have an X-ray halo, AGN activity, a kinematically decoupled core.
- Fainter ($M_V > -20.5$ mag) rotationally flattened, discy Es do not in general have X-ray haloes, nor AGNs, nor a decoupled core.
- Es of intermediate brightness (-22.0 mag $< M_V < -20.5$ mag) may belong to either of the two classes.

Recent observations of galactic cores with the HST have confirmed this division into two classes [Lauer et al. 1995; Faber et al. 1997].

Since SO galaxies also consist of the superposition of a "bulge" on a stellar disc they are morphologically similar to "discy" ellipticals.

iii) Interesting questions concerning the age and the nature of the stellar populations in ellipticals can be tackled by comparing models of stellar populations of different age and metallicity with observations of real spectra. A problem here is the age–metallicity degeneracy: the spectrum of a model stellar population does not change, if e.g. the metallicity is doubled, and the age is reduced by one-third.

Bright E galaxies have an overabundance of the Mg to Fe ratio

$$[\mathrm{Mg/Fe}] \equiv \log(\mathrm{Mg/Fe}) - \log(\mathrm{Mg/Fe})_\odot \qquad (2.34)$$

over the solar value. [Mg/Fe] = 0.4 is reached, while faint E and SO galaxies have solar abundances [Mg/Fe] $\simeq 0$ [Davies et al. 1993].

There is generally an enrichment of so-called "α-elements" (O, Mg, Si, Ca, Ti) in bright E galaxies. This indicates a dominance of SNII explosions (which enrich α-elements) over SNIa which produce mainly Fe. This can be explained by star formation time-scales less than 10^9 years, since SNIa need a somewhat longer time for the evolution of the lower-mass progenitor stars.

iv) From these arguments we may conclude that the large elliptical galaxies formed from the merging of massive progenitor objects that consisted partly of stars and partly of gas. The bulk of the stars in a large fraction of massive ellipticals formed at redshifts above 3, and the formation time-scale for most of the stars was about 10^9 years.

v) One of the cosmologically most relevant properties of elliptical galaxies is the existance of a relation between the three global parameters of elliptical galaxies – the effective radius r_e, the central velocity dispersion σ_0, and the mean effective surface brightness SB_e – given by

$$\log r_e = \alpha \log \sigma_0 + \beta \mathrm{SB}_e + \gamma, \tag{2.35}$$

with constants α, β, γ. This equation defines the "fundamental plane" of elliptical galaxies. SB_e is a logarithmic quantity describing the surface brightness or specific intensity

$$I_e \equiv 10^{-0.4(SB_e - 27)}$$

within r_e.

A short note on the velocity dispersion σ is in order. Consider the motion of N stars in a galaxy with a mean velocity $\langle \boldsymbol{v} \rangle = \frac{1}{N} \sum \boldsymbol{v}_i$. Projection of these velocities along the line of sight, i.e. some fixed direction $\boldsymbol{n}$, leads to the definition of a one-dimensional velocity dispersion σ:

$$\sigma \equiv \frac{1}{N} \left(\sum (\boldsymbol{v}_i \cdot \boldsymbol{n} - \langle \boldsymbol{v} \rangle \cdot \boldsymbol{n})^2 \right)^{1/2} .$$

This quantity cannot be measured, of course, since in general the motion of individual stars cannot be observed. But the velocity dispersion leads to a broadening of the spectral lines ("Doppler broadening"). The real spectrum is thus the result of a convolution of the velocity distribution along the line of sight with a model spectrum which would be the addition of all the stellar spectra with the stars at rest. Such a spectrum is computed from a distribution function of the spectral classes of the individual stars. The velocity distribution is assumed to be Gaussian,

$$f(v) = \frac{1}{\sqrt{2\pi}\sigma} \exp\left(-\frac{v^2}{\sigma^2} \right) .$$

A comparison of the convolution integral with the measured spectrum leads to a determination of the free parameter σ, the velocity dispersion. The central velocity dispersion is just the value of σ obtained when a small region around the centre of the galaxy is considered; for details see [Dressler 1984]. The effective radius r_e is determined from fits of surface brightness profiles to de Vaucouleurs' $r^{1/4}$ law. These fits are commonly done outside of the innermost 3″ of a galaxy image, and within an outer bound of typically 1′.

One form of the fundamental plane is given as

$$r_e \quad \propto \quad (\sigma_0^2)^{0.7} \quad I_e^{-0.85} \tag{2.36}$$

[Bender et al. 1992].

Since I_e is connected to the distance D of the object measured (the luminosity $L \propto I_e\, r_e^2$, and the flux $F \propto L/D^2$), (2.36) can be used as a distance indicator. As it turns out, the scatter around the fundamental plane is quite small ($\leq$ 20% in r_e), such that very reliable distances can be derived from the fundamental plane relations. If we use the virial theorem in the form $M \propto \sigma_0^2\, r_e$, and also $L \propto r_e^2 I_e$, we find from (2.36)

$$M/L \quad \propto \quad L^{0.2} \tag{2.37}$$

a modest increase of the M/L ratio with luminosity. The fundamental plane relation has been found empirically; its theoretical foundation certainly lies in the virial theorem. The detailed derivation from a theoretical model remains an unsolved problem.

vi) The mass-to-light ratios M/L found for E galaxies vary for visible matter between $5h$ and $10h$ [Bender et al. 1992; Djorgovski and Davis 1987].

b) Disc Galaxies

i) The flat disc of spiral galaxies has a surface-brightness distribution

$$I(r) = I_0 \exp(-\alpha r). \tag{2.38}$$

The central surface brightness is constant, $I_0 \simeq 145 L_\odot pc^{-2}$ [Freeman 1970]. A spherical component (bulge) occurs in spiral galaxies with an intensity distribution following the $r^{1/4}$ law (2.33). The ratio of the luminosity in the disc (L_d) to the luminosity in the bulge (L_b) is one of the classification parameters of the different types of disc galaxies. For instance, SO galaxies have been found to have a larger L_b/L_d ratio than spiral galaxies [Burstein 1979]. Further observations have more or less established the brightness distribution (2.38) for old stellar populations. Exceptions occur, as in the case of M83, owing to the presence of young blue stars.

ii) The random velocities of nearby stars in our own galaxy are small ($\sim 30\,\mathrm{km\,s^{-1}}$) compared with the rotational velocity of the galactic disc of about 250 km s^{-1}. We expect similar behaviour in other galaxies – especially as discs are thin, and therefore large random motions would have to be very anisotropic, confined to the plane of the disc.

One of the most interesting, cosmologically relevant results came from measurements of the rotational velocity (i.e. the component of the rotational velocity along the line of sight) as a function of the radial distance from the centre, the so-called rotation curve [Rubin et al. 1985]. The rotation curve of a disc with an internal mass distribution that follows the brightness law (2.38) is expected to show a Keplerian $r^{-1/2}$ behaviour at large radii (cf. Fig. 2.34).

As can be seen from Fig. 2.34, the measurements of the rotation curves [Carignan and Freeman 1985] show a quite different behaviour: $v(r)$ stays constant out to the visible edge in many cases. Radio observations of the 21 cm line have established that a constant rotational velocity appears out to several times the optical radius. This indicates that the luminous matter does not give a true picture of the mass distribution. Indeed, if $v \equiv v_m =$ const. then the mass contained within a radius r, $M(r)$, varies in proportion to r,

$$M(r) = v_m^2 G^{-1} r. \tag{2.39}$$

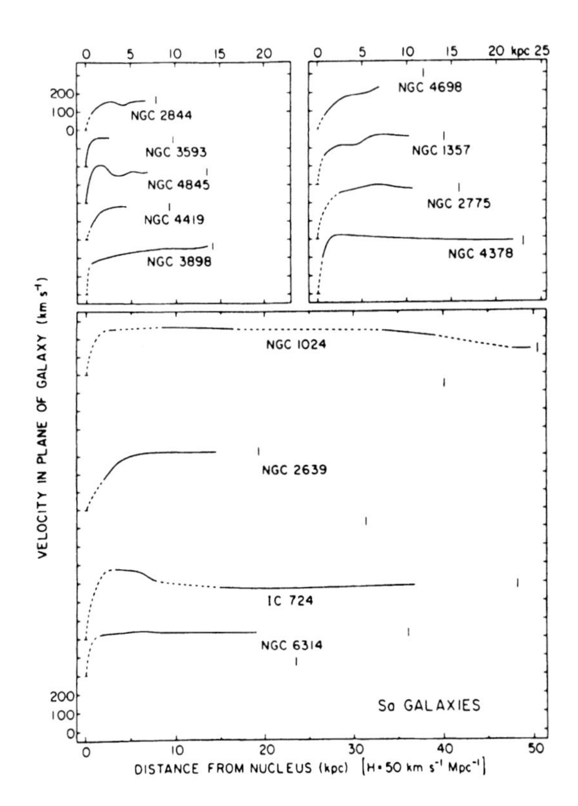

Fig. 2.34. Rotation curves of spiral galaxies [after Carignan and Freeman 1985]

If all the matter is distributed in a disc, then the surface density μ_d must follow

$$\mu_d = v_m^2/(2\pi G r), \tag{2.40}$$

but if the matter at large radii is distributed in a sphere, the density behaves as

$$\varrho(r) = v_m^2/(4\pi G r^2) \tag{2.41}$$

for constant v_m.

In any case, the law $M(r) \propto r$, implies that a large fraction of the total mass of a galaxy is in the form of a non-luminous, dark component located at large radii. The mass per unit luminosity increases with radius.

iii) A striking observational example is the polar ring galaxy A0136-0801 (Fig. 2.35) [Schweizer et al. 1983]. It consists of a flattened rotating disc – the surface brightness profile and velocity gradients have been measured and confirm this physical description. In addition, a thin ring of normal galactic matter – gas, stars, and dust – surrounds the disc, lying almost perpendicular to it. The matter in the ring is also rotating, with its axis of rotation almost in the plane of the disc.

The ring is about three times as large as the disc; in the ring orbital velocities are equal to the velocities at the edges of the disc. It can be safely inferred that in this object mass continues to increase with radius, at least out to a distance three times the disc's radius. The detailed analysis [Schweizer et al. 1983] leads to the conclusion that the distribution of the matter is not flattened but more nearly spherical.

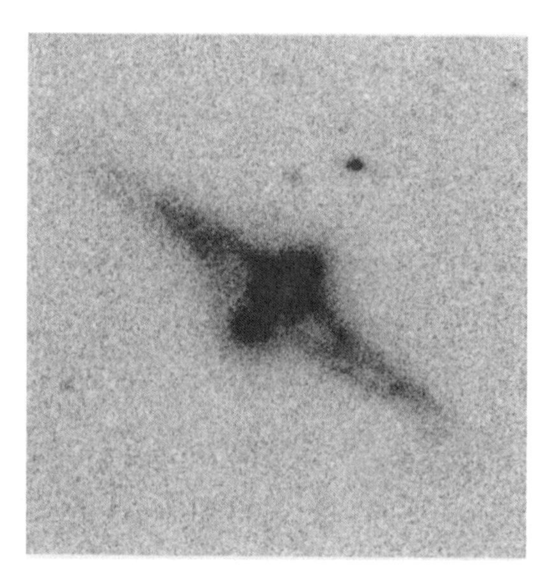

Fig. 2.35. The polar ring galaxy A0136-0801 (Cerro Tololo 4 m plate, courtesy of [F. Schweizer et al. 1983])

The formation of such a system probably proceeded in two steps: first, the formation of the disc galaxy; and second, matter being captured by the existing galaxy to form the ring. A0136-0801 is just one example of a class of about 50 similar objects [Rubin 1984; Whitmore et al. 1990].

The peculiar ring system gives direct observational evidence of a spherical halo of dark matter extending far beyond the optical disc. The non-luminous matter inferred from the rotation curves must be distributed quite differently from the luminous material – in a spherical halo with a much larger extent than the optically luminous disc. It is not clear what the constituents of this "dark matter" are; it may be concentrated in massive black holes – the remnants of burnt-out massive stars – or in Jupiter-like, small-mass stars ($M \leq 0.1 M_{\odot}$). The dark matter may also consist of a non-baryonic component, one of the possible – albeit very speculative – massive relic particles discussed in Chap. 7.

iv) The mass-to-light ratio within the Holmberg radius r_H (defined by $I(r_H) = I_0/e$, i.e. $r_H \propto 1/\alpha$) may be obtained from measurements of the rotation curve via (2.39) if the luminosity can be determined.

There is an empirical relation between the mean rotational velocity v_m – determined by a measurement of the 21 cm line profile – and the total luminosity L_t (corrected for internal absorption and absorption in our galaxy), the TF relation [Tully and Fisher 1977; Giovanelli et al. 1997]. There are actually different TF relations depending on the spectral band. For the total blue luminosity L_B one finds

$$L_B \propto v_m^3, \tag{2.42}$$

whereas for the infrared bands some observers find a steeper relation,

$$L_K \propto v_m^4. \tag{2.43}$$

For thin discs

$$I(r_H) 2\pi r_H^2 = L.$$

Hence $L^{1/2} \propto r_H$. Thus (2.42) together with (2.43) or (2.42) lead to

$$M/L \quad \propto (v_m^2 r_H/L) \propto L^{\frac{4-\beta}{2\beta}}. \tag{2.44}$$

The mass-to-light ratio depends on the parameter β which we use to parametrize the TF relation $L \propto v_M^{\beta}$.

For $\beta = 3$, we find $M/L \propto L^{1/6}$, a slight increase of M/L with luminosity.

v) The Tully–Fisher relation is the equivalent to the fundamental plane for spirals. A naive application of the virial theorem for a disc of size R proceeds from

$$v^2 \propto M/R \quad \text{i.e.} \quad v^4 \propto M^2/R^2$$

or

$$v^4 \propto (M/L)^2 \left(\frac{L}{R^2}\right) L. \tag{2.45}$$

Thus $L \propto v^4$ if $M/L = $ const. and if the surface brightness L/R^2 is constant. This assumption is definitely not satisfied, because galaxies of low surface brightness have been observed. Therefore the TF relation must have a deeper foundation. The fundamental plane relation $r_e \propto (\sigma^2)^{0.7} I_e^{-0.85}$ leads to

$$L \propto I_e r_e^2 \propto (\sigma^2)^{1.4} I_e^{-0.7} \tag{2.46}$$

which is $L \propto \sigma^{2.8}$ for $I_e = $ const., close to $L \propto \sigma^3$.

An interesting idea is to try to explain the TF relation as a consequence of the adopted star formation law [Silk 1995]: Approximating the luminosity by $L \propto M v_{\mathrm{gas}}/R$, and applying the virial theorem $M = Rv^2G^{-1}$, one finds

$$L \quad \propto \quad v^3 \frac{M_{\mathrm{gas}}}{M}, \tag{2.47}$$

where the star formation rate has been set equal to the rotational frequency v/R times the surface mass density $M_{\mathrm{gas}}/2\pi R^2$. This in turn is proportional to the surface brightness $\frac{L}{2\pi R^2}$; (2.47) follows.

vi) The flat rotation curves of spirals seem to establish the existence of an extended halo of dark matter. One can try to build models of galaxy formation which make use of such a dark matter component. In these models the disc forms when the gas cools and settles into a rotationally supported configuration at the centre of a dark matter halo. It has recently been shown [Mo et al. 1998] that such a picture can reproduce some of the general properties of observed disc galaxies. The TF relation in particular can be interpreted quite simply within such models.

There are now several experiments which test for massive dark objects in the halo of our galaxy. The idea is to observe changes in the light of a distant star due to the gravitational lens effect of a massive dark halo object. This so-called "microlensing" occurs when the halo object passes close to the line of sight from us to the star. An increase and a corresponding later decrease in the brightness of the star is expected on a time-scale of days to months, depending on the mass and velocity of the halo objects.

If the light-ray from a source passes within the Einstein radius θ_E of a point mass lens (see below for a definition of θ_E) the image is magnified. The surface brightness is preserved in gravitational light deflection, but the apparent solid angle of a source is changed. The total flux received is changed in proportion to the solid angle, and there is a magnification equal to the ratio of image area to source area. The peak magnification for a source lying on the Einstein radius is 1.34, and higher for a closer approach [see e.g. Bartelmann and Narayan 1999; Schneider et al. 1992].

Paczyński has proposed monitoring millions of stars in the LMC to detect such magnification events [Paczyński 1986]. The Einstein radius for a star in the LMC lensed by an object in the galactic halo is of the order of milliarcseconds. The magnification of 1.34 corresponds to a brightening by 0.32 magnitudes. Such changes in the light of a star can easily be detected. The expected time-scale for the light curve change depends on the typical angular scale θ_E, the relative velocity v between source and lens, and the distance to the lens D_d:

$$\Delta t = \frac{D_d \theta_E}{v}.$$

When the formula for θ_E is used:

$$\Delta t = 0.214\,\mathrm{y} \left(\frac{M}{M_\odot}\right)^{1/2} \left(\frac{D_d}{10\,\mathrm{kpc}}\right)^{1/2} \left(\frac{D_{ds}}{D_s}\right)^{1/2} \left(\frac{v}{200\,\mathrm{km\,s^{-1}}}\right)^{-1}.$$

$D_{ds}/D_s \sim 1$ for lenses in the galactic halo and sources in the LMC. Massive halo objects in the mass range $10^{-6} M_\odot$ to $10^2 M_\odot$ lead to Δt between about an hour and a year.

The Massive Compact Halo Objects (MACHO) project is one of these observation programmes with a particularly endearing name [Alcock et al. 1993, 1997]. Other projects are EROS [Aubourg et al. 1993] which also monitors stars in the LMC, and DUO and OGLE which observe microlensing events towards the Galactic bulge [Alard 1995 (DUO); Udalski et al. 1992 (OGLE)].

All the monitoring projects detect a huge number of variable stars, a nice reward, but paid for by the difficulty of distinguishing these from the few stars where variability is caused by microlensing. Two very helpful properties are that microlensing events are time symmetric with magnification independent of wavelength, while intrinsically variable stars have typically asymmetric light curves and change their colour.

More than 100 microlensing events towards the Galactic bulge and about 10 towards the LMC have been identified.

The estimated optical depth due to the LMC events is

$$\tau_{\mathrm{LMC}} = \frac{1}{\Delta\Omega} \int dV n(D_d) \pi \theta_E^2 = \left(2.1 \pm \begin{matrix} 1.1 \\ -0.7 \end{matrix}\right) \times 10^{-7},$$

($dV \equiv \Delta\Omega D_d^2 dD_d$) about four times larger than the contribution from known populations of lenses in the disc and halo of the Galaxy, as well as "self-lensing" within the LMC. The microlensing observations have apparently revealed the existence of a previously undetected population of dark objects. Can these be the dominant contribution to the dark mass of the Galactic halo?

At first sight this seems to be the obvious interpretation. But some doubts have been cast on it. The lens mass inferred is $\sim 0.5 M_\odot$, and it is puzzling why such objects have not been seen in searches for white dwarfs and subdwarfs [Graaf and Freese 1996].

There are also arguments that most of the baryonic dark matter in the universe exists in the form of diffuse hot gas, and not in the form of massive objects. At present it seems that $\sim 30\%$ of the halo of the Milky Way can be supplied by MACHOs, while the rest are just unknown dark matter particles. The mean dark matter density in the halo is $0.3\,\mathrm{GeV/cm^3}$.

vii) Great interest in the TF relation as a distance indicator [Giovanelli et al. 1997] has led to many observations of disc galaxies, and large samples are now available [Mathewson and Ford 1996; Courteau 1997]. It is thus possible to place constraints on the standard picture of disc formation from these observations [Syer et al. 1999]. So, for example, the total M/L in the infrared I-band has been derived within such models, and by comparison with observations – it has been found that

$$M/L = 43h. \tag{2.48}$$

2.3.2 Estimates of Ω_m from Galaxies

The estimates for Ω_m obtained from the M/L ratios for various objects will be presented here. The basis for these estimates is (2.32),

$$\Omega_m = (1450)^{-1}(M/L)h^{-1}.$$

An M/L ratio for a particular class of cosmic sources will result in a value of Ω_m.

a) Stars

Limits on the mass-to-light ratio of the stars in the Galactic disc in the solar neighbourhood can be derived from a combination of kinematic measurements and star counts. A local surface mass density in the disc of $40\, M_\odot pc^{-2}$ and the V-band luminosity density of $15\, L_\odot pc^{-2}$ result in a mass-to-light ratio in the visual band of

$$M/L_v = 2.67 \tag{2.49}$$

[Gould et al. 1996].

The mass-to-light ratio of extragalactic discs can also be measured directly from kinematic studies [Bottema 1993, 1997]. The M/L measurement relies on the relation between vertical velocity dispersion, surface density, and vertical scale height. A value of M/L in the I-band is derived:

$$(M/L)_I = (1.7 \pm 0.5)h \tag{2.50}$$

[Bottema 1997].

Mass-to-light ratios can also be derived from stellar population synthesis models. The main uncertainty lies in the poorly known initial stellar mass function. The predictions depend on the metallicity and age of the stellar population. $(M/L)_I$ is found to be between 0.9 and 1.8, in agreement with other values.

Thus from starlight

$$\Omega_{m,\star} = 0.0014h^{-1}, \tag{2.51}$$

much lower than previous estimates which have relied on the maximum disc hypothesis, i.e. on the assumption that the inner part of the rotation curve is completely due to the luminous disc. It seems that this assumption does not hold in general. $\Omega_{m,\star}$ is really far away from the critical value $\Omega_m = 1$.

b) Rotation Curves and FP Relations

It is interesting to note that for dynamical measurements of M/L the uncertainty in H_0 does not influence the Ω_m estimate. $\langle L \rangle$ is proportional to H_0, and M/L is also proportional to H_0. Thus Ω_m is independent of H_0.

As we have seen, the optical parts of E and spiral galaxies have $M/L \simeq (5 \text{ to } 10)h$, and therefore

$$\Omega_{m,G} = 0.0035 \quad \text{to} \quad 0.007. \tag{2.52}$$

c) Models for Disc Galaxies

Modelling the dark matter halo of disc galaxies and comparing it to observations resulted in $M/L = 43h$ [Syer et al. 1999]. Taking this as typical, we can derive a maximal contribution of galaxies to the matter density of

$$\Omega_{m,G} = 0.03. \tag{2.53}$$

This is still far below the critical density, but we are not necessarily done. There may be a lot of matter not clustered in galaxies, and this can help to increase Ω_m substantially, as we shall see.

d) Binaries and Groups of Galaxies

Binary galaxies and galaxies with satellites have been used to determine the mass of a galaxy. The Kepler relation $\frac{GM}{r} = kv^2$ is used, and for binary galaxies the characteristic radius is the projected separation, and the velocity is the redshift difference. The dimensionless parameter k depends on model assumptions such as the shape of galaxy orbits, on the relative mass distribution, on selection and projection effects. The uncertainties are large, the values given $M/L \simeq (50 - 100)h$, i.e. $\Omega = (0.03 - 0.06)$.

The situation is better for groups of galaxies because a larger number of members reduces the scatter in k. While binaries are typically separated by $\leq 100\,\mathrm{kpc}$, the typical size of groups is $\sim 1\,\mathrm{Mpc}$. Their dynamics allow the mass distribution to be inferred on a much larger scale not tightly connected to an individual galaxy. There is, however, one great difficulty: groups are loose associations, and deciding whether a galaxy belongs to a group or not is a problem. The collapse times of groups are very long, and they may be far from equilibrium (Henriksen and Mamon 1994). These uncertainties make M/L determinations for groups quite unreliable. In the literature one can find a M/L between $50h$ and $200h$, i.e.

$$\Omega_m = (0.03 - 0.12). \tag{2.54}$$

2.3.3 Estimates of Ω_m from Clusters of Galaxies

Rich clusters of galaxies offer several possibilities for estimating their mass, and therefore M/L.

1. They are the largest virialized systems, and the virial theorem can be applied to the motion of the galaxies in the cluster. For a relaxed stationary object the virial theorem states that $2T + \Omega = 0$, where T is the kinetic energy and Ω the potential energy:

$$T = \frac{1}{2}\Sigma m_i \dot{r}_i^2 \equiv \frac{1}{2}\langle v^2\rangle M;$$

$$\Omega = -\frac{GM^2}{R_v}, \qquad \text{i.e. } \langle v^2\rangle = \frac{GM}{R_v}. \tag{2.55}$$

M is the total mass of the cluster; R_v is the effective virial radius; $\langle v^2\rangle$ is actually a non-observable quantity, the sum of the mass-averaged velocity squares. Only the velocity in the direction of the line of sight is measureable. Assumptions are necessary to get an estimate of $\langle v^2\rangle$. Often a luminosity weighted mean is taken. Often one uses $\langle v\rangle^2 = 3\sigma^2$, where σ^2 is the line-of-sight velocity dispersion, but this may be quite wrong – by a factor as high as 2.

The virial radius R_v is determined from the projected number of galaxies. The mass determinations can be roughly represented by

$$M = 1.1 \times 10^{14} \left(\frac{\sigma}{10^3\,\mathrm{km\,s^{-1}}}\right)^2 \left(\frac{\theta}{30\,\mathrm{s}}\right) \left(\frac{D}{1\,\mathrm{Gpc}}\right) M_\odot \tag{2.56}$$

(θ is the angular extent analysed; D the distance). The method has been applied to various rich clusters. The Coma cluster in particular has a large number of measured redshifts, and various authors find $(M/L_B)_{\mathrm{Coma}} = 150h$ to $400h$, i.e. $\Omega_m = 0.1$ to 0.27, if this is a typical object. For the cluster MS1224 one finds $M/L = 250h$. The errors are just guesses of the uncertainties involved, and at present one usually takes

$$\Omega_0 = 0.1 \quad \text{to} \quad 0.3$$

as the result derived from the application of the virial theorem to clusters. The fact that reliable M/L values can be obtained only for the cores of clusters, where the orbital periods are sufficiently short, gives rise to another cautionary point: the outer regions of clusters have orbital periods too long to make similar estimates, but the dense cores contribute only 10% of the light in the universe [Salpeter 1984]. Thus there is room for dramatically different models of the distribution of the dark matter.

2. The mass contained in a cluster can act as a gravitational lens on the light of distant galaxies. Sometimes the image appears as a giant arc, and the radius of the arc determines the mass within. Very often the galaxies appear as tiny elliptical images called "arclets". Typically there are about 70 arclets per $(\text{arcmin})^2$. The statistical analysis of these arclets permits a reconstruction of the cluster mass distribution.

This interesting approach deserves a few more words of explanation. For details, see [Schneider et al. 1992; Narayan and Bartelmann 1999].

Consider first the case of a point mass M and a light ray which would, if unperturbed, pass M at a distance b. The deflection angle due to the gravitational field is

$$\alpha = \frac{4GM}{c^2 b}. \tag{2.57}$$

The light deflection at the edge of the Sun is $\alpha = 1.7$ arcsec. This was first measured by Arthur Eddington's famous expedition to the solar eclipse of 1919. The measurement confirmed a prediction of General Relativity, and marked the beginning of the immense popularity of Albert Einstein.

In Fig. 2.36 a schematic diagram of the light rays in a gravitational lens is shown.

The true angular distance between a distant galaxy S and the centre of the gravitational lens is $\boldsymbol{\beta}$ (a 2-component vector in the plane of the sky). Because of the light deflection the source S appears to be at the angular distance $\boldsymbol{\theta}$. The lens can be considered thin compared to the total lenght of the light path. Its mass distribution can be projected along the line of sight and be replaced by a mass sheet orthogonal to the line of sight. This flat lens approximates the real situation quite well, and it can be characterized by its surface mass density

$$\Sigma(\boldsymbol{\xi}) \equiv \int \rho(\boldsymbol{\xi}, z) dz \tag{2.58}$$

(ρ is mass density, z is distance along the unperturbed ray). The deflection angle at position $\boldsymbol{\xi}$ is the sum of the deflections due to all the mass elements in the lens plane. For the special case of a symmetric lens this 2-component angle can be written as

$$\hat{\alpha} \equiv \mid \hat{\boldsymbol{\alpha}} \mid = \frac{4GM(\xi)}{c^2 \xi} \tag{2.59}$$

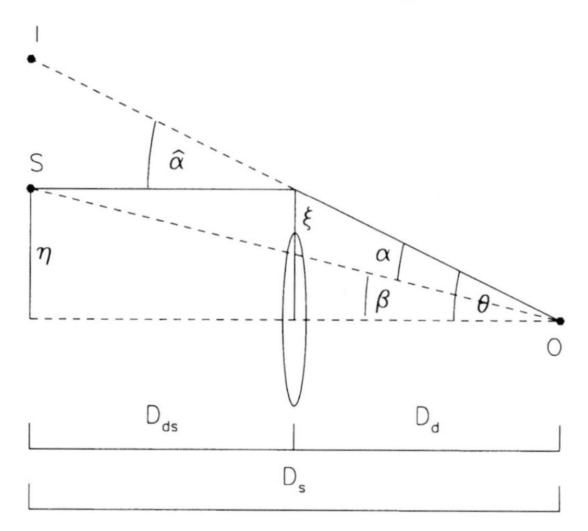

Fig. 2.36. Schematic drawing of a gravitational lens system [from Bartelmann and Narayan 1999]. A light ray from the source S at transverse distance η from the optical axis to the observer O passes the lens at a transverse distance ξ. It is deflected by an angle $\hat{\alpha}$. The angular separations of the source and the image from the optical axis are β and θ, respectively, as seen by the observer. The distance between O and S, O and the lens, and the lens and S are D_s, D_d and D_{ds} respectively. The reduced deflection angle α is $\alpha = D_{ds}/D_s\hat{\alpha}$

where $M(\xi)$ in the mass within radius ξ:

$$M(\xi) = 2\pi \int_0^{\xi} \xi' \Sigma(\xi') d\xi' .$$

The distances between source and observer, observer and lens, as well as lens and source are denoted as D_s, D_d and D_{ds}, respectively, in Fig. 2.36. The position of source and image on the sky are related by the simple equation (the "lens equation")

$$\boldsymbol{\beta} = \boldsymbol{\theta} - \frac{D_{ds}}{D_s} \hat{\boldsymbol{\alpha}} . \tag{2.60}$$

The derivation of the lens equation requires only that the relation between the angle enclosed by two lines and their separation is:

$$\text{separation} = \text{angle} \times \text{distance}.$$

In curved space-times distances D_{ds}, D_d, D_s are defined such that this relation still holds; this corresponds to defining them as angular diameter distances r_A (see Chap. 1).

A point source which is exactly on the optical axis appears as a circular ring with radius

$$\theta_E = \left[\frac{4GM(\theta_E)}{c^2} \frac{D_{ds}}{D_d D_s} \right]^{1/2} . \tag{2.61}$$

This "Einstein" radius is, of course, an ideal case which is never seen in reality. But parts of Einstein rings, namely giant arcs in the sky, have been seen as images of distant galaxies.

As instructive examples take the case of a star in the Galaxy, for which $M \sim 1 M_\odot$ and $d_{\rm eff} = \frac{D_d D_s}{D_{ds}} \sim 10\,{\rm kpc}$, and the case of a galaxy cluster with $M \sim 10^{14} M_\odot$, and $d_{\rm eff} \sim 1\,{\rm Gpc}$.

The corresponding Einstein radii are:

$$\begin{aligned}\theta_E &= 0.0009''(M/M_\odot)^{1/2}(d_{\rm eff}/10\,{\rm kpc})^{-1/2},\\ \theta_E &= 9''(M/10^{14}M_\odot)^{1/2}(d_{\rm eff}/1\,{\rm Gpc})^{-1/2}.\end{aligned}$$

Consider a lens with a constant surface mass density. Using $\xi = D_d\theta$ and $\alpha \equiv \hat{\alpha} D_{ds}/D_s$, we obtain

$$\alpha = \frac{D_{ds}}{D_s}\frac{4G}{c^2\xi}(\Sigma\pi\xi^2) = \frac{4\pi G}{c^2}\Sigma\frac{D_d D_{ds}}{D_s}\theta.$$

In this case $\beta \propto \theta$. We can write

$$\alpha = \Sigma/\Sigma_{\rm cr}\theta \tag{2.62}$$

with a critical surface mass–density

$$\Sigma_{\rm cr} = \frac{c^2}{4\pi G}\frac{D_s}{D_d D_{ds}} = 0.35 \quad {\rm gcm}^{-2}\left(\frac{d_{\rm eff}}{1\,{\rm Gpc}}\right) \tag{2.63}$$

where $d_{\rm eff} \equiv \frac{D_d D_{ds}}{D_s}$. For a lens with $\Sigma = \Sigma_{\rm cr}$, we have $\alpha(\theta) = \theta$, i.e. $\beta = 0$ for all θ. A "critical" gravitational lens focuses perfectly.

A giant arc in a cluster traces a part of a circle which encloses the mass responsible for the gravitational lens effect. One can show that $\Sigma = \Sigma_{\rm cr}$ for a circularly symmetric lens with $\theta_{\rm arc} \simeq \theta_E$. Thus

$$\langle\Sigma(\theta_{\rm arc}\rangle = \langle\Sigma(\theta_E\rangle = \theta_E,$$

and the mass enclosed within $\theta_{\rm arc} = \theta_E$ is:

$$M = \Sigma_{\rm cr}\pi(D_d\theta)^2 \simeq 1.1 \times 10^{14} M_\odot \left(\frac{\theta}{30''}\right)^2\left(\frac{D}{1\,{\rm Gpc}}\right) \tag{2.64}$$

here D is another effective distance $\frac{D_d D_s}{D_{ds}}$.

Mass-to-light ratios in the blue for three clusters with giant arcs are:

This shows good agreement with the virial estimates.

The lens equation is in general non-linear with several images $\boldsymbol{\theta}$ of a single source position $\boldsymbol{\beta}$. Therefore it cannot be simply inverted to derive properties of the source from observations. Light from a circular source will be focused isotropically – leading to an amplification – as well as being distorted into

Cluster	(M/L_B)
A370	$\simeq 200h$
A2390	$\simeq 120h$
MS2137-23	$\simeq 250h$

[Narayan and Bartelmann 1999 and ref. therein].

an elliptical image. These features of a lens can be described by quantities called "convergence" $\kappa(\theta)$ and "shear" $(\gamma_1(\theta), \gamma_2(\theta))$. The surface density of the lens is obtained from $\Sigma(\theta) = \Sigma_{\rm cr}\kappa(\theta)$. Both quantities can be derived from an effective lens potential $\psi(\boldsymbol{\theta})$:

$$\psi(\boldsymbol{\theta}) = \frac{D_{ds}}{D_d D_s} \frac{2}{c^2} \int \phi(D_d \boldsymbol{\theta}, z) dz \tag{2.65}$$

(ϕ being the Newtonian potential of the mass distribution of the lens). For completeness' sake:

$$\kappa(\boldsymbol{\theta}) = 1/2(\psi_{,11} + \psi_{,22}), \tag{2.66}$$

$$\gamma_1(\boldsymbol{\theta}) = 1/2(\psi_{,11} - \psi_{,22}), \tag{2.67}$$

$$\gamma_2(\boldsymbol{\theta}) = \psi_{,12}, \tag{2.68}$$

$$\left(\psi_{,ij} \equiv \frac{\partial^2 \psi}{\partial \theta_i \partial \theta_j}\right). \tag{2.69}$$

There is a connection between κ and γ which can be expressed as a convolution integral in θ-space:

$$\kappa(\boldsymbol{\theta}) = 1/\pi \int d^2\theta' \,\mathrm{Re}[D^*(\boldsymbol{\theta} - \boldsymbol{\theta}')\gamma(\boldsymbol{\theta}')]$$

where D is a komplex kernel

$$D(\theta) = \theta^{-4}[\theta_1^2 - \theta_2^2 - 2i\theta_1\theta_2]$$

and $\gamma = \gamma_1 + i\gamma_2$; the asterisk denotes complex conjugation. The shear field γ can be measured, and thus the reconstruction of $\kappa(\boldsymbol{\theta})$ becomes a typical inverse problem.

$\gamma(\boldsymbol{\theta})$ can be obtained from the elliptical distortion of background galaxies. (The experts' jargon for this phenomenon is "weak lensing".) The ellipticity of an image can be defined as:

$$\epsilon = \epsilon_1 + i\epsilon_2 = \frac{1-r}{1+r} e^{2i\phi}; \quad r = b/a,$$

ϕ is the angle in the (r, ϕ) description of an ellipse, b is the semimajor, a the semiminor axis. The mean ellipticity produced by a gravitational lens is

$$\langle \epsilon \rangle = \left\langle \frac{\gamma}{1-\kappa} \right\rangle \tag{2.70}$$

where $\langle \cdots \rangle$ denotes averages over a finite area of the sky. When the case of weak lensing is considered, e.g. the imaging of distant background galaxies by a massive foreground cluster, then in general $\kappa \ll 1$, and $|\gamma| \ll 1$ (Fig. 2.37). In this limit the shear is directly given by the mean ellipticity:

$$\langle \gamma_1 \rangle \simeq \langle \epsilon_1 \rangle \; ; \qquad \langle \gamma_2 \rangle \simeq \langle \epsilon_2 \rangle \, .$$

Thus the integral for $\langle \kappa(\theta) \rangle$ can be directly evaluated from the measured ellipticity. Actually, to obtain $\langle \epsilon_i \rangle$ with an acceptable signal-to-noise ratio, one has to average over many slightly deformed galaxy images [Kaiser and Squires 1993; Seitz and Schneider 1996; Bartelmann and Narayan 1995]. Clearly this method is a very promising way to obtain masses and density profiles of clusters, no matter whether they are virialized or not, because the lensed images provide a momentary snapshot of the current mass distribution.

Some M/L ratios (for blue light) from weak lensing (arclets) are given in Table 2.6.

First measurements of the cosmic shear field produce results more or less in agreement (see [Van Waerbeke et al. 2002]).

3. Cluster masses can also be obtained from X-ray data with the additional bonus that the mass in baryons contained in the cluster can also be determined. The X-rays are emitted from a hot gas in the cluster at a temperature of $\sim 10^8$ K. In Fig. 2.38 the X-ray image of a galaxy cluster is shown. The hot gas must be kept in equilibrium and the mass $M(r)$ within the radius r must be large enough to hold it. Both the density $\rho(r)$, and the temperature $T(r)$

Fig. 2.37. The cluster A2218 acts as a gravitational lens and distorts the images of distant galaxies into elliptical shapes

Table 2.6. Mass-to-light ratios of several clusters derived from weak lensing (from [Bartelmann and Narayan 1999])

Cluster	M/L	Remark	Reference
MS1224	$800h$	Virial mass ~ 3 times smaller	[Fahlmann et al. 1994]
Cl1455	$520h$	DM more concentrated than galaxies	[Smail et al. 1995]
A2218	$440h$	Gas mass fraction $< 4\% h^{-3/2}$	[Squires et al. 1996]
A851	$200h$	Mass distrib. agrees with galaxies and X-rays	[Seitz et al. 1996]

Ω_m for clustered matter derived from these measurements is in the range $0.13 \leq \Omega_m \leq 0.5$.

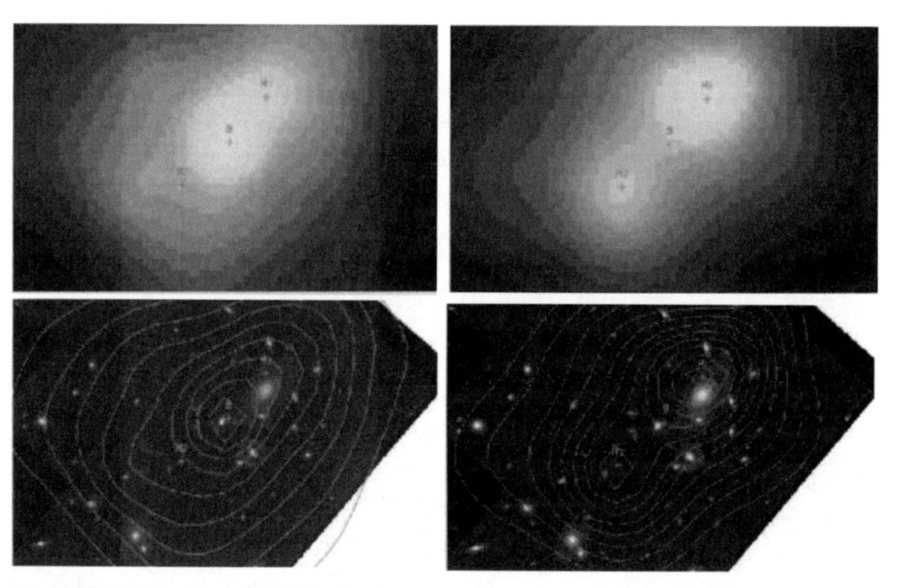

Fig. 2.38. The cluster A2218 also shows extended X-ray emission as recorded here in this observation from the Chandra X-ray observatory

of the gas can be obtained directly from X-ray observations. The modeling of the X-ray emission is done under various simplifying assumptions on the cluster shape and its gravitational potential. The estimates of the gas density from the X-ray emission measure seem quite reliable, wheras temperatures can only be determined for spectra of extremely good resolution.

Often the cluster mass determination by X-ray observations is based on the assumption that the hot gas emitting the X-rays is kept in hydrostatic equilibrium by the gravitational mass of the cluster.

The mass profile for a spherically symmetric cluster is

$$M(r) = -\frac{kT(r)r^2}{m_n \mu G}\left(\frac{d\log T(r)}{dr} + \frac{d\log \varrho(r)}{dr}\right) \tag{2.71}$$

where $\varrho(r)$ and $T(r)$ are the density and temperature profile of the gas.

ROSAT observations (Böhringer 1995; David et al. 1994; Schindler 1996) lead to the conclusion that the hot gas producing the X-rays has about 5 times the mass of the galaxies in the cluster. For rich clusters of galaxies on finds typically

$$M_{\text{grav}} \sim (5 \times 10^{14} - 5 \times 10^{15})\, M_\odot$$

Galaxies are about 1–4% of the total mass

$$M_{\text{Gal}}/M_{\text{grav}} = (0.01 - 0.04)h^{-1}.$$

The results are consistent with the M/L estimates derived from weak lensing [Squires et al. 1996].

The X-ray observations of clusters offer in addition the opportunity to derive an estimate of the gas mass, and hence of the mass in baryons contained in a cluster.

The X-ray luminosity L_X of a spherical cluster of radius R is

$$L_X \simeq n^2 T^{1/2} R^3;$$

here n is the electron (and thus also the proton) density, T the X-ray temperature. For a virialized cluster the total mass M is proportional to TR. Since the gas mass M_g is proportional to nR^3, we have the following dependence on the Hubble constant

$$L_X \propto h^{-2}, \text{ hence } \quad n^2 R^3 \propto h^{-2}, \text{ and } \quad n \propto h^{-2.5}.$$

The ratio M_g/M is then $\propto h^{-1.5}$. From the X-ray data we have

$$\frac{M_g}{M} = (0.1 - 0.4)h^{-1.5}. \tag{2.72}$$

This can be translated into a limit on Ω_m, if we assume that the ratio M_g/M has a universal value and if we take into account the limit on the baryonic density parameter $\Omega_B < 0.025h^{-2}$ (see Chap. 3). Then

$$\frac{\Omega_B}{\Omega_m} > 0.1h^{-1.5}, \quad \text{hence} \quad \Omega_m < 10\Omega_B h^{1.5} < 0.25h^{0.5} \tag{2.73}$$

(cf. [White et al. 1993]).

4. Three conclusions can be drawn immediately:

Firstly, the value of Ω_m derived from the M/L ratio of cluster is between 0.1 and 0.3. Thus it is more than 10 times higher than the density parameter for visible light. The universe is dominated by dark matter.

Secondly, the limits on the baryonic density Ω_B (these will be discussed in Chap. 3) indicate that the main component of the dark matter is non-baryonic.

Thirdly, the matter clumped on the scale of clusters is not enough to reach the critical density $\Omega_m = 1$.

2.3.4 The Cosmic Virial Theorem

Typical clusters of galaxies at a scale r will have an overdensity with respect to the mean $\langle\varrho\rangle$ of

$$\frac{\delta\varrho}{\langle\varrho\rangle} \sim \xi(r). \tag{2.74}$$

Clumps of galaxies with a high excess over the background density might be bound and stable – the virial theorem may be applicable. The mass of a typical clump is

$$M \simeq \langle\varrho\rangle\, \xi r^3;$$

$\xi(r)$ is actually the two-point correlation function [Peebles 1980] defined by $n(r) = (1 + \xi(r))\langle n\rangle$, where $n(r)$ is the number density of galaxies at a distance r from any given galaxy and $\langle n\rangle$ is the general average density. Note that this definition of $\xi(r)$ makes sense only if there exists an average density $\langle n\rangle$ whose value is independent of the volume considered. As we have remarked above, there is only rather weak observational evidence for such a behaviour. The following discussion rests, however, on this assumption. An interpretation challenging this fairly generally accepted approach is contained in [Pietronero 1987]. These arguments will be discussed in some detail in Chap. 10. For a wide range of distances ($10\,\mathrm{kpc} \le r \le 10\,\mathrm{Mpc}$) ξ seems to follow a simple power law $\xi \propto r^{-1.8}$ [Peebles 1980]. The velocities should therefore scale as

$$\langle v^2\rangle \sim \frac{GM}{r} \sim \langle\varrho\rangle\, G\xi r^2 \propto r^{0.2} \tag{2.75}$$

[Efstathiou and Silk 1983]. A more rigorous argument makes use of the three-point correlation function $\zeta(r_{12}, r_{23}, r_{31})$ [cf. Peebles 1980]. If the velocity distribution is isotropic and the cluster is in gravitational equilibrium, the mean-square relative peculiar velocity between pairs of galaxies at separations r_{12} is a one-dimensional quantity $\langle v_{12}^2\rangle$ which can be related to the sum of the accelerations of any three galaxies

$$\langle v^2\rangle = \frac{6G\langle\varrho\rangle}{\xi(r)} \int_r^\infty \frac{dr}{r} \int d^3z \frac{\boldsymbol{r}\cdot\boldsymbol{z}}{z^3} \zeta(r, z|\boldsymbol{r}\cdot\boldsymbol{z}|). \tag{2.76}$$

Inserting $\xi(r)$ and an analytic expression for ζ [Peebles and Groth 1975] in terms of two-point correlation functions,

$$\zeta(r_{12}, r_{23}, r_{31}) = Q[\xi(r_{12})\xi(r_{23}) + \xi(r_{23})\xi(r_{31}) + \xi_{(r_{31})}\xi(r_{12})]; \qquad (2.77)$$

$$Q = 1.3 \pm 0.2$$

yields the simple formula

$$\left\langle v_{12}^2 \right\rangle^{1/2} = 800 Q^{1/2} \Omega_m^{1/2} \left(\frac{r_0}{8}\right)^{\gamma/2} \left(\frac{r}{h^{-1}Mpc}\right)^{0.2} \mathrm{km\,s^{-1}}. \qquad (2.78)$$

This relation is sometimes called the "cosmic virial theorem". It is consistent with the result of the dimensional analysis given above. The cosmological importance of (2.78) is the inherent possibility of measuring the density parameter Ω_m from a determination of the relative peculiar velocities of pairs of galaxies. This method assumes that the galaxy two-point correlation function is determined by the actual mass distribution – as we have remarked, the presence of dark matter sheds some doubt on this.

A simple way to take into account a different distribution of galaxies and dark matter clumps is the introduction of a "bias", a factor b^2 relating the galaxy correlation function $\xi_g(r)$ to the matter correlation function $\xi_m(r)$:

$$\xi_g(r) = b^2 \xi_m(r). \qquad (2.79)$$

Theoretical and observational results seem to indicate that b is not a constant, but depends on the separation r.

Then instead of Ω_m, the ratio Ω_m/b is determined by (2.78). Earlier applications of the cosmic virial theorem gave low values

$$\left\langle v_{12}^2 \right\rangle^{1/2} = (300 \pm 50)\,\mathrm{km\,s^{-1}} \quad \text{at} \quad r \sim 1h^{-1}\,\mathrm{Mpc} \qquad (2.80)$$

and correspondingly low values of $\Omega_{m/b} \simeq 0.1$. [Davis and Peebles 1983b; Bean et al. 1983; Efstathiou and Silk 1983].

Later investigations have shed doubt on these earlier estimtes. It was shown that the pairwise velocity dispersion depends on selection effects quite strongly. Especially the inclusion of rich cluster is crucial – if there are not enough rich clusters in the sample, then $\langle v_{12}^2 \rangle$ cannot be measured reliably [Mo et al. 1993, 1997]. Recently a large redshift survey, the LCRS [Shectman et al. 1996], has been completed. This sample is large enough to measure $\langle v_{12}^2 \rangle$ accurately. The value obtained was

$$\left\langle v_{12}^2 \right\rangle^{1/2} = (570 \pm 80) \quad \mathrm{km\,s^{-1}} \quad \text{at} \quad r \sim 1h^{-1}\,\mathrm{Mpc}, \qquad (2.81)$$

larger than early estimates.

The LCRS is a data set good enough to measure the three-point correlation function too. The result is that the "hierarchical" model of (2.77) does not hold. Rather Q is a function of the separation of the galaxies [Jing and Börner 1998]. The cosmic virial theorem in its simplified form does not hold either – which means that the structures measured are in general not gravitationally relaxed. A reasonable fit to the observations can be produced by a Cold Dark matter (CDM) model with $\Omega_m = 0.2$, and $\Omega_\Lambda = 0.8$. [Jing and Börner 1998]. (More on LCRS and correlation functions can be found in Chap. 12).

The best estimate for Ω_m using this approach is [Mo et al. 1993]

$$\Omega_m = 0.28b^2 \quad (\langle v_{12}^2 \rangle^{1/2} / 400\,\mathrm{km\,s^{-1}})^2.$$

2.3.5 Infall to the Virgo Cluster

The observed anisotropy of the local velocity field has been ascribed to the overdensity of the Virgo cluster. Our local group is falling towards the centre of Virgo because of the gravitational attraction of this overdensity. The measurements of the infall velocity allow a determination of Ω_0 according to the simple formula

$$\Omega_0 = \left(\frac{3v_v}{\delta(v_v + v_0)} \right)^{1.5}$$

derived for a spherical flow model [Davis and Peebles 1983a]. In one analysis the value $v_v = 220 \pm 50$ has been derived for the infall velocity (in $\mathrm{km\,s^{-1}}$) [Tammann and Sandage 1985]. Taking the overdensity $\delta = 2.8 \pm 0.5$ and v_0 the normal Virgo cluster member velocity (~ 970), one arrives at

$$\Omega_0 = 0.09 \pm 0.04.$$

Thus Ω_0 is well within the range of the values derived from the virial theorem applied to clusters.

2.3.6 Large-Scale Peculiar Motions

Besides virialized parts of the cosmos, we can also look at large-scale density fluctuations which are not in gravitational equilibrium. The difference in mass in large regions leads to large-scale peculiar velocity patterns which can give direct information on the underlying density field.

Multipole expansions of the light distribution in a large volume of space are used to infer properties of the peculiar velocity field, as well as the method of locally reconstructing the gravitational potential from direct distance and redshift determinations of a large number of individual galaxies.

The cosmic microwave background exhibits a dipole anisotropy (cf. Sect. 2.5) which cannot be explained just by the motion of the Local Group towards the Virgo cluster.

There is another component of this motion which must be caused by the superposition of all the mass concentrations around our position. It is probably not possible to ascribe this motion to a few attractors, evern if they are "Great". It is very likely that an N-body problem must be solved to find the true explanation.

There are some encouraging results already. The dipole moments of the galaxy brightness distribution in optical and infrared whole-sky catalogues have been estimated. A combination of three optical diameter-limited catalogues was used to estimate the optical dipole [Lahav 1987; Lynden-Bell et al. 1989]. It was found to lie only 7–15° away from the CMB dipole. About 80% of it is produced within a radius of $4000\,\mathrm{km\,s^{-1}H_0^{-1}}$. The IRAS (IRAS, the Infrared Astronomical Satellite observes sources in the infrared light) dipole has been newly determined by using the QDOT survey, a redshift catalogue of IRAS galaxies, where one out of six galaxies was randomly selected to have its redshift measured. This way 2163 randomly sampled galaxies brighter than 0.6 Jy covering all the sky at galactic latitudes $|b| > 10°$ have been available to determine the peculiar velocity field [Rowan-Robinson et al. 1990; Strauss 1989]. The IRAS dipole is found to originate within $5000\,\mathrm{km\,s^{-1}}$ to $10\,000\,\mathrm{km\,s^{-1}}$, and to agree in direction with the CMB dipole.

In the manner described earlier, these results can be applied to a derivation of a value for Ω_0. For such estimates, however, the effect of biasing should be taken into account. The acceleration

$$g = \frac{1}{2} H_0^2 r \Omega_m \frac{\delta n}{n}$$

is deduced from the distribution of galaxies, not from the matter distribution. Thus a relation between the fluctuation in galaxy density $\delta n/n$ and in matter density $\delta \varrho/\rho$ of the form

$$\frac{\delta n}{n} = b \frac{\delta \varrho}{\varrho}$$

may be used. The peculiar velocity is given by

$$v = \frac{2 f(\Omega_m)}{3 H_o \Omega_m} g(r),$$

where $f(\Omega_m) \simeq \Omega_m^{0.6}$ is a useful analytical approximation [Peebles 1980, Sect. 14]. Finally,

$$v_p = \frac{1}{3} \delta H_0 r \Omega_m^{0.6} / b,$$

where we have written $\delta \equiv \delta\rho/\rho$. One finds

$$\Omega_m^{0.6}/b = 0.4 \quad \text{for optical galaxies,}$$

$$\Omega_m^{0.6}/b = 0.8 \quad \text{for IRAS galaxies.}$$

Different biasing parameters must be assumed to obtain the same Ω_m, i.e. $b_{\rm opt} = 2b_{\rm IRAS}$. For the critical value $\Omega_m = 1$, e.g. $b_{\rm opt} = 2.5$, and $b_{\rm IRAS} = 1.3$. Thus optical and IRAS galaxies are different tracers of the mass distribution.

An alternative zero curvature model with baryonic matter $\Omega_{0,B} = 0.1$, and a cosmological constant $\Omega_\Lambda = \Lambda/3H_0^2 = 0.9$, would lead to the conclusion that the galaxies are less clustered than the matter. The antibiasing parameters would then be $b_{\rm opt} = 0.6$, and $b_{\rm IRAS} = 0.3$. The anisotropy limits of the microwave background (see Sect. 2.5) are a great concern for such models.

The measurements of a bulk motion for a sample of 200 elliptical galaxies [Dressler et al. 1987] and the subsequent ("undeserved"?) publicity for the "seven samurai" and their "Great Attractor" have motivated several groups to use various distance indicators to get more accurate peculiar velocities. The real uncertainties in the use of semi-empirical relations such as the infrared Tully–Fisher relation for spirals (luminosity-line width) or the fundamental plane relation for ellipticals are difficult to access. The prediction of the "Great Attractor" [Lynden-Bell et al. 1988] depended crucially on the calibration of the D_n–σ relation $\log D_n = x \log \sigma + f(r)$ (where $r = zc/H_0$, D_n is the photometrically defined angular size, σ the velocity dispersion). A higher accuracy in the distance determination can be achieved in a statistical sense, by measuring the distance to groups of galaxies, and then reducing the error by dividing by the square root of the number of galaxies in the group. Systematic errors in this procedure, like the dependence of the intrinsic quantity on a third parameter, may really bring trouble. In particular, the algorithm defining the different groups of galaxies is of crucial importance. It has been shown that a slightly different calibration [Kates and Weigelt 1990] makes the model of a "Great Attractor" statistically less favourable (in a χ^2 test) than even simple models of a nearly constant velocity field varying linearly with distance.

Other attempts to improve the modelling involve a reconstruction of the velocity field with a specific smoothing length [Bertschinger and Dekel 1989]. The authors claim that by this reconstruction, named "POTENT", one can achieve from the few data points a higher accuracy than the actual observation. This claim seems to me doubtful, since the smoothing procedure must have a great influence on the results.

The aplication of this procedure to a sample of spirals (Tully–Fisher distances) and ellipticals (D_n–σ distance) roughly confirmed the Great Attractor model [Bertschinger et al. 1990].

This potential reconstruction method ("POTENT") relies on the central assumption of a curl-free peculiar velocity field. There is no proof for this assumption from observations since the radial velocity components along the line of sight which can be measured do not contain sufficient information to allow a decision whether curl $\boldsymbol{v}$ is zero or not.

As we shall see in Chap. 11 the rotational motion of an incompressible fluid as a linear perturbation decays proportionally to the inverse of the expansion factor, $R^{-1}(t)$. Numerical simulations have demonstrated [Bertschinger and

Dekel 1989] that smoothing of the velocity field on large scales results in a curl-free field, since collapse processes which can produce large circulations are then smoothed out. If one accepts the assumption curl $\boldsymbol{v} = 0$, then the velocity field can be written as the gradient of a scalar potential $\psi(r)$:

$$\boldsymbol{v}(\boldsymbol{r}) = \nabla\psi(\boldsymbol{r}). \tag{2.82}$$

If in addition the radial component of $\mathbf{v}$ can be determined within a sphere of radius r_0, then the velocity potential ψ within that sphere can be calculated by integration of $v_r(r, \Theta, \varphi)$ along radial rays

$$\psi(\mathbf{r}) = \int_0^r dr' v_r(r', \Theta, \varphi); \tag{2.83}$$

r, Θ, φ are spherical coordinates, and the observing astronomer (e.g. A. Dressler) is at $r = 0$. $\boldsymbol{v}(\boldsymbol{r})$ is simply the gradient of ψ.

The tricky part of the POTENT method is to find a good approximation to the radial component v_r of the continous peculiar velocity field by interpolating between a few discrete measurements which are not completely without errors either.

Avishai Dekel has written a commendable review on the general strategy, and various applications [Dekel 1994].

An application of the POTENT method to the newest catalogue of groups of galaxies, the Mark III catalogue [Willick et al. 1997], and its comparison to the IRAS 1.2 Jy redshift survey [Fisher et al. 1994] has resulted in an estimate of

$$\Omega_m^{0.6}/b_I = 0.89 \pm 0.12 \tag{2.84}$$

[Sigad et al. 1998], consistent with high as well as with low estimates of Ω_m. b_I is the linear biasing parameter for IRAS galaxies. For $b_I > 0.75$, as theoretical models of biasing indicate, $\Omega_m > 0.3$ at a 95% confidence level. This is the only claim so far that observations lead to a value of Ω_m close to the critical one, $\Omega_m = 1$. It seems to be only marginally consistent with most other determinations which indicate Ω_m below 0.3.

2.3.7 An Upper Limit to $\Omega_{0,B}$

In Chap. 3 the predicted abundances from hot big-bang nucleosynthesis are compared with the observed abundances of the light elements. Deuterium and lithium provide the most stringent limits, requiring the density parameter for the baryons that have participated in nucleosynthesis – all baryons, if exotic scenarios such as quark nuggets and hot quark bubbles are not involved – to be less than 0.025. Thus an upper limit

$$h^2\Omega_B \leq 0.025$$

seems to be quite a safe estimate.

2.3.8 Conclusion

We must conclude that observations give a value of Ω_m between 0.1 and 0.3 for matter bound in clusters, but that $\Omega_m = 1$ or even $\Omega_m \geq 1$ cannot be excluded, because any background density distributed more homogeneously than the large clusters is not accounted for in these measurements. Therefore a decision whether we live in a closed or open universe cannot yet be made.

There must be at least 10 times the amount of luminous matter present in the form of non-luminous matter. What kind of matter? Black holes, neutrinos of non-zero rest mass, dark stars of small mass, and rocks have all been mentioned. We see that the value of the mean density is extremely uncertain, and that part of this uncertainty is due to a lack of knowledge of local physics.

The CMB data of the Boomerang and Maxima experiments (cf. Chaps. 11, 12) indicate that $\Omega = 1$. That value is also preferred by some people for philosophical reasons. The simplest model is the $K = 0$ FL universe; only for $\Omega_0 = 1$ is $\Omega = \Omega_0$ always. For $\Omega < 1$ we have Ω as a function of t, which for small t must nevertheless be very close to 1. This "fine-tuning of the initial condition" seems unaesthetic to many physicists. In Chap. 9 we shall comment on this aspect of the standard model.

2.4 Ω_Λ Estimates

The big puzzle of the cosmological constant is the fact that $\frac{\Lambda}{8\pi G}$ is determined by astronomical observations to be comparable to ρ_c:

$$\Omega_\Lambda \equiv \frac{\Lambda}{3H_0^2} \cong 1.$$

This is absolutely tiny from the particle physics point of view: the vacuum energy density of "Grand Unified Theories (GUT)" is $\rho_{\mathrm{GUT}} \simeq 10^{108} \varrho_c$, and one would naively estimate a cosmological constant Λ_{GUT} from it, $\Lambda_{\mathrm{GUT}}/8\pi G \simeq \rho_{\mathrm{GUT}}$, which is 10^{108} times larger than the value that can be tolerated at the present epoch in the universe (see Chap. 9 for another view on this dicrepancy).

If we forget about this deep problem, and take the pragmatic view that Λ is just one more cosmological parameter to be determined by observations, we know that Ω_Λ is around 0.7.

Gravitational lensing can be used to obtain upper limits on Ω_Λ. The argument is statistical in nature, basically a comparison of the expected number of lensed sources with observations. The volume per unit redshift $\frac{dV}{dz}$ increases with increasing Ω_Λ, and consequently the relative number of lensed sources for a fixed number density of galaxies increases also with Ω_Λ. For example [Maoz and Rix 1993], observations of images of quasars with a narrow separation of less than 2 arcseconds, and the statistics of such images produced by a conserved, non-evolving population of elliptical galaxies in a cosmological model with $\Omega_{tot} = 1$, lead to the following result: the best fit is obtained for $\Omega_\Lambda < 0.7$, and $\Omega_\Lambda < 0.9$ can be excluded with high significance. Another analysis [Kochanek 1996] gives a limit

$$\Omega_\Lambda \leq 0.66.$$

This value is consistent with $\Omega_\Lambda = 0.7$ derived frum SNIa (cf. Sect. 2.2). For a_0, we simply have $a_0 = \Omega_{m/2} - \Omega_\Lambda$.

2.5 The 3 K Cosmic Black-Body Radiation

In 1964 Penzias and Wilson discovered an excess radio background at $\lambda = 7.35$ cm corresponding to a black-body temperature of 2.5 to 4.5 K. Since its publication in 1965 this discovery has been interpreted as a cosmological background signal indicating the existence of a hot, dense early state of the universe, when radiation and ionized matter were in equilibrium. The history of this discovery is told in Weinberg's book *The First Three Minutes* [Weinberg 1977].

In the years following the discovery balloon and rocket measurements covered a wider frequency range, and confirmed the Planck distribution of the radiation with a temperature $T \sim 3$ K [Woody and Richards 1979].

The peak of the radiation at a wavelength of 2 mm is absorbed by the Earth's atmosphere, and can only be measured from space. Then, on November 18, 1989, the satellite COBE ("Cosmic Background Explorer") was launched by NASA to measure the cosmic microwave background.

First observational results from this satellite of the cosmic microwave background are displayed in Fig. 2.39 [Mather et al. 1990]. The data were taken by the Far Infrared Absolute Spectrometer (FIRAS) aboard the satellite in 9 minutes during scans centred on the Galactic North Pole. Shown in the figure are small error boxes around the individual measured points. The wavelengths between 1 cm and 0.05 cm are covered, and over this range the agreement with a theoretical Planck spectrum at a temperature of 2.728 K (solid line in Fig. 2.39) is truly spectacular. The universe seems to be the ultimate black body! This smooth radiation field strongly supports the idea of a hot big bang. It is becoming more and more difficult to find other explanations than the simple hypothesis that the cosmic microwave background (CMB) is the remnant of a homogeneous and structureless primeval explosion.

Attempts to explain the radiation as the result of the superposition of the microwave emissions of many individual sources have all crumbled beneath the weight of evidence produced in recent years (see Sect. 2.5.1 below). The measurements satisfy all the tests for a cosmic background radiation that have been proposed up to the present.

Another interesting result was obtained by the observation of radiation from CN molecules in a distant ($z = 1.78$) galaxy. The observations allow us

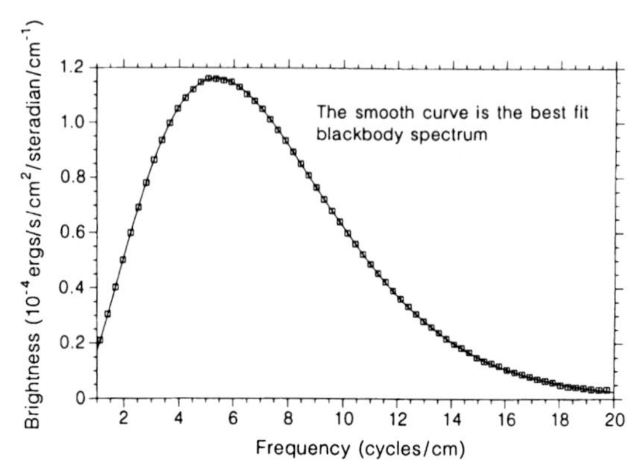

Fig. 2.39. The spectrum of the cosmic microwave radiation as obtained by COBE during a 9 min observation which included the Galactic North Pole. The error boxes of the individual measurements are all in perfect agreement with a theoretical black-body radiation curve of a temperature $T = 2.728\,\mathrm{K}$ (*solid line*). This perfect fit finds a natural explanation in the hypothesis of a "hot big bang": The universe has gone through a hot early phase of thermodynamic equilibrium which was almost structureless (Courtesy of the COBE collaboration [Mather et al. 1990])

to deduce that the excited states of the CN molecules are populated with an abundance which corresponds to a radiation field of a temperature of $(7.6 \pm 0.8)\,\mathrm{K}$. This fits in very nicely with the standard cosmological interpretation of the CMB which, at $z = 1.78$, has a temperature of $T = (1+z)T_{\gamma 0} = 7.58\,\mathrm{K}$.

Besides excellent earlier reviews of both the experimental and theoretical situation [Weiss 1980; Sunyaev and Zeldovich 1980], there is a huge number of more recent reviews and books [Partridge 1995; White et al. 1994].

2.5.1 The Spectrum

One important fact about a Planck radiation field in a homogeneous FL universe is that its spectrum keeps the Planck shape during the expansion: for $\kappa = 4/3$ (photons), the adiabatic expansion law $T_\gamma (R^3)^{\kappa-1} = \text{const.}$ gives $T_\gamma R = \text{const.}$ Thus the Planck spectrum

$$\varrho_\gamma d\nu = 8\pi h \left(\frac{\nu}{c}\right)^3 d\nu \left[\exp\left(\frac{h\nu}{k_B T}\right) - 1\right]^{-1} \qquad (2.85)$$

remains a Planck spectrum, since $T_\gamma = T_{\gamma 0}(1+z)$, and $h\nu = h\nu_0(1+z)$. Integration over frequency yields

$$\varrho_\gamma = a\,T_\gamma^4 \qquad \text{with} \qquad a = \frac{\pi^2}{15}\frac{k_B^4}{h^3 c^3} = 7.564 \times 10^{-15}\,\mathrm{erg\,cm^{-3}\,deg^{-4}};$$

the number density of photons is

$$n_\gamma = \frac{1}{\pi^2}\frac{(k_B T_\gamma)^3}{(2\pi\hbar c)^3} 2\zeta(3),$$

where ζ is Riemann's ζ function. Thus

$$n_\gamma = \frac{30\zeta(3)}{\pi^4}\frac{a}{k_B}T_\gamma^3 \approx 0.37\frac{a}{k_B}T_\gamma^3.$$

After observing for two years, and measuring many spectra, the observers have determined the temperature of the CMB within narrow limits:

$$T_\gamma = 2.728 \pm 0.002\,\mathrm{K}.$$

[Fixsen et al. 1996].

The energy density is therefore

$$\varrho_{\gamma 0} c^2 = a T_{\gamma 0}^4 = (4.19 + 0.01, -0.02) \times 10^{-13}\,\mathrm{erg\,cm^{-3}} = 0.26\,\mathrm{eV\,cm^{-3}}.$$

The number of photons is

$$n_{\gamma 0} = 412 + 1(-2)\,\mathrm{cm^{-3}}.$$

The CMB is the dominating radiation energy density, but its mass equivalent is unimportant at the present epoch:

$$\varrho_{\gamma 0} = a T_{\gamma 0}/c^2 = 4.7 \times 10^{-34}\,\mathrm{g\,cm^{-3}}.$$

An interesting quantity derived from the CMB is the ratio of the number of photons to baryons,

$$\frac{n_{\gamma 0}}{n_{B0}} \simeq 10^{10}, \tag{2.86}$$

which is constant in FL cosmologies, except for epochs when photons were produced by heating processes.

The number $n_\gamma/n_B = 10^{10}$ is a characteristic of the present universe. At early times $n_\gamma R^3$ was not fully conserved (e.g. during the e^+e^- annihilation epoch (see Chap. 3), but even then n_γ/n_B did not change by orders of magnitude. Thus a large value of n_γ/n_B has to be set down as an initial condition in any classical FL cosmology. Modern particle theories applied to the early universe may find a way of deriving this number without having to introduce a net baryon assymmetry at the beginning (cf. Chap. 8).

Such a background radiation field has important consequences: Since the energy density of matter $\varrho_m = \varrho_{m0}(1+z)^3$, and of radiation $\varrho_\gamma = \varrho_{\gamma 0}(1+z)^4$, we have $\varrho_\gamma/\varrho_m = (1+z)\varrho_{\gamma 0}/\varrho_{m0}$. Matter dominates now, but at some time in our past, radiation was the dominating energy density. Before a redshift z_γ, between $z_\gamma = 10^3$ and $z_\gamma = 10^4$ (because of the uncertainty in ϱ_{m0}!), the universe was dominated by hot radiation.

Before the time $t_{\rm dec}$ of the recombination of hydrogen ($T \approx 3000\,$K), i.e. $z_{\rm dec} \approx 1000$, matter and radiation were in thermodynamic equilibrium. It is a remarkable coincidence that these two times may be equal: $z_\gamma \approx z_{\rm dec}$.

At $t < t_\gamma$ the universe must have been completely different from its present appearance: this radiation-dominated phase of the early universe was a hot and dense gas of radiation and matter. Structures such as galaxies and stars did not yet exist.

Although at $z_{\rm dec}$ the mean free path of photons increased dramatically, there was still a small interaction with free electrons via Thomson scattering. Therefore the photons that we observe come from a surface at which they were last scattered, which has a smaller redshift than ≈ 1000.

No deviations from a Planckian spectrum were found in the COBE observations, and the experimental limits are such that any deviation larger than 0.005% of the peak intensity can be excluded [Fixsen et al. 1996].

This high-precision result can, of course, be used to set limits on other cosmic parameters. The most important elementary mechanism of interaction of radiation and matter in the early universe is the scattering of photons on electrons ("Compton scattering"). The optical depth

$$\tau_T = \int_0^{t_{\rm max}} \sigma_T c N_e(z) dt = \int_0^{z_{\rm max}} \sigma_T N_e(z) c \frac{dt}{dz} dz,$$

is $\tau_T = 0.024(\Omega h^2)^{1/2} z_{\rm max}^{3/2}$ if the primeval hydrogen–helium plasma is fully ionized for $0 < z < z_{\rm max}$ ($\sigma_T = 0.665 \times 10^{-24}\,{\rm cm}^2$; $N_e(z)$: number of electrons along the line of sight between z and $z + dz$). Before recombination this optical depth was very high. Under such conditions the most important mechanism of energy transfer between plasma and radiation is Compton scattering which involves a frequency change of the photons ("Comptonization"). If there are hot electrons in the intergalactic medium, they can influence the spectrum of the CMB [Sunyaev and Zel'dovich 1980]. A typical spectrum arises which depends only on a parameter y given by

$$y = \int_{t_{\rm min}}^{t_{\rm max}} \frac{k(T_e - T_\gamma)}{m_e c^2} \sigma_T N_e(z) c dt,$$

where T_e is the electron temperature, m_e the electon mass, and $t_{\rm min}, t_{\rm max}$ characterize a time interval within

$$0 < z < 4 \times 10^4 (\Omega h^2)^{-6/5}.$$

During that epoch the scattered photons are redistributed over frequency. The number of photons is conserved, but their average energy increases:

$$n_\gamma = n_{\gamma 0} = \frac{a}{2.7 k_B} T_{\gamma 0}^3,$$

$$E_\gamma = E_0 e^{4y} = a\, T_{\gamma 0}^4 e^{4y}.$$

The energy input to the radiation field is

$$\frac{\Delta E}{E_0} = e^{4y} - 1.$$

The observations set a limit (with 95% confidence)

$$|y| < 1.5 \times 10^{-5}.$$

The fractional energy input $\frac{\Delta E}{E_0}$ to the CMB between $z \simeq 10^5$ and the present epoch is therefore strongly limited:

$$\frac{\Delta E}{E_0} < 6 \times 10^{-5}.$$

This bound is good enough to exclude or reduce to insignificance several of the less standard ideas in cosmology. To mention just a few examples:

1. The model of a smooth, hot intergalactic medium suggested to explain the X-ray background can be ruled out by this limit, since the intensity of the predicted X-ray background scales as y^2. Even for a very late heating time, the contribution of the hot, intergalactic medium would be limited to less than 1% of the observed value.
2. A model for galaxy formation by repeated supernova explosions can be ruled out.
3. The model advocated by Sir Fred Hoyle, in which an appropriate distribution of needle-like iron dust grains produces the CMB by absorption and re-emission of starlight runs into difficulties too. First of all, the absorption and emission processes must model a black-body spectrum almost perfectly in the range of $\frac{1}{2}$ mm to 1 cm wavelength. Secondly, there are distant objects (e.g. a galaxy with redshift $z = 2.2$) observed in the mm wavelength range. This further limits the opacity that the iron grains might have.

Even a possible energy input at epochs $z > 4$ can be constrained by these precise observations. The cooling of the electrons by interaction with the radiation background was very short compared to the Hubble time. So T_e was close to T_γ, and any energy input increased both T_e and T_γ.

Multiple Compton scatterings lead to a Bose–Einstein spectrum for the number density of photons with frequency between ν and $\nu + d\nu$: $n = (e^{x+\mu} - 1)^{-1}$, where $x \equiv h\nu/kT$ and μ is chemical potential. The relaxation time to a black body

$$n = (e^x - 1)^{-1}$$

for the radiation is slow, because $n_\gamma \simeq 10^{10} n_e$. Therefore the effects of an energy release might be detected as far back as $z \simeq 10^7$.

For small μ [Sunyaev and Zel'dovich 1980]

$$\frac{\Delta E}{E_\gamma} = 0.714\mu,$$
$$|\mu| < 9 \times 10^{-5},$$

i.e. an energy input $\frac{\Delta E}{E_\gamma}$ larger than 6×10^{-5} during $10^4 < z < 10^7$ would have been detected by the COBE experiment.

2.5.2 Isotropy of the Background

The last scattering surface of the CMB is not necessarily sharply defined: photons seen along one line of sight may have been last scattered in various uncorrelated regions. One expects to see inhomogeneities in the matter distribution reflected in a variation of the temperature (cf. [Sunyaev and Zeldovich 1980]).

The largest anisotropy signal found is due to the proper motion of the Earth (Fig. 2.40).

The CMB can be viewed as an almost perfectly isotropic medium, defining a preferred frame of reference. Any peculiar motion with respect to this frame can be measured. The expected dipole anisotropy due to the Earth's motion has a cos θ behaviour:

$$T(\theta) = T_0 \left(1 - \frac{v^2}{c^2}\right)^{1/2} \left(1 - \frac{v}{c}\cos\theta\right)^{-1} = T_0 \left(1 + \frac{v}{c}\cos\theta\right). \tag{2.87}$$

Observations at 33 MHz with a U2 plane [Smoot et al. 1977; Muller 1978, 1980] indeed showed such a cosine behaviour of T, as indicated in Fig. 2.41.

This large-angle anisotropy feature has been remeasured with high precision by the differential radiometers aboard COBE. The overall sky pattern is shown in Fig. 2.42. The velocity derived for the barycentre of the solar system is

$$v/c = 0.001236 \pm 2 \times 10^{-6}$$

or

$$v = 371 \pm 0.5\,\mathrm{km\,s^{-1}}$$

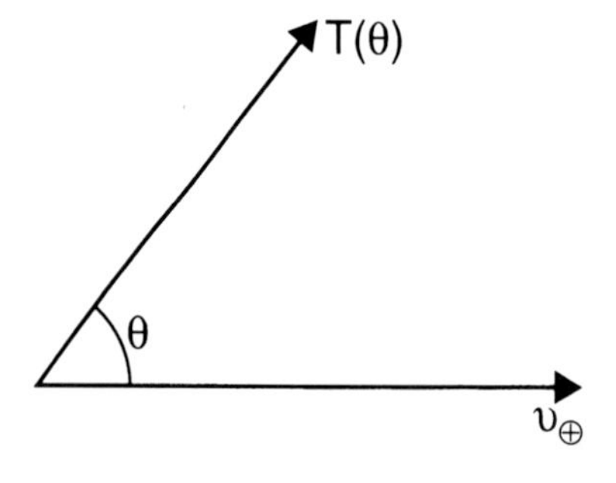

Fig. 2.40. Schematic plot of the surface of last scattering, and the decoupling surface $t = t_R$

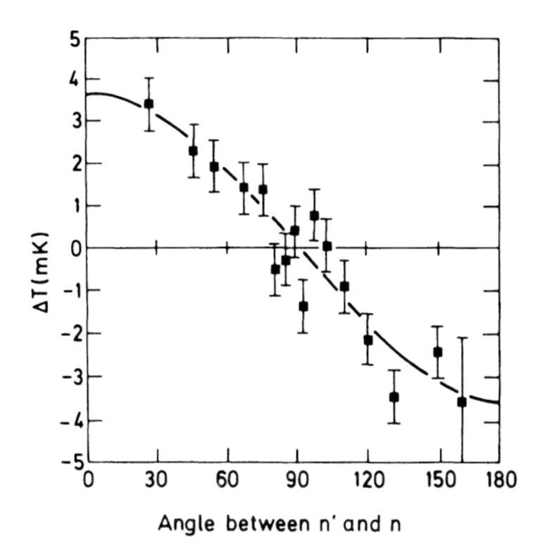

Fig. 2.41. The dipole anisotropy – proportional to $(v/c)\cos\theta$ – caused by the motion of the Earth in the CMB rest-frame

towards

$$(\alpha, \delta) = (11.16^h \pm 0.01^h, -6.7^0 \pm 0.15^0),$$
$$(l, b) = (264.14^0 \pm 0.15^0, 48.26^0 \pm 0.15^0).$$

This in turn implies a velocity for the Local Group of galaxies relative to the CMB:

$$V_{\mathrm{LG}} = 627 \pm 22\,\mathrm{km\,s^{-1}}$$

toward $(l, b) = (276^\circ \pm 3^\circ, 30^\circ \pm 3^\circ)$. The large magnitude of this velocity is interesting, since the peculiar velocities of nearby galaxies with respect to the frame of the Virgo cluster – from optical observations – are small, $\sim 80\,\mathrm{km\,s^{-1}}$ to $200\,\mathrm{km\,s^{-1}}$. The galactic motion implied by the dipole anisotropy has a component toward Virgo $\sim 200\,\mathrm{km\,s^{-1}}$, and components of $400\,\mathrm{km\,s^{-1}}$ tangential and normal to the plane of the local supercluster. There must be another large-scale velocity field which explains the CMB dipole.

The search for smaller anisotropies produced by matter fluctuations is not easy. The true cosmic signal is always mixed up with internal (receiver) and atmospheric noise. In the case of anisotropy measurements, for example, the expected signal is only a fraction of the receiver system noise. To cope with the formidable difficulty of disentangling the data the observers use sophisticated beam-switching set-ups, and low-noise receivers – in the COBE DMR instruments cryogenically cooled receivers are used with an overall system noise corresponding to a temperature in the range 80 to 100 K.

Nevertheless, in 1992 the COBE team announced the discovery of small anisotropies in the CMB , with an rms variation of the temperature of $(30 \pm 5)\,\mu\mathrm{K}$ over patches of $\sim 10^\circ$, i.e. $\frac{\Delta T}{T} = 1.1. \times 10^{-5}$ [Smoot et al. 1992].

Theorists heaved a sigh of relief because for 28 years the hunt for fluctuations in the CMB had been in vain. Such fluctuations had been expected

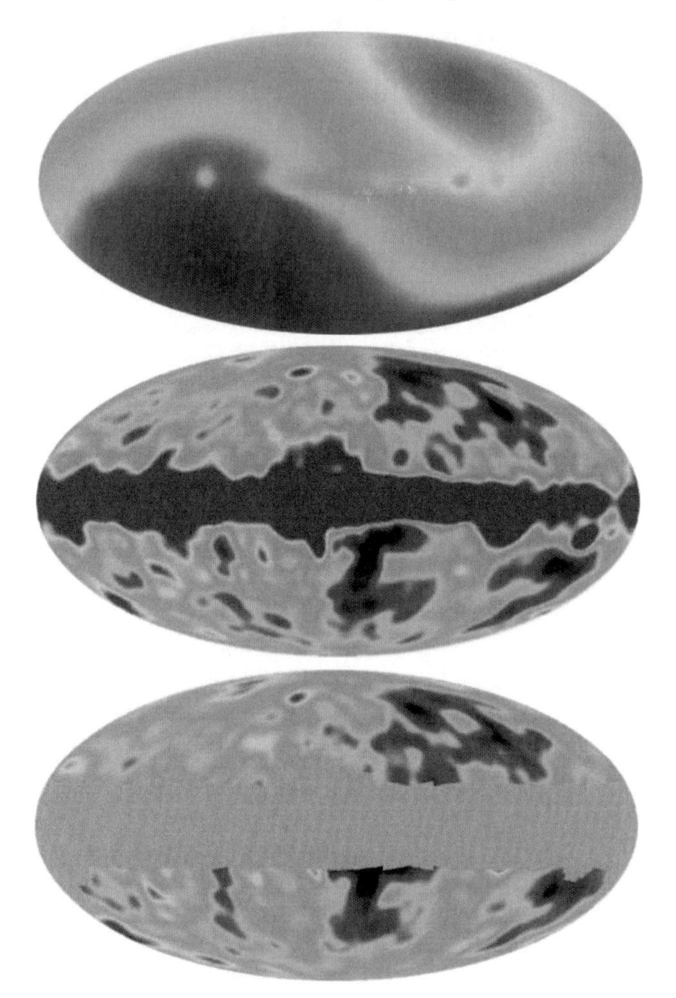

Fig. 2.42. The pattern of the COBE observations of the CMB sky is shown here. At a sensitivity of $\Delta T \sim 3\,\mathrm{mK}$ a dipole pattern is seen (*top*). After subtraction of the dipole and an increase of the sensitivity to $\Delta T \sim 30\,\mu\mathrm{K}$ a patchwork of hot and cold spots appears with a prominent band of emission from the Milky Way (*middle*). After subtraction of the Milky Way, the cosmic signal of hot and cold patches appears

on the basis of models for structure formation. The galaxies, the clusters and superclusters, and the large-scale patterns in the sky are supposed to evolve during the expansion of the universe from initially small-amplitude perturbations. Their density contrast grows due to the action of gravity, until they reach an overdensity which causes them to collapse by gravitational instability. Since the discovery of the CMB it had been realized that these seed fluctuations should be visible as small temperature anisotropies. The

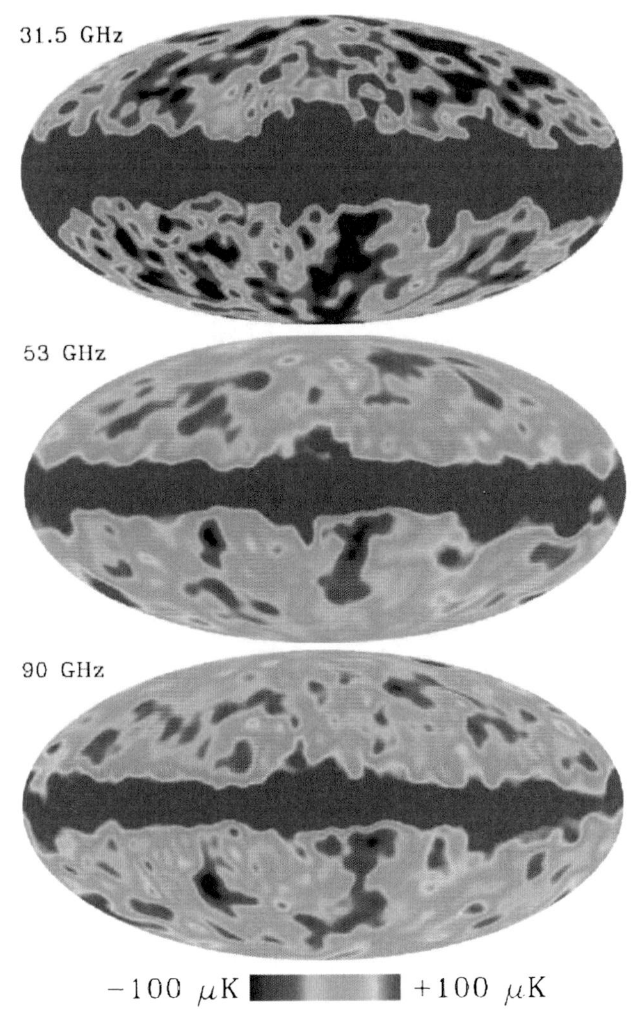

Fig. 2.43. The COBE DMR measurements for 31.5 GHz, 53 GHz and 90 GHz are dis-played as sky maps

COBE result shows that this notion has a good chance of being somewhere near the truth.

The COBE sky maps are shown in Figs. 2.42 and 2.43. There are two DMR instruments A and B on the satellite measuring temperature differences of small regions separated by 60° on the sky. The instruments gather data at three different freqencies: 90 GHz (3.3 mm), 53 GHz (5.7 mm), 31 GHz (9.5 mm). They provide sky maps of 2.5° × 2.5° pixels, i.e. 6144 pixels on the celestial sphere. The pixel size is determined from a smoothing of the measurement points with the 7° beam width of the antenna. As can be seen from Fig. 2.43, the signals from the Milky Way are very prominent. It is necessary to remove this galactic contribution from the data. This is done by first cutting out an area of $|b| \leq 20°$ in the sky. The remaining 4038 pixels are then analysed statistically. The 31 GHz map is taken as a model for the galactic emission, and the estimated contribution from dust and galactic radio sources is subtracted from the 53 GHz and 90 GHz maps. The receiver noise is estimated from sky maps of (A–B)/2, i.e the signal of instrument B is subtracted from the signal received by A. Thus any cosmic signal is removed, and only internal noise patterns remain. A few years ago there was a cute colour poster announcing a workshop on COBE results, but the picture displayed was inadvertently the (A–B) map. After these subtractions what remains on the (A+B)/2 maps is the cosmic signal – hopefully. The temperature in a certain direction $\boldsymbol{n}$ (the direction of the incoming photon) is

$$T(\boldsymbol{n}) = T_0(1 + \Delta T(\boldsymbol{n})),$$

where T_0 is the background temperature and $\Delta T(\boldsymbol{n})$ is the possible deviation dependent on direction. Usually the average value of the product $\Delta T(\boldsymbol{n})\Delta T(\boldsymbol{n}')$, the autocovariance function, is considered:

$$C(\theta) \equiv \langle \Delta T(\boldsymbol{n})\Delta T(\boldsymbol{n}')\rangle$$

for directions $\boldsymbol{n}$ and $\boldsymbol{n}'$ enclosing an angle $\theta, \boldsymbol{n}\cdot\mathbf{n}' \equiv \cos\theta$. The distribution of the radiation field on the celestial sphere can be analysed most conveniently when it is expanded in terms of spherical harmonics. The corresponding power spectrum of $C(\theta)$ is given by an expansion in terms of Legendre polynomials $P_l(x)$:

$$T(\boldsymbol{n}) = T_0 \sum_{l,m} a_l^m Y_l^m(\boldsymbol{n}),$$

$$C(\theta) = \frac{1}{4\pi} \sum (2l+1) c_l P_l(\cos\theta).$$

For Gaussian random fields,

$$(2l+1)c_l = \sum_{-l}^{+l} |a_l^m|^2.$$

Monopole ($l = 0$), and the well-established dipole ($l = 1$) contributions are substracted out before the statistical analysis begins. The summation over l thus starts with a quadruple term $l = 2$ (for a detailed exposition see [Bond and Efstathiou 1987]). The resolving power of the telescope, the size and geometry of the beams, must also be taken into account in the data analysis. A simple Gaussian approximation for the resolution function (RF) may be written as

$$RF(\boldsymbol{n} - \boldsymbol{n}') = \frac{1}{2\pi\Theta_S^2} \exp\left(-|\boldsymbol{n} - \boldsymbol{n}'|^2/2\Theta_S^2\right),$$

where $\boldsymbol{n}$ is the direction to the center of the beam and Θ_S describes the charcteristic dispersion of the instrument. The resolution function folded with the true sky distribution gives the data. To test for significant non-random temperature variations one has adopted a maximum likelihood method [Davis et al. 1987], where the contours are constructed from varying the parameters of a single Gaussian $C(\theta)$ folded with the beam resolution function.

The expansion of the radiation field into multipoles compresses the 4038 pixels which would in principle be available for a statistical analysis, and one might worry that some information is lost. It has been shown, however, that this is good enough, because the low angular resolution of COBE does not give reliable constraints on c_l beyond $l = 10$ or so. The physical (angular) size Θ_0 corresponding to a certain value of l can be estimated as $\Theta_0 = \pi/l$, although a physical feature would not in general correspond to a single multipole. Another problem is the region $|b| < 20°$ cut out of the celestial sphere: the spherical harmonics are then no longer a complete, orthogonal system of functions, and different multipoles may influence each other in the analysis. The COBE results give a flat power spectrum $n = 1.11 \pm 0.29$ and a quadrupole amplitude

$$\langle Q \rangle = (20.2 \pm 4.6)\,\mu\mathrm{K}.$$

The rms temperature variations across the sky are

$$\Delta T = (30 \pm 5)\,\mu\mathrm{K} \quad \text{at} \sim 10°$$

i.e.

$$\frac{\Delta T}{T} = 1.1 \times 10^{-5}.$$

We should appreciate the high degree of isotropy of the universe exhibited by this result! This is strong support, indeed, for the standard FL models. At higher l there are peaks expected in the anisotropy graph, when mass concentrations of a certain scale oscillate as sound waves just before recombination. The precise location and amplitude of these peaks depend on cosmological parameters and structure formation models (Fig. 2.44). In Chap. 11 we will look in detail at this interesting aspect; that measurements of the anisotropy graph can determine the complete set of cosmological parameters to high accuracy.

For adiabatic fluctuations the ratio of entropy $s = 4aT_\gamma^3/3k_B$ and baryon density n_B is constant, i.e.

$$\frac{\delta n}{n} = +3\frac{\Delta T_\gamma}{T_\gamma}.$$

The angular scale associated with a typical galactic mass of $M = 10^{11} M_\odot$ at z_R is 40 arcseconds, and the scale is proportional to $(M)^{1/3}$ (cf. Chap. 10), i.e. for a galaxy cluster, $M = 10^{14} M_\odot$, the angle is $\sim 6'$. Values of $l \geq 1000$ must

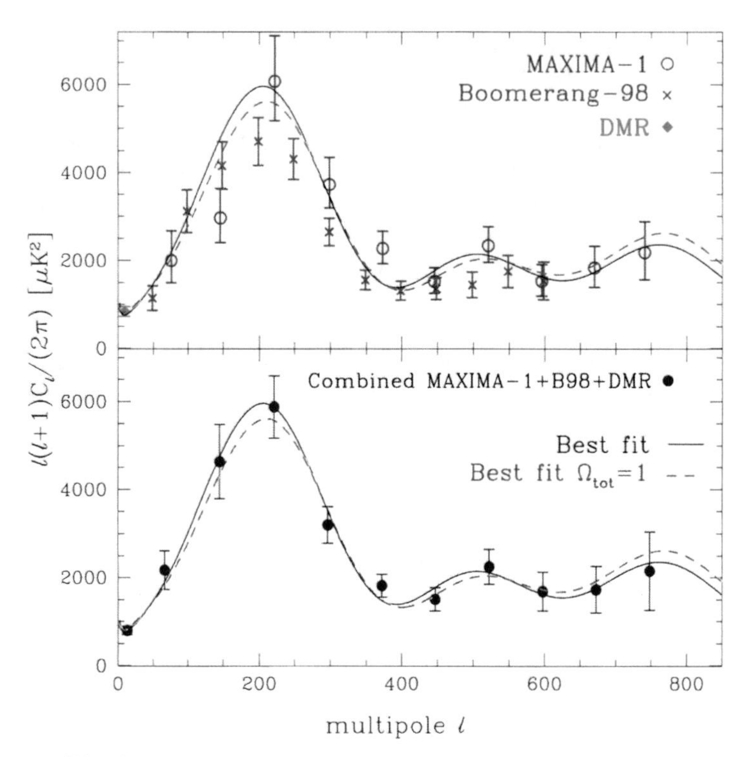

Fig. 2.44. The Boomerang experiment reveals the structure of the $\sim 10°$ COBE spots with its $0.5°$ angular resolution .

be probed to measure the fluctuations corresponding to galactic or cluster mass scales.

Recent balloon and interferometer experiments (Boomerang, Maxima, Dasi, Fig. 2.44) have achieved this (cf. Chap. 11).

Two new space missions are at the planning stage. Both will have much higher angular resolution than COBE. NASA has launched MAP (Microwave Anisotropy Probe) in the year 2000, and ESA has scheduled the launch of PLANCK for the year 2006. Both satellites will make full-sky maps of the CMB temperature with angular resolutions of $0.2°$ (MAP) and $0.1°$ (PLANCK). Both satellites will be stationed at the remote, stable L2 Lagrange point, 1.5 million kilometers antisunward from the Earth. If everything goes according to plan, by the year 2010 we shall have all the cosmological parameters measured to a few percent accuracy, as well as the initial conditions for structure formation determined precisely. Then the final book on cosmology can be written, and as a field of research cosmology will have come to an end ... unless something unexpected turns up.

2.5.3 A Limit to Cosmic Ray Energies

Very high-energy cosmic rays are thought to be individual protons or neutrons. They produce air showers in the Earth's atmosphere. The highest observed energies are around 10^{20} eV [Takeda et al. 1999].

The Larmor radius $r_L \equiv pc/eB$ of a proton with $pc = 10^{20}$ eV moving in the interstellar field $B \sim 10^{-6}$ eV is $r_L \simeq 100$ kpc. Since this is much larger than the distance to the centre of the galaxy, these cosmic rays, if produced locally, should point back to their sources. From their distribution no preference for the plane of the Milky Way can be seen, and thus they are taken to be extragalactic. But for extragalactic ultrahigh-energy cosmic rays the CMB photons appear as gamma rays in their rest-frame, reducing the mean free path by e^+e^- energy losses and pion production. The mean free path for 10^{20} eV protons is only 2 Mpc. Cosmic rays coming from greater distances are slowed down by their interaction with the CMB medium. The extragalactic component of cosmic rays should have a peak and a cut-off at energies just below 10^{20} eV. The observations so far seem to agree.

2.5.4 The Sunyaev–Zel'dovich Effect

X-ray satellites have established the presence of a large amount of hot ($T_e \sim 10^8$ K), dilute ($n_e \sim 10^{-2}$ to $10^{-3}\,\mathrm{cm}^{-3}$) gas in clusters of galaxies. Compton scattering of microwave photons from hot electrons might lead to a distortion of the spectrum [Sunyaev and Zel'dovich 1980; Sunyaev 1978]. Scattering by electrons shifts the photons' frequency by a mean value

$$\frac{\langle \Delta\nu \rangle}{\nu} \sim 4\frac{k_B T_e}{m_e c^2} . \tag{2.88}$$

In a black-body spectrum of photons the first-order effect of a scattering by hot electrons cancels out, but a second-order effect remains ("Comptonization"). For $T_\gamma \ll T_e$ there is a decrease in the intensity and in the brightness temperature

$$\frac{\Delta T_\gamma}{T_\gamma} = -2\frac{k_B T_e}{m_e c^2}$$

in the Rayleigh–Jeans part ($h\nu < 3.83\ k_B T_\gamma$) of the spectrum. In the Wien region ($h\nu > 3.83\ k_B T_\gamma$) the intensity increases. The effect is proportional to the fraction τ_T of photons that have been scattered at least once. The brightness temperature in the direction of the hot gas cloud decreases (Fig. 2.45). The magnitude of the effect can be estimated: with $\tau_T = 0.1$ and $kT_e = 5$ keV one obtains $\Delta T_\gamma/T_\gamma = 2\tau_T(k_B T_e/m_e c^2) \sim 2 \times 10^{-3}$.

Radio and X-ray observations may determine the gas density and the temperature distribution inside the cluster. Then this effect can be used for a determination of the Hubble constant. Consider as a simple model for the

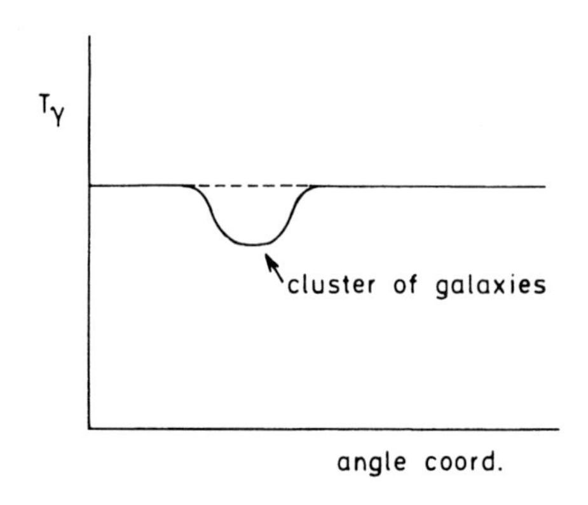

Fig. 2.45. For $h\nu < 3.83 k_B T_\gamma$ the brightness temperature decreases in the direction of a cluster of galaxies [Sunyaev and Zel'dovich 1980]

cluster a spherically symmetrical isothermal cloud of uniform density. The surface brightness of the X-rays is

$$I_x = 2ARn_e^2 T_e^{-1/2} \exp(-h\nu/k_B T_e)\, g \tag{2.89}$$

(g = Gaunt factor; A = const.; R = radius of the cloud, n_e = electron density). This can be supplemented by

$$\frac{\Delta T_\gamma}{T_\gamma} = -\frac{4k_B T_e}{m_e c^2} \sigma_T n_e R. \tag{2.90}$$

X-ray observations of the spectrum determine T_e. Then, using (2.89) and (2.90), both n_e and R can be found. Knowledge of the absolute dimension R, together with the angular extent, implies knowledge of the distance. Redshifts of galaxies in the cluster then immediately give H_0 – without the use of intermediate distance-scale steps.

More realistic cluster models – e.g. $n_e(r)$ and $T_e(r)$ profiles – require more detailed observations of small-scale structures. The right-hand side of (2.85) is then replaced by a line integral $-4y \equiv \int (k_B T_e/m_e c^2) \sigma_T n_e(r) dr$, where y is once more the Comptonization parameter. An SZ temperature decrement has been reported for several clusters [e.g. Rephaeli 1995; Kobayashi et al. 1996], and even determinations of the Hubble constant have been reported. H_0 seems to be on the low side, between 50 and 60 km s^{-1} Mpc^{-1}.

New satellite missions may allow us to observe clusters of galaxies via the SZ effect in the mm and sub-mm range, i.e. near the maximum or in the Wien part of the CMB's Planck spectrum. The intensity of the SZ effect does not suffer from the $(1+z)^{-4}$ intensity decrease with redshift of the optical or X-ray brightness, and therefore, especially at high z such an observational window can be a great advantage.

The increase in intensity due to the SZ effect is

$$\Delta I_\nu = i_0 y g(x) \tag{2.91}$$

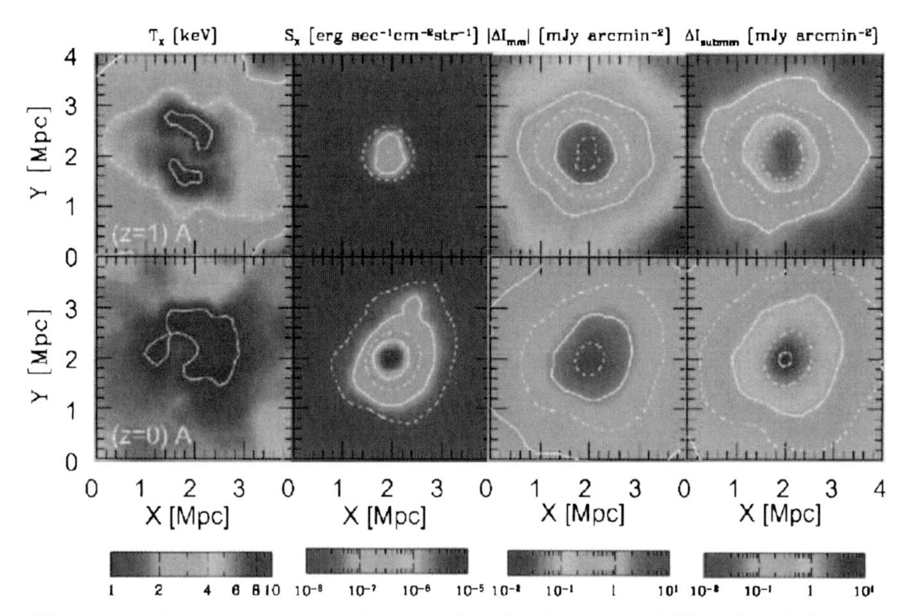

Fig. 2.46. X-ray temperature, X-ray surface brightness, and SZ surface brightness contours at 150 GHz (mm) and 350 GHz (sub-mm) are shown for a simulated cluster at $z = 1$ (top) and $z = 0$ (see text for further explanation; courtesy of [Yoshikawa et al. 1998])

where $i_0 \equiv 2(k_B T_\gamma)^3/h^2c^2, y$ is the Comptonization parameter

$$y = \int_{-\infty}^{+\infty} \frac{k_B T_e}{m_e c^2} \sigma_T n_e dl,$$

and g(x) is the function

$$g(x) \equiv x^4 e^x (e^x - 1)^{-2} [x \cot h(x/2) - 4]; \qquad x \equiv h\nu/k_B T_\gamma.$$

In Fig. 2.46 ΔI_ν contours are shown for a simulated cluster at $z = 1$ and $z \simeq 0$ [Yoshikawa et al. 1998]. The X-ray emission-weighted temperature T_x, the X-ray surface brightness S_x and the SZ surface brightness contours $\Delta I_{\rm mm}$ (at 150 GHz) and $\Delta I_{\rm submm}$ (at 350 GHz) are plotted. This projected view of the plane perpendicular to the line of sight shows quantities which have been integrated along the line of sight. Since S_x is an integral over $n_e^2(l)$ it appears less extended than the integrals over $n_e(l)$, at ΔI_ν. Notice also the difference between the cluster at $z = 0$ and at $z = 1$ in S_x, whereas the SZ surface brightness is practically unchanged. It seems that high-z clusters will be much easier to identify by the SZ increment than by the actual X-ray brightness.

2.6 The X-ray Background

The IR and UV backgrounds seem to be entirely produced by well-observable discrete sources. For the X-ray background there is a growing evidence that it also can be accounted for in terms of discrete sources.

In the soft (0.5–2 keV) band more than 90% of the X-ray background flux has been resolved by ROSAT, Chandra, and XMM-Newton observations. In the harder (2–10 keV) band a similar fraction of the X-ray background has been resolved by the Chandra and XMM-Newton surveys (cf. [Hasinger 2002]). So far the X-ray background observations are completely consistent with models of accretion onto supermassive black holes, integrated over cosmic time-scales. Population synthesis models based on a mixture of obscured and unobscured AGN have so far been consistent with the observations, although these models rest on a number of additional assumptions, such as the existence of a substantial number of high-luminosity obscured QSOs which have not been observed yet [Gilli et al. 2001]. It seems very likely that the complete explanation of the X-ray background will be in terms of such models of discrete sources and their cosmic evolution.

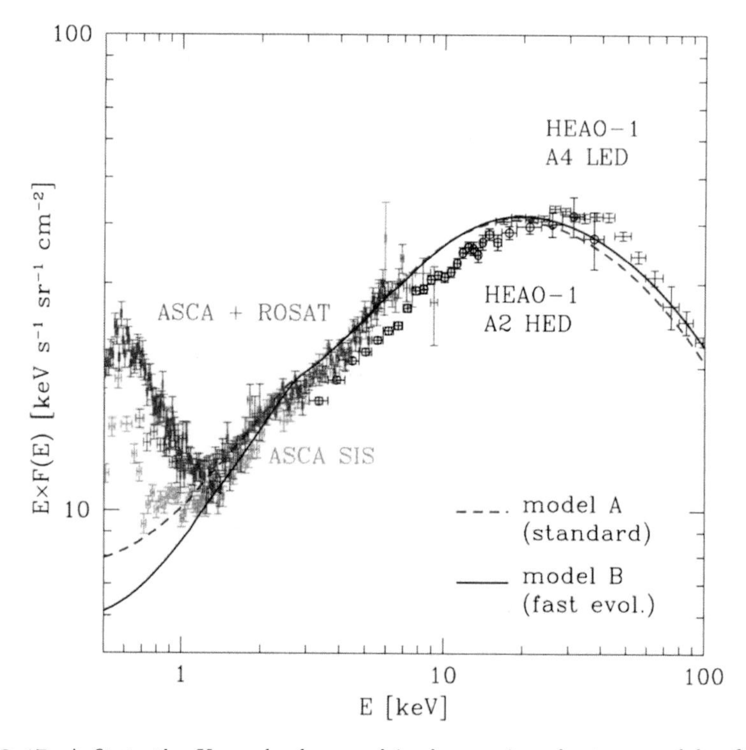

Fig. 2.47. A fit to the X-ray background is shown given by two models of AGN evolution [Gilli et al. 2001]. The data are from [Gendreau et al. 1995; Miyaji et al. 2001; Gruber et al. 1999] (courtesy of R. Gilli et al.)

In Fig. 2.47 the fit to the X-ray background given by two models of AGN evolution (A and B) [Gilli et al. 2001] is shown. Since the X-rays can penetrate a column density of $1\,\mathrm{g/cm^2}$, they could, in principle, propagate towards us from extremely large distances, covering a sizable fraction of the Hubble sphere. Fluctuations in the background are thus essentially measurements of fluctuations of the column density, and therefore measurements of density fluctuations $\delta\varrho/\varrho$ on scales corresponding to the angular scales of the measurements, of ~ 50 to 500 Mpc.

2.7 Evolutionary Effects

Counts of radio sources with flux above a certain limit S, or counts of optical sources with redshifts less than a certain limit z, have played an important part in establishing the presence of evolutionary effects in the source populations. The "steady-state" model of cosmology (cf. e.g. [Weinberg 1972]), which was prominent in the 1950s, first came under fire from these classical cosmological tests. In the meantime evolutionary effects can be seen and imaged directly by the large, newly active telescopes.

2.7.1 Counts of Radio Sources

The observational material on radio sources has increased steadily in quantity and quality. Now large samples of sources are available at frequencies below 1415 MHz, but also at 2700 MHz and 5000 MHz (cf. [Wall et al. 1980, 1981]). The flux limits obtained at 408 MHz correspond to a density of 10^5 sources per steradian. The searches at 2700 MHz have revealed the existence of two distinct classes of radiosources: the steep spectrum sources (SS) whose flux S follows a power law in frequency $S \propto \nu^{-\alpha}$ with $\alpha \sim 0.75$. They comprise very luminous sources with extended radio structure (radio galaxies and quasars) which make up about 90% of the sources counted at 408 MHz. The non-steep spectrum sources (NSS) have appeared as a second population in the 2700 MHz surveys; at high powers they have been identified with quasars of a compact radio structure, and a flatter ($\alpha < 0.5$) and often self-absorbed spectrum. It has turned out that these two classes of sources show quite a difference in their behaviour as the number of sources per steradian above a certain flux level is considered.

The usual approach to the data is to extract a generalized radio luminosity function ϱ, which describes the space density of radio sources (the comoving density in cosmological models). ϱ may depend on the radio luminosity P, the redshift z, and the type of source population – in the analysis discussed here, essentially NSS and SS sources.

For a constant space density ϱ_0 in Euclidean space, the number of sources $N(\geq S)$ above a certain flux level S is

$$N(\geq S) = \int_0^\infty dP \varrho_0(P) \frac{4\pi}{3} d^3, \qquad P \equiv \frac{L}{4\pi}.$$

Here $d = \frac{1}{4\pi}(P/S)^{1/2}$, and thus

$$N(\geq S) = S^{-3/2} \int_0^\infty dP P^{3/2} \varrho_0(P) \frac{4\pi}{3}. \tag{2.92}$$

Quite early on, when this analysis was carried out, it became clear that the source counts did not follow this $N \sim S^{-3/2}$ law. Figure 2.48 shows the famous $\log N$–$\log S$ plot of the first counts at 408 MHz [Ryle 1968]. The number counts deviate strongly from a constant, Euclidean, space density – there are many more sources at low fluxes. This has always been interpreted as an evolution of the radio source population – these sources were more common in the past than they are now.

The source counts are usually done at one particular frequency (408 MHz, 1415 MHz, etc), and the assumption is made that there are but two source populations each with definite spectral index $P_\nu \propto \nu^\alpha$.

Therefore $\varrho(P_{\nu'}, \nu', z) = \varrho\left(\left(\frac{\nu}{\nu'}\right)^\alpha P_\nu, \nu, z\right)$, whence it is sufficient to consider $\varrho(P_\nu, \nu', z)$ for only one frequency ν, and to write $\varrho(P_\nu, z)$ for it. The function $\varrho(P_\nu, z)$ is written in the form

$$\varrho = F(P_\nu, z)\varrho_0(P_\nu),$$

where $\varrho_0(P\nu)$ is the present intrinsic, local luminosity function. One tries to determine this function from the data by choosing a complete sample

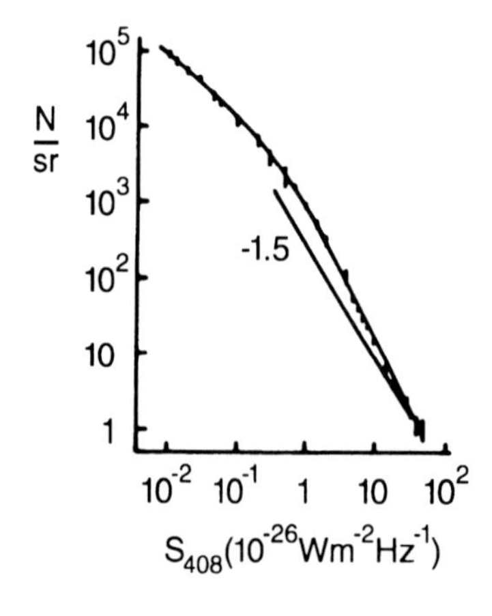

Fig. 2.48. An early $\log N$–$\log S$ plot of radio sources at 408 MHz [after Ryle 1968]. The straight line represents the Euclidean distribution $N \propto S^{-1.5}$

of sources for which redshifts are known, and using the Hubble law to determine distances and hence intrinsic luminosities. Such a sample seems to exist for the SS sources at 408 MHz and at flux levels above 10 Jy ($1\,\mathrm{Jy} = 10^{-26}\,\mathrm{Wm^{-2}Hz^{-1}}$) [Wall et al. 1980], and thus $\varrho_0(P_{408})$ can be determined from the data. In other cases a model function has to be assumed, and tested together with the other fits.

The function $F(P_\nu, z)$ is the "evolution function" – changes in the number density with cosmic history should turn up in the shape of this function. $F = 1$ corresponds to a uniform space distribution considering only expansion, and no evolution.

As a next step the choice of a particular FL model provides the framework for the interpretation of the data. The area distance r_A referred to the present time ($K = 0$ model) (Chap. 1 and Sect. 2.2) is chosen as a measure of the distance

$$d_0 = \frac{2c}{H_0}\left(1 - \frac{1}{\sqrt{1+z}}\right);$$

$z(S)$ is the redshift at which a source of power P_ν has a flux density S_ν. The expression

$$S_\nu = \frac{P_\nu}{r_A^2(1+z)^{1+\alpha}}$$

holds generally. Then the number of sources with a flux density above S_ν is given by

$$N(\geq S_\nu;\nu) = \int_0^\infty dP_\nu \int_0^{z(S_\nu,P\nu)} \varrho_0(P)F(P_\nu,z)dV(z).$$

The comoving volume element $dV(z)$ is given as

$$dV(z) = 4\pi r_A^2 \frac{c}{H_0}(1+z)^{1/2}dz.$$

This involves a projection from an interval dt_e on our past light-cone to the hypersurface $t_e =$ const. (corresponding to z), on which the sources are placed: $dV(z)$ is the volume element on this hypersurface. Again, $F = 1$ does not represent the data well, as is clearly shown even in older data from 20 years ago (Fig. 2.49). More detailed information about the nature of the sources' evolution can, of course, be derived from abundant new data.

Considering counts from 408 MHz and 2700 MHz, it can be shown conclusively [Wall et al. 1980, 1981], that strong evolutionary effects are present ($F = 1$ is excluded). In addition, it seems possible to obtain information about the nature of this evolution: simple power-law models $F \sim (1+z)^\beta$

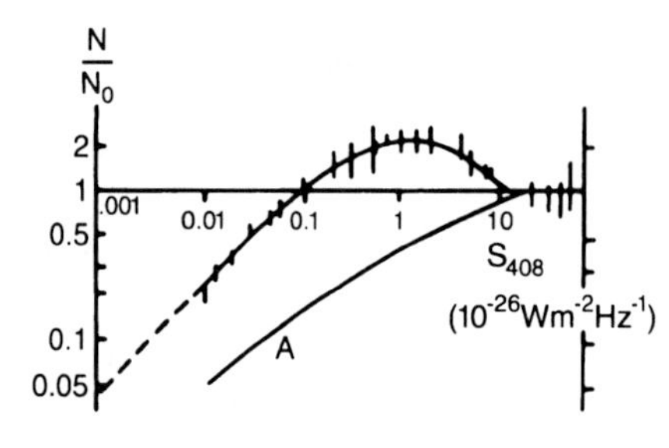

Fig. 2.49. The counts of radio sources expressed as the ratio N/N_0, where N_0 is the number from the Euclidean $S^{-1.5}$ behaviour. The *solid line* (A) shows the variation predicted for an FL world model with $q_0 = -3$ [after M. Ryle 1968]

could be eliminated, as well as exponential functions $F \sim \exp M(1 - t/t_0)$ with constant M for the SS sources ($t/t_0 = (1+z)^{-3/2}$). Satisfactory fits to the data for these sources could be obtained for an exponential F with M depending on the power P, and a redshift cut-off at $z_c = 3.5$, i.e. $F = 0$ for $z \geq z_c$. The distribution of the SS sources is shown in Fig. 2.50. The best fit parameters are

$$M = 0 \quad \text{for} \quad P_1 > P$$

$$F = \exp[M(1 - t/t_0)];\, M = M_{\max} \log(P/P_1)/\log(P_2/P_1);$$

$$M = M_{\max} \quad \text{for} \quad P > P_2$$

$$M_{\max} = 11.5; \quad \log{(P_1/WHz^{-1})}\,\mathrm{sr}^{-1} = 25.0$$

$$\log{(P_2/WHz^{-1})}\,\mathrm{sr}^{-1} = 27.3.$$

There is a strong evolutionary effect for the bright sources, which disappears for the very faint sources; little evolution is found for small redshift. The

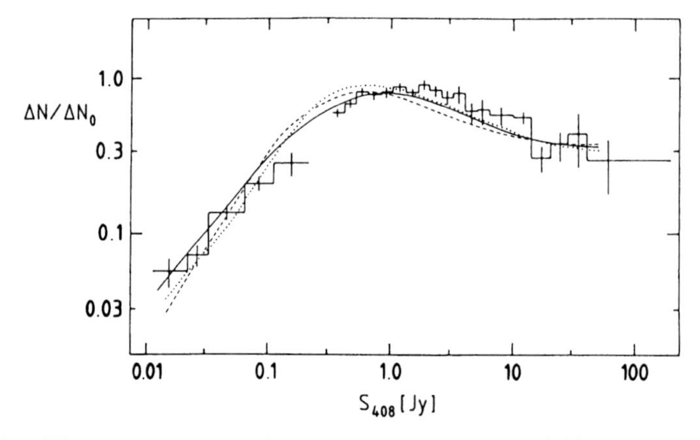

Fig. 2.50. The source counts of steep-spectrum sources (SS) at 408 MHz [after Wall et al. 1980]. The *solid* lines are model fits. ΔN is the number of sources per steradian. The counts are normalized to the Euclidean prediction $\Delta N_0 = 100/(S/\mathrm{Jy})^{-1.5}$

NSS sources show a less pronounced evolution; a good fit for these sources can be obtained with a function

$$F(P, z) \equiv F(P) \exp M(1 - t/t_0)$$

with $M = 5$ to 7 [Wall et al. 1981].

This ties in with another analysis [Peacock and Gull 1981], where the luminosity function is expanded as a power series both in luminosity P_ν and in redshift z for each of the two spectral classes (NSS and SS). The expansion coefficients are then found iteratively. The result (Fig. 2.51) is again that strong evolution is present for the higher flux sources, but that this seems to fade away at the very faint end. The interpretation is that few of these sources have turned on before the epoch corresponding to about $z \sim 4$. This time must then be considered as the time of formation of these sources. The general conclusion is that the number of sources in a comoving volume was greater by a factor of $\sim 10^3$ when epochs at redshift between 1 and 3 are

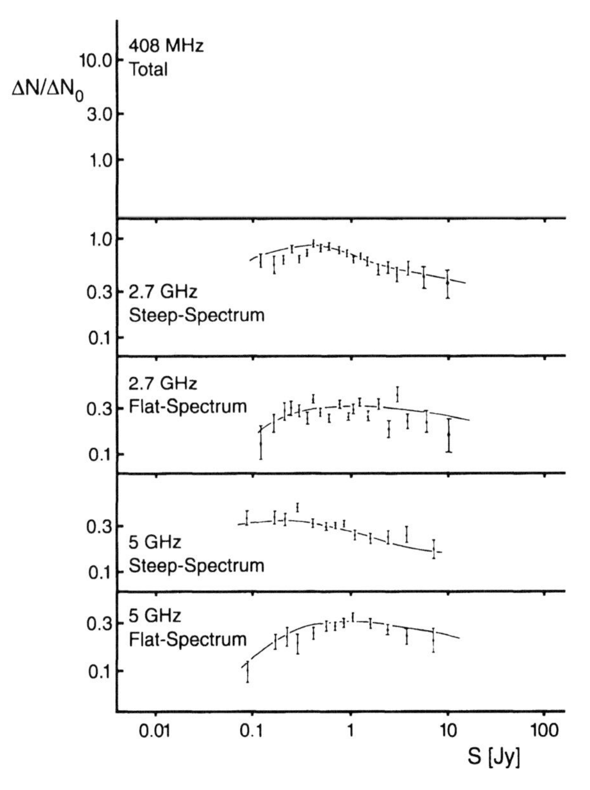

Fig. 2.51. Source counts at 408, 2700 and 5000 MHz together with the model predictions (normalized to $\Delta N_0 = 100 S^{-1/5}$ [after Peacock and Gull 1981]. Models give a typical redshift z between 1.0 and 3.0. The range of z in evolution models is 0.03 to 3 for SS sources, and 0.3 to 10 for NSS sources

considered. Back beyond redshifts of ~ 3 there is good evidence that there is no further increase in source numbers.

2.7.2 The Luminosity–Volume Test for Quasars

The conclusions just given are well supported by deep surveys looking for quasars with very high redshift [Osmer 1982]. It seems that beyond $z = 5$ few quasars are found. Therefore it seems a safe conclusion that the quasars started their activity around $z = 5$ or 6. This epoch must then also coincide with the epoch of galaxy formation, in accordance with the general belief that quasars are the active nuclei of newly formed galaxies.

The luminosity–volume test for quasars is another indication for evolutionary effects [Kafka 1968; Schmidt 1968; Osmer 1982]. Consider a sample of sources which is complete up to a known apparent-brightness limit, and for which redshifts are well determined. Then one can compute for each source the ratio of the volume V corresponding to the redshift z, and the volume $V_{\max}$ corresponding to a redshift $z_{\max}$ up to which the source would have been detected and included in a complete sample.

For a uniform space density of sources, the expected mean value in a complete sample is

$$\left\langle \frac{V}{V_{max}} \right\rangle = 0.5.$$

Various different observations [Osmer 1982] yield average values consistently above 0.5,

$$\left\langle \frac{V}{V_{max}} \right\rangle = 0.558 \quad \text{to} \quad 0.69,$$

depending on the redshift intervals considered. The QSOs seem to be concentrated near the edge of their observable volume. This means that there is an evolution of their number density with redshift.

2.7.3 Direct Evidence for Cosmic Evolution

The Hubble Space Telescope, as well as the Keck telescope and other large ground-based telescopes, have increased the ability to study directly the evolution of galaxies at high redshift. Number counts versus magnitude and redshift surveys are used. The HST provides information on the morphology of distant galaxies [Schade et al 1995].

Faint Blue Galaxy Excess. The number counts of blue galaxies increase with magnitude $B > 22$ more rapidly than for red galaxies, and more rapidly than expected, if the local population at $z = 0$ had just undergone passive evolution with comoving density and mixture of types preserved for $z > 0.3$

[Koo and Kron 1992]. The interpretation is difficult because of uncertainties in the faint end slope and normalization of the local luminosity function. There is evidence from ground-based photometry that the luminosity function of red galaxies shows little evolution for $z < 1$, while the blue luminosity function brightens by about one magnitude and steepens at $z > 0.5$ [Lilly et al. 1995].

Faint blue galaxies might be a population of dwarf galaxies undergoing starbursts and then fading away, or they might be star-forming fragments of galaxies that merged into bright galaxies at $z < 1$. Their abundance now seems quite consistent with galaxy formation models based on hierarchical clustering. The morphology of the excess faint blue galaxies has been revealed by the HST [Schade et al. 1995], especially in the Hubble Deep Field images: a high percentage are late-type spirals, irregulars and peculiar (possibly merging) galaxies. Blue irregular galaxies seem to be an intermediate step in the galaxy formation process, since they appear to be missing in the nearby universe. Figure 2.12 is the $1' \times 1'$ section of the sky containing the ~ 2500 galaxies of the Northern Deep Field.

Lyman Break Galaxies. The UV Lyman continuum emission shortward of the wavelength of 912 Å is absorbed by the gas and stars in a galaxy. At high redshifts this absorption edge moves into the optical, and by a clever arrangement of filters a very efficient way of finding high-z galaxies has been worked out. Redshifts are now available for more than 700 objects at $z \geq 3$ [Steidel et al. 1998], with a record holder above $z = 6$. The number density and clustering properties of these galaxies indicate that they might be the central galaxies of the most massive DM halos present at $z \sim 3$ (see Chaps 11, 12). Lyman break galaxies probably have low angular momentum and small size for their mass in contrast to damped *Ly*α systems (see below) which probably have large angular momentum [Mo et al. 1998].

Absorption Lines. Absorption lines along the line of sight to background quasars probe the gas clouds and/or protogalaxies present at high redshift. The "Lyman-alpha forest" consists of the numerous optically thin absorption lines (with HI column densities $\leq 10^{17}\,\mathrm{cm}^{-2}$) seen at wavelengths shorter than rest-frame Lyα (1216 Å) in the spectrum of a quasar.

"Lyman limit systems" are absorbers with $n(\mathrm{HI})$ exceeding $10^{17}\,\mathrm{cm}^{-2}$, so that the medium is optically thick in the Lyman continuum (rest-frame $\lambda \leq 912$ Å), resulting in strong continuum extinction. "Damped Lyα systems" have $n(\mathrm{HI}) \geq 10^{20.2}\,\mathrm{cm}^{-2}$ and thus show damping wings in the Lyα line, and opaque Lyman continuum. They are probably early-type disc galaxies with large angular momentum.

High HI column densities are associated with young galaxies [Wolfe et al 1995]; lower HI column density systems may show Mg II absorption [Steidel et al. 1994], and other metal lines [Pettini et al. 1994]. The total amount of baryons in the damped Lyα systems at $z \sim 3$ is consistent with the amount of baryons seen in galaxies today [Lanzetta et al. 1995].

The Lyα forest can be identified in numerical simulations as the web of structure forming in hierarchical clustering models. Seminal work on this model was done in the early nineties [Bi et al. 1992; Bi 1993]. but it took the community some time to accept and rediscover these models [Haehnelt et al. 1996]. The observed $n(\mathrm{HI})$ distribution is close to a power law from 10^{12} to $10^{21}\,\mathrm{cm}^{-2}$ [Wolfe et al. 1995].

Lyα absorbers with $n(\mathrm{HI}) > 3\times10^{14}\,\mathrm{cm}^{-2}$ also have measurable absorption lines from 3-fold ionized carbon (CIV) and a medium metallicity of 1% of the solar. This has led to the speculation of star formation in subgalactic clumps. All this supports the view that we see in those low-metallicity, high-z Lyα absorbers the hydrogen clumps from which galaxies are forming.

Exercises

This is a long chapter, so there are plenty of difficult exercises.

2.1 The darkness of the night sky. Show that a uniform distribution of stars in infinite Euclidean space will lead to a bright night sky.

For an expanding universe the apparent brightness of a source I is related to its intrinsic brightness I_0 by

$$I = \frac{I_0}{(1+z)^4}.$$

Show that the finite lifetime of the stars limits I_0 and I. Discuss the CMB brightness. Show that the brightness of the night sky can be approximated by

$$I(\nu) \simeq \frac{n_0}{4\pi H_0}\int_\nu^\infty \frac{L(\nu')}{\nu'}d\nu'.$$

Here n_o is the number density of galaxies, $L(\nu)$ is the mean luminosity of a galaxy at frequency ν. Look up numbers to find that at $\nu = c/5500\,\text{Å}$, $\nu I(\nu) \simeq 2.5\times10^{-6}\,\mathrm{erg\,cm^{-2}s^{-1}sr^{-1}}$.

2.2 For a sample of objects the data (m_i, z_i) can be used to determine the Hubble law. Assume $cz = H_0 r^p$ as a "generalized" law (in FL models $p = 1$). The relation

$$m_i = -\frac{5}{p}\log H_0 + \frac{5}{p}\log cz_i + 25 + M_i$$

can be evaluated by linear regression to find p and H_0.

Show that the estimator $E(m/V)$ of equation (2.18) with $V = \log cz$ reduces to the simple linear regression only if $M_i = M_0$, i.e. the objects are standard candles! Show that a magnitude cut-off ($m \leq m_g$) is harmless if $M_i = M_0$, but introduces errors if the magnitudes are distributed according to some luminosity function.

Compute the correction term to $E(m/V)$ if the luminosity function is a Gaussian with mean M_0 and variance σ, and $m \leq m_g$.

2.3 Consider classical electromagnetic waves. What is the intensity I of light at an arbitrary point which is illuminated by N sources? Take the electromagnetic field of source j as a $\exp(i\varphi_j)$.

Take the phase average $\langle I \rangle$ and show that $\langle (I - \langle I \rangle)^2 \rangle \simeq \langle I \rangle^2$ for large N. Why is the CMB so much smoother?

2.4 The system of photons and matter in thermal contact can be described as an ideal gas in equilibrium with black-body radiation. The continuity equation can be used to find $T(R)$. Derive the differential equation, introduce the abbreviation

$$\sigma \equiv \frac{4}{3}\frac{a\,T^3}{kn};$$

then show that for $\sigma \ll 1, T \propto R^{-1}$ and $\sigma = \text{const.}$

2.5 Use the SAHA equation to estimate the temperature of recombination. Show that

$$\frac{n_e n_p}{n_{\mathrm{H}}} = e^{-\Delta/kT} \frac{(2\pi m_e kT)^{3/2}}{h^3};$$

n_e, n_p, n_{H} are electron, proton, hydrogen atom densities, and Δ is the ionization energy of hydrogen $\sim 13.6\,\mathrm{eV}$.

Evaluate the formulae for the degree of ionization $x \equiv \frac{n_P}{n_P + n_{\mathrm{H}}} = \frac{1}{2}$. Why is T_R, obtained this way, higher than the real recombination temperature (see text)?

2.6 Consider a finite number N of atoms in a background of black-body radiation with temperature T. Discuss the final state of this system. What is the time it takes to reach it?

Why is recombination in the universe nevertheless possible?

Look at [Bach et al. 2000] to get an idea of what is required for a rigourous treatement of this problem.

2.7 The evidence for dark matter from the rotation curves of galaxies, from the dynamics of galaxies in clusters, depends on the assumption that Newton's law of gravitation holds universally. One could think of modifying the gravitational force law (see [Bekenstein and Milgrom, 1984, Milgrom 1983]).

The ansatz for the Poisson equation for the modified potential φ for a spherical mass distribution might be

$$\nabla\left(\mu\left(\frac{|\Delta\varphi|}{a_o}\right)\nabla\varphi\right) = -4\pi G\varrho;$$

where a_o is a fixed acceleration, and the function μ might be

$$\mu(x) = \begin{cases} 1 \text{ for } x \gg 1, \\ x \text{ for } x \ll 1, \end{cases}$$

i.e. for small values of the gradient of the potential, far from the sources, the gravitational force is modified.

Take $a_o = 1.21 \times 10^{-8}\,\mathrm{cm^2 s^{-2}}$, $\mu(x) = x(1+x^2)^{-1/2}$, and show that the rotation curves of galaxies are flat.

The non-linear equation for φ is not easy to solve. So, one takes the Newtonian potential φ

$$\nabla^2\varphi_N = -4\pi G\varrho,$$

and writes

$$\mu(F/a_o)\boldsymbol{F} = \boldsymbol{F}_\mathrm{N} + \nabla\times\boldsymbol{h};$$

with arbitrary $\boldsymbol{h}$. $\nabla\times\boldsymbol{h}$ can be neglected for large distances. To solve the equation, just calculate the Newtonian force $\boldsymbol{F}_\mathrm{N}$, and finally solve for $\boldsymbol{F}$.

2.8 Take the modified gravitational law of exercise 2.7, and apply it to the virial theorem for clusters, assuming a spherically symmetric mass distribution. Use $\mu(x) = \sqrt{x}$ for $x \ll 1$, and derive $M/L = Y \equiv \left(\frac{3J}{\sqrt{Ga_o}}L^{-3/2}\right)^2$ for clusters. Here L is the luminosity,

$$J \equiv 3\pi\int_o^\infty I(r)\sigma_p^2(r)rdr,$$

$\sigma_P(r)$ is the velocity dispersion, $I(r)$ the surface brightness.

$K \equiv \frac{1}{2}\Sigma m_i v_i^2 = M/LJ$ [cf. Binney and Tremaine 1987]. For constant σ_P the result is

$$Y = \frac{81}{4}\frac{1}{L}\frac{\sigma_P^4}{Ga_o}.$$

Derive this, and taking $a_o = 1.21 \times 10^{-8}\,\mathrm{cm\,s^{-2}}$, find some M/L ratios for clusters.

2.9 Compare the change with redshift of the apparent angular extent of the X-ray image of a cluster with the Sunyaev–Zel'dovich image obtained from the CMB decrement.

2.10 The rate of the total energy gain $\varepsilon(z)$ of an intergalactic plasma is given by the average energy input per unit time $\bar{\varepsilon}_Q$ of a quasar. Then

$$\varepsilon(t) = (n_Q(z)\bar{\varepsilon}_Q)/\rho_{Pl}(z),$$

where $n_Q(z) = n_Q(0)(1+z)^3 n(z)$ is the number density of quasars (or active galaxies); $n(z)$ is a cosmic evolution factor for quasars ($n(z) = (1+z)^6$); $\rho_{Pl}(z)$ is the plasma mass density.

Use $T(t)dS(t) = \epsilon(t)dt$ to compute the temperature T and entropy S. For $S(z)$ use the Sackur–Tetrode formula

$$S(z) = \frac{R_g}{\mu}\frac{3}{2}\ln\left(\frac{T(z)}{\rho_{Pl}(z)}\right)^{3/2} + \text{const.}$$

Show that $T(z) = T_0(1+z)^2\frac{F(z_k)-F(z)}{F(z_k)-F(0)}$, where z_k is the redshift for which $T = 0$. Compute $F(z)$ and T_0 for $n(z) = (1+z)^6$, and plot $T(z)$ for $\Omega_o = 1$. [Field and Pierrenod 1977].

2.11 Carry out a detailed calculation for the luminosity–volume test for quasars. Consider a sample of sources complete up to an apparent brightness limit with well-determined redshift. One can compute for each source the ratio of the volume V corresponding to the redshift z, and the volume $V_{\max}$, corresponding to the redshift $z_{\max}$, up to which the source would have been detected and included in a sample.

Show that for a uniform space density of sources, the expectation value of the ratio is 0.5. What is the result if an evolution of the luminosity function as in Sect. 2.7.1 is assumed?

3. Thermodynamics of the Early Universe in the Classical Hot Big-Bang Picture

"The present universe is something like the old professor nearing retirement with his brilliant future behind him."

A. Sandage (1994)

The hot big-bang picture emerges from an FL model, where the expansion factor $R(t)$ approaches zero towards the limit ($t \to 0$) of the cosmic time t. The presence of a uniform radiation field implies a "hot" singularity, i.e. an ever-increasing temperature as $t \to 0$. (This is true as long as the constituent particles of the radiation background can be treated as non-interacting. There are model theories of elementary particles where the interactions limit the temperature to a maximum, finite value, which is, however, large when compared with the energy scales discussed in this chapter; cf. also Chap. 7.)

A lot of effort is required to obtain secure facts about the hot big bang

A scheme that avoids the hot big bang can be acceptable only if an explanation of the origin of the CMB from specific astronomical sources is given. Attempts at such unorthodox scenarios do not look very promising, and have received little attention so far. During the radiation-dominated phase ($z > z_R$) the energy density in singular FL models evolves proportionally to R^{-4} or, equivalently, proportionally to the fourth power of the temperature. The curvature term K/R^2 and the cosmological constant Λ become negligible compared with the "potential energy" term $\propto \varrho$ in the Friedmann equation, which can be written approximatly as

$$\dot{R}^2 = \frac{8\pi}{3} G\varrho R^2. \tag{3.1}$$

Since ϱR^4=const., the expansion factor $R(t)$ satisfies $\dot{R}R^2$ = const., hence $R \propto t^{1/2}$, and finally

$$\frac{32\pi}{3c^2} G\varrho t^2 = 1. \tag{3.2}$$

At sufficiently early times all homogeneous and isotropic solutions with a "hot" singularity evolve like the $K = 0$ Einstein–de Sitter universe.

A crucial assumption in all attempts to describe the early universe is the existence of a phase of complete thermodynamic equilibrium at some very early epoch (e.g. around $T = 10^{11}$ K, or earlier).

3.1 Thermodynamic Equilibrium

3.1.1 Statistical Equilibrium Distributions

The existence of a state of complete thermodynamic equilibrium implies an enormous simplification of the physics in the early universe; all the quantities of interest depend only on the temperature and the chemical potentials. The previous history of the thermodynamic system need not be known, except for the interesting complication that the values of some chemical potentials may derive from particle reactions earlier on.

In thermal equilibrium the number density n_i of particles of type i – and their energy density ϱ_i – at temperature T can easily be derived [Weinberg 1972; Steigman 1979; Jüttner 1928]:

$$n_i = \left(\frac{g_i}{2\pi^2}\right)\left(\frac{k_B T}{\hbar c}\right)^3 \left\{\exp\left(\frac{E_i - \mu_i}{k_B T}\right) \pm 1\right\}^{-1} z^2 dz, \tag{3.3}$$

$$\varrho_i c^2 = \left(\frac{g_i}{2\pi^2}\right)\left(\frac{k_B T}{\hbar c}\right)^3 k_B T \int_0^\infty \left\{\exp\left(\frac{E_i - \mu_i}{k_B T}\right) \pm 1\right\}^{-1} z^2 \frac{E_i - \mu_i}{k_B T} dz. \tag{3.4}$$

Here g_i is the number of spin degrees of freedom; i.e. the number of helicity states.

Correct initial conditions are absolutely necessary

$$E_i \equiv \left[(\mathbf{p}c)^2 + (m_i c^2)^2\right]^{1/2},$$

and

$$z \equiv \frac{|\mathbf{p}|c}{k_B T}.$$

The $+(-)$ sign in the integral applies to fermions (bosons). Antiparticles are treated as independent particle species.

The chemical potentials μ_i are unknown parameters in this state of thermodynamic equilibrium. In the standard model their values have to be set as an initial condition. In GUT theories they can be computed from particle reactions. The convergence of the integrals in (3.3) and (3.4) requires, for bosons

$$|\mu_i| \leq m_i c^2.$$

For fermions there are no restrictions.

Photons are bosons with $g_i = 2$ and rest mass $m_i = 0$. Hence also $\mu_i = 0$, and from (3.4) we recover the Planck formula:

$$n_\gamma = \frac{2\zeta(3)}{\pi^2}\left(\frac{k_B T}{\hbar c}\right)^3 = \frac{2.404}{\pi^2}\left(\frac{k_B T}{\hbar c}\right)^3, \tag{3.5}$$

$$\varrho_\gamma c^2 = \frac{2\zeta(4)}{\pi^2}\left(\frac{k_B T}{\hbar c}\right)^3\left(\frac{k_B T}{c^2}\right) \approx \left(\frac{k_B T}{c^2}\right) n_\gamma. \tag{3.6}$$

Here $\zeta(x)$ is Riemann's zeta function:

$$\zeta(3) \equiv 1/2 \int_0^\infty x^2(e^x - 1)^{-1} dx,$$

$$\zeta(4) \equiv 1/6 \int_0^\infty x^3(e^x - 1)^{-1} dx.$$

The electromagnetic and weak interaction processes between the particles, which establish equilibrium below $T \simeq 10^{12}$ K (the particles involved are photons, leptons, and nucleons), conserve various additive quantum numbers; (Table 3.1 lists some of the properties of these particles).

The total electric charge, Q, is obtained by adding up the charges of the individual particles participating in a process. Charge conservation means that the total charge of the initial state is equal to the total charge of the final state. For example, the process

$$e^- + \mu^+ \to \nu_e + \overline{\nu}_\mu \qquad \text{has} \qquad Q = -e + (+e) = 0$$

initially, and $Q(\nu_e) + Q(\overline{\nu}_\mu) = 0$ for the final state.

The total baryon number, B, is ascribed to the particles participating in the strong interactions; the proton and the neutron have a baryon number $B = +1$. The antiparticles have $B = -1$. Leptons (neutrino, muon, electron) have $B = 0$. The quarks, three of which form a proton, have a baryon number of 1/3. The total baryon number B is obtained by adding up the B of the individual particles. Baryon number conservation means the equality of the initial and final baryon numbers in a process. For example, $e^- + p \to \nu_e + n$ has $B = 1$ initially, and $B = 1$ in the final state.

Table 3.1. Basic parameters of a few elementary particles

Particle	Mass [MeV]	Mean life [s]	Bose (B) Fermi (F)
γ photon	$< 3 \times 10^{-33}$	stable	B
ν_e electron neutrino	$< 46 \times 10^{-6}$	stable($> 3 \times 10^8 m_{\nu e}$)	F
e^- electron	0.511	stable($> 2 \times 10^{22}$ y)	F
ν_μ muon neutrino	<0.17	stable($> 10^5 m_{\nu\mu}$)	F
$\mu^\pm$ muon	105.7	2.2×10^{-6}	F
ν_τ^* tau neutrino	< 18		F
τ tau lepton	1784±3	3×10^{-13}	F
$\pi^\pm$ pi-meson	139.6	2.6×10^{-8}	B
π^0 pi-meson	135	0.78×10^{-16}	B
p proton	938.3	stable($> 10^{31}$ to 10^{32} y)	F
n neutron	939.6	887±3	F

*not observed

Baryon number is conserved in all experimentally known processes. Only the electroweak phase transition and some theoretical schemes of the GUTs (see Chap. 8) allow a violation of baryon number.

The total electron–lepton number, L_e, is $+1$ for the electron and the electron–neutrino, and -1 for their anitparticles. For example, $e^- + p \to \nu_e + n$ has $L_e = +1$ in the initial and final states.

The total μ-lepton number, L_μ, is $+1$ for the μ^- and $\nu +_\mu$ particles and -1 for their antiparticles. For example, $\mu^- + p \to \nu_e + n$ has $L_\mu = +1$ in the initial and final states.

The Salam–Weinberg model of electroweak interactions describes processes which violate L_e or L_μ conservation; but the total lepton number $L = L_e + L_\mu$ is conserved in that theory (cf. Chap. 5) and only violated in special GUTs.

None of these quantities is conserved in the GUTs, which we shall describe in Chap. 6. Therefore in the spirit of GUTs all chemical potentials should be set equal to zero.

If we consider, in a more classical spirit, Q, B, L_e and L_μ as conserved additive quantum numbers, there are correspondingly four independent chemical potentials. Since particles and antiparticles can by annihilation processes transform into photons, their respective chemical potentials have opposite signs,

$$\mu_{e^-} = -\mu_{e^+}$$

etc.

The reactions

$$\begin{aligned} e^- + \mu^+ &\to \nu_e + \overline{\nu}_\mu , \\ e^- + p &\to \nu_e + n , \\ \mu^- + p &\to \nu_e + n \end{aligned}$$

lead to relations for the corresponding chemical potentials in equilibrium:

$$\mu_{e-} - \mu_{\nu_e} = \mu_{\mu-} - \mu_{\nu_\mu} = \mu_n - \mu_p \tag{3.7}$$

Therefore the four quantities $\mu_p, \mu_{e-}, \mu_{\nu_e}, \mu_{\nu_\mu}$ can be chosen as the independent chemical potentials, corresponding to the conservation laws for Q, B, L_e, L_μ. It is a reasonable approximation to set these chemical potentials equal to zero. Let us denote by n_Q, n_B, n_{L_e} and n_{L_μ} the densities corresponding to the four different quantum numbers. Each chemical potential is then a function of $n_Q, n_B, n_{L_e}, n_{L_\mu}$; if in a system all particles are replaced by their antiparticles, the four densities change sign, and the chemical potentials also change sign – they are odd functions of these four densities.

Now, because of charge neutrality in any small volume, one has $n_Q \simeq 0$. Furthermore, since $n_B \ll n_\gamma$ one can also neglect (to a good approximation) the small value of n_B – small as compared to the entropy density ($\propto n_\gamma$), and set $n_B \simeq 0$. Then

$$\begin{aligned} n_{Le} &= n_{e-} + n_{\nu e} - n_{e+} - n_{\overline{\nu} e}, \\ n_{L\mu} &= n_{\mu-} + n_{\nu\mu} - n_{\mu+} - n_{\overline{\nu} e}. \end{aligned}$$

It is reasonable to expect these lepton numbers to be small in the same sense as the baryon number is small. This is equivalent to assuming that there is, at most, a small excess of particles or antiparticles.

It should be mentioned here that cosmological models with a large excess of neutrinos – or antineutrinos – for example, have been discussed. In such models there is a large neutrino background behaving like a degenerate Fermi gas. The equilibrium distributions (3.3) show that for $\mu_i \neq 0$, and $|\mu_i| \gg k_B T$, there will be a considerable excess of particles – which have a factor $\exp(|\mu_i| \gg k_B T)$ in their favour over antiparticles, which have their number reduced by a factor $\exp(-|\mu_i| \gg k_B T)$. We shall discuss the relevance of such theories a bit more in Chap. 7.

The reasonable expectation is, however, that $n_{Le} = n_{L\mu} \approx 0$. Then the chemical potentials are all equal to zero too. (Since they are odd functions of $n_Q, n_B, n_{Le}, n_{L\mu}$ their value at the zero point is zero.) With all the chemical potentials equal to zero only those particles can be present in appreciable numbers for which $mc^2 \ll kT$. These particles can all be treated to a good approximation as ultrarelativistic, $c|p| \gg mc^2$.

Then for bosons

$$n_B = (g_B/2) n_\gamma;\ \varrho_B = (g_B/2) \varrho_\gamma \tag{3.8}$$

and for fermions

$$n_F = \frac{3}{8} g_F n_\gamma;\ \varrho_F = \frac{7}{16} g_F \varrho_\gamma. \tag{3.9}$$

The total energy density in the radiation-dominated phase is then

$$\varrho = \varrho_B + \varrho_F \equiv \frac{1}{2} g(T) \varrho_\gamma \tag{3.10}$$

where

$$g(T) \equiv g_B + \frac{7}{8} g_F.$$

To determine $g(T)$, all particles in thermal equlibrium at temperature T are taken into account, plus additional relativistic particles that have dropped out of equilibrium.

At temperatures below 10^{12} K (corresponding to energies $\simeq$ 100 MeV) only the leptons $\mu^\pm, e^\pm, \nu_\mu, \overline{\nu}_\mu, \nu_e, \overline{\nu}_e$ and the photons γ exist in appreciable numbers in equilibrium. The nucleons are an unimportant "contaminant". The function $g(T)$ is then equal to 57/4.

The electromagnetic interactions guarantee equilibrium distributions for the photons and the charged particles.

The neutrinos and antineutrinos take part in weak interaction processes such as

$$e^+ + e^- \leftrightarrow \nu + \overline{\nu}; e^\pm + \nu \to e^\pm + \nu;$$
$$e^\pm + \overline{\nu} \to e^\pm + \overline{\nu};$$

$$\begin{aligned}
&e^- + \mu^+ \leftrightarrow \nu_e + \overline{\nu}_\mu; && e^- + \mu^+ \leftrightarrow \overline{\nu}_e + \nu_\mu; \\
&\nu_e + \mu^- \leftrightarrow \nu_\mu + e^-; && \overline{\nu}_e + \mu^+ \leftrightarrow \overline{\nu}_\mu + e^+; \\
&\nu_\mu + \mu^+ \leftrightarrow \nu_e + e^+; && \overline{\nu}_\mu + \mu^- \leftrightarrow \overline{\nu}_e + e^-.
\end{aligned}$$

The so-called standard model of Salam, Weinberg, and Glashow (cf. Chap. 5) describes these weak interactions correctly. Since the muons – because of their larger mass – are around in appreciable numbers only at temperatures above 100 MeV, these reactions are mainly responsible for maintaining the coupling of neutrinos to matter. The interaction cross-sections are (on dimensional grounds) all of the order of ($\hbar = c = k_B = 1$)

$$\sigma_W \simeq G_F^2 T^2. \tag{3.11}$$

Here G_F is the universal Fermi constant

$$G_F \simeq 10^{-5} m_p^{-2}\,[\mathrm{GeV}^{-2}] \tag{3.12}$$

and m_p is the mass of a proton. In the standard electro-weak model

$$G_F = \frac{\alpha\pi}{\sqrt{2}} \sin^{-2}\Theta_W m_W^{-2}.$$

Here $\sin^2\Theta_W = 0.215 \pm 0.014$ from experiments, and $\alpha = \frac{1}{137}$, the fine-structure constant. m_W is the mass of the W-boson,

$$m_W \simeq 80\,\mathrm{GeV} \quad \text{(see Sect. 5.5)}.$$

On the other hand, the equilibrium distributions for the electrons (positrons) of (3.3) give

$$n_e \simeq T^3 \tag{3.13}$$

for $T \gg m_e c^2, n_e = (2/\pi^2)T^3$.

Therefore the reaction rates for ν-scattering and ν-production through $e^\pm$ scattering all are of magnitude

$$\sigma_W n_e c \simeq G_F^2 T^5 \tag{3.14}$$

The total energy density

$$\varrho = aT^4 g(T)/2$$

gives rise to an expansion rate of an FL cosmos

$$H \equiv \frac{\dot{R}}{R} \simeq (G\varrho)^{1/2} \simeq G^{1/2} g^{1/2} T^2. \tag{3.15}$$

Thus the ratio of the weak interaction rate to H is

$$\frac{\sigma_W n_e}{H} \simeq G^{-1/2} G_F^2 g^{-1/2} T^3 \simeq g^{-1/2} T_{10}^3 \tag{3.16}$$

where T_{10} is temperature in units of 10^{10} K.

This ratio is larger than 1 for $T > g^{1/2} 10^{10}\,\mathrm{K} \simeq 3 \times 10^{10}\,\mathrm{K}$ ($g = \frac{43}{4}$ at this epoch). The neutrino interactions can maintain a thermal equilibrium for temperatures above 10^{10} K. At lower temperatures one expects the neutrinos to drop out of thermal equilibrium. Even after their decoupling, the neutrinos follow a thermal Fermi distribution

$$n_\nu(p)dp = \frac{4\pi}{h^3} p^2 dp \left[\exp(p/k_B T_\nu) + 1\right]^{-1}. \tag{3.17}$$

The entropy of relativistic particles is constant: $S/k_B = \frac{4}{3}(R^3/T)\varrho_{eq}$; as $\varrho_{eq} \propto R^{-4}$ it follows that $TR = \mathrm{const}$. The neutrino temperature T_ν decreases, after decoupling, proportionally to R^{-1}. (The distribution (3.17) is maintained, if $T_\nu \propto R^{-1}$ since $n_\nu \propto R^{-3}$ and $p \propto R^{-1}$.)

3.1.2 The Neutrino Temperature

After the annihilation of the muons the universe is made up of highly relativistic particles ($e^\pm, \gamma, \nu_e, \overline{\nu}_e, \nu_\mu, \overline{\nu}_\mu$ and possibly $\nu_\tau, \overline{\nu}_\tau$) and of negligibly small contributions from massive nucleons. This is the epoch in the history of the universe where it appears in its simplest state. All these particles are initially in thermal equilibrium with a temperature $T \propto R^{-1}$. The neutrino temperature shows this behaviour even after decoupling.

The photon temperature has a more complicated evolution: below $T \sim 5 \times 10^9$ K, electrons and positrons start to annihilate into photons, and this leads to a change in the behaviour of T_γ. Let us first of all discuss the ratio T_γ/T_ν after the annihilation of e^+, e^-. The entropy of relativistic particles is

$$S = sR^3 = R^3 \frac{(\varrho_{eq} + p_{eq})}{T} = \frac{4}{3}\frac{R^3}{T}\varrho_{eq}, \tag{3.18}$$

$$\varrho_{\nu_e} = \varrho_{\overline{\nu}_e} = \varrho_{\nu_\mu} = \varrho_{\overline{\nu}_\mu} = \varrho_{\nu_\tau} = \varrho_{\overline{\nu}_\tau} \equiv \varrho_\nu = \frac{7}{16}\varrho_\gamma. \tag{3.19}$$

For $k_B T \gg m_e c^2$ the radiation density contributed by electrons and positrons is

$$\varrho_{e-} = \varrho_{e+} = 2\varrho_\nu = \frac{7}{8}\varrho_\gamma \tag{3.20}$$

For the photons, of course,

$$\varrho_\gamma = aT^4. \tag{3.21}$$

Then the requirement that the entropy (3.18) is constant, leads to a relation between the system of $e^\pm$, neutrinos and photons when $k_B T \gg m_e c^2$(BF) and the system of neutrinos and photons when $k_B T \ll m_e c^2$(A): $S(kT \gg m_e) = S(kT \ll m_e)$ is equivalent to

$$(T_\gamma R)^3_{BF}(1 + \frac{2 \times 7}{8}) + (T_\nu R)^3_{BF} = (T_\gamma R)^3_A + (T_\nu R)^3_A.$$

As $(T_\nu R)_{BF} = (T_\nu R)_A$, we find

$$(T_\gamma R)^3_{BF} \frac{11}{4} = (T_\gamma R)^3_A$$

or

$$\frac{(T_\gamma R)_A}{(T_\gamma R)_{BF}} = \left(\frac{11}{4}\right)^{1/3}. \tag{3.22}$$

For $kT \gg m_e$ the photon and neutrino temperatures are equal, and hence

$$\left(\frac{T_\gamma}{T_\nu}\right)_A = \left(\frac{11}{4}\right)^{1/3} = 1.401. \tag{3.23}$$

For all lower temperatures T_γ and T_ν are again proportional to R^{-1}. the decoupling of photons from matter at $T \sim 10^4$ K does not appreciably change this law, since there are so few hydrogen atoms: $n_H \simeq 10^{-10} n_\gamma$. Thus at the present epoch we expect a thermal neutrino radiation with a temperature

$$T_{\nu 0} = \left(\frac{4}{11}\right)^{1/3} T_{\gamma 0} \simeq 1.9\,\mathrm{K}. \tag{3.24}$$

After the annihilation of the $\mu^\pm$ and before the $e^\pm$-pair annihilation, i.e. for $T > 1\,\mathrm{MeV}$, the density ϱ is given (considering ν_μ, ν_e, ν_τ) by:

$$\varrho_{eq} = \varrho_\gamma + 2\varrho_{e-} + 6\varrho_\nu = aT^4 \frac{43}{8}. \tag{3.25}$$

Inserting this expression for ϱ into (3.2) we arrive at

$$t = \left(\frac{24c^2}{43 \cdot 32\pi G a T^4}\right)^{1/2} = 0.99 T_{10}^{-2}\,\mathrm{s}. \tag{3.26}$$

After the $e^\pm - e^-$ annihilation ($T < 0.5\,\mathrm{MeV}$)

$$\varrho = \varrho_\gamma(T_\gamma) + 6\varrho_\nu(T_\nu) = aT_\gamma^4[1 + 6 \cdot \frac{7}{16}(\frac{4}{11})^{4/3}] = aT_\gamma^4\, 1.68.$$

For this temperature-range,

$$t = 3.16\, T_{10}^{-2} s. \tag{3.27}$$

Again $RT = \mathrm{const}$.

During the period of the annihilation of electrons and positrons, the behaviour of $T(R)$ is somewhat more complicated, but it can be determined easily by the requirement that the entropy in a volume $R^3(t)$ stays constant:

$$
\begin{aligned}
S = \frac{R^3}{T}(\varrho + p) &= \frac{R^3}{T}(\varrho_{e-} + \varrho_{e+} + \varrho_\gamma + \varrho_e + p_{e-} + p_{e+} + p_\gamma) \\
&= \frac{4}{3}a(RT)^3 J\left(\frac{m_e}{T}\right) = \text{const.}
\end{aligned} \tag{3.28}
$$

where

$$
\begin{aligned}
J\left(\frac{m_e}{T}\right) &\equiv J(x) \\
&= 1 + \frac{45}{2\pi^4}\int_0^\infty dy \frac{y^2}{\exp\sqrt{x^2+y^2}+1}\left\{\sqrt{x^2+y^2} + \frac{1}{3}\frac{y^2}{\sqrt{x^2+y^2}}\right\}.
\end{aligned}
$$

The constant in (3.28) must be equal to the entropy of the radiation after the annihilation

$$
\frac{R^3}{T}(\varrho_\gamma + p_\gamma) = \frac{4}{3}a(RT_\gamma)^3_A. \tag{3.29}
$$

Since $T^3_{\gamma,A} = \frac{11}{4}T^3_\nu$, this is also equal to $\frac{4}{3}a(\frac{11}{4})(RT_\nu)^3$. Therefore

$$
T_\nu = \left(\frac{4}{11}\right)^{1/3} T\left\{J\left(\frac{m_e}{T}\right)\right\}^{1/3}. \tag{3.30}
$$

The energy density is

$$
\begin{aligned}
\varrho &= \varrho_\gamma + 6\varrho_\nu(T_\nu) + \varrho_{e-} + \varrho_{e+} \\
&= aT^4 + \frac{42}{16}a\,T_\nu^4 + \frac{2}{\pi^2}\int_0^\infty (q^2+m_e^2)^{1/2} \\
&\qquad \times \left\{\exp[(q^2+m_e^2)^{1/2}/T]+1\right\}^{-1} q^2 dq.
\end{aligned} \tag{3.31}
$$

Use of (3.25) leads to

$$
\varrho = aT^4\varepsilon\left(\frac{m_e}{T}\right), \tag{3.32}
$$

where

$$
\begin{aligned}
\varepsilon(x) = 1 &+ \frac{42}{16}\left(\frac{4}{11}\right)^{4/3} J(x)^{4/3} \\
&+ \frac{30}{\pi^4}\int_0^\infty \sqrt{x^2+y^2}y^2\left\{\exp\left(\sqrt{x^2+y^2}/T\right)+1\right\}^{-1} dy.
\end{aligned} \tag{3.33}
$$

Because the entropy (3.31) can also be expressed, in terms of the temperature of the present epoch, as

$$\frac{4}{3}a(R_0 T_{\gamma 0})^3,$$

we find

$$\frac{R}{R_o} = \left(\frac{T}{T_{\gamma 0}}\right)^{-1} J^{-1/3}\left(\frac{m_e}{T}\right). \quad (3.34)$$

As

$$dt = \left(\frac{8\pi G\varrho}{3}\right)^{-1/2} \frac{dR}{R}$$

and

$$\frac{dR}{R} = -\frac{dT}{T} + \frac{1}{3}\frac{dJ}{J},$$

$$t = -\int \left\{\frac{8\pi G}{3} aT^4 \epsilon\left(\frac{m_e}{T}\right)\right\}^{-1/2} \left\{\frac{dT}{T} + \frac{dJ\left(\frac{m_e}{T}\right)}{3J\left(\frac{m_e}{T}\right)}\right\}. \quad (3.35)$$

In this way the temperature T of the photons is determined as a function of time. The neutrino temperature $T_\nu(t)$ – and hence $R(t)$, because $T_\nu R = \text{const.}$ – is then fixed by (3.33).

Figure 3.1 [Harrison 1973] shows the density as a function of temperature and time for various epochs. The relation of the temperature of other weakly interacting particles, such as "photinos, gravitinos, or axions" – particles which may exist according to certain versions of unified field theories – to the photon temperature can be computed in a similar way.

The simple case, where a particle species X remains relativistic from its decoupling epoch to some final annihilation process which heats up the photon gas, can be written as follows. Until decoupling

$$\rho_x = \frac{1}{2} g_x a T_\gamma^4,$$

where g_x for bosons is the number of helicity states, and for fermions 7/8 times the number of helicity states.

Let $g_A(T)$ be the function $g(T)$ for the interacting particles at and before decoupling, and call it $g_E(T)$ after a series of annihilation processes. After the final annihilation, the photons and other interacting particles have a temperature T_γ. The X-particles which have not been heated up after decoupling have a temperature lowered by

$$(T_x/T_\gamma)^3 = g_E/g_A.$$

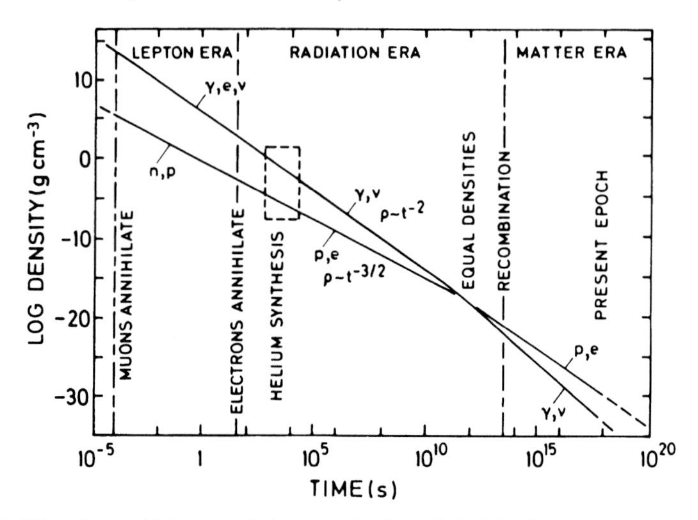

Fig. 3.1. The thermal history of the standard model. The densities of protons, electrons, photons, and neutrinos are shown at various stages of cosmological evolution [after Harrison (1973)]

The contribution to the energy density is then

$$\rho = a(g_E/2)T_\gamma^4(1 + g_x/g_E(g_E/g_A)^{4/3}).$$

More complicated processes can be treated within the same scheme of entropy conservation, leading to more complicated formulae. One can imagine, for example, that the decoupled particles have new interactions allowing them to annihilate and create other non-interacting particles. There may also be earlier phases, where the photon gas is heated up by the annihilation of exotic particles, but such stages leave traces only with respect to the temperature of other "-inos". More about this in Chap. 7.

In general – neglecting such effects – we can use (3.2) to write

$$t_s = 2.42 \times 10^{-6} g^{-1/2} T_{\mathrm{GeV}}^{-2}. \tag{3.36}$$

This equation should hold for $T > 1\,\mathrm{MeV}$. The function $g(T)$ increases from 43/4 at $T = 1\,\mathrm{MeV}$ to 423/4 at $T = 100\,\mathrm{GeV}$, if presently favoured theories of particle behaviour (e.g. [Olive et al. 1979]) are accepted. For the minimal SU(5) GUT (cf. Chap. 6) the function $g(T)$ has the value

$$g = \frac{643}{8} \quad \text{at} \quad 10^{15}\,\mathrm{GeV}.$$

If we stretch our belief in modern particle theories to the limit of energy scales around 10^{15} GeV, then we approach the initial singularity very closely: the time is $\simeq 10^{-35}$ s, and the place is very hot ($T \simeq 10^{15}$ GeV) and dense

($\simeq 10^{75}$ g cm^{-3}). The hot and dense early stage of the universe was a unique laboratory for high-energy particle and nuclear physics. Unfortunately nobody was there to observe, and we have to be content with the faint possiblity of finding a few vestiges of the original chaos. Monopoles, massive stable charged leptons, heavy hadrons, or free quarks might all be produced in the high-temperature medium. Some of these relics may have survived until now. In Chap. 7 we shall cover some aspects of this scenario.

One important conclusion has been derived from the assumption of particle–antiparticle symmetry, namely that the ratio of surviving nucleons to photons at present would be [Steigman 1979]

$$\left(\frac{n_N + n_{\overline{N}}}{n_\gamma}\right) = 7 \times 10^{-19}. \tag{3.37}$$

This number is about 9 orders of magnitude smaller than the observed value. In the standard model, the ratio n_B/n_γ of baryon number density to photon number density is approximately a constant. The value observed at the present epoch,

$$\frac{n_\gamma}{n_B} \simeq 2 \times 10^{10},$$

must be imposed on the specific cosmological model as an initial condition.

The consideration of baryon-non-conserving interactions in a fundamental theory should enable one to calculate this number from an initially symmetric state, making theory and cosmological model subject to one more observational test.

3.2 Nucleosynthesis

> "The numerical results show that (...) the n-p ratio in the beginning of the element formation is nearly 1:4 almost irrespective of its initial values, as long as initial temperatures are so high ($T \geq 2 \times 10^{10}$ K) that equilibrium has once been attained."
>
> *C. Hayashi (1950)*

The "hot big-bang" model of the universe predicts the synthesis of the light elements helium, deuterium, and lithium with abundances which are quite close to the estimates obtained from observations. This is another great success for this simple model. How does nucleosynthesis proceed in the early universe? [Hayashi 1950; Weinberg 1972; Zel'dovich and Novikov 1983].

3.2.1 The Neutron-to-Proton Ratio

The ratio of neutrons to protons determines the amount of heavier elements that can be formed. Therefore this ratio has to be computed as accurately as possible as a function of time. For $T > 1\,\mathrm{MeV}$ the weak interactions (Fig. 3.2)

$$p + e^- \leftrightarrow n + \nu_e; \quad n + e^+ \leftrightarrow n + \overline{\nu}_e; \quad n \leftrightarrow p + e^- + \overline{\nu}_e \tag{3.38}$$

have rates per nucleon that can be written down in integral form [Weinberg 1972; Dicus et al. 1982]. For the process $\nu + n \to p + e^-$, for example,

$$\lambda = \int \left(1 - \left[\exp\left(\frac{E_e}{k_B T}\right) + 1\right]^{-1}\right) v_\nu \times \sigma(\nu + n \to p + e^-) \frac{4\pi}{(2\pi)^3} \frac{k^2 dk}{\exp(E_\nu / k_B T) + 1}. \tag{3.39}$$

Here E_e, E_ν are the electron and neutrino energies, v_ν the velocity, T_ν the neutrino temperature, and the energy difference is given by

$$E_e - E_\nu = Q \equiv m_n - m_p = 1.293\,\mathrm{MeV}. \tag{3.40}$$

The cross-section σ can be computed from the (V-A) charged-current interaction of the weak interactions as

$$\sigma(\nu + n \to E^- + p) = \frac{G_V^2 + 3G_A^2}{\pi} v_e E_E^2(p). \tag{3.41}$$

Then

$$\lambda = \frac{G_V^2 + 3G_A^2}{\pi} \int v_e E_E^2 p_\nu dp_\nu \left(1 + \exp\frac{E_\nu}{k_B T_\nu}\right)^{-1} \times \left(1 + \exp -\frac{E_e}{k_B T}\right)^{-1}. \tag{3.42}$$

The rates for the other processes can be obtained in a similar manner. The total rate for the transition $n \to p$ can now be written as follows [Weinberg 1972, p. 547]:

$$\lambda(n \to p) = \frac{G_V^2 + 3G_A^2}{\pi} \int \left(1 - \frac{m_e^2}{(Q+q)^2}\right)^{1/2} \times (Q+q)^2 q^2 \left(1 + \exp\frac{q}{k_B T \nu}\right)^{-1} \left(1 + \exp\left(-\frac{Q+q}{k_B T}\right)\right)^{-1} dq. \tag{3.43}$$

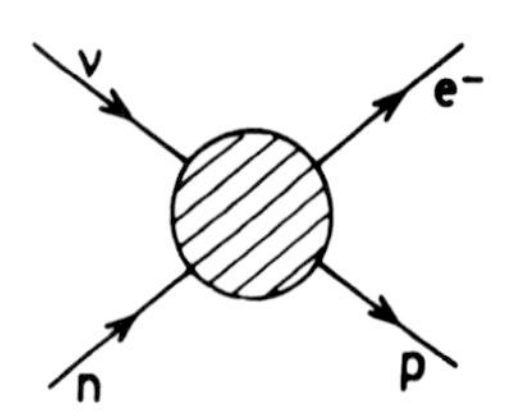

Fig. 3.2. The transition $n \to p$ involves processes such as $n + \nu \to p + e^-$ pictured here

By detailed balance

$$\lambda(p \to n) = \lambda(n \to p) \times e^{-Q/k_BT}. \tag{3.44}$$

The integrals are to be carried out over the real axis $-\infty < q < +\infty$ with the exception of the interval $(-Q - m_e < q < -Q + m_e)$.

The ratio X_n of the neutrons N_n to all other nucleons $N_n + N_p$,

$$X_n = \frac{N_n}{N_n + N_p},$$

is determined from the balance of $p \to n$ and $n \to p$ reactions

$$-\frac{d}{dt}X_n = \lambda(n \to p)X_n - \lambda(p \to n)(1 - X_n). \tag{3.45}$$

In general this equation has to be solved numerically; e.g. [Peebles 1966]. Some qualitative remarks can be made at this point. For $k_BT \gg Q$, we have $T = T_\nu$ and Q and m_e can be neglected in (3.43). Then the rates are

$$\lambda(n \to p) \simeq \lambda(p \to n) = \frac{7}{15}\pi^4 \frac{G_V^2 + 3G_A^2}{\pi}(k_BT)^5 = 0.36 \quad T_{10}^5 \quad s^{-1}. \tag{3.46}$$

The expansion time (see (3.26)) is

$$t = 0.99\, T_{10}^{-2}\,\mathrm{s} \tag{3.47}$$

and thus $\lambda t \geq 10$ for $T > 3 \times 10^{10}\,\mathrm{K}$. Therefore the system is in thermodynamic equilibrium for temperatures above $T = 3 \times 10^{10}\,\mathrm{K}$.

As long as $T \simeq T_\nu$ (i.e. $T > 10^{10}\,\mathrm{K}$) the ratio of the two rates

$$\frac{\lambda(p \to n)}{\lambda(n \to p)} = e^{-Q/k_BT}. \tag{3.48}$$

In this case the equilibrium solution of (3.45) $\dot{X}_n = 0$ can be written down simply as

$$X_n = \frac{\lambda(p \to n)}{\lambda(p \to n) + \lambda(n \to p)} = \left(1 + \exp\frac{Q}{k_BT}\right)^{-1}. \tag{3.49}$$

We see that

$$X_n = 1/2 \qquad \text{for} \qquad T \to \infty. \tag{3.50}$$

The hot big-bang model does not require an initial condition for the abundance ratio of neutrons and protons. The conditions of thermal equilibrium at very high temperatures fix the initial abundance ratio as $X_n = 1/2, X_p = 1/2$. Thus the results of the primeval nucleosynthesis are real and honest predictions of the cosmological model.

The ratio X_n is

$$X_n = 0.38 \qquad \text{for} \qquad T = 3 \times 10^{10}\,\mathrm{K}. \tag{3.51}$$

As the temperature drops below 1 MeV ($T \sim 1.2 \times 10^{10}\,\mathrm{K}$) the weak interactions can no longer establish thermal equilibrium against the expansion rate $H(t)$:

$$\frac{\sigma_W n_e c}{H(t)} \ll 1.$$

X_n will start to deviate strongly from the thermal equilibrium value (3.49). The ratio of neutrons to protons "freezes out" and at lower temperatures is changing only by the β-decay of the neutrons. Then the time evolution from $T = 1\,\mathrm{MeV}$ till the start of the nucleosynthesis is described by

$$X_n(t) = N \exp\left[-\frac{t}{\tau_n}\right], \tag{3.52}$$

with $\tau_n \simeq 887 \pm 3\,\mathrm{s}$ [Schreckenbach and Mampe 1992; Mampe et al. 1989].

The "freeze-out" occurs, when the expansion rate $H \simeq g^{1/2} T^2 G^{1/2}$ is equal to the reaction rate per nucleon

$$\sigma_W n_e c \simeq G_F^2 T^5.$$

Hence the "freeze-out" temperature

$$T_f \simeq g^{1/6} G^{1/6} G_F^{-2/3} \tag{3.53}$$

depends on the statistical degrees of freedom $g(T)$. $T_f \simeq 1\,\mathrm{MeV}$, and the time, when this occurs, $t_f \simeq 1\,\mathrm{s}$. The neutron abundance at this time can be approximated by its equilibrium value

$$X_n(t_f) \simeq [1 + \exp Q/k_B T_f]^{-1}. \tag{3.54}$$

Since the exponent Q/T is of order one, a substantial fraction of neutrons survive when the chemical equilibrium between protons and neutrons is broken. The neutron–proton mass difference Q is determined by the strong and electromagnetic interactions, while the freeze-out temperature is fixed by the weak and gravitational interactions. Thus all four fundamental interactions conspire to produce a sufficient number of surviving neutrons. Additional relativistic species of particles, if they are neutrino-like, such as gravitinos, and other types of light neutrinos increase g to a value above the canonical $g = \frac{43}{4}$. Then the freeze-out temperature will increase, and $X_n(t_f)$ as well.

The asymptotic abundance X_n for $Q/T \gg 1$ must be evaluated carefully by considering all the weak interaction processes which change neutrons into protons and vice versa [Bernstein et al. 1989; Sarkar 1996], finally yielding

$$X_n(Q/T \to \infty) \simeq 0.150$$

which is already reached around $Q/T \simeq 5$, i.e. at an epoch $t \simeq 20\,\mathrm{s}$, or at a temperature $T \simeq 0.25\,\mathrm{MeV}$.

It seems remarkable that only comparatively recently (since 1989) has sufficient precision been achieved in the measurement of the neutron lifetime to determine this and other nucleosynthesis numbers accurately. Before that time the neutron lifetime was the main uncertainty in big-bang nucleosynthesis (cf. [Boesgaard and Steigmann 1985]).

3.2.2 Nuclear Reactions

At temperatures above $10^{10}\,\mathrm{K}$ various kinds of complex nuclei are also present, with an abundance determined by thermodynamic equilibrium. The number density of a nucleus of type i, with chemical potential μ_i, is given by the Maxwell–Boltzmann distribution

$$n_i = g_i \left(\frac{2\pi m_i T k_B}{h^2}\right)^{3/2} \exp\left(\frac{\mu_i - m_i}{k_B T}\right), \tag{3.55}$$

where m_i is the mass of the nucleus and g_i the statistical weight of the ground state.

The equilibrium condition for the chemical potential is

$$\mu_i = Z_i \mu_p + (A_i - Z_i)\mu_n. \tag{3.56}$$

If we set

$$X_i \equiv \frac{n_i A_i}{n_N}; \quad X_n \equiv \frac{n_n}{n_N}; \quad X_p \equiv \frac{n_p}{n_N}, \tag{3.57}$$

with n_N the total number of nucleons per unit volume and A_i the mass number of nucleus i, then (3.55), together with (3.56) and (3.57), results in

$$X_i = \frac{g_i}{2} X_p^{Z_i} X_n^{A_i - Z_i} A_i^{1/2} \epsilon^{A_i - 1} \exp\left(\frac{B_i}{T}\right). \tag{3.58}$$

In this equation B_i is the binding energy ($k_B = 1$)

$$-B_i = m_i - Z_i m_p - (A_i - Z_i) m_n,$$

and ϵ is the dimensionless number

$$\epsilon \equiv \frac{1}{4} n_N (2\pi M_n T)^{-3/2}.$$

Particle density conservation

$$n_N = n_{N_0} \left(\frac{R_0}{R}\right)^3 = \frac{\varrho_{N_0}}{m_N} \left(\frac{R_0}{R}\right)^3,$$

together with the approximation $m_N \simeq m_n \simeq m_p$, allows to write ϵ as

$$\epsilon = 1.61 \times 10^{-12} \frac{\varrho_{N0}}{10^{-30}\,\mathrm{gcm}^{-3}} \left(\frac{R}{10^{-10} R_0}\right)^{-3} T_{10}^{-3/2}. \tag{3.59}$$

During the epochs that interest us here, ϵ is a very small number. Nuclei of type i occur in appreciable amounts only if the temperature has dropped to a value T_i where

$$T_i \simeq \frac{|B_i|}{(A_i - 1)|\ln \epsilon|}. \tag{3.60}$$

Typical values for T_i are between 1 and 3×10^9 K. But at such temperatures the densities are already much too small to maintain equilibrium by direct many-body reactions, such as

$$2\mathrm{n} + 2\mathrm{p} \to {}^4\mathrm{He}.$$

Thus the only way to synthesize complex nuclei is to proceed via a network of 2-body reactions. The most important reactions use deuterium (D) as a link:

$$\mathrm{p} + \mathrm{n} \leftrightarrow \mathrm{D} + \gamma, \tag{3.61}$$

$$\mathrm{D} + \mathrm{D} \leftrightarrow {}^3\mathrm{He} + \mathrm{n} \leftrightarrow {}^3\mathrm{He} + \mathrm{p}, \tag{3.62}$$

$${}^3\mathrm{He} + \mathrm{D} \leftrightarrow {}^4\mathrm{He} + \mathrm{n}. \tag{3.63}$$

The deuterium production $\mathrm{p} + \mathrm{n} \to \mathrm{D} + \gamma$ proceeds very rapidly with a rate per free neutron of

$$\lambda_0 = (4.55 \times 10^{-20}\,\mathrm{cm^3 s^{-1}}) n_p = 27.4\,\mathrm{s}^{-1} \left(\frac{R}{10^{-9} R_0}\right)^3 \left(\frac{\varrho_{N0}}{10^{-30}\,\mathrm{gcm}^{-3}}\right) X_p. \tag{3.64}$$

The expansion rate $t^{-1} = \frac{1}{1.09} T_{10}^2$ is much smaller. Therefore deuterium nuclei are produced with their equilibrium abundance. According to (3.58)

$$X_D = \frac{3}{\sqrt{2}} X_p X_n \epsilon \exp\left(\frac{B_D}{T}\right). \tag{3.65}$$

When this abundance is at a sufficiently high level, then heavier elements can be built up at a sufficient rate. The temperature for which $\epsilon \exp(B_D/kT) \simeq 1$ is about $T_d \simeq 0.8 \times 10^9$ K for deuterium (with $B_D = 2.23$ MeV, this is $T \simeq B_d/32$). At higher temperatures deuterium is immediately photo-dissociated by the hot radiation field.

Again we should be aware at this point of the influence of the large number of CMB photons per nucleon, $n_\gamma \sim n_B \times 10^{10}$. There are enough high-energy photons in the Wien part of the thermal spectrum to photodissociate deuterium far beyond the point, where the average blackbody photon energy of $2.7k_BT$ falls below B_D.

A careful estimate gives $T_d \simeq \frac{B_D}{26} \simeq 0.086\,\text{MeV}$ [Bernstein et al. 1989; Sarkar 1996], and thus nucleosynthesis will begin at an epoch $t_{ns} \simeq 180\,\text{s}$.

By this time the neutrons surviving at freeze-out have been somewhat reduced by β-decay:

$$X_n(t_{ns}) \simeq X_n(\frac{Q}{T} \to \infty)e^{-t_{ns}/T_n} \simeq 0.122. \tag{3.66}$$

As soon as $T \leq T_d$ deuterium can survive, and a chain of nuclear reactions sets in, building up helium nuclei very rapidly; cf. (3.61) to (3.63). Heavier elements are not produced in significant amounts, because at the mass numbers $A = 5$ and $A = 8$ there is no stable nucleus. A small amount of ^{7}Li and ^{7}Be is produced through the reaction chains

$$^4\text{He} + {}^3\text{H} \to {}^7\text{Li} + \gamma, \tag{3.67}$$

$$^4\text{He} + {}^3\text{He} \to {}^7\text{Be} + \gamma, \tag{3.68}$$

$$^7\text{Be} + \text{e}^- \to {}^7\text{Li} + \nu_e. \tag{3.69}$$

The ^{7}Be electron capture is effective at densities $\Omega_B > 0.04$ in producing ^{7}Li, while the fusion of ^{3}He and ^{4}He is effective at lower densities. As a result, the abundance of ^{7}Li does not change very much as Ω_B increases from 0.01 to 0.1. Nucleosynthesis is completed when all neutrons present at $T = 0.086\,\text{MeV}(X_n = 0.122)$ have been cooked into deuterium – which is only a small fraction – and ^{4}He – which predominates. This is completed within $\sim 1000\,\text{s}$ (cf. *The First Three Minutes* by S. Weinberg). The abundance of ^{4}He by mass is about twice the neutron abundance X_n at the beginning of the nucleosynthesis

$$Y_P \equiv X(^4\text{He}) = 2X_n(t_{ns}) \simeq 0.24. \tag{3.70}$$

Detailed calculations of D and He synthesis have led to very precise predictions. The first numerical calculations were by [Hayashi 1950]. Other early numerical solutions were by [Wagoner et al. 1967]. Wagoner's code has been significantly improved and updated by [Kawano 1992]. L. Kawano has made his code publicly available, and it is now the standard tool for big-bang nucleosynthesis studies.

Interestingly, an analytic approach to these calculations has also been developed [Esmailzadeh et al. 1991]. The abundances of the light elements can be obtained to good accuracy from the fixed points of the corresponding rate equations. Quite generally, the abundance of a nucleus i is determined by the balance of source terms $I(t)$ and sink terms $\Gamma(t)$:

$$\frac{dX_i}{dt} = I(t) - \Gamma(t)X_i. \tag{3.71}$$

$I(t)$ and $\Gamma(t)$ depend in general also on the abundances of the other elements. Their change with time is determined by the expansion time-scale $H(t)$. As long as $\Gamma \gg H$, an equilibrium solution $\dot{X}_i \simeq 0$ exists with $X_i = I(t)/\Gamma(t)$.

It is shown in [Esmailzadeh et al. 1991] that X_i approaches the equilibrium value if $|\dot{I}/I - \dot{\Gamma}/\Gamma| \ll \Gamma$. This condition is somewhat more stringent than $\Gamma \gg H$, since $\dot{I}/I \simeq \dot{\Gamma}/\Gamma \simeq H$. This state is called "quasi-static" equilibrium (QSE). As the universe expands, the nuclear reaction rates slow down rapidly due to the decrease of particle densities, and the fact that Coulomb barriers inhibit most reactions. Source and sink terms are rapidly falling functions of time, and at some epoch t_f the value of X_i "freezes out", i.e. no longer changes appreciably after that time:

$$\int_{tf}^{\infty} \dot{X}_i dt \ll X_i(t_f).$$

Generally freeze-out occurs when $\Gamma \simeq H$; then an asymptotic value can be defined

$$X_i(t \to \infty) \simeq X_i(t_f) = \frac{I(t_f)}{\Gamma(t_f)}. \tag{3.72}$$

The largest source and sink term for each species are identified from a careful examination of the reaction network. Then $\Gamma \simeq H$ defines the freeze-out temperature, and $I(t_f)/\Gamma(t_f)$ the QSE abundance at this epoch. The abundances of D, ^{3}He, ^{7}Li, ^{4}He are predicted correctly with sufficient accuracy by this method. The great advantage is the clarification of important features of the underlying physics, especially when more exotic paths to nucleosynthesis are considered. But normal nuclear physics can already be understood better, such as the effect that the synthesis of ^{4}He is delayed until enough ^{3}H has been synthesized, since the main process for making ^{4}He is

$$\mathrm{D} + {}^3\mathrm{H} \to {}^4\mathrm{He} + \mathrm{n}.$$

Readers interested in these analytic approaches can obtain more insight from [Esmailzadeh et al. 1991; Bernstein et al. 1989]. Reviews of big-bang nucleosynthesis are numerous, and the following are just a few examples which I have consulted myself [Schramm and Wagoner 1977; Sarkar 1996; Walker et al. 1991; Boesgaard and Steigmann 1985]. You need not read them all, but if you do, it is like an exercise in Buddhist meditation: the eternal recurrence of everything. Every small change in an abundance, to the third decimal place, prompts the experts to write a new review – I hope the attentive reader will be able to do so after finishing this chapter.

The review by Subir Sarkar [Sarkar 1996] is highly recommandable – informative, well balanced, and complete. In Fig. 3.3 the results of a numerical calculation (using the Wagoner code upgraded by Kawano) are shown [taken from Sarkar 1996].

The resulting abundance of ^{4}He – plotted and listed always as the ratio of the ^{4}He mass density to the overall mass density – varies very little with nucleon density, and hence with the baryon density ϱ_{B0} of the present epoch (as $\varrho_B \propto R^{-3}$). For $3 \times 10^{-31}\,\mathrm{gcm}^{-3} \leq \varrho_{B0} \leq 10^{-30}\,\mathrm{gcm}^{-3}$ one finds a variation of X (^{4}He) between 0.23 and 0.25. These values are much larger than the abundance that would be expected from fusion processes in stars. Models of galactic evolution estimate the amount of helium produced in stars to be a few percent: $Y_* = 0.01$ to 0.04 from stellar fusion [Tinsley 1977a,b].

The observations (cf. below) indicate, however, a large cosmic abundance $Y \simeq 0.24$. This agreement between the prediction of big-bang models and observations is another impressive point in favour of such simple cosmological models. We shall discuss qualifications and uncertainties of these abundance determinations in Sect. 3.3.

The abundances in Fig. 3.3 are plotted as functions of η, the ratio of nucleon to photon number density. This is simply connected to the baryon density Ω_B:

$$\Omega_B h^2 = 3.65 \times 10^{-3} (T_\gamma/2.726)^3 \eta_{10}$$

where $\eta_{10} = \eta \times 10^{10}$.

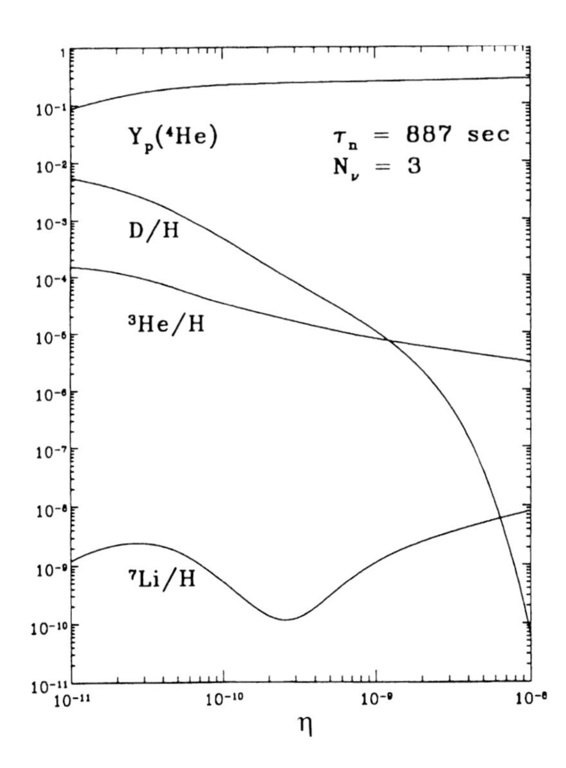

Fig. 3.3. Calculated abundances of elements from big-bang nucleosynthesis as a function of present-day baryon density [from Sarkar 1996]

Table 3.2. Cosmic primeval abundances of H, D, ^{3}He, ^{4}He, ^{7}Li as predicted from hot big-bang nucleosynthesis. Y_p gives the ^{4}He abundance by mass for three neutrino generations. All other abundances are by number relative to hydrogen

$Y_p(^4\mathrm{He})$	=	$0.2462 \pm 0.01 \ln(\eta_{10}/5)(\frac{\eta_{10}}{5})^{-0.2} \pm 0.0012$; over $\eta_{10} = 1$ to 10
D/H	=	$3.6 \times 10^{-5 \pm 0.06}(\eta_{10}/5)^{-1.6}$
^{3}He/H	=	$1.2 \times 10^{-5 \pm 0.06}(\eta_{10}/5)^{-0.63}$
^{7}Li/H	=	$1.2 \times 10^{-11 \pm 0.2}[(\eta_{10}/5)^{-2.38} + 2.17(\eta_{10}/5)^{2.58}]$

In contrast to ^{4}He, the predicted deuterium abundance is strongly dependent on the baryon density. This is quite plausible, because deuterium can only survive if the reactions leading to He formation are not rapid enough to use up all the D nuclei. With increasing density 2-body collisions are more frequent, and fewer D nuclei can escape fusion into helium. The graphs of Fig. 3.3 show that a deuterium abundance is so sensitive a function of the mean density, that a determination of the cosmic abundance of deuterium could fix the uncertain baryon density of the present epoch – or rather the density of all the baryons that have participated in the primeval nucleosynthesis.

Some fits to the numerical calculations are represented in Table 3.2 (from [Sarkar 1996]). The accuracy of these calculations for D, ^{3}He, ^{4}He should be within a few percent if the presently used reaction rates can be trusted. Various important reactions involved in the production of ^{7}Li ($^4\mathrm{He} + {}^3\mathrm{He} \to \gamma + {}^7\mathrm{Be}$; $^7\mathrm{Li} + p \to 2\,{}^4\mathrm{He}$; $^7\mathrm{Be} \to p + {}^7\mathrm{Li}$) still have rather uncertain reaction rates, and therefore the final abundance may be accurate only to within a factor of 2 or 3.

3.3 Observations of Cosmic Abundances

The reviews mentioned in Sect. 3.2 also cover the subject of cosmic abundances. Since helium and lithium are produced, deuterium is destroyed in stars, it is not an easy task to find the primeval abundance of these elements. The processes of galactic and stellar evolution contaminate the measurements. Observers therefore look for sites with low metallicity – high-redshift hydrogen clouds to measure deuterium, and highly ionized hot gas in blue compact galaxies to determine the helium content. Some aspects of the light element observations are discussed in the following sections (see Table 3.3).

Table 3.3. Limits on primaeval abundance ratios from observations

	Remark	Range of η_{10} ($N_\nu = 3, \tau_n = 10.6$ min)
$X(^4\mathrm{He}) = 0.238 \pm 0.005$ (by mass)	Blue compact gal.	
$(\mathrm{D/H})_p > (1$ to $2) \times 10^{-5}$	Interstellar	η_{10} >7 to 10
$[(\mathrm{D} + {}^3\mathrm{He})/\mathrm{H}]_p <(6$ to $10) \times 10^{-5}$	[a]	η_{10} >3 to 4
$(^7\mathrm{Li/H})_p = (0.7$ to $1.8) \times 10^{-10}$	Pop. II stars	$1.5 < \eta_{10} <5$

[a] The survival of ^{3}He and the conversion of D to ^{3}He is taken into account; the small abundance of newly synthesized ^{3}He is ignored.

3.3.1 The Abundance of ^{4}He

The abundance of ^{4}He has been determined by several different methods in H II regions, in various types of stars, in planetary nebulae, and in the Solar System; the derivation of a primeval abundance from these measurements generally requires a considerable reliance upon theory. The recombination lines He II $\rightarrow$ He I give the best way to determine the ^{4}He abundance, but the proper degree of He ionization is often difficult to assess. Measurements of line strengths in stellar atmospheres have to be supplemented by a complete theory of stellar evolution to find the amount of enrichment and the initial helium abundance.

Evolved systems like our Solar System, or our galaxy, have certainly changed the initial abundances considerably by enrichment processes, star formation, contamination by heavier elements, redistribution by stellar explosions into the interstellar medium, fractionation in the planets, and so on. Thus a strong dependence on local conditions must be expected in the values of the helium abundance. In general one tries to estimate the degree of chemical evolution by measuring abundances of heavier elements, which are mainly produced in stars. (The so-called "metal" abundance usually stands for the abundance of all elements heavier than oxygen (O); the mass fraction Z of heavy elements is often approximated by $Z = 25[\mathrm{O/H}]$ i.e. 25 times the ratio of oxygen to hydrogen.)

The most favoured method is to measure helium in extragalactic H II regions of low metal content. The helium abundance Y is then plotted against the [O/H] or [N/H] abundance and extrapolated linearly to the zero value:

$$Y = Y_p + Z\frac{dY}{dZ}.$$

The gradient dY/dZ depends on details of the initial mass distribution of the stars and on stellar evolution. Its empirical determination is quite difficult, but at present there are more than 70 extragalactic H II (ionised hydrogen) regions where the helium abundance can be extrapolated reliably to the value $Z = 0$, corresponding (hopefully) to the primeval Y_p [Olive et al. 1997; Skillman et al. 1994; Pagel et al. 1992]. The overall result is

$$Y_p = 0.232 \pm 0.003.$$

In principle any H II region gives an upper limit to Y_p, since some stellar processing has taken place. The most reliable results should therefore be the measurements of the lowest-metallicity objects. The irregular galaxy IZw 18 has some very low-metallicity H II regions, and a total average for this galaxy has been determined as

$$Y_p = 0.230 \pm 0.003$$

[Olive 1998], with 2σ errors of 0.006.

There is the claim that some new data lead to an estimate of $Y_p \geq 0.24$ [Izotov et al. 1994, 1997]. This discrepancy is still under discussion, but the general conclusion seems rather safe that

$$Y_p = 0.234 \pm 0.002 \qquad \pm 0.005 \tag{3.73}$$

(the first error ($\pm$ 0.002): random; the second ($\pm$ 0.005): systematic); i.e. Y_p between 0.227 and 0.241.

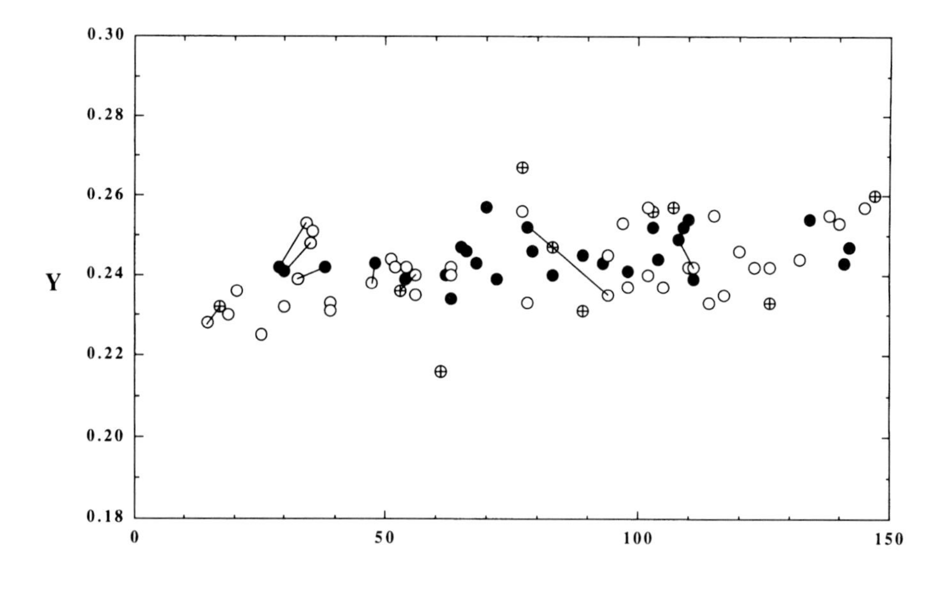

Fig. 3.4. The helium (Y) and oxygen (O/H) abundances from different observations (*open* and *filled circles*; *crossed circles* are regions excluded by the observers). *Lines* connect the same regions observed by different groups (from [Olive 1998])

In Fig. 3.4 [Olive 1998] the helium and oxygen abundances obtained from different groups are shown. Especially instructive are the lines which connect the values for the same H II regions observed by different groups. The differences may give an indication of the systematic errors.

3.3.2 The Deuterium Abundance

Deuterium has been observed in the solar system, the interstellar medium (ISM), and recently in absorption systems towards distant quasars. Since the element is destroyed in stars by the reaction

$$\mathrm{p} + \mathrm{D} \leftrightarrow {}^3\mathrm{He} + \gamma,$$

as soon as it is formed by the fusion of two protons,

$$\mathrm{p} + \mathrm{p} \leftrightarrow \mathrm{D} + \mathrm{e}^+ + \nu_e,$$

the abundance may be depleted by large factors from primeval to present D/H. As we can see from Fig. 3.3 the D/H abundance varies quite strongly with the nucleon density, and therefore can provide strong constraints. The primeval abundance is difficult to determine, since deuterium is affected by galactic evolution most drastically. Because D is destroyed in stars, and no production mechanism is known, any measured value is a lower bound to the primeval abundance.

The solar abundance of D/H is inferred in two steps:

i) Solar wind measurements of ^{3}He, as well as the stepwise heating of meteorites (the low-temperature component) yield the presolar ratio $(\mathrm{D} + {}^3\mathrm{He})/\mathrm{H}$, since D was efficiently burnt to ^{3}He in the pre-main sequence phase of the Sun.
These measurements give

$$(\mathrm{D} + {}^3\mathrm{He})/\mathrm{H} = (4.1 \pm 0.6 \pm 1.4) \times 10^{-5};$$

(the first error (± 0.6) being random; the second (± 1.4) systematic). The high-temperature component in meteorites is believed to yield the true solar abundance of ^{3}He/H:

$$({}^3\mathrm{He}/\mathrm{H})_\odot = (1.5 \pm 0.2 \pm 0.3) \times 10^{-5}.$$

The difference obviously must be the presolar D/H:

$$\mathrm{D}/\mathrm{H} = (2.6 \pm 0.6 \pm 1.4) \times 10^{-5} \tag{3.74}$$

[Scully et al. 1996; Geiss 1993].
There are claims that on Jupiter the surface abundance of HD leads to $\mathrm{D}/\mathrm{H} = (5 \pm 2) \times 10^{-5}$. Such discrepant values may be due to chemical fractionation in various molecules.

ii) D in the interstellar medium has been measured by the HST. Absorption lines towards the star Capella reveal a value

$$(\mathrm{D/H})_{\mathrm{ISM}} = (1.60 \pm 0.09^{+0.05}_{-0.10}) \times 10^{-5} \tag{3.75}$$

[Linsky et al. 1995].

iii) In absorption line systems of a few high redshift quasars the Ly-α line of DI has been identified. This would, of course, be the best way to find the primeval abundance of D, by measuring it directly at high redshift. The results, so far, are still under discussion. Certainly a theory of line formation in a turbulent medium is required to derive D/H from the line observations. It seems that earlier claims of a high value of D/H = $(1.8 - 3.1) \times 10^{-4}$ [Webb et al. 1997] are gradually falling out of favour, and a convergence towards lower values has set in.
The ratio D/H = $(3.3 \pm 0.3) \times 10^{-5}$ with a robust upper limit of 3.9×10^{-5} [Burles and Tytler 1998] probably has its accuracy overstated. A reanalysis of the line formation process with a mesoturbulent model – where a finite-velocity correlation length is considered in the stochastic velocity field – gives better fits to different data sets [Levshakov et al. 1998]. The result is

$$(\mathrm{D/H})_{\mathrm{ISM}} = (2.9 \text{ to } 4.8) \times 10^{-5}. \tag{3.76}$$

Comparison with Fig. 3.3 indicates a corresponding range of

$$3.9 \leq \eta_{10} \leq 6.2. \tag{3.77}$$

This result clearly contradicts the limits obtained from the low ^{4}He abundance $Y_p \leq 0.24$ which correspond to

$$1.6 \leq \eta_{10} \leq 0.3.$$

Only the somewhat higher result of $Y_p = (0.246 \pm 0.003)$ [Izotov et al. 1997; Olive 1997] has a small overlap with the D/H measurements:

$$3.9 \leq \eta_{10} \leq 5.3.$$

Observations and fits to the DI absorption line for the quasar Q 1937–1009 (emission $z = 3.78$) are shown in Fig. 3.5. The absorption system has $z = 3.57$, and you should appreciate the high resolution of the relative velocity features in the absorption lines.

3.3.3 The ^{3}He Abundance

As we have seen, the solar wind and gas-rich meteorites give an abundance of ^{3}He/H = $(1 \text{ to } 2) \times 10^{-5}$. The interstellar ^{3}He abundance has been estimated [Balser et al. 1994] by observation of the $^3\mathrm{He}^+$ line at 3.46 cm in different galactic H II regions. The values obtained lie in the range $(1 - 5) \times 10^{-5}$, indicative of a bias or of pollution. Observation of ^{3}He in planetary nebulae gave high values $^3\mathrm{He/H} \sim 10^{-3}$ [Rood et al. 1992].

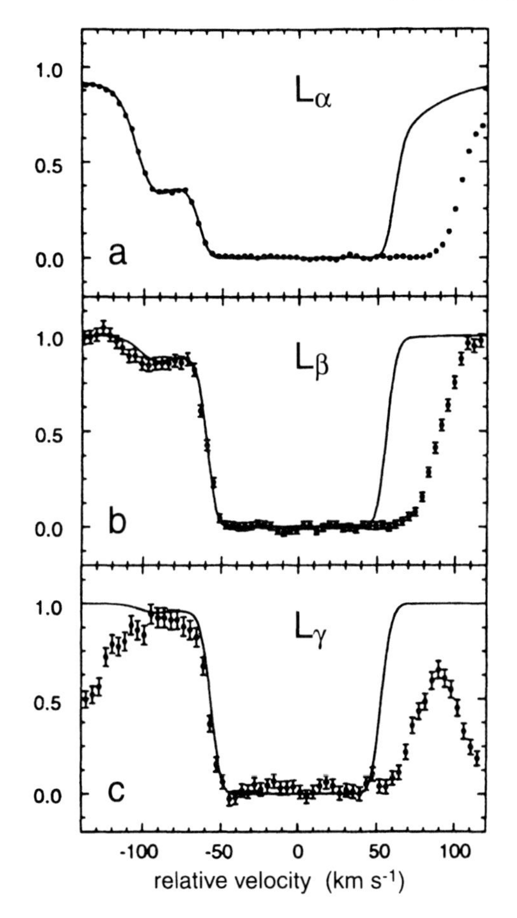

Fig. 3.5. Observations and fits for absorption lines of QSO 1937–1009. Velocity plots of D Ly-α, Ly-β, and Ly-γ obtained from the Keck/Hires echelle data are shown. The instrumental resolution of $9\,\mathrm{kms}^{-1}$ is the basis for such a clear signal at the redshift of $z = 3.57$. In panel a) error bars are too small to be distinguished [after Levshakov et al. 1998]

3.3.4 The ^{7}Li Abundance

Observations of several halo stars – which are supposed to be very old – in the Li line at $\lambda = 6702$ Å [Spite and Spite 1982] have led to a determination of the Li/H ratio as $\approx 10^{-10}$ (by number). At that time this was about 10 times lower than values quoted for the Solar System and for the interstellar medium [Audouze 1984; Steigman and Boesgaard 1985]. As a consequence one has to assume that ^{7}Li is enriched during galactic evolution – by nova explosions or by stellar winds from Li-rich giants – to account for the large Li abundance observed in the Solar System and in most of the main-sequence F stars. In stars with convective zones the surface lithium is mixed into the interior and partially destroyed. This effect should, however, be less pronounced in old stars. Nevertheless, it seems difficult to derive a primeval Li abundance from the observations.

Further observations of old, hot population II stars strengthened the result of a nearly uniform abundance.

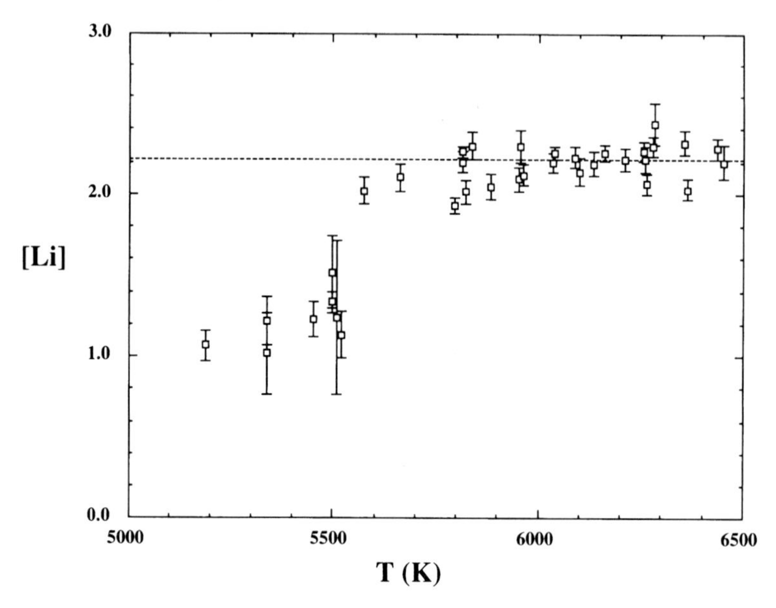

Fig. 3.6. The Li abundance in halo stars with [Fe/H] < -1.3, as a function of surface temperature

There are now ^{7}Li data for $\sim$ 100 halo stars. Especially for metal-poor stars with [Fe/H] $= -1.3$, the Li abundances as a function of surface temperature reach an approximately constant value at temperatures $T > 5500$ K (Fig. 3.6). Since at high temperatures no surface convection is expected, the existence of this low dispersion plateau is evidence that this abundance is indeed primordial.([Fe/H] is defined as the logarithm of Fe/H vs. $(\text{Fe/H})_\odot$: $[\text{Fe/H}] \equiv \log \text{Fe/H} - \log (\text{Fe/H})_\odot$).

The depletion of ^{7}Li by stellar processes would be accompanied by a severe depletion of ^{6}Li. The observation of ^{6}Li in hot population II stars with an abundance consistent with its origin in cosmic ray reactions, and a small amount of depletion as expected from standard stellar evolution shows that this is not the case.

The ^{7}Li abundance is

$$\text{Li/H} = \left(1.6 \pm 0.1 \begin{array}{c} +\,0.4 + 1.60 \\ -\,0.3 - 0.5 \end{array}\right) \times 10^{-10} \tag{3.78}$$

[Bonifacio and Molaro 1997]. These are a fancy set of errors [Olive 1997], and their meaning is as follows

i) ± 0.1: random;
ii) $(+0.4, -0.3)$: systematic, covering the range of abundances derived by different methods;
iii) $(+1.6, -0.5)$: possible depletion and production in stars. The allowed range of η_{10} is large: $\eta_{10} \simeq 1$ to 6.

3.3.5 Conclusion

The observed abundances of the light elements can be compared to the big-bang nucleosynthesis predictions. The calculations depend only on the parameter η_{10}, i.e. on the baryon density at the present epoch. In Fig. 3.7 boxes indicate the observational results in comparison to the theoretical curves for ^{4}He, D, ^{3}He, ^{7}Li production. Only a generous tratment of the error bars can still produce a narrow overlap between the measeurements of Y_p and ^{7}Li on one hand, and the D/H values from QSO absorption lines on the other.

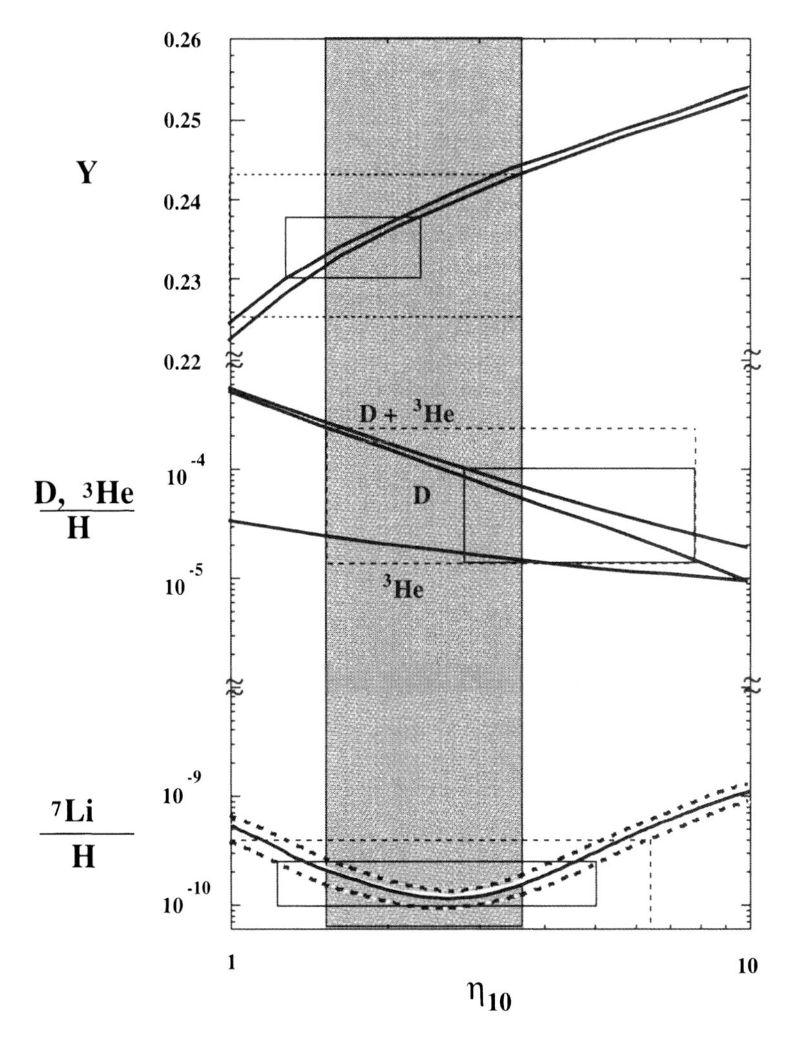

Fig. 3.7. The observed light element abundances compared to big bang nucleosynthesis results are plotted as functions of η_{10} (from [Olive 1997])

The narrow interval for η_{10} required to find agreement between theory and observation is viewed by some as a triumph of big-bang nucleosynthesis. The worry was, of course, that a very narrow margin might eventually give way to a real discrepancy. Even now, the best fit values already create a problem. For Y_p and ^{7}Li data

$$1.6 \le \eta_{10} \le 3.0 \qquad \text{with a best fit} \qquad \eta_{10} = 1.8 \tag{3.79}$$

[Olive 1997]. From D/H data

$$3.9 \le \eta_{10} \le 6.2 \tag{3.80}$$

[Levshakov et al. 1998] But care is needed. It seems too early to rejoice that the big-bang model has been refuted. A reasonable procedure in my opinion would be to allow for slightly larger systematic uncertainties in the experimental data, and to accept larger values of Y_p, up to 0.249. Then $3.9 \le \eta_{10} \le 5.3$ is a common overlap.

Alternatively, we may choose Y_p and ^{7}Li as the most secure abundance determinations. This leads to η_{10} between 1.6 and 3.0, and predicts a primeval $D/H = (4.7 - 28) \times 10^{-5}$ (within 95% confidence limits).

Differences in D/H and ^{3}He/H (which is predicted as $\simeq (1.3 - 2.7) \times 10^{-5}$) measurements from the predictions would then be attributed to the effects of chemical or stellar evolution [Olive 1997].

The intervals for η_{10} translate into constraints on

$$\Omega_B h^2 (\Omega_B h^2 = 3.65 \times 10^{-3} \eta_{10}) :$$

$$0.006 \le \Omega_B h^2 \le 0.011 \qquad (\text{for } 1.6 \le \eta_{10} \le 3.0), \tag{3.81}$$

$$0.015 \le \Omega_B h^2 \le 0.02 \qquad (\text{for } 3.9 \le \eta_{10} \le 5.3). \tag{3.82}$$

We can compare this with the analysis of CMB anisotropy measurements by balloon experiments (see Chap. 11): $\Omega_B h^2 = 0.022(^{+0.004}_{-0.003})$ is on the high side of the intervals acceptable from nucleosynthesis. This result seems to favour the D/H limits.

Even if we accept the complete interval $0.006 \le \Omega_B h^2 \le 0.02$, as the realistic uncertainty, we may conclude that baryonic matter alone is not enough to account for the estimates of the mean cosmological density. Dark non-baryonic matter must clearly be present, unless the homogeneous big-bang nucleosynthesis model can be changed appreciably. Furthermore, the lower limit on $\Omega_B h^2$ shows that there is more baryonic matter than the amount estimated from starlight, i.e. there also is baryonic dark matter.

The discussion given here and the analytic fits of Table 3.2 should enable the attentive reader to write his own nucleosynthesis review as soon as new data appear.

3.4 Helium Abundance and Neutrino Families

Another interesting conclusion may be drawn from the ^{4}He abundance [Schramm and Wagoner 1977; Sarkar 1996; Copi et al. 1995]: the weak reaction rate Γ for n $\leftrightarrow$ p processes decreases with temperature, and when $\Gamma < H(t)$, where $H(t)$ is the expansion rate, the n/p ratio freezes out. Practically all the neutrons present at freeze-out time combine to make helium. Thus the expansion rate $H(t)$ around $T \simeq 1\,\mathrm{MeV}$ determines the freeze-out ratio, and hence the ^{4}He abundance. Now $H^2(t) \propto \varrho$, and all families of relativistic particles at $T = 1\,\mathrm{MeV}$ contribute to ϱ. The helium abundance therefore depends sensitively on the number of types of relativistic particles present at $T \simeq 1\,\mathrm{MeV}$.

If only 2-component ν_e and ν_μ neutrinos (and their antiparticles) are present, then before the annihilation of e^+e^- pairs

$$\varrho = \varrho_\gamma + \varrho_{e^-} + \varrho_{e^+} + 4\varrho_\nu = \frac{9}{2}aT^4. \tag{3.83}$$

A number ΔN of additional massless, two-component neutrino families increases the density to

$$\varrho' = \varrho + 2\Delta N \varrho_\nu. \tag{3.84}$$

The ratio $\epsilon^2 = \varrho'/\varrho$ is therefore

$$\epsilon^2 = 1 + \frac{7}{36}\Delta N. \tag{3.85}$$

As the expansion time $t \propto \varrho^{-1/2}$, we have

$$\frac{t(T)}{t'(T)} = \left(\frac{\varrho'}{\varrho}\right)^{1/2} = \epsilon = \left(1 + \frac{7}{36}\Delta N\right)^{1/2}. \tag{3.86}$$

The expansion is more rapid when $\epsilon > 1$ i.e. $\Delta N > 0$. For $\epsilon > 1$, $t'(T) < t(T)$, i.e. the same temperature is reached at an earlier time. This leads to an earlier He synthesis, and to a higher neutron–proton ratio at the onset of the nuclear reactions. The net result is that the abundance of ^{4}He increases.

Roughly speaking, the primordial abundance of helium increases by 1% for each additional neutrino family. Accelerator results on the decay of the Z^0 particle (L3 Collab. 1989) lead to another limit on the number of different kinds of neutrinos. The width of the Z^0 together with a thorough analysis of all decay channels allows the limit $N_\nu(Z^0) \leq 3$ to be derived for all neutrinos that couple maximally to the Z^0 with mass $m_\nu \leq m(Z^0)/2 = 46\,\mathrm{GeV}$.

An analytic fit to Y_p [Sarkar 1996] shows in detail the dependence on N_ν:

$$Y_p = 0.244 + 0.014(N_\nu - 3) + 0.0002\Delta\tau_n + 0.009\ln\left(\frac{\eta_{10}}{5}\right).$$

The observational limits on Y_p lead to $N_\nu \leq 3.4$.

Very few possibilities for additional non-baryonic types of particles are left by these results. One possiblility would be a light supersymmetric particle, the "neutralino", if the τ-neutrino was heavy. The coupling of this light particle must be arranged such that it does not influence the Z^0 decay. Then it can be the third species allowed by the ^{4}He results, while the τ-neutrino would be the third neutrino-type permitted by the Z^0 decay.

We can also turn the argument around, and state that the existence of three neutrino species and the results on the ^{4}He abundance give us some confidence that the evolution with time of the scale factor, as described by the Friedmann equations – namely $R(t) \propto t^{1/2}$ – is experimentally verified for $t \sim 10\,\mathrm{s}$ after the initial singularity. As the ^{4}He production is so sensitive to a change in the expansion rate, a deviation from the law $R(t) \propto \sqrt{t}$ would be clearly imposed on the ^{4}He abundance.

We can venture the courageous statement that the evidence that the Friedmann equations correctly describe the evolution of the universe with time is better around $t = 10\,\mathrm{s}$, than it is for the present epoch.

3.5 Non-standard Scenarios

Finally, it should be mentioned that during the last few years there has been some activity on the possible effects of inhomogeneous nucleosynthesis (for reviews see [Thielemann and Wiescher 1990; Jedamzik and Fuller 1995; Jedamzik 1998]). One can imagine that after the decoupling of weak interactions density inhomogeneities form which cause different neutron and proton density distributions. Such inhomogenities may form if the quark hadron transition was a first-order phase transition. The mean free path of neutrons is much larger than that of protons. Thus in the initial high-density regions a very proton-rich environment appears, while the low-density regions are almost entirely filled with diffused neutrons. The overall picture becomes more complex and a number of new free parameters can be introduced, even if the high- and low-density regions are assumed to have constant density, respectively. These are, besides Ω_B, the density ratio between high- and low-density regions and the distance between high-density zones. The hope was to find a way to increase the baryonic density Ω_B to a value $\Omega_B = 1$, because one could expect to have weaker constraints than in the standard model. Thus the abundance of ^{4}He would be reduced – in the high-density regions because of the efficient neutron diffusion leading to smaller n/p ratios, in the low-density regions because of the scarcity of protons. Detailed calculations have shown that it is very difficult to avoid high ^{7}Li production. It seems possible to weaken the standard homogeneous nucleosynthesis limits somewhat – by a factor of 2 or 3.

A possible final outcome is also that the permitted parameter space coincides with the standard big bang. If the separation between high-density regions was smaller than the proton diffusion length or larger than the neutron diffusion length the resulting abundances are very close to the standard predictions. Instabilities of the boundaries of high-density regions may lead to a development of finger-like high-density structures, leading to efficient mixing, and to a very good approximation back to the standard, homogeneous big bang nucleosynthesis (e.g. [Freese and Adams 1989]). In general the models contain many *ad hoc* parameters which are – for the astrophysically relevant range of values – disfavoured by the standard model of strong interactions.

Other non-standard big-bang nucleosynthesis scenarios include the presence of antimatter domains, stable or unstable massive neutrinos, neutrino oscillations, neutrino degeneracy, and massive decaying particles. The influence of many of these effects is on a change of the expansion rate. The abundances of ^{4}He, ^{7}Li are used to constrain these more exotic nonstandard processes([Jedamzik 1998] gives a nice summary; see also [Sarkar 1996] and Chap. 7 below).

Exercises

3.1 Derive the equilibrium distributions of (3.3) starting from the quantum statistics of Bose and Fermi particles. What would be the difference in particle number density between neutrinos and antineutrinos, if you introduced a positive chemical potential μ_ν for the neutrinos, but kept the chemical potential zero for the antineutrinos? What would be the maximum energy for such a cosmic neutrino background?

3.2 Compare the statistical factor $g(T)$ for various epochs in the early universe, considering the particles in and out of equilibrium: for the electroweak phase, before and after the quark–hadron phase transition, for the $SU(5)$ GUT, SUSY GUT and superstring theory (see Chaps. 5 and 6).

3.3 Calculate the changes introduced in the neutron–proton ratio by a nonzero chemical potential of neutrinos. What are the limits that can be put on μ_ν/kT from the abundance of ^{4}He?

3.4 Derive the limits that can be set on a cosmological constant from the cosmic abundance of ^{4}He.

3.5 F. Hoyle and G. Burbidge argue for an origin of the CMB from local sources [Hoyle and Burbidge 1998], starting from the observation that the energy density set free in helium synthesis is equal to the CMB energy density. Can you find arguments against this suggestion?

4. Can the Standard Model Be Verified Experimentally?

"From our home on the Earth, we look into the distances and strive to imagine the sort of world into which we are born. Today we have reached far out into space. Our immediate neighbourhood we know rather intimately. But with increasing distance our knowledge fades, and fades rapidly, until at the last dim horizon we search amongst ghostly errors of observations for landmarks that are scarcely more substantial."

E. Hubble [Hubble 1958]

So far, we have considered a standard cosmological model, and have shown how the data can be accommodated within it. In this chapter we want to ask to what extent the data actually determine the features of our universe [Kristian and Sachs 1966; G. Ellis 1980, 1984; Ellis et al. 1985]. As before, we take GR as the correct theory of gravitation.

One basic problem with experimental verification of large-scale features of the universe is the very large size of all cosmic models. It is not possible for us to travel any cosmologically significant distance within our lifetime. Even the whole history of the science of astronomy corresponds to only a tiny interval in the evolution of the universe. We can describe the observational situation by stating that all astronomical cosmological evidence is gained from one space-time point.

In all cosmological investigations it is tacitly assumed that the local laws of physics hold everywhere, and that the astrophysics and astronomy of specific objects – the organization of matter into stars, the formation of stellar associations, etc. – is well understood, and can, at least in principle, be explained by known physical laws. The hypothesis is, of course, supported by the observation and analysis of the spectra of distant galaxies. Taking these assumptions for granted, how much can we learn about the structure of the universe from ideal observations?

4.1 Ideal Galactic Observations

The propagation of electromagnetic radiation from distant objects can probably be described quite well by geometrical optics. In a specific cosmological model the light rays are thus null-geodesics of the space-time. The question is

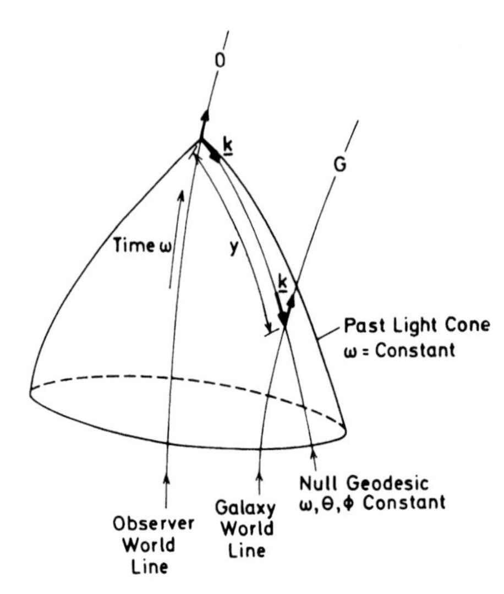

Fig. 4.1. Coordinates based on the world-line of an observer O characterize the time of observation ω; the projected position in the sky θ, ϕ; and the distance y to the galaxy observed [Ellis 1980]

how the local space-time geometry can be determined by observations [Kristian and Sachs 1966; Ellis et al. 1978]. Here we shall follow the description given by Ellis [Ellis 1980].

Consider the past light-cone of an observer O, and set up observational coordinates $(\omega, y, \theta, \phi)$ in the following way (Fig. 4.1): ω is the time of observation of a galaxy as measured by an observer O; y is a measure of the distance of the galaxy; θ, ϕ the coordinates of this galaxy in the sky. Then the surface ω = const. is the past light-cone of the observer O at time ω. The null geodesics generating this light-cone have the tangent vector field $k_\alpha = \omega_\alpha$ and are given by $\omega, \theta, y = const.$; ϕ measures "distance" along the null cone.

Care must be taken in avoiding problems with the regularity of the fluid flow and of space-time on the central world-line $y = 0$ of this coordinate system. To this end conditions have to be chosen for the fluid velocity, the null-geodesic tangent vector, and the metric components as $y \to 0$. The central conditions also give the relation between coordinate directions θ, ϕ at different times ω along $y = 0$ [Ellis et al. 1985].

If we choose $y = 0$ as the observer's world-line, then the space-time metric in these coordinates can be brought into the form

$$g_{\mu\nu} = \begin{pmatrix} a & b & \nu_1 & \nu_3 \\ b & 0 & 0 & 0 \\ \nu_1 & 0 & h_{22} & h_{23} \\ \nu_3 & 0 & h_{23} & h_{33} \end{pmatrix} . \tag{4.1}$$

The scale for y may be chosen such that one component is equal to 1; e.g. $b = 1$, if y is an affine parameter for the null-geodesic vector field $\boldsymbol{k}$.

From the observer's position (corresponding to the point $\omega = y = 0$), an ideal observational programme may now be carried out:

1) The metric components $h_{AB}(A, B = 2, 3)$ are determined by the intrinsic size and shape of the observed object. The image can be measured, and compared with the size and shape of the object – if they are known. The metric components h_{AB} just relate the actual size of the object to the corresponding observed angles through the relationship

$$dl^2 = h_{AB} dx^A \, dx^B; dy = d\omega = 0 \quad (x^A \equiv \theta, \phi). \tag{4.2}$$

Distortion effects caused by anisotropic curvature (transforming a spherical galaxy into an elliptical image, for example) as well as the focusing effect caused by isotropic curvature may be measured in this way.
The focusing effect can be conveniently expressed by employing an area distance r_A (cf. Chap. 1)

$$r_A^4 \sin^2 \theta = \det(h_{AB}), \quad A, B = 2, 3. \tag{4.3}$$

The knowledge of h_{AB} implies a knowledge of the intrinsic geometry of the null cones ($\omega = \text{const.}$).
2) The components $u^\omega, u^\theta, u^\phi$ of the matter's 4-velocity are observable. (If τ is the proper time along the world line of a galaxy, its 4-velocity is $dx^a/d\tau \equiv u^a$).
In principle, u^θ, u^ϕ can be measured directly from the proper motions of the galaxy. u^ω is obtained from the redshift, e.g. [Ellis 1971]

$$1 + z = \frac{(u^a k_a)_{\text{Gal}}}{(u^\beta k_\beta)_{\text{Obs}}} \tag{4.4}$$

(normalization $\boldsymbol{u} \cdot \boldsymbol{k} = 1$ at $y = 0$).
The metric component u^ϕ cannot be measured, but it can be found in terms of $g_{\mu\nu}$ of (4.1) from the normalization condition $u^\nu u_\nu = -1$.
3) The matter density ρ can be derived if the metric component b is known. The number-count dN of galaxies between y and $y + dy$ along the null geodesics determines bn, the product of b and the number density n of sources at y. This is observable if y is an observable quantity (e.g. by its relation to r_A). If b is known, n can be determined, and then, by using a mass-to-light ratio, the mass density ϱ can be derived.
The quantities listed above – h_{AB}; u^a $(a = 0, 2, 3)$; bn – are the only detailed properties of the cosmological model that can, in principle, be measured on the past light-cone of the space-time position of the observer.
4) The metric components a, b, ν_A $(b = 1)$ are not measurable directly through observations on our past light-cone. In general only their limiting values at $y = \omega = 0$ are known – determined by the central conditions at $y = 0$.
We have to conclude that the space-time metric or the matter distribution on the past null cone cannot be determined by observations alone. Thus, for instance, when no dynamical equations are imposed, the implication of all cosmological observations being spherically symmetric is *not* that the space-time is spherically symmetric around the observer.

5) The use of the field equations of GR, however, allows us to determine a, ν_2, ν_3 from the known quantities by integration of four of Einstein's equations down the null geodesics from the origin.

The remarkable result emerges that the quantities that are, in principle, observable on our past light-cone are precisely sufficient to determine the space-time and its matter content on that light-cone when Einstein's field equations are used. A priori, the geometry could either have been over- or under-determined. As a specific example, consider an equation of state for a perfect pressure-free fluid. The equations of GR show that spherically symmetric observations do imply that the space-time is spherically symmetric on our past null cone.

These data on the past light-cone are sufficient to determine the geometry in the interior by integration of the field equations. As is indicated in Fig. 4.2, one cannot make predictions into the future of the light-cone. New information can come from the outside and produce unexpected effects – we cannot be sure that the Sun will rise tomorrow unless we can be sure that there will not be a strong incoming gravitational field affecting the stability and motions of the Solar System. In practice, however, we are fairly certain that such drastic disturbances will not occur and that the predictions made from data in our past light-cone will not be invalidated, at least at short time-scales. The "non-interference" is owing to the fact that for small time-steps into the future nearly all the data needed to make predictions are known (the surface N in Fig. 4.2 is very small), and that the effects of uncertainties in the data are weakened by redshift and distance effects. Also the fact that up to now we have been successful in predicting motions within the Solar System increases our trust in a "no-interference" hypothesis.

We can never feel completely secure, of course, that there isn't a black hole waiting to meet our world-line. Modern astronomical methods would, however, be able to give us a warning a few thousand years before the ultimate collision.

This "observational approach" to cosmology should eventually be completed by examining the consistency of the results obtained with other data, such as the properties of element formation and of the background radiation. Within this framework – accepting a "no-interference" hypothesis – we can then formulate which observational results are necessary and sufficient to establish the universe as an FL model (here, as before and in what follows, we are simply reproducing the ideas presented by G. Ellis over the years [Ellis 1980; Ellis et al. 1985]).

i) There must be no proper motions, i.e. $u^\theta = u^\phi = 0$.
ii) There is no distortion effect; h_{AB} takes the form

$$h_{AB} = \begin{pmatrix} r_A^2 & 0 \\ 0 & r_A^2 \sin^2\theta \end{pmatrix} . \tag{4.5}$$

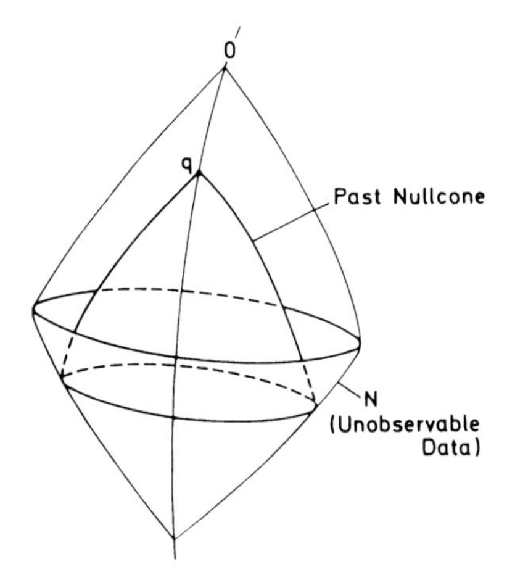

Fig. 4.2. Observational data on the past null cone of the point q must be supplemented by data on a surface such as the surface N shown, in order to make predictions (unless the past null cone already contains the whole universe, as it does in the Einstein static universe, or in some versions of the small universe hypothesis) [after Ellis 1980]

iii) All relations between observable quantities (e.g. relations between m and z, or between dN and z) are isotropic. Relations between the fundamental quantities r_A and $\varrho(z)$ are isotropic; we may use z instead of y, and then $r_A(z, \theta, \phi) = r_A(z)$ and $\varrho(z, \theta, \phi) = \varrho(z)$.
iv) The derived relations take the FL form; i.e. (for $p = \Lambda = 0$, for example)

$$\begin{aligned} r_A(z) &= H_0^{-1} q_0^{-2} (1+z)^{-2} [z q_0 + (q_0 - 1)(\sqrt{1 + 2 q_0 z} - 1)], \\ \varrho(z) &= (l+z)^{-3} \varrho_0; \quad \varrho_0 = 6 q_0 H_0^2. \end{aligned} \tag{4.6}$$

v) In addition, the observations must be compatible with the background radiation measurements, with local information on element and particle abundances, age estimates, and local density measurements.

In principle, an observational programme systematically verifying points (i)–(v) could establish the FL nature of the universe.

4.2 Real Observations

The observational reality is in stark contrast to the splendid picture painted above. In the real world the ideal observational programme can be carried out only to a limited extent. First of all, the inhomogeneity of the distribution of matter tells us that only by overaging over sufficiently large volumes can the FL form of the distribution be found. Since superclusters seem to form a network on scales $\sim 100\,\mathrm{Mpc}$, even over such scales deviations from the precise FL relations may exist. The expectation is that a convergence to the true FL relations appears as larger and larger scales are considered.

a) Proper motions on cosmological scales cannot be detected. We can observe practically only on one light-cone. Transverse velocities are too small to be measured directly, although they may be large enough to be important dynamically.

b) The distortion effect cannot be detected reliably. The reason is that there are not any objects with standard shapes such that observations of their apparent shapes could reveal a distortion effect. For example, an apparently elliptical galaxy cannot be shown (e.g. by a specific spectral or morphological feature) to be intrinsically spherical. Apparent alignments of the axes of clusters of galaxies could be ascribed to physical processes effective during galaxy formation rather than to a distortion effect. Even if for some objects, or for some direction in the sky, the distortion effect could be definitely measured, it seems quite impossible to achieve this for all redshifts and for every direction in the sky.
c) The isotropy of the distribution of matter and radiation around us can in principle be established. Possible departures from isotropy, however, cannot be given a well-defined interpretation unless criteria for isotropy or for the lack of isotropy have been laid down.
d) The estimates of $r_A(z)$ and $\varrho(z)$ from observations are beset by a variety of uncertainties, including observational limits, astrophysical uncertainty and statistical problems, as well as by the redshift and the dilution of the information by screening – through receiver bandwidth for example.

As it is, the observational situation requires "an entire package of information" [Ellis 1980] before the observed (m, z) or (N, z) curves can be understood properly. The interpretation of observational curves has to include assumptions about:

i) the form of the spectrum and surface-brightness distribution for the objects in the class considered;
ii) the distribution of objects in the (luminosity, diameter) plane at a certain redshift z;
iii) the luminosity, diameter, and spectral evolution of the class of objects – where luminosity and spectral evolution can in principle be determined from the observations, if the morphological evolution is known;
iv) the limits of the detection systems used;
v) a particular cosmological model which prescribes a specific area distance–redshift relation $r_A(z)$. In addition, a detailed investigation would have to take into account absorption, seeing effects, and effects of local inhomogeneities in the universe that give an $r_A(z)$ relation different from the FL form. A few examples serve to illustrate these remarks [G. Ellis 1980, 1984].

4.2.1 Observational Methods

The methods of observation have an important influence on the nature of the objects considered. The impression is that in many cases "the data tell

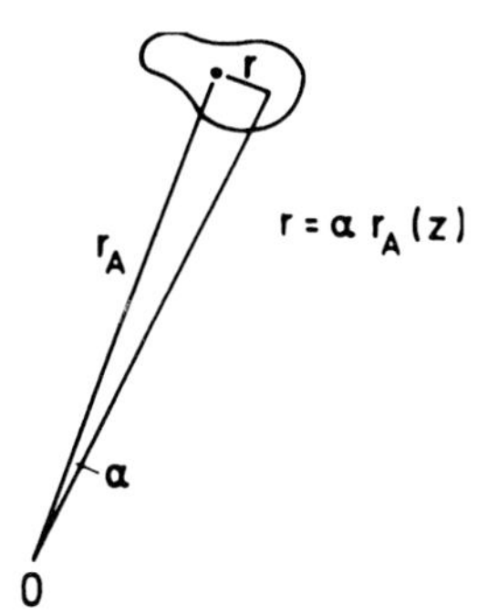

Fig. 4.3. An idealized observation of an extended object at a distance $r_A(z)$

us more about the threshold of detection than the object involved" [Baum 1962]. This quotation refers to the increase of the observed diameter of the galaxy M32, between 1920 and 1960, from 3 to 30 arcsec. Similar problems may arise in the determination of galactic magnitudes.

Consider a simple model (Fig. 4.3) of the observation of an extended object [Ellis et al. 1984]. Let $r = \alpha r_A(z)$ be the distance from the centre of a point on the object, taken perpendicular to the line of sight. The brightness distribution across the surface perpendicular to the line of sight is a product of the central surface brightness B_0, the spectrum $J(\nu)$ and a structure function $f(r/a)$ with $f(0) = 1$ (examples are $f(r/a) = (1 + r/a)^{-2}$ (Hubble profile) or $f(x) = \exp(-x^{1/\beta})$ (de Vaucouleurs), with $\beta = 1$ for spiral galaxies and $\beta = 4$ for ellipticals (see Chap. 2)):

$$B = B_0 J(\nu) f\left(\frac{r}{a}\right). \tag{4.7}$$

The observed specific intensity at an angle α from the centre of the image is then given by

$$I_\nu(\alpha) = B_0(1+z)^{-3} J(\nu(1+z)) f\left(\frac{\alpha r_A}{a}\right). \tag{4.8}$$

Let us idealize the situation further by neglecting absorption and seeing effects. Then the following observable quantities can be defined [Ellis and Perry 1979]:

1) The magnitude $m_v(A)$ measured through an aperture of semi-angle A centred on the galactic image. This can be determined from (4.8) by integrating up to the angle A, and changing to a magnitude scale:

$$m_v(A) = -2.5 \log \left\{ B_0(1+z)^{-3} J(\nu(1+z)) \left(\frac{a}{r_A}\right)^2 g\left(\frac{A r_A}{a}\right) \right\},$$

where

$$g(\beta) = \int_0^\beta 2\pi\delta f(\delta) d\delta. \tag{4.9}$$

2) The limiting specific intensity $S_L(\nu)$ to which the galactic image can be detected defines a detection threshold. The angle at which the observed specific intensity has dropped to this limit is the apparent angle A_{ap} measured for this galaxy,

$$I_\nu(A_{\mathrm{ap}}) = S_L(\nu). \tag{4.10}$$

Using an aperture A_{ap} gives the apparent magnitude

$$m_{\mathrm{ap}} \equiv m_v(A_{\mathrm{ap}}). \tag{4.11}$$

If the observed specific intensity never exceeds the detection limit $S_L(\nu)$, then $A_{\mathrm{ap}}=0$, $m_{\mathrm{ap}} = \infty$.

4.2.2 Observational Limits

The common assumption that observations of galaxies are magnitude-limited, i.e. that m_{ap} is the detection limit, is incorrect – average image brightness and apparent angle are clearly important in determining detection limits and selection effects. Consider a set of sources with the same intrinsic magnitude but different diameters – for very large diameters the average surface brightness will drop below the detection limit, while for very small diameters the apparent angle will be very small, and the object will appear "quasi-stellar". Clearly, the actual selection effects are determined by the image size and intensity, and thus, in addition to aspects intrinsic to the objects, there are aspects connected with the measuring situation.

It seems appropriate to mention here a few salient points – all well known to observers, of course. For a given class of objects Ellis introduces an object plane – intrinsic magnitude vs. diameter – and an image plane – apparent magnitude vs. apparent angle. The "observational map" (4.11) for objects at a specific distance from the observer in a given space-time maps each point in the object plane into the observed point in the image plane. It depends on (i) the spectrum $J(\nu)$ and the radial brightness distribution function f, (ii) the detection limit $S_L(r)$, (iii) the space-time properties as specified by the distance $r_A(z)$, and (iv) the distance between object and observer as measured by the redshift z.

An example is shown in Fig. 4.4 [from Ellis 1980], where for a high-density ($\Omega_0 = 2$) FL universe, galaxies with a spectral index ($J(\nu)\alpha\nu^{-\lambda}, \lambda = 0.7$) of 0.7, and a $\beta = 1$ de Vaucouleurs profile are plotted. The objects (1–4) are mapped onto images, which move along the curves labelled (1)–(4) as the redshift is varied.

The observational limits are placed in the image plane. Typical examples are shown in Fig. 4.4: images to the right of the brightness limit are difficult to detect because of their faintness. Optical images to the left of the angular limit can easily be confused with stars, and the magnitude limit might also play a role.

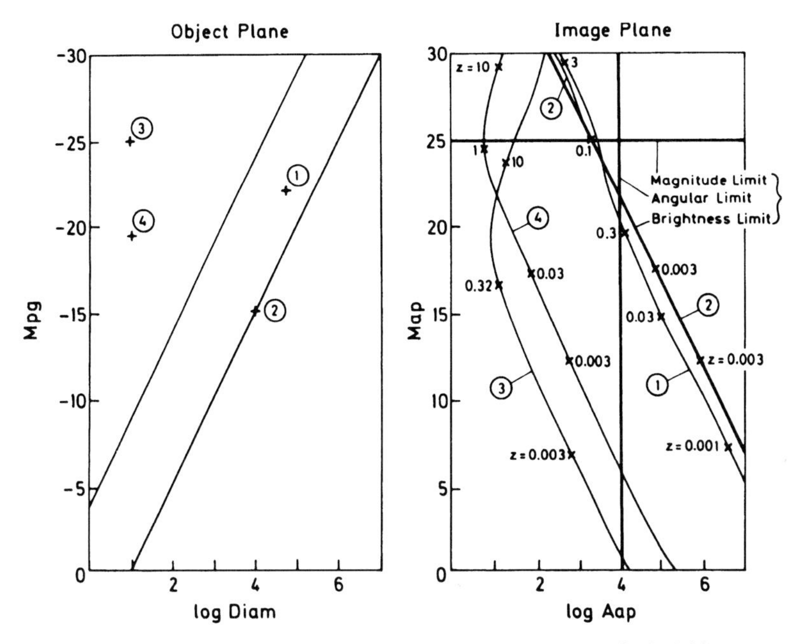

Fig. 4.4. The observational map for galaxies with an $f \propto \exp\{-(r/a)\}$ profile and spectral index 0.7 in an FL universe with $q_0 = 1$. The image points move along the curves 1 to 4 as the redshift at which the objects 1 to 4 have been placed varies [after Ellis 1980]

As the redshift increases the image of each object moves into a region beyond the observational limits.

The projection of the map leads to (apparent angle, redshift) or (apparent magnitude, redshift) relations for specific objects. These relations establish the existence of a limiting redshift for the observation of any particular object. The specific intensity received will decrease with redshift, and disappear below the detection limit for a high enough redshift. The $(A_{\rm ap}, z)$ and $(m_{\rm ap}, z)$ curves asymptotically approach this limiting redshift value for small angles and large magnitudes. (cf. curves marked "no evolution" in Figs. 4.5 and 4.6).

4.2.3 Effects of Evolution

The objects in Fig. 4.4 are assumed to keep the same magnitude and diameter as the redshift changes. But there are indications from theoretical models of galactic evolution [Tinsley 1977a,b] that both luminosity and spectral evolution may be expected in an FL universe for high redshifts. Studies of "galactic cannibalism", i.e. the merging processes of two galaxies into one [Hausmann and Ostriker 1978] make an appreciable radial evolution seem very plausible. In Figs. 4.5 and 4.6 (for the graphs of $m_{\rm ap}$ vs. z) the effects of radial evolution $a(z)$ and of luminosity evolution $L(z)$ are shown. It can

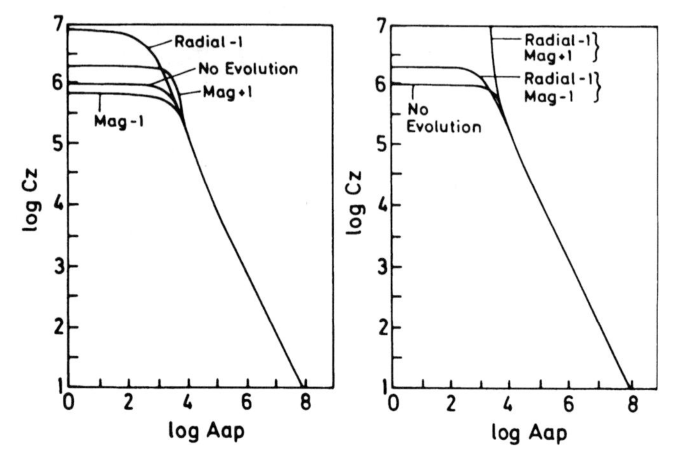

Fig. 4.5. The graphs of redshift vs. apparent angle for galaxies as in Fig. 4.4. Evolutionary effects are also evident: intrinsic luminosity and radius are assumed to change in proportion to $(1+z)$ or $(1+z)^{-1}$. Thus "radial-1" means a length $R \propto (1+z)^{-1}$ and "Mag+1" means $L \propto (1+z)$

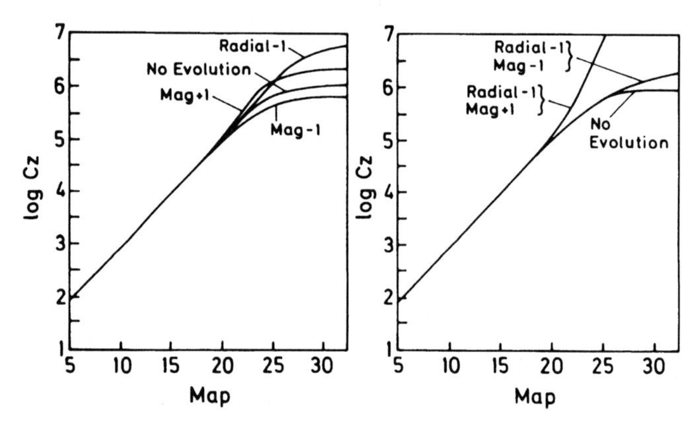

Fig. 4.6. Apparent magnitude m_{ap} versus redshift for the objects as in Fig. 4.4. Evolution is shown in the same way as in Fig. 4.5

be seen that such an evolution changes the limiting redshift value, and thus introduces a considerable spread in the curves.

This evolution of luminosity and radius has to be known from astrophysical considerations before the cosmologically relevant information can be extracted from the (m, z) and (N, z) relations.

4.2.4 Selection Effects

It is also clear that the observed curves for number-counts within a certain distance, or for the luminosity function of a cluster of galaxies, represent a selection of objects, namely those that are seen and counted, while it does not necessarily represent all the objects of a specific class that are actually there. Let us illustrate this by two examples:

1) A line of constant magnitude in the object plane is mapped into a line in the image plane that is approximately at a constant apparent magnitude at small angles, but has a sharply defined turnover point at large angles – the surface brightness drops towards the limiting value (cf. Fig. 4.7). The detection probability is greatest at a given magnitude for the largest apparent angle, and therefore the observations will preferentially pick out galaxies close to the turnover point [Disney 1976]. Many other galaxies of the same magnitude could be present but would not be detected.
2) Consider a cluster of galaxies in which the galaxies exhibit a linear relation between magnitude and log diameter (Fig. 4.8). They lie on a line in the object plane. The situation in the image plane now depends on the slope of this line; if it is equal to or steeper than the slope of lines of constant magnitude, then (Fig. 4.9) the image points lie on a line that is close to a line of constant observed brightness. At a definite redshift z there will always be some of the large galaxies with large radii which pass undetected.

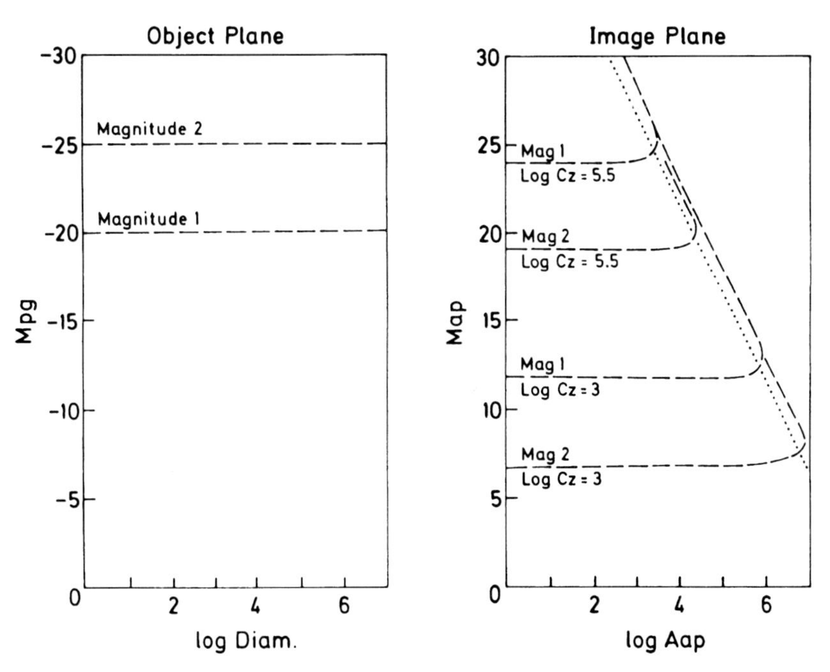

Fig. 4.7. The images of families of galaxies that are all of the same intrinsic magnitude, but of varying diameters are plotted. The image lines turn over when the brightness limit (*dotted line*) is approached. Two redshift values are chosen, $\log cz = 3$ and $\log cz = 5.5$, in a $q_0 = 1$ FL universe

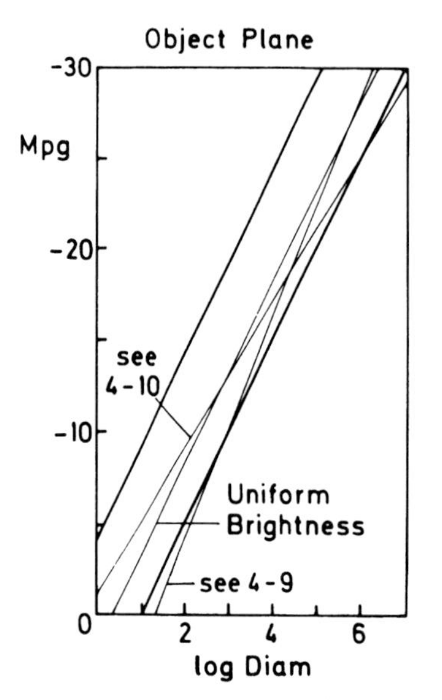

Fig. 4.8. Object plane (FL universe, $q_0 = +1$) for galaxies with a linear relation between magnitude and log (diameter); all galaxies of nearly the same intrinsic brightness

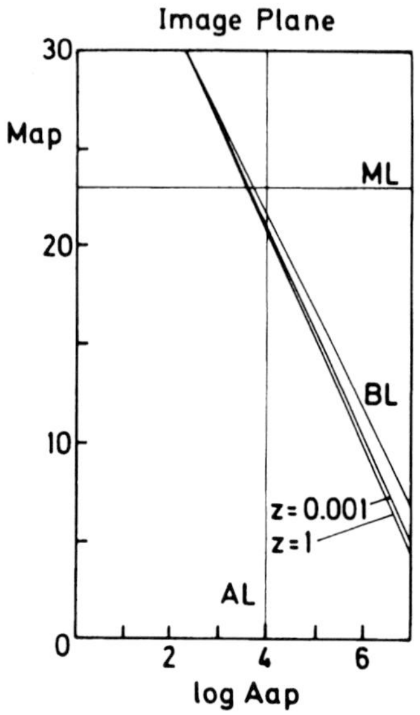

Fig. 4.9. Image plane for galaxies on a line with a slope greater than for a family of constant brightness ($q_0 = 1$, FL)

If the slope of the line in the object plane is less than the slope of a constant brightness line, then the galaxies at the bottom exceed both angular and apparent magnitude limits, and the galaxies at the top exceed brightness limits.

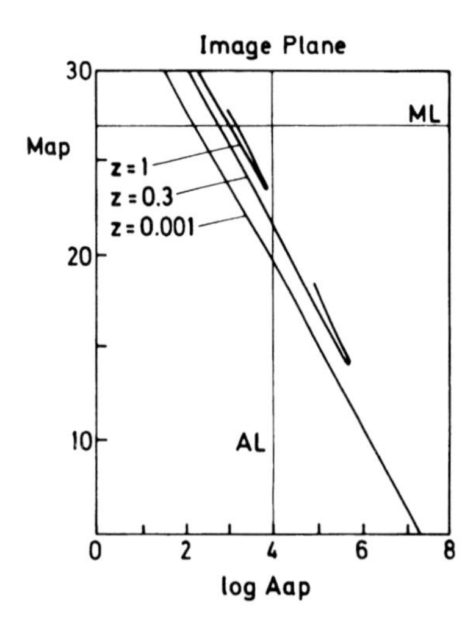

Fig. 4.10. Image plane for galaxies on a line with a slope less than that for a family of constant brightness ($q_0 = 1$, FL)

The result is a sharp turnover at a certain apparent magnitude and angle (Fig. 4.10). As the redshift increases, the turnover point moves towards smaller A_{ap} and larger m_{ap}. Observations of a family of such galaxies would detect those near the turnover point as the most luminous objects.

The turnover point corresponds to the cross-over of the brightness limit and the line of the cluster of galaxies in the object plane. The existence of such a sharp turnover could lead to selection effects, which may produce the rather precise constant value for the apparent magnitudes of the "brightest galaxies" in a cluster [Sandage et al. 1976].

4.2.5 Uncertainty Increases with Redshift

The measurable information becomes very scarce with increasing redshift – "Uncertainty increases with redshift" – the flux received decreases, the angles subtended by the objects decrease. The possible evolution effects of the sources also increase with growing redshift. The information received from distant regions is further reduced by galactic cores intervening between us (the observer) and the source. Most of the intervening galaxies would not be detected because they would be below the brightness detection limit. But the medium made up of galactic cores may be rather opaque at early times. Cores of a diameter of about 4 kpc can obscure over 40% of the sky in a high-density universe at $z = 10$ (60% in a low-density universe), and 70% (and 90%, respectively) at $z = 20$. Radiation originating at higher redshifts must pass through this "forest of galaxies" [Ellis 1980]. The existence of the Planck spectrum of the 3 K background is evidence that the universe must become opaque – i.e. the optical depth must reach unity – for some finite redshift z^*. In an FL universe the maximum distance of this cosmic photosphere is $z^* = 1500$, but z^* may be much smaller in high-density models.

"The whole world is made of sand" – the scientifically sound extrapolation of an ant living in the Gobi desert

4.3 Outline of a Procedure to Verify the FL Universe

To examine a particular FL universe by observations we first of all have to establish the isotropy and homogeneity of the universe around us.

4.3.1 Isotropy

The best evidence of isotropy seems to be revealed in the near-isotropy of the microwave background. Certainly the measurements cannot directly prove absolute isotropy – a necessary condition for an FL universe – but they do produce upper limits on the anisotropy that is observed from a suitable 4-velocity. In fact, we have seen in Chap. 2 that a dipole anisotropy is prominent in the CMB observations, which can be subtracted out by ascribing a definite proper motion to the Earth. This implies large-scale motions in our vicinity – and thus problems of definition arise: how large a peculiar motion of our local group of galaxies can be accepted without leading to a violation of the isotropy demanded by the FL models? At present we do not have a valid operational criterion that would tell us how large a deviation from isotropy in any specific test could still be reconciled with the hypothesis of an FL model.

It seems that the upper limits obtained after subtracting the dipole effect make it reasonable to accept the isotropy of the radiation around us. Even then inferences about the isotropy of space-time around our world-line have to be treated with caution, because the measurements do not give any direct evidence about the emission redshift of the radiation. There are many strange possibilities: the emission redshifts might vary with direction, or anisotropic radiation and matter distributions might be arranged around our world-line to produce isotropic radiation observations [Ellis 1980]. The dipole anisotropy may partly be owing to the large-angle signature of an anisotropic world [Barrow et al. 1983].

Such arrangements seem rather contrived, but they illustrate the point that without further assumptions – such as the "Copernican principle" that our position should not be special – very little can be inferred from the measurements.

Further tests of isotropy are, in principle, possible – the proper motions and distortions must be zero, and the (n, z) and (dN, z) curves must be isotropic. In practice, proper motions and distortions can only be checked in our immediate neighbourhood, and the observations show that peculiar motions are present. Again the problem arises of tolerance limits for local deviations from isotropy. It seems, nevertheless, that the overall evidence is compatible with an isotropic model at epochs after the decoupling of the CMB radiation. To extend this assumption back to very early times is, of course, a daring extrapolation, since the observational results can be compatible with a very early universe that is very different from an FL model.

Sometimes a dissenting opinion leads to new insights

4.3.2 Homogeneity

To verify a spatially homogeneous FL universe model within the class of spherically symmetrical models, one has to show that the (r_A, z) and (ϱ, z) relations have the FL form. In principle this can be tested by the observed (r_A, z) and (dN, z) or (dN, r_A) relations. The problems discussed above of selection effects, source evolution and observational limits have prevented, so far, the determination of even the quadratic term in a power-series expansion of $r_A(z)$. The conversion of the information on number-counts of galaxies and radio sources into knowledge about $\varrho(z)$ is very difficult. The mass-to-light ratio has not been determined directly even in our immediate neighbourhood. At large distances, then, it is virtually impossible to discriminate between a time variation and a spatial variation of source properties – both leading to an apparent deviation from homogeneity.

It seems difficult to distinguish between an FL universe and a spatially inhomogeneous universe that is like an FL model near us. A particular model [Ellis 1980; Ellis et al. 1978] for such a situation can be described by an inhomogeneous, spherically symmetric space-time:

$$ds^2 = -g^2(r)dt^2 + dr^2 + f^2(r)d\Omega^2. \tag{4.12}$$

There is a Killing vector $\partial/\partial t$, i.e. the space-time is static for observers with 4-velocity $u^\mu = g^{-1}\delta^\mu_0$. The model is spherically symmetric around $r = 0$, if f and g satisfy

$$\begin{aligned} &f(0) = 0, \quad f'(0) = 1, \quad (f''/f)|0 \quad \text{finite;} \\ &g(0) = 1, \quad g'(0) = 0. \end{aligned}$$

A "cosmic black-body" radiation in this space-time has a temperature varying with r:

$$T(r) = T_0(1 + z) = T_0 g^{-1}(r), \tag{4.13}$$

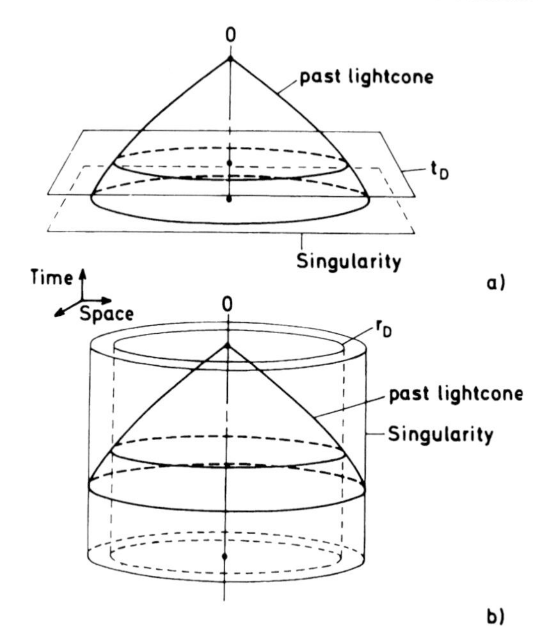

Fig. 4.11. In an FL universe a spatially homogeneous primeval fireball occurs (**a**), whereas in an inhomogeneous, spherically symmetric universe the hot big bang corresponds to a time-like singularity (**b**). To a central observer both situations can appear quite similar

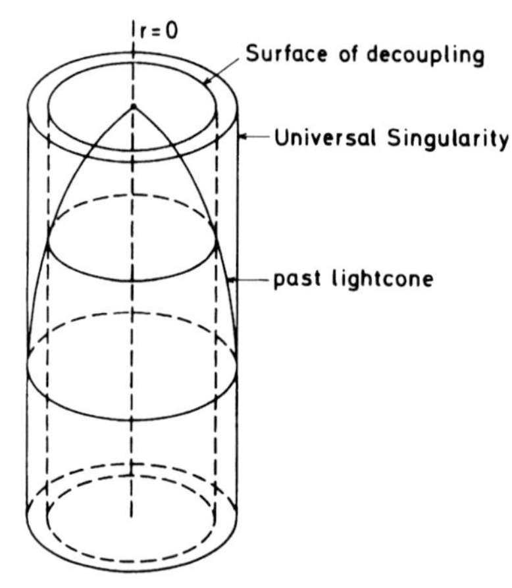

Fig. 4.12. The static, spherically symmetric cosmological model surrounded by a time-like singularity

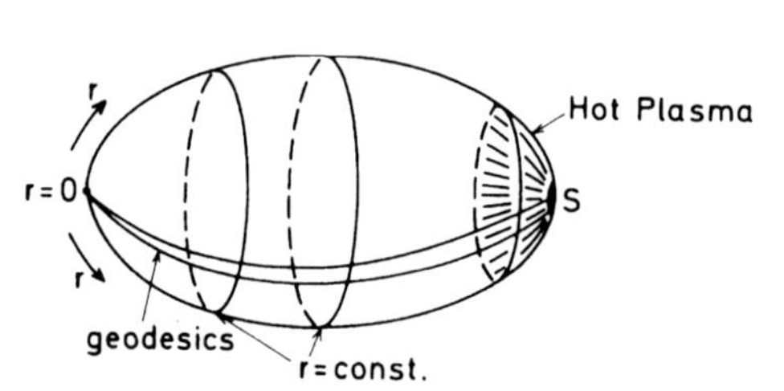

Fig. 4.13. The t = const. surface of the static spherical model. The time-like singularity S is surrounded by a hot plasma

where T_0 is the temperature along $r = 0$ $(T_0 \simeq 3\,\mathrm{K})$. Let $g(r)$ decrease monotonically to zero in the interval $(0, R)$ with finite R. Then $T(r)$ will grow as $r \to R$, and, for some value r_d, reach the decoupling temperature $T_d = 3000\,\mathrm{K}$. The region beyond r_d would be occupied by a hot plasma $(r_d < r < R)$ – the source of the black-body radiation.

The radial variation of the radiation and matter content of the space-time would closely correspond to the dependence on cosmic time t in the standard models. Successive values of r corresponding to decoupling, to nucleosynthesis, to pair production could be determined, as $T(r)$ increases. Each variation would here be ascribed to a spatial variation. The situation is that all observations which are made on our past light-cone are extended off the light-cone on timelike surfaces. (In contrast, in FL cosmologies the extensions are made on spacelike surfaces; see Fig. 4.11.)

If R is finite, then $g(r) \to 0$ in a finite proper distance from $r = 0$, and $T(r) \to \infty$, i.e. there is a singularity S at $r = R$ (time-like). All past radial null geodesics intersect S. All space-like radial geodesics intersect it too. Therefore the singularity surrounds the central world-line (Fig. 4.12). The universe is spatially finite and bounded; each space section (t = const.) has a diameter $2R$. If $f(r) \to 0$ as $r \to R$ then the singularity can be regarded as a point singularity on each space section. The space-time is spherically symmetric about $r = 0$, and about S. In Fig. 4.13 a space section (t = const., $\theta = \pi/2$) is displayed.

Models of this sort have not been considered, because the assumption of spatial homogeneity has been taken as a cosmological principle ("Copernican principle"). This assumption has not been tested by observations, but it is introduced into any cosmological data analysis, because it is believed to be unreasonable to assume that we are near the centre of the universe. However, in such a spherical, static model, conditions for life are most favourable near the central line $r = 0$, where the universe is cool. This "anthropic principle" would give a reason why we should happen to live near the centre of our static universe.

The specific model presented above [Ellis et al. 1978] can accommodate almost all the data, except for one crucial point. When the field equations of GR are used on the line element (4.12), together with a plausible equation of state, then it is not possible to obtain a good fit to the observed (m, z) relation. This difficulty of obtaining Hubble's law in a static universe gives an idea for a proof that the universe is expanding – a proof without resort to cosmological principles. Many more investigations of this kind should be carried out, to explore various different possible ways of explaining the measurements.

Another possible approach to the homogeneity problem [Ellis 1979; Ellis and Schreiber 1986] is the investigation of an expanding universe with small, finite space sections.

Such a "small universe" should be small enough for us to have time to see round it many times since decoupling. The topology of the compact spatial sections can be quite complicated; in general they will not be simply connected.

Spatially homogeneous and isotropic small universes are identical locally to an FL universe, since they are the same solution of the Einstein field equations. These differential equations do not specify the global topology of space-time. Small universes can be obtained for each sign $K = +1, 0$ or $K = -1$ of the spatial curvature in FL universes. Thus the low-density, $K = -1$, FL universe can be constructed with compact, i.e. finite, 3-spaces.

Small-universe models are found from the large FL model by suitable identifications within a discrete group of isometries. For instance, choosing a rectangular basic cell with opposite faces identified gives a toroidal topology for the small universe. The never-ending identical repetition of the basic cell (the build-up of its "universal covering space") reproduces the original FL universe.

Observations in the small universe will be exactly the same as in a simply connected universe covering space-time with a set of galaxies that repeats itself exactly again and again. Thus many images of each galaxy will be seen.

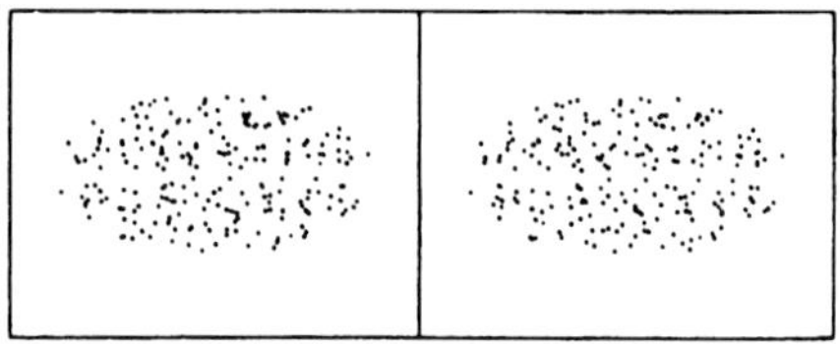

```
STEREO PAIR SHOWING THE DISTRIBUTION OF GALAXIES
IN 3 DIMENSIONS
TOPOLOGY 3.
( REDSHIFT DISTRIBUTION )

ID. PARAMETERS (IN MPC) :800.00 , 800.00 , 800.00
CUT-OFF REDSHIFT= 0.50
NO. OF GALAXIES IN BASIC CELL=10.
TOTAL NO. OF GALAXY IMAGES = 217.
== RANDOM DISTRIBUTION OF GALAXIES IN BASIC CELL ==
ORIGINAL COORDINATES (IN MPC) :

258.68 , 326.71 , 201.36 ,
176.70 , 297.36 , 393.02 ,
338.32 , 153.87 , -298.33 ,
-154.14 , -30.72 , -215.99 ,
377.24 , 376.34 , -97.33 ,
-178.40 , 26.11 , 247.54 ,
144.21 , 179.49 , 345.96 ,
140.65 , -311.45 , 123.28 ,
60.23 , -310.63 , -344.55 ,
-86.61 , -353.53 , 224.49

HISTOGRAM OF        REDSHIFTS
----------------------------


0. - 10. %|
10. - 20.%| ■
20. - 30.%| ■■■
30. - 40.%| ■■■■■■■
40. - 50.%| ■■■■■■■
50. - 60.%| ■■■■■■■■■
60. - 70.%| ■■■■■■■■■■■■■■■
70. - 80.%| ■■■■■■■■■■■■■■
80. - 90.%| ■■■■■■■■■■■■■
90. - 100%| ■■■■■■■■■■■■■
ONE POINT CORRESPONDS TO 3. GALAXIES
ABOVE PERCENTAGE REFERS TO PERCENTAGE
OF MAX.        REDSHIFT : 0.50
```

Fig. 4.14. Galaxy distribution in a small universe with a toroidal topology [figure obtained by courtesy of G. Ellis]

Can such a set-up be distinguished from a slightly perturbed FL model, where a basic cell is repeated in a statistically average fashion only? Of basic importance is the identification length which must be larger than the scale on which distinct structures are still observed. The $\sim 10°$ structures in the CMB may require the basic scale to be $\sim 10^3$ Mpc, i.e. close to c/H_0 already. Then the small universe will almost be like the real, big one. Nevertheless, it is interesting to speculate about such more complicated global topologies.

Consider, for example, a $K = 0$ FL universe. The spatial sections are flat, and a rectangle may be chosen as a basic cell. The identification of opposite faces leads to a toroidal topology. Then all images of a galaxy up to a certain cut-off redshift z^* are seen. The different images correspond to the identical repetitions of the basic cell. But each time the galaxy is seen at different times in its history, and so at a different evolutionary stage. The redshifts, area distances, and absorptions (i.e. selection effects) are different for each image. The galaxy is seen each time from a different direction. Different distortion and lensing effects will affect each image. The identification of different images is a difficult problem. This is illustrated in Fig. 4.14 (courtesy of [Ellis 1986]), where the distribution in the sky of the images of galaxies is shown. The topology is that of a three-dimensional torus (opposite faces of the basic cell have been identified, but one pair has been rotated by 180°). There is no obvious pattern in the distribution, and the visual impression indicates an approximate isotropy.

This is an interesting proposal of how the approximate homogeneity and isotropy of the universe may be explained! The idea should also be attractive to observers, since in that case all the matter in the universe can be seen, even many times over.

4.3.3 The Mixmaster Universe

A theoretical investigation of the reasons for an FL universe, apart from appealing to initial conditions, is the suggestion [Misner 1968] that a viscous fluid model should be used for the description of matter in the early universe. For a large class of homogeneous and anisotropic models it can then be shown that in such "mixmaster" universes the anisotropy decays because of the action of the viscosity. This result is, however, not of general validity, because there are other classes of anisotropies that survive even damping by viscosity.

Another theory is the model of the "inflationary universe" (cf. Chap. 9), which attempts to explain the observed isotropy and homogeneity by the fact that within a generally quite chaotic universe, the past observable by us appears smooth and isotropic, because it is the interior of a bubble of a low-energy phase of the GUT vacuum. The probability of observing a boundary of such a bubble is very low.

4.4 The Classical Cosmological Tests

The normal procedure in classical cosmology, which we have followed so far, is to parametrize the cosmological solutions of Einstein's equations in terms of a few fundamental quantities. Then observations are used to obtain fits to these few parameters – for FL models with matter and radiation these are the quantities $(t_0, H_0, \Omega_0, \Omega_\Lambda, T_{\gamma,0}, K, R_0)$, with relations between them from the dynamical equations. This seems the best way to proceed, especially because there are so few cosmological observations. It is, however, of interest in principle to investigate how far the cosmological model is determined by observational relations. In order to get anywhere with that approach the models considered have to be considerably restricted. One approach which we have mentioned already in (4.1) starts out from the isotropic and homogeneous metric of the FL models, and examines in detail which observable relations could determine such a kinematic model. The dynamical equations of Einstein's field theory are not used, and thus their validity may even be tested in this approach. The main objective is to bring out clearly the information contained in the various observations, and the logical connections between them. Here I shall follow [Ehlers and Rindler 1987].

4.4.1 Observational Relations

The Robertson-Walker metric

$$ds^2 = dt^2 - R^2(t)(d\chi^2 + f^2 d\Omega^2)$$

$$\text{with} \quad f = \begin{cases} \sin\chi & \text{for } K = +1, \\ \chi & \text{for } K = 0, \\ \sinh\chi & \text{for } K = -1 \end{cases}$$

$$\text{and } d\Omega^2 = d\theta^2 + \sin^2\theta d\varphi^2$$

is used as in Chap. 1. This metric is isotropic, and therefore the information reaching us along the past light-cone is essentially determined by the information on one incoming ray of light (Fig. 4.15) Let us define $\chi = 0$ as the world-line of our galaxy, t_0 as the present time. The area of a sphere $f =$ const. at time t is $4\pi R^2(t) f^2(\chi)$. This leads to the definition of an area distance

$$r_A = R(t) f(\chi).$$

The calculation of $r_A(z)$ in terms of q_0, H_0, and z can be done analytically for $\Lambda = 0$ (see Chap. 1)

$$r_A = \chi_1 R_1 = H_0^{-1} q_0^{-2} (1+z)^{-2} \left\{ z q_0 + (q_0 - 1)\left[-1 + \sqrt{1 q_0 z + 1}\right]\right\}.$$

On the incoming null ray

$$\frac{d\chi}{dt} = -1/R.$$

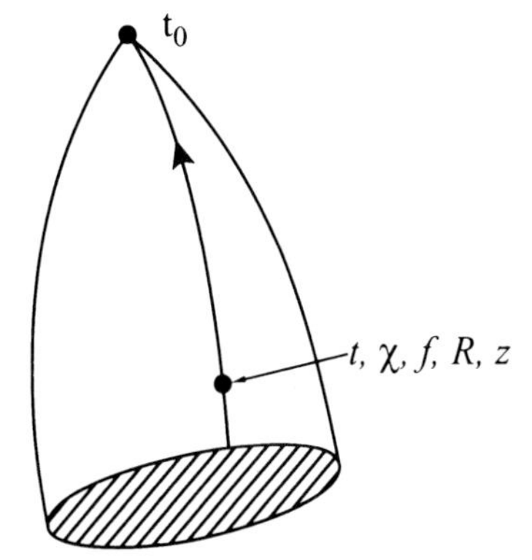

Fig. 4.15. An incoming ray of light on our past light-cone carries sufficient information, if the cosmological model is FL

Assume that the universe was expanding, $\dot{R}(t) > 0$, for some time $t \leq t_0$ at least. Then t, χ, R, z are strictly monotonic functions of each other in that range. $f(\chi)$ can be expressed as a function of any one of these variables.

Of these variables only $z = R_0/R - 1$ is given directly by observations of distant galaxies. Other directly observable quantities are the bolometric flux S, the solid angle ω subtended by the object, and $N(z), N(S)$, the number of galaxies with redshift $\geq z$ (or flux $\geq S$, respectively).

Each of the observables is related to a quantity describing an intrinsic property of the objects observed. S is related to the (bolometric) luminosity L of a standard galaxy, ω to the cross-sectional area A of a galaxy orthogonal to the line of sight, and N to the number n of galaxies per unit volume of space. L, A, and nR^3 change because of the evolution of galaxies. We consider $L(t), A(t)$, and $n(t)$ as unknown functions of cosmic time t.

Within the FL models the following relations hold:

$$S = L(1+z)^{-2} R_0^{-2} f^{-2}/4\pi; \tag{4.14}$$

$$\omega = A(1+z)^2 R_0^{-2} f^{-2}; \tag{4.15}$$

$$dN = nR_0^3 f^2 (1+z)^{-3} d\chi. \tag{4.16}$$

These relations are the basis for all the different empirical tests which we will mention briefly in Sect. 4.4.2.

From a given model we can determine $t(z), f(z)$. Then, if $L(t)$ were known, $L(t(z))$ can be calculated, and (4.14) gives a relation between S and z predicted by the model.

Similarly $N(z)$ and $\omega(z)$ relations may be derived. Let us reverse the question, and ask, whether observed relations $S(z), \omega(z)$, and $N(z)$ determine the cosmological model. The answer is no. Any three functions $S(z), \omega(z)$, and $N(z)$ are compatible with a given function $R(t)$, and a given curvature index K: the model $[R(t); K]$ determines $z(t), f(t)$, and finally $S(z)$ gives $L(t)$, ω gives $A(t)$, and $N(z)$ leads to $n(t)$. The given model is completed by the observable functions, but not restricted.

It is an interesting result that the situation is quite different if some detailed predictions from astrophysical theory for the evolution of galaxies (their sizes and luminosities) and for the change in their number density are accepted. $N(z)$ and $S(z)$ taken together do determine the metric, if $L(t)$ – the evolution with time of the intrinsic luminosity of a standard galaxy – and $n(t)$ – the true number density of galaxies in the past – are given by theory. Actually one of these functions is sufficient to determine the metric, and the other can be found from the model. The curvature K/R_0^2 can be found from the formula

$$\frac{K}{R_0^2} = \frac{20\pi}{3} \lim_{A \to 0} \frac{36\pi V^2 - A^3}{A^4}$$

where the volume V and surface area A of a geodesic sphere around the observer are calculated from the pairs $N(z), n(t)$ or $S(z), L(t)$.

If we can find K_0 independently, e.g. by measuring the parallax of one galaxy, then $[S, L]$ together with the observables $\omega(z), N(z)$ are sufficient to determine the model *and* the intrinsic properties $A(t), n(t)$.

This line of argument may also be an interesting way to test models of galaxy formation, once they become sufficiently detailed.

4.4.2 Cosmological Tests

The hope is to determine one favoured FL model from these tests. High-redshift objects are most suitable for this purpose.

The assumptions usually made are: a) a conservation law for the galaxies

$$n_G R^3 = \text{const.},$$

and b) the existence of a luminosity function

$$\Phi(L)dL = \Phi^*(L/L_*)^\alpha \exp(-L/L_*)d(L/L_*),$$

giving the number of galaxies with luminosity between L and $L + dL$. It is assumed that α is a constant, but a z-dependence of L_*. is taken into consideration: $L_* = L_*(z)$.

But the effects of galaxy evolution, the uncertainties in the assumptions made about star formation, as well as merging effects lead to $n_G R^3$ not being constant, to luminosity evolution, to $\alpha = \alpha(z)$. The tests are spoiled by these effects.

a) First we show the relation of time and redshift $t(z)$ for various FL models $(\Omega_0, \Omega_\Lambda)$; see Fig. 4.16.
b) The angle $\theta(z)$ under which a source of linear extent D at redshift z appears, is given by

$$\theta(z) = D(1 + z)R_0^{-1} f^{-1}.$$

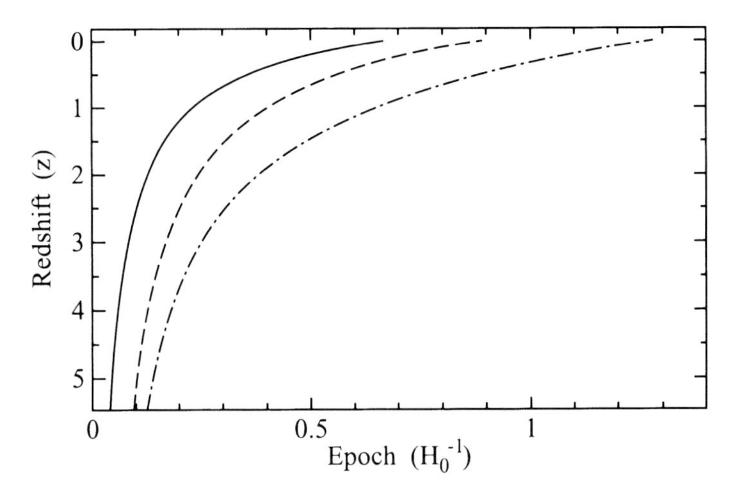

Fig. 4.16. The relation between cosmic time and redshift, $t(z)$, is shown for three FL models: $(\Omega_m, \Omega_\Lambda) = [1, 0]$ – *solid line*; $[0, 1, 0]$ – *dashed line*; $[0.1, 0.9]$ – *dot-dashed line*

This is really the test of (4.15) with $D = \sqrt{A}$, etc. For sources of standard size D, the angle $\theta(z)$ is plotted in Fig. 4.17. for three different FL models. To obtain $\theta(z), R_0 f(\chi)$ has to be computed in terms of z, H_0, q_0, Ω_0.
It has been claimed that certain short-lived jet structures in sources measured with very long baseline interferometry (VLBI) may define such a standard size. The claim was that the fit would lead to $\Omega_0 = 1, \Omega_\Lambda = 0$. Subsequent observations have failed to substantiate this.

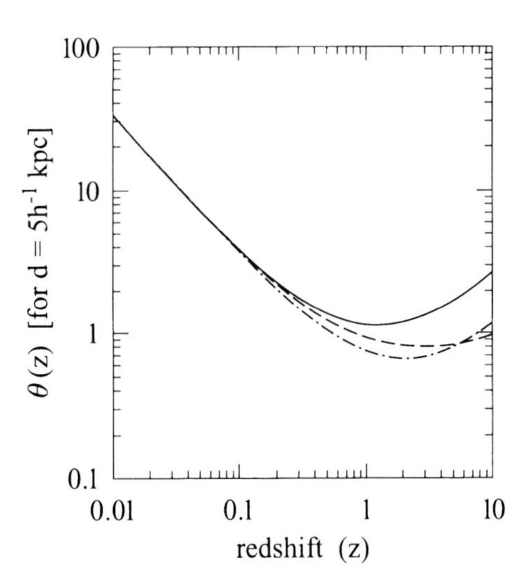

Fig. 4.17. The angle $\theta(z)$ for an object of linear extant D is shown for the same three FL models as in Fig. 4.16

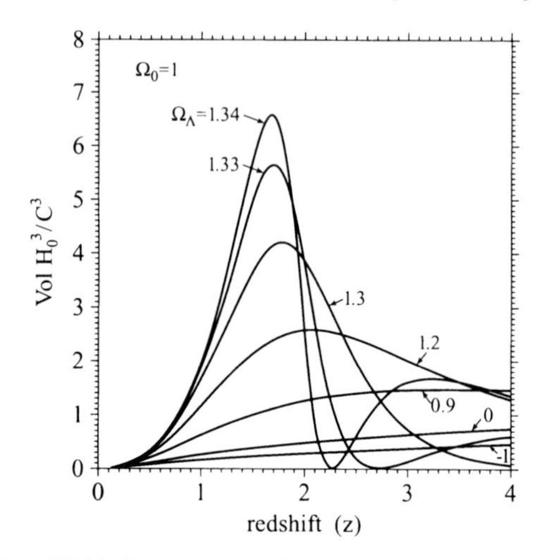

Fig. 4.18. dN/dz is shown for the same models as in Fig. 4.16

A similar test was tried for E galaxies. A photometrically defined radius r_E satisfies

$$\log \mathrm{r}_E = -0.58M_B + \text{const.}$$

To get r_E within a factor of 2, one needs to determine M_B to an accuracy of $\Delta M_B = 0.5$. So far there is no reasonable result from this test. For radio sources the unexpected behaviour of $\theta(z) \propto 1/z$ has been observed even for very high redshifts. This is the flat-space Newtonian result. It is strange that evolution and selection effects should conspire in a way which leads to $\theta(z) \propto 1/z$.

c) There is a strong dependence of dN/dz or dV/dz on the cosmological constant Ω_Λ, as can be seen from Fig. 4.18. There are claims that the counts of faint galaxies can be fitted much better by an $\Omega_\Lambda \neq 0$ model. Again it is not yet clear whether this result stands up to further investigation. Normalization problems and evolution, and in particular the assumption of constant $n_G R^3$, have to be clarified.

4.5 Concluding Remarks

Whereas the isotropy of the universe – at least around us – is established fairly well by the CMB observations, it is impossible to test the spatial homogeneity without precise knowledge of the intrinsic properties and structure of all astrophysical objects such as radio galaxies, quasars, etc. At present we have things the other way round: the assumption of an FL cosmology is used to derive limits on the physical parameters of radio sources. Even with local physical laws we find a similar situation: uncertainties in the microphysics reflect themselves in cosmological uncertainties, but the theory of element formation in FL universes (for example), together with the observations of light-element abundances (see Chap. 3) have been used to give limits on various local physical parameters such as the possible types of heavy leptons, neutrino masses and variations in the coupling constants. Elsewhere the abundances have been used to provide constraints for FL universes. But the same small set of numbers cannot be used to elucidate both local and cosmological physics. "One data point is one data point", and squeezing more than one bit of information out of it requires all the artistic skill of modern cosmologists.

One basic feature is the increasing uncertainty with increasing redshift, both because of source evolution, and because of the scarcity of measurable information (owing to the decrease of flux received and the decrease in the angle subtended by the objects).

The determination of the parameters of one particular FL model is also made difficult by these uncertainties. H_0, q_0, Λ and ϱ_0 are only known to within rather wide limits today. The major problem here is the determination of ϱ_0. Dynamical methods (see (2.4)) give estimates of galactic and cluster masses but they do not determine the density of a uniform mass-distribution, which could have gone unobserved. Therefore a lack of knowledge of the fundamental matter component in the universe adds to the uncertainty in ϱ_0.

Using only observations, therefore, we do not get very far. It is quite remarkable, on the other hand, that very far-reaching conclusions can be drawn, if the existence of the cosmological radiation background is supplemented by some reasonable hypotheses. I want to mention two such results.

The first is connected with the shape of the spectrum of the radiation [Hawking and Ellis 1973]. It is assumed that the radiation has been at least partially thermalized by repeated scattering, as indicated by the black-body nature of the spectrum, and by the exact isotropy at small angular scales. Therefore the amount of matter on each past-directed null geodesic from us must be sufficient to cause a sufficiently high opacity. It can then be shown that this amount of matter is enough to make our past light-cone reconverge. This then satisfies the conditions of a theorem that proves that there should be a singularity somewhere in the universe, provided Λ is not too large (see Chap. 1).

The second result [Ehlers et al. 1968] involves the assumption of the Copernican principle that we do not occupy a privileged position in space-time. It is assumed that the 3 K background would appear equally isotropic to any observer moving on a time-like geodesic, and that the photons have been freely propagating for a long distance – this is supported by observations. Then the universe has the Robertson–Walker metric.

But we should always remember that the detection threshold of cosmic signals limits our insight considerably. Just imagine the universe at a somewhat older stage, with an expansion factor $R(t)$ larger by a factor of 10.

Observations with present-day sensitivity would then fail to detect the CMB and the distant galaxies. Only a local system of galaxies could be detected. Surely in that case quite a different "standard model of cosmology" would be constructed.

Exercises

4.1 Calculate the Gaussian curvature K/R_0^2 from the pair $N(z), n(z)$ or from $L(z), S(z)$. Hint: Use the formula giving K/R_0^2 in terms of the volume and surface area of a geodesic sphere.

4.2 How can K/R_0^2 for an FL model be determined from the parallax of a single galaxy?

4.3 Consider possible equations of state for the matter in the spherically static model. Take a perfect fluid form for the energy–momentum tensor. Investigate the consequences of dust, radiation, or of collisionless particles as the matter content. Derive the relation between redshift and pressure changes

$$\frac{dz}{1+z} = \frac{dp}{\rho + p}.$$

Estimate the decoupling redshift at $T \sim 10^3$ K for dust and relativistic particles.

4.4 What cosmological test can be done with a set of (m_i, z_i) data to distinguish between the spherical static model described in this chapter, and a standard FL model?

4.5 Observations in a small-universe model can be described most conveniently by performing the relevant calculations in an associated FL model and then making the appropriate identifications on a fundamental cell. Show that for the simple case of a flat FL model, reduced to a small universe by appropriate identifications, the past light-cone of a given point intersects each world-line many times.

4.6 Compute the CMB anisotropy signal expected from a small universe model. Is there a difference from the standard flat FL model ($\Omega_m = 0.3$, $\Omega_\Lambda = 0.7$)?

Part II

Particle Physics and Cosmology

5. Gauge Theories and the Standard Model

"... the best reason for believing in a renormalizable gauge theory of the weak and electromagnetic interactions is that it fits our preconceptions of what a fundamental field theory should be like."

S. Weinberg

According to our present understanding of elementary particle physics four fundamental forces are responsible for the appearance of the physical world: electromagnetic, strong, weak, and gravitational force. These four interactions exhibit a remarkable hierarchical pattern.

The size of atoms, about 10^{-8} cm in the case of hydrogen, is in quantum mechanics fixed by the balance of electromagnetic forces between the positively charged protons in the atomic nucleus, and the negatively charged electrons in the hull. This is a universal property: hydrogen atoms in the ground state are of equal size. The nucleons, neutrons and protons, have their dimensions defined by the strong interaction which acts between the quarks, the fundamental constituents of the nucleons. The radius of the nucleons is determined to be $\sim 10^{-13}$ cm. On the scale of the atomic nucleus the electromagnetic interaction is unimportant, while the strong interaction controls the behaviour on these and on smaller scales, but is irrelevant for the structure of the atom.

The weak interaction is just a correction to these interactions in processes such as radioactive decay, or in all reactions where neutrinos take part. Up to the size of macroscopic solid bodies these interactions dominate. The gravitational force is negligibly small on these scales. Although the proton and the electron in the hydrogen atom attract each other by gravity too, this force is weaker by a factor of 10^{-39} than the electromagnetic force. Gravitation, on the other hand, is an attractive force for all particles, and it decreases slowly with distance, while strong and weak interactions are limited to nuclear dimensions and decay exponentially outside. The electromagnetic force is completely screened on average, because positive and negative charges attract each other, while equally charged particles repel each other. Therefore macroscopic bodies are electrically neutral, and the structure and motion of heavenly bodies is determined by the force of gravitation although it is negligible in laboratory experiments.

Particle physicists have succeeded to some extent in their attempts to understand this hierarchy of interactions. The idea that the different forces observed at present merge into one fundamental force at sufficiently high energies has led to a very successful derivation of the weak and electromagnetic interactions from a common origin.

This standard model can be considered as a first step towards a unified treatment of the different interactions of the elementary particles: a kind of unification is achieved for the electromagnetic and weak interactions [Weinberg 1967; Salam 1968], whereas the strong as well as the electroweak interactions are formulated as gauge theories.

The following sections attempt to explain the basic concepts involved in this scheme. Although there is no immediate cosmological implication of the electroweak model it is presented in some detail, since it exhibits many features that are important in more speculative and, if true, cosmologically more relevant theories (cf. Chap. 6).

The experiments carried out at CERN in 1983 gave support to the standard model and also to the chain of reasoning leading to it. There are, however, various fundamental assumptions which have not yet been tested by experiments.

In the following sections various simple examples serve to illustrate the concepts of gauge invariance, Yang–Mills theories and the Higgs mechanism of spontaneous symmetry-breaking. The description is semi-classical throughout – the arguments remain valid in each order of perturbation theory, but non-perturbative phenomena cannot be discussed. The Salam–Weinberg theory of the electroweak interactions and the gauge theory of the strong interactions (quantum chromodynamics) are briefly sketched. Quantum chromodynamics is a theory with an unbroken gauge invariance. The fundamental constituents, the quarks, behave like non-interacting particies (asymptotic freedom) at very high energies, and they also seem to have the interesting property of "confinement", i.e. the interaction force increases linearly with distance.

Finally, some shortcomings of the standard model are pointed out, which indicate that despite its successes this scheme cannot be the ultimate theory.

5.1 Introduction – The Concept of Gauge Invariance

The formulation of General Relativity by Einstein, a theory which gave a new and deep geometrical insight into the force of gravitation, was followed by attempts to find a unified geometrical description of electromagnetism and gravitation. In General Relativity the gravitational field is made into a geometrical object by the introduction of an affine connection, which defines the parallel transport of vectors, and therefore determines the relative orientation of local frames in space-time. As early as 1918, Hermann Weyl considered the question of whether this concept of geometrization could be generalized to other forces of nature. He proposed a general connection that defined a path-dependent comparison of the length of vectors at different space-time points [Weyl 1919]. This idea became known as scale or "gauge" invariance. However, because it seemed to contradict the absolute mass scales present in the real world, this approach fell into disrepute. The central idea of the local character of gauge symmetry has, however, proved very fruitful in recent years. (The resurrection was initiated by Weyl himself, and continued in a seminal paper by [Yang and Mills 1954].)

Many well-known symmetries, such as the Lorentz invariance of special relativity, are "global", in the sense that the same transformation is carried out at all space-time points. In contrast, the requirement of the invariance of a theory under transformations of a group G depending on the space-time point – "gauge transformations" – is more restrictive, and usually requires the introduction of new fields into the theory.

For example, the equivalence principle of Einstein's GR allows the introduction of local inertial frames, i.e., locally, at each point, gravity can be transformed away. GR has the property of local Lorentz invariance. We could speak of GR as a gauge theory with the homogeneous Lorentz group as the gauge group. We can also take the point of view that the theory of GR arises from the special theory of relativity by making the Lorentz group into a local gauge group. Similarly, any globally invariant theory can be generalized to a locally invariant theory. In general, however, this requires the introduction of new fields which participate in the dynamics – the so-called gauge fields. For GR those "gauge fields" are the "Vierbein fields". The coupling of the gauge fields to the matter fields is almost uniquely determined by the requirement of local invariance.

Let me illustrate these remarks with the example of electrodynamics (taking only the classical theory): the matter fields ϕ – electrons and positrons – satisfy the Euler equation of a Lagrange function) $L_M(\phi, \phi_{,\mu})$:

$$\frac{\partial L_M}{\partial \phi} = \partial_\mu \frac{\partial L_M}{\partial \phi_{,\mu}}$$

(this is just the Dirac equation for classical electrons and positrons if L_M is the Lagrangian for non-interacting particles). L_M is a Lorentz-invariant function of the fields ϕ, and in addition it is invariant under the global phase transformation

$$\phi(x) \rightarrow e^{-iq\epsilon}\phi(x) \tag{5.1}$$

(the integer q is the charge in units of the electric charge e). This global symmetry is given by the set of all phase factors $e^{-i\alpha}$. The phase factors can be multiplied and inverted, and they form a group, the one-dimensional unitary group $U(1)$. The group elements can be represented by numbers $U(\alpha) = e^{-i\alpha}$, where α is any real number between 0 and 2π.

There are infinitely many group elements, but only a one-dimensional parameter space: the interval of real numbers between 0 and 2π.

$U(1)$ is the simplest example of a Lie group (invented by Sophus Lie, 1842–1899), a continuous group with the property that the parameters of a product must be analytic functions of the parameters of each factor.

The elements of a Lie group are differentiable. Take $U(1)$, for example:

$$\frac{dU}{d\alpha} = \frac{d}{d\alpha}e^{-i\alpha} = ie^{-i\alpha} = e^{-i[\alpha+(\pi/2)]} = U[\alpha + (\pi/2)].$$

$dU/d\alpha$ is also an element of $U(1)$. The $U(1)$ symmetry leads to a conservation law $\partial_\mu J^\mu = 0$ for the Noether current

$$J^\mu = -ie\frac{\partial L_M}{\partial \phi_{,\mu}} q\phi.$$

A local gauge transformation, where ϵ depends on the space-time point x,

$$\phi(x) \to e^{iq\epsilon(x)}\phi(x), \tag{5.2}$$

will in general not leave the Lagrange function $L_M(\phi, \phi_{,\mu)}$ invariant. The natural way to obtain a locally gauge-invariant theory is to introduce an additional vector field A_μ, which transforms in such a way that the "covariant derivative" $\mathrm{D}_\mu\phi$,

$$\mathrm{D}_\mu\phi \equiv (\partial_\mu - ieqA_\mu)\phi, \tag{5.3}$$

transforms as ϕ. That is,

$$(\partial_\mu - ieqA'_\mu)e^{iq\epsilon(x)}\phi(x) = e^{iq\epsilon(x)}(\partial_\mu - ieqA_\mu)\phi(x) \tag{5.4}$$

and from this we find immediately

$$A'_\mu = A_\mu - \frac{1}{e}\partial_\mu\epsilon.$$

The vector potential A_μ, the photon field appears as the new gauge field, when $U(1)$ is made into a local gauge group. The new Lagrange function $L_M(\phi, \mathrm{D}_\mu\phi)$ is, by construction, invariant under the local gauge group $U(1)$. This Lagrange function also describes automatically the coupling of A_μ and ϕ through the $\mathrm{D}_\mu\phi$ terms. To complete the description, the "free Lagrangian"

$$L_F \equiv -\frac{1}{4}F_{\mu\nu}F^{\mu\nu} \tag{5.5}$$

with the fields

$$F_{\mu\nu} \equiv \partial_\mu A_\nu - \partial_\nu A_\mu \tag{5.6}$$

is added to L_M to give the total Lagrange function as $L = L_\mathrm{F} + L_\mathrm{M}$. $F_{\mu\nu}F^{\mu\nu}$ is (up to total divergences) the only Lorentz-invariant, gauge-invariant function that contains only first derivatives of A_μ.

When the representations of the fields with respect to $U(1)$ are given, there is only one free coupling constant, namely e (free, because it is not assigned a value within the theory). A mass term proportional to $m^2 A_\mu A^\mu$

would not be gauge-invariant. Hence as a classical object the gauge field A_μ has to be massless. We shall see shortly how this constraint of local gauge-invariance – a feature which for a long time shed doubt on the usefulness of gauge theories in the description of the weak or the strong interactions – can be loosened at the quantum level.

Besides the phase transformations typical of electrodynamics, it is possible to formulate theories where more complicated internal symmetries are made into local gauge groups.[1] An important example are the iso-spin transformations of nuclear physics. Here proton and neutron are viewed as two different states in an abstract "charge space" – an idea that is due to Heisenberg, who proposed considering the proton and the neutron as the "up" and "down" states of an abstract quantity, the isotopic spin, in close analogy with the ordinary spin states of the electron. The group operating in this isotopic spin space is $SU(2)$ (isomorphic to the usual rotation group), and this group leaves the strong nuclear force invariant. The different possible iso-spin states can thus be classified by analogy with the familiar mathematics of angular momentum. The identification of the π-meson as the particle that carries the nuclear force proved to be crucial to a deeper understanding: the charged and neutral π-mesons are the three components of a state with isotopic spin $T = 1$ (proton, neutron: $T = \frac{1}{2}$). The isotopic spin group could be used to determine the coupling of the π-meson to the nucleons, and thus nucleon-scattering processes could be calculated as exchange processes of a π-meson (Fig. 5.1).

In 1954 C.N. Yang and R. Mills [Yang and Mills 1954] made use of this analogy between the strong nuclear interactions and electromagnetism and proposed a gauge theory of nuclear interactions with $SU(2)$ as the local gauge group.

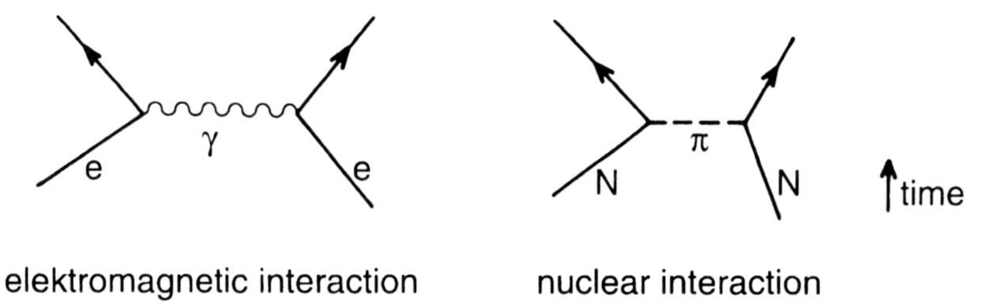

Fig. 5.1. The electromagnetic interaction between two electrons can be pictured as arising from the exchange of a photon. Similarly, the strong force between two nucleons is due to a meson. The Feynman graphs drawn here are a shorthand notation for the quantum-theoretical matrix elements describing such processes. We can also view these graphs just as pictures of certain reactions, keeping in mind that a computational rule is also involved (for a precise definition see any textbook on quantum field theory)

[1] The groups used in gauge theory, $U(n)$, $SU(n)$ $(n = 1, 2, \dots)$ are all "compact" Lie groups, i.e. they have a compact parameter space.

This idea changed the point of view of the identity of an elementary particle: the local isotopic spin symmetry allows one to choose arbitrarily a reference direction in the internal isotopic-spin space at each space-time point. With a given labelling at one point, it is clear that a rule is needed defining how to make a comparison with the choice at any other position. In electromagnetism the vector potential provides a connection between the phase values of the wave function at different positions. With $SU(2)$ as the gauge group instead of $U(1)$ the phase is replaced by the internal orientation of the isotopic spin. We shall describe in Sect. 5.2 some principal features of the Yang–Mills theory. The reason why these theories did not gain immediate acceptance is the fact that gauge invariance invariably implies that the quanta of the gauge fields have zero mass in a direct unsophisticated interpretation of perturbation theory. This is, of course, alright for quantum electrodynamics, but apart from the photon – and perhaps a hypothetical graviton – there are no mass-zero gauge particles in nature. On the other hand, a "non-abelian" gauge group such as $SU(2)$ brings into the theory a number of independent gauge fields equal to the dimension (3 for the rotation group). All these massless fields then lead to a long-range $1/r$ potential for the interactions that are described by the exchange of these particles. Thus, for instance, a gauge theory of electromagnetic and weak interactions would lead to a long-range weak interaction too, contrary to the experimental evidence, which gives an upper limit of 10^{-15} cm for the range of the weak forces. These difficulties were overcome in the mid-sixties when it was discovered that the "spontaneous breaking" of the gauge symmetry could lead to massive gauge bosons without destroying the gauge invariance of the Lagrangian. In the next sections I want to discuss some principal features of Yang–Mills theories, of spontaneous symmetry-breaking, and of unification schemes.

5.2 Yang–Mills Theory

The qualitative remarks of the last paragraph will now be presented in a bit more detail. Let the internal symmetry group be a Lie group G (e.g. $SU(2), SU(2) \times U(1), SU(3)$). The different matter fields (leptons, quarks, etc.) have transformation laws according to various irreducible representations of the group G. Different components ϕ_A are collected in one many-component field $\phi \equiv (\phi_A)$ transforming, in general, according to a reducible representation $U(g)$ of G. Global invariance of the Lagrange function $L_M(\phi, \partial_\mu \phi)$ under the group, i.e.

$$L_M(U(g)\phi), \partial_\mu(U(g)\phi) = L_M(\phi, \partial_\mu \phi)$$

for all elements g of G, leads to conserved currents in the usual way (Noether theorem).

The construction of a local gauge theory – which is invariant under transformations

$$\phi(x) \to (U(g)(x)\phi(x))$$

depending on the space-time points x – proceeds similarly to the electrodynamic case. Let $T_a \in \mathcal{G}$ be a basis of the Lie algebra $\mathcal{G}$ of G, and introduce a vector field

$$A_\mu(x) = \sum_a A_\mu^a(x) T_a;$$

$A_\mu(x)$ is a $\mathcal{G}$-valued vector field.[2,3]

Then replace $\partial_\mu \phi$ by the covariant derivative

$$\mathrm{D}_\mu \phi = \partial_\mu \phi + U_*(A_\mu)\phi, \tag{5.7}$$

where $U_*(A_\mu)$ denotes the representation of the Lie algebra corresponding to $U(g)$. As before, it is required that under gauge tranformations $\mathrm{D}_\mu \phi$ changes exactly as ϕ itself:

$$\mathrm{D}_\mu (U(g(x)))\phi(x) = (U(g(x)))\mathrm{D}_\mu \phi(x). \tag{5.8}$$

Here $\mathrm{D}\mu = \partial_\mu + U_*(A'_\mu)$ with the tranformed field A'_μ. Equation (5.8) can be translated into a transformation law for (A_μ):

$$\begin{aligned} &\partial_\mu (U(g(x)))\phi(x) + U_*(A'_\mu(x)_\mu)(U(g(x)))\phi(x) \\ &= (U(g(x)))\partial_\mu \phi(x) + U_*(A_\mu(x)_\mu)\phi(x). \end{aligned}$$

For this to hold for arbitrary $\phi(x)$ one must have

$$\partial_\mu U(g) + U_*(A'_\mu)U(g) = U(g)U_*(A_\mu).$$

Multiplying on the right by $U^{-1}(g)$,

$$U_*(A'_\mu) = U(g)U_*(A_\mu)U^{-1}(g) - (\partial_\mu U(g))U^{-1}(g).$$

For the identity representation $g \to g$ this reads

$$A_\mu(x) = g(x)A_\mu(x)g^{-1}(x) - (\partial_\mu g(x))g^{-1}(x). \tag{5.9}$$

For infinitesimal transformations $g(s,x) \equiv g(0,x) + s(d/ds)g(s,x)|_{s=0} = (e + sT(x))$, hence

$$\delta A_\mu^a T_a \equiv A'_\mu(x) - A_\mu(x) = [T(x), A_\mu(x)] - \partial_\mu T(x), \quad T \in G. \tag{5.10}$$

---

[2] The Lie algebra $\mathcal{G}$ is the tangent space of G at the unit element $e \in G$. In a parametrization $g(s) \in G$ with $g(0) = e$, the elements X of $\mathcal{G}$ are the derivatives of the group elements $X \equiv (d/ds)g(s)|_{s=0}$.

[3] The Lie algebra $SU(n)$ has a basis of $n^2 - 1$ (equal to the dimension of the group) linearly independent elements T_a.

If we write out (5.10) in component form, we find – using $T(x) \equiv \Sigma\epsilon^a(x)T_a(x)$ – that

$$\delta A_\mu^a\, T_a = [\epsilon^b T^b, A_\mu^c T_c] - \partial_\mu \epsilon^a T_a = [T_b, T_c]\epsilon^b A_\mu^c - \partial_\mu \epsilon^a\, T_a .$$

For the basis elements, the generators T_a of the Lie algebra, the commutation relations are

$$[T_b, T_c] = f_{bc}^a\, T_a$$

with the structure constants f_{bc}^a.[4] Then

$$\delta A_\mu^a(x) = f_{bc}^a \epsilon^b A_\mu^c(x) - \partial_\mu \epsilon^a(x). \tag{5.11}$$

The indices a, b, c refer to the components in the Lie algebra; μ is the space-time index.

Since ϕ and $\mathrm{D}_\mu\phi$ have identical transformation properties the global invariance of $L_M(\phi, \partial_\mu\phi)$ implies the invariance of $L_M(\phi, \mathrm{D}_\mu\phi)$ under local gauge transformations. This Lagrangian contains via $\mathrm{D}_\mu\phi$ the coupling of the gauge fields to the matter fields.

The coupled system of gauge fields A_μ and matter fields ϕ is completely described when a free Lagrangian for the A_μ has been specified.

It can be shown that a Lorentz- and gauge-invariant form of L_F is

$$L_F = -\frac{1}{4} F_{\mu\nu} \cdot F^{\mu\nu} . \tag{5.12}$$

The difference between this and the electromagnetic L_F is that the "·" represents an invariant inner product with respect to the Lie algebra $\mathcal{G}$ and that $F_{\mu\nu}$ is defined as

$$F_{\mu\nu} = \partial_\mu A_\nu - \partial_\nu A_\mu + [A_\mu, A_\nu] \tag{5.13a}$$

or, in component form,

$$F_{\mu\nu}^a = \partial_\mu A_\nu^a - \partial_\nu A_\mu^a + f_{bc}^a\, A_\mu^b A_\mu^c . \tag{5.13b}$$

The total Lagrangian is thus

$$L = L_M(\phi, \mathrm{D}_\mu\phi) + L_F .$$

Even for $L_M \equiv 0$ a non-trivial theory results; as G is in general a non-abelian group the product $F_{\mu\nu} \cdot F^{\mu\nu}$ contains cubic $(A_\mu)^3$ and quartic $(A_\mu)^4$ self-coupling terms. Interesting exact solutions for such a pure Yang–Mills theory have been found.

Since terms such as $m^2 A_\mu A^\mu$ are not gauge-invariant, there is no possibility of introducing a mass term for the gauge bosons into the Lagrangian L_M

[4] The structure constants in turn uniquely determine the generators.

without losing the gauge invariance. The gauge bosons are thus expected to be mass-zero fields. We shall, however, see in Sect. 5.3 that notwithstanding the absence of mass terms in the Lagrangian, the physical gauge-boson states can be massive if the gauge invariance is "hidden" (or spontaneously broken).

An explicit example might be instructive here; take the Yang–Mills theory with the non-abelian group $SU(2)$ ($2^2 - 1 = 3$ free parameters) [Chaichian and Nelipa 1984]. The global invariance under $SU(2)$ transformations can be read off immediately from a Lagrange function for the matter fields ψ^a,

$$L_M = \frac{i}{2}(\overline{\psi}_a \gamma_\mu \partial^\mu \psi^a - \partial_\mu \overline{\psi}_a \gamma_\mu \psi^a) - M\overline{\psi}_a \psi^a. \tag{5.14}$$

ψ^a ($a = 1, 2$) are the components of a spinor iso-doublet with respect to $SU(2)$, $\psi \equiv (\psi^1, \psi^2)$, where ψ^1 and ψ^2 may be viewed as describing a neutron and a proton respectively. ψ^1, for instance, is a Dirac spinor, not a complex number (γ^μ are the Dirac matrices, cf. [Chaichian and Nelipa 1984. p. 11]).

The functions ψ transform under $SU(2)$ transformations as

$$\begin{aligned} \psi^a \to \psi'^a &= exp[(-i/2)\epsilon_k \tau_k]^a_b \psi^b, \\ \overline{\psi}_a \to \overline{\psi}'_a &= \psi_b \exp[(-i/2)\epsilon_k \tau_k]^a_b, \end{aligned} \tag{5.15}$$

where ϵ_k denotes the (constant) parameters of the group (three free parameters), and the matrices $(-i/2)\tau_k$ are the infinitesimal generators of $SU(2)$; in the (2×2) representation considered here, τ_k are the Pauli matrices

$$\tau_1 = \begin{pmatrix} 0 & 1 \\ 1 & 0 \end{pmatrix}; \quad \tau_2 = \begin{pmatrix} 0 & -i \\ i & 0 \end{pmatrix}; \quad \tau_3 = \begin{pmatrix} 1 & 0 \\ 0 & -1 \end{pmatrix}. \tag{5.16}$$

The law for infinitesimal transformations is

$$\begin{aligned} \delta\psi^a &= \frac{-i}{2}\epsilon_k (\tau_k)^a_b \psi^b, \\ \delta\overline{\psi}_a &= \frac{i}{2}\epsilon_k \overline{\psi}_b (\tau_k)^b_a. \end{aligned}$$

Therefore the generators of the transformations are represented as

$$T^a_{kb} = -\frac{i}{2}(\tau_k)^a_b$$

with commutation relations

$$[T_k, T_i] = -\frac{1}{4}[\tau_k, \tau_i] = \epsilon_{klm} T_m$$

and structure constants

$$f_{klm} = \epsilon_{klm}.$$

(ϵ_{klm} is the totally antisymmetric tensor, $\epsilon_{123} = +1$.)

With these properties of the global symmetry in mind, the local gauge invariance can now easily be implemented; the derivative $\partial_\mu \psi^a$ (according to (5.7)) has to be replaced by

$$\mathrm{D}_\mu \psi^a \equiv \partial_\mu \psi^a + \frac{i}{2}(\tau_k)^a_b A^k_\mu \psi^b. \tag{5.17}$$

The triplet of vector fields A^k_μ is the gauge field. The Lagrangian for the gauge fields has the form.

$$L_F = -\frac{1}{4} F^a_{\mu\nu} \cdot F^{a\mu\nu},$$

where

$$F^a_{\mu\nu} \equiv \partial_\mu A^a_\nu - \partial_\nu A^a_\mu - \frac{1}{2}\epsilon_{abc}(A^b_\mu A^c_\nu - A^b_\nu A^c_\mu).$$

The gauge fields in L_F are self-interacting – $(A^k_\mu)^3, (A^k_\mu)^4$ terms appear. The A^k_μ terms transform as follows:

$$\delta A^k_\mu = \epsilon_{klm} A^m_\mu \epsilon_l(x) + \partial_\mu \epsilon_k(x).$$

The total, locally invariant, Lagrangian is

$$L \equiv \frac{1}{g^2} L_F + \frac{i}{2}\overline{\psi}\gamma^\mu D_\mu \psi - M\overline{\psi}\psi = L_F + \frac{i}{2}(\overline{\psi}\gamma^\mu \partial_\mu \psi - \partial_\mu \overline{\psi}\gamma^\mu \psi) - M\overline{\psi}\psi$$
$$-g\frac{i}{2}\overline{\psi}_a \gamma^\mu T^a_{kb}\, A^k_\mu \psi^b. \tag{5.18}$$

The coupling constant g has been introduced by multiplying the free Yang–Mills Lagrangian by $1/g^2$. Replacing A_μ by gA_μ transforms L into the form generally in use.

5.3 Spontaneous Symmetry-Breaking

A consequence of formulating locally gauge-invariant theories is the appearance of massless gauge fields. But for a realistic description of the short-range interactions massive gauge fields are needed. The explicit gauge symmetry of the Lagrangian ensures that the theory is renormalizable, i.e. that well-defined non-divergent perturbation calculations can be done. Therefore the gauge invariance must not be destroyed by adding non-symmetric terms to the Lagrangian. This dilemma seemed for many years like a fatal flow in Yang–Mills theories, until in the 1960s a scheme was developed which in semi-classical approximation can be described as follows: the exact invariance of the Lagrangian is kept, but the symmetry is not realized in the ground state of an additional (Higgs) field of the system. This phenomenon has been called "spontaneous symmetry-breaking "[Higgs 1964,1965].

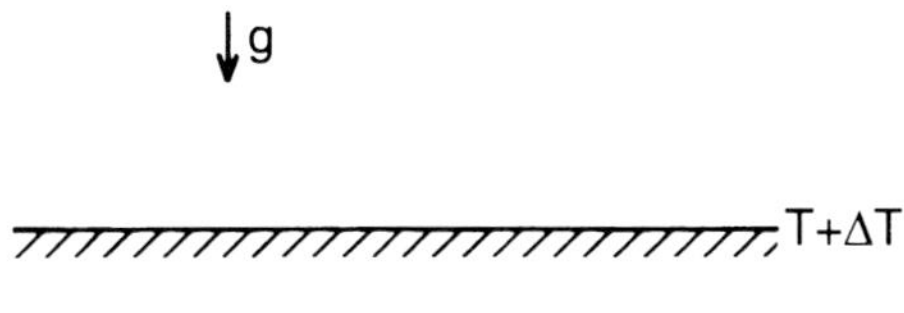

Fig. 5.2. Two horizontal plates at different temperatures T and $T + \Delta T$, respectively, in a vertical gravitational field

Examples of spontaneous symmetry-breaking already exist in classical physics: consider a stationary, rigidly rotating, ideal fluid under its own gravitational field (perhaps an idealized "planet"). Since the problem is symmetrical with respect to the rotational axis of the fluid, all the equations show this symmetry too. One would expect the same symmetry to show up in the equilibrium configurations. But, as Jacobi showed in 1834, there are exceptions: all equilibrium configurations are ellipsoids, but besides the axially symmetrical ellipsoids (described by MacLaurin) there are triaxial ellipsoids – asymmetrical solutions – which branch off from the sequence of the symmetrical solutions at a critical rotation period. Beyond the critical point the "asymmetrical" configurations are stable, the symmetrical ones not [Chandrasekhar 1968].

Another instructive example is the Benard problem. Consider a fluid contained between two plane, horizontal plates in the Earth's gravitational field (Fig. 5.2). The obvious idealization is to make the plates infinitely extended, the gravitational force acting in the vertical direction. Let the two plates have different temperatures, with the lower plate at a higher temperature $T + \Delta T$. For sufficiently small values of ΔT the fluid stays at rest, only heat conduction being important. If, however, ΔT increases beyond a critical value, then the translationally symmetrical state of the fluid at rest becomes unstable, and convection sets in. This convection has a peculiar pattern: in horizontal directions the motion is periodic, the space between the two plates is divided up into regular cells, and in each cell the fluid streams in the same way. The horizontal cross-sections of these cells form a 2-dimensional horizontal lattice (a spontaneous breakdown of translation invariance!). Experimentally one sees a hexagonal pattern (Fig. 5.3).

In the centre of the hexagonal prisms the fluid is streaming upwards, at the walls downwards. Further increase in ΔT produces an instability of the stationary cell pattern, and a transition to turbulence.

These classical examples illustrate the general aspects and properties of a spontaneously broken (or hidden) symmetry. The next step is to introduce part of the vocabulary that has become common in quantum theory with respect to spontaneous symmetry breaking. A quantum mechanical system, described by the Hamiltonian H – or equivalently by the Lagrangian L – can be in various states Ψ_n with energies E_n:

$$H\Psi_n = E_n\Psi_n.$$

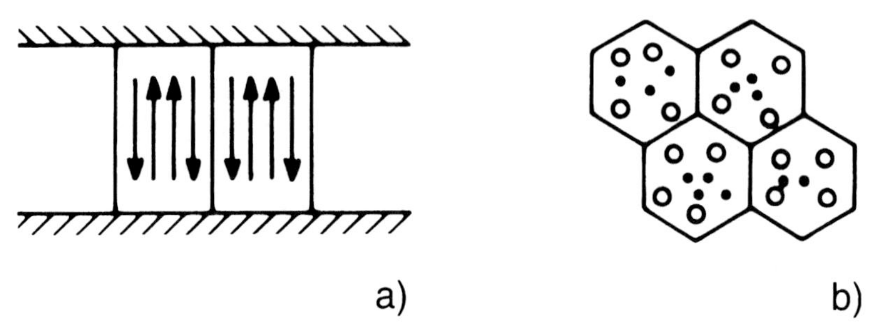

Fig. 5.3. Schematic cell pattern for Benard convection: (**a**) side view; (**b**) horizontal cross-section

Each state is specified by the wave function Ψ_n and by the value of the energy E_n. The state of minimum energy E_0 with wave function Ψ_0 is called the ground or vacuum state. If a single, unique state Ψ_0 exists, the vacuum is non-degenerate, otherwise it is called degenerate.

The vacuum state is invariant under a symmetry group G, if it is transformed into itself under the group G; it is non-invariant otherwise. In the frame work of local relativistic quantum-field theory there exists a connection between the invariance of the vacuum state under a group of transformations and the invariance of the Lagrangian under the same group (Fig. 5.4).

If the vacuum state is invariant, the Lagrangian must necessarily be invariant too ("the invariance of the vacuum state is the invariance of the universe"). This is the case of an "exact symmetry". If the vacuum state is non-invariant, L may be either invariant or non-invariant. In both cases an exact symmetry is not realized – the symmetry as a whole is broken. The term "explicit symmetry-breaking" is used in the case of a non-invariant vacuum state and a non-invariant Lagrangian. When only the vacuum state is non-invariant, but the Lagrangian still shows the symmetry, the expression "spontaneous symmetry-breaking" is used.

A simple field theory model will now be used to demonstrate the spontaneous breaking of a symmetry ([Chaichian and Nelipa 1984]). Consider a massive, self-coupled neutral scalar field with the Lagrangian

$$L = \frac{1}{2}\partial^\mu\phi\partial_\mu\phi - \frac{1}{2}\mu^2\phi^2 - \frac{1}{4}\lambda\phi^4, \tag{5.19}$$

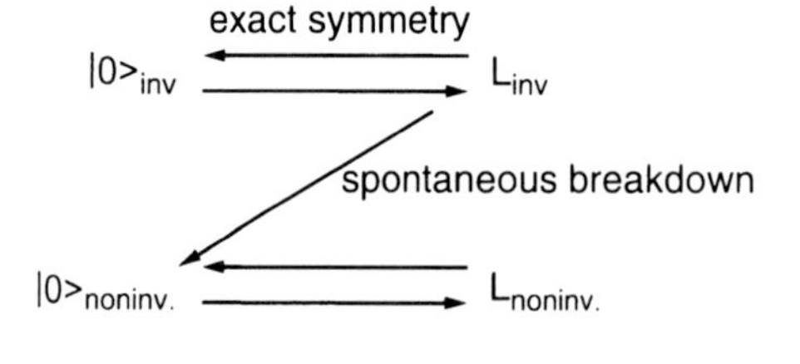

Fig. 5.4. Schematic diagram illustrating the relations between symmetries of the Lagrangian and symmetries of the vacuum $|0\rangle$

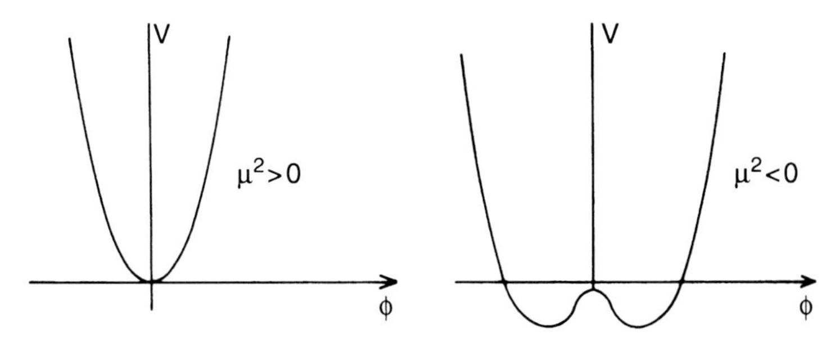

Fig. 5.5. The self-interaction $V(\phi)$ for $\mu^2 > 0$ and $\mu^2 < 0$. At the critical parameter value $\mu^2 = 0$ a degenerate state of lowest energy appears

symmetric under the action $\phi \to -\phi$ of the group Z_2. The "potential"

$$V(\phi) = \frac{1}{2}\mu^2\phi^2 + \frac{1}{4}\lambda\phi^4 \tag{5.20}$$

has different shapes in the cases $\mu^2 > 0$ and $\mu^2 < 0$ (Fig. 5.5). For $\mu^2 > 0$ there is only one minimum of $V(\phi)$, namely at $\phi = 0$. This corresponds to the case of a stable, non-degenerate, unique ground state $\phi = 0$. In the case of $\mu^2 > 0$, on the other hand, the minima of $V(\phi)$ are at

$$\phi = \pm\phi_0, \left(-\frac{\mu^2}{\lambda}\right)^{1/2} = \phi_0 > 0. \tag{5.21}$$

It is possible to choose either one of these values as the ground state, but not both at the same time. To emphasize: different vacuum states correspond to different worlds.

Each one of the two ground states breaks the symmetry $\phi \to -\phi$. This really is a classical field theory model for Buridan's ass. If the symmetry

Buridan's ass

is preserved, the ass will starve. Choosing one pile of hay to feed on is a spontaneous breaking of the symmetry.

We proceed to study the solutions near one ground state – say $\phi_0 > 0$. For this purpose we write $\phi' = \phi - \phi_0$, and use ϕ' as the new, physically relevant variable. $\phi = 0$ is now unstable, $\phi' = 0$ is the new, stable minimum. The theory restricted to $+\phi_0$ is no longer symmetrical with respect to the transformation $\phi \to -\phi$. Solutions near ϕ_0 are not mapped into solutions close to the same ϕ_0. Rewrite the Lagrangian (5.19) in terms of ϕ':

$$L = \frac{1}{2}(\partial^\mu \phi')(\partial_\mu \phi') + \mu^2 \phi'^2 - \lambda \phi_0 \phi'^3 - \frac{1}{4}\lambda \phi'^4. \tag{5.22}$$

Here a mass term for ϕ' has appeared which has the correct signature ($(\text{mass})^2 = -2\mu^2$). The symmetry $\phi \to -\phi$ of the Lagrangian can no longer be immediately read off from (5.22); it is still present – hidden, but not lost.

In general the global symmetry is a continuous group of transformations, not just the discrete reflection $\phi \to -\phi$ as in the example above. If such a general global symmetry is spontaneously broken, particles of spin zero and mass zero occur, the so-called Goldstone bosons [Ezawa and Swieca 1967]. (The simple example of a Lagrangian of two real fields σ and π with self-interaction $V \equiv \frac{1}{2}\mu^2(\sigma^2 + \pi^2) + \lambda(\sigma^2 + \pi^2)^2, \lambda > 0$, can be used to illustrate the Goldstone theorem on the semi-classical level [Chaichian and Nelipa 1984].)

These massless bosons do not seem to show up in particle physics experiments. They are considered an undesirable consequence of spontaneous symmetry-breaking. But in solid-state physics the occurrence of Goldstone bosons is a well-known phenomenon: the spontaneous symmetry-breaking in a ferromagnet, for example, is connected with the occurrence of spin waves – these so-called "magnons" are the Goldstone bosons (e.g. [Goldstone 1961]; also [Kittel 1976]).

There is an important exception to the Goldstone theorem: if the symmetry of a local gauge group is broken spontaneously the Goldstone bosons partially acquire mass. This so-called Higgs phenomenon will be discussed in the next section.

5.4 The Higgs Mechanism

Consider the simple example of a complex scalar field $\phi = \phi_1 + i\phi_2$ with a Lagrangian

$$L = \phi^* \partial^\mu \phi - V(\phi), \tag{5.23}$$

$$V(\phi) = \frac{\mu^2}{2}(\phi^* \phi) + \frac{\lambda}{4}(\phi^* \phi) \tag{5.24}$$

("*" denotes complex conjugation). This Lagrange function is invariant under global $U(1)$ transformations $\phi \to -e^{-i\epsilon}\phi$. If we "gauge" this symmetry, i.e. require L to be invariant under the local $U(1)$ group,

$$\phi(x) \to e^{-i\epsilon(x)}\phi(x), \tag{5.25}$$

we have to replace $\partial_\mu \to \partial_\mu + ieA_\mu$ with a gauge field $A_\mu(x)$. Under $U(1)$ gauge transformations

$$A_\mu(x) \to A_\mu(x) - \frac{1}{e}\partial_\mu \epsilon(x), \tag{5.26}$$

and the invariant Lagrangian is of the form

$$L = -\frac{1}{4}F_{\mu\nu} \cdot F^{\mu\nu} + (\partial_\mu + ieA_\mu)\phi^*(\partial^\mu - ieA_\mu)\phi - V(\phi), \tag{5.27}$$

$$F_{\mu\nu} \equiv \partial_\mu A_\nu - \partial_\nu A_\mu. \tag{5.28}$$

For $\mu^2 > 0$ the theory is just scalar electrodynamics (photons and massive scalar particles). The spontaneously broken $U(1)$ symmetry occurs for $\mu^2 < 0$ (Fig. 5.6).

The minima of $V(|\phi|)$ lie on the circle $|\phi| = \phi_0$,

$$\phi_0 = \left(\frac{-\mu^2}{\lambda}\right)^{1/2}.$$

A specific choice of a minimum $\phi = \phi_0$ defines a ground state, and a "physical" field $\phi' = \phi - \phi_0$ can be introduced. Furthermore, to obtain a particle interpretation a special gauge is chosen, namely $\phi' = \rho$, where ρ is a real field ($\rho > 0$). (This is possible, as the Lagrangian (5.27) is locally gauge invariant.)

Up to now different steps have been taken: first the local gauge invariance has been spontaneously broken by choosing a particular local minimum $\phi = \phi_0$. The Lagrangian $L(A_\mu, \phi')$ is still gauge-invariant. But in a second step, a special gauge has been chosen (namely $\phi' = \rho$) to give an interpretation to the various terms in the Lagrangian. In this special gauge the Lagrangian reads

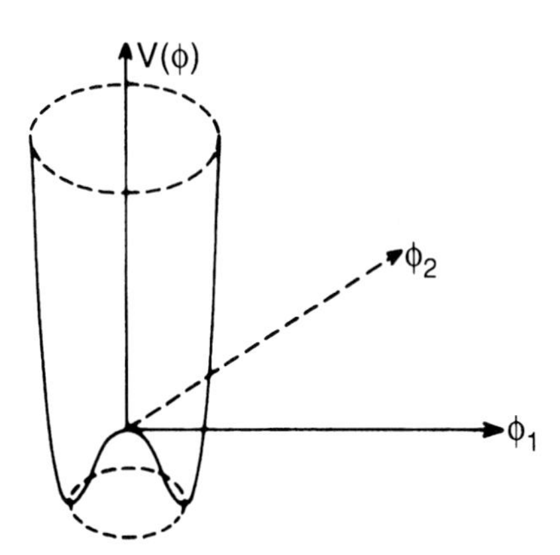

Fig. 5.6. The potential $V(\phi^*\phi)$ for $\mu^2 < 0$

$$L = \frac{1}{4}F_{\mu\nu}F^{\mu\nu} + \frac{1}{2}\partial_\mu\varrho\partial^\mu\varrho + \frac{1}{2}e^2\phi_0^2A_\mu A^\mu$$
$$+\frac{1}{2}e^2A_\mu A^\mu\varrho(2\phi_0 + \varrho) - \frac{1}{2}\varrho^2(3\lambda^2\phi_0^2 + \mu^2)$$
$$-\lambda\phi_0\varrho^3 - \frac{1}{4}\lambda\varrho^4. \tag{5.29}$$

The quadratic terms in (A, ρ) are diagonal in this gauge, and it may be read off from (5.29) that there is a real, scalar particle ρ with $(\text{mass})^2 = 3\lambda^2\phi_0^2 + \mu^2$, and a massive vector boson A_μ with mass $= |e|\phi_0$.

Masses of particles can only be identified rigorously by an analysis of the n-point functions (LSZ formalism, see e.g. [Bjorken and Drell 1965]). In the semi-classical picture presented here, however, the fundamental fields correspond to particles, and the coefficients of the quadratic terms in the Lagrangian to particle masses.

One of the two degrees of freedom ϕ_1', ϕ_2' in the original complex field ϕ has been removed by the specific choice of a gauge. (One has derived a new theory with a fixed gauge.) But this special choice has provided an additional component for A_μ: the system ϕ_1' (1 degree of freedom), $\phi_2'(1)$, A_μ (2 transverse components, massless) has been changed in the unitary (physical) gauge into $\rho(1)$, A_μ (3, massive). The particle physicist's jargon for this is "the would-be Goldstone particle has been eaten by A_μ, and therefore A_μ has become massive". The non-physical Goldstone boson has been exchanged for the physical state of the vector boson with longitudinal polarization. This remarkable property of gauge fields is called the "Higgs mechanism"; the scalar particles described by the field ϕ_1' or ρ are referred to as "Higgs bosons".

We should again stress that this discussion is on the classical level. If the Higgs field is treated as a quantum object, it may no longer be possible to fix the ground state by $\phi = \phi_0$. Quantum fluctuations of the field can be expected to make all the other ground states $|\phi| = \phi_0$ accessible (cf. the Mermin–Wagner theorem [Mermin and Wagner 1966; Mermin 1967]).

Within the semi-classical picture, two essential points should have become clear with these examples:

i) There exists a critical point (here $\mu^2 = 0$) which determines whether or not the symmetry is spontaneously broken.
ii) The change of the sign of the term $-\frac{1}{2}\mu^2\phi^*\phi$ in the Lagrangian (5.27) may appear as a manipulation which introduces particles with imaginary mass. A term $\frac{1}{2}m^2\phi^*\phi$ in the Lagrangian corresponds, however, to a mass term only if the state $\phi = 0$ is the stable equilibrium position, i.e. the global minimum of the potential energy. For $\mu^2 < 0$ the potential energy $V(|\phi|)$ has a local maximum at $\phi = 0$. At $\phi = 0$ the system is unstable, whereas at $\phi = \phi_0$ it is stable. Thus $\phi - \phi_0$ is the physically relevant field, rather than ϕ itself.

The example described above (which is due to Higgs) has a close analogy in the theory of superconductors. Many of the salient features of superconductivity, such as the Meissner effect and flux quantization, can be understood by describing the electrons in the superconductor as a single coherent system which behaves like a self-interacting quantum-mechanical particle. The microscopic, rigorous description of how the electrons form a coherent macroscopic system was given by Bardeen, Cooper, and Schrieffer (the BCS theory [Bardeen et al. 1957]). (A very readable qualitative, semi-classical account can be found in [Weisskopf 1981a]). Loosely speaking, the interaction of the conduction electrons with the atomic lattice of the superconductor produces an attractive force between two electrons. For sufficiently small electron energies this attractive force becomes larger than the normal Coulomb repulsion and binds electrons together into "Cooper pairs". In a pair, the spins of the two electrons are oriented in opposite directions so that a Cooper pair acts like a spin-zero boson with two units of negative charge. The attractive, binding force is small, and therefore the effective size of a Cooper pair is very large, about 10^{-4} cm. Every Cooper pair overlaps with approximately 10^6 other Cooper pairs in a typical superconductor. Those superpositions produce strong correlations and coherence effects over macroscopic scales. The interaction of the electrons in the superconductor, with an external magnetic field, has been successfully described by the phenomenological Landau–Ginzburg theory. Within this theory there is an order parameter $\phi(\boldsymbol{x})$, a complex function of the space point $\boldsymbol{x}$. In the fundamental BCS theory this order parameter is proportional to the ground-state expectation value of the electron-pair field $\phi(x) \equiv \langle \Psi_\uparrow(\boldsymbol{x})\Psi_\downarrow(\boldsymbol{x})\rangle$ ($\downarrow\uparrow$: opposite spins):

$$
\begin{aligned}
L &= \frac{1}{4}F^2 + \frac{1}{2}|\mathrm{D}\phi|^2 + V(\phi), \\
V &\equiv \frac{\lambda}{8}(|\phi|^2 - v^2)^2 + \text{const.}\,; \qquad V(0) = 0,
\end{aligned}
\tag{5.30}
$$

hence

$$
\text{const.} = -\frac{\lambda}{8}v^4.
$$

L is the free energy of the system with external magnetic field $\boldsymbol{B} = \nabla \times A$ and the current $\boldsymbol{J}$ is given by

$$
\boldsymbol{J} = \frac{e\hbar}{im}\left\{\phi^*\left(\nabla - \frac{2ie}{\hbar}\boldsymbol{A}\right)\phi - \phi\left(\nabla - \frac{2ie}{\hbar}\boldsymbol{A}\right)\phi^*\right\}. \tag{5.31}
$$

This semi-phenomenological theory [Ginzburg and Landau 1950] describes the free energy density of the superconductor as an effective potential V, depending on two temperature-dependent parameters $\lambda(T)$ and $v^2(T)$. The Landau–Ginzburg equations (Euler–Lagrange equations of (5.30)) are

$$\nabla \times \boldsymbol{B} = 4\pi \boldsymbol{J},$$

$$\frac{1}{2m}\left(\frac{\hbar}{i}\nabla - \frac{2e}{m}\boldsymbol{A}\right)^2 \phi - \frac{\lambda}{4}v^2\phi + \frac{\lambda}{4}|\phi|^2\phi = 0. \tag{5.32}$$

The superconducting phase is the non-trival solution of (5.32) for $\boldsymbol{B} = 0$. Looking for a solution $\phi = \phi_0 = \text{const.} \neq 0$ the equations (5.32) require $\lambda(\phi_0 - v^2)/4 = 0$. The potential $V(\phi)$ therefore has exactly the same form as the Higgs potential of the preceding example, for the case of spontaneous symmetry-breaking. The ground state of the superconducting phase is asymmetrical with respect to the $U(1)$ abelian gauge group of the Lagrangian (5.30).

The first equation (5.32) leads, for $\boldsymbol{B} \neq 0, \phi_0 = \text{const.}$, to

$$\nabla \times (\nabla \times \boldsymbol{A}) = 4\pi \boldsymbol{J} = 4\pi \left(-\frac{4e^2}{m}\right) \boldsymbol{A}|\phi_0^2|,$$

$$\nabla \times (\nabla \times \boldsymbol{A}) = -\frac{1}{\Lambda^2}\boldsymbol{A}.$$

The introduction of appropriate boundary conditions leads to an exponential fall-off of the solution in the superconductor with a characteristic scale

$$\Lambda = \left(\frac{mc^2}{16\pi e^2}|\phi_o|^{-2}\right)^{1/2} = \left(\frac{mc^2}{16\pi e^2 v^2}\right)^{1/2}. \tag{5.33}$$

Λ is the so-called "London penetration depth". Its value is typically $(3-5) \times 10^{-6}$ cm.

Thus the magnetic field decreases rapidly within the superconductor. This is the Meissner effect – the generation of a short-range field without violating the gauge invariance of the equations. The quantity $1/\Lambda$ corresponds to a "mass" of the vector potential, acquired in the superconducting phase, i.e. in the case of spontaneous symmetry-breaking.

For readers at home with group theory, I shall give a few short remarks on the general case of the Higgs mechanism [Straumann 1984b]. For a general gauge theory let the gauge group be $G = G_1 \times G_2 \times \ldots \times G_n$ with Lie algebra $\mathcal{G}$ and coupling constants $g_1, \ldots, g_n$. The Lagrangian has the structure

$$L = -\frac{1}{4}\frac{1}{g^2}F_{\mu\nu}F^{\mu\nu} + \frac{1}{2}(\mathrm{D}_\mu\phi, \mathrm{D}^\mu\phi) - V(\phi) + \text{ terms with other fields.}$$

As usual $\mathrm{D}_\mu\phi$ is the covariant derivative

$$\mathrm{D}_\mu\phi = (\partial_\mu - iA_\mu^a\, T_a)\phi, \qquad T_a \in \mathcal{G}$$

(the coupling constants g_i are absorbed in the gauge fields A_μ^i).

Let the potential $V(\phi)$ be G-invariant, and v be a local minimum of $V(\phi)$:

$$DV(v) = 0 \qquad \mathrm{D}^2V(v) > 0.$$

$H \in G$ is the stabilizer of v ($Hv = v$) with corresponding Lie algebra $\mathcal{H}$. Without the gauge fields A_μ we would expect $(\dim \mathcal{G} - \dim \mathcal{H})$ massless Goldstone bosons to occur. The change induced in a gauge theory can be seen from a rewriting of the Lagrangian in terms of the field $\phi' = \phi - v$:

$$L = -\frac{1}{4g^2}F^2 + \frac{1}{2}(\mathrm{D}_\mu\phi', \mathrm{D}^\mu\phi') + (\mathrm{D}^\mu\phi', \mathrm{D}_\mu v) + \frac{1}{2}(\mathrm{D}_\mu v, \mathrm{D}^\mu v) - V(\phi' + v) + \ldots$$

Expanding the potential V around v gives

$$\begin{aligned} L = &-\frac{1}{4g^2}F^2 + \frac{1}{2}(\mathrm{D}_\mu\phi', \mathrm{D}^\mu\phi') - \underline{(\mathrm{D}_\mu\phi, T_a\ v)A^{a\mu}} \\ &-\frac{1}{2}(T_a\ v, T_b v)A^b_\mu, A^a_\mu - V(v) - \frac{1}{2}\mathrm{D}^2V(v)(\phi', \phi') + \ldots ; \end{aligned}$$

the underlined terms are "non-diagonal" interaction terms in the fields (mixtures of A_μ, v and ϕ) and prevent a simple particle interpretation. The choice of a special gauge can solve this problem. It can be shown that a gauge exists where $(\phi', T_a v) = 0$ for all a, i.e.

$$(\phi', U_*(X)v) = 0 \qquad \text{for } X \in \mathcal{G}.$$

Since the Goldstone particles correspond exactly to the components of ϕ' in the sub-space $U_*(X)v, X \in \mathcal{G}$, there are no such particles in this gauge.

The mass terms of the gauge fields are contained in a matrix $M_{ab} \equiv (T_a\ v, T_b v)$. When we diagonalize this matrix, we find the masses of the gauge bosons. Since the rank of the matrix equals the dimension of the vector space spanned by $T_a\ v$, it is $\dim \mathcal{G} - \dim \mathcal{H}$. Hence one can conclude that $\dim \mathcal{H}$ gauge bosons stay massless, while $\dim \mathcal{G} - \dim \mathcal{H}$ gauge bosons become massive. This is the general statement of the Higgs phenomenon.

All these examples have been set up in a classical framework. When quantized versions of these theories are considered they have the beautiful property of allowing a consistent treatment of the divergences occurring in quantum field theory. They are renormalizable [t'Hooft 1971; Lee and Zinn-Justin 1972] just as quantum electrodynamics is, and therefore they are subject to perturbation calculations. The parameters of the theory are the coupling constants g, the parameter λ of the self-interaction, and the minima v of the potential (in a quantized theory the v correspond to the vacuum expectation values of the Higgs fields). In specific models – such as the Salam–Weinberg model of electroweak interactions discussed in Sect. 5.5 – there are usually more masses and interaction terms than free parameters. Therefore one finds relations between masses and coupling constants, and in that sense a "unification" of the various interactions is achieved. These relations are also important with respect to the renormalizability and the behaviour at high energies of such theories.

These perturbative computations essentially keep the classical picture unchanged. Non-perturbative effects such as tunnelling require a different framework (lattice calculations). The situation then changes drastically: typically there is no spontaneous breaking of gauge invariance [Elitzur 1975; Fröhlich et al. 1981; Borgs and Nill 1986a, b; Kennedy and King 1985; De Angelis et al. 1978].

The scheme presented above, generally in use in perturbation approaches, is therefore probably only a phenomenological description, and the classical Higgs field in this picture must be viewed as a complicated average of more fundamental quantum objects.

5.5 The Salam–Weinberg Theory of Electroweak Interactions

The Salam–Weinberg theory [Salam and Weinberg 1962; Weinberg 1967; Salam 1968; Schwinger 1956; Glashow 1958] is an attempt to unify the electromagnetic and the weak gauge symmetry. The requirements set by experiments have been satisfied by this theory in a very economical way, and the discovery of the W and Z intermediate bosons (e.g. [Darriulat 1984]) has verified predictions of this theory. Nevertheless it will become clear in the following that this theory still contains many arbitrary and artificial elements, and leaves many fundamental questions – such as the origin of the parity violation of weak interactions – unanswered.

As we have seen above, the construction of a spontaneously broken gauge symmetry involves the following steps:

1) The gauge group has to be chosen. The number of the gauge fields is equal to the dimension of G. If only leptonic processes are considered, then there are the three known lepton pairs of the electron and its neutrino $\nu_e(e^-\nu_e)$, the μ^- lepton and its neutrino $\nu_\mu(\mu^-\nu_\mu)$, and the τ^- lepton and its neutrino $\nu_\tau(\tau^-\nu_\tau)$. The study of charged-current experiments indicates that these pairs must be considered as "doublets" of isotopic spin like the neutron and the proton. As the known weak processes seem to require three intermediate vector-bosons, mediating processes such as β-decay, or the scattering of e^- and ν_μ (Fig. 5.7), one needs at least three gauge fields $(W^\pm, Z^0)$. The photon field A_μ must also be incorporated. The natural and simplest choice for a gauge group is thus the direct product $SU(2) \times U(1)$. Suppose the particles are placed in the lowest representation of the group $SU(2)$ as the following doublets:

$$\begin{pmatrix}\nu_e \\ e^-\end{pmatrix}_L , \begin{pmatrix}\nu_\mu \\ \mu^-\end{pmatrix}_L , \begin{pmatrix}\nu_\tau \\ \tau^-\end{pmatrix}_L .$$

Since the electromagnetic current $j_\mu^{\text{em}} \equiv \overline{\psi}\gamma_\mu\psi$ is a vector, whereas the weak current

$$j_W^\mu \equiv \sum_{l=e,\mu'} \overline{\nu}_l\gamma^\mu(1-\gamma_5)l$$

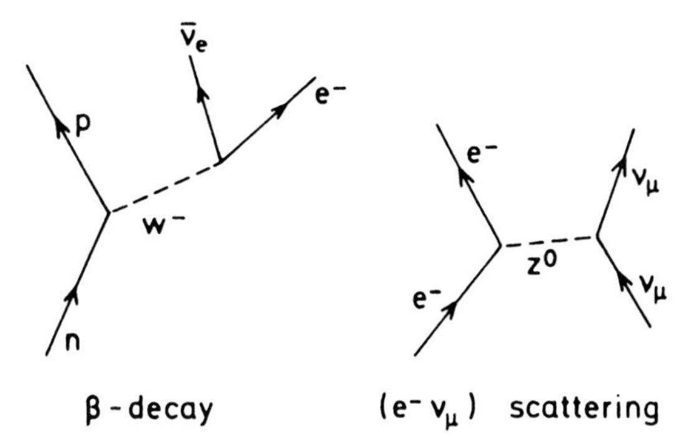

Fig. 5.7. Feynman diagrams for β-decay and $e - \nu_\mu$, scattering involving the exchange of electroweak gauge bosons

(γ^μ, γ_5 are Dirac's gamma-matrices; see e.g. [Bjorken and Drell 1965].) contains vector and axial vector terms (l characterizes the different types of leptons; it is the "flavour index"), it is convenient to separate the left- and right-handed spin polarization states of the spinors. The left-handed doublet is

$$L \equiv \frac{1}{2}(1-\gamma_5)\begin{pmatrix} \nu_e \\ e^- \end{pmatrix} = \begin{pmatrix} \nu_e \\ e^- \end{pmatrix}_L \equiv L^\alpha . \tag{5.34}$$

The right-handed neutrinos and the doublets of right-handed leptons are not observed in experiments, and thus the right-handed state contains only the electron

$$R \equiv \frac{1}{2}(1+\gamma_5)e^- \equiv e_R^- . \tag{5.35}$$

Similarly for the μ and τ.

2) These fundamental Fermi fields form a multiplet ψ_f, giving rise to a fermionic part of the Lagrangian

$$L_F = \overline{\psi}_f[i\gamma^\mu \mathrm{D}_\mu - m_f]\psi_f$$

where

$$\mathrm{D}_\mu \psi_f \equiv [\partial_\mu + U_*^f(A_\mu)]\psi_f .$$

The Yang–Mills part is given by

$$L_{\mathrm{YM}} = -\frac{1}{4}F_{\mu\nu}\cdot F^{\mu\nu}; \quad F_{\mu\nu} \equiv \partial_\mu A_\nu - \partial_\nu A_\mu + [A_\mu, A_\nu].$$

The gauge bosons still have mass zero so far. The four gauge bosons $A^a_\mu (a = 1, \ldots, 4)$ are renamed in the following way. The three gauge fields belonging to the $SU(2)$ gauge group are called W^i_μ:

$$W_\mu \equiv W_\mu^i \frac{\tau_i}{2};$$

τ_i are the Pauli matrices. The field belonging to the $U(1)$ factor is called B_μ:

$$\begin{array}{ccc} SU(2) & \times & U(1) \\ \downarrow & & \downarrow \\ W_\mu \equiv W_\mu^i \frac{\tau_i}{2} & & B_\mu \quad \text{(field strengths} \quad W_{\mu\nu}, B_{\mu\nu}) \end{array}$$

There must be combinations of W_μ^i belonging to $SU(2)$ but with definite charge $\pm e$.

Setting

$$W_\mu^\pm \equiv \frac{1}{\sqrt{2}}(W_\mu^1 \pm iW_\mu^2) \tag{5.36}$$

takes care of these states. The electrodynamic potential A_μ is taken to be a combination of W_μ^3 and B_μ:

$$\begin{pmatrix} A_\mu \\ Z_\mu \end{pmatrix} \equiv \begin{pmatrix} \cos\theta_w & \sin\theta_w \\ -\sin\theta_w & \cos\theta_w \end{pmatrix} \begin{pmatrix} B_\mu \\ W_\mu^3 \end{pmatrix}. \tag{5.37}$$

The photon and the Z-boson are obtained by a rotation by the angle θ_w (the "Weinberg" angle) from the $SU(2)$ gauge field W_μ^3 and the $U(1)$ field B_μ. The representation of the leptons as in (5.34) and (5.35) has the strange property (required by experiments) that the electron appears as a component of the left-handed doublet $(\binom{\nu}{e})_L$, and as a right-handed singlet e_R!

In this representation the fermionic part of the Lagrangian contains coupling terms like

$$\frac{g}{\sqrt{2}}(l_+ W_\alpha^+ + l_- W_\alpha^- + l_0 W_\alpha^3),$$

where

$$l = \sum_l \overline{\nu}_l \gamma^\alpha l; \quad l = e, \mu, \tau \quad \text{as above.}$$

The $SU(2)$ invariance is characterized by isotopic spin-like quantum numbers T_i ("weak isotopic spin"), whereas the $U(1)$ factor is characterized by a quantum number called "weak hypercharge" Y. The electric charge Q is a linear combination of T_3 and Y. To have the right Q values for the particles in L and R representations, the convention is to set

$$Q = T_3 + \frac{1}{2}Y \tag{5.38}$$

with

$$Y = \begin{cases} -1 & \text{for } L \text{ doublets,} \\ -2 & \text{for } R \text{ singlets.} \end{cases}$$

If the coupling constants are chosen as

$$\begin{array}{cc} SU(2)_L \times & U(1) \\ \downarrow & \downarrow \\ g & \frac{1}{2}g' \end{array}$$

then the Yang–Mills part of the standard model is

$$L_{\text{YM}} = -\frac{1}{4} W^{\mu\nu} W_{\mu\nu} - B_{\mu\nu} B^{\mu\nu}$$

and the leptonic part can be written as

$$L_{\text{Lept}} = \overline{R} i \gamma^\mu (\partial_\mu + i g' B_\mu) R + \overline{L} i \gamma^\mu [\partial_\mu + i(g'/2) B_\mu - i g W_\mu] L. \qquad (5.39)$$

There are no mass terms invariant under $SU(2)_L$.

The neutral part of the Lagrangian is written in terms of the fields A_μ and Z_μ:

$$\begin{aligned} L_{\text{int}}^{\text{neutr}} = & -A_\mu \left\{ g' \cos\theta_w \left(\bar{e}_R \gamma^\mu e_R + \frac{1}{2} \overline{\nu}_L \gamma^\mu \nu_L + \frac{1}{2} \bar{e}_L \gamma^\mu e_L \right) \right. \\ & \left. - \frac{1}{2} g \sin\theta_w \left(\overline{\nu}_L \gamma^\mu \nu_L - \bar{e}_L \gamma^\mu e_L \right) \right\} \\ & + Z_\mu \left\{ g' \sin\theta_w \left(\bar{e}_R \gamma^\mu e_R + \frac{1}{2} \overline{\nu}_L \gamma^\mu \nu_L + \frac{1}{2} \bar{e}_L \gamma^\mu e_L \right) \right. \\ & \left. + \frac{1}{2} g \cos\theta_w \left(\overline{\nu}_L \gamma^\mu \nu_L - \bar{e}_L \gamma^\mu e_L \right) \right\} . \end{aligned}$$

Additional terms for the μ and τ leptons should be added mentally to each expression. Possible mixings in the case of non-zero ν masses are ignored for the moment. Since the photon should couple only to the electromagnetic current

$$j_\mu^{\text{em}} = -(\bar{e}_R \gamma^\mu e_R + \bar{e}_L \gamma_\mu e_L)$$

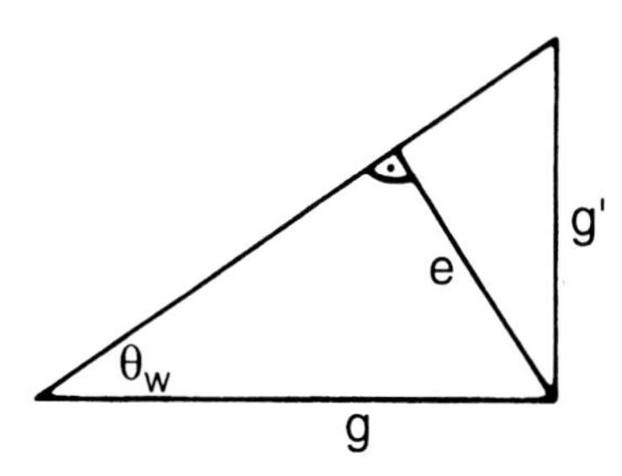

Fig. 5.8. Geometrical picture of the relation between g, g', e, and θ_w

there must be a relationship between g, g' and θ_w (Fig. 5.8):

$$g' \cos\theta_w = g \sin\theta_w = e. \tag{5.40}$$

Hence $\tan\theta_w = g/g'$ and $e = gg'(g^2 + g'^2)^{-1/2}$. The relation (5.40) also fixes the coupling of Z_μ:

$$\begin{aligned} L_{\text{int}}^{\text{neut}} &= \frac{g}{\cos\theta_w}[l_\mu^3 - \sin^2\theta_w j_\mu^{\text{em}}]Z^\mu + e j_{\text{em}}^\mu A_\mu; \\ l_\mu^3 &\equiv \frac{1}{2}\bar{\nu}_L\gamma_\mu\nu_L - \frac{1}{2}\bar{e}_L\gamma_\mu e_L \end{aligned} \tag{5.41}$$

(third component of the iso-spin current).

3) The next step is to introduce a (reducible) multiplet of Higgs fields chosen appropriately such that the spontaneous symmetry–breakdown produces the desired results. In the case of the standard model the aim is to obtain masses for the W_μ and Z_μ bosons – since the weak interaction is of short range – while the photon A_μ has to remain massless.

Therefore the stabilizer H must be chosen as the subgroup $U(1)_{\text{em}}$ of $SU(2)_L \times U(1)_Y$ generated by the charge $Q = T_3 + \frac{1}{2}Y$. This can be achieved by introducing two complex Higgs fields ϕ^+ and ϕ^0 – with $Q\phi^+ = 1; \quad Q\phi^0 = 0$ and arranging them in a pair

$$\phi = \begin{pmatrix}\phi^+ \\ \phi^0\end{pmatrix}, \quad Y\phi = \phi. \tag{5.42}$$

The Higgs potential is introduced in the same form as above:

$$V(\phi) = \mu^2\phi^*\phi + \lambda(\phi^*\phi)^2; \quad \lambda > 0, \mu^2 < 0. \tag{5.43}$$

If the minimum is chosen at

$$\langle\phi\rangle = \frac{1}{\sqrt{2}}\begin{pmatrix}0 \\ v\end{pmatrix}$$

it is left invariant by $U(1)_{\text{em}}(Q\binom{0}{v} = 0)$. Choose v to be real, $v = \sqrt{(-\mu^2/\lambda)}$ and fix the gauge as

$$\phi = \frac{1}{\sqrt{2}}\begin{pmatrix}0 \\ v + \eta\end{pmatrix} \tag{5.44}$$

with η a real field.

The Higgs–field Lagrangian reads

$$L_\phi \equiv \mathrm{D}_\mu\phi^*\mathrm{D}^\mu\phi - \mathrm{V}(\phi^*\phi).$$

In terms of v and η,

$$L_\phi = \frac{1}{2}\partial^\mu \eta \partial_\mu \eta + \frac{(v+\eta)^2}{8}\chi_-^*[(g'B_\mu + 2gW^\mu)(g'B^\mu + 2gW^\mu)]\chi_- \\ -V\left(\frac{(v+\eta)^2}{4}\right) \tag{5.45}$$

where $\chi_- \equiv \binom{0}{1}$. The neutral scalar field η (Higgs field) has the mass

$$m_\eta^2 = -2\mu^2 \tag{5.46}$$

The terms of L_ϕ that are quadratic in the vector mesons can be read off as

$$\frac{1}{8}v^2[(g'B_\mu - gW_\mu^3)(g'B^\mu - gW^{\mu 3}) + g^2((W_\mu^1)^2 + (W_\mu^2)^2)] \\ \equiv m_\text{w}^2 W_\mu^+ W^{-\mu} + \frac{1}{2}m_Z^2 Z_\mu Z^\mu. \tag{5.47}$$

Hence (using $\tan\theta_w = g'/g$ and (5.37)) we can conclude that

$$m_\text{w} = \frac{1}{2}gv, \\ m_Z = \frac{1}{2}\sqrt{g^2 + g'^2}v. \tag{5.48}$$

The photon mass remains zero according to this construction. The relation between m_W and m_Z that can be derived from (5.48) [Weinberg 1967] is

$$m_Z = \frac{m_\text{w}}{\cos\theta_w}. \tag{5.49}$$

The mass of the Z-bosons is greater than the mass of the W.

In the matrix element M of the processes caused by the weak charged currents there appears a factor (the propagator)

$$M \propto \frac{(g/2\sqrt{2})^2}{q^2 - m_\text{w}^2},$$

where q is the momentum, and m_w the mass of the intermediate bosons. In the limit of small energies $|q^2| \ll m_\text{w}^2$ this reduces to

$$M \propto = \frac{g^2}{8m_\text{w}^2}. \tag{5.50}$$

In this limit the exchange of the virtual W-bosons can be reduced to a description by an effective 4-Fermi Lagrangian $L_\text{eff} = (-4/\sqrt{2})G_F l_W^\lambda l*_{\lambda W}$, describing the interaction of four fermions at one point. Consider, for example, the process $\mu^- + e^+ \rightarrow \nu_e + \nu_\mu$ (Fig. 5.9). Comparing the two couplings gives

$$\frac{G_F}{\sqrt{2}} = \frac{g^2}{8m_\text{w}^2}. \tag{5.51}$$

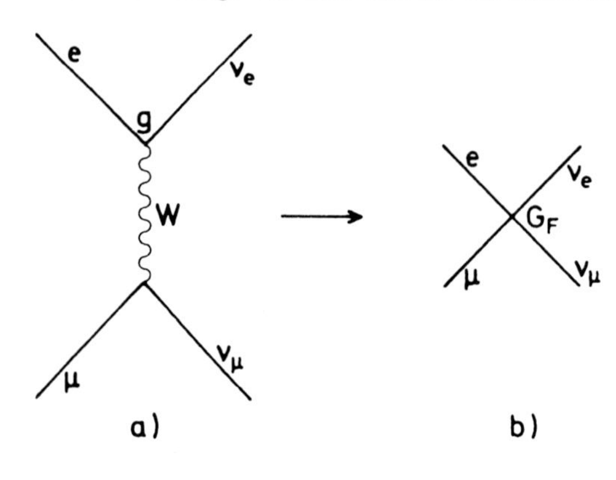

Fig. 5.9. Comparative diagrams for the electroweak theroy and the old 4-Fermi theory for descriptions of the process $\mu^- + e^+ \to \nu_e + \nu_\mu$

Here G_F is the Fermi constant of weak interactions; with m_P the mass of the proton, it is given by

$$G_F m_p^2 \simeq 10^{-5}. \tag{5.52}$$

Equations (5.51) and (5.40), taken together, lead to an expression for the mass of the W bosons in terms of G_F, e and θ_w

$$m_{\mathrm{w}}^2 = \frac{g^2\sqrt{2}}{8G_F} = \frac{e^2}{4\sqrt{2}G_F \sin^2\theta_w} = \frac{\alpha\pi}{\sqrt{2}} G_F^{-1} \sin^{-2}\theta_w. \tag{5.53}$$

Inserting $\alpha = \frac{1}{137}$ and (5.52) sets m_{w} at

$$m_{\mathrm{w}} = \frac{37.3}{\sin\theta_w} \quad [\mathrm{GeV}] \tag{5.54}$$

and correspondingly

A good theory must fit the experimental facts

$$m_Z = \frac{74.6}{\sin 2\theta_w} \quad [\text{GeV}]. \tag{5.55}$$

One can see that the lower bounds for the masses of W bosons and Z bosons are quite large. The experimental study of neutral current processes leads to a value for the mixing angle θ_w [Fayet 1984 and ref. therein]

$$\sin^2\theta_w = 0.215 \pm 0.014. \tag{5.56}$$

All results on ν-nucleon ν-electron, e^+e^- scattering, and parity-violation effects in atomic physics are in agreement with the above value of θ_w. Thus a prediction for the masses of the W and Z bosons emerges:

$$\begin{aligned} m_{\mathrm{W}} &= 83.0 \pm 2.9 \quad \mathrm{GeV}/c^2, \\ m_Z &= 93.8 \pm 2.4 \quad \mathrm{GeV}/c^2. \end{aligned}$$

It is a great success of the standard model of electroweak interactions that in 1983 in proton–antiproton collision experiments at CERN the $\mathrm{W}^{\pm}$ and Z particles were observed ([Darriulat 1984]; see also Fig. 5.10). Their masses were measured, and agree well with the predictions:

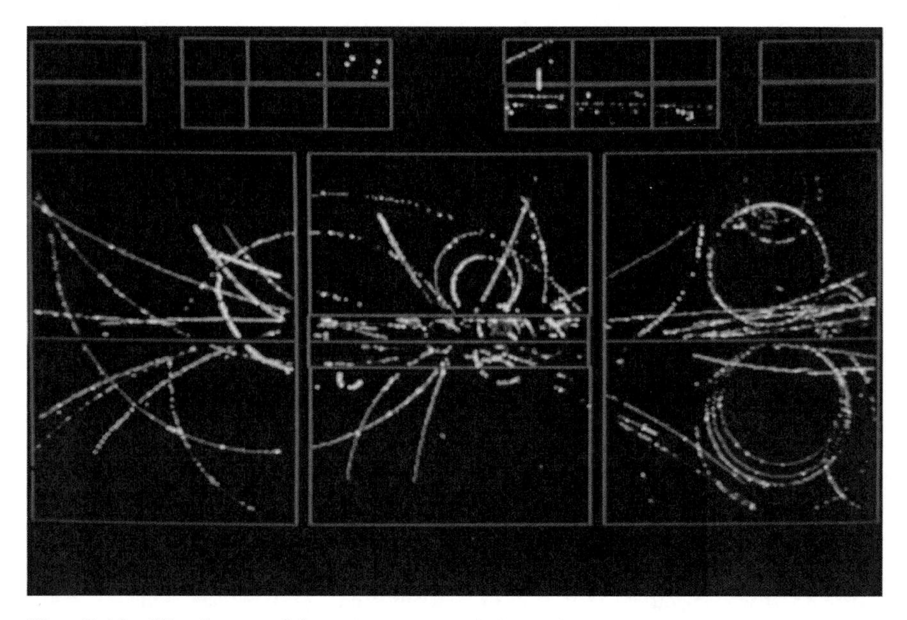

Fig. 5.10. The famous May 1983 event of the UA1 experiment at CERN which revealed the signature of the Z^0 particle as it decayed into a very high-energy electron–positron pair (the two-prong yellow lines in the lower half of the right-hand part of the picture). The Z^0 particle is the gauge boson carrying the neutral weak current. These experiments firmly establish the gauge theory of the electroweak interactions (courtesy of CERN documentation office)

$$m_{\mathrm{W}} = 81 \pm 2 \quad \mathrm{GeV}/c^2,$$
$$m_Z = 94 \pm 2 \quad \mathrm{GeV}/c^2. \tag{5.57}$$

This detection (which led to the Nobel prize in 1984 for C. Rubbia and J. van der Mer) has put the standard model on a firm footing. It has also increased confidence in the concept of spontaneously-broken gauge symmetries as a description of elementary particle interactions. It would, of course, be extremely convincing if the Higgs boson were also detected. Unlike the mass of W bosons and Z bosons the mass of the Higgs boson is not fixed in the standard model, which raises difficulties for detection.

Ultimately, however, the symmetry-breaking Higgs field may be the most important feature of the Weinberg–Salam theory. The Higgs field is involved in two of the most revolutionary aspects of this mode. It shows how a mass for the gauge bosons is generated through an interaction, and it plays the role of a new type of vacuum in the gauge theory. It acts like an all-pervading background medium even at very short distances. But what is its origin? In the case of a superconductor, the "Higgs" field was a composite system of electrons bound into Cooper pairs. The Higgs field in relativistic field theories is not at all well understood in this respect of composite or elementary quantity.

Non-perturbative calculations on a lattice in the Weinberg–Salam model indicate that a well-defined continuum limit does not exist (i.e. the limit where, for example, the electromagnetic coupling $\alpha \to 0$, instead of $\alpha = 1/137$) [e.g. Hasenfratz 1986]. If this result is accepted, the standard model must be viewed as an effective theory where the regularization depends on a cut-off energy Λ_c (with the limit $\Lambda_c \to \infty$ not being possible). Λ_c depends (among other factors) on the Higgs mass m_H in a way that allows an upper limit to be derived: $m_H \leq 10 m_W$ [Lindner 1986].

5.6 The Colour Gauge Theory of Strong Interactions – Quantum Chromodynamics

The emergence of a gauge theory for the strong interactions is another promising step in the unified description of particle interactions. The theory is based on a new hypothetical charge-like quantum number, called "colour", which is carried by the quarks, the fundamental constituents of normal nucleons. Thus the new gauge theory for strong interactions has been called "colour gauge theory" or "chromodynamics". Unlike earlier theories of the strong interactions involving meson exchange, chromodynamics is postulated to be an exact local gauge theory based on the colour gauge group $SU(3)_C$. This statement should puzzle us, because previously we have pointed out that in gauge theories without spontaneous symmetry-breaking the gauge potential fields necessarily have zero mass, and therefore can only describe long-range interactions. Obviously the understanding of the structure of the hadrons (baryons and mesons – from Greek "hadros" meaning heavy) and of their interactions is intimately tied to this idea. First of all there is the great success of the quark model in describing the huge number of hadrons as bound states of a small number of elementary quarks (see e.g. [Gell-Mann and Ne'eman 1964]). This suggests that we should consider the quarks as the true sources of the strong interactions. However, the production of free quarks in high-energy collisions has not been observed. In deep inelastic lepton–hadron scattering experiments it has been established, on the other hand, that at high momentum transfer or at short distances, hadrons behave like systems consisting of weakly interacting particles, which, to a first approximation, may be considered free. The same result follows from the quantum-chromodynamic theory in which all interactions between the quarks are mediated by massless gauge fields, the so-called "gluons": at small distances the effective coupling constant is quite small. Such theories are said to have the property of "asymptotic freedom". This is a general property of non-abelian gauge theories [Politzer 1973; Gross and Wilczek 1973; Peterman 1979]. Thus the qualitative picture given by quantum chromodynamics (QCD) is the following: hadrons have an internal structure, and the interaction of the constituent particles tends to zero at distances much smaller than the hadron size.

One expects that at large distances (of the order of the hadron size) the interaction becomes strong, and the constituent particles cannot escape the hadron; the quarks and gluons are "confined". Contrary to the property of asymptotic freedom, however, this confinement has not been proved generally and rigorously. Numerical computations in the context of QCD-like models in a lattice give strong indications that there is confinement of the quarks.

Despite many impressive results neither the theoretical predictions nor the experimental tests of "chromodynamics" have yet achieved the status of certainty of either quantum electrodynamics or the Weinberg–Salam theory. The difficulty, on the theoretical side, lies in the treatment of the non-linear terms in the Yang–Mills equations. On the experimental side, the energies reached by accelerators are not yet large enough to allow crucial tests.

The new internal quantum numbers in QCD are three colours – they are rather like the charge in quantum electrodynamics – which are conventionally called red, green, and blue. Each quark can occur in three colour states, which form a fundamental basis for the QCD gauge group $SU(3)$. $SU(3)$ is the three-parameter unimodular group with $(3)^2 - 1 = 8$ independent generators of the Lie algebra – thus there are eight gauge fields, called the gluons.

Some of the mathematical tools needed for the $SU(3)$ colour group can be taken over directly from the well-known $SU(3)$ classification of the hadrons in the old quark model (e.g. [Han and Nambu 1965]). The three colours are assigned to a fundamental triplet, and the complex conjugate anti-triplet forms another base for the fundamental representation of $SU(3)$ (by complex 3×3 matrices) (Fig. 5.11).

The gluons, as gauge fields of the colour $SU(3)$, act as operators in the internal colour space. They carry colour themselves, and they change the

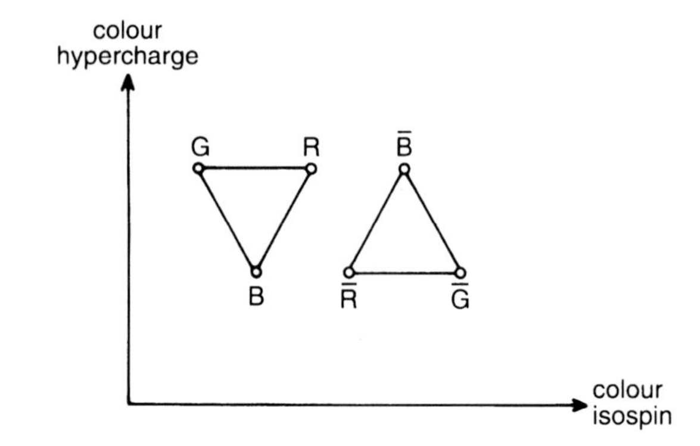

Fig. 5.11. The three quark colours red, green, and blue belong to a fundamental triplet of $SU(3)$. The anti-colours belong to an anti-triplet. The colours red and green form a "colour isotopic spin" pair

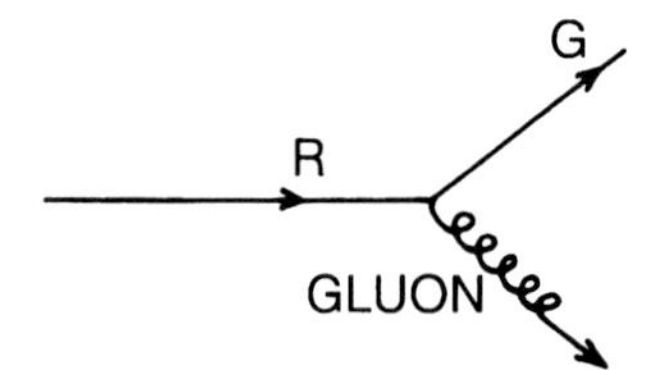

Fig. 5.12. "QCD bremsstrahlung": a red quark changes to green by emitting a gluon. The gluon carries off the red and anti-green colours

colour of the quarks during the interaction. Consider a red quark, which emits a gluon and changes to green, as shown in Fig. 5.12.

No free, coloured quarks or gluons have been observed in experiments. This constraint on QCD means that quarks must be bound in mesons and baryons as (respectively) quark or antiquark, three-quark states with a zero net colour charge. Combining the colour theory of the strong interactions $SU(3)_C$ with the Weinberg–Salam $SU(2)_L \times U(1)_Y$ electroweak theory gives a standard model of weak, electromagnetic, and strong interactions as a gauge theory with symmetry group $SU(3)_C \times SU(2)_L \times U(1)_Y$. The particle content at this stage can be represented as in Table 5.1.

The quarks and the τ-neutrino ν_τ have been observed only indirectly through their influence on certain collision processes. The spin-zero-Higgs boson has not hitherto been detected, either directly or indirectly.

The different quark "flavours" u (up), d (down), c (charm), s (strangeness), t (top), and b (bottom) have been introduced to take into account certain selection rules observed in the strong interactions (the discovery of the strange particles; the J, ψ meson decays, etc.; for a review see [Langacker 1981]). Both the left-handed (L) and right-handed (R) parts of the quarks are triplets under $SU(3)_C$, while L quarks are doublets (2) and R quarks are singlets (1) under $SU(2)$. The leptons do not participate in the strong interactions, i.e. they are colour ($SU(3)_C$) singlets (1), but as we have seen they do participate in electroweak interactions with a structure as described in Sect. 5.4. Thus

Table 5.1. Particle content of the standard model $SU(3)_C \times SU(2) \times U(1)$

Spin 1 gauge bosons:				
γ, $W^{\pm}$, Z (electro-weak)				
8 Gluons (indirectly observed strong interactions)				
Spin $\frac{1}{2}$ matter fields: families of leptons and quarks				
				charge
leptons $\begin{pmatrix} \nu_e \\ e \end{pmatrix}$	$\begin{pmatrix} \nu_\mu \\ \mu \end{pmatrix}$	$\begin{pmatrix} \nu_\tau \\ \tau \end{pmatrix}$		$\begin{pmatrix} 0 \\ -1 \end{pmatrix}$
quarks $\begin{pmatrix} \text{up} \\ \text{down} \end{pmatrix}$	$\begin{pmatrix} \text{charm} \\ \text{strange} \end{pmatrix}$	$\begin{pmatrix} \text{top} \\ \text{bottom} \end{pmatrix}$		$\begin{pmatrix} +2/3 \\ -1/3 \end{pmatrix}$
Spin-zero-Higgs boson:				
1 neutral, spin 0 Higgs				

one generation of fermions from Table 5.1 under $SU(3)_C \times SU(2)_L \times U(1)_Y$ contains 15 members.

$$\begin{matrix} u_L^{\text{red}} & u_L^{\text{green}} & u_L^{\text{blue}} & & u_R^r, & u_R^g, & u_R^b & \nu_e & \\ & & & ; & & & & & ; e_R. \\ d_L^{\text{red}} & d_L^{\text{green}} & d_L^{\text{blue}} & & d_R^r, & d_R^g, & d_R^b & e_L & \end{matrix}$$

There are at least two more generations (c, s, v_μ, μ) and (t, b, ν_τ, τ) with the mass of the constituents increasing in this sequence. The observed baryons must be colour singlets, i.e. they are three-quark (qqq) states completely antisymmetric in the three possible colours. Since the quarks are fermions, the wave function for any baryon must be antisymmetric with interchange of any two of its quarks.

The well-known baryons are made up of just d, u, and s quarks. Thus the well-known octet and decaplet assignments remain, with the proviso that all states are colour singlets (Fig. 5.13). The mesons are colour singlets of quark–antiquark states.

The masses of the quarks cannot be determined directly from scattering experiments, because they do not exist as free particles. They appear always "dressed", never "naked". From resonances in electron–positron scattering one can extrapolate to the asymptotic state, where the momentum transfer becomes infinite ($q^2 \to \infty$), and derive masses for the naked d, u, s, c quarks. These estimates require a fair amount of theoretical input. The values obtained are

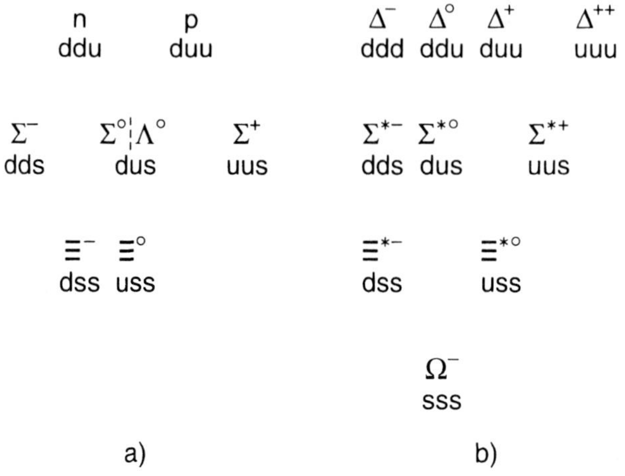

n ddu p duu

Σ^- dds Σ^0 ¦ Λ^0 dus Σ^+ uus

Ξ^- dss Ξ^0 uss

a)

Δ^- ddd Δ^0 ddu Δ^+ duu Δ^{++} uuu

Σ^{*-} dds Σ^{*0} dus Σ^{*+} uus

Ξ^{*-} dss Ξ^{*0} uss

Ω^- sss

b)

Fig. 5.13. The octet (**a**) and decaplet (**b**) assignments of $SU(3)$

$(5-15)$ MeV	for the d quark;
$(2-8)$ MeV	for the u quark;
$(100-300)$ MeV	for the s quark;
$(1-1.6)$ GeV	for the c quark;
$(4.1-4.5)$ GeV	for the b quark.

The top-quark has been discovered in proton–antiproton collisions (900 GeV energy in the lab system, i.e. 1.8 TeV in the centre-of-mass system) [Abe et al. (CDF collab.) 1994]. Its mass is given as $m_{\text{top}} = (180 \pm 14)$ GeV.

These schemes have had several impressive successes in analysing the structure of hadrons. An essential argument for using gauge theories to describe the fundamental interactions is the possibility of renormalizing such theories, just as one can renormalize QED. It is then possible to compute processes for any order of perturbation theory and obtain finite results with only a finite number of undetermined parameters. It has been shown [t'Hooft 1971] that the Higgs mechanism of spontaneous symmetry-breaking does not spoil this property.

Let us finally add a few more remarks on "asymptotic freedom". The experiments suggesting this new feature have been carried out at the 2-mile-long linear accelerator at Stanford. High-energy electrons from this machine were scattered on proton targets in order to study the structure of the proton. The interaction is mediated by the exchange of a high momentum q. According to the uncertainty relation ($\Delta p \Delta q \geq \hbar$) the value of q is inversely proportional to the size of the region that is seen. For $q \geq 1\,\text{GeV}/c$, one is probing dimensions smaller than the radius of the proton, hence these reactions are named "deep-inelastic" (Fig. 5.14). The scattering results showed – similar to the way in which Rutherford's experiments probed the atom and found it almost empty except for a tiny nucleus – that (down to probing scales of 10^{-17} cm), the proton consists of point-like charged constituents, the quarks.

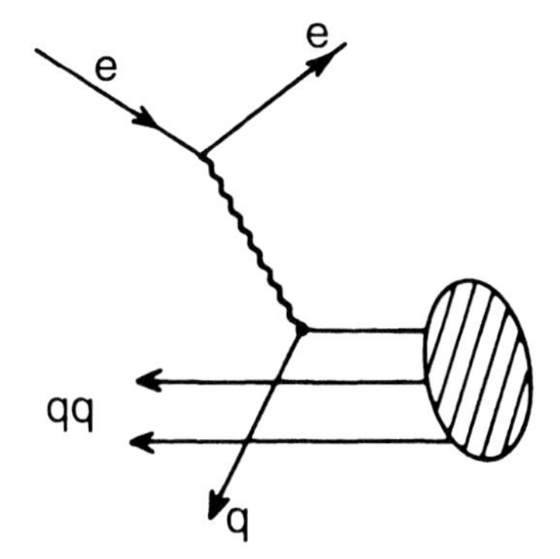

Fig. 5.14. Diagram of the deep-inelastic scattering of high-energy electrons by a proton

As q was increased, i.e. as the size of the volume being probed decreased, the effective strength of the interaction between the quarks appeared to become weaker. The quarks act as if they were essentially free particles in the proton. This behaviour and its theoretical extrapolation to very high energies became known as "asymptotic freedom".

The reason for this asymptotic freedom is the peculiar polarization property of the QCD vacuum. For a discussion of this effect we shall compare it to the familiar properties of vacuum polarization in quantum electrodynamics: an electron is thought to be surrounded by a cloud of virtual (i.e. not really existing) electron–positron pairs of the vacuum. These virtual pairs contribute to the Lamb shift in the energy spectrum of the hydrogen atom. The virtual positrons are attracted slightly closer to the electron while the virtual electrons are repelled. The polarization of the vacuum is such that there is a shielding effect of the "bare" electron charge by the virtual positrons (Fig. 5.15). At large distances the effective charge of the electron appears to be smaller.

In QCD a similar vacuum polarization from pairs of virtual quarks and antiquarks occurs. There is in addition an "anti-screening" effect from the virtual gluons. Since the gluons can carry colour charge, there is – as detailed calculations show – a decrease in the effective colour charge as the quark is approached more closely. The situation may be depicted as in Fig. 5.16.

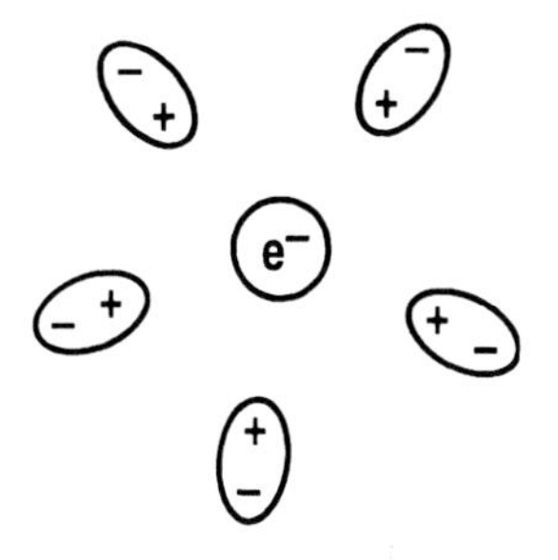

Fig. 5.15. A "bare" electron is surrounded by virtual electron–positron pairs

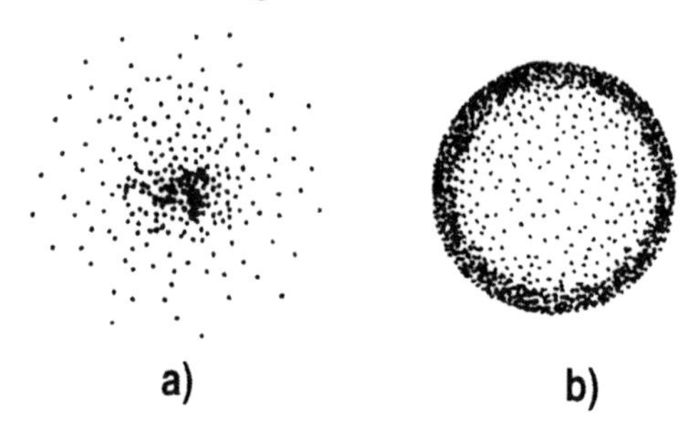

Fig. 5.16. The effective charge density in quantum electrodynamics (**a**) and in quantum chromodynamics (**b**)

The calculation of the effective colour charge from vacuum polarization in QCD by perturbation techniques can be supplemented by a derivation in which the QCD vacuum is treated semi-classically like a dielectric medium [Lee 1979; Weisskopf 1981b]. The potential between two charges is written in the form

$$V(r) = \frac{q_1 q_2}{4\pi\epsilon(r)r}$$

and all of the vacuum polarization effects are thought to be absorbed into an effective dielectric function of the vacuum, $\epsilon(r)$, which depends on the distance between the two charges [Nielsen and Olesen 1978]:

$$\epsilon = \left[1 + \frac{(33 - 2N_f)}{24\pi^2} g^2 \log\left(\frac{E_2}{E_1}\right)\right]^{-1} . \tag{5.58}$$

The energies E_2 and E_1 are cut-off parameters. If E_1 is kept fixed, and $E_2 \to \infty$, then $\epsilon \to 0$. N_f is the number of different types of "flavours" of quarks. If $N_f < 16$, then $\epsilon < 1$, corresponding to anti-screening.

5.7 Successes and Problems of the Standard Model

We have mentioned above that the $SU(3)_C \times SU(2) \times U(1)$ model – or, as abbreviated, the $(3,2,1)$ model – provides a potentially complete description of the strong, weak, and electromagnetic interactions. The total Lagrangian, as before, consists of a Yang–Mills and a fermion part, where the covariant derivative for the fermions D_μ^f now includes the gluon gauge fields A_μ^a and the generators of $SU(3)$, $\lambda^a (a = 1, \dots, 8)$:

$$\mathrm{D}_\mu^f = \left(\partial_\mu - ig_3 A_\mu^a \frac{\lambda^a}{2} - ig_2 W_\mu^b \frac{\sigma^b}{2} - ig_1 B_\mu Y\right) . \tag{5.59}$$

Here g_j are the coupling constants as before, σ^b $(b = 1, 2, 3)$ are the generators of $SU(2)$ (Pauli matrices) and

$$Y = -\frac{\sigma_3}{2} + Q \tag{5.60}$$

as in (5.4). The Yang–Mills part includes a non-linear gluon term $-\frac{1}{2}F^a_{\mu\nu}F^{\alpha\mu\nu}$, with

$$F^a_{\mu\nu} = \partial_\mu A^a_\nu - \partial_\nu A^a_\mu + g_3 f^{abc} A^b_\mu A^c_\nu \quad a, b, c = 1, \dots, 8.$$

Fermions are grouped in three families (or generations), the e, μ, and τ families. Thus we may write

$$L(1, 2, 3) = L_f + L_{\mathrm{YM}} + L_H + L_{\mathrm{YUK}}.$$

The Higgs term $L_H = -V(\phi)$, with V as in (5.5). The term $L_{\mathrm{YUK}} \equiv \sum f\overline{\psi}_f(\Gamma, \phi)\psi_f$ contains all kinds of couplings of matter fermions to the scalar Higgs fields ϕ. The spontaneous symmetry-breaking leads – via $\langle\phi\rangle \neq 0$ – to mass terms for the fermions. In the standard model, the Lagrangian L_f does not contain mass terms, and all the quark and lepton masses are generated with the Higgs pair $\binom{\phi^+}{\phi_0}$ described in (5.5). This has the consequence that in the standard model, the chirality (left- or right-handedness of quarks and leptons) is preserved in the terms $L_{\mathrm{YM}} + L_f$. Or, in other words, only the spontaneous symmetry-breakdown introduces mass terms, and hence mixes chirality states.

The standard model has been formulated in such a way that the symmetry of CP (CP stands for the combined operation of charge conjugation C, and parity transformation P) is exact at the classical level. It is known, however, that this symmetry of a mirror reflection of all processes and of replacing all particles by their antiparticles is only an approximate symmetry. It has been observed that the long-lived neutral K-meson K^0_L, which is its own antiparticle, decays slightly more often into $\pi^- + e^+ + \nu_e$ than into $\pi^+ + e^- + \overline{\nu}_e$, whereas CP invariance would require both decay modes to be equal.

Higher-order terms in perturbation theory lead to violations of the CP invariance that are too large to be reconciled to the experiments. Some ideas concerning this "strong CP problem" are discussed in Sect. 6.5.

There are several impressive successes of the standard model:

1) The low-energy weak interactions are well described with two parameters, θ_w, the Weinberg angle, and the Fermi constant G_F. Taking $\sin^2\theta_w = 0.215 \pm 0.015$ from experiment, the masses of the W and Z bosons are predicted in good agreement with experiments.
   Even precision tests of the radiative correction calculations within the standard model showed no deviations, to the disappointment of experimentalists, who had hoped that the first signs of "new physics" would emerge. The measurement of the mass of the top quark has established the consistency of these precision tests, and has even led to a prediction of the Higgs mass, $m_H = 92^{+141}_{-67}$ GeV, with a 95% confidence upper limit $m_H < 255$ GeV [Marciano 1999].
2) There is a conservation of the fermion chirality (right- and left-handedness of fermions is preserved) in QCD. The theory is asymptotically free, in agreement with results from large momentum-transfer processes.

3) The theory is renormalizable and unitary for each order of perturbation theory. It may therefore be used to calculate all quantities in terms of the fundamental parameters of the theory.

A number of questions are left unanswered:

i) The mixing parameter $\sin^2\theta_w$ is arbitrary, and there are three arbitrary coupling constants corresponding to the three factors of the gauge group.
ii) There is no reason why the matter multiplet structure should be as chosen with three families of fermions. Quarks and leptons are not related.
iii) Charge quantization is not explained since Y in (5.60) is arbitrary.
iv) The number of Higgs bosons and their interactions are practically arbitrary.

These limitations seem to suggest that there must be a more fundamental theory, of which the $(3, 2, 1)$ model is only a manifestation at low energies. The underlying theory, if it exists, can in principle turn out in two different ways: the first possibility is the existence of more fundamental, elementary building blocks, which generate the fields for the standard model as composite structures. The second possibility is that the fields of the standard model are fundamental, but they are related by further symmetries, broken at some large energy scale M_X. This leads to schemes of unification, such as grand unified theories (GUTs) or supersymmetric theories (SUSY GUTs). In the next chapter some properties of grand unification schemes will be discussed. I shall not discuss consequences of a possible sub-quark structure, because this field of speculation is not strongly suggested by observations. and also does not seem to be in a shape suitable for cosmological exploitation.

Exercises

5.1 A simple system from classical mechanics can serve as an analogue for the Higgs mechanism. Consider a rotating whoop with a freely moving ball (mass m) inside. The whoop (radius a) rotates around an axis (frequency ω) parallel to the direction of the homogeneous gravitational field (g). Write down the Lagrangian, and the potential energy of the ball (q is the angle between the rotation axis and the radial line to the position of the ball). Show that there exists a critical frequency $\omega_o^2 = g/am$ such that for $\omega < \omega_o$ the stable equilibrium is at $q = 0$, but for $\omega > \omega_o$ a new stable equilibrium is at $\cos q_o = \left(\frac{\omega_o}{\omega}\right)^2$, and $q = 0$ becomes unstable.

5.2 Follow the arguments after (5.14) and show explicitly the local gauge invariance of the Lagrangian of (5.18).

5.3 Carry out explicitly the calculations of (5.24) to (5.29), but do not choose a special gauge in the beginning. Rather keep $\phi' \equiv \phi_1' + i\phi_2' = \phi - \phi_o$, and show explicitly how the gauge fields and Higgs fields mix to produce a Higgs boson, and a massive gauge vector boson.

5.4 Consider the model of Fig. 5.7 and (5.23) of a complex scalar field with potential $V(\phi)$. The spontaneous symmetry-breaking produces a Goldstone particle. What is the Goldstone particle in this model? Show that it has zero mass.

5.5 Show that neutrinos are massless in the standard model.

5.6 The Higgs field for spontaneous symmetry-breaking must be a scalar. Why are vectors and spinors not possible with a non-zero vacuum expectation value?

6. Grand Unification Schemes

"To the inventor the products of his imagination appear necessary and natural, such that he does not look at them as objects of the mind, but rather as things belonging to reality, and thus he wishes them to be seen like that."

A. Einstein

"There is something fascinating about science. One gets such wholesale return of conjecture out of such a trifling investment of fact."

Mark Twain, Life on the Mississippi (1874)

There is no experimental evidence so far that could not be accommodated within the standard model. There are, however, several theoretical shortcomings which must be remedied by a more fundamental theory. The unification of forces has not really found a satisfactory formulation, since three different gauge groups $U(1), SU(2)$, and $SU(3)$ are used to describe the electromagnetic, weak, and strong interactions.

As a simple first attempt to improve upon the standard model in this aspect of fundamental symmetries one can introduce a larger gauge group that contains the factors $SU(3) \times SU(2) \times U(1)$ as a subgroup. This larger group G should generate all gauge interactions that relate particles of the same spin. It should have only a single coupling constant g. Since the rank of the (1, 2, 3) symmetry group is 4 – its Lie algebra has four commuting operators – the rank of a simple group G must be at least 4. The simplest extension is thus the group $SU(5)$, which will be described in Sect. 6.1. Several of the results obtained for $SU(5)$ in Sect. 6.1 are of general validity, and we will point these out in Sect. 6.2. It is worthwhile to consider this simple case, because several of the guiding principles can be seen in it, although it has become clear that $SU(5)$ cannot be the right unification group.

The general idea suggests itself intuitively. Combine a quark field, say d (which exists in three colour states d_1, d_2, d_3), and the electron and electron–neutrino to form a five-component spinor. The transformations of this 5-dimensional quantity under $SU(3)$ and $SU(2) \times U(1)$ simultaneously can be represented by a 5×5 matrix:

$$\begin{pmatrix} d_1 \\ d_2 \\ d_3 \\ e \\ \nu \end{pmatrix} \to \begin{pmatrix} \times & \times & \times & \circ & \circ \\ \times & \times & \times & \circ & \circ \\ \times & \times & \times & \circ & \circ \\ \circ & \circ & \circ & \bullet & \bullet \\ \circ & \circ & \circ & \bullet & \bullet \end{pmatrix} \begin{pmatrix} d_1 \\ d_2 \\ d_3 \\ e \\ \nu \end{pmatrix} .$$

The first three rows and columns represent an element of the colour gauge group $SU(3)_C$ acting only on d, and in the lower right-hand corner there is a complex 2×2 matrix acting on e and ν_e, an element of $SU(2) \times U(1)$. The generalization then consists simply in allowing this 5×5 matrix to be a general element of $SU(5)$ (i.e. unitary and with determinant $= 1$), and assuming this symmetry for the theory. Obviously this implies that not only transformations between $SU(3)$ states and $SU(2) \times U(1)$ states occur separately, but also transformations that mix these states.

As a further step one may be tempted to incorporate the basic ideas of an $SU(5)$ GUT into a supersymmetric theory (SUSY GUT) where particles of different spin can be related (see [Ross 1984] for a review of GUT schemes).

If the supersymmetry is implemented as a local symmetry, a "supergravity" theory results, which in some sense includes the gravitational interaction in its unification scheme. In Sect. 6.4 some consequences of a SUSY GUT scheme will be pointed out. An additional basic ingredient is, of course, the spontaneous breaking of the big symmetry G at some exorbitantly high-energy scale – so as to avoid interference with the successes of the standard model.

The recent development of superstring theories leaves behind all these modest attempts to construct a unified theory using as basic building-blocks the fundamental particles of the standard model. The basic element is instead an extended object of 10 space-time dimensions, the string (a short discussion of these concepts will be found in Sect. 6.8).

6.1 *SU*(5) GUT

6.1.1 The Group Structure

The group $SU(5)$, consisting of 5×5 unimodular matrices of determinant 1, has $5^2 - 1 = 24$ free parameters. Therefore there will be 24 gauge boson fields associated with $SU(5)$ as a local gauge group. Clearly 12 of them must be identified with those of the standard model: $W_\mu, Z_\mu, A_\mu, A_\mu^a$ (gluons). The other 12 are new fields, usually designated as X and Y bosons. How should the fermion basis of $SU(5)$ be set up? One requires, of course, that the known particles of the standard model are taken over as elementary building blocks into the $SU(5)$ representations. Usually one finds it more convenient to work with fermions of a fixed helicity, since the gauge interactions conserve the helicity of fermions (the chirality symmetry). The left-handed fields f_L, and their charge conjugates f_L^C – describing the same particles with opposite charge – are often used. To relate the irreducible representations of $SU(5)$ to the known generations of fermions, one investigates their content with respect to the subgroup $SU(3)_C \times SU(2)_L \times U(1)_Y$. These are designated by three numbers of the form $(3,1)_{-2/3}$, for example. This denotes a 3-dimensional representation (triplet) with respect to $SU(3)_C$, a $SU(2)_L$ singlet, and a hypercharge $Y = -2/3$. (Y is defined as in (5.38): $Q = (\sigma_3/2) + Y$). For example,

$$Y \begin{pmatrix} d_1^C \\ d_2^C \\ d_3^C \\ 0 \\ 0 \end{pmatrix}_L = -\frac{2}{3} \begin{pmatrix} d_1^C \\ d_2^C \\ d_3^C \\ 0 \\ 0 \end{pmatrix}_L ; Y \begin{pmatrix} 0 \\ 0 \\ 0 \\ e \\ -\nu \end{pmatrix}_L = 1 \begin{pmatrix} 0 \\ 0 \\ 0 \\ e \\ -\nu \end{pmatrix}_L .$$

We have made use of the obvious imbedding of $SU(3)_C \times SU(2)_L \times U(1)_Y$ in a 5-dimensional representation of $SU(5)$ as mentioned above:

$$SU(3)_C \times SU(2)_L \to \left(\begin{array}{c|c} SU(3)_C & 0 \\ \hline 0 & SU(2)_L \end{array} \right) .$$

The subgroup $U(1)_Y$ is generated by

$$Y = \mathrm{diag}\left(-\frac{2}{3}, -\frac{2}{3}, -\frac{2}{3}, 1, 1 \right) .$$

A short excursion into the representation structure of these groups may be permissible here. A basic reference is [Itzykson and Nauenberg 1966]. The 5-dimensional representation $\underline{5}$ has the $SU(3)_C, SU(2)_Y$ content

$$\underline{5} : (3,1)_{-2/3} \oplus (1,2)_1$$

and the conjugate representation

$$\underline{5}^* : (3^*,1)_{2/3} \oplus (1,2)_{-1} .$$

The (1,2,3) content of higher-dimensional representations can be evaluated by considering tensor products of $\underline{5} \otimes \underline{5}$, etc. For example, using diagrams of Young tableaux,

$$\begin{array}{ccccccc} 5 & & 5 & & 15 & & 10 \\ \square & \otimes & \square & = & \square\square & \oplus & \begin{smallmatrix}\square\\\square\end{smallmatrix} \end{array}$$

(The use of the Young tableaux and the multiplication rules for tensor products are explained in [Itzykson and Nauenburg 1966].)

$$\underline{5} \otimes \underline{5} \equiv [(3,1)_{-2/3} \oplus (1,2)_1] \otimes [(3,1)_{-2/3} \oplus (1,2)_1] .$$

To carry out the multiplication, remember that $U(1)$ is abelian, i.e. the Y values just have to be added up. Furthermore, in $SU(3)$ products

$$\overset{3}{\square} \otimes \overset{3}{\square} = \overset{6}{\square\square} \oplus \overset{3^*}{\boxminus}$$

and in $SU(2)$

$$\underline{2} \otimes \underline{2} = \overset{1}{\boxminus} \oplus \overset{3}{\square\square}$$

Therefore one can split up $\underline{5} \otimes \underline{5}$ in the following way:

$$\underline{5} \otimes \underline{5} = (6,1)_{-4/3} \oplus (3^*,1)_{-4/3} \oplus (3,2)_{1/3} \oplus (3,2)_{1/3} \oplus (1,1)_2 \oplus (1,3)_2 .$$

Because of symmetry properties (6,1) and (1,3) have to belong to $\underline{15}$ and then, from dimensional arguments,

$$\underline{10} : (3,2)_{1/3} \oplus (3^*,1)_{-4/3} \oplus (1,1)_2$$

and

$$\underline{15} : (6,1)_{-4/3} \oplus (3,2)_{1/3} \oplus (1,3)_2 .$$

Table 6.1 gives the (1,2,3) content of various irreducible representations of $SU(5)$.

We have seen that one generation of fermions in the standard model has 15 different basic states, e.g. $u_L, d_L, d_L^C, u_L^C, \nu_L, e_L, e_L^C$ for the first generation. (As mentioned, by convention, one groups fermions of a fixed helicity together.) The (1,2,3) irreducible representation reads

$$\begin{aligned} u_L, d_L &: (3,2)_{Y=1/3} \\ d_L^C &: (3^*,1)_{2/3} \\ u_L^C &: (3^*,1)_{-4/3} \\ \nu_L, e_L &: (1,2)_{-1} \\ e_L^C &: (1,1)_2 \end{aligned}$$

(u_L^C, etc. are the charge conjugates of u_L etc.). A look at Table 6.1 shows that these basic states cannot be put into the 15-dimensional irreducible representation, but they can be accommodated in the direct sum $\underline{5} \oplus \underline{10}$ in the following way:

$$\begin{aligned} \underline{5}^* &: \nu_L, e_L, d_L^C, \\ \underline{10} &: e_L^C, u_L, d_L, u_L^C . \end{aligned}$$

Table 6.1. Irreducible representations of $SU(5)$

Young diagram dimension		$(SU(3), SU(2))_Y$
□ $\underline{5}$		$(3,1)_{-2/3} \oplus (1,2)_1$
□ $\underline{10}$ □		$(3,2)_{1/3} \oplus (3^*,1)_{-4/3} \oplus (1,1)_2$
□□ $\underline{15}$		$(6,1)_{-4/3} \oplus (3,2)_{1/3} \oplus (1,3)_2$
□□ $\underline{24} = \underline{24}^*$ □ □ □		$(8,1)_0 \oplus (3,2)_{-5/3} \oplus (3^*,2)_{5/3}$ $\oplus (1,3)_0 \oplus (1,1)_0$
Tensor products:	$\underline{5} \oplus \underline{5} = \underline{15} \oplus \underline{10}$ $\underline{5} \oplus \underline{5}^* = \underline{24} \oplus \underline{1}$	

Because $\underline{5}^* = (3^*,2)_{2/3} \oplus (1,2^*)_{-1}$, the 5-component spinor has to be chosen as

$$(\psi_{5^*})_L \equiv \begin{pmatrix} d_1^C \\ d_2^C \\ d_3^C \\ e^- \\ -\nu_e \end{pmatrix}_L . \tag{6.1}$$

This spinor transforms according to the 5-dimensional representation $\underline{5}^*$ of $SU(5)$. The choice of the component $\binom{e^-}{-\nu}_L$ derives from the fact that $\binom{\nu}{e}_L$ transforms according to $\underline{2}$ of $SU(2)$: $\underline{2} \simeq \underline{2}^*$. Therefore $\binom{e^-}{-\nu}_L$ transforms according to 2^*. For $U \in SU(2)$ and $(\varepsilon_{\alpha\beta})$: $\left\{ {\varepsilon_{12}=+1 \atop \varepsilon_{21}=-1} \right.$, zero otherwise, one has $\varepsilon \overline{U} \varepsilon^{-1} = U$, and therefore $\overline{a}_\alpha$ transforms as $\varepsilon_{\alpha\beta} a_\beta$ for 2-component spinors $a_\alpha : \varepsilon_{\alpha\beta}\binom{\nu}{e} = \binom{e}{-\nu}$ The 10-dimensional representation ⊟ can be written down as an antisymmetric matrix χ_{10} as follows:

$$(\chi_{10})_L \equiv \frac{1}{\sqrt{2}} \begin{pmatrix} 0 & u_3^C & -u_2^C & -u_1 & -d_1 \\ -u_2^C & 0 & u_1^C & -u_2 & -d_2 \\ +u_2^C & u_1^C & 0 & -u_3 & -d_3 \\ u_1 & u_2 & u_3 & 0 & -e^C \\ d_1 & d_2 & d_3 & e^C & 0 \end{pmatrix}_L .$$

1,2,3 is the colour index. χ_{10} and Ψ_{5^*} contain just the members of one family of the (1,2,3) model. The other families (or generations) may be included by inducing additional $\underline{5}^* \oplus \underline{10}$ representations. The $SU(5)$ scheme does not give any idea as to why there should be three generations. There is evidently some progress in the unification, since just a $\underline{5}^* \oplus \underline{10}$ representation is certainly an improvement over the structure of the $SU(3)_C \times SU(2)_L \times U(1)_Y$ model for a single family of fermions.

We should note that the 15 elementary fields do not include a right-handed neutrino. Therefore, in this scheme, the neutrinos have to be mass-zero particles. For a massive fermion, the right-handed helicity state is also present. A right-handed neutrino ν_R – equivalently a ν_L^C state – could, however, be accommodated in the singlet representation $(1,1)_0$.

The $SU(5)$ multiplets consist of both quarks and leptons, i.e. $SU(5)$ transformations will mix quark and lepton states. In addition, the $SU(5)$ gauge invariant interactions are the same for quarks and leptons.

The product $SU(2) \times U(1)$ is a subgroup of $SU(5)$, and therefore the charge operator Q of (5.38) belongs to the generators of $SU(5)$. The trace of any generator of a simple Lie group is zero in any representation. (For any $A \in \mathcal{G}$ there are $B, C \in \mathcal{G} : \mathcal{A} = [\mathcal{B}, \mathcal{C}]$, thus $TrA = TrBC - TrCB \equiv 0$.) Therefore $TrQ = 0$. The sum of the charges of the particles in a multiplet must be zero. Take, for example, the $TrQ = 0$ condition for Ψ_{5^*}, the $SU(5)$ quintet:

$$TrQ = q_{d_1^C} + q_{d_2^C} + q_{d_3^C} + q_{e^-} + q_\nu = 0. \tag{6.2}$$

It follows that

$$\begin{aligned} 3q_{d^C} + (-e) &= 0, \\ q_{d^C} &= \frac{1}{3}e. \end{aligned} \tag{6.3}$$

Placing the particles in $SU(5)$ multiplets thus immediately explains the quantization of the electric charge, as well as the relation of quark to lepton charges. The factor of 1/3 in (6.3) clearly follows from the existence of three different colours.

In the fundamental representation of $SU(5)$ (e.g. [Ross 1983]), the charge operator Q is equal to

$$Q \equiv \mathrm{diag}\left(-\frac{1}{3}, -\frac{1}{3}, -\frac{1}{3}, 1, 0\right). \tag{6.4}$$

The gauge bosons – as we have seen in Chap. 5 – correspond to the generators of the Lie algebra. There are 24 independent generators for $SU(5)$. They can be placed in the 24-dimensional representation. Table 6.1 suggests that one may identify the representations:

- $(8,1)_0$ with the gluon fields of $SU(3)_C$
- $(1,3)_0$ with the gauge bosons W^+, W^- W_3
- $(1,1)_0$ with the B-field (remember that B and W_3 are just rotations of A – the photon field – and Z).

In addition, there are 12 new gauge fields X_j, Y_j with charge $X(-4/3)$, $Y(-1/3)$ – and $X(4/3)$, $Y(+1/3)$, respectively – belonging to $(3,2)_{-5/3}$ and $(3^*,2)_{5/3}$ in the decomposition of $\underline{24}$.

With the $SU(5)$ gauge field in component form

$$V_\mu \equiv \sum_{a=1}^{24} V_\mu^a \, T_a \tag{6.5}$$

(the base T_a is normalized: $TrT_aT_b = \delta_{ab}$), the covariant derivative is

$$D_\mu \equiv \partial_\mu - i\frac{g_5}{\sqrt{2}}V_\mu . \tag{6.6}$$

Here g_5 is the universal coupling constant of $SU(5)$, and the factor $1/\sqrt{2}$ is chosen to give the conventional normalization.

The field strength is

$$F_{\mu\nu} = \partial_\mu V_\nu - \partial_\nu V_\mu - i\frac{g_5}{\sqrt{2}}[V_\mu, V_\nu]. \tag{6.7}$$

The requirement of a single coupling constant leads to a definite value of the Weinberg angle. (The following discussion follows [Straumann 1984a].) In the limit, when the $SU(5)$ symmetry is valid the coupling constants g_S, g, g' of the standard model must all be expressible through g_5. It turns out that g_5 and g are already normalized such that $g_S = g = g_5$. Only the coupling constant g' of the $U(1)_Y$ group is slightly different: Y is not an element of the Lie algebra normalized with the same constant as the $SU(3)_C$ or $SU(2)$ generators.

$$T_a = \frac{1}{2}\tau_a \quad \text{for} \quad SU(2) \quad a = 1,2,3 \quad \text{and}$$
$$T_a = \frac{1}{2}\lambda_a \quad \text{for} \quad SU(3) \quad a = 4,\ldots,11.$$

For both we have $Tr(T_a\, T_b) = \frac{1}{2}\delta_{ab}$.

Set $\frac{1}{2}Y = CT_0$, and write the charge operator $Q = T_3 + CT_0$. Since for a simple group G there exists a basis of generator such that $Tr(T_a\, T_b) =$ const. $\times\, \delta_{ab}$ in any representation, we have the relation

$$TrQ^2 = (1 + C^2)Tr(T_3^2). \tag{6.8}$$

Since the covariant derivative $\mathrm{D}_\mu = \partial_\mu - igT_a\, W_\mu^a - ig'(\frac{1}{2}Y)B_\mu$ is to be replaced by $\mathrm{D}_\mu = \partial_\mu - ig_5 T_a\, W_\mu^a - ig_5 T_0 B_\mu$, we find that $g' = g_5/C$ in the symmetry limit. In the standard model the following relation between the Weinberg angle and the coupling constants is valid:

$$\sin^2\theta_w = g'^2/(g^2 + g'^2).$$

In the $SU(5)$ limit this should read

$$\sin^2\theta_w = \frac{g_5^2/C^2}{g_5^2 + g_5^2/C^2} = \frac{1}{C^2+1} = \frac{Tr(T_3^2)}{Tr(Q^2)}. \tag{6.9}$$

With $(\Psi_L)_{5^*}$ from (6.2) this can be easily evaluated as

$$\sin^2\theta_w = \frac{2\times 1/4}{1+3\times 1/9} = \frac{3}{8} = 0.375.$$

The manifest disagreement with the experimental value of ≈ 0.2 brought $SU(5)$ into disfavour for some time, until Georgi, Quinn and Weinberg [Georgi et al. 1974] suggested that (6.10) could be correct only at high energy – at a large scale m_X – where $SU(5)$ is an exact symmetry. At lower energies, where $SU(5)$ is spontaneously broken, the value of $\sin^2\theta_w$ has to be computed by including radiative corrections (renormalization).

The fact that $\sin^2\theta_w = \frac{3}{8}$ in the symmetry limit does not depend on the specific GUT group $SU(5)$. This result can be proved for any simple Lie group G containing $SU(3)\times SU(2)\times U(1)$ as a subgroup. The general proof has to make use only of the theorem from group theory that in any representation of the Lie algebra of a simple group G there is a basis T_a such that

$$Tr(T_a\, T_b) = \text{const.}\ \delta_{ab}.$$

The computation of TrQ^2 etc. in the calculation of the Weinberg angle is, of course, only possible with the implicit assumption that the known fermions generate complete representations of the symmetry group.

The coupling of the fermion fields to the gauge fields within $SU(5)$ follows the usual procedure: replace $\partial_\mu\psi_f$ by the covariant $\mathrm{D}_\mu\psi_f$:

$$\mathrm{D}_\mu\psi_f \equiv \left(\partial_\mu - i\frac{g_5}{\sqrt{2}}U_*^f(V_\mu)\right)\psi_f.$$

Here U_*^f is the representation of $\mathcal{G}$ for a Fermi field representation U^f. Consider the representation 5^* with the spinor $(\psi_{5^*})_L$. It proves convenient to introduce, in addition, $(\psi_R)_5 \equiv \begin{pmatrix} d \\ e^C \\ -\nu^C \end{pmatrix}_R$ which transforms according to $\underline{5}$. The interaction term read off from the Lagrangian

$$L_f = (\overline{\psi}_f, i\gamma^\mu D_\mu\psi_f)$$

is

$$L_{\text{int}} = \frac{g_5}{\sqrt{2}} Tr\overline{\psi}_R\gamma^\mu V_\mu\Psi_R. \tag{6.10}$$

Insertion of $(\psi_R)_5$ and of V_μ, which in terms of the physical fields reads

$$V = \left(\begin{array}{ccc|cc} & \textit{Gluons} & & X_1 & Y_1 \\ & & & X_2 & Y_2 \\ & & & X_3 & Y_3 \\ \hline X_1^C & X_2^C & X_3^C & 0 & W^+ \\ Y_1^C & Y_2^C & Y_3^C & W^- & 0 \end{array}\right) + \frac{1}{\sqrt{12}} \begin{pmatrix} -A & & & & \\ & -A & & & 0 \\ & & -A & & \\ & & & 3A & \\ & 0 & & & 0 \end{pmatrix} \tag{6.11}$$

$$+\frac{1}{\sqrt{20}} Z \begin{pmatrix} 1 & & & & \\ & 1 & & 0 & \\ & & 1 & & \\ 0 & & & 1 & \\ & & & & -4 \end{pmatrix},$$

leads to a contribution from the neutrino field (for example) of $(g_5/2\cos\theta_w) \times \overline{\nu}_L \gamma^\mu \nu_L Z_\mu$. This agrees with the coupling of the Z in the standard model – for $g_S = 2\sqrt{2/3}e$ and $\sin^2\theta_w = \frac{3}{8}$, i.e. $\cos\theta_w = \sqrt{5/8}$. In a similar way the 10-dimensional representation of $SU(5)$ contributes various interaction terms, and finally the Lagrangian for the gauge-boson–Fermi-field interactions is

$$\begin{aligned} L_{\text{int}} &= L_{\text{int}}^{\text{SM}}(\text{gauge couplings of the standard model with} \\ &\quad g = g_5, \text{and} \quad \sin^2\theta_w = \tfrac{3}{8}) \\ &\quad + \frac{g_5}{\sqrt{2}} \left[X_{\mu;\alpha}^{-} \left\{ \overline{d}_R^\alpha \gamma^\mu e_R^C + \overline{d}_L^\alpha \gamma^\mu e_L^C + \epsilon_{\alpha\beta\gamma} \overline{u}_L^{C\gamma} \gamma^\mu u_L^\beta \right\} + \text{h.c.} \right] \\ &\quad - \frac{g_5}{\sqrt{2}} \left[Y_{\mu;\alpha} \left\{ \overline{d}_R^\alpha \gamma^\mu \nu_R^C + \overline{u}_L^\alpha \gamma^\mu e_L^C + \epsilon_{\alpha\beta\gamma} \overline{u}_L^{C\beta} \gamma^\mu d_L^\gamma \right\} + \text{h.c.} \right] \end{aligned} \tag{6.12}$$

(α, β, γ = 1,2,3 are the colour indices).

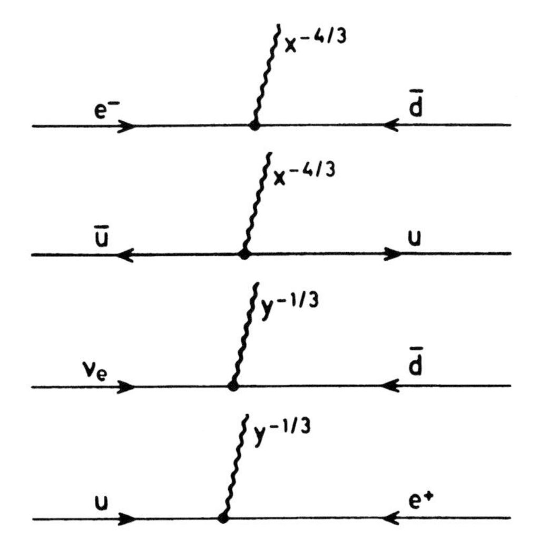

Fig. 6.1. *B*- and *L*-violating vertices

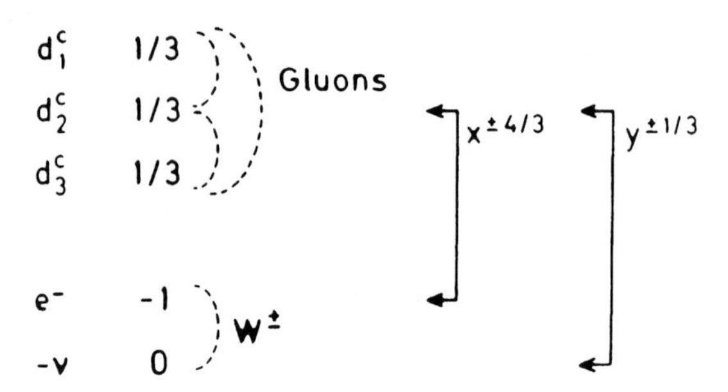

Fig. 6.2. Couplings of the components of ψ_{5^*}

The new terms in (6.12) lead to couplings between X, d, and e particles, u and X particles etc. All these processes violate the conservation of baryon number B and lepton number L (Figs. 6.1 and 6.2).

6.1.2 Spontaneous Symmetry-Breaking

The $SU(5)$ symmetry is spontaneously broken at high energies by inserting a number of scalar Higgs fields. As in other cases too, the Higgs fields is chosen in the adjoint representation $\underline{24}$:

$$\phi = \sum \varphi^a \, T_a$$

where T_a is a basis of the Lie algebra of $SU(5)$. The Higgs potential for the 24-plet of Higgs fields is written in the well-known form

$$V(\phi) = -\frac{1}{2}\mu^2 Tr(\phi^2) + \frac{1}{4}a(Tr\phi^2)^2 + \frac{1}{2}bTr\phi^4. \tag{6.13}$$

The constants can be arranged in such a way that the minimum lies at

$$\phi_0 \equiv v_{24} \left(\begin{array}{ccc|cc} 1 & & & & \\ & 1 & & & 0 \\ & & 1 & & \\ \hline & & & -3/2 & \\ & 0 & & & -3/2 \end{array}\right). \tag{6.14}$$

The parameter v_{24} is connected with a, b, and μ^2:

$$(15a + 7b)v_{24}^2 = 2\mu^2. \tag{6.15}$$

The group leaving ϕ_0 invariant is the gauge group of the standard model $SU(3)_C \times SU(2)_L \times U(1)_Y$. Since all the masses of X and Y bosons are equal, the trace of the mass matrix of the gauge bosons is [Ross 1983]

$$Tr(M^2) = 5g_5^2 Tr(\phi_0^2) = \frac{5 \times 15}{2} v_{24}^2 g_5^2 \tag{6.16}$$

and

$$m_X^2 = m_Y^2 = \frac{1}{12} Tr(M^2) = \frac{25}{8} g_5^2 v_{24}^2. \tag{6.17}$$

The 12 X and Y bosons acquire a mass, and their longitudinal degrees of freedom are supplied by the 12 scalar fields contained in the adjoint representation of the Higgs fields ϕ_a.

The remaining 12 Higgs scalars correspond to the generators of $SU(5)$ which leave the vacuum of (6.15) invariant. These should be identified with the Higgs sector carrying the same quantum numbers as the gluons, W, Z, and the photon. All the gauge bosons of the standard model remain massless after this first stage of symmetry breaking.

The 24-plet of Higgs fields cannot be used to generate fermion masses. These can only come through couplings with Higgs multiplets which transform according to irreducible representations which appear in the product decomposition of the fermion representations $\overline{\psi}_R \chi_L \simeq \overline{(\psi_L)^C} \chi_L$. Since one fermion generation

$$\underline{5}^* \oplus \underline{10}$$

we decompose

$$(\underline{5}^* \oplus \underline{10}) \otimes (\underline{5}^* \oplus \underline{10}) = \underline{10}^* \oplus \underline{15}^* \oplus \underline{5} \oplus \underline{45}^* \oplus \underline{5}^* \oplus \underline{45} \oplus \underline{50}.$$

Clearly the adjoint 24-dimensional representation does not occur. But there are possible couplings to $\underline{5}^*$ or $\underline{45}$, and one can choose a 5-plet H in addition. The fact that the 24-plet Higgs fields cannot be coupled to the fermions is, of course, fortunate, since otherwise the fermion masses would certainly have turned out too large – of the order of the GUT breaking scale M_X, M_Y. The 5-plet H can contain the Higgs doublet of the standard model. The symmetry-breaking scheme is thus introduced in two steps: first $SU(5)$ is broken down to the standard model by the 24-plet,

$$SU(5) \underset{\underline{24}}{\rightarrow} SU(3)_C \times SU(2)_L \times U(1)_Y; \tag{6.18}$$

then secondly, the usual breakdown of the electroweak symmetry to $SU(3)_C \times U_{\text{em}}(1)$ is achieved by the 5-plet H:

$$SU(3)_C \times SU(2)_L \times U(1)_Y \underset{\underline{5}}{\rightarrow} SU(3)_C \times U_{\text{em}}(1). \tag{6.19}$$

The minimum of the effective potential can be chosen at

$$H_0 \equiv \frac{v_5}{\sqrt{2}} \begin{pmatrix} 0 \\ 0 \\ 0 \\ 0 \\ 1 \end{pmatrix}. \tag{6.20}$$

The mass matrix for the gauge bosons has a part derived from H:

$$M_{ab}^2 V^a V^b = \frac{g_5^2}{2} H_0^T V V H_0 = \frac{g_5^2}{2}\frac{v_5^2}{2}\left(W^+W^- + \frac{4}{5}Z^2\right). \tag{6.21}$$

The mass of the W and Z bosons can be read off from (6.22) as

$$m_W = \frac{1}{2} g_5 v_5, \qquad \text{and} \tag{6.22}$$

$$m_Z = \frac{m_W}{\cos\theta_w}; \qquad \text{for} \qquad \sin^2\theta_W = \frac{3}{8} \tag{6.23}$$

This $SU(5)$ model with 29 Higgs fields is called the minimal model. Extensions that involve more Higgs fields can be easily made. The formulae are exactly the same as in the standard model – except that all couplings are evaluated at the $SU(5)$ scale. But, if we assume that the coupling constant g_2 of the electroweak model changes into g_5 in the symmetry limit of $SU(5)$ (see Sect. 6.2 for a discussion of these effects) – i.e. for high enough energies – then we may conclude that

$$v_5 = (\sqrt{2}G_F)^{-1/2} \simeq 246\,\text{GeV}. \tag{6.24}$$

Besides the "physical" Higgs particles H_4, H_5 corresponding to the complex doublet (ϕ^+, ϕ_0) of Sect. 5.5, there is a triplet of complex fields H_1, H_2, H_3 which transform according to $(3,1)_{-2/3}$ and $(3^*,1)_{2/3}$. It is a straightforward necessity to provide masses of the order of the GUT scale for these particles, because otherwise they might induce rapid proton-decay reactions:

$$u + d \to H^{1/3} \to \begin{matrix} e^+ + \overline{u} \\ \overline{\nu}_e + \overline{d} \end{matrix};$$

$\frac{1}{3}$ is the value of $Y = Q + \frac{1}{3}T_3$.

Masses originate from interaction terms of the ϕ and H fields:

$$V_{H,\phi} \equiv \alpha|\overline{H}^*|^2 Tr\phi_{24}^2 + \beta\overline{H}^*\phi_{24}^2\overline{H}. \tag{6.25}$$

The minimum of the effective potential then involves (6.25) as well as (6.13), and a similar $V(H)$ for the 5-plet. Despite the mixing of the two parameters v_{24} and v_5 in the relations for the minima, the ratio of $v_5/v_{24} \approx 10^{-13}$ of the electroweak and the $SU(5)$ scale has to be preserved. This so-called "hierarchy problem" is generally viewed as an artificial, "unnatural" prescription. Radiative corrections will, in general, not preserve such a fine-tuning. In perturbation calculations divergencies of the self-energies require, at each order, a rearrangement to keep the small v_5/v_{24} ratio [Haber and Kane 1985]. Such corrections would come from graphs such as those in Fig. 6.3.

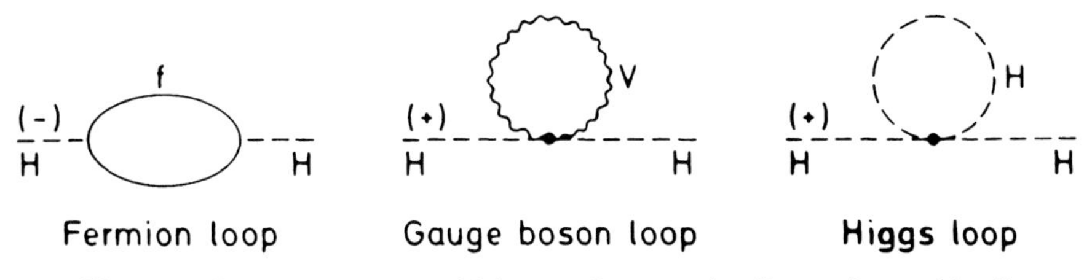

Fig. 6.3. Loop corrections which contribute to the "hierarchy problem"

The ways out of this hierarchy problem are either to make the Higgs particles into composite structures – the so-called "Technicolor theories" – or to introduce a supersymmetric GUT (see Sect. 6.3).

The Higgs 5-plet H generates fermion masses. The conjugate representation 5^* may be used to write the Yukawa couplings

$$L_{\mathrm{YUK}} = \frac{1}{\sqrt{2}}(\chi_i^+)^{\alpha\beta}\gamma_0 M_{ij}^D [H_\alpha^* \psi_\beta H_\beta^* \Psi_\alpha]_j \\ -\frac{1}{4}\epsilon^{\alpha\beta\gamma\delta\epsilon}(\chi_i)_{\alpha\beta} M_{ij}^U (\chi_j)_{\delta\epsilon} H_\gamma^* . \tag{6.26}$$

Here i and j are generation indices and $\alpha\beta\gamma\delta\epsilon$ are $SU(5)$ indices. The mass terms have the consequence that

$$m_d = m_e . \tag{6.27}$$

Similarly, it can be shown that

$$m_S = m_\mu , \qquad m_b = m_\tau \tag{6.28}$$

for the other generations. The mass of the up quark is contained in the second term of (6.26) – this mass is a free parameter of the model.

The relations (6.27) and (6.28) should be viewed as equalities for large momentum-transfer processes $q^2 \geq m_X^2$, and their continuation to small values of q^2 requires the techniques of the renormalization group [Buras et al. 1978].

6.2 Evolution of the Coupling Constants

We have seen that the gauge theory for $SU(5)$ requires the following relations:

$$\sin^2\theta_w = \frac{3}{8} ,$$
$$m_d = m_e ; \qquad m_s = m_u ; \qquad m_b = m_\tau ,$$
$$g_3 = g_2 = g_5 .$$

All these results are in conflict with laboratory measurements – the strong coupling constant g_3, for example, is much larger than g_2, the coupling constant of the weak interaction. One has to realize, however [Georgi et al. 1974], that the above relations can be valid only at an energy scale $O(m_X)$, where $SU(5)$ is a good symmetry, and that for comparison with experiments it is necessary to include radiative corrections. Then the masses and coupling "constants" will no longer be constants, but functions of energy. They can be continued to an energy scale, where measurements are made.

It is a well-known fact in electrodynamics that the charge appearing in the Lagrangian has to be corrected for virtual e^+e^- loops (the vacuum polarization, as shown in Fig. 6.4).

The whole scheme of renormalization group techniques was developed originally for quantum electrodynamics: the photon propagator is modified in the 1-loop approximation (Fig. 6.5) as follows:

$$\frac{\eta_{\mu\nu}}{q^2} \to \frac{\eta_{\mu\nu}}{q^2}(1 - \Pi(q^2)) \tag{6.29}$$

where

$$\Pi(q^2) = \frac{\alpha}{3\pi} q^2 \int_{4m^2}^{\infty} ds \frac{(1 + 2m^2/s)\sqrt{1 - 4m^2/s}}{s(s - q^2 - i\epsilon)}; \qquad \Pi(0) = 0. \tag{6.30}$$

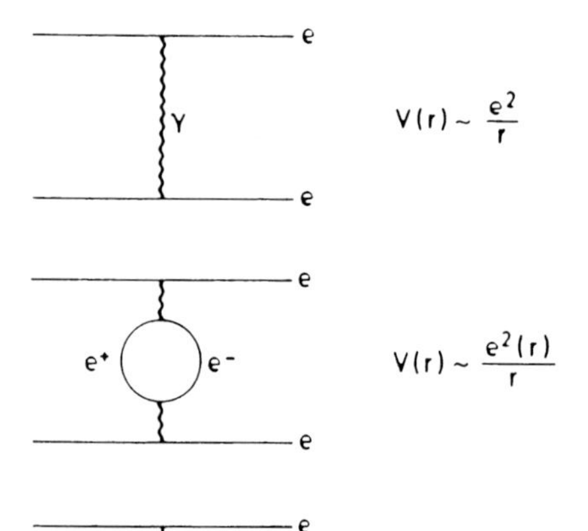

Fig. 6.4. The effective Coulomb potential has an additional $e(r)$ dependence caused by radiative corrections. The "loop correction" consists of the contribution, by creation and subsequent annihilation, of "virtual" pairs of e^+ and e^-. The particles are "virtual" because their 4-momentum $p^2 \neq m^2$

Fig. 6.5. One-loop correction to the photon propagator

For $|q^2| \gg m^2$,

$$\Pi(q^2) \simeq -\frac{\alpha}{3\pi} ln(+q^2/m^2). \tag{6.31}$$

Here α is the fine-structure constant ($\alpha = e^2/\hbar c$).

Equation (6.29) has the consequence that the fine-structure constant in scattering amplitudes has to be replaced by an energy-dependent quantity

$$\alpha(q^2) = \alpha(0)\left[1 - \Pi(q^2)\right].$$

Using the approximation of (6.31) yields

$$\alpha(q^2) = \alpha(m^2)\left[1 + \frac{\alpha(m^2)}{3\pi^2}\ln\left(\frac{+q^2}{m^2}\right) + O(\alpha^2)\right]. \tag{6.32}$$

In general the energy dependence of a coupling constant $g(\mu)$ appears through a solution of the equation (e.g. [Ross 1984])

$$\mu\frac{d}{d\mu}g(\mu) = \beta[g(\mu)]. \tag{6.33}$$

Here $\mu^2 = q^2; g(\mu)$ is the effective coupling constant, and $\beta(g)$ is a function that can be evaluated, for small values of g, by perturbation calculation.

From the lowest-order term the integration of (6.33) yields the result

$$\frac{1}{g^2(\mu)} = \frac{1}{g^2(m)} + b\ln\frac{m^2}{\mu^2}. \tag{6.34}$$

Taking m_X to be the grand unification scale, where $g_1(m_X) = g_2(m_X) = g_3(m_X) = g_5$, we find that

$$\frac{1}{\alpha_i(\mu)} = \frac{1}{\alpha_5} + b'_i\ln\frac{m_X^2}{\mu^2}, \tag{6.35}$$

where $\alpha_i = g_i^2/4\pi$ and $b'_i = 4\pi b_i$. The lowest-order contributions to the coefficients b'_i come from graphs of the type shown in Fig. 6.6.

This leads to the following behaviour of the effective couplings of the (1,2,3) model below the scale m_X

$$\frac{1}{\alpha_3(\mu)} = \frac{1}{\alpha_5} + \frac{1}{6\pi}(4N_G - 33)\ln\left(\frac{m_X}{\mu}\right),$$

$$\frac{1}{\alpha_2(\mu)} = \frac{\sin^2\theta(\mu)}{\alpha\mu} = \frac{1}{\alpha_5} + \frac{1}{6\pi}\left(4N_G - 22 + \frac{1}{2}\right)\ln\left(\frac{m_X}{\mu}\right), \tag{6.36}$$

$$\frac{1}{\alpha_1(\mu)} = \frac{3}{5}\frac{\cos^2\theta(\mu)}{\alpha\mu} = \frac{1}{\alpha_5} + \frac{1}{6\pi}\left(4N_G + \frac{3}{10}\right)\ln\left(\frac{m_X}{\mu}\right).$$

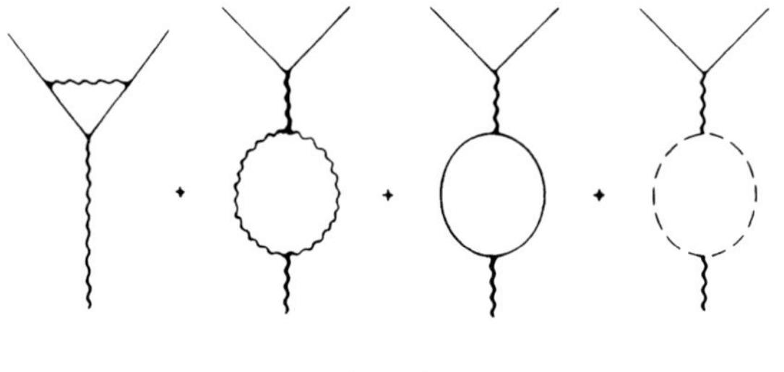

Fig. 6.6. Loop contributions to $g(\mu)$

N_G is the number of generations; in each generation of fermions there are four left-handed doublets: one lepton doublet and a quark doublet in three different colours. The contribution from scalar Higgs particles has been neglected in α_3; it is 1/2 in $1/\alpha_2$ and 3/10 in $1/\alpha_1$ of (6.36). If we remember that the number of flavours $N_F = 2N_G$, and that the condition for asymptotic freedom is $N_F < 16$ or $N_F < 8$, then we see that $\alpha_3(\mu)$ decreases with increasing energy μ. The same is true for the $SU(2)$ coupling $\alpha_2(\mu)$. This expresses the property of asymptotic freedom of non-abelian gauge groups. $\alpha_1(\mu)$, the $SU(1)$ coupling constant, increases with energy. In addition, α_3 decreases faster than α_2: $|4N_G - 33| > |4N_G - 22 \pm \frac{1}{2}|$.

Any GUT theory must be aware of obstacles preventing a smooth extrapolation to high energies

All this leads to the conclusion that eventually all three couplings will meet approximately at same large energy $\approx m_X$. The precision measurements of standard model radiative correction effects indicate that the couplings meet almost, but not quite, at one point (see Fig. 6.7, where the situation is sketched). Some consider this slight non-convergence as a first hint of the effects of supersymmetry. That theory introduces corrections which make the running couplings meet at one point.

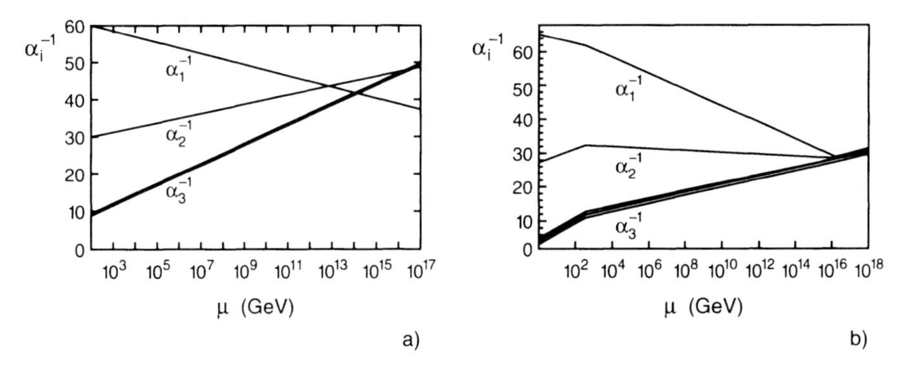

Fig. 6.7. The low-energy couplings measured at LEP (**a**) do not quite converge at a GUT energy scale, but (**b**) do so, when supersymmetry is included [Amaldi et al. 1991]

It is easily seen that m_X can be determined in terms of α_3 and α_2 from (6.36). Moreover, a prediction for α_1 (or $\sin^2\theta_w$) can be extracted:

$$\frac{3}{5\alpha_2(\mu)} + \frac{1}{\alpha_1(\mu)} = \frac{3}{5\alpha(\mu)}, \tag{6.37}$$

$$\frac{3}{5\alpha(\mu)} - \frac{8}{5\alpha_3(\mu)} = \frac{201}{30\pi}\ln\frac{m_X}{\mu}, \tag{6.38}$$

or equivalently,

$$\frac{1}{\alpha(\mu)} - \frac{8}{3\alpha_3(\mu)}\frac{33.5}{6\pi}\left(\frac{m_X}{\mu}\right)^2 \simeq \frac{11}{2\pi}\left(\frac{m_X}{\mu}\right)^2. \tag{6.39}$$

This combination of coupling constants, $1/\alpha - 8/3\alpha_3$, is independent of the number of fermion generations N_G.

For the Weinberg angle θ_w we obtain

$$\frac{1}{\alpha_2} - \frac{1}{\alpha_1} = \frac{\sin^2\theta_w}{\alpha} - \frac{1}{5}\frac{\cos^2\theta_w}{\alpha} = \frac{1}{6\pi}\left(\frac{-1}{10} - 22\right)\ln\left(\frac{m_X}{\mu}\right). \tag{6.40}$$

Therefore,

$$\sin^2\theta_w(\mu) = \frac{3}{8} - \frac{54.5}{18\pi}\alpha(\mu)\ln\left(\frac{m_X}{\mu}\right)^2. \tag{6.41}$$

To determine m_X we can use the values of the coupling constants at the value $\mu^2 = m_W^2$ of the electroweak unification energy:

$$\alpha(m_W^2)^{-1} = 127.8, \tag{6.42}$$

$$\alpha_3(m_W^2) = 0.11. \tag{6.43}$$

These values lead to

$$m_X = 2 \times 10^{15}\,\text{GeV}, \tag{6.44}$$

and

$$\sin^2\theta_w(m_W) \simeq 0.21. \tag{6.45}$$

The very large mass obtained for m_X follows from the logarithmic evolution of the coupling constants with μ. The scale m_X has to be large in any case to inhibit the decay of the proton (see Sect. 6.3). The agreement of $\sin^2\theta_w$ with the experimental results is very impressive. Remember that the initial value of $\frac{3}{8}$ – derived in the symmetry limit – was in strong disagreement. These two results indicate that the GUT theory concept may have some basis in reality.

It is also a gratifying consistency check that m_X is still a few orders of magnitude below the Planck mass $M_{Pl} = 10^{19}$ GeV. Thus one is still below the scale at which quantum-gravitational interactions become important.

The derivation of the change in the coupling constants with energy has been improved several times. For example, the scenario hinted at in Fig. 6.7 supposes that above m_X they coincide, whereas in reality they will only asymptotically and approximately be equal. Such refinements tend to reduce somewhat the value of m_X.

Detailed calculations show rather large uncertainties. Because of the logarithmic dependence of α_i on m_X, the errors in, for example, α_3, appear exponentially magnified in these calculations.

6.3 Nucleon Decay in $SU(5)$ GUT

GUTs combine quarks and leptons in common multiplets and therefore introduce interactions which couple these particles. This does not necessarily imply the violation of baryon number (or lepton number). For example, the terms coupling the $\underline{5}^*$ multiplet of $SU(5)$ to the X and Y bosons have a structure such as $d_L^C\gamma^\mu\nu_L$, i.e. they all have a baryon number $B = +1/3$ and a lepton number $L = +1$. Thus if $\underline{5}^*$ was the only fermion representation, these quantum numbers could be ascribed to the $\overline{X}$ and $\overline{Y}$ fields and thus B and L would be separately conserved. But a combination of this and the couplings involving the $\underline{10}$ multiplet, with q and $\overline{q}$ in the same representation, leads to a separate violation of B and L: terms involving $\overline{u}_L^C\gamma^\mu u_L$ have $B = 2/3$, $L = 0$ and B and L cannot be assigned to X and Y in a way that conserves B or L separately.

In $SU(5)$ the difference $B - L$ is conserved. This is an accidental global symmetry of the minimal $SU(5)$ model which is lost in extensions of this model.

In the minimal $SU(5)$ model the various B- and L-violating interactions have been represented in Figs. 6.1 and 6.2. These fundamental processes give

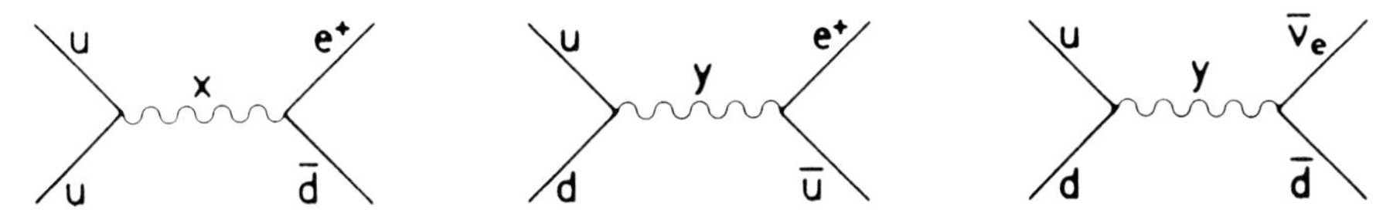

Fig. 6.8. Graphs contributing to nucleon decay in $SU(5)$

rise to the interactions of Fig. 6.8. These lead to graphs for the decay of the proton (Fig. 6.9). The large mass of the X boson makes the proton decay very slow. The simplest diagrams indicate that the proton lifetime

$$\tau_p \propto m_X^4/m_p^5. \tag{6.46}$$

The diagrams are similar to the diagrams for μ-decay in the standard model. Since

$$\tau_\mu \propto m_W^4/m_\mu^5, \tag{6.47}$$

we can obtain a simple estimate

$$\tau_p = \tau_\mu \left(\frac{m_\mu}{m_p}\right)^5 \left(\frac{m_X}{m_W}\right)^4. \tag{6.48}$$

For $m_X \simeq 3 \times 10^{14}$ GeV, i.e. $m_X/m_W \sim 3 \times 10^{12}$, $\tau_p \simeq 10^{32}$ years.

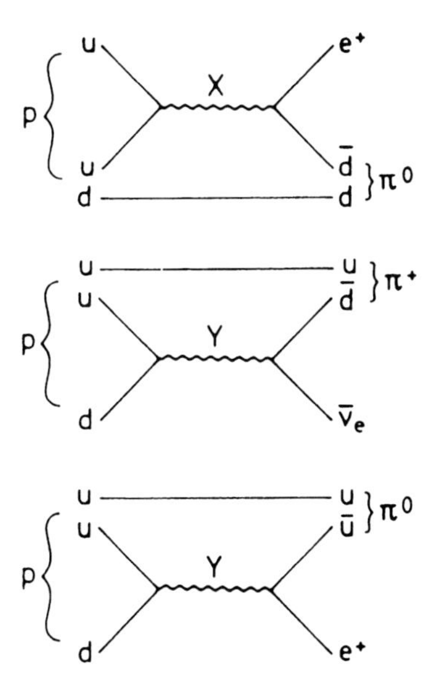

Fig. 6.9. Some diagrams of proton decay

For a better estimate of τ_p one needs:

i) to know m_X as exactly as possible, since $\tau_p \sim m_X^4$;
ii) reliable operator-matrix elements for physical hadron states to compute processes such as $p \to e^+ + \pi^0$ (Fig. 6.10).

Both estimates are liable to rather large uncertainties. The results [Ross 1983] are in the range:

$$m_X = (1 \text{ to } 6) \times 10^{14}\,\text{GeV}, \tag{6.49}$$

$$\tau_p = (0.25 \text{ to } 10) \times 10^{30} \left(\frac{m_X}{4 \times 10^{14}}\,\text{GeV} \right)^4 \text{ years } = 10^{29 \pm 2} \text{ years.} \tag{6.50}$$

The calculations also predict the decay modes expected. The favoured decay mode for the proton is $\mathrm{p} \to e^+ + \pi^0$.

$$\tau(\mathrm{p} \to e^+ + \pi^0) \sim 4.5 \times 10^{29 \pm 1.7} \text{ years.} \tag{6.51}$$

A lower limit was obtained from first experiments (IMB, Frejus, Kamiokande combined) [Fiorini 1984],

$$\tau(\mathrm{p} \to e^+ + \pi^0) > 4 \times 10^{32} \text{ years.} \tag{6.52}$$

Another experimental limit is $\tau(\mathrm{p} \to \nu + K^+) > 6 \times 10^{31}\,\mathrm{y}$ (IMB, Kamiokande). The prediction of the minimal $SU(5)$ model (6.50) already seems outside the error interval of this experimental result.

Because of the large uncertainties in the theoretical estimate of m_X which exponentiates errors in α_3, one should, at this point, be cautious with such very definite conclusions. The situation does not immediately discredit the GUT concept anyway, because a more complicated GUT can be employed (see Sect. 6.4).

The experiments looking for proton decay all tried to detect the secondaries from nucleon decay. The main background effect is due to neutrinos. One set-up, the so-called "calorimetric" approach, used a sandwich structure of horizontal or vertical plates of massive material (normally iron) interspaced

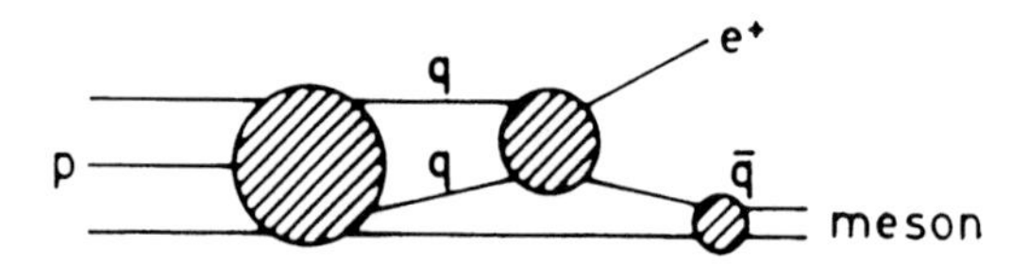

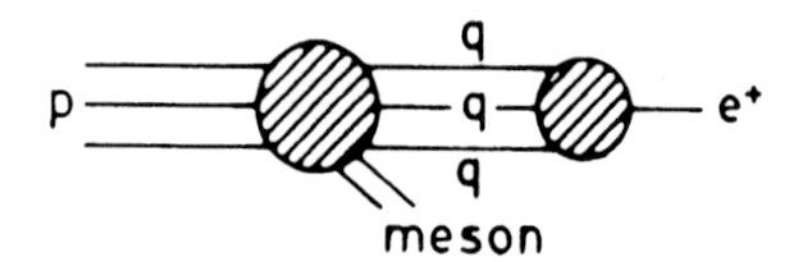

Fig. 6.10. Operator-matrix elements of proton decay

with planes of detectors. An example of this set-up was the Kolar gold-mine experiment, a detector of dimensions 6 × 4 × 3.7 m^3 and with a mass of 140 tonnes.

Another approach is based on the detection of the Čerenkov light emitted in a transparent medium by high-velocity charged secondaries of the nucleon decay ($p \to e^+\pi^0$, for example). The source for nucleon decay can be a large amount of highly purified water. The Čerenkov light is collected by photomultipliers on the walls of, or inside, the water tank.

The largest Čerenkov detector operates in Japan (Kamiokande and Super-Kamiokande). These detectors and others in the USA and in Europe are now used for neutrino observations from supernovae, the Sun, and the Earth's atmosphere.

The experimental results obtained up to now are negative, except for a few candidate events that are disputed and not of reliable significance [Perkins 1984; Fiorini 1984; Meyer 1986].

All experiments find events that can be attributed to interaction with atmospheric neutrinos. In fact, within the bounds of statistical significance all events found so far can be explained by neutrino interactions (such as $\nu_\mu + p \to \mu^+ + \pi^0$) in agreement with estimated neutrino background properties.

6.4 Beyond $SU(5)$

6.4.1 General Remarks

In the previous sections of this chapter we have seen that the minimal $SU(5)$ model has several attractive features:

a) It contains the (3,2,1) standard model with all its successes.
b) The number of gauge coupling constants is reduced to one.
c) The prediction for the value of the Weinberg angle θ_w is in good agreement with experiment.
d) The electrical charge is quantized, and the 1/3 nature of the quark charges is explained – a very remarkable result.
e) There are some reasonably good predictions for the quark masses.
f) There are baryon-number-violating reactions which give the possibility of eventually explaining the baryon excess of the universe.

But there are also a number of less enjoyable properties:

i) The number of generations, and the whole family structure, are not explained. There is no connection between the vector, fermion, and scalar representations.
ii) The CP violation is not incorporated naturally.

iii) The appearance of the large mass scale m_X explains, on the one hand, why the proton decay should be slow. On the other hand, it introduces the disastrous hierarchy problem of the ratio $m_X/m_w \simeq 10^{13}$ of the $SU(5)$ and the weak interaction-breaking scale.
iv) No experimental evidence exists as yet for the decay of the proton.
v) No experimental evidence exists for the new particles.
vi) There is no asymptotic freedom of the Higgs self-coupling λ.

The general consensus is that the minimal $SU(5)$ model cannot be the final theory. But the hope is that all of its nice properties can be preserved in a more complete theory, perhaps an extension of $SU(5)$.

6.4.2 Larger Gauge Groups

Various larger groups, such as $SO(10)$, $E6$ have also been considered in GUT models. $SO(10)$ is a group of rank 5, which contains $SU(3) \times SU(2) \times SU(1)$ as a subgroup. Its complex 16-dimensional spinor representation has the $SU(3) \times SU(2)$ decomposition

$$\underline{16} = (3,2) \oplus 2 \times (3^*,1) \oplus (1,2) \oplus 2 \times (1,1). \tag{6.53}$$

This representation accommodates one generation of fermions plus a neutrino state ν_R or $\overline{\nu}_L$.

In $SU(5)$ the unification scale m_X as well as $\sin^2\theta_w$ are uniquely determined, because there is no larger subgroup of $SU(5)$ that contains $SU(3) \times SU(2) \times U(1)$. This is no longer the case for $SO(10)$. Since $SU(5)$ is contained in $SO(10)$ the quantization of the electric charges works the same way, and $\sin^2\theta_w = 3/8$ in the symmetry limit, as derived previously. The predictions due to the renormalized coupling constants vary according to the scheme for breaking the $SO(10)$ symmetry. Several different mass scales may be introduced; and it is easy to change the prediction of the proton decay rate by a factor of 10^3. The minimal version for breaking $SO(10)$ has a pattern

$$\begin{array}{c} SO(10) \\ \underline{16} \text{ of Higgs} \end{array} \rightarrow \begin{array}{c} SU(5) \\ \underline{45} \text{ of Higgs} \end{array} \rightarrow \begin{array}{c} SU(3) \times SU(2) \times U(1) \\ \underline{10} \text{ of Higgs} \end{array} \rightarrow SU(3) \times U(1)\,. \tag{6.54}$$

In general, fermion masses can be generated by Higgs fields transforming as the irreducible representations that are present in the direct product decomposition

$$D_R \times D_L \sim \underline{16} \times \underline{16} = 10_S \oplus \underline{126}_S \oplus \underline{120}_A \tag{6.55}$$

(S: symmetric, A: antisymmetric).

As under $SU(5)$, $\underline{10} = \underline{5} \oplus \underline{5}^*$, with ϕ_{10} one obtains again the 10-plet of Higgs fields, the $SU(5)$ relations

$$m_d/m_e = m_s/m_\mu = m_b/m_\tau = 1 \tag{6.56}$$

plus new relations for the Dirac masses

$$m_u = m_{\nu_e}; \qquad m_c = m_{\nu_\mu}; \qquad m_t = m_{\nu_\tau}. \tag{6.57}$$

Neutrino masses therefore appear naturally in an $SO(10)$ scheme. These neutrino masses correspond to Dirac terms of the form $\overline{\nu}_L \nu_R$ + h.c. Diagonal mass terms for the neutrinos (Majorana terms) can only be generated via a ϕ_{126} of Higgs fields.

We should remind ourselves here that the neutrinos in the standard model have to be massless. Since only ν_L appears, only Majorana terms $m\nu_L\nu_L$ (i.e. $m\nu_L^T C\nu_L$ + h.c.) would be allowed. Since they transform with $SU(2)_L$ as $\underline{T} = 1$, $\underline{T}_3 = 1$ objects, their masses cannot be generated by couplings to a Higgs doublet. The same conclusion holds for the minimal $SU(5)$ model. $m_\nu = 0$, as only a Majorana term is possible, which transforms according to $\underline{15}$, and therefore cannot be generated by Yukawa couplings to ϕ_{24} or H_5.

6.4.3 Neutrino Masses

In $SO(10)$, as in most other extensions of $SU(5)$, a right-handed neutrino ν_R can be accommodated. Then a Dirac mass term $\overline{\nu}_L \nu_R$ may arise, as well as a Majorana term from couplings to the Higgs fields $(H\nu_L)^2$.

If both ν_L and ν_R fields are allowed, the mass matrix involving both Dirac and Majorana masses has the form

$$(\nu_L, \overline{\nu}_R) \begin{pmatrix} m_1 m_2 \\ m_2 m_3 \end{pmatrix} \begin{pmatrix} \nu_L \\ \overline{\nu}_R \end{pmatrix}. \tag{6.58}$$

m_2 is chosen to be of order m_q (m_q: a typical quark mass). m_1 may derive from a Yukawa coupling to a Higgs field $(H\nu_L)^2$ of order v_0^2/m_X, where m_X is the very large grand-unification scale. Since ν_R is a singlet, the expected value of m_3 is of the scale m_X. Therefore a mass matrix of the form

$$(\nu_L, \overline{\nu}_R) \begin{pmatrix} -v_0^2/m_X & -m_q \\ -m_q & m_X \end{pmatrix} \begin{pmatrix} \nu_L \\ \overline{\nu}_R \end{pmatrix} \tag{6.59}$$

must be diagonalized.

The mass eigenstates are $\nu_L + \epsilon\nu_R$ and $\nu_R + \epsilon\nu_L$ with $\epsilon = v_0^2/m_X^2$ and with mass eigenvalues $\sim v_0^2/m_X^2$ and $\sim m_X$ respectively. Since m_q is a typical scale for $\langle H \rangle \approx v_0$, we have $m_{\nu_L} \approx m_q^2/m_X$.

Because m_X is so large, the neutrino masses expected in any particular GUT scheme are very small: 10^{-6} to 10^{-3} eV for the left-handed neutrinos that are observed. The right-handed ν_R becomes superheavy $\sim m_X$.

Somewhat larger masses are possible in particular models. One can in a general, qualitative way argue as follows [Weinberg et al. 1980]: the ex-

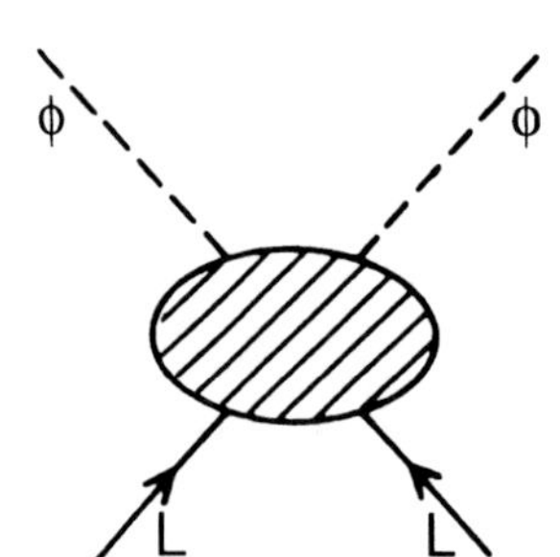

Fig. 6.11. Schematic diagram of an interaction that violates lepton–number conservation

change of a supermassive particle can lead to an effective interaction which violates lepton number conservation (Fig. 6.11).

$$L_{\text{eff}} = \frac{f}{M}\left(\phi^T \epsilon l\right)^2 = \frac{f}{M}\langle\phi_0\rangle^2 \nu_L \nu_L + \dots . \tag{6.60}$$

This leads to

$$m_\nu = \frac{f}{M}\langle\phi_0\rangle^2 . \tag{6.61}$$

Now $\langle\phi_0\rangle \simeq 300\,\text{GeV}$, and for $M \sim 10^{14}\,\text{GeV}$ one obtains

$$m_\nu = \frac{f}{M}\langle\phi_0\rangle^2 = (1\,\text{eV}) \cdot f . \tag{6.62}$$

There is also the possibility [Witten 1980] of generating, via higher-order corrections, an effective Majorana mass for ν_R of the order of $\alpha^2 M_X$. This leads to a mass for the light neutrino of

$$m_\nu = m_q \frac{m_w}{\alpha^2 m_{\text{GUT}}} . \tag{6.63}$$

(m_q is the up quark of the same generation.) Then, for example,

$$m_{\nu_\tau} \simeq 10^{-9} m_{\text{top}} \simeq 180\,\text{eV} \qquad \text{for} \qquad m_{\text{top}} = 180\,\text{GeV} .$$

In general, neutrino masses in GUT theories are small $m_\nu \sim m_q^2/m_X$, but they are not very precisely determined within present-day frameworks.

Neutrinos will be discussed further in Chap. 7.

6.5 Axions

Axions are an ingredient – some would say an unwelcome one – of theoretical attempts to deal with the CP problem of the strong interactions. The CP symmetry, i.e., invariance against an interchange of particles and antiparticles, seems to be preserved in most experiments, but attempts to embody it in the theoretical formulation meet with difficulties.

The only experimental evidence for a violation of CP invariance is found in the decay of the K_0 meson. For the strong interactions, especially QCD, this symmetry should hold, at least to a very good approximation. For some time the invariance of QCD was indeed taken for granted, because the QCD fermion Lagrangian can be brought into a form explicitly invariant under CP transformations.

There are, however, different quantum configurations that correspond to the same classical Lagrangian. In general, these configurations are no longer CP–invariant. They are classified by an additional term in the Lagrangian, proportional to a parameter θ:

$$L_\theta = \theta \frac{g_s^2}{32\pi^2} G \cdot \widetilde{G}. \tag{6.64}$$

Here G is the gluon field strength

$$G^a_{\mu\nu} = \partial_\mu A^a_\nu - \partial_\nu A^a_\mu + g_s f^{abc} A^b_\mu A^c_\nu \tag{6.65}$$

and

$$\overline{G}^a_{\mu\nu} \equiv \frac{1}{2} \epsilon_{\mu\nu\varrho\sigma} G^{a\varrho\sigma}$$

is the dual tensor. The L_θ term destroys the CP and $U(1)$ invariance of the Lagrangian for the quark fields (in QCD, considering only one type of quark and setting the quark mass equal to zero)

$$L = i\bar{q}\gamma^\mu D_\mu q + L_\theta.$$

The chiral transformations

$$\begin{aligned} q &\to e^{i\alpha\gamma_5} q, \\ \bar{q} &\to \bar{q} e^{-i\alpha\gamma_5} \end{aligned} \tag{6.66}$$

change L_θ. The reason is that the generator of the chiral transformations cannot be identified with the charge

$$Q_A \equiv \int J^0_A \, d^3x$$

of the axial current

$$J^\mu_A = \bar{q}\gamma^\mu\gamma_5 q,$$

since the divergence is non-zero:

$$\partial_\mu J^\mu_A = 2G^a_{\mu\nu}\overline{G}^{\mu\nu}_a/32\pi^2.$$

The additional term L_θ is a total divergence, and thus is unimportant within the classical theory. It is, however, relevant for non-linear instanton solutions of QCD.

The product $G \cdot \overline{G}$ changes sign under a parity transformation (similar to the product $\mathbf{E} \cdot \mathbf{B}$ of electrodynamics) or a time inversion, but remains invariant under a charge conjugation. Thus it induces CP-violating processes with a strength depending on the parameter θ. The similarity to the Aharonov–Bohm effect is obvious: this quantum effect is also due to a divergence term, which has no physical relevance in classical electrodynamics.

To restore the symmetry by setting $\theta = 0$ does not help, because the contributions of the weak interactions will renormalize θ to a non-zero quantity. L_θ would contribute to a hypothetical electric dipole moment d_n of the neutron via its coupling to the electromagnetic current with an amplitude

$$\langle n | i \int d^4x L_\theta J_\mu^{\mathrm{em}}(0) | n \rangle$$

One finds [Baluni 1979]

$$\left| \frac{d_n}{e} \right| \simeq 3 \times 10^{-16} \theta \quad \mathrm{cm}. \tag{6.67}$$

A comparison with the experimental upper limit of

$$\left| \frac{d_n}{e} \right| \leq 6 \times 10^{-25}\,\mathrm{cm} \tag{6.68}$$

[Dress et al. 1977] leads to a bound on the CP-violating parameter

$$|\theta| \leq 2 \times 10^{-9}. \tag{6.69}$$

How can such a small parameter be incorporated into the theory in a natural way?

The approach of [Peccei and Quinn 1977] consists of a transmutation of the parameter θ to a dynamically determined, minimum-energy value of a scalar field $\theta(x)$. This is achieved by the introduction in the standard model of additional Higgs fields leading to a Higgs potential for $\theta(x)$ which has a minimum at $\theta = 0$. These additional Higgs fields are used to break an additional chiral symmetry spontaneously (defined as in (6.66); called $U_{PQ}(1)$): The mass terms in the Lagrangian originate through the interactions with the Higgs fields ϕ, and are typically of the form $\sim (\phi \overline{q_R} q_L + \phi^* \overline{q}_L q_R)$. Invariance under $U_{PQ}(1)$ requires a transformation law

$$\phi_L \rightarrow e^{-2i\alpha} \phi.$$

The doublet of Higgs fields introduced in the standard electroweak model (see Chap. 5) leads to mass terms that violate the $U_{PQ}(1)$ symmetry. In the original scheme, therefore, the Higgs sector is enlarged by an additional doublet. Then, one doublet ϕ_u induces mass terms for the (u, s, t) quarks, and a doublet ϕ_d induces masses for the (d, c, b, ...) quarks. These Higgs fields transform as

$$\phi_u \to\to e^{-2i\alpha}\phi_u; \qquad \phi_d \to\to e^{2i\alpha}\phi_d.$$

A Higgs potential $V(\phi_u, \phi_d)$ leads to expectation values at minimum as usual:

$$\langle\phi_u\rangle = \frac{1}{\sqrt{2}}\begin{pmatrix}0\\ v_u\end{pmatrix}e^{i\vartheta_u}; \langle\phi_d\rangle = \frac{1}{\sqrt{2}}\begin{pmatrix}0\\ v_d\end{pmatrix}e^{i\vartheta_d}.$$

The phases ϑ_u and ϑ_d are undetermined.

In the full standard model with N_f families of quarks, the parameter θ in (6.64) is changed into $\theta + N_f(\vartheta_u - \vartheta_d)/2$. The Peccei–Quinn mechanism arranges the explicit breaking by $G\overline{G}$ terms such that $\theta + N_f(\vartheta_u - \vartheta_d)/2 = 0$ at the minimum of the Higgs potential. Thus the original, naively derived, form (6.64) of the QCD Lagrangian is recovered.

As a consequence of the spontaneous breaking of the $U_{PQ}(1)$ symmetry a new particle appears, the "axion". In contrast to the usual "Goldstone bosons" of a spontaneously-broken symmetry, the axion is not massless. It acquires a mass owing to the explicit breaking of $U_{PQ}(1)$ by the QCD instanton effects of the $G \cdot \overline{G}$ terms.

The properties of this new particle, which has been called the "standard axion", are determined by the expectation values v_u and v_d. The weak interactions fix the absolute value $v = \sqrt{v_u^2 + v_d^2}$ at

$$(\sqrt{2}G_F)^{-1/2} \simeq 250\,\text{GeV}. \tag{6.70}$$

As the couplings to the fermions are $\sim m_q/v$ (with m_q a typical quark mass), the "standard axion" was expected to participate in various processes through decays into two photons or electron–positron pairs. Experiments looking for such features in nuclear reactions, solar γ-rays, and monochromatic γ-transitions have failed to detect any trace of a "standard axion" (the only exception is an experiment done in Aachen, Germany).

To rescue the Peccei–Quinn idea a further enlargement of the Higgs sector was proposed. A typical example is the model where an additional Higgs singlet ϕ is introduced with an expectation value $\langle\phi\rangle = v_{PQ}/\sqrt{2}$, which is a free parameter not restricted by the structure of the standard model. The axion field is then essentially the imaginary part of this new Higgs field with small admixtures of ϕ_u and ϕ_d [Dine et al. 1981] (the "DFS" axion). Masses and couplings are proportional to v_{PQ}^{-1}, and therefore if v_{PQ} is very large, and consequently m_a very small, the axions remain "invisible".

This axion is a pseudo-scalar particle, just like the π-meson, and it has analogous couplings to the fermions. Therefore the relation

$$\frac{m_a}{m_\pi} \sim \frac{f_\pi}{v_{PQ}} \tag{6.71}$$

may be expected to hold to a good approximation. (f_π = π-meson decay constant $\approx 0.1\,\mathrm{GeV}$.)

There are some constraints which limit the range of v_{PQ}:

i) Experiments require $v_{PQ} \geq 300\,\mathrm{GeV}$.
ii) The energy loss of white dwarfs [Raffelt 1986] would be in conflict with observations, unless

$$v_{PQ} \geq 10^9\,\mathrm{GeV}. \tag{6.72}$$

(Models that do not couple the axions directly to electrons [Raffelt 1986] lead to a weaker limit of $v_{PQ} \geq 10^7$ GeV.)

The solar neutrino problem would be made worse by axions, because they provide an additional cooling mechanism for the solar interior. To maintain the photospheric luminosity the temperature in the core must be higher, and as a consequence more solar neutrinos would be produced. Some astrophysical limits on axions are listed in Table 6.2 [Raffelt 1986].

Table 6.2. Limits on axion parameters [Raffelt 1986]

Axion-electron coupling[a]			
$g <$	$m_a \cdot b <$	$v_{PQ}/b >$	Mechanism
5×10^{-11}	3.2 eV	10^7 GeV	Axion luminosity of the Sun must be smaller than photon luminosity
8×10^{-13}	0.06 eV	6×10^8 GeV	Axion luminosity of read giants smaller than photon luminosity
2×10^{-13}	0.01 eV	3×10^9 GeV	Cooling of white dwarfs
Axion-photon coupling[b]			
$G \cdot m_e <$	$m_a <$	$v_{PG}/N >$	Mechanism
10^{-12}	17 eV	4×10^5 GeV	Solar luminosity
6×10^{-14}	0.7 eV	8×10^6 GeV	Red-giant luminosity

[a] The coupling of axions to electrons $L_{aee} = ig\overline{\psi}_e\gamma_5\psi_e\varphi_a$ gives rise to axions emitted from the interior of stars by axion bremsstrahlung and by Compton photoproduction $\gamma + e^- \to e^+ + a$. Limits on g can be translated into limits on m_a and v_{PQ} for DFS axions. $b = 2\cos^2\beta$ is a free parameter, $g = (m_e/v_{PQ})b$ for DFS axions and $g^2/4\pi = 1.6 \times 10^{-23}(m_a/\mathrm{eV})^2 b^2$.

[b] The coupling of axions to photons $L_{a\gamma\gamma} = (G/4)F_{\mu\nu}\overline{F}_{\mu\nu}\varphi_a$ gives rise to a direct transformation of photons into axions. Usually electrostatic screening suppresses this process. It is of importance only for specific axions which couple only very weakly to electrons. Limits on $G \cdot m_e$ can be translated into limits on m_a and v_{PQ}, if $G \simeq 8.7 \times 10^{-4}/v_{PG}/N$, $(Gm_e)^2/4\pi \simeq 4.37 \times 10^{-28}(m_a/\mathrm{eV})^2$, with N of order 1, is used [Dine et al. 1981].

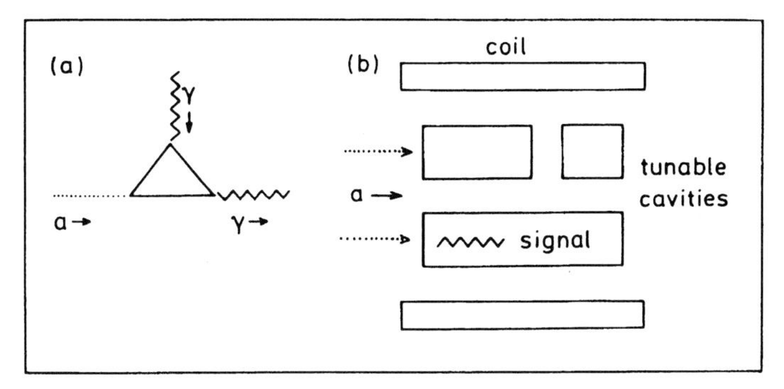

Fig. 6.12. The Feynmann graph for the decay of an axion into two photons is shown (**a**), as well as an experimental set-up to detect this process (**b**) [Smith 1986]

An experimental set-up to detect axions is shown in Fig. 6.12 [see also Raffelt 1986]. Attempts to derive models in the framework of GUTs, e.g. in a $SU(5)$ model [Wise et al. 1981], lead to

$$\begin{aligned} v_{PQ} &\sim 10^{15}\,\mathrm{GeV}, \\ m_a &\sim \frac{f_\pi m_\pi}{v_{PQ}} \sim 10^{-8}\,\mathrm{eV}. \end{aligned} \tag{6.73}$$

These parameter values are in conflict with cosmological considerations:

A background of axions in the universe may become the dominating component under certain conditions. The number density of axions created in the early universe cannot be estimated by assuming that there are about as many axions as there are background photons or neutrinos. If this were done, the low mass of the axion of $m_a \leq 10^{-2}\,\mathrm{eV}$ would render them completely unimportant. The cosmic density due to axions would be

$$\Omega_a \leq 10^{-3}.$$

Cosmic axions do not originate in thermal processes, but through the twofold symmetry-breaking of the Peccei–Quinn mechanism. An investigation of the cosmological effects of this speculation might lead to the following picture: as the universe expands and cools to temperatures $T \simeq v_{PQ}$, the Higgs field settles down in one of the infinitely many ground states

$$\phi(x) = \frac{v_{PQ}}{\sqrt{2}} e^{i(\vartheta_0 + \vartheta(x))} \tag{6.74}$$

(ϑ_0 is the phase which brings the CP violation parameter to zero). At a much lower temperature, around $T \sim 1\,\mathrm{GeV}$, the explicit breaking of $U_{PQ}(1)$ by instanton effect occurs. A simplified description containing the essential features of this twofold symmetry breaking can be given with the Lagrangian for the Higgs sector

$$H = \partial_\mu \phi^* \partial^\mu \phi - \frac{1}{2}\lambda(|\phi|^2 - v^2)^2 + 2m^4(\cos\vartheta - 1); \tag{6.75}$$

here $\phi = |\phi| \exp\,[i(\vartheta_0 + \vartheta(x))]$, and the axion field can be identified with $a \equiv v\vartheta$; m corresponds to the QCD transition scale $m \sim 1\,\mathrm{GeV}$.

Below $T \sim m$ oscillations of the phase around the minimum of the energy at $\vartheta = 0$ will occur. For $T \ll v_{PQ}$ we can derive an effective Lagrangian for $\vartheta(x)$, by inserting $\phi = v_{PQ} \exp\,[i(\vartheta_0 + \vartheta(x))]$:

$$L_{\mathrm{eff}} = v_{PQ}^2 \partial_\mu \vartheta \partial^\mu \vartheta + 2m^4(\cos\vartheta - 1). \tag{6.76}$$

This effective Lagrangian leads to the Sine–Gordon equation for ϑ:

$$\Box\vartheta + m_a^2 \sin\vartheta = 0. \tag{6.77}$$

$m_a = m^2/v_{PQ}$ can be identified with the axion mass.

In an expanding FL model the evolution of ϑ can be computed from

$$\ddot{\vartheta} + 3H(t)\dot{\vartheta} + m_a^2(t)\sin\vartheta = 0. \tag{6.78}$$

This equation describes oscillations of the axion field around the minimum $\vartheta = 0$. Assuming m_a and H to vary slowly over a "period" $1/m_a$, the energy density of the coherent axion oscillations $\phi(x) = v_{PQ}\exp(ia/v)$ is $\sim m_a^2 a^2 \sim m_a^2 v_{PQ}^2 \vartheta^2$. Since $\vartheta = 0(1)$ initially, this is of the order of $m_a^2 v_{PQ}^2$. These coherent oscillations correspond to a number density of axions at rest $n_a \sim m_a^2 v_{PQ}^2$.

Model calculations which take the time dependence of $m_a(t)$ into consideration give the following formula for the axion density parameter [Preskill et al. 1983]:

$$\Omega_a = 10^7 \left(\frac{v_{PQ}}{M_{Pl}}\right) \left(\frac{200\,\mathrm{MeV}}{T_i}\right) h_0^{-2}. \tag{6.79}$$

Here T_i depends on v_{PQ} and $T_i \sim 800\,\mathrm{MeV}$ for $v_{PQ}/M_{Pl} \sim 4 \times 10^{-7}$. T_i increases slowly with increasing v_{PQ}.

We see that axions can be of cosmological importance, $\Omega_a \sim 1$, if $v_{PQ} \simeq 10^{12}\,\mathrm{GeV}$. For $\Omega_a \leq 1$, we have

$$v_{PQ} \leq 10^{12}\,\mathrm{GeV}. \tag{6.80}$$

The interpretation of v_{PQ} as a GUT-breaking scale is, however, excluded by these cosmological limits.

6.6 SUSY GUT

6.6.1 Supersymmetry

It is the aim of supersymmetric theories to combine bosons and fermions in multiplets. To this end a "supersymmetric" operator Q is defined which changes the spin of particles by 1/2 units. It thus transforms bosons into fermions and fermions into bosons:

$$Q|\text{boson}\rangle = |\text{fermion}\rangle, \qquad (6.81)$$
$$Q|\text{fermion}\rangle = |\text{boson}\rangle.$$

Q must satisfy the anti-commutation relation

$$\{Q, \overline{Q}\} = -2\gamma^\mu P_\mu \qquad (6.82)$$

(P_μ: momentum operator of the Poincaré group) [Nicolai 1983 for a nice introduction; Wess and Zumino 1974; Fayet and Ferrara 1977].

The operator Q is considered as one of the generators of the supersymmetry. It satisfies commutation relations with boson fields, and anti-commutation relations with fermion fields

$$[Q, \phi_B] = \Psi_f; \qquad (6.83)$$
$$\{Q, \Psi'_f\} = \phi'_B.$$

This leads to the relation

$$[\{Q, \overline{Q}\}, \phi] = \quad 2i\gamma^\mu \partial_\mu \phi \equiv -i\gamma^\mu P_\mu \phi. \qquad (6.84)$$

These equations, plus the relation

$$[Q, P^\mu] = 0, \qquad (6.85)$$

which expresses the fact that the result of a supersymmetry transformation does not depend on the space-time point where it is performed, complete the supersymmetry algebra. The generator P^μ of the Minkowski space-time translations appears on the right-hand side of (6.84), and this indicates a basic connection between supersymmetry and space-time transformations.

An instructive, basic example is the system of a harmonic oscillator and a spin without interactions [Nicolai 1983]. Creation and annihilation operators a^+ and a for the harmonic oscillator (corresponding to bosons) satisfy canonical commutation relations

$$[a, a^+] = aa^+ - a^+a = 1. \qquad (6.86)$$

The spin states (corresponding to fermions) are described by creation (Ψ^+) and annihilation operators (Ψ) with canonical anti-commutation relations

$$\{\Psi, \Psi^+\} = 1. \qquad (6.87)$$

In addition one requires $[a, \Psi] = 0$, $[a^+, \Psi^+] = 0$, etc. The Hamiltonian is then

$$H = a^+a + \Psi^+\Psi.$$

Application of a^+ and Ψ^+ to the vacuum states yields states with Bose and Fermi particles

$$\begin{aligned}|n_b, 0\rangle &= \frac{(a^+)^{n_b}}{\sqrt{n_b!}}|0,0\rangle \\ |n_b, 1\rangle &= \Psi^+|n_b, 0\rangle\end{aligned} \tag{6.88}$$

where n_b is the number of Bose states. A fermionic charge Q may be introduced by

$$Q \equiv a^+\Psi, \qquad \text{with} \qquad Q^+ = a\Psi^+. \tag{6.89}$$

Then the Hamiltonian is

$$H = \{Q, Q^+\}, \tag{6.90}$$

and H commutes with Q and Q^+:

$$[H, Q] = [H, Q^+] = 0. \tag{6.91}$$

Q generates a symmetry of the system. The operator Q destroys one spin quantum and creates one oscillator mode. Since it commutes with H one finds the following relations

$$\begin{aligned}\langle n_b, 1|H|n_b, 1\rangle &= (n_b+1)^{-1/2}\langle n_b, 1|HQ^+|n_b+1, 0\rangle \\ &= (n_b+1)^{-1/2}\langle n_b, 1|Q^+H|n_b+1, 0\rangle \\ &= \langle n_b+1, 0|H|n_b+1, 0\rangle.\end{aligned} \tag{6.92}$$

This model shows a degeneracy between Bose and Fermi states. The energy is unchanged if one fermion is removed and one boson is added. There is only one non-degenerate state, the vacuum $|0,0\rangle$, for which

$$Q|0,0\rangle = Q^+|0,0\rangle = 0. \tag{6.93}$$

The vacuum is the only state invariant under the supersymmetry transformation. The positivity of the Hamiltonian,

$$H = QQ^+ + Q^+Q \geq 0,$$

ensures that this is also the state of lowest energy. This rather simple example nevertheless exhibits one important aspect of supersymmetry theories: the boson–fermion degeneracy (which leads to the cancellation of divergencies in quantum field theory).

A linear representation of a supersymmetric theory describes equal numbers of bosonic and fermionic states – because the generator Q in any representation produces a supersymmetric boson for any fermion, and vice versa. If supersymmetry was an exact symmetry, the fermions and bosons in one multiplet would all have the same mass. Since this clearly contradicts the real world, supersymmetry can at best be realized as a broken symmetry. The spontaneous breaking of supersymmetry is made difficult by the properties of the operator Q: it turns out that any supersymmetric vacuum state is necessarily stable [Nicolai 1983; Witten 1981].

In an ordinary gauge theory the situation is completely different. As we have seen (Chaps. 5 and 6), the gauge-symmetric state can easily be made unstable. This is the Higgs trick of producing a spontaneously-broken gauge symmetry. Nevertheless, devious theorists have been able to cook up a mechanism for the spontaneous symmetry-breaking (SSB) of supersymmetry [Fayet 1975; O'Raifeartaigh 1975]. The result is an analogy to the Goldstone theorem: in broken supersymmetric theories massless neutral spin-half particles appear – corresponding to the spin-zero bosons of the usual Goldstone theorem. These Goldstone fermions are called "goldstinos".

A rather undesirable aspect of supersymmetry is its tendency to introduce large numbers of new particles. Originally one hoped to associate the known fermions and bosons in a supersymmetric theory, to have, for example, the photon and the neutrino, the W^- and the electron, as partners (Fig. 6.13). This idea has proved unsuccessful owing to the fact that there exist several families of leptons and quarks. If the photon is associated with one neutrino (or with several, as in the framework of extended supersymmetry), the $e - \mu - \tau$–quark universality cannot be maintained.

How can the real world be supersymmetric? Where are the bosons and fermions that can be related? How can lepton and baryon numbers be conserved?

The simplest solution of these problems has turned out to be the introduction of new particles as the supersymmetric partners of the ordinary, well-known particles [Fayet 1977]. In such a scheme the superpartners of the photon and the W would be interpreted as new particles (called the "photino" and "Wino").

In Fig. 6.14 the simplified diagram of Fig. 6.13 is repeated, now showing the photon, $W^{\pm}$ and their superpartners. The electroweak symmetry is

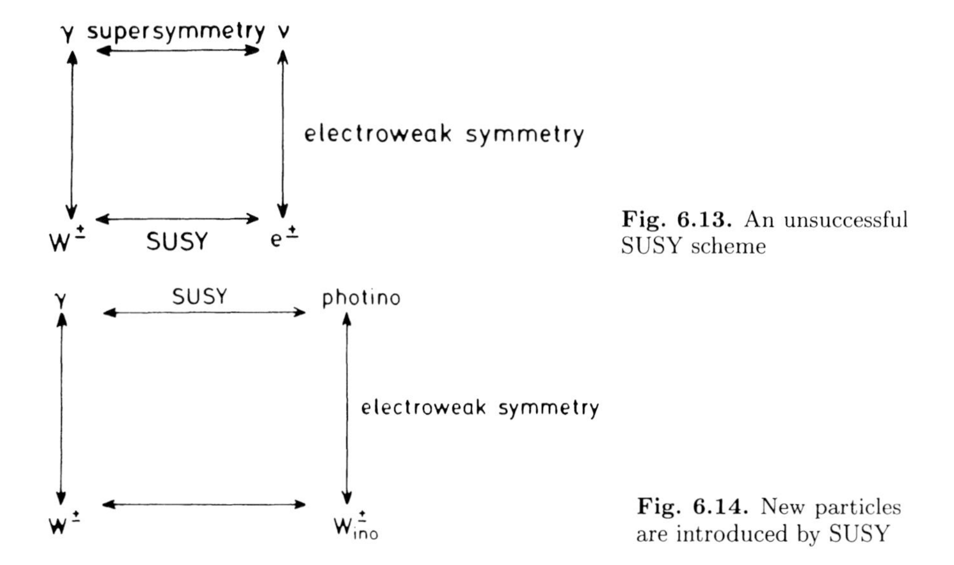

Fig. 6.13. An unsuccessful SUSY scheme

Fig. 6.14. New particles are introduced by SUSY

spontaneously broken, and therefore the $W^{\pm}$ is associated with 2 Winos$^{\pm}$ and a spin-zero Higgs boson $\overline{H}_W^+$ (similarly for W^-).

The octet of gluons of $SU(3)_C$ is associated with an octet of spin-half selfconjugate fermions called "gluinos".

Leptons and quarks are associated with spin-zero "s"-leptons and "s"-quarks. Thus the simplest supersymmetric theory requires twice as many particles as found in the ordinary theory.

Supersymmetry allows one to obtain relations between particles with different spins (1, $\frac{1}{2}$, 0), and also between particles with different electroweak properties: $W^{\pm}$ and Z belong to a triplet and a singlet; Higgs bosons $\overline{H}_W^+$ and $\overline{H}_Z$ to doublets; the Winos and Zinos are a mixture of triplets, singlets, and doublets. Supersymmetry also relates the couplings of the gauge bosons (g or g') to the couplings of the Higgs bosons

$$\sim g \frac{m(\text{fermions})}{\mathrm{m_W}}.$$

In a supersymmetric scheme the spontaneous breaking of the $SU(2)\times U(1)$ electroweak gauge symmetry requires two doublet Higgs superfields instead of only one doublet (cf. Sect. 5.5): This leads to a minimal content of a supersymmetric gauge theory as listed in Table 6.3 – these structures are common to all supersymmetric theories.

The spin-2 graviton and its superpartner, the spin-$\frac{3}{2}$ gravitino, are present when the supersymmetry is realized as a local gauge group. In locally supersymmetric theories the spontaneous symmetry-breaking by a super-Higgs mechanism avoids the occurrence of a spin-half goldstino – instead the superpartner of the spin-2 graviton, the spin-$\frac{3}{2}$ gravitino acquires a mass [Deser and Zumino 1977]. The gravitino mass $m_{3/2}$ is the parameter that introduces mass splittings between the ordinary particles and their supersymmetric partners.

Table 6.3. Minimal particle content of a supersymmetric gauge theory

spin 1	spin $\frac{1}{2}$	spin 0
Gluons	Gluinos	
photon	photino $\overline{\gamma}$	
$W^{\pm}$	2 (Dirac) Winos $\overline{W}$	$\overline{H}_{W^\pm}$
Z	2 (Majorana) Zinos $\overline{Z}$	$\overline{H}_Z$
	1 (Majorana) Higgsino $\overline{H}$	standard $\overline{h}^0$
		pseudoscalar $\overline{h}^0$
	leptons	spin-0 leptons $\overline{e}, \overline{\nu}$
	quarks	spin-0 quarks $\overline{u}, \overline{d}$

In addition there are the spin-2 graviton and spin-$\frac{3}{2}$ gravitino $\overline{g}$.

6.6.2 Particle Masses

Basically there is no experimental evidence as yet either in support of supersymmetry, or against it. Some limits on the masses of the supersymmetric particles can be placed by considering specific reaction channels (see [Haber and Kane 1985] for a review of supersymmetry phenomenology). For instance, from searches for the process

$$e^+ + e^- \to \gamma + 2\overline{\gamma}$$

(see Fig. 6.15) one finds that the mass of the scalar electron $m(\overline{e})$ is

$$\begin{array}{lll} m(\overline{e}) > 22\,\mathrm{GeV} & \text{for} & m(\overline{\gamma}) < 19\,\mathrm{GeV}, \\ m(\overline{e}) > 33\,\mathrm{GeV} & \text{for} & m(\overline{\gamma}) \sim 10\,\mathrm{GeV}, \\ m(\overline{e}) > 51\,\mathrm{GeV} & \text{for} & m(\overline{\gamma}) < 100\,\mathrm{eV}. \end{array}$$

This reaction could allow one to detect the effects of a spin-zero electron with a mass below $50\,\mathrm{GeV}/c^2$. For larger masses the signal will get lost in a competing background from

$$e^+ + e^- \to \gamma\nu\overline{\nu}.$$

Similar $e^+ + e^-$ processes give limits on the scalar muon and scalar tau mass:

$$m(\overline{\mu}) > 21\,\mathrm{GeV},$$

$$m(\overline{\tau}) > 18\,\mathrm{GeV}.$$

Quite generally one calls the lightest, neutral SUSY particle a "neutralino". This is probably stable – it could be the photino or the gravitino, or a mixture of both.

Even if it were not the lightest stable particle, the gravitino would be very long-lived. Therefore both the photino and the gravitino may show up us a supersymmetric relic from the hot big bang. In what follows we want to discuss a few effects of the photino and gravitino in astrophysics.

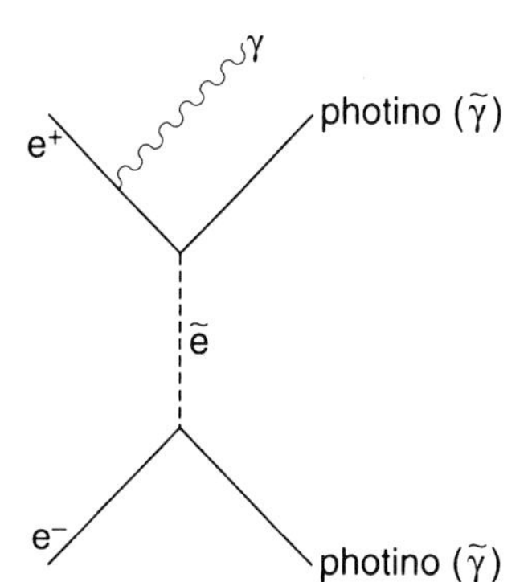

Fig. 6.15. Radiative production of a photino pair in e^+e^- annihilation

6.6.3 Effects of the Photino in Astrophysics

Photinos ($\overline{\gamma}$) interact with charged leptons and quarks, as in Fig. 6.16. The corresponding cross-sections are roughly comparable with neutrino cross-sections: for example [Fayet 1979, 1984],

$$\begin{aligned} \sigma(\overline{\gamma}) + N \rightarrow \overline{\gamma} + \text{hadron} &= 0.7 \times 10^{-38}\,\text{cm}^2\text{E}\,[\text{GeV}]\left[\frac{40\,\text{GeV}}{m_{\overline{q}}}\right]^4 , \\ \sigma(\overline{\gamma} + N \rightarrow \text{gluino} + \text{hadron}) &= 50 \times 10^{-38}\,\text{cm}^2\text{E}\,[\text{GeV}]\left[\frac{40\,\text{GeV}}{m_{\overline{q}}}\right]^4 . \end{aligned} \tag{6.94}$$

Compare this with standard neutrino cross-sections (elastic scattering)

$$\begin{aligned} \sigma(\nu - N) &= 0.7 \times 10^{-38}\,\text{cm}^2\text{E}\,[\text{GeV}], \\ \sigma(\overline{\nu} - N) &= 0.3 \times 10^{-38}\,\text{cm}^2\text{E}\,[\text{GeV}] \end{aligned}$$

(N here stands for nucleon). If it has a small mass, the photino has the same effect as an additional neutrino on the abundance of the primordial helium. The mass limit from cosmological energy density limits is then, as for the neutrinos (cf. Chap. 3),

$$m_{\overline{\gamma}} \leq 100\,\text{eV}. \tag{6.95}$$

In the case of a heavy photino its mass may vary between 2 and 20 GeV, depending on the mass of the s-leptons and s-quarks [Goldberg 1983]. The photino may also be an unstable particle decaying into (e.g.) a lighter gravitino, or neutralino.

This decay is important in suppressing the abundance of photinos when $m_{\overline{\gamma}} \leq 100\,\text{eV}$, because otherwise there may already be conflicts with the number of neutrino flavours – $\Delta N_\nu = 3$ seems to be required by the helium abundance (cf. Chap. 3), and there is no room for an additional light "-ino".

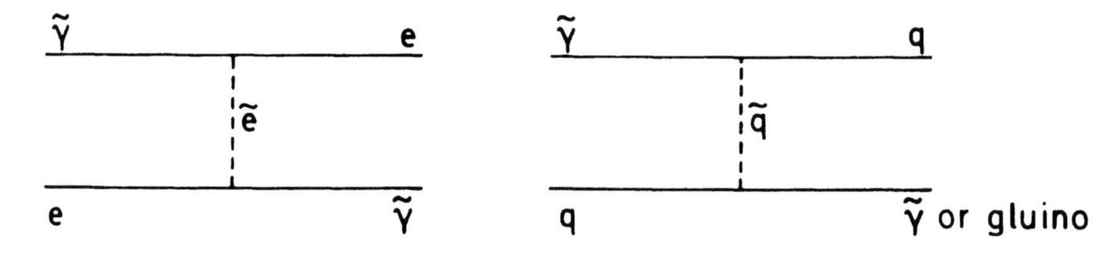

Fig. 6.16. Photino–electron and photino–quark interactions

6.6.4 Effects of the Gravitino

A heavy gravitino $\overline{g}$ can decay into lighter boson–fermion pairs at a rate

$$\Gamma(\overline{g} \to bf) \simeq G m_{3/2}^3 \tag{6.96}$$

[Weinberg 1982]. A rough estimate of the lifetime of the gravitino gives

$$\begin{array}{lll} \text{for} & m_{3/2} \sim 1\,\text{GeV}; & \tau_{\overline{g}} \sim 10^{15}\,\text{s} \\ \text{for} & m_{3/2} \sim (10^4 \text{ to } 10^5)\,\text{GeV}; & \tau_{\overline{g}} \sim 10^{-2}\,\text{s to } 10\,\text{s}. \end{array} \tag{6.97}$$

If gravitinos were more massive than 1 GeV they would be cosmologically unimportant, because they would all have decayed. The expansion rate of the early universe might, however, have been modified by them, with a consequent increase in the primordial helium abundance. Also their decays might have produced too many high-energy photons. Therefore a heavy gravitino should satisfy

$$m_{3/2} \geq 10^4\,\text{GeV} \tag{6.98}$$

or the gravitinos should be diluted by some mechanism, such as cosmological inflation.

The gravitino may also be light [Fayet 1984] in some models,

$$m_{3/2} \simeq \frac{G^{1/2} m_W^2}{g''} \tag{6.99}$$

where g'' is the coupling constant of an extra $U(1)$ symmetry. If $g'' \approx 0(1)$, then

$$m_{3/2} \simeq 10^{-6} \text{ to } 10^{-5}\,\text{eV}. \tag{6.100}$$

The interaction strength of a light gravitino is fixed by the ratio $G m_{3/2}^{-2}$.

A very light gravitino would interact too strongly and should already have been detected in particle physics experiments. From the non-observations of the decay

$$\Psi' \to \pi^+ + \pi^- + \overline{g} + \overline{\gamma},$$

a limit on $m_{3/2}$ follows:

$$m_{3/2} > 1.5 \times 10^{-8}\,\text{eV}. \tag{6.101}$$

A better limit $\sim 10^{-6}$ eV will soon be available from the search for the process

$$e^+ + e^- \to \gamma + \overline{g} + \overline{\gamma}. \tag{6.102}$$

Light gravitinos have interactions similar to neutrino interactions:

$$\sigma(\overline{g}+\mathrm{e}\to\overline{\gamma}+\mathrm{e}) = 0.4\sigma(\nu_\mu+\mathrm{e}\to\nu_\mu+\mathrm{e})\left(\frac{m_{3/2}}{10^{-5}\,\mathrm{eV}}\right)^{-2}. \tag{6.103}$$

In a cosmological context a light gravitino in a $\pm 1/2$ polarization state would have a decoupling temperature of roughly

$$T_d \simeq \left(\frac{m_{3/2}}{10^{-5}\,\mathrm{eV}}\right)^{2/3} \qquad [\mathrm{MeV}]. \tag{6.104}$$

Gravitinos in a $\pm 3/2$ polarisation state would have an interaction $\approx G$, such that they decouple very early, or never are in equilibrium. A gravitino lighter than $\simeq 10^{-2}\,\mathrm{eV}$ would decouple at $T_d \leq 100\,\mathrm{MeV}$. This would have similar effects to an additional 2-component neutrino species and might produce conflicts with the helium production. Thus a lower limit of $m_{\overline{g}} \geq 10^{-2}\,\mathrm{eV}$ should be imposed. For $m_{\overline{g}} \sim 0.1\,\mathrm{eV}$, for example, the decoupling temperature is $T_d \simeq 200\,\mathrm{MeV}$. Below T_d the gravitinos cool as $T_{\overline{g}} \propto R^{-1}$, whereas the photon–neutrino gas is heated up repeatedly by the annihilation of various particle species, down to $T \sim 1\,\mathrm{MeV}$ (decoupling of neutrinos). Thus gravitinos are colder than neutrinos and their energy density is too small at the time of helium abundance to be of importance.

For stable gravitinos an upper limit to the mass is obtained, in a similar way as for massive neutrinos, by comparing their density with the critical density. Then

$$m_{3/2} \leq 100\,\mathrm{eV}\,\frac{g_I(< T_d)}{g_I(=\frac{43}{3})} \tag{6.105}$$

where $g_I(< T_d)$ is the effective number of interacting degrees of freedom just after gravitinos decouple. Equation (6.105) gives a limit of 1 to 100 keV. Finally, these cosmological aspects, taken together, would confine the gravitino mass to the following interval:

$$10^{-2}\,\mathrm{eV} \leq m_{3/2} \leq (1\text{ to }100)\,\mathrm{keV} \tag{6.106}$$

for a stable or very long-lived gravitino, and

$$m_{3/2} \geq 10^4\,\mathrm{GeV}$$

for a short-lived gravitino.

6.6.5 A Few Comments

It is a curious fact that despite the lack of experimental evidence in favour of supersymmetry it has become exceedingly popular. The reasons are connected with the attractive algebraic properties of supersymmetry: There are relations between the masses and couplings of gauge bosons and of Higgs bosons. The hope of an automatic connection with gravity when the supersymmetry algebra is realized locally, pervades many discussions (the $N = 8$ "supergravity" theory; N refers to the number of SUSY charges).

There is also hope for a (partial) solution of the hierarchy problem in supersymmetric grand unified theories. Remember that the question is two-fold. Why is $m_W/m_X \leq 10^{-13}$ so small? Second, how can it remain so small, despite radiative corrections that lead to $\delta m_W^2 \approx \delta m_H^2 \simeq O(\alpha)m_X^2$? Supersymmetry may answer the second question by the special relations present between fermions and bosons which lead to a correction term

$$\delta m_W^2 = O(\alpha)|m_B^2 - m_F^2|.$$

This "solution" can work if

$$\frac{O(\alpha)|m_B^2 - m_F^2|}{m_X^2} \leq 10^{-26} \quad \text{or} \quad |m_B^2 - m_F^2| \leq (10^3\,\text{GeV})^2.$$

For many theorists this possibility provides the strongest motivation for considering SUSY GUTs. The experimental situation requires patience, because if SUSY partner particles have masses much larger than 10^3 GeV it will be a long time before they are found.

The $N = 8$ "supergravity" theory is the realization of the supersymmetry algebra (for $N = 8$) as a local gauge group – with an internal "gauge" symmetry group which may be as large as $SO(8)$. But $SU(3) \times SU(2) \times U(1)$ is not contained as a subgroup.

In minimal SUSY models the mass of the X bosons increases and the proton becomes more stable – it is easy to reach agreement with experiments. These corrections have a negligible effect on the successful $SU(5)$ predictions of $\sin^2\theta_W$ and m_b/m_t since the changes are of the order of $\Delta m_X/m_W$. It is interesting to note that various new baryon decay modes become possible

$$B \to \overline{\nu} + K; \qquad B \to \mu^+ + K; \qquad B \to e^+ + \pi^0. \tag{6.107}$$

Even the largest supergravity theory does not include among its fundamental fields all the known elementary particles [Ellis et al. 1980]. Therefore it seems necessary to explain all the known particles as composite entities.

6.7 Monopoles, Strings, and Domain Walls

In this section we shall discuss structures that may appear as a result of the spontaneous breaking of a global or of a gauge symmetry. The Higgs field in its new ground state can have various types of defects in its topological arrangement. Zero-dimensional, point-like defects correspond to monopoles; one-dimensional, vortex-like defects are strings; two-dimensional defects are domain-walls; 3-dimensional defects are textures which decay in time.

The whole concept of spontaneous symmetry-breaking with the help of Higgs fields is a somewhat speculative procedure well understood only at the level of perturbation theory. Therefore doubts may be entertained as to whether such structures really exist. Fortunately these topological objects have a secure basis for their existence as solutions, with finite energy, of the Euclidean Yang–Mills–Higgs equations in one, two, or three dimensions.

Furthermore, a close analogy can be drawn with the magnetic flux vortices in type II superconductors. (I am indebted to [Straumann 1982] for the following description.) These are solutions to the Ginzburg–Landau equations of superconductivity [Abrikosov 1961].

Topological defects are everywhere

In Chap. 5 the close correspondence between the Meissner effect and the Higgs phenomenon was outlined. The electron two-point function

$$\langle \Psi \uparrow (x) \Psi \downarrow (x) \rangle$$

defines a complex function $\phi(x)$, which is an order parameter for the phase transitions that occur. The Lagrangian can be written (as in (5.30)) as

$$\begin{aligned} L &= \tfrac{1}{4}F^2 + \tfrac{1}{2}|D\phi|^2 + V(\phi) \qquad \text{where} \\ V(\phi) &= (\lambda/8)(|\phi|^2 - v^2)^2 - (\lambda/8)v^4, \\ F_{\mu\nu} &= \partial_\mu A_\nu - \partial_\nu A_\mu, \\ D_\nu\phi &= \partial_\nu\phi - 2ieA_\nu\phi. \end{aligned} \tag{6.108}$$

Formally, the order parameter $\phi(x)$ appears just as the Higgs field in a simple abelian Higgs model [Nielsen and Olesen 1973], with the difference, of course, that here it is composed of more fundamental quantities, the electron-field operators.

The parameter $v^2(T)$ depends on the temperature, and changes sign at the critical temperature T_c. Below T_c, the minimum of the free energy occurs at $|\phi|^2 = v^2$, corresponding to the superconducting state. The thermodynamic critical magnetic field H_c (for external fields $H > H_c$ the superconducting state is destroyed), is given by

$$V(|\phi| = v) - V(\phi = 0) = -\frac{H_c^2}{2}, \tag{6.109}$$

$$H_c = \frac{\sqrt{\lambda}}{2}v^2(T).$$

In the superconducting state the coherence length ξ corresponds to a "Higgs mass" $m_H = \sqrt{\lambda}v$:

$$\frac{\xi}{\sqrt{2}} = m_H^{-1}. \tag{6.110}$$

The London penetration depth δ corresponds to a "gauge boson" mass $m_A = 2\,\mathrm{eV}$:

$$\delta = m_A^{-1}. \tag{6.111}$$

Thus the critical field can be written as

$$H_c = \frac{1}{4e}m_H m_A. \tag{6.112}$$

The microscopic theory allows ξ and δ to be expressed by fundamental quantities. Since $m_H \le 2m_A$ always [Plohr 1981] it follows that $\xi/\sqrt{2} \ge \delta/2$. Both are proportional to $|T_c - T|^{-1/2}$, and therefore the parameter

$$\chi \equiv \frac{\delta}{\xi} = \frac{1}{\sqrt{2}}\frac{m_H}{m_A} \tag{6.113}$$

is independent of temperature.

There are two types of superconductors, as

$$\begin{aligned} \chi &< 1/\sqrt{2} \qquad \text{type I}, \\ \chi &> 1/\sqrt{2} \qquad \text{type II}. \end{aligned}$$

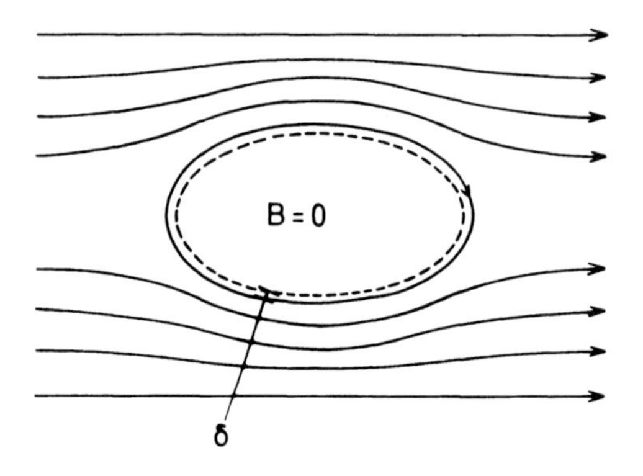

Fig. 6.17. Meissner–Ochsenfeld effect. δ is the penetration depth; typically $\delta \sim 5 \times 10^{-6}$ cm

Superconductors exclude all external magnetic fields, if they are less than some critical field value H_{c_1} (Meissner–Ochsenfeld effect; Fig. 6.17). In type II superconductors, however, for $H > H_{c_1}$ the external field penetrates in tube-like structures (Fig. 6.18). The magnetic flux of these vortices is quantized, equal to integer multiples of $2\pi/2e$. For type II superconductors the vortices repel each other, and a regular lattice array is formed. This lattice was predicted by Abrikosov in 1957 and was observed 10 years later. These magnetic-flux vortices correspond precisely to the string solutions of Yang–Mills–Higgs field theoretical models.

The two types of superconductors are distinct in the nature of the phase transition. In the case of type II the superconducting state is reached below H_{c_2}, by a second-order phase transition (with $|\phi|$ increasing continuously from zero). The critical field H_{c_2} is given by

$$\frac{H_{c_2}}{H_c} = \frac{m_H}{m_A}. \tag{6.114}$$

The phase-transition leads to the formation (far from the surface) of superconducting domains (s-phase) within the normal matter. The lower critical field H_{c_1} – below which flux is expelled – can be approximated by

$$\frac{H_{c_1}}{H_c} \simeq \frac{m_A}{m_H} \ln \frac{m_H}{m_A}. \tag{6.115}$$

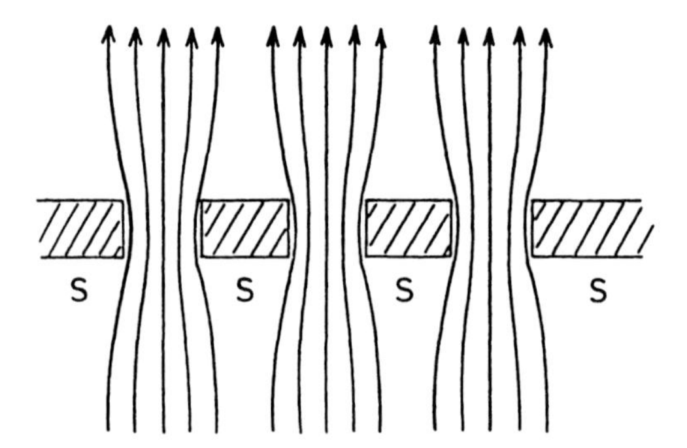

Fig. 6.18. Formation of vortices in a type II superconductor for $H_{c_1} < H < H_{c_2}$; H_{c_1} is the critical field below which superconductivity sets in

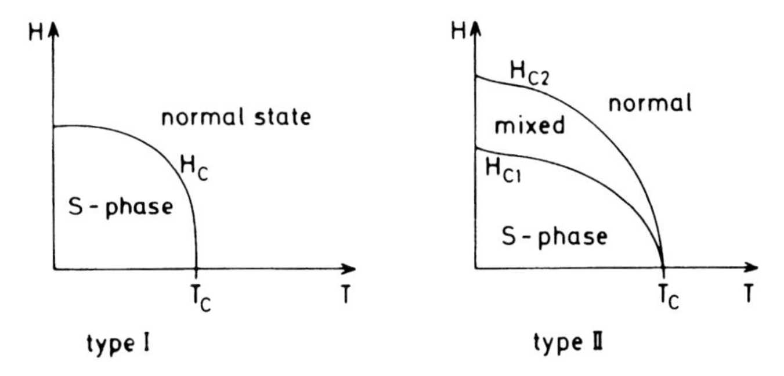

Fig. 6.19. Phase diagrams for type I and type II superconductors

The two critical fields depend on T, and become zero for $T = T_c$. The phase diagram for both types of superconductors is shown in Fig. 6.19.

Let us consider the behaviour of such systems if the temperature is increased for a given external field $H > H_{c_1}$. At low temperatures in both types a complete flux expulsion occurs (the Meissner effect). Type I superconductors show a first-order phase transition at $H = H_c(T)$. If the superconductor is of type II the occurrence of small domains of the n-phase is thermodynamically favoured; the increase in the volume energy is compensated by the surface energy, which is negative in this case. In the "mixed state" between H_{c_1} and H_{c_2} a triangular array of magnetic-flux vortices is formed.

Similarly, if the system is cooled down from an initially high temperature (n-phase), bubbles of the s-phase will form below H_{c_2}. The transition is second-order for type II superconductors. But in the case of type I "supercooling" phenomena may occur. The system remains in the normal state even below H_c, ideally until $H_{c_2} < H_c$ is reached. Then nucleation occurs, accompanied by a discontinuous and irreversible jump of $|\phi|^2$ to v^2.

To transfer some of these concepts to the universe, we replace the superconductor by the hot big-bang model. Various phase transitions may then be possible due to the spontaneous breaking of GUT symmetries as the universe expands and cools. Various topological defects may occur as a result of these transitions, by analogy with the vortex solutions for the superconductor.

The exotic structures of our imagination such as monopoles, strings, and domain walls in the early universe are thus analogous to the configurations found in laboratory experiments. The question whether the topological defects in cosmology really exist cannot, unfortunately, be decided by an appeal to this analogy.

6.7.1 Magnetic Monopoles

All ordinary magnetic substances have paired regions of opposite polarity, i.e. all ordinary magnets are dipoles. These observations led Maxwell to formulate electrodynamics in such a way that single particles carrying a specific magnetic charge did not occur. This gives electrodynamics its curious asymmetry between electric and magnetic fields. Establishing a symmetry between electricity and magnetism was the motivation for P. Dirac inventing magnetic monopoles in 1931.

Although we are mainly interested in monopoles as they occur in non-abelian gauge theories (such as $SU(5)$ GUT), a few remarks are appropriate here on the properties of Dirac monopoles. This is also interesting because GUT monopoles – such as the t'Hooft–Polyakov monopole – look asymptotically like Dirac monopoles.

Dirac managed to show that the existence of at least one free magnetic charge implies the quantization of the electric charge. Dirac's quantization condition established a relation between the elementary electric charge e and the elementary magnetic charge g:

$$eg = \left(\frac{\hbar c}{2}\right) n \tag{6.116}$$

with $n = 1, 2, 3 \ldots$. Thus the minimum unit of magnetic charge g is about 70 times as large as the unit of electric charge (in c.g.s. units $e = 5 \times 10^{-10}$; $g = 3.5 \times 10^{-8}$).

A monopole that passes through a conducting loop induces a flux

$$\Delta\phi = \int \boldsymbol{E} \cdot ds dt = \frac{4\pi g}{c}. \tag{6.117}$$

Using the quantization condition (6.116)

$$\Delta\phi = 2n \frac{hc}{2e}; \tag{6.118}$$

$hc/2e$ can be viewed as an elementary flux quantum. This induction property is used in a class of experiments to detect monopoles.

Several groups have searched for monopoles, for example in cosmic rays and in paramagnetic substances deep under the sea; Moon rocks and meteorites have been passed through superconducting coils in the hope of finding a change $\Delta\phi$ in the quantized flux. All these experiments have produced zero results, and hence upper limits for a cosmic flux of monopoles F have been obtained. Typical numbers are $F < 10^{-18}\mathrm{Mon\,cm^{-2}s^{-1}}$ for monopoles with kinetic energies $T < 10^{10}$ GeV.

The next chapter in the story of monopoles unfolded in 1974, when t'Hooft and Polyakov [t'Hooft 1974; Polyakov 1974] showed that certain non-perturbative solutions of the Yang–Mills–Higgs equations exist, which are regular with finite energy, and which behave like monopoles. Such solutions occur when a large, local gauge group is spontaneously broken down to a subgroup containing the abelian group $U(1)$. For example, the breaking of $SU(5) \rightarrow SU(3) \times SU(2) \times U(1)$ necessarily implies the existence of a monopole.

The classification of topological defects proceeds via homotopy groups. If the vacuum manifold M_0 of the Higgs field,

$$M_0 \equiv \{\phi : V(\phi) = 0\},$$

is such that $\Pi_2(M_0) \neq 0$, then monopole solutions are present. Such a GUT monopole can be considered as a zero-dimensional defect in the GUT Higgs field – a "hedgehog" pattern, as illustrated in Fig. 6.20. The GUT monopole has mass

$$m_M \simeq \frac{m_X}{g^2/4\pi} \simeq 10^{16}\,\text{GeV}. \tag{6.119}$$

This mass is extremely large, about 10^{-8} g, similar to the mass of a bacterium. These large masses indicate that monopoles could have been produced only close to the big bang in the very early universe. For instance the $SU(5)$ GUT symmetry should be exact for temperatures above 10^{15} GeV, i.e. during the first 10^{-35} seconds. Then the spontaneous breaking of the symmetry $SU(5) \rightarrow SU(3) \times SU(2) \times U(1)$ – with a $U(1)$ factor occurring for the first time – produced monopoles which should still be around as relic particles.

We should note here that the further breaking of $SU(2) \times U(1) \rightarrow U(1)$ does not lead to additional monopoles connected with the electroweak breaking scale of ≈ 100 GeV (cf. e.g. [Jaffe and Taubes 1980]).

Since monopoles correspond to topological defects in the Higgs field, their density is determined by the coherence of these fields: if ξ is a typical scale of the spatial variations – a correlation length of the Higgs field – then the density of monopoles $\sim (\xi)^{-3}$.

For GUT physics monopoles are extremely interesting objects: they have an onion-like structure (schematically shown in Fig. 6.21) which contains the whole world of grand unified theories. Near the centre ($\sim 10^{-29}$ cm) there is a GUT symmetric vacuum. At $\approx 10^{-16}$ cm, out to the Yukawa tail $\sim \exp(-m_W r)$, the field is the electroweak colour field of the (3,2,1) standard model, and at 10^{-13} cm it is made up of photon and gluons, while at the edge there are fermion–antifermion pairs. Far beyond nuclear distances it behaves as a magnetically charged pole of the Dirac type.

This view of the GUT monopole raises the possibility that it may catalyse the decay of the proton (Rubakov–Callan effect [Rubakov 1981; Trowers

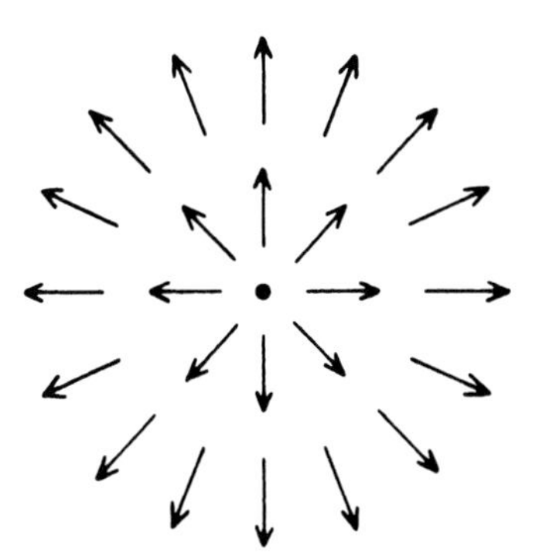

Fig. 6.20. A pictorial representation of a zero-dimensional defect. The directions of the arrows indicate the orientations of the Higgs fields in internal space, while their location represents ordinary space

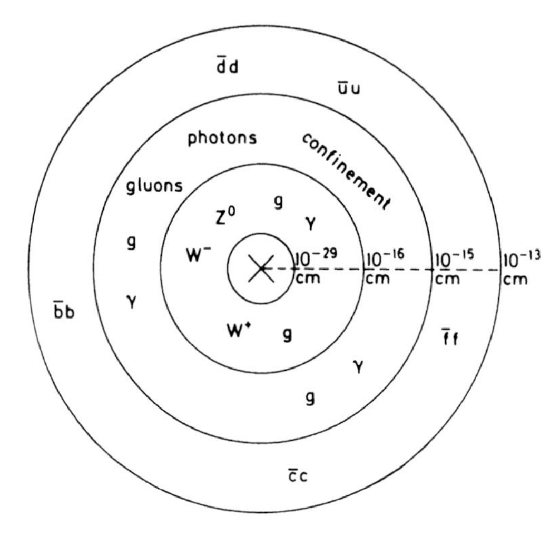

Fig. 6.21. The onion structure ot the GUT monopole. From the grand unification core outwards there is the electroweak unification region, the confinement region, and the fermion–antifermion condensate

1983]). In a collision with a proton the GUT region may be overlapped, producing a rapid decay of that particular proton. A monopole passing through a proton-decay detector could induce of the order of ten decays during its passage, which typically lasts 50 μs [Fiorini 1983; Giacomelli 1984]. Since such decays have not been observed, the experiments produce upper limits on the flux of monopoles near the detector.

Neutron stars should be good collectors of monopoles. The catalysing monopoles inside a star would transform nucleons at such a rate that the resulting X-ray luminosity would be considerable. The measured upper limits of neutron-star X-ray luminosities imply limits on the monopole flux – assuming reasonable cross-sections for catalysis – [Ritson 1982] of

$$\phi_M < 5 \times 10^{-22}\,\mathrm{cm}^{-2}\mathrm{s}^{-1}\mathrm{sr}^{-1}, \tag{6.120}$$

much lower than any other limit.

A more conventional astrophysical limit can be obtained by requiring the decay time of the galactic magnetic field to be larger than the regeneration time $\tau_r \simeq 10^8$ years. Free monopoles will be accelerated along the field lines, and the current density $j_M = n_M g u$ (g = pole strength, u = monopole velocity) will be mainly along $\boldsymbol{B}$. The work done by the magnetic field on the monopoles is $\boldsymbol{B} \cdot \boldsymbol{j}_M \sim B n_M g u$. This is also the dissipation rate of the field energy $B^2/8\pi$, and hence the field is neutralized on a time-scale

$$\tau = \frac{B}{8\pi n_M g u} = \frac{B}{8\pi g \phi_M} .$$

Here B is the galactic magnetic field $B \sim 3\,\mu\mathrm{G}$; $\phi_M = n_M u$. From $\tau_r \le \tau$, the so called "Parker limit" [Parker et al. 1982; Parker 1970] is derived

$$\phi_M < 3 \times 10^{-15}\,\mathrm{cm}^{-2}\mathrm{s}^{-1}. \tag{6.121}$$

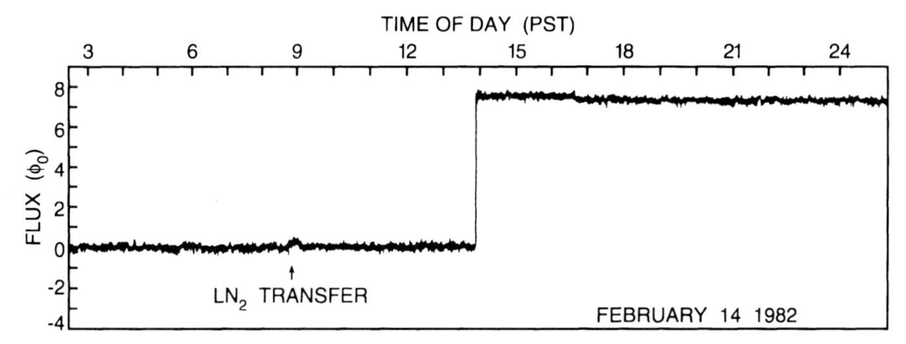

Fig. 6.22. The current change induced by the candidate monopole event (after [Cabrera 1982])

Another limit can be obtained from the requirement that the cosmic monopole density $m_M n_M$ should not exceed the critical mass density ϱ_c in the universe. This can be translated into a flux limit

$$\phi_M < 3 \times 10^{-15}(\beta 10^3)\left(\frac{10^{16}\,\mathrm{GeV}}{m_M}\right)\mathrm{cm}^{-2}\mathrm{s}^{-1}, \tag{6.122}$$

where $\beta = u/c$ is a typical monopole velocity. It seems reasonable to assume that $\beta \simeq 10^{-2}$ to 10^{-3} as monopoles are so heavy that they certainly will be non-relativistic.

In view of these limits it was quite a surprise when in 1982 the observation of an interesting event was announced [Cabrera 1982]. This can be interpreted as a monopole passing through a superconducting coil (Fig. 6.22). The flux limit $\phi < 10^{-9}\,\mathrm{cm}^{-2}\,\mathrm{s}^{-1}\,\mathrm{sr}^{-1}$ for this candidate event is larger than all the other limits discussed above. Additional experiments are under way, but to date no further candidate events have been found.

The cosmological aspects of monopoles will again be discussed in Chaps. 7 and 9.

6.7.2 Strings

Strings are solutions, with finite energy of the 2-dimensional Euclidean Yang–Mills–Higgs equations, just as monopoles are solutions of the 3-dimensional equations. These objects really are the exact analogy of the vortex-lines in a type II superconductor.

A simple model of a complex scalar field ϕ with self-interactions exhibits this analogy explicitely. The Lagrangian of the abelian Higgs model [Nielsen and Olesen 1973] is

$$L = \mathrm{D}_\mu \phi^* \mathrm{D}^\mu \phi - \frac{1}{4} F_{\mu\nu} F^{\mu\nu} - \frac{1}{2}\lambda(\phi^*\phi - v^2)^2,$$

where

$$\begin{aligned} \mathrm{D}_\mu &\equiv \partial_\mu - ieA_\mu \\ F_{\mu\nu} &\equiv \partial_\mu A_\nu - \partial_\nu A_\mu . \end{aligned} \tag{6.123}$$

The potential $V(\phi) \equiv \frac{1}{2}\lambda(|\phi|^2 - v^2)^2$ is minimal on the circle $M_0 \equiv \{\phi : |\phi|^2 = v^2\}$, the "vacuum manifold" of the Higgs field ϕ.

String solutions of this model will be fields that are static and constant in one spatial direction. Consider such a string in cylindrical coordinates (r, ϑ, z) at $r = 0$, constant in the z-direction. At large distance $(r \to \infty)$ the field behaves as

$$\phi \simeq v e^{in\vartheta}, \tag{6.124}$$

where n is an integer. The gauge field is

$$A_\mu = \frac{1}{ie}\partial_\mu \ln \phi . \tag{6.125}$$

For $r \to \infty$ the energy density vanishes outside the string core, ($F_{\mu\nu} \to 0$, $|D\phi| \to 0$ for $r \to \infty$). The Higgs field defines a mapping $f : S^1 \to M_0$, where each direction ϑ in ordinary space is mapped to the asymptotic value of the Higgs field $\phi \sim v \exp(in\vartheta)$ in the vacuum manifold. The integer n is the winding number (or degree) of the map f. This characterizes the homotopy class, in our case an element of $\pi_1(M_0)$. This is a simple example of a topological "quantum number".

The "magnetic" flux of the string is n times the elementary quantum $2\pi/e$. Values of $n = \pm 1$ correspond to stable, elementary strings; $|n| \geq 2$ probably relates to unstable objects that can decay into elementary ones.

The dimensions of the string core are given by the Compton wavelengths of the Higgs particle m_ϕ^{-1}, and the vector mesons m_A^{-1}:

$$m_\phi = \sqrt{2\lambda} v, \qquad m_A = \sqrt{2} e v. \tag{6.126}$$

When $m_A < m_\phi$, an inner core of "false vacuum" $\sim m_\phi^{-1}$ is surrounded by a tube of magnetic field $\sim m_A^{-1}$. Regardless of the shape of the Higgs potential, the maximal decay of the inner core is $\sim (2m_A)^{-1}$ [Plohr 1981]. The mass per unit length of the string is

$$\mu \sim v^2 (\sim 10^{-6} M_{Pl}^2 \quad \text{in GUT theories}).$$

For a macroscopic description of strings the internal structure is unimportant, and one can carry out integrations over cross-sections. Thus, for a static, straight string along the z-axis, one finds the averaged energy-momentum tensor to be

$$\int T_{\mu\nu} dx dy = \mu(1, 0, 0, -1). \tag{6.127}$$

The mass density per unit length in macroscopic units is

$$\mu = G\mu \frac{M_{Pl}}{L_{Pl}} = 2.2 \left(\frac{G\mu}{10^{-6}} \right) 10^{10} M_{\odot}/\text{kpc}, \tag{6.128}$$

quite a high value. Such macroscopic strings, if they occur at all, define regions of extremely high mass density. They may act as seeds for galaxy formation (cf. Chap. 11).

String solutions are expected whenever the homotopy group $\pi_1(M_0)$ is non-trivial. In our example $\pi_1(M_0) = Z$. $\pi_1(M_0)$ is non-trivial in the case of a symmetry-breaking of a group G to a smaller symmetry group H, if H contains a discrete symmetry. For example the symmetry-breaking scheme $SO(10) \to SU(5) \times Z_2$ contains string solutions. Strings form in many GUTs [Vilenkin 1985; Zel'dovich 1980]. One expects them in the low-energy limit of superstring theories [Witten 1985]. Because of their topological quantum number, strings can have no ends, and must be infinite or closed.

The dynamics of macroscopic strings can be discussed in terms of a 2-dimensional world-sheet $x^\mu(\tau, \sigma)$, where τ is a time-like and σ is a space-like parameter. The action S is a functional of $x^\mu(\tau, \sigma)$,

$$S = -\mu \int \sqrt{-g^{(2)}} d\tau d\sigma, \tag{6.129}$$

where $-g^{(2)}$ denotes the determinant of the induced metric on the world-sheet. Several properties of moving strings, and of strings in an FL model, have already been worked out (see e.g. [Vilenkin 1985]). These exercises suffer from a serious deficiency, however: an exact solution to the Einstein field equations has been found only for the case of a straight, static string.

For closed loops, dimensional considerations allow the conjecture that a circular loop will rapidly collapse due to gravitational radiation, with a velocity reaching the velocity of light as it shrinks to a point. Other radiation (due to Yang–Mills interactions) may be important too, but has not yet been considered. The energy loss due to gravitational radiation can be estimated from the quadrupole formula (the loop mass is $M \sim \mu R$) (cf. Fig. 6.23):

$$\dot{M} = -G(MR^2\omega^3)^2 = -G\mu^2. \tag{6.130}$$

The lifetime

$$\tau = \frac{M}{|\dot{M}|} \simeq \frac{R}{G\mu}, \tag{6.131}$$

and the energy-loss fraction in one oscillation period

$$\frac{|\dot{M}|}{M} \frac{2\pi}{\omega} \simeq G\mu \sim 10^{-6} \tag{6.132}$$

for typical GUT scales of $v \sim 10^{16}$ GeV.

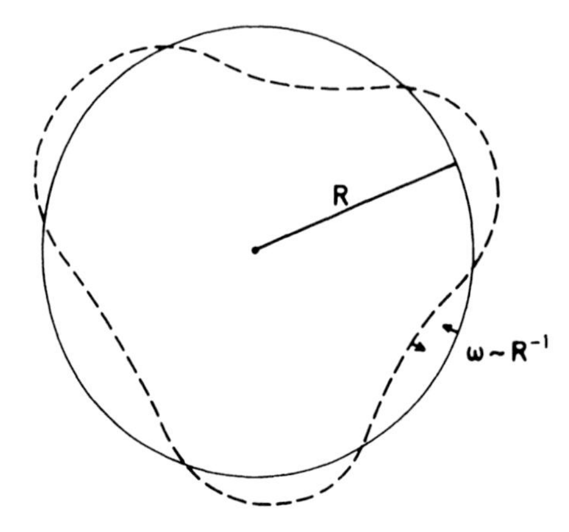

Fig. 6.23. A loop of dimension R has a typical oscillation frequency $\sim R^{-1}$

Again, on dimensional grounds, we can argue in a cosmological context that loops surviving the gravitational energy loss at an epoch t must have a size $R \sim G\mu t$. For $G\mu \sim 10^{-6}$ this scale is a few kpc at the present time t_0. At the time of galaxy formation, closed loops are of much smaller size, in general, than the galaxy condensing around them. Accretion of matter onto these loops would produce massive, compact objects such as quasars and active galactic nuclei. Strings may act as seeds for the formation of galaxies, by laying out a definite pattern of density enhancements (more about this in Chap. 11).

It seems a good idea to look earnestly for possible observational tests of such a speculative scenario. In that respect there are a few interesting suggestions:

(i) Oscillating loops emit gravitational waves and thus can produce a stochastically fluctuating background. This will induce a "timing noise" in the period of pulsars [Hogan and Rees 1984; Taylor 1986]. This effect would become detectable within the next few years if $G\mu \sim 10^{-6}$. A non-detection would mean $G\mu < 10^{-6}$, and then the string scenario would lose its importance for cosmology – no significant influence on galaxy clustering would be expected (see Chap. 11 for further discussion).

(ii) A string can act as a gravitational lens for a quasar. This can most easily be seen by considering the exact solution of a static, straight string. The metric outside the string core (taken along the z-axis) is, in cylindrical coordinates (Fig. 6.24),

$$ds^2 = dt^2 - dz^2 - dr^2 - (1 - 4G\mu)^2 r^2 d\varphi^2. \qquad (6.133)$$

The coordinate transformation $\varphi' = (1 - 4G\mu)\varphi$ results in a Minkowski metric. Thus the space outside the string core is locally flat. Since φ' varies only from 0 to $(1 - 4G\mu)2\pi < 2\pi$, it is a conical space, where a flat wedge of angular size $8\pi G\mu$ is cut out, and its faces are identified.

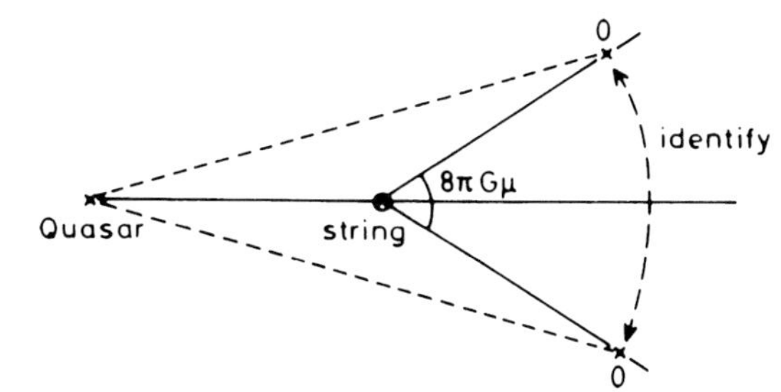

Fig. 6.24. The conical space outside a straight string core. The two edges of the wedge $\varphi \leq 8\pi G\mu$ must be identified

Thus an observer O will see a double image of a quasar when the line of sight passes across the string. The separation angle between the two images is $8\pi G\mu \sim 3$ to 30 arcsec for $G\mu \sim 10^{-6}$ to 10^{-5}. Thus in principle such observations, as well as the lensing effect on distant galaxies, could be carried out by the Hubble Space Telescope. The probability of detection is, however, quite small. Expectations are that every 10^4th quasar is doubled by a string [Vilenkin 1984; Paczyński 1986].

(iii) Rapidly moving strings can produce step-like discontinuities in the 3 K background with $\delta T/T \sim 10\,G\mu$ [Kaiser and Stebbins 1984]. Present limits require $G\mu < 10^{-5}$.

6.7.3 Domain Walls

Domain walls are two-dimensional defects in the Higgs field's topological arrangement (Fig. 6.25). A typical example is the "kink" solution (Fig. 6.26) which occurs when the vacuum expectation value $\langle\phi\rangle$ is $+v$ for $x \to +\infty$ and $-v$ for $x \to -\infty$.

A simple model with a real scalar field ϕ, with the Lagrangian

$$L = \frac{1}{2}(\partial_\mu \phi)^2 - \frac{\lambda}{2}(\phi^2 - v^2)^2 \tag{6.134}$$

[Vilenkin 1985] can be used to derive this solution. There is a discrete symmetry $Z_2 : \phi \to -\phi$; the minima of $V(\phi)$ are at $\phi = \pm v$. Whenever a domain

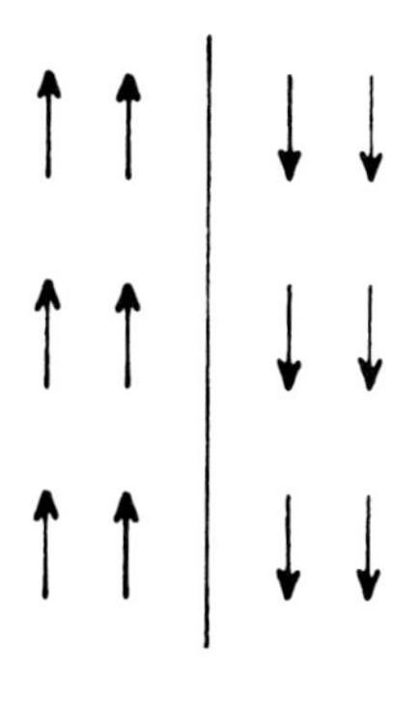

Fig. 6.25. A domain wall is indicated by the line separating two regions of space-time where the Higgs fields point in different directions

with $\langle\phi\rangle = +v$ is separated from a domain with $\langle\phi\rangle = -v$, one has to pass a region with the "false vacuum" $\langle\phi\rangle = 0$ in between. The two domains are separated by a domain wall of false vacuum. The "kink" solution is given as the classical solution of

$$\Box\phi + 2\lambda\phi(\phi^2 - v^2) = 0. \tag{6.135}$$

depending only on x,

$$\phi_0(x) = v\tanh(x/\delta), \tag{6.136}$$

where $\delta = \lambda^{-1/2}v^{-1}$ is the wall's thickness. The vacuum energy density $\varrho_v \sim \lambda v^4$ gives rise to a surface energy density of

$$\sigma \sim \varrho_v\delta \sim \lambda^{1/2}v^3.$$

The energy–momentum tensor can be written as

$$\begin{aligned} T^\nu_\mu &= f(z)\mathrm{diag}(1,1,1,0) \qquad \text{with} \\ f(z) &= \lambda v^4[\cosh(z/\delta)]^{-4}. \end{aligned} \tag{6.137}$$

This "kink" solution is displayed in Fig. 6.26.

In my opinion domain walls are less firmly rooted in reality than strings or monopoles. There is no surface energy for domain walls even on the classical level. Even in simple models it is not clear that they will survive a consistent treatment of the Higgs field as a quantum object. Because of the local gauge symmetry it does not cost any energy to reorientate the Higgs field. By making the Higgs fields parallel on both sides of a domain wall, this entity would disappear from the realm of existing objects – mathematical or otherwise.

Monopoles and strings are protected from such a fate by their topological quantum numbers. In terms of homotopy classification, domain walls correspond to $\pi_0(M_0) \neq 0$. This is equivalent to a vacuum manifold that is not connected.

Domain walls in the early universe must be avoided. They carry enormous energy densities, which would lead to a catastrophic collapse situation. Any theory predicting domain walls runs into cosmological difficulties (cf. Chap. 7).

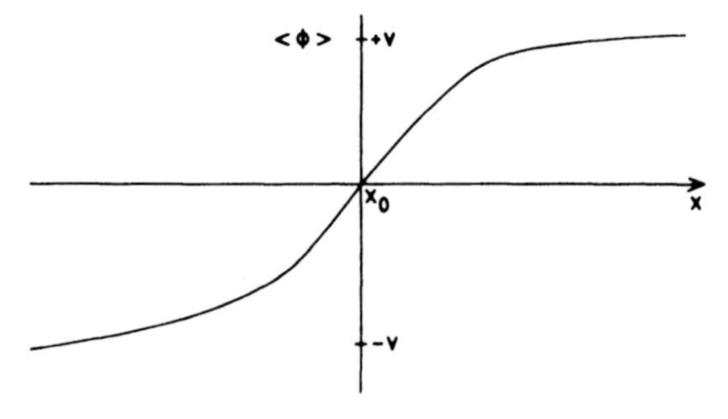

Fig. 6.26. The "kink" solution for the vacuum manifold of ϕ leads to a domain wall located at $x = 0$

6.7.4 Textures

Explicitly broken non-abelian symmetries lead to another kind of topological defect: a texture. A global texture occurs, for instance, in the simple case of the breaking of an $SU(2)$ symmetry by a complex-valued scalar field ϕ. The Lagrangian

$$L = \frac{1}{2}\phi,_{\mu}\,\phi,^{\mu} - \lambda(\phi \cdot \phi - \eta^2)^2 \tag{6.138}$$

is specified by a spherically symmetric ansatz

$$\phi = \eta\left(\frac{\mathbf{r}}{r}\sin\chi, \cos\chi\right), \tag{6.139}$$

where χ is an angular variable depending only on r and t.

Defining a coordinate $y \equiv (t - t_c)/r$ in flat space with an arbitrary t_c, the exact solution of the equations of motion is

$$\chi = 2\arctan(\pm y) \pm n\pi. \tag{6.140}$$

A collapsing texture with winding number 1 for $t < t_c$ and winding number 0 for $t > t_c$ can be patched together from

$$\chi = \begin{cases} 2\arctan y + \pi, & -\infty \le y \le 1, \\ 2\arctan(1/y) + \pi, & 1 \le y \le \infty. \end{cases} \tag{6.141}$$

The winding number

$$n = \int \frac{1}{12\pi^2}\epsilon_{abcd}\phi^a\, d\phi^b d\phi^c d\phi^d = \begin{cases} 1, & \text{if } t < t_c, \\ 0, & \text{if } t > t_c. \end{cases} \tag{6.142}$$

The kink at $y = 1$ reflects the singularity of this model approach (the so called σ-model) at the unwinding event $t - t_c = r = 0$. There the gradient energy of the solution diverges, i.e. becomes bigger than the symmetry-breaking scale η. The Higgs field leaves the vacuum manifold at $t \simeq t_c$, unwinds, and the kinetic energy can be radiated away in massless Goldstone bosons [Turok 1989].

6.8 Further Developments

Quantum gravity, the theory of superstrings, and quantum cosmology are rapidly evolving subjects, a detailed description of which I have excluded from this book. We restrict the description of the universe to the case of a given classical cosmological model, which serves as a background for different models of elementary particles. Such an approach can make sense only at energies below the Planck scale. All these new theories, on the other hand, are thought to be valid at energy scales beyond the Planck mass $\sim 10^{19}$ GeV; thus they will affect a possible quantum phase of the early universe. The significance for cosmology of these new concepts has not been worked out in detail. A discourse on these speculations would definitely require another book. Here only a few cursory remarks on these lively subjects will be given.

Fundamental theories and their contact with reality

Theories with a local supersymmetry must include a spin-two particle, i.e. they contain some aspects of gravity.

The largest 4-dimensional $N = 8$ supergravity theory can be formulated as an $N = 1$ supergravity theory in 11 dimensions. This has revived the old ideas of Kaluza and Klein [Kaluza 1921; Klein 1926] of a unified description of spin-one gauge fields and gravity. For example [J. Ellis 1984], in a geometrical unification of a 5-dimensional theory, the 5×5 metric tensor g_{MN} might be decomposed into the usual 4×4 metric tensor $g_{\mu\nu}$, a gauge field A_μ, and a scalar field ϕ:

$$g_{MN} = \begin{pmatrix} g_{\mu\nu} & A_\mu \\ A_\mu & \phi \end{pmatrix} .$$

Isometries of the extra dimension correspond to gauge symmetries in the effective 4-dimensional theory.

In a similar way, Kaluza–Klein theories in higher dimensions should have isometries of some internal space that appear as gauge symmetries in the effective 4-dimensional theory (e.g. [Salam and Strathdee 1982; Wetterich 1985; Kolb 1986]). As at present we seem to live in a 4-dimensional space-time, there is the common assumption that the extra dimensions somehow curl up to a very small size of the order of the Planck length, $L_{Pl} \sim 10^{-33}$ cm. In the 11-dimensional theory, each space-time point x_μ, for example, would thus become a small compact space $(x_\mu, x_5, \dots, x_{10})$ sitting at location x_μ. The "curling up" of the extra dimensions is a rather mysterious process.

"Dimensional reduction" can be an everyday experience

In a cosmological context, it seems natural to look at the early universe as the site where extra dimensions may become dynamically important. Clearly, if R_D is the physical size of the D extra dimensions, then at temperatures higher than R_D^{-1} these structures may break up. Above a critical temperature $T_c = R_D^{-1}$ the space time will be $(4 + D)$-dimensional, and below T_c the D extra dimensions retract into a size $\sim R_D$ and the universe becomes effectively 4-dimensional. After this "dimensional reduction" the extra space dimensions are dynamically unimportant.

The condensed state of the D dimensions must be imagined as an equilibrium between forces tending to expand and to collapse the space. Away from equilibrium there should be an unbalanced force which will lead to a huge positive cosmological constant in the 4-dimensional "dimensionally reduced" real world. This positive constant could drive an inflationary phase of the universe in a similar way to that proposed for GUT theories in a cosmological model (cf. Chap. 9 in this book, and [Wetterich 1985]). I do not think, however, that the use of classical cosmological models beyond the Planck length is easy to justify.

Another possible effect might be a direct production of entropy in the transition of the universe when three space dimensions expand and grow big, while D dimensions contract and stay small [Kolb 1986].

This type of investigation within the Kaluza–Klein theories has not found answers to the fundamental question of why there are three expanding, large spatial dimensions with a size comparable to the Hubble radius at present, and why D dimensions are small and compact with size $\sim L_{Pl}$. Why is there a distinction between large and small dimensions at all? Why are three dimensions large? The Einstein field equations would couple the dynamics of the evolution of the spatial dimensions. There is then the vague hope that such questions can be attacked in a consistent cosmological model.

Supergravity theories are an attempt at all or nothing: as they are unrenormalizable, they can make sense only when they are finite. Unfortunately it has not been possible to formulate a consistent perturbation theory for supergravity theories because the quantum corrections have uncontrollable infinities and violate the gauge symmetry of the theory, i.e. "anomalies" occur.

6.9 Superstrings

It is hoped that the infinities and anomalies of supergravity theories are absent from another attempt at unification which has become very popular in recent years. Superstring theories [Green and Schwarz 1984, 1985; Green 1986; Ellis 1986] propose a radical change in the fundamental objects of the theory. They are based on the dynamics of a one-dimensional object – a string – in one time and nine space dimensions. The fundamental quanta are the infinitely many excited states of the string, instead of the point-like quanta of the Yang–Mills gauge theories. Typical energy differences between these states are multiples of the string tension T (of dimension $(\mathrm{mass})^2$ in natural units).

Superstring theories incorporate supersymmetry and the lowest excited states of zero mass as a first approximation and have all the interactions and quantum numbers of $N = 1$ supergravity in ten dimensions. It is conjectured that the superstring tension is characterized by the Planck scale

$$T^{1/2} \leq M_{Pl}.$$

Usually one assumes that the string tension is somewhat smaller than the Planck mass; similarly the string scale L_s is often taken to be somewhat larger than the Planck length.

This makes the excited states so massive that the usual GUT theories will appear as an effective point-field theory of the massless ground state. The excitation of the higher mass states is, however, a crucial aspect of the short-distance behaviour of the theory.

A typical feature of most string theories is that they are defined in more than four dimensions. The extra dimensions must then be compactified to very small scales, just as the extra dimensions are rolled up in Kaluza–Klein theories.

Originally the big motivation for string theories was that in a perturbation scheme there were no infinities or anomalies in a calculation of one-loop amplitudes, provided that either $O(32)$ or $E_8 \times E_8$ was used as the internal gauge group.

The non-singular behaviour was expected, because the fundamental length scale L_S provides an ultraviolet cut-off, such that perturbation theory is divergence-free. The surprise was that there seemed to be only one or two basic theories – the dream of a uniquely determined "theory of everything" seemed to come true. Meanwhile the higher-order perturbation terms had divergences, and the number of compactified 4-dimensional string theories has increased steadily because of the many possible ways in which the extra dimensions could curl up. So the hope of finding a unique description of the world dwindled.

In recent years the discovery of the duality symmetry has given a new boost to the hopes of the followers of string theory. Roughly this means that there is a simple relationship between perturbative and non-perturbative solutions of string theory. So a way to obtain non-perturbative solutions for string theory has opened up. In perturbation calculations many string theory solutions have been found.

In addition, the existence of a fundamental "M-theory" in 11 dimensions, the mother of all the different string theories, has been glimpsed.

How our world, with its standard model and big-bang cosmology, can be found among the millions of solutions of M-theory, remains, of course, a big open problem.

One amusing cosmological aspect has emerged from a consideration of the gauge group $E_8 \times E_8$ [Kolb et al. 1985]. If the symmetry between the two factors persisted even in the dimensionally reduced 4-dimensional theory, then two forms of matter might exist: one for each gauge factor E_8. The only interaction between these two limits of matter would be by gravity. There might exist two kinds of worlds side by side, "shadows" of each other, but not interacting other than by gravity. Certain restrictions would be imposed on the amount and distribution of "shadow" matter in our neighbourhood. But even within the galaxy, other shadow Solar Systems might exist with shadow planets and shadow cosmologists (σκιᾶς ὄναρ ἄνθρωπος ... (Pindar): "A shadow's dream (is) man ... ").

Most of the physics of superstrings happens at energies above $M_{Pl} \sim 10^{19}$ GeV, and from there it is a long way to every-day accelerator physics. Let us hope that from its phase of floating around in the imagination, superstring theory will be able to find some connection with reality. Let us wait and see whether it succeeds or whether it ends up as the kind of thing that Pauli would have characterized as being "not even wrong" [Ginsparg and Glashow 1986].

The connection to Einstein's theory of General Relativity has not yet been achieved on a fundamental level. It is rather the occurrence of spin-two fields in perturbation solutions which leads to the claim that GR is contained in string theory. Many relativists would like to see a deeper link concerning the fundamentally non-linear features of GR such as the singularities. The linearized version of GR in quantized form with a few gravitons jogging around in empty space is not what they expect from quantum gravity.

String cosmologist about the universe like beaver about a dam: "I did not build it myself, but it was based on my idea."

6.10 Quantum Gravity

The singularities of GR inside black holes or in the big bang show very clearly that a new theory is needed to describe such phenomena. This theory already has a name, "quantum gravity", although nobody knows as yet what it will look like.

But we can make some guesses, or formulate some expectations for such a theory. Since all non-gravitational interactions must be described by quantum theories, it seems compelling that a unified theory including gravitation must also be a quantum theory.

The approaches treating gravity as a classical theory on a fundamental level – such as using "semi-classical" Einstein equations with the energy momentum tensor as the expectation value of quantum field operators – have not been successful so far. That drastic changes are required can be seen when one considers the problem of defining the space-time concept on scales smaller than the Planck length. We can see this in a simple example [Nicolai and Niedermaier 1989]:

Try to measure the position of a particle of mass m by scattering a photon of frequency ω from it. The direction of the photon is uncertain up to an angle ϵ. The transverse position of the particle (in the x-direction, say) has an uncertainty

$$\Delta x \geq (\omega \sin \epsilon)^{-1}, \tag{6.143}$$

which can be reduced by increasing ω. There is also a very weak gravitational shock wave accompanying the photon, and the particle experiences an acceleration by that wave which moves it by a distance $\sim G\omega$ [e.g. Padmanabhan 1988]. The uncertainty in this shift is $\Delta x \sim G\Delta\omega$. Together with (6.143), this leads to

$$(\Delta x)^2 \geq G = L_{Pl}^2. \tag{6.144}$$

The frequency of the photon disturbs the particle position, and thus the resolution is limited. Neither GR nor quantum mechanics (QM) can cope with such a situation. Not because their validity is limited, but they just do not have a conceptual framework for such situations. In QM it is unimportant whether spatial coordinates are discrete or continuous. In GR the concept of a differentiable manifold, and therefore the existence of arbitrarily small separations, is a fundamental ingredient. This is denied by (6.144). How can a theory be formulated which dissolves the classical space-time below the Planck length? We expect that a fundamental theory exists which links GR and QM (superstring theory?). This should be a finite theory, and the status of renormalizable field theories within that framework should be clarified. The validity of treating quantum fields in a classical background gravitational field, such as Hawking radiation from black holes, should be understandable in a semi-classical approximation. The predictions of such a theory must be in agreement with cosmological conditions.

The operational limitations imposed on the space-time structure should be derivable within the theory, similar to the derivation of the uncertainty principle within QM.

In the following remarks only gravity will be considered, without any interaction with matter or quantum fields. Since the true, fundamental theory of everything is not in sight yet, this may at least give some indication of what to expect from quantum gravity. (The discussion follows [Nicolai and Niedermaier 1989; Kiefer 1997].) The quantization of GR might follow the canonical rules. The configuration space of GR consists of the set of metric tensor fields $g_{\mu\nu}(x)$ which define distance measures in 4-dimensional space-time. The Hamiltonian formalism needs a slicing of the 4-dimensional space-time M into a time sequence of 3-dimensional, space-like hypersurfaces Σ. The components $h_{ij}(\boldsymbol{x})(i, j = 1, 2, 3)$ of $g_{\mu\nu}(\boldsymbol{x})$ on Σ and the time derivatives of $g_{\mu\nu}$ on Σ are the initial conditions. The time evolution of these "Cauchy" data follows uniquely from the Einstein equations. The variables $h_{ij}(\boldsymbol{x})$ have the canonically conjugate momenta $i\delta/\delta h_{ij}(\boldsymbol{x})$ which are functional differential operators, and quantization proceeds in the usual manner, replacing Poisson brackets by (equal time) commutators. The quantum states of the gravitational field are described by "wave functionals" $\psi(h)$. The requirement of general (3-dimensional) coordinate invariance on Σ leads to the equation

$$H(h)\psi(h) = 0, \tag{6.145}$$

where $H(h)$ is the Hamilton operator of the gravitational field, a non-linear functional differential operator. Equation (6.145) is called the "Wheeler–de Witt equation", although it is just the stationary Schrödinger equation of gravity. This equation is non-linear to such a degree that the construction of realistic solutions is out of the question. There is no "hydrogen atom" of quantum gravity. A transformation to new variables has led to sufficient simplification for the Hamiltonian of the Einstein equations that formally a large class of exact solutions has been obtained [Ashtekar 1995]. The status of these solutions with respect to quantum gravity is not completely clear, but progress is being made.

An essential difficulty is the concept of renormalization itself. The elimination of divergencies in successful quantum field theories depends crucially on the careful study of operator products at small space-time separations. If space-time itself is the quantum object the usual scheme of renormalization cannot be carried out. In addition, we must ask how meaningful observables can be defined, what the configuration space of the theory is like, and how suitable expectation values can be calculated.

Finally, there are enormous difficulties in understanding the physical meaning of quantum gravity. What is meant by the "wave function of the universe"? Who is the classical observer who can interpret the wave functional $\psi(h)$?

Time does not appear in (6.145). On the other hand, there are three degrees of freedom in h_{ij} (six independent components minus three coordinate transformations possible in three dimensions), while the gravitational field has only 2 physical degrees of freedom (two helicity states $s = \pm 2$).

The missing time variable can be approximated semi-classically by a scalar function $T(h)$. Only when such a scalar function of the variables h_{ij} is identified with time is a time evolution of a 2-component field contained in (6.145).

Many investigations have been undertaken within a path integral formalism by analogy with Feynman's path integral method in QM. The transition amplitude between two spatial geometries is represented by an expression such as

$$\langle 1|2\rangle = \int_1^2 DgD\phi e^{iS[g\Phi]}, \tag{6.146}$$

where S is the Einstein action

$$S[g,\phi] = \int R(-\det g)^{1/4} d^4x,$$

with R the Riemann scalar; $Dg, D\phi$ signifies functional integration. The functional integration involves all suitable metrics g and all matter fields ϕ which have prescribed values at the two ends of the transition process. The expression (6.146) is purely formal, but some possible effect of quantum gravity can nevertheless be studied. There might be a tunnelling between two spatial geometries which are not connected by a solution of the equations of GR.

A formal solution of (6.145) can be written down:

$$\psi[h,\phi] = \int_{(h,\varphi)} DgD\phi e^{iS[g,\phi]}. \qquad (6.147)$$

This looks rather similar to (6.146), but the boundaries of the integration are quite different. Instead of prescribing the initial and final configuration, one integrates in (6.147) over all g and ϕ which take the values h and φ on the boundary of the manifold $\Sigma \equiv \partial M$. These functional integrals are not at all well-defined mathematical expressions. The divergencies occurring are of a similar nature to those in the canonical formalism. A typical difficulty shows up when the oscillating integrand $\exp(iS)$ is replaced by an exponentially damped one $\exp(-S_E)$. This change is achieved by changing from space-time metrics with Lorentz signature $(+,-,-,-)$ to a 4-dimensional Euclidean, positive definite metric. This change is not one-to-one, and the consequence is that we loose the space-time interpretation of the expressions under the integral in (6.147). We cannot be sure that this prescription ends up anywhere close to the correct theory.

Another serious problem is the structure of the action S_E. The summation over all Euclidean metrics includes all metrics differing just by a conformal factor, say $\Omega(x)$. A suitably, or rather unsuitably, chosen $\Omega(x)$ can give a negative value to S_E, as large as you want. Thus the action S_E, is unbounded from below, and the integral (6.147) is a sum over exponentially growing terms.

This is a serious difficulty because the Einstein action is practically uniquely determined by the requirements of general covariance and the absence of derivations higher than second order. The usual recipe – just use another model, if this one does not work – is not acceptable in this case.

Do these problems have a deep significance, or will they just dissolve in the context of a fundamental theory?

A final remark concerns some of the more speculative possibilities seen in connection with so-called "worm-holes". If the "transition amplitude" (6.146) is seen as a tunnelling process between geometries which can be interpreted as space-times, the typical configurations which link the different space-times are called "worm-holes" – configurations with a complicated topology.

A simple example of a worm-hole is the following. Write x^2 for the Euclidean scalar product

$$x^2 = x_1^2 + x_2^2 + x_3^2 + x_4^2.$$

The metric

$$ds^2 \equiv \left(1 + \left(\frac{a^2}{x^2}\right)^2\right) dx^2$$

seems to have a singularity at $x_\mu = 0$. But this is just a coordinate effect. Since ds^2 is invariant under the transformation

$$x^\mu \to x^\mu (a^2/x^2),$$

the region $x^2 \leq a^2$ has the same structure as the region $x^2 \geq a^2$.

The metric space is not equivalent topologically to the flat Euclidean space, but it splits asymptotically into two such spaces (for $x^2 \to 0$, and $x^2 \to \infty$). These are connected by a worm-hole of circumference $4\pi a$. This can be interpreted as the creation and annihilation of a 4-dimensional sphere. (Poetically minded colleagues call this a "baby universe".)

The action for this type of worm-hole is

$$S_E = 3\pi a^2/G,$$

and all such configurations are exponentially suppressed (because of the factor $\exp(-S_E)$) except those confined to extensions comparable to the Planck length $L_{Pl} = \sqrt{G}$. All spaces with an extent larger than L_{Pl} should be connected by worm-holes in the integral (6.147).

Possible consequences of such a worm-hole interpretation of the path integral are discussed with respect to the values of the physical constants [Weinberg 1989], to time travel, to faster-than-light travel and to the origin of the universe. This is fascinating, but still outside the horizon of physics as an empirical science. Nevertheless, as Albert Einstein said, "you cannot build a theory only on observable quantities. In reality it is quite the opposite. Only theory can tell us what is observable."

Exercises

6.1 Show that in GUT schemes like $SU(5)$ the Weinberg angle is quite generally given by $\sin^2\theta_w = 3/8$.

6.2 Show that neutrino masses are still zero in $SU(5)$ GUT. Why are non-zero masses possible in GUTs with larger symmetry groups, such as $SO(10)$?

6.3 Show that $B - L$ is still conserved in $SU(5)$ GUT, although B and L are violated.

6.4 The monopole solution can be described by taking the Higgs field ϕ, which is a space-time scalar, to be composed of iso-vectors $\phi = (\phi^1, \phi^2, \phi^3)$ in some 3-dimensional abstract iso-spin space.

The Lagrangian

$$L = \frac{1}{2}(\mathrm{D}_\mu \phi^a)\mathrm{D}^\mu \phi^a + \frac{m^2}{2}\phi^a\phi^a - \lambda(\phi^a\phi^a)^2 - \frac{1}{4}F^a_{\mu\nu}\ F^{a\mu\nu},$$

where

$$F^a_{\mu\nu} = \partial_\mu A^a_\nu - \partial_\nu A^a_\mu + e\epsilon^{abc}A^b_\mu A^c_\nu;$$
$$\mathrm{D}_\mu \phi^a = \partial_\mu \phi^a + e\epsilon^{abc}A^b_\mu \phi^c.$$

After symmetry-breaking, the vacuum state consists of ϕ^a iso-vectors with length

$$(\phi^1)^2 + (\phi^2)^2 + (\phi^3)^2 = \frac{m^2}{2\lambda}.$$

The 3D vectors ϕ^a in iso-space can be drawn in normal space as a hedgehog pattern

$$\phi^a = \left(\frac{m^2}{2\lambda}\right)^{1/2} r^a/r.$$

Show that the gauge fields A^a_μ can be chosen such that for large r

$$A^a_i = \epsilon^{iab}\frac{r^b}{er^2}, \qquad A^a_0 = 0,$$

and that consequently $\mathrm{D}_\mu\phi^a$ goes to zero as r goes to infinity.

Show also that the energy

$$H = \frac{1}{2}(\mathrm{D}_\mu\phi)^2 + V(\phi) + \frac{1}{4}F^2$$

goes to zero as $r \to \infty$.

Finally, show that $F_{\mu\nu}$ does not vanish as $r \to \infty$, but has the form of a radial magnetic field.

6.5 Try to derive a domain wall as a solution of Einstein's equations of General Relativity.

6.6 Show that a straight cosmic string is a solution of Einstein's field equations of General Relativity.

Show that a straight cosmic string does not exert a gravitational force.

6.7 A spherical shell of radius r with homogeneous charge and mass density of charge e and mass m_0 has a self-energy

$$m(r) = m_0 + \frac{e^2}{r} - \frac{Gm_0^2}{r}.$$

A sequence of such shells with radius r going to zero has a diverging $m(r)$. But one central idea of General Relativity is that all energies act gravitatively. Thus the gravitational energy is $\frac{Gm^2}{r}$ with $m(r)$ replacing m_0. Show that this renormalizes the self-energy, and for $r \to 0$ a finite result is obtained.

Show that this is a non-perturbative result, i.e. an expansion in G would lead term by term to a divergent $m(r)$ for $r \to 0$.

7. Relic Particles from the Early Universe

"The evolution of the world can be compared to a display of fireworks that has just ended: some few red wisps, ashes, and smoke. Standing on a cooled cinder, we see the slow fading of the suns, and we try to recall the vanishing brillance of the origin of the world."

G. Lemaitre

"Unifit to mend the azure sky,
I passed some years to no avail.
My life in both worlds written here,
whom can I ask to pass it on?"

(*Inscription on the jade stone rejected by the goddness Nu Wa when she repaired the sky* [from *A Dream of Red Mansions* by Tsao Hsueh-Chin; Foreign Language Press, Beijing 1987]).

Relics of a previous epoch bring puzzles and surprises

7.1 Introductory Remarks

We have seen in Chap. 3 that during the radiation-dominated epoch the cosmic time t is proportional to the inverse square of the temperature, $t \propto T^{-2}$, except during the annihilation processes which heat up the gas of photons. We can write

$$t = 2.42 \times 10^{-6} g^{-1/2}(T) T_{\mathrm{GeV}}^{-2}, \tag{7.1}$$

where kT is measured in GeV, k is set equal to one, and t is measured in seconds.

There are fundamental units that can be expressed in terms of the gravitational constant G, of Planck's quantum of action $\hbar$, and of the velocity of light c. Thus a fundamental unit of mass, the so-called "Planck mass" can be defined as

$$M_{Pl} \equiv \left(\frac{\hbar c}{G}\right)^{1/2} = 2.2 \times 10^{-5}\,\mathrm{g}. \tag{7.2}$$

The Compton wavelength of this mass $\hbar/M_{Pl}c$ defines a fundamental length, the "Planck length"

$$L_{Pl} \equiv \frac{\hbar}{M_{Pl}c} = 1.5 \times 10^{-33}\,\mathrm{cm}. \tag{7.3}$$

This length is equal to half the Schwarzschild radius of the Planck mass

$$L_{Pl} = \frac{GM_{Pl}}{c^2}. \tag{7.4}$$

Finally, L_{Pl}/c can be used to define a fundamental unit of time, the "Planck time"

$$t_{Pl} \equiv \frac{L_{Pl}}{c} = \left(\frac{G\hbar}{c^5}\right)^{1/2} = 5 \times 10^{-44}\,\mathrm{s}. \tag{7.5}$$

These natural units were introduced by Max Planck, as early as 1899 [Planck 1899]. He commented:

"These quantities have their natural significance as long as the laws of gravitation and of the propagation of light in a vacuum, and the first and second laws of thermodynamics, remain valid. They must therefore always turn out the same – even if measured by various intelligent beings with various methods."

All physical quantities become pure numbers when measured in these units. For example, $1\,\mathrm{g} = 5 \times 10^4; 1\,\mathrm{cm} = 6 \times 10^{32}; 1\,\mathrm{s} = 2 \times 10^{43}; \hbar = c = G = 1$. In the following we retain G in the equation as a constant with dimensions of its own, and we set $\hbar = c = 1$. Then all quantities can – as is usually done in elementary particle physics – be expressed in terms of an energy scale:

$$M_{Pl} = 1.36 \times 10^{19}\,\text{GeV} \equiv G^{-1/2},$$
$$t_{Pl} = \frac{1}{1.36} \times 10^{-19}\,\text{GeV} = G^{1/2} = L_{Pl}.$$

There is a widespread opinion that for processes with energy scales close to M_{Pl}, the classical theory of general relativity breaks down, and must be replaced by "quantum gravity" – i.e. quantum effects will somehow change the metric tensor into a kind of fluctuating quantum object. Although the concepts of quantum gravity seem sufficiently vague to allow for many different avenues of speculation, it seems reasonable to take M_{Pl} as an upper limit of these energy scales that allow a description of their gravitational effects in terms of a classical theory. There may be doubts whether, at a scale $M_{\text{GUT}} \simeq 10^{-4} M_{Pl}$, one is still safely in the classical realm. But we shall follow here the procedure generally in use: the effects of elementary particles up to and beyond scales $\approx M_{\text{GUT}}$ are set up in a classical background solution of Einstein's equations. Usually the homogeneous and isotropic Friedmann space-times are taken as cosmological models.

We have seen in Chap. 3 that, in thermal equilibrium, the number density of relativistic particles ($T > m$) in the ideal gas approximation is given as

$$n \simeq g(T)\zeta(3)\pi^{-2}T^3.$$

The concept of "asymptotic freedom" of the strong interactions (cf. Chap. 5) implies that interaction cross-sections decrease at high energies as

$$\sigma \sim \alpha^2 T^{-2} \tag{7.6}$$

where α describes the strength of the coupling. The reaction rate can be approximated by

$$(n\sigma) \sim \alpha^2 T$$

at energies where the rest masses of the particles are negligible. The expansion rate is

$$H \sim G^{1/2}T^2$$

and the ratio of the two rates

$$H^{-1}n\sigma \approx \alpha^2 \frac{G^{-1/2}}{T} \approx \alpha^2 \frac{M_{Pl}}{T}. \tag{7.7}$$

Take α to be the GUT coupling constant, $\alpha \sim 10^{-2}$; then $H^{-1}n\sigma > 1$ for $T < 10^{-4}T_{Pl}$. Below $\alpha_x^2 T_{Pl}$ the strong, weak, and electromagnetic interactions should all be in thermal equilibrium, since the particles mediating these interactions have masses that are small compared to the GUT mass scale. This thermal equilibrium should be maintained until thermal energies are far below the electroweak breaking scale of $\sim 300\,\mathrm{GeV}$. Without this stage of thermal equilibrium the physics of the universe would have the complicated structure of a single, strongly interacting many-body state.

Another important feature of particle physics in the early universe is the view that the fundamental objects – quarks and leptons – are point-like particles. Extended particles, on the other hand, would be in trouble in the early stages of the cosmological evolution: the particle horizons that exist in FL space-times have a size $\approx t$, and this is, for example, smaller than the nuclear radius of $10^{-13}\,\mathrm{cm}$ for times t earlier than $10^{-23}\,\mathrm{s}$. It is a blessing that the fundamental particles are not like extended nucleons. The causality problem is avoided by point-like quarks and leptons.

The concept of thermal equilibrium still can be used within each horizon only, and then for small horizon sizes new problems appear [Barrow 1983; Ellis and Steigman 1979].

In this respect superstring theories can be accommodated, because the "strings", as fundamental, extended objects, are in any case smaller than the Planck length L_{Pl}. For $t \geq t_{Pl}$ the strings may be considered as point-like objects.

7.2 Production, Destruction, and Survival of Particles

The general situation we want to describe here is concerned with an equal number of particles and antiparticles – produced thermally – which interact with a given, temperature-dependent cross-section. Let us picture the system under discussion as embedded in a heat bath of temperature $T(t)$ and occupying a volume $V(t)$ (Fig. 7.1).

In the radiation-dominated phase

$$V \propto T^{-3}, \qquad t \propto T^{-2}.$$

Let $n(\overline{n})$ be the particle (or antiparticle) densities of a certain particle species (for example e^+ and e^-, or nucleons and antinucleons). We make the

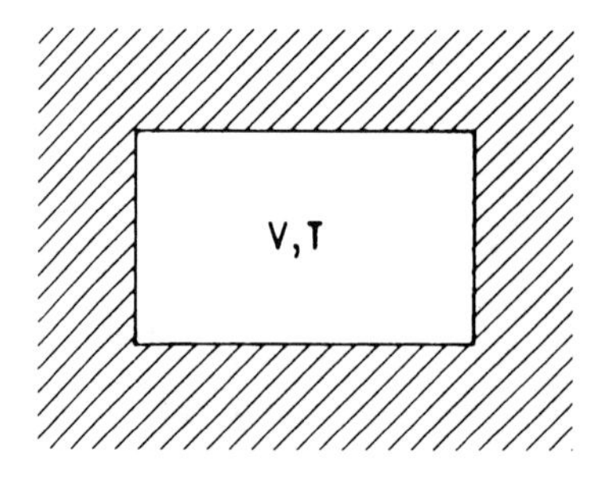

Fig. 7.1. Schematic picture of a system of particles occupying a volume $V(t)$ in a heat bath $T(t)$

assumption $n = \overline{n}$ – this is satisfied for nucleons to an accuracy of 10^{-10} in our almost baryon-symmetric universe. The time dependence of the particle number $N \equiv nV$ can be written as follows [Steigman 1979; Barrow 1983]:

$$\frac{dN}{dt} = \psi V - \beta n^2 V. \tag{7.8}$$

Here ψ describes the production rate and βn^2 the annihilation rate of the particles.

The production term ψ can be related to β by the principle of detailed balance (time reversal invariance). The equilibrium condition $dN/dt = 0$ leads to

$$\psi(T) = \beta(T) n_{eq}^2. \tag{7.9}$$

Equation (7.9) can be rewritten as

$$dN/dt = \beta(n_{eq} + n)(N_{eq} - N). \tag{7.10}$$

Numerical integrations of (7.10) are presented in [Wolfram 1979; Lee and Weinberg 1977; Kolb and Turner 1983].

Let us discuss (7.10) in some detail (following [Steigman 1979]). We replace the temperature by the dimensionless variable

$$x(t) \equiv \frac{mc^2}{k_B T} = \frac{m(\mathrm{GeV})}{T(\mathrm{GeV})}. \tag{7.11}$$

The equilibrium distributions n_{eq} were derived in Chap. 3, and it was found that $n_{eq} \sim T^3$ for relativistic particles, and $n_{eq} \sim T^3 x^{3/2} e^{-x}$ for non-relativistic particles $T \leq m$. Therefore ($N_{eq} \propto T^{-3} n_{eq}$)

$$N_{eq} = \begin{cases} \text{const.} & \text{for } x \leq 1, \\ x^{3/2} e^{-x} & \text{for } x > 1. \end{cases} \tag{7.12}$$

Thus at high temperatures $x \leq 1$,

$$\frac{dN_{eq}}{dt} = 0. \tag{7.13}$$

We see that $N = N_{eq}$ is a solution for $x \leq 1$ – there is equilibrium in this regime, and the number of particles in a comoving volume is constant. For massless particles ($x = 0$) there is always the solution $N = N_{eq}$. Independent of interactions, the relativistic particles in a comoving volume are conserved – as long as heating caused by the annihilation of other particles can be neglected.

For $x \geq 1, N = N_{eq}$ is no longer a solution $[(d/dt)N_{eq} \neq 0]$ – equilibrium is lost when the expansion rate $H \equiv \dot{R}/R$ is larger than the annihilation rate βN^2. Dividing (7.10) by N_{eq}, we can write

$$N_{eq}^{-1} \frac{dN}{dt} = \beta n_{eq}(1 + n/n_{eq}) \left(1 - \frac{N}{N_{eq}}\right). \tag{7.14}$$

It is conventient to define a "freeze-out" temperature x_f, which marks the point where the deviations from equilibrium begin, but where they are still small: $\Delta \equiv (N - N_{eq})/N_{eq} \leq 1$. Then

$$N_{eq}^{-1} \frac{dN}{dt} = -\Delta(2+\Delta)\beta n_{eq}. \tag{7.15}$$

Let $dN/dt \approx dN_{eq}/dt$, and choose $\Delta_f \equiv \Delta(x_f)$. Then x_f is fixed by the condition

$$N_{eq}^{-1} \frac{dN_{eq}}{dt} = \beta n_{eq} \Delta_f (2+\Delta_f). \tag{7.16}$$

Since $x \propto T^{-1} \propto \sqrt{t}$, we have for $x \geq 1$:

$$\frac{d \ln N_{eq}}{d \ln t} = \frac{1}{2} \frac{d \ln N_{eq}}{d \ln x} = \frac{1}{2}\left(\frac{3}{2} - x\right). \tag{7.17}$$

From (7.16) and (7.17), x_f is given as

$$x_f - \frac{3}{2} \approx x_f = 2 t_f \beta n_{eq,f} \Delta_f (2+\Delta_f). \tag{7.18}$$

Now

$$t_f = \left(\frac{45}{16\pi^3 g}\right)^{1/2} T_f^{-2} \tag{7.19}$$

and

$$n_{eq,f} = \frac{g_i}{(2\pi)^{3/2}} T_f^3 x_f^{3/2} e^{-x_f}. \tag{7.20}$$

Hence

$$x_f^{1/2} e^{x_f} = 2(2\pi)^{-3}\left(\frac{45}{2}\right)^{1/2} \frac{g_i m_i}{g^{1/2}} \beta(\Delta_f + 2)\Delta_f. \tag{7.21}$$

x_f depends logarithmically on the choice of Δ_f. For $\Delta_f(2+\Delta_f) \approx 1$ one finds

$$x_f^{1/2} e^{x_f} = 4.0 \times 10^{19} g_i Z_i, \tag{7.22}$$

where

$$Z_i = m_i \overline{\beta} g^{-1/2}, \tag{7.23}$$

$$\overline{\beta} = 10^{15} \beta\,[\mathrm{cm^3\,s^{-1}}]. \tag{7.24}$$

In (7.23), m_i is in units of GeV (cf. [Steigman 1979]). For $x > x_f$ the annihilation processes dominate:

$$\frac{dN}{dt} = -\beta n^2 V = -\frac{\beta N^2}{V}. \tag{7.25}$$

Here $V = V_f(t/t_f)^{3/2}$; the volume V_f is equal to the ratio $N_{\gamma f}/n_{\gamma f}$ for the photons. Integration of (7.25) can easily be carried out to give ($N_0 = N(t \gg t_f)$)

$$\int_{N_f}^{N_0} \frac{dN}{N^2} = -\beta \int_{t_f}^{t} \frac{dt}{V_f(t/t_f)^{3/2}}.$$

We may also set $N_f = N_{eq,f}$ to obtain

$$N_0 \simeq \frac{N_{eq,f}}{2\beta n_{eq,f} t_f} = \frac{N_{eq,f}}{x_f}, \tag{7.26}$$

where we have also made use of the relation (7.18). This is true if $2\beta n_{eq,f} t_f \gg 1$, i.e. if the expansion rate is less than reaction rates.

We can compare N_0 with the number of photons per comoving volume at the present epoch $N_{\gamma 0}$ – or equivalently compare the number densities –

$$\left(\frac{n}{n_\gamma}\right)_0 = \frac{n_{eq,f}}{x_f} \frac{N_{\gamma f}}{n_{\gamma 0}} \frac{1}{N_{\gamma 0}}.$$

For the equilibrium distributions

$$\frac{n_{eq,f} t_f}{n_{\gamma f}} = \frac{g_i}{(2\pi)^{3/2}} T_f^3 x^{3/2} e^{x_f} (2.404)^{-1} \pi^2 T_f^{-2},$$

and therefore

$$\left(\frac{n}{n_\gamma}\right)_0 = x^{-1/2} e^{x_f} x_f g_i (2\pi)^{3/2} \frac{\pi^2}{2.404} \frac{N_{\gamma f}}{N_{\gamma 0}}. \tag{7.27}$$

With the help of (7.23) and (7.24) the final outcome is

$$\left(\frac{n}{n_\gamma}\right)_0 = 6.5 \times 10^{-21} \frac{x_f}{Z_i} \frac{N_{\gamma f}}{N_{\gamma 0}}. \tag{7.28}$$

This gives the present-day ratio of relic particles to photons (as well as of antiparticles to photons).

The photon number ratio $N_{\gamma_f}/N_{\gamma 0}$ can be calculated from the condition of entropy conservation,

$$g(T_f)(RT)_f^3 = g(T_>)(RT)_>^3$$

for any $T_> > m_e$. Around $T \sim m_e$ the photon gas is heated by the annihilation of the electron–positron pairs (cf. Chap. 3). If we assume that from

$T \sim m_e$ to T_f the expansion was adiabatic, we have

$$g(T_f)(RT)_f^3 = \left(2 + 3 \times \frac{7}{4}\left(\frac{T_\nu}{T_\gamma}\right)_0^3\right)(RT)_0^3 = \frac{43}{11}(RT)_0^3,$$

and then it follows trivially that

$$\frac{N_{\gamma 0}}{N_{\gamma f}} = \frac{11}{43} g(T_f). \tag{7.29}$$

For the sum $(n = \overline{n})_0$ this leads to

$$\left(\frac{n + \overline{n}}{n_\gamma}\right)_0 = 5.1 \times 10^{-20} \frac{x_f}{m_i \overline{\beta} g^{1/2}}, \tag{7.30}$$

$$x_f + \frac{1}{2} \ln x_f = 45.2 + \ln(g_i Z_i), \qquad Z_i \simeq \frac{m_i \overline{\beta}}{g^{1/2}}, \tag{7.31}$$

where $\overline{\beta} = 10^{15} \langle \sigma v_{th} \rangle$, σ being the cross-section, v_{th} the velocity.

In Fig. 7.2 the general properties of these results are illustrated. Several simple conclusion can be drawn:

i) For $T > m_i$, particles and antiparticles are relativistic and approximately as abundant as photons. Even such massless particles as freeze out of equilibrium (weakly interacting or massless particles) stay roughly as abundant as photons; their number in a comoving volume is conserved.
ii) Massive, strongly interacting particles can maintain equilibrium by collisions even for some time below $T \leq m_i$. During this equilibrium phase, their number decreases exponentially $\propto \exp(-m_i/T)$ until the freeze-out temperature T_f is reached. Below T_f creation of new particles is suppressed, and eventually the annihilation processes also cease. This way a finite number of such particles in a comoving volume survives – eventually $N = \text{const}$.

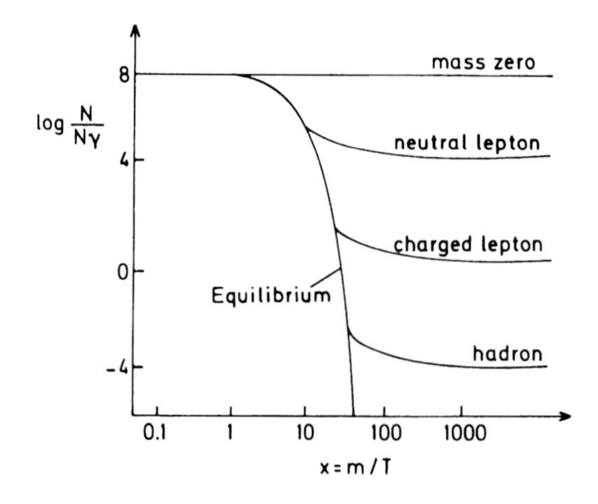

Fig. 7.2. The change of particle number in the early universe as a function of $x = m/T$. The various particles are distinguished by the value of $Z \sim m\langle \sigma v \rangle$ (after [Steigman 1979])

Table 7.1. Estimates for stable relic particles

Particle	T_f	$n_0\,[\mathrm{cm}^{-3}]\ (n_{\gamma 0} = 400)$	$\rho_0\,[\mathrm{g\,cm}^{-3}]$
Light neutrinos	1 MeV	100	$10^{-31} m_{\mathrm{eV}}$
Heavy neutrinos	$0.05 m_\nu$	$10^{-4} m_{\mathrm{GeV}}^{-2}$	$10^{-28} m_{\mathrm{GeV}}^{-2}$
Charged leptons	$0.03 m_L$	$10^{-10} m_{\mathrm{GeV}}$	$10^{-34} m_{\mathrm{GeV}}^{2}$
Heavy hadrons	$0.02 m_H$	$10^{-16} m_{\mathrm{GeV}}^{-1}$	10^{-40}

iii) The strongly interacting particles remain in equilibrium longer, and fewer survive ($\sim e^{-x}$) than the neutral or charged leptons. In Table 7.1 we list some order of magnitude estimates of stable, relic particles [from Steigman 1979].

The application to nucleons (and quarks) for the case of baryon symmetry $n = \overline{n}$ is treated in early papers by Zel'dovich and Chiu [Chiu 1965; Zel'dovich et al. 1965].

The cross-section is roughly

$$\langle \sigma v \rangle \simeq \frac{c}{m_\pi^2} \approx 3 \times 10^{10} \times 2 \times 10^{-26}\,\mathrm{cm^3 s^{-1}} \simeq 10^{-15}\,\mathrm{cm^3 s^{-1}}. \tag{7.32}$$

Experimentally [Particle Data Group 1996] one finds

$$v \sigma_{n\overline{n}} = 1.5 \times 10^{-15}\,\mathrm{cm^3 s^{-1}} \tag{7.33}$$

for energies $0.4\,\mathrm{GeV} < E < 7\,\mathrm{GeV}$. This results in $x_f \simeq 44$, therefore $T_f \simeq 22\,\mathrm{MeV}$ and

$$\left(\frac{n + \overline{n}}{n_\gamma} \right)_0 \simeq 10^{-18}. \tag{7.34}$$

This is almost 10 orders of magnitude smaller than the observed value (cf. Chap. 3). If the universe is symmetric between baryons and antibaryons, then the annihilation is so efficient that baryons do not survive. Therefore a significant initial baryon asymmetry must already exist for $T \gg T_f$, and the observed value is determined not by the annihilation processes, but by the baryon excess imposed at $T \geq T_f$. If nucleons contain a quark substructure, then at temperatures above the quark–hadron phase transition ($T_c \simeq 200\,\mathrm{MeV}$), a quark–antiquark asymmetry must be realized of the order of

$$\frac{n_q - n_{\overline{q}}}{n_q + n_{\overline{q}}} \approx 10^{-10}. \tag{7.35}$$

In Chap. 8 we shall present the arguments in favour of the generation of such a tiny asymmetry within the framework of grand unified theories.

7.3 Massive Neutrinos

> "Dear radioactive ladies and gentlemen, ...as a desperate remedy to save the principle of energy conservation in beta-decay, ... I propose the idea of a neutral particle of spin 1/2 and with mass not larger than 0.01 proton masses."
>
> *W. Pauli (1930)*

7.3.1 Experimental Limits

The Particle Data Group [Groom et al. 2000] gives the following limits for the masses of the stable neutrinos:

$$m_{\nu_e} \leq 2.2\,\text{eV} \qquad m_{\nu_\mu} \leq 1.7\,\text{MeV}. \tag{7.36}$$

The mass limit for ν_e is found from β-decay experiments. The limit on ν_μ comes from investigations of π-meson decays ($\pi^+ \to \mu^+ + \nu_\mu$) at the PSI μ-spectrometer facility. Preliminary limits on the mass of the τ-neutrinos are gained from e^+e^- collider experiments at DESY and CERN (using the reactions ($e^+e^- \to \tau^+\tau^- \to 5\pi\nu_\tau$). An upper limit of

$$m_{\nu_\tau} \leq 18.2\,\text{eV} \tag{7.37}$$

is derived. All these limits are, hitherto, compatible with $m_{\nu_i} = 0$.

The only experiment claiming a definite value for the mass of the electron antineutrino has been the tritium decay measurement by the ITEP group under Lyubimov in Moscow [Lyubimov et al. 1980; Boris et al. 1985]. This experiment measured the electron spectrum of the decay reaction

$$^3\text{He} \to {}^3\text{He}^+ + e^- + \overline{\nu}_e. \tag{7.38}$$

The mass m_ν of the electron antineutrino reveals itself in a distortion of the beta spectrum near the maximum energy $E_{\text{max}} \simeq E_0 - m_\nu c^2$. The process of (7.38) has a rather low end-point energy of $E_0 = 18.6\,\text{keV}$. The ITEP group claimed a mass for the electron anitneutrino of

$$m_{\overline{\nu}_e} = 31.3 \pm 0.3\,\text{eV}. \tag{7.39}$$

An effect in this range is, however, beset with many experimental uncertainties. Energies of several eV are typical of excitations in solid-state physics, and it is a difficult undertaking to take all those effects properly into account. Also the resolution function must be very accurately known to determine the real effect properly. The consensus is now that this difficult experiment can hardly probe the range of $\overline{\nu}_e$ masses below 10 eV. The result (7.39) is not accepted.

In 1985 an independent experiment [Fritschi et al. 1986] was succesful in measuring the end-point region of the tritium beta-spectrum with a 27 eV

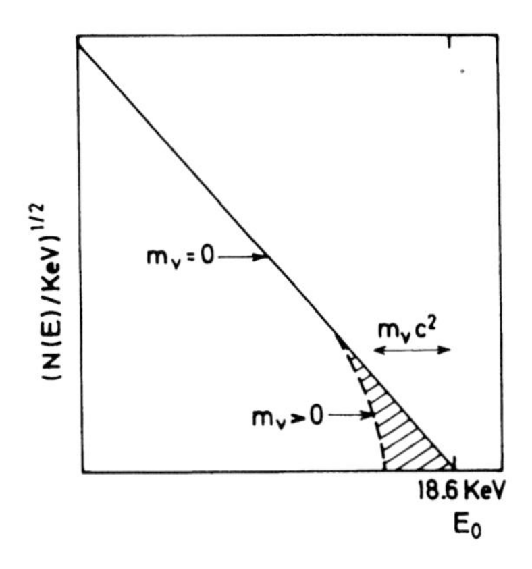

Fig. 7.3. Schematic Kurie plot of the β-spectrum of tritium decay indicating the effect of a non-zero neutrino mass (after [Winter 1986])

resolution. The result is a mass of $\overline{\nu}_e$ consistent with zero, with an upper limit of

$$m_{\overline{\nu}_e} \leq 18\,\mathrm{eV}. \tag{7.40}$$

Criticism of the Lyubimov experiment is supported by this measurement. In Fig. 7.3 we show a Kurie plot for tritium decay (schematic).

The best tritium decay experiments [Lobashev et al. 1999; Weinheimer et al. 1999] give a limit $m_{\overline{\nu}_e} \leq 2.2\,\mathrm{eV}$. Other experiments on massive neutrinos involve the study of double beta decay processes without neutrino emission:

$$(A, Z) \rightarrow (A, Z+2) + e^- + e^-. \tag{7.41}$$

Besides such a neutrinoless β-decay, there is a two-neutrino transition of the kind

$$(A, Z) \rightarrow (A, Z+2) + 2e^- + 2\overline{\nu}_e. \tag{7.42}$$

Here A is the mass number of a nucleus and Z the number of charges.

At this point it may be worthwhile saying a few words about the nature of neutrinos. Are they Dirac or Majorana particles (cf. [Winter 1986])? Let us consider the transformation properties under CPT (particle into antiparticle and time inversion). A neutrino ν_- with negative helicity is transformed by CPT into the antineutrino $-\overline{\nu}_-$ with positive helicity. If, in addition, the neutrino is massive, ν_- can be turned into ν_+ by an appropriate Lorentz transformation. (The massive neutrino travels with a velocity v less than light and can be overtaken by an observer travelling with $v_0 \geq v$. In his frame the neutrino travels in the opposite direction while spinning in the same way as in the original frame.) This ν_- neutrino may have its own CPT image $\overline{\nu}_+$. The final outcome is four states $(\nu_-, \overline{\nu}_+, \nu_+, \overline{\nu}_-)$ with the same mass. Such an object is described by a Dirac field.

There is also the possibility that the states ν_+ and $\overline{\nu}_+$, as well as ν_- and $\overline{\nu}_-$, are identical. Such a two-state neutrino is described by a Majorana field.

Since a Majorana neutrino is its own antiparticle, its lepton number is zero. Majorana neutrinos lead to L-violation.

For zero-mass neutrinos the distinction between Dirac and Majorana neutrinos is artificial. The two Dirac states $(\overline{\nu}_-, \nu_+)$ may not exist in nature, and $(\nu_-, \overline{\nu}_+)$ can be viewed as a Dirac neutrino or as the two helicity states of a Majorana neutrino.

Clearly the occurrence of a neutrinoless (0ν) double β-decay requires two things:

i) There must be no difference between the neutrino emitted and the antineutrino absorbed, i.e. the neutrinos must be 2-component ("Majorana") particles.
ii) A helicity change must happen to the neutrino between emission and reabsorption – either by a finite neutrino mass (giving a probability $\propto (m_\nu/E_\nu)^2$ for the helicity transition) or through the direct introduction of a right-handed weak current into the theory.

Experimental studies have so far not provided evidence that neutrinoless beta decay actually occurs in nature. Such a result would have far-reaching consequences for particle physics: in a gauge theory of weak interactions a non-vanishing $\beta\beta0\nu$ decay rate has the consequence that the neutrinos are Majorana particles, regardless of the mechanism inducing the decay.

Since the neutrino with a definite mass will in general not be the state corresponding to a specific flavour (such as ν_e), the mass limits obtained from studies of the $\beta\beta0\nu$ process refer to an average mass $\langle m_\nu \rangle_e$, where the state $|\nu_e\rangle$ is a superposition

$$|\nu_e\rangle = \sum_i U_{ei} \, |\nu_i\rangle \, ,$$

and

$$\langle m_\nu \rangle_e = \sum_i m_i \, |U_{ei}|^2 .$$

The effective mass can even be smaller than m_i for all types of neutrinos. But in many GUT models the relation $\langle m_\nu \rangle_e \simeq m_{\nu_e}$ holds.

The experimental limits on the average mass are very strong

$$\langle m_\nu \rangle_e \leq 0.2 \, \mathrm{eV} .$$

[Klapdor-Kleingrothaus et al. 2001].

If the neutrinos have mass, then the proper weak interaction eigenstates are in general not identical with the states corresponding to a definite mass. In a Lagrangian formulation this corresponds to non-diagonal terms quadratic in the neutrino fields – the diagonalization gives the neutrino states with a definite mass as linear combinations of the original fields.

In that case there are non-zero transition probabilities between different neutrino species – "neutrino oscillations".

In the case of two species of neutrino with non-zero mass (e.g. ν_e, ν_μ) the mixing is determined by just one mixing angle ϑ:

$$\underbrace{\begin{pmatrix} \nu_e \\ \nu_\mu \end{pmatrix}}_{\text{weak interaction eigenstates}} = \begin{pmatrix} \cos\vartheta & \sin\vartheta \\ -\sin\vartheta & \cos\vartheta \end{pmatrix} \underbrace{\begin{pmatrix} \nu_1 \\ \nu_2 \end{pmatrix}}_{\text{mass eigenstates}}. \tag{7.43}$$

The mass eigenstates ν_1, ν_2 have a time dependence

$$\begin{pmatrix} \nu_1(t) \\ \nu_2(t) \end{pmatrix} = \begin{pmatrix} \nu_1(0)e^{-iE_1 t} \\ \nu_2(0)e^{-iE_2 t} \end{pmatrix}$$

where

$$E_i^2 = m_i^2 + p_i^2 .$$

Weak eigenstates following the mixing (7.43) will then show an oscillating behaviour (similar to the Faraday rotation of light) such that, for example, the probability of finding ν_μ at time t, when at $t = 0$ we started with ν_e, is $P(\nu_e \to \nu_\mu) = \frac{1}{2}\sin^2 2\vartheta \sin^2[(E_e - E_\mu)t]$. For ultra-relativistic neutrinos $E_e - E_\mu = \frac{1}{2}\Delta m^2/p$, and

$$P(\nu_e \to \nu_\mu) = (\sin 2\vartheta)^2 \left(\sin\left[\frac{1.27\Delta m^2 L}{E_\nu}\right]\right)^2 , \tag{7.44}$$

where $L = ct$ is the distance between the source and the detector in [m], E_ν the neutrino energy (in [MeV]), and $\Delta m^2 \equiv m_1^2 - m_2^2$ is the difference of the squares of the neutrino masses. Examples of oscillation lengths are listed in Table 7.2 (from [Mössbauer 1984]).

To determine the oscillation parameters Δm^2 and $\sin^2 2\vartheta$, one measures the probability given in (7.44) as a function of E_ν/L. Nuclear fission reactors are appropriate terrestrial sites for such measurements. They produce

Table 7.2. Neutrino oscillation lengths

$\Delta m^2 = m_1^2 - m_2^2$ [eV^2]	Oscillation length [cm]			
(E_ν)	(1 MeV)	(10 MeV)	(1 GeV)	(100 GeV)
1	250	2.5×10^3	2.5×10^5	2.5×10^7
10^{-3}	2.5×10^5	2.5×10^6	2.5×10^8	
2.5×10^{-7}		Sun		
1.7×10^{-10}		1 au		

Sun: 10^{10} cm: effective diameter of the solar fusion core
1 au $= 1.5 \times 10^{13}$ cm: average distance between Sun and Earth

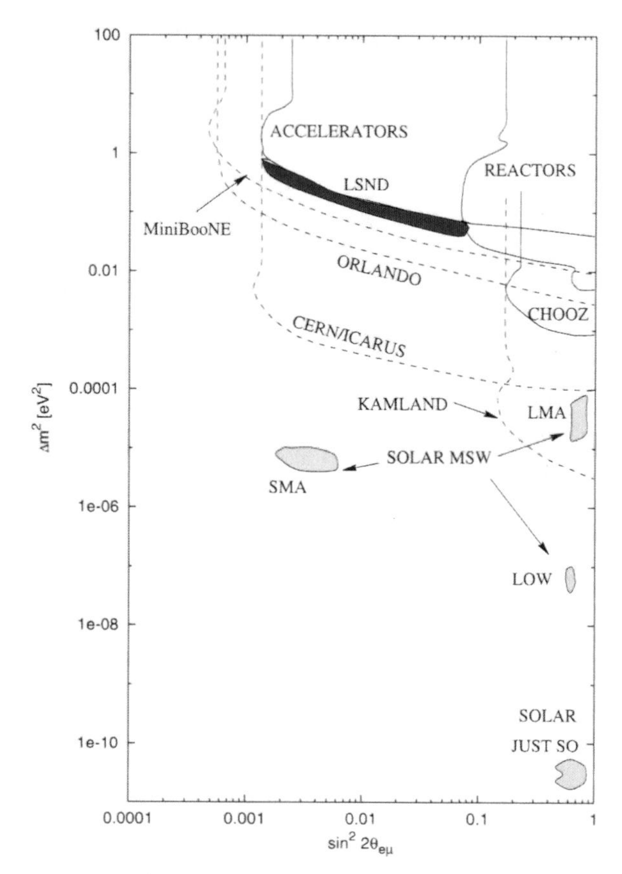

Fig. 7.4. Present bounds (*solid lines*), projected sensitivities of future experiments (*dashed lines*), and values suggested by the LSND experiment

abundant $\overline{\nu}_e$ with maximum energies of $\approx 8\,\mathrm{MeV}$ from the decay of neutron-rich fission produced isotopes. The oscillation experiments at reactors check for the disappearance of $\overline{\nu} : \overline{\nu}_e \leftrightarrow X$. These experiments are very sensitive to small values of the mass difference Δm^2; they are less sensitive to small mixing angles.

For instance, setting up a detector at 100m from a reactor allows us to test oscillations of ν_e into other flavours to $\Delta m^2 \simeq 10^{-2}\,\mathrm{eV}^2$, if $\sin^2 2\theta \leq 0.1$. The CHOOZ experiment in France has reached $\Delta m^2 \sim 10^{-3}\,\mathrm{eV}^2$ (for not too small $\sin^2 2\theta \geq 0.1$) [Apollonio et al. 1999] by putting a large detector at a distance of 1 km. Combining the CHOOZ result with mass limits from other experiments shows that $\sin^2 2\theta \geq 0.1$. The KAMLAND experiment which is planned in Japan will be sensitive to neutrinos arriving from 100 km, and hence will probe the range $\Delta m^2 \sim 10^{-5}\,\mathrm{eV}^2$ (some bounds are shown in Fig. 7.4).

Very long baseline experiments similar to KAMLAND are planned, to demonstrate convincingly an oscillation pattern (a web page with links to the experiments is the Neutrino Industry homepage: http://www.hep.anl.gov/ndk/hypertext/nuindustry.html). Another excellent source is John Bahcall's homepage: http://www.sns.ias.edu/~jnb/.

7.3.2 The Solar Neutrino Puzzle

R. Davis started a programme to measure neutrinos from the Sun in the Homestake mine by using the capture rate of the reaction

$$\nu_e + {}^{37}\mathrm{Cl} \to e^- + {}^{37}\mathrm{Ar}. \tag{7.45}$$

He performed many experiments from 1970 to 1983 and found a rate

$$R_{\mathrm{exp}} = 3.2 \pm 0.3\,\mathrm{SNU} \tag{7.46}$$

[Davis 1980] where $1\,\mathrm{SNU} = 10^{-36}$ captures per ^{37}Cl atom per second.

In the Sun's interior the basic fusion reaction

$$4\mathrm{p} \to {}^{4}\mathrm{He} + 2e^+ + 2\nu_e \tag{7.47}$$

produces a flux of neutrinos. 27 MeV are liberated by this fusion reaction, of which 3% is emitted as neutrinos, 97% eventually as photons. The neutrino flux from the Sun at the Earth can thus be estimated as

$$\sim 6 \times 10^{10} \nu_e/\mathrm{cm}^2\,\mathrm{s}.$$

Davis's first results were a factor of 3 lower than the rate expected from a standard solar model

$$R_{\mathrm{th}}({}^{37}\mathrm{Cl}) = 7.6 \pm 1.5\,\mathrm{SNU}.$$

Further experiments with gallium (SAGE, GALLEX, GNO), with water Čerenkov detectors (Kamiokande, Super-Kamiokande), and more recently the heavy water Sudbury Neutrino Observatory (SNO) have all looked for neutrinos from the Sun in different energy ranges. The puzzling result, first found by Davis, is still with us, that only 1/3 to 1/2 of the expected fluxes are observed [Cleveland et al. 1998; GNO collab. 2000; SAGE collab. 1999; Fukuda et al. 2001]. Figure 7.5 gives an impression of the observational status (after [Bahcall et al. 2001; courtesy of Schlattl and Weiss 1999]).

The oscillation length for solar neutrinos with energies between 0.1 and 10 MeV is of the order of the distance to the Earth for $\Delta m^2 \simeq 10^{-10}\,\mathrm{eV}^2$. Even tiny neutrino mass differences can have an observable effect if the mixing angles are large.

Even more remarkable is the idea that oscillations may be enhanced resonantly when the neutrinos propagate outwards through the Sun. The so-called

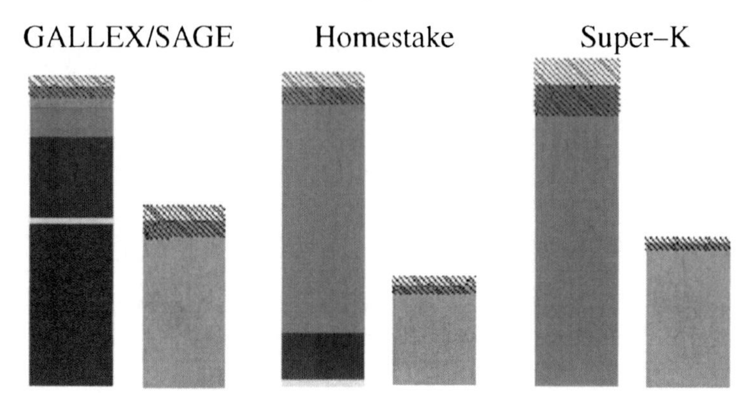

Fig. 7.5. The predicted count rates for solar neutrinos in the three experiments, GALLEX/SAGE, Homestake, Super-Kamiokande (coloured bars) in comparison with the actual detections (grey histograms). The different neutrino energies are colour-coded (Super-Kamiokande, for example, detects only the highest-energy neutrinos). The ratio of observed to predicted counts differs between the different experiments, but is $\simeq 50\%$

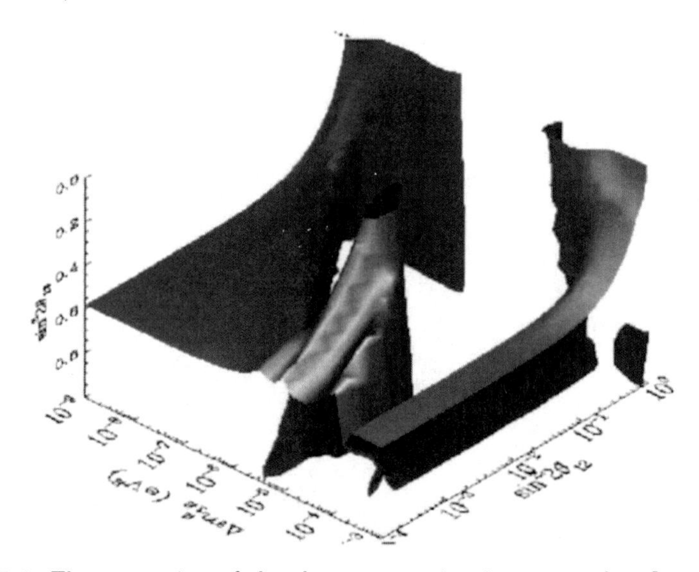

Fig. 7.6. The conversion of the electron neutrino into two other flavours 2 and 3 (μ and τ) depends on four parameters. The dependence on $(\Delta m)^2_{13}$ is omitted. Regions of solutions are surrounded by the red surfaces (from [Schlattl 2001])

MSW (Mikheyev–Smirnov–Wolfenstein [Mikheyev and Smirnov 1986]) effect generates large conversion effects of electron neutrinos into μ-neutrinos.

For solar neutrinos this will happen according to the model for Δm^2 between $10^{-8}\,\mathrm{eV}^2$ and $10^{-4}\,\mathrm{eV}^2$, and for $\sin^2\theta \geq (E_\nu/10\,\mathrm{MeV})6\times10^{-8}\,\mathrm{eV}^2/\Delta m^2$. Large suppression of the flux of ν_e seems easily possible. Figure 7.6 shows a model calculation [Schlattl 2001] for the conversion of electron neutrinos

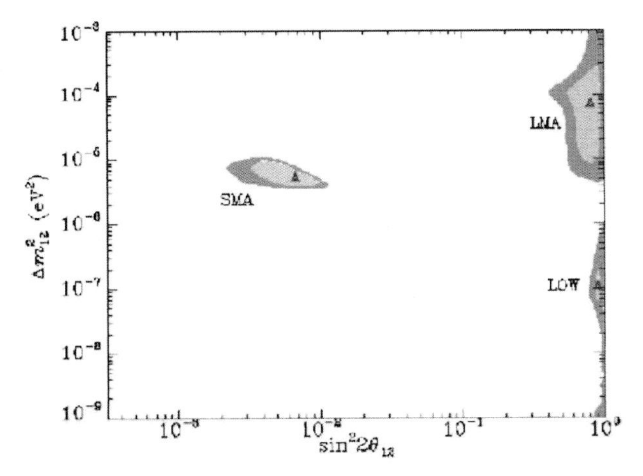

Fig. 7.7. Conversion into only one other flavour may be sufficient. Then Fig. 7.6 reduces to a plane with the solutions shown as coloured islands. The best solutions are marked by *triangles*

into two other flavours which reproduces approximately the number of measured neutrinos in all three types of experiments. It is actually sufficient to consider the conversion into only one other flavour, since recent measurements of Super-Kamiokande revealed a more or less flat energy spectrum of the solar neutrinos. Then the solutions due to MSW are represented by three coloured islands in the $(\Delta m^2, \sin^2 2\theta)$ plane:

$\Delta m^2 \simeq 10^{-5}\,\mathrm{eV}^2$ with small mixing angles $\sin^2 2\theta \geq 10^{-2}$ (SMA);
$\Delta m^2 \simeq 10^{-4}\,\mathrm{eV}^2$ with $\sin^2 2\theta \simeq 1$ (LMA);
$\Delta m^2 \simeq 10^{-7}\,\mathrm{eV}^2$ with $\sin^2 2\theta \simeq 1$ (LOW) (Fig. 7.7).

Besides the overall fluxes, other important tests are measurements of the neutrino spectrum in the water Čerenkov detectors, a day–night effect due to matter effects in the Earth, and an annual modulation of the signal. No conclusive evidence for any of these signals has yet been found.

The SNO has started to produce data, and the results [SNO collab. 2002] from the registration of solar neutrionos generated in the decay of ^{8}B provide convincing evidence for neutrino flavour oscillations. The deuterium-filled tanks at SNO allow electron neutrinos to register via the charged current (CC) reaction

$$\nu_e + D \to \mathrm{p} + \mathrm{p} + e^-,$$

and the total flux from neutral current (NC) reactions, where electron, muon, and tau neutrinos are registered

$$\nu + D \to \mathrm{p} + \mathrm{n} + \nu.$$

The results are: for the flux from CC,

$$\phi_{\nu_e} = (1.76 \pm 0.14) \times 10^6 \, \mathrm{cm}^{-2} \, \mathrm{s}^{-1};$$

and for the NC flux,

$$\phi_{NC} = (5.09 \pm 0.9) \times 10^6 \, \mathrm{cm}^{-2} \, \mathrm{s}^{-1}.$$

The excess of the NC flux over the CC implies neutrino flavour transformations. The total flux of ^{8}B neutrinos is determined to be $(5.44 \pm 0.99) \times 10^6 \, \mathrm{cm}^{-2} \, \mathrm{s}^{-1}$, in close agreement with the predictions of solar models [Schlattl et al. 1997; Schlattl 2001; Bahcall and Pinsonneault 2000].

The future Borexino experiment will be sensitive to the neutrino energies (spectral lines) from e^- capture on Be in solar fusion reactions. These ν_e seem to be strongly suppressed, as present data indicate.

At present the experiments favour the LMA solution.

7.3.3 Atmospheric Neutrinos

When a cosmic ray hits the atmosphere, its collision with a nucleon initiates a hadronic and electromagnetic shower in which many π-mesons are produced.

Charged pions decay through the chain

$$\pi \to \mu \nu_\mu \to e \nu_e \nu_\mu \nu_\mu .$$

Observations of atmospheric neutrinos by the Super-Kamiokande (and other detectors) [Okada et al. 2000] showed a dependence of the flux of μ-neutrinos on zenith angle, ν_μ from above (pathlengths ~ 20 km) being about twice as numerous as ν_μ from below (passing through the Earth, with pathlengths $\sim 13\,000$ km). The most plausible explanation for this difference is an oscillation $\nu_\mu \to \nu_\tau$ with a large mixing angle, and $\Delta m^2 \simeq (2-7) \times 10^{-3} \, \mathrm{eV}^2$ [Fukuda et al. 2000, Learned 2001]. Oscillations $\nu_\mu \to \nu_e$ cannot explain this effect, because the electron neutrino flux is completely according to expectations.

Another possibility is the oscillation into a fourth type of neutrino which is postulated to have no weak interactions at all. This so-called "sterile" neutrino has been invoked by inventive model builders to give them more freedom in the parameters that can be fitted to experiments, and now the experimentalists are having a hard time chasing it away again.

A sterile neutrino would not be affected by the Z^0-width limit, and it can just be accommodated within the uncertainties of the helium abundance (cf. Sect. 7.3.4). The experiments show that the angular dependence of the oscillations does not give support to this idea.

An experiment (named K2K) is under way in which a ν_μ beam is sent over 250 km to the Super-Kamiokande detector. First results indicate a deficit of

ν_μ at the detector (40 events expected and only 27 observed) consistent with $\nu_\mu \to \nu_\tau$ oscillation models (see http://neutrino.kek.jp/ for updates). Unfortunately towards the end of 2001 the photomultipliers in Super-Kamiokande were destroyed by some mysterious accident. So it will take some time before it can operate again.

These experiments prove conclusively that neutrinos have a mass, but they determine only $\sqrt{(\Delta m)^2}$, not the absolute scale. If the neutrino masses are different, such that, e.g., $m_{\nu_e} < m_{\nu_\mu} < m_{\nu_\tau}$, then their masses are $\sim \sqrt{(\Delta m)^2}$, and they are unimportant for cosmology. A degenerate mass spectrum $m_{\nu_e} \simeq m_{\nu_\mu} \simeq m_{\nu_\tau}$ with small Δm may still be a possibility. $\beta\beta 0\nu$ experiments may be able to set a mass scale in the future.

7.3.4 Neutrinos from Supernova 1987A in the Large Magellanic Cloud

On the morning of February 23, 1987, between 2 and 7 o'clock universal time (UT) a star in the Large Magellanic Cloud (LMC) at the position of the B31 supergiant Sanduleak-69202 began to brighten rapidly. By 10 o'clock UT it had brightened by a factor of 200. Then during the next 24 hours it increased by a further factor of 10 (see also Figs. 2.5 and 2.6). This is the first very bright supernova in the modern era. Because of the proximity of LMC (at a distance of 55 kpc) the supernova could be seen with the naked eye – but only from the southern hemisphere.

The fact that the progenitor of the supernova seems to be a B-type supergiant came as a surprise; it had been thought that only M-type giants or supergiants could have an evolved core where collapse would trigger a supernova. But the astronomical community soon recovered, and produced many models of collapsing and exploding B stars. The key point here is the low-metallicity environment. Metal-poor B stars do have an evolution all of their own, and they do not branch off to become M supergiants prior to the onset of carbon burning. The explosion of a B-type supergiant can also explain the rather low peak luminosity of the supernova (2 to 5 magnitudes fainter than expected for a type II supernova) [Hillebrandt et al. 1987]. The spectra showed the hydrogen lines indicating a type II supernova. The Hα line shows structure with a width of 1000 Å. This corresponds to an outflow velocity of 25 000 km s^{-1}.

Type II supernovae are thought to leave a collapsed remnant – a neutron star or a black hole. We shall only have to wait a little to obtain evidence on this question.

An immediate and fascinating aspect of SN 1987A has been the detection of neutrino bursts correlated with this event. At 7:35 UT the water Čerenkov detectors at Kamiokande (Japan) and IMB (USA) each registered a neutrino burst within 1 minute (the time uncertainty is due to the lack of precision timing at the Kamiokande experiment). The Kamiokande experiment detected a neutrino burst lasting 12 s, containing 12 events in the energy range between 7 and 50 MeV [Hirata et al. 1987]. The IMB experiment is in complete agreement with this measurement, finding eight events within 6 seconds in an energy range of 20 to 40 MeV.

These detections proved that our ideas about the physical processes in a collapsing stellar core are very likely to be correct.

Intrinsic properties of the neutrino bursts can be used to obtain limits on various supernova models, neutrino properties, and even (model-dependent) mass estimates.

7.3.5 Theoretical Possibilities

The various possibilities for massive neutrinos have already been discussed in Chap. 6. I will briefly repeat some of the results here.

Neutrinos can have Dirac as well as Majorana mass terms, since they are electrically neutral particles. Dirac mass terms would be of the form $m\overline{\psi}_L\psi_R$ in the Lagrangian; Majorana mass terms of the form $m\psi_L C\psi_L$. The minimal $SU(5)$ model does not contain right-handed neutrinos ν_R, and no $\underline{15}$-dimensional Higgs – therefore neutrinos are exactly of zero mass in this model. In an $SO(10)$ GUT the neutrinos can have a non-zero mass. The 16-dimensional representation of $SO(10)$ contains

$$\underline{5} \oplus 10 \oplus \underline{1} \quad \text{of} \quad SU(5) \quad \text{representations.}$$

The singlet (1) can be a right-handed neutrino ν_R. The spontaneous symmetry-breaking of $SU(5)$ produces a mass for this ν_R proportional to the GUT scale $M \sim 10^{15}$ GeV. The diagonalization of the mass matrix

$$(\nu_L \nu_R)\begin{pmatrix} m & m_q \\ m_q & M \end{pmatrix}\begin{pmatrix} \nu_L \\ \overline{\nu_R} \end{pmatrix} + \text{h.c.}$$

gives

$$m_{\nu_L} \approx g\frac{m_q^2}{M}$$

with $M = 10^{15}$ GeV, m_q a typical quark mass. Taking $m_q \simeq 100$ GeV,

$$m_{\nu_L} = g10^{-2}\,\text{eV}. \tag{7.48}$$

This rather small value is characteristic of many GUT models, since most of them contain a right-handed, heavy neutrino and have a sufficiently rich Higgs structure to allow for Majorana mass terms. Masses of this typical magnitude fit quite nicely to the oscillation experiments, but they are not directly accessible by experiments.

We see also that a cosmologically favoured range of neutrino masses is not easily reached in theoretical models, and also is not favoured by experiments.

7.3.6 Cosmological Limits for Stable Neutrinos

Since observational limits on the mean density exist (the age of the universe and the Hubble constant limit the mean density, for instance), there is also a cosmologically permitted range for the masses of stable neutrinos.

The neutrinos decouple from other particles (eγ) at a temperature $T_{\nu d}$, where their reaction rate becomes less than the expansion rate (cf. Chap. 3)

$$T_{\nu d} = 1\,\mathrm{MeV}. \tag{7.49}$$

Light neutrinos with $m_\nu < 1$ MeV are therefore still relativistic when they decouple, and according to Sect. 7.1 have their density n_ν equal to the photon density n_γ – up to statistical factors

$$\frac{n_\nu}{n_\gamma} = \frac{3}{8} g_\nu . \tag{7.50}$$

At $T < 1$ MeV the annihilation of e^+ and e^- pairs raises n_γ by a factor $\frac{11}{4}$ (cf. Chap. 3) and therefore at present

$$\left(\frac{n_\nu}{n_\gamma}\right)_0 = \frac{3}{22} g_\nu . \tag{7.51}$$

For each type of neutrino (ν and $\overline{\nu}$) the helicity counting factor is $g_\nu = 2$. The contribution of all the various light neutrinos – they become non-relativistic at $t \gtrsim 10^{12}$ s – to the mass density is thus

$$\varrho_{\nu 0} = \frac{3}{11} \sum m_\nu n_{\gamma 0} \tag{7.52}$$

(adding all neutrino masses in $\sum m_\nu$). The ratio $\varrho_{\nu 0}/\varrho_c \equiv \Omega_{\nu 0}$ accordingly is

$$h_0^2 \Omega_{\nu 0} \simeq \left(\frac{T_{\gamma 0}}{2.7\,\mathrm{K}}\right)^3 \frac{\sum m_\nu}{100\,\mathrm{eV}} . \tag{7.53}$$

With a zero cosmological constant, the observations limit Ω_0 to be less than 1 (Chap. 2), and therefore

$$\sum m_\nu < 100\,\mathrm{eV} h_0^2 \tag{7.54}$$

for an FL universe undergoing a hot big bang. The reader can convince himself that for three types of neutrinos of equal mass this corresponds to

$$m_\nu < 33 h_0^2\,\mathrm{eV}. \tag{7.55}$$

Heavy neutrinos with $m_\nu > 1$ MeV will be non-relativistic beforethey decouple. Then according to (7.27) their density is given by

$$n_\nu = 2T^3 \left(\frac{m_\nu}{2\pi T}\right)^{3/2} \exp(-m_\nu/T). \tag{7.56}$$

The decoupling temperature can be derived according to the scheme presented in Sect. 7.1.

The quantity $\langle\sigma \cdot v\rangle$ can be approximated by a constant, since for exothermal processes the cross-section $\sigma(\nu + \overline{\nu} \to X)$ is proportional to $1/v$. Hence

$$\langle\sigma v\rangle = \frac{N_A}{2\pi} G_F^2 m_\nu^2, \tag{7.57}$$

where G_F is the universal Fermi constant, and N_A is a constant depending on open reaction channels.

Setting the annihilation rate $n_\nu\langle\sigma v\rangle$ equal to the expansion rate gives the decoupling temperature $T_{\nu d}$. It turns out that, for $m_\nu \simeq m_p$, the parameter

$$X_{\nu d} \equiv m_\nu / T_{\nu d} \simeq 20.$$

The condition that $\Omega_{\nu 0} < 1$ leads, for massive neutrinos, to the limit

$$\left(\frac{m_\nu}{m_p}\right) \geq 2. \tag{7.58}$$

There should be no stable neutrino in the interval between 100 eV and ~ 2 GeV. The importance of massive neutrinos for the problem of galaxy formation will be discussed in detail in Chaps. 10 to 12.

7.3.7 Asymmetric Neutrinos

The radiation-dominated phase of the early universe was discussed in Chap. 3, and there the assumption was made that initially there are equal numbers of neutrinos and antineutrinos – the chemical potentials of the neutrinos were set equal to zero.

A non-zero chemical potential μ_ν can give rise to an excess of antiparticles over particles or vice versa. The mass-zero antineutrinos (or neutrinos) would then form a Fermi sea with degenerate particles filling the Fermi levels up to $E_\nu \leq \mu_\nu$.

For light neutrinos the tritium-decay experiments imply the limits of (7.36) for μ_ν. If there is an asymmetry for massive neutrinos with $m_\nu >$ 1 MeV, then this asymmetry will determine the final abundance [Hut and Olive 1979]. The mass density ϱ_ν will not decrease with increasing mass, because the annihilation reactions will stop when annihilation partners have been used up. This kind of a lepton asymmetry arises naturally in GUT theories. Its magnitude may be expected to be equal to the baryon asymmetry of $\sim 10^{-10}$. For the excess neutrinos, then, the requirement not to exceed the critical density of the universe leads to a further mass limit: the neutrino mass must be less than

$$m_\nu \leq (\Omega_0 h_0^2) 10^4\,\text{GeV}. \tag{7.59}$$

7.3.8 Unstable Neutrinos

So far, we have discussed limits on stable neutrinos. For unstable neutrinos these considerations change considerably. But this additional freedom does not just weaken the limits derived previously, it also introduces new astrophysical effects: [Glashow and de Rujula 1980]. The decay of unstable neutrinos deposits energy into the radiation background, and thus new limits exist from the observation of the background at different wavelengths (Fig. 7.8 gives a graphic description of such limits).

A heavy neutrino ν_H decaying into a lighter one ν_L at rest (i.e. at an epoch where $T \leq m_{\nu H}$),

$$\nu_H \to \nu_L + \gamma, \tag{7.60}$$

emits a photon with energy

$$E_\gamma = \frac{(m_{\nu H}^2 - m_{\nu L}^2)}{2m_{\nu H}} \simeq \frac{1}{2} m_{\nu H} \tag{7.61}$$

for $m_{\nu H} \gg m_{\nu L}$. For example, a neutrino mass between 10 and 100 eV results in E_γ between 5 and 50 eV, i.e. the photon energy is in the UV range.

It has been pointed out [e.g. Stecker 1980] that one might hope for the detection of two types of signals from ν-decay:

i) A line feature centred on E_γ, with a narrow Doppler broadening resulting from the decaying ν_H in the galactic halo or in nearby clusters, would occur.
ii) The integrated cosmological background of heavy neutrinos decaying at various redshifts would give rise to a power-law spectrum.

In case (i), Doppler velocities of $\sim 270\,\mathrm{km\,s^{-1}}$ broaden the line by no more than 1 Å typically (in the UV). In case (ii) the flux (per eV) is [Stecker 1980]:

$$I_\lambda = \frac{10^{29}}{\tau} \left(\frac{\lambda_0^3}{\lambda^5} \right)^{1/2} \tag{7.62}$$

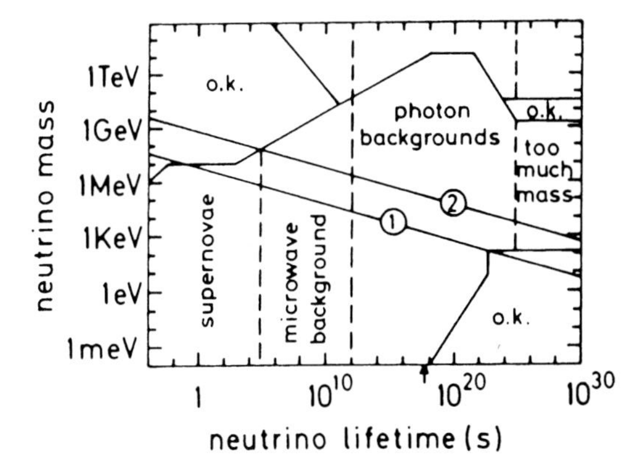

Fig. 7.8. Limits on the allowed mass and range of unstable neutrinos (after [Gunn et al. 1978]) (see text for explanations)

where $\lambda_0 = E_\gamma^{-1}$, τ is the lifetime of ν_H, and H_0 has been assumed to be $\sim 70\,\mathrm{km\,s^{-1}\,Mpc^{-1}}$.

As $I_\lambda \propto \lambda^{-5/2}$, the best limits can be obtained from observations at short wavelengths. The observational limit [Barrow 1983; Stecker 1980] of

$$I(1250\,\text{Å}) \leq 200\,\mathrm{cm^{-2}s^{-1}sr^{-1}}\text{Å}^{-1} \tag{7.63}$$

can be used to set a limit on the neutrino lifetime. If $m_\nu \geq 20\,\mathrm{eV}$ is assumed (i.e. $E_\gamma > 10\,\mathrm{eV}, \lambda_0 > 250\,\text{Å}$), then

$$\tau \geq 4 \times 10^{22}\,\mathrm{s}. \tag{7.64}$$

An assessment of the constraints that can be placed upon neutrino masses m_ν and lifetimes τ, when these two quantities are uncorrelated has been given by various authors [Gunn et al. 1978; Sato and Kobayashi 1978].

7.3.9 Neutrino Generations

In Chap. 3 we presented an argument that limits the number of neutrino flavours to less than 4, from considerations concerning the cosmological helium abundance. Additional neutrino types or other weakly interacting particles – the number of species exceeding 4 – would lead to ^{4}He abundances that were too large to be compatible with observations. These limits, however, depend crucially an the assumption of a Friedmann–Lemaître model for the early universe. In fact, as cosmologists, we would like to reverse the argument, and draw the conclusion that the strong dependence of the ^{4}He production on the expansion rate of the early universe provided a test of the behaviour of the scale factor $R(t)$ at a cosmic time $t \simeq 20$ seconds. The rather good agreement of the observations of the cosmic helium abundance with the predicted value gives confidence that the simple FL model is a good approximation to reality even at $t \simeq 20$ seconds.

A limit on the neutrino flavours has been set by an analysis of the decay of the Z^0 intermediate boson [Groom et al. 2000]:

$$N_\nu < 2.9835 \pm 0.0083. \tag{7.65}$$

The neutrinos included here have masses small enough that the decay $Z^0 \to \bar{\nu}\nu$ can proceed freely.

It is very remarkable that (7.65) and the helium limit are so close. But it is also important to note that these two limits do not necessarily refer to the same number N_ν. The helium limit includes all other light, stable particles that were relativistic at $T \sim 1\,\mathrm{MeV}$, even if they interact only superweakly (as the photino or gravitino or a sterile neutrino, for example). Very massive neutrinos ($\sim 1\,\mathrm{GeV}$) are not included in the cosmological limit. The Z^0 experiments, on the other hand, give a limit for all kinds of neutrinos, even heavy ones, provided they couple to the intermediate bosons with full strength. These limits can eventually restrict phenomenological models (motivated by the superstring theory) which contain additional neutral particles, such as light, right-handed neutrinos [Candelas et al. 1985].

7.4 Axions

Axions were discussed in Chap. 6; for reviews see [Raffelt 1986, 1990, 1996, 2000]. They are not generated in thermal equilibrium, but by an intricate twofold symmetry-breaking process. Here I want to discuss a few more aspects of the axion in cosmology.

Below a temperature of $\sim 1\,\text{GeV}$ quarks start to combine to form hadrons, and coherent axion waves appear $-\phi(x) = \vartheta e^{i\vartheta(x)}$ – with a typical correlation length of the order of the horizon size $t(1\,\text{GeV}) \sim 10^{-4}$ seconds. The characteristic momenta of the particle excitations corresponding to these waves are therefore $p = O(1/t) = O(10^{-20})\,\text{GeV}$; much less than the axion masses (see Chap. 6). The axions are extremely non-relativistic particles with $v/c = O(10^{-6})$.

The oscillatory modes can therefore be treated as an almost stationary classical field. The oscillation energy density is $\frac{1}{2}m_a^2 v^2 \vartheta^2$. Because of the axions' small couplings this energy dissipates very slowly. The coherent axion waves can serve as an ideal model for cold, dark matter in the universe.

As quoted in Chap. 6 the axion contribution to the cosmic density is

$$\Omega_\alpha \equiv \frac{\varrho_a}{\varrho_c} = 10^7 \left(\frac{v}{M_{Pl}}\right) \left(\frac{200\,\text{MeV}}{T_i}\right) h_0^2; \tag{7.66}$$

$T_i \sim 800\,\text{MeV}$ for $v/M_{Pl} \sim 4 \times 10^{-7} T_i$ decreases slowly with v/M_{Pl}. Then for $v \simeq 10^{12}\,\text{GeV}$ the axions rise to cosmological importance, $\Omega_\alpha \simeq 1$. Their field energy on scales $> 1/m_a \sim 1\,\text{cm}$ (for $m_a \simeq 10^{-5}\,\text{eV}$) has the properties of dust – no pressure, no dissipation. Since axions couple only extremely weakly to ordinary matter, fluctuations in the axion density can grow on all scales, after the axion field energy begins to dominate the energy density of the universe.

$\varrho_a = \varrho_\gamma$ for $T_\gamma/T_{\gamma 0} = 10^4 \Omega_{a,0}\ 2\ h_0$, and for $\Omega_{a,0} \simeq 1, H_0 = 50$, we find $T_\gamma = 2.7 \times 10^4\,\text{K} = (2.7 \times 10^4/1.16 \times 10^4)\,\text{eV} = 2.3\,\text{eV}$. This is larger than the recombination temperature $T_R \simeq 0.3\,\text{eV}$. Thus axions could be a candidate for cold dark matter.

To plan and carry out realistic searches for axions, the limits on v and m_a must be considered. The best limit to the coupling of photons to axions comes from globular cluster stars

$$G_{a\gamma} \cdot m_e \leq 0.3 \times 10^{-13} \tag{7.67}$$

[Raffelt 1996]. The limit is stronger than the red-giant limit (Table 6.2). Relating this to models, one arrives at a solid limit of

$$m_a \leq 0.01\,\text{eV}. \tag{7.68}$$

for the axion mass.

Cosmic axion strings are also a theoretical possibility. After axions acquire a mass at the QCD phase transition these strings could quickly become non-relativistic, and form a cold dark matter component. String loops are the dominant contribution, and the contribution to the cosmic axion density is [Battye and Shellard 1994]

$$\Omega_a \, h_0^2 \simeq 88 \times 3^{\pm 1} \left[(1 + \alpha/\kappa)^{3/2} - 1 \right] \left(10^{-6} \, \text{eV}/m_a \right)^{1.175} . \tag{7.69}$$

Probably $0.1 < \alpha/\kappa < 1.0$, and axions can be the dark matter, if their mass in 10^{-6} eV is between 6 and 2500. The high-mass end of the string scenario overlaps with the sensitivity planned for galactic dark matter axion search experiments in the US (Livermore [Hagman et al. 1998]) and Japan (Kyoto [Ogawa et al. 1996]). In Fig. 7.9 (from [Raffelt 2000]) an overview of the limits on axion mass and couplings is shown (this supplements Table 6.2). It appears that the only practical chance to discover these "invisible" particles is with the search experiments for galactic axions.

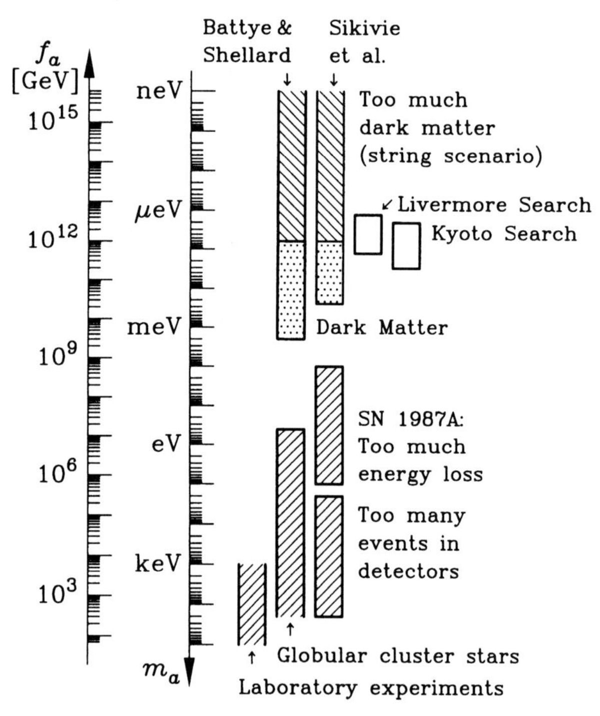

Fig. 7.9. Limits on axion masses and couplings [Raffelt 2000]

7.5 Domain Walls, Strings, Monopoles

The topological defects described in Chap. 6 can all be created in phase transitions in the early universe. They can be consistent with a standard cosmological model only under severe restrictions – as we shall see in the following discussion.

7.5.1 Domain Walls

Domain walls have disastrous cosmological consequences. According to (6.136) the surface energy density of a domain wall is

$$\sigma \simeq \lambda^{1/2} v^3. \tag{7.70}$$

A domain wall that extends over the dimension of the horizon $d_H = 2t$ has an energy content of $(\pi t^2)\sigma$. As the energy in relativistic particles is $[(4\pi/3)t^3]\varrho$, and as long as ϱ dominates the energy density of a Friedmann model, one has $\varrho = (\frac{3}{8}\pi G)(t^2/4)(R \sim \sqrt{t})$. Therefore the ratio of the two energies is

$$\frac{\pi t^2 \sigma 8\pi}{\frac{4\pi}{3} t^3 \frac{3}{4t^2 G}} = 8\pi G \sigma t. \tag{7.71}$$

The domain wall dominates energetically for $t > t_{dw}$:

$$t_{dw} = \frac{1}{8\pi G\sigma}. \tag{7.72}$$

In axion models the thickness of a domain wall is $\delta \sim m_\alpha^{-1}$, the surface energy density $\delta = m_\alpha v^2$. Hence

$$t_{dw} = 10^{-4}\,\mathrm{s} \left(\frac{m_a}{10^{-5}\,\mathrm{eV}}\right)^{-1} \left(\frac{10^{10}\,\mathrm{GeV}}{v}\right)^2. \tag{7.73}$$

The domain wall dominance occurs when the temperature has dropped below

$$T \le \lambda^{1/4} \left(\frac{v}{M_{Pl}}\right)^{3/2} M_{Pl}. \tag{7.74}$$

For example, when $v \sim 10^{15}\,\mathrm{GeV}, \lambda \simeq 10^{-4}$, the wall dominates for $T \le 10^{11}\,\mathrm{GeV}$.

Under the dominance of domain-wall structures, the universe expands as $R(t) \propto t^2$. The energy density of a medium with a cell structure is proportional to $R^{-1}(t)$; $\varrho = k/R(t)$; with the Friedmann equation the solution [Vilenkin 1985] is then

$$R(t) = R_1 t^2; \qquad R_1 = \frac{2\pi G}{3k}; \qquad \varrho = \frac{3}{2\pi G t^2}. \tag{7.75}$$

Equation (7.75) shows that domain walls would quickly dominate the energy density of the universe in a way that is not consistent with present-day structures. Therefore these objects have to be removed from the observable universe. Several ways out of these problems have been suggested. Cosmological inflation (cf. Chap. 9) just after the spontaneous breaking of the Peccei–Quinn symmetry removes the domain walls from the observable universe. Alternatively, the different vacua can be continuously connected in certain axion models – this lets the domain walls disappear.

7.5.2 Strings

The evolution of a network of cosmic strings depends crucially on the way the strings interact. Qualitative discussions [Vilenkin 1985] and numerical studies [Turok 1986] paint the following picture. The network of strings has a scale ϵ below which strings are straight, and above which their nature becomes like a random walk. The self-interaction of strings on scales less than the horizon produces loops, which decay by the emission of gravitational radiation. At scales larger than the horizon, strings are conformally stretched by the cosmic expansion. As a result ϵ becomes about equal to the horizon scale t. Thus the energy density in strings evolves as

$$\varrho_s \simeq (\epsilon\mu)t^{-3} \approx \mu t^{-2}. \tag{7.76}$$

Then in the radiation-dominated phase the ratio of ϱ_s to the total energy-density $\varrho \sim (Gt^2)^{-1}$ is constant

$$\frac{\varrho_s}{\varrho} \simeq G\mu. \tag{7.77}$$

Loops surviving decay by gravitational radiation up to time t have a length $L \sim G\mu t$. The density of loops larger than L is $n \sim L^{-3/2}t^{-3/2}$ [Turok 1986]. Then the energy density is

$$\varrho_s \simeq (\mu L)L^{-3/2}t^{-3/2} \simeq (G\mu)^{1/2}(Gt^2)^{-1}.$$

The ratio of the densities for loops turns out to be

$$\frac{\varrho_s}{\varrho} \simeq (G\mu)^{1/2}. \tag{7.78}$$

The vacuum strings do not dominate the energy density ($G\mu \sim 10^{-5}$ or 10^{-6}), but they do not disappear either, and may act as seeds for galaxy formation.

7.5.3 Monopoles

The limits for the density of monopoles can be derived from the limits given in Chap. 6. The Parker limit for the flux $gn_M c\beta$ of monopoles can be translated into a limit on the density n_M. The velocity $c\beta$ of galactic monopoles comes from their acceleration in the interstellar field ($B \simeq 3 \times 10^{-6}$ G), $g \cdot B = 6 \times 10^{-2}$ eV cm^{-1}. Since these fields extend over distances $L > 10^{21}$ cm, any monopole acquires at least an energy of 10^{11} GeV, and a β of 10^{-3}. Then

$$n_M \leq 10^{-23}\,\mathrm{cm}^{-3}; \tag{7.79}$$

or the density parameter of monopoles

$$\Omega_M \leq 0.04 \left(\frac{m_M}{10^{16}\,\mathrm{GeV}}\right) h_0^{-2}. \tag{7.80}$$

This requirement is stronger than the cosmological limit $\Omega_M \leq 4h_0^{-2}$.

All these limits are many orders of magnitude lower than the expected values from a first- or second-order phase transition. Typically one monopole per horizon length t at the phase transition ($T \sim 10^{15}$ GeV) would be expected. This leads to

$$\Omega_M \approx 10^{12} \frac{m_M}{(10^{16}\,\mathrm{GeV})} h_0^{-2}, \tag{7.81}$$

and a terrible consistency problem emerges. The proposal of the inflationary universe promises to solve the dilemma. We shall discuss this model in Chap. 9.

7.5.4 Textures

Textures appear typically in the same way as long strings – one per horizon volume. Thus their density $\varrho_T \propto t^{-2}$, and the ratio to the radiation energy density

$$\frac{\varrho_T}{\varrho} \simeq \mathrm{const.}$$

Textures do not dominate the energy density because initially their density is $\sim G\eta$, several orders of magnitude below 1.

7.6 Gravitinos, Photinos, and Neutralinos

Some astrophysical aspects of the super-symmetric particles, the spin-$\frac{3}{2}$ gravitino and the spin-half photino, were mentioned in Chap. 6. We have seen that these particles – if they are stable – can contribute a large fraction to the cosmological density. For the spin-$\frac{3}{2}$ gravitinos we have found the following limits (6.105):

For a stable or very long lived gravitino $10^{-2}\,\mathrm{eV} \leq m_{3/2} \leq (1$ to $100)$ keV (cf. (6.105)).

The upper limit is derived from the requirement that the density in gravitinos should be less than the critical density in an FL model. This limit is larger than the corresponding mass limits on massive neutrinos, because gravitinos may decouple very early, and then the subsequent annihilation reactions of other massive particles will heat up the neutrinos and photons still in equilibrium, but not the gravitinos. Thus the gravitinos will be diluted, and their number density will be less than that of the neutrinos and photons. A lower limit for short-lived gravitinos can be obtained from restrictions derived from the helium synthesis, as $m_{3/2} \geq 10^4$ GeV (cf. (6.105)).

Concerning the photino, we can only repeat the remarks of Chap. 6 that the photino is a candidate for dark matter in the universe. The mass of the photino must be below 100 eV or above 2 to 20 GeV depending on the masses of the lepton and quark superpartners. The same remarks apply to the neutralino, which by definition is the lightest stable supersymmetric particle.

7.7 QCD Transition Relics – The Aborigines of the Nuclear Desert

Symmetry restorations within the standard model may also lead to phase transitions in the course of cosmic evolution [de Rujula 1985; Witten 1984]. Around the Fermi scale $G_F^{-1/2} \simeq 300$ GeV one expects a phase-transition owing to the temperature dependence of the effective Higgs potential. For $T \gg G_F^{-1/2}$ the stable minimum occurs for a vanishing expectation value $\langle \phi \rangle$ of the Higgs field ϕ. In the high-temperature phase then all gauge bosons are massless, and additional scalar bosons are present. But there does not seem to be any important cosmological effect connected with this electroweak transition. We shall see in the next two chapters how GUT phase transitions leave their mark on the early cosmological evolution.

Another, potentially more influential, effect can come from the QCD transition of hadrons to quarks. Below a critical temperature of $T_c \sim -200$ MeV quarks are bound in hadrons, but above T_c a plasma of quarks and gluons constitutes the matter component in the universe.

Lattice calculations in QCD predict this deconfinement transition [Satz 1985]. For a pure gluon gas a first-order phase transition occurs. For a model calculation with quarks of zero chemical potential one can show only that the entropy density changes rapidly within a narrow temperature range (Fig. 7.10).

The possibility of such a first-order phase transition associated with QCD effects at a temperature of a few hundred MeV has led to speculation about a stable form of "strange quark matter": perhaps not all quarks have ended up as ordinary protons and neutrons.

In particular, if there is a more stable form of baryon conglomerates than ordinary nuclear matter, a large fraction of the baryons may eventually be

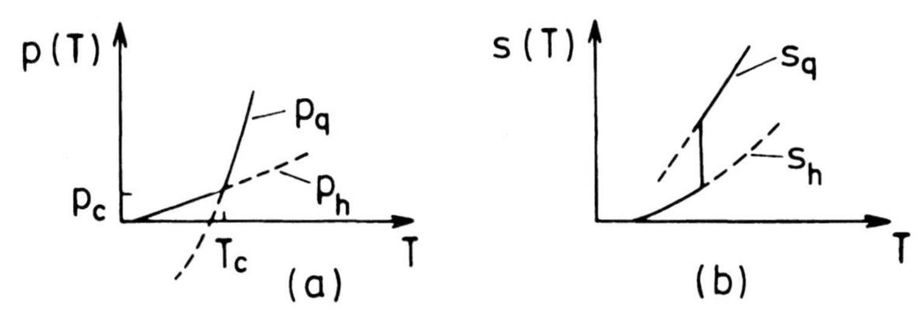

Fig. 7.10. Schematic course of the entropy density for hadron matter ($s_h(T)$), and for a quark gluon plasma ($s_q(T)$). The QCD transition corresponds to an abrupt change in $s(T)$

contained in quark nuggets of varying mass. Arguments can be found [de Rujula 1985] in favour of such unusual systems. Nuclei consist of protons and neutrons, and they are not a single bag of u and d quarks. For the same value of the mass number A, a (p, n) mixture has a lower energy than a hypothetical state of dissolved nuclear constituents held in a single bag. If one replaces some of the up and down quarks in the hypothetical bag by strange quarks, then – at first sight – at fixed A, the relation $M(\mathrm{p+n}) < M(\mathrm{u+d})$ should become an even stronger inequality $M(\mathrm{p+n}) < M(\mathrm{u+d+s})$, since $m_s > m_u, m_d$.

But the introduction of another species of quarks relaxes the requirements of the Pauli exclusion principle. In Fermi statistics it is more favourable to populate levels with three rather than two different quark types. (In a degenerate gas of massless quarks at zero temperature the energy per quark for three different types is 0.9 times the energy per quark for two types). It is a delicate balancing of large energies that finally decides whether or not, for fixed $A, M(\mathrm{p+n})$ is bigger or smaller than $M(\mathrm{u+d+s})$. If Fermi statistics are the main effect then

$$M(\mathrm{u+d+s}) < M(\mathrm{p+n}), \qquad \text{at fixed } A, \tag{7.82}$$

and the ground state of matter may be single bags with approximately equal numbers of u, d, and s quarks. Calculations [Farhi and Jaffe 1984] on the stability of strange quark bags in the context of the MIT bag model, and of a Fermi gas model, have been carried out. The results depend on the strange quark mass m_s, the QCD fine-structure constant α_s, and the bag parameter B. The conclusion of these investigations was that within the allowed range of the parameters B, m_s, α_s it is impossible to decide whether strange quark balls are more – or less – stable than ordinary nuclei. For certain acceptable ranges of the parameters B, α_s, m_s the strange quark matter is stable for essentially any value of A. The estimated density of this hypothetical new ground state of matter is [Farhi and Jaffe 1984]

$$\varrho \sim 3.6 \times 10^{14}\,\mathrm{g\,cm^{-3}}, \tag{7.83}$$

a value quite close to the density of ordinary nuclear matter.

The decay of ordinary nuclei into this more stable state of matter is, however, suppressed, because for a nucleus of mass-number A this decay would be of order $(G_F)^A$ in the weak interaction coupling constant. (Weak interactions of the type ud $\rightarrow$ su; d $\rightarrow$ se$^+\nu$, would allow the nucleus to decay into a bag containing up to A strange quarks, emitting photons, protons, α-particles, or $(e^+\nu)$ pairs in the process.) But these strange quark balls may have formed in the early universe at the time of the quark–hadron transition.

Another speculative suggestion [Witten 1984] for the existence of strange quark matter makes use of the probable occurrence of a first-order phase transition at the epoch of the change from a state of quasi-free quarks to a state of mesons and baryons at $T_c \sim 100 - 200\,\mathrm{MeV}$.

Thermal equilibrium may perhaps be maintained during this phase transition, and it has the consequence that there will be coexisting phases. The low-temperature phase – composed of hadrons – and the high-temperature phase – composed of quark matter – can coexist for a long time. Are there possible observable remnants of this epoch of phase coexistence? Witten [1984] argues that there may be such remnants if neutrino losses are the main energy loss of the high-temperature phase. Generally speaking, as the universe expanded and cooled, the high-temperature-phase regions shrank further and further in size. But eventually the excess baryons in these regions will exert a pressure that will help in stabilizing against further shrinkage. The final outcome will then be lumps of hot quark matter of density $\sim 10^{15}\,\mathrm{g\,cm^{-3}}$, and a mass between 10^9 and 10^{18} g. About 90% of the baryonic mass of the universe can be contained in these strange quark balls.

As the transition from the quark plasma to the cooler nuclear state occurred at a higher temperature (an earlier epoch) than the primordial synthesis of nuclei, the baryons hidden in the quark plasma are not subject to the constraints on the baryon density derived from the deuterium abundance (cf. Chap. 3). Thus the presence of quark balls allows a critical density $\Omega_0 = 1$ in a universe made up of baryons – 90% of which have not participated in nucleosynthesis! It seems worthwile following up the consequences of such speculations. Possible ways of searching for these lumps of quark matter have been suggested [de Rujula 1985].

Energy loss in matter is mainly by collisions with atoms – lumps with $A \leq 2 \times 10^{14}$ might come to rest in the Earth's crust. Lumps of quark matter with velocities greater than $30\,\mathrm{km\,s^{-1}}$ can be detected in experiments searching for proton decay.

Also, some experiments set up to catch magnetic monopoles are sensitive detectors of this matter. Quark balls may appear as very fast meteors, they may cause scars on the surface of the Earth – tracks that might be etched out. Any quark system more massive than 2.4×10^{-10} g can leave an observable track at great depth – an effect that cannot be produced by meteorites, cosmic rays or natural radioactivity. Objects more massive than 1 tonne will cause epilinear earthquakes in their passage through the Earth.

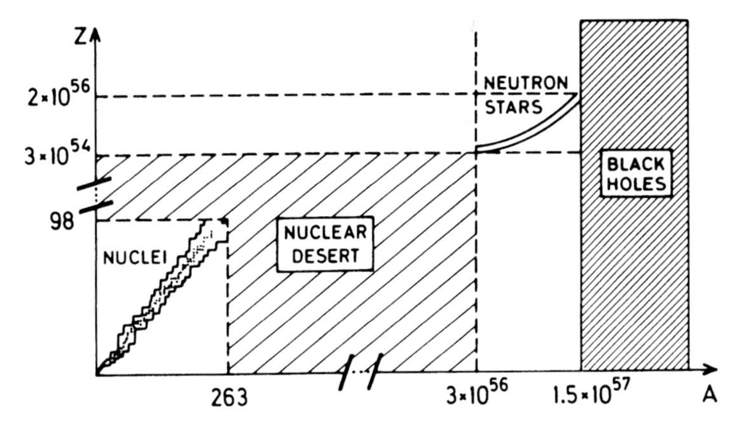

Fig. 7.11. Chart of nuclides with the nuclear desert inhabited by the aboriginal quark balls (after [de Rujula 1985])

It is perhaps appropriate to conclude this chapter with an unusually complete chart of the nuclides (Fig. 7.11). The lower left-hand corner contains the known nuclei, a narrow band extending to $(Z, A) \sim (98, 263)$. Higher up on the far right, around $A \sim 10^{57}$, is an island of stability: the neutron stars. Their Z is a few percent of A. "Nuclides" of larger mass number are unstable against gravitational collapse. The region in between, $263 < A < 3 \times 10^{56}$ is traditionally thought to be a "barren land", where no nuclei are stable. Compared with the mere 13 orders of magnitude – from 100 GeV to 10^{15} GeV – of the particle physics "desert", this "nuclear desert" looks formidable indeed: 54 orders of magnitude. It is some consolation that this desert may be populated by the remnants of the quark–hadron transition in the early universe. Quark balls may be the aborigines of the nuclear desert.

Exercises

7.1 Study [Mikheyev and Smirnov 1986] to see how the medium influences the oscillations of neutrino states.

7.2 The neutrino burst from the supernova SN 1987A in the LMC can be used to obtain limits on the neutrino mass (electron-antineutrinos were registered in the Kamiokande experiment). A massive neutrino with energy E_ν has a certain velocity below c. Assume an energy distribution $f(E_\nu) = (\exp E_\nu/T + 1)^{-1}$ for the emitted neutrinos with $T = 2.8\,\mathrm{MeV}$, compute the mean and dispersion, and derive an upper limit to the mass from the condition that the time difference between the different neutrino events should be less than 12.4 s, the duration of the burst in the Kamiokande detectors. Read [Spergel and Bahcall 1988] to see what a more careful analysis requires.

7.3 The neutrino burst from SN 1987A was observed within 3 hours of the first sight of the optical event. This observation can be used to test the weak equivalence principle: the Shapiro time delay in a parametrized post-Newtonian (PPN) formalism is given by

$$\Delta t = \frac{1+\gamma}{2}2\int \phi(x(t))dt;$$

ϕ is the Newtonian potential along the radiation path which can be replaced by a straight line. If the weak equivalence principle were violated, $1+\gamma$ would be different for photons and neutrinos. Remember that 1987A occurred in the LMC, assume an upper limit $\Delta t \simeq 10^4$ s, and compute the accuracy to which the weak equivalence principle for photons and neutrinos is satisfied.

7.4 Compute the cosmological upper limit on the mass of the gravitino using the formalism outlined here and in Chap. 3.

7.5 Derive the expansion law for a cosmological model dominated by domain walls.

7.6 Take a closed, $K = +1$, FL model with radiation, matter, and vacuum energy density (i.e. the potential energy $V(\varphi)$ of a Higgs field), and assume that early on, say at $t_c = 10^{-35}$ s, a phase transition which produces monopoles occurs. Let $V(\varphi)$ be completely transformed into monopoles at t_c, and describe the evolution of the model. Consider the case where $V(\varphi)$ is smaller than the radiation energy density at t_c, and also the case where it is larger. Take the matter density to have the value now of $(n_B/n_\gamma)_0 = 10^{-10}$. How does the total mass–energy content of the model behave with time?

8. Baryon Synthesis

"The most serious uncertainty affecting the ultimate fate of the universe is the question whether the proton is absolutely stable against decay into lighter particles. If the proton is unstable, all matter is transitory and must dissolve into radiation."

F.J. Dyson (1978)

"To be or not to be . . . "

[Shakespeare: "Hamlet"]

One of the fundamental questions of cosmology is how the small baryon-asymmetry of

$$n_B/s \simeq 10^{-10} \tag{8.1}$$

that we observe in our present universe was created. In classical cosmology – within the standard big-bang model – this number is imprinted on the universe as an initial condition: Heaven-sent and inexplicable. Another answer is given by the GUT theories of elementary particle interactions: the B-violating interactions – i.e. interactions that do not conserve the baryon number – in those theories can produce a small net baryon asymmetry in an expanding universe that was initially completely symmetric in baryons and antibaryons. The number n_B/s turns out to be small, because the masses of the particles responsible for B-violating interactions are extremely large.

There is a widespread belief that B-violation need not to be tied to a specific GUT. It is quite generally expected that B is not completely conserved, since there is no compelling theoretical reason for a B-conservation law. On the other hand, the consideration of primordial mini black holes, which are built from baryons and radiate themselves away into $B = 0$ particles by the Hawking radiation, suggests that B-violation may be quite commonplace at very high energy scales [Hawking 1974; Barrow 1980; Dolgov 1980].

Another consequence of such interactions is the decay of the proton. As discussed in Chap. 6, there are no positive results as yet from experiments looking for proton decay. Until such experiments are successful, with a positive answer, any ideas on baryon synthesis will remain highly speculative. But we should take our speculations seriously, too. The best evidence for processes that do not conserve the baryon number is the existence of experimental physicists looking for decaying protons.

The speculations are, nevertheless, great fun – and there is the hope of landing on firmer ground in the not-too-distant future. The theory eventually established will almost certainly be quite different from the considerations presented in the following discussion.

8.1 Evidence for *B*-Asymmetry

The Solar System does not contain large quantities of antimatter (such as "anti-asteroids") – interaction with the solar wind would transform such objects into very luminous sources of γ-radiation. This would be in conflict with observations.

Cosmic rays with energies above 100 MeV are mostly of extra-solar origin – they can provide information on the abundance of antimatter in our galaxy, and perhaps beyond. The most abundant nuclei of antimatter in cosmic rays are antiprotons and antihelium nuclei with abundances relative to ordinary nuclei of

$$\bar{\mathrm{p}}/\mathrm{p} \simeq 3 \times 10^{-4}, \quad \overline{{}^4\mathrm{He}}/{}^4\mathrm{He} \leq 10^{-5}; \tag{8.2}$$

cf. [Steigman 1976]. These antinuclei may have been produced by collisions, e.g. $\mathrm{p} + \mathrm{p} \to 3\mathrm{p} + \bar{\mathrm{p}}$. Such reactions can explain the spectrum and the flux of antiprotons.

For clusters of galaxies and their intercluster gas one can derive limits for the amount of antimatter present from the γ-flow expected from π^0-decay (the matter–antimatter annihilation would produce π^0 mesons, and those decay into γ-rays). The observations exclude, for example, the existence of antimatter in the Virgo cluster.

Very distant islands of antimatter separated by empty space from regions containing normal matter would, however, not signal their presence to us. Electromagnetic radiation cannot give precise information about its matter or antimatter origin. Neutrinos carry the signal of their matter or antimatter production in their helicity – but only large neutrino detectors could, in principle, find direct evidence of the matter–antimatter structure of the universe.

We shall take it for granted then that our present-day universe contains almost only matter (and possibly very small amounts of antimatter) with a baryon-to-photon ratio of

$$n_B/n_\gamma \simeq (4 \div 7) \times 10^{-10} \tag{8.3}$$

[cf. Chap. 3].

For the early universe, where the number of baryons and antibaryons was roughly equal to the number of photons, this indicates the presence of a tiny matter–antimatter asymmetry: for every 10^{10} antibaryons there existed $(10^{10} + 1)$ baryons. It is to that one part in 10^{10} excess of ordinary matter that we owe our existence!

8.2 Some Qualitative Remarks

In Chap. 7 we pointed out that a locally B-symmetric universe would leave as a relic abundance of baryons only small fractions of the order of

$$\frac{n_B}{n_\gamma} = \frac{n_{\overline{B}}}{n_\gamma} \simeq 10^{-18}. \tag{8.4}$$

Even if a separation of matter and antimatter could be achieved, this value would be about 10 orders of magnitude too small. Since the annihilation processes after the "freeze-out" temperature $T_f \simeq 20\,\mathrm{MeV}$ are so efficient, different macroscopic regions of matter or antimatter must be formed before the temperature has dropped to T_f.

Statistical fluctuations of the number of baryons in a certain volume are too small to achieve a noticeable separation effect. Consider the comoving volume that contains our galaxy at the present epoch. It contains about $10^{12} M_\odot$, about 10^{69} baryons, and about 10^{79} photons. At times for which $T \geq 1$ GeV this comoving volume enclosed about the same number of photons, and approximately an equal number of baryons and antibaryons. Statistical fluctuations can produce a net baryon number of $|N_B - N_{\overline{B}}| \sim N_B^{1/2} \sim 10^{39}$. There is no way here to reach the required number of $\sim 10^{69}$.

Another limit on baryon–antibaryon separation can be obtained from the horizon structure of the standard model. Separate regions formed by causal processes must fit into one horizon. Now the size of a horizon at time t (for $T > 20\,\mathrm{MeV}$) is

$$d_H(t) = 2t = \left(\frac{45}{4\pi^3 g(T)}\right)^{1/2} T^{-2}. \tag{8.5}$$

The baryon mass within one horizon is thus

$$\begin{aligned} M_H(t) &= \frac{4\pi}{3} d_H(t)^3 \varrho_B \\ &= \frac{4\pi}{3} d_H^3 m_p \left(\frac{n_B}{s}\right) s \\ &= \frac{4\pi}{3} \left(\frac{45}{4\pi^3}\right)^{3/4} \frac{2\pi^2}{45} g(T)^{1/4} (2t)^{3/2} \left(\frac{n_B}{s}\right). \end{aligned} \tag{8.6}$$

Putting in numbers, one finds

$$M_H(t) = 0.7 M_\odot g^{1/4} t_s^{3/2} (\Omega_B h^2). \tag{8.7}$$

For $t \simeq 3 \times 10^{-3}$ s corresponding to $T \simeq 20\,\mathrm{MeV}$, $M_H \simeq 10^{-5} M_\odot$.

This estimate changes drastically when an inflationary phase precedes the baryon–antibaryon separation. The horizon constraint essentially disappears, because the whole observable universe at the present epoch is contained within one causal length.

Within the standard big-bang model, however, there seems to be little chance of achieving a physical separation of baryon and antibaryon phases in an initially baryon-symmetric cosmological model. If baryon number was exactly conserved – as it is assumed to be in the standard model – the small asymmetry necessary for our existence must be postulated initially. Grand unified theories offer the possibility of creating this small asymmetry from physical processes.

During the 1960s Sakharov looked into the conditions necessary to create a baryon excess in an initially B-symmetric universe [Sakharov 1967]. Clearly there must be B-violating interactions; in addition, the symmetry between particles and antiparticles must be disturbed, i.e. a violation of the CP symmetry (cf. Chap. 6) must occur. Besides these symmetry violations there must be epochs in the early universe where thermodynamic equilibrium was not established. Loosely speaking, the CPT-invariance of local, relativistic field theories and thermodynamic equilibrium imply the invariance under CP, because in thermodynamic equilibrium there is no arrow of time.

A more quantitative argument would run as follows [Straumann 1984a]: the expected value of the baryon number $\langle B\rangle$ can be calculated with the help of the density matrix ϱ:

$$\langle B\rangle = Tr\{B\rho\}/Tr\{\varrho\}. \tag{8.8}$$

Now in thermodynamic equilibrium the density matrix has the form

$$\varrho \propto \exp\{-\beta(H - \Sigma\mu_i Q_i)\}. \tag{8.9}$$

where Q_i are the conserved additive quantum numbers (H is the Hamiltonian of the system). This precise form maximizes the entropy for given average values of Q_i. Since B (baryon) and L (lepton number) are no longer considered as conserved quantities, only the electric charge and perhaps colour charges remain in (8.9). Because of charge neutrality, however, the corresponding chemical potentials are zero, and then ϱ is the canonical ensemble

$$\varrho \propto \exp(-\beta H). \tag{8.10}$$

But, under the CPT transformation, the Hamiltonian function remains invariant – always true for local, relativistic field theories – while the baryon number B changes sign: CPT: $B \to -B$, because particles are transformed into antiparticles under CPT. Since all vacuum expectation values of the theory are also invariant under CPT, we have

$$\langle B\rangle = Tr\{Be^{-\beta H}\}/Tr\{e^{-\beta H}\} = -\langle B\rangle, \tag{8.11}$$

hence $\langle B\rangle = 0$. Therefore, any net baryon or antibaryon gain will be damped out, if there is enough time to reach complete thermodynamic equilibrium.

8.3 GUTs and Thermodynamic Equilibrium

Since the typical energy or mass scales of grand unified theories are of the order of $10^{-4}M_{Pl}$, the times we are considering here are $t_{\rm GUT} \approx 10^8 t_{Pl} \approx 10^{-35}$ s. Even that close to the big bang, the universe is assumed to be an expanding, homogeneous, and isotropic FL space-time that provides the classical background on which the particles and their interactions are described. Some readers may be reluctant to accept such extreme excursions from the known regime of physics accessible in the laboratory. The following is a speculative adventure, but perhaps a speculation from which new insights can be derived.

The GUT particles in the hot early universe experience collisions, decays, and inverse decays. In Fig. 8.1 several typical collision processes are pictured, and in Fig. 8.2 the decay processes and inverse decays of heavy bosons are shown. Without taking into account the detailed structure of a particular GUT – such as minimal $SU(5)$ – various general properties can be pointed out (e.g. [Weinberg 1964; Nanopoulos 1981]).

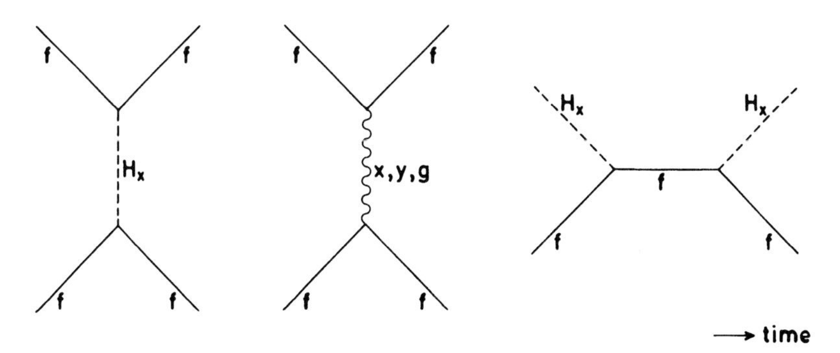

Fig. 8.1. Typical collision processes between fermions involving vector bosons (X,Y). Higgs particles (H_X) or supersymmetric objects (g = gravitino)

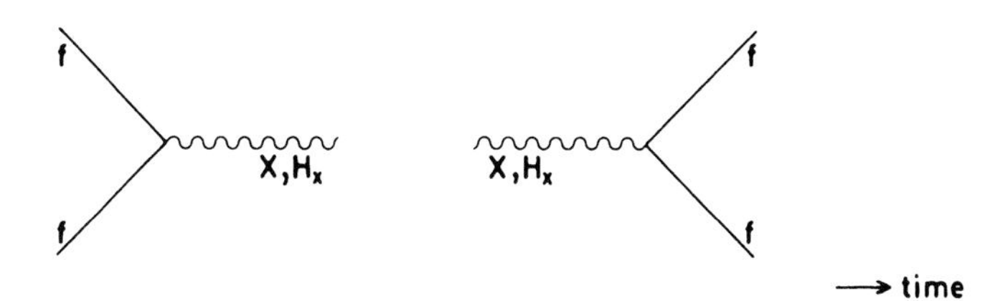

Fig. 8.2. Decays and inverse decays of vector bosons (X) and Higgs bosons (H_X)

The decays, as well as the inverse decays of a massive particle X, will be characterized by the free decay time of a movingparticle (a time dilatation occurs because of the thermal motion)

$$\tau_D \simeq \frac{(T^2 + m_X^2)^{1/2}}{\alpha_x m_X^2} . \tag{8.12}$$

This can be rather a long time for $T \gg m_X$. In a local gauge theory – as we have seen – particles acquire their mass from spontaneous symmetry-breaking. Therefore above an energy $T_c \geq m_X$ they can be treated as essentially massless particles, which are very long-lived. At low temperatures, $T \ll m_X$, the density of X-particles is reduced by a Boltzmann factor $\propto \exp(-m_X/T)$ – and the probability of two light particles coming together to form a massive X-particle is similarly reduced. Thus a characteristic mean free time for the decay and inverse decay processes is

$$\tau_{DiD} = \tau_D \exp\left(-\frac{m_X}{T}\right) . \tag{8.13}$$

Scattering via an intermediate boson of mass m_X gives a collision cross-section

$$\sigma_c = \alpha_X^2 T^2 (T^2 + m_X^2)^{-2} . \tag{8.14}$$

In the ultrarelativistic limit $T \gg m_X$:

$$\sigma_c = \frac{\alpha_X^2}{T^2} ; \tag{8.15}$$

$\alpha_X \simeq \frac{1}{45}$ in $SU(5)$ GUT.

The scattering is strongly reduced at temperatures $T \ll m_X$. The cross-sections can be used to define a mean free time between collisions

$$\tau_c = (n\sigma_c)^{-1} . \tag{8.16}$$

From (8.14), together with a formula for a relevant number-density n,

$$n \simeq A\, T^3 ,$$

one can derive (roughly)

$$\tau_c = (T^2 + m_X^2)^2 \alpha_X^{-2} T^{-5} . \tag{8.17}$$

At very high temperatures, then,

$$\tau_c = \alpha_X^{-2} T^{-1} \tag{8.18}$$

(approximately) and at very low temperatures

$$\tau_c = \alpha_X^{-2} m_X^4 T^{-5} . \tag{8.19}$$

The epochs during which various interactions are in thermal equilibrium can now be determined by comparing the expansion time of the cosmological model to the various mean free times τ_D and τ_c.

The expansion rate

$$H = \frac{\dot{R}}{R} = \left(\frac{8\pi G}{3}\rho\right)^{1/2}$$

depends on the energy density

$$\rho = \frac{1}{2}g(T)\frac{\pi^2}{15}T^4. \tag{8.20}$$

For the minimal $SU(5)$ model $g(T) \simeq \frac{643}{4}$, and therefore

$$H = 1.66G^{1/2}g^{1/2}T^2 \simeq 1.66g^{1/2}\frac{T}{T_{Pl}}T. \tag{8.21}$$

If $\tau > H^{-1}$, the corresponding interaction is out of equilibrium, and if $\tau < H^{-1}$, then the interaction is in thermal equilibrium. Several different cases are illustrated in Fig. 8.3.

For strong, weak, and electromagnetic interactions the masses of the particles mediating these interactions are negligible on the energy scales considered around the GUT mass, and therefore (8.18) should be applicable.

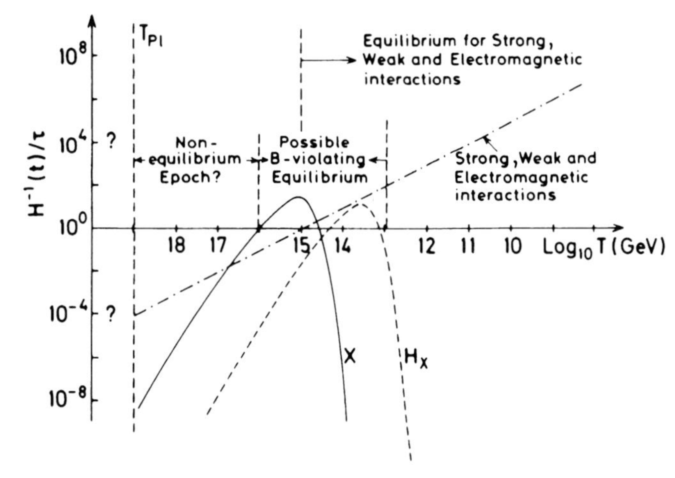

Fig. 8.3. Qualitative picture of the various possible interactions in the early universe. The *solid line* denotes gauge boson interactions, the *dashed line* Higgs boson interactions. Between T_{Pl} and $T_{\text{GUT}} \sim 10^{15}$ GeV there seems to be an epoch of non-equilibrium. Around $T = 10^{15}$ GeV there is a period of equilibrium for B-violating interactions. The ratio of $H^{-1}(t)$ to τ, the mean free time, is plotted. Equilibrium requires $H^{-1} > \tau$ (adopted from [Nanopoulos 1981])

Then

$$\begin{aligned}(H\tau)^{-1} &= (\alpha_X^{-2}T^{-1}T^2G^{1/2})^{1/2} \\ &= \alpha_X^2 t^{-1}G^{-1/2} = \alpha_X^2 \tfrac{T_{Pl}}{T}.\end{aligned} \tag{8.22}$$

As $\alpha_X^2 \simeq 10^{-4}$ one expects these interactions to be out of equilibrium from $T \simeq T_{Pl}$ to $T \simeq \alpha_X^2 T_{Pl} \simeq 10^{15}$ GeV. Below $\alpha_X^2 T_{Pl}$, the strong, weak, and electromagnetic interactions are in thermal equilibrium. This equilibrium should be realized down to and below the electroweak breaking scale, perhaps as far as the quark–hadron transition temperature.

For GUT interactions which involve exchanges of (or describe the decay of) the superheavy gauge bosons or Higgs bosons we expect for collisions at temperatures $T \gg m_X$, a behaviour as in (8.22):

$$(H\tau_c)^{-1} \ll 1 \qquad \text{for} \qquad T > \alpha_X^2 t_{Pl}. \tag{8.23}$$

At temperatures $T \ll m_X$ we would accept an estimate

$$(H\tau_c)^{-1} = \alpha_X^2 T^3 m_X^{-4} G^{-1/2} \tag{8.24}$$

for the collision processes with a smooth interpolation between $T \gg m_X$ and $T \ll m_X$. Therefore it seems plausible that $(H\tau)^{-1}$ can be greater than one only around

$$\frac{T}{T_{Pl}} \simeq \alpha_X^2 \simeq \frac{m_X}{M_{Pl}} \simeq 10^{-4}. \tag{8.25}$$

For the decays an estimate may be extracted from (8.12) and (8.13):

$$(H\tau_D)^{-1} \simeq \alpha_X^2 m_X^2 (T^2 + m_X^2)^{-1/2} T^{-2} e^{-m_X/T} G^{-1/2}. \tag{8.26}$$

This function is clearly $\ll 1$ at $T \ll m_X$ and at $T \gg m_X$, with a maximum near $T \simeq m_X$ at a value of the order of $(\alpha_X M_{Pl}/m_X) \simeq 10^2$.

The various ratios of $(H\tau)^{-1}$ are plotted in Fig. 8.3 (after [Nanopoulos 1981]). For an epoch where the temperature T was of the order of m_X, the interactions involving X-bosons may have been in thermodynamic equilibrium.

Similar estimates can be made for the interactions of the superheavy Higgs bosons H_X expected in GUTs. All the arguments are the same as above, except that there is more uncertainty about the mass of the Higgs bosons and their coupling parameter α_H than there is about the vector gauge bosons. Preferred values are $m_H \simeq 10^{14}$ GeV and $\alpha_H \simeq 10^{-3}$.

There are two interesting conclusions that we can draw from these discussions. First, it seems likely that during the epochs in the early universe with temperatures between T_{Pl} and $T \approx m_X$ – i.e. between 10^{-43} s and 10^{-35} s after the big bang – thermodynamic equilibrium was not established. Second, it seems likely that around $T \approx m_X$ a state of complete thermodynamic equilibrium existed. If this equilibrium state was dominated by B-violating interactions, then any initial baryon asymmetry would have been strongly diluted. As a consequence the baryon synthesis by GUT interactions is not only a possibility, but even a necessity in order to come up with the observed ratio.

8.4 A Mechanism for Baryon Synthesis

The appearance of supermassive gauge bosons or Higgs bosons is a natural outcome of any scheme of grand unified theories. As we have seen in Chap. 6, the collisions and decays of these particles violate B and L conservation. We have discussed these reactions previously and in detail, in the framework of a minimal $SU(5)$ GUT. It seems likely that in simple GUT schemes the dominant mechanisms generating a baryon–antibaryon asymmetry are the decays and inverse decays of such supermassive bosons – just as in the $SU(5)$ scheme. The generation of a net baryon number may have been quite complicated, with several different baryon-number-violating processes contributing or competing. (For reviews see [Kolb and Turner 1983, 1990].)

Nevertheless a basic mechanism in simple GUTs is the decay of X-particles into pairs of quarks or into an antilepton and an antiquark (Fig. 8.4):

$$X \text{ or } H_X \to (q+q) \text{ or } (\overline{q}+\overline{l})$$

[Nanopoulos 1981; Ellis et al. 1979]. The total decay rates of particles and antiparticles must be the same because of CPT:

$$\begin{aligned} \Gamma_{\text{tot}}(X) &= \Gamma(X \to qq) + \Gamma(X \to \overline{q}\overline{l}) \\ \Gamma_{\text{tot}}(\overline{X}) &= \Gamma(\overline{X} \to \overline{q}\,\overline{q}) + \Gamma(\overline{X} \to ql) . \end{aligned}$$

Let

$$\begin{aligned} r &\equiv \frac{\Gamma(X \to qq)}{\Gamma_{\text{tot}}} , \\ 1-r &\equiv \frac{\Gamma(X \to \overline{q}\overline{l})}{\Gamma_{\text{tot}}} \end{aligned} \qquad (8.27)$$

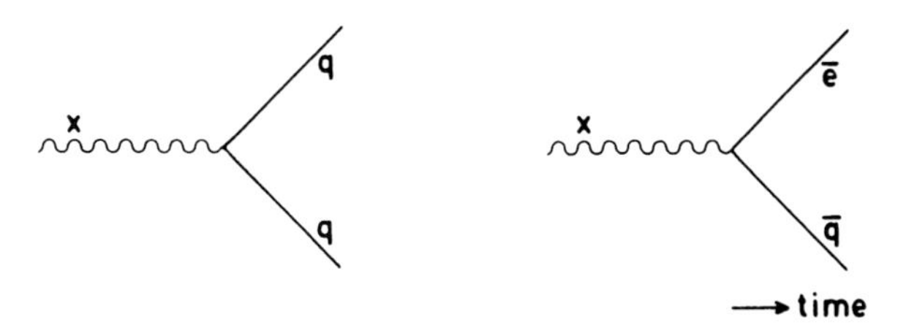

Fig. 8.4. Decay modes of superheavy vector bosons X

and similarly for the antiparticle decays

$$\overline{r} \equiv \frac{\Gamma(\overline{X} \to \overline{q}\,\overline{q})}{\Gamma_{\text{tot}}}$$

$$1 - \overline{r} \equiv \frac{\Gamma(\overline{X} \to ql)}{\Gamma_{\text{tot}}}.$$

The baryon numbers for the endproducts are, in this example,

$$r : (qq), \qquad B = \frac{2}{3}$$

$$1 - r : (\overline{q}\overline{l}), \qquad B = -\frac{1}{3}.$$

A pair consisting of an X-boson and an $\overline{X}$-antiboson has $B = 0$; the decay into the two channels qq and $\overline{q}\,\overline{l}$, produces a net baryon number per decay:

$$\Delta B = \left[r\frac{2}{3} + (1-r)\left(-\frac{1}{3}\right)\right] + \left[-\overline{r}\frac{2}{3} + (1-\overline{r})\frac{1}{3}\right] = (r - \overline{r}). \qquad (8.28)$$

$\Delta B \neq 0$ is only possible if $r \neq \overline{r}$, i.e. if the decays of particle and antiparticle are different. Therefore a violation of the CP symmetry is necessary to have such an effect. The CP-violating terms in GUT theories have been discussed to a certain extent in Chaps. 6 and 7. The CP violation is rather an uncertain quantity: its magnitude is only limited by certain experiments (cf. Chaps. 6 and 7). All theoretical estimates for CP-violating terms are strongly model-dependent.

The baryon number produced per decaying $(X\overline{X})$ pair is, in principle, completely determined by particle physics. In the first-order Born approximation (Fig. 8.5) $r = \overline{r}$. This is due to the fact that in this approximation the processes are invariant under time reversal T. The CPT-invariance then implies invariance under CP, hence $r = \overline{r}$ [Weinberg 1982].

Second- and higher-order terms contribute to a ΔB-producing effect, as in Fig. 8.6. The decay rates for collisions and decays can now be compared with the expansion rate H – as in Sect. 8.3. The decoupling temperature T_D, defined by $\Gamma_X = H$, can either be $> m_X$ or $< m_X$.

The case $T_D < m_X$ can be treated rather simply: in that case the ratio of decay rate and expansion rate is

$$K \equiv (\tau H)^{-1}\,|_{T=m_X} < 1. \qquad (8.29)$$

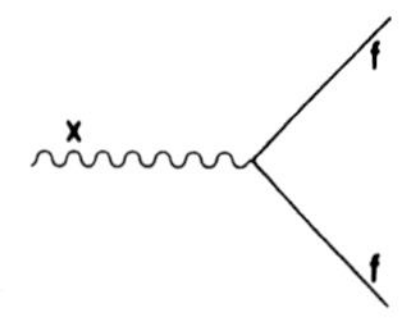

Fig. 8.5. Graph for the first-order Born approximation for the decay X → ff

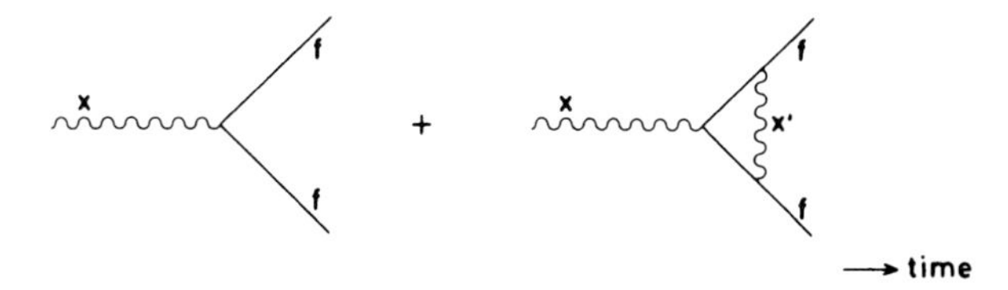

Fig. 8.6. Higher-order graph contributing to ΔB production

Annihilation reactions and decays will occur rapidly enough, even for $T < m_X$, to keep up a thermal distribution $\propto \exp(-m_X/T)$. The X- and $\overline{X}$-bosons are initially almost as abundant as photons $n_X \simeq n_{\overline{X}} \simeq n_\gamma$. For $T < T_D$ they decay freely, and the decay products are no longer energetic enough to reproduce X, $\overline{X}$ pairs by the inverse decay reaction. In that case a $\Delta B \neq 0$ may be generated. The situation is pictured in Fig. 8.7.

For $T_D > m_X(K > 1)$ the decay rate $1/\tau_D$ is larger than H in the regime $T < m_X$. Therefore, one would expect an equilibrium-distribution of the X-bosons to be maintained by rapid decay reactions. No appreciable B-excess can therefore be produced this way. The situation is depicted in Fig. 8.8.

This washing-out effect owing to a long maintenance of equilibrium has been computed numerically by various groups [Kolb and Wolfram 1980; Treiman and Wilczek 1980]. The Boltzmann equation has been integrated using decay and collision cross-sections.

The effect is a dependence of the efficiency of generating baryon number from a given ΔB, on the coupling α_X and on the mass m_X (Fig. 8.9): the efficiency is 100% only for $m_X \geq 10^{17}$ GeV, if $\alpha_X \approx \frac{1}{45}$ – as in $SU(5)$. For Higgs bosons with $\alpha \sim 10^{-3}$ complete efficiency is guaranteed down to $m_H = 10^{15}$ GeV.

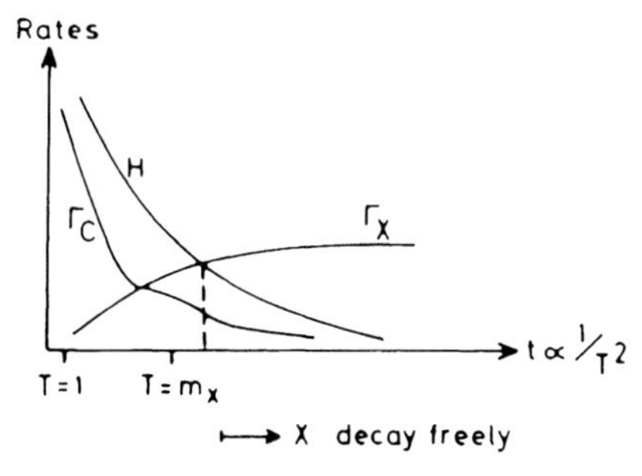

Fig. 8.7. Schematic plot of expansion rate H, and decay and collision rates, for $K \equiv (\Gamma/H)_{T=m_X} < 1$

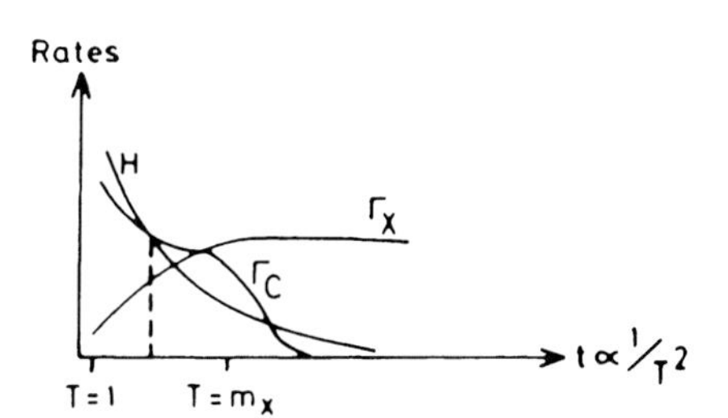

Fig. 8.8. As Fig. 8.7, except that $K > 1$. B-synthesis is suppressed, because X-bosons are too long in thermal equilibrium

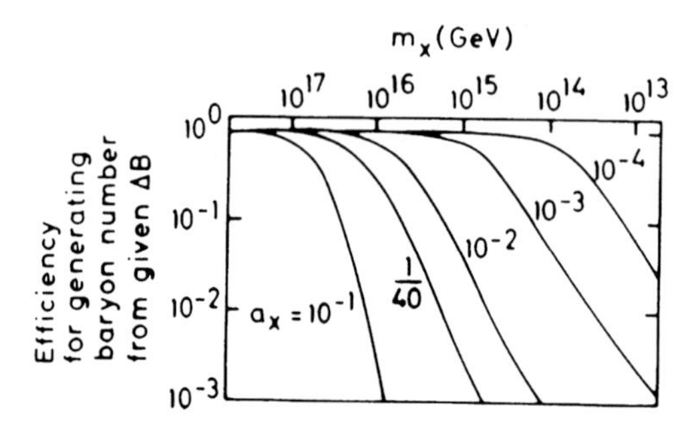

Fig. 8.9. The decrease in efficiency by collisions in the B-violating process (from [Kolb and Wolfram 1980])

For $K \geq 1$ a large enough ratio n_B/n_γ can only be obtained if

$$m_X \geq 10^{15}\,\text{GeV} \quad \text{or} \quad m_H \geq 10^{13}\,\text{GeV}. \tag{8.30}$$

The numerical calculations show that the resulting asymmetry for $K \geq 1$ is not washed out very rapidly. Only for $K \gg 1$ is the baryon excess completely damped away. For a certain regime between $K \simeq 1$ and $K \gg 1$ one finds a dependence

$$\frac{n_B}{s} = \frac{3 \times 10^{-2}}{(1 + 4.5K)^{1.2}} \Delta B \tag{8.31}$$

[Fry et al. 1980a,b]. For $K \geq 1$ there is also the possibility that the ratio n_B/s varies from point to point for anisotropic perturbations.

The case $K < 1$ can be evaluated analytically. For $K \equiv (H\tau)^{-1}\,|_{T=m_X}$ we have

$$K \simeq \alpha_X \frac{M_{Pl}}{m_X} \frac{1}{g^{1/2}} \quad \text{at} \quad T \simeq m_X. \tag{8.32}$$

The case $K < 1$ is equivalent to

$$m_X > g^{-1/2} \alpha_X M_{Pl}. \tag{8.33}$$

For gauge bosons $\alpha_X \simeq \frac{1}{45}$, and therefore $m_X > 10^{16}$ GeV. The vector gauge bosons in simple GUTs are not that massive. For Higgs bosons the mass can be smaller. The coupling constant α_X is not fixed for Higgs bosons, but in specific models α_X can be estimated. If the supermassive Higgs boson is in the same representation as the light Higgs particles responsible for the fermion masses, then one would expect $\alpha_X \simeq (m_f/m_W)^2 \simeq 10^{-3}$ to 10^{-6} (here m_f is an average fermion mass). Because of the smaller α_X, the mass m_H of the Higgs boson has a smaller lower limit from (8.33); $m_H \geq (10^{12}\text{–}10^{15})$ GeV.

Until the temperature has dropped down to the decoupling temperature T_D, the density of the heavy bosons n_X is proportional to T^3, and

$n_X \simeq n_\gamma$. Then at $T = T_D$,

$$n_{XD} \simeq N_X \frac{1}{\pi^2} \zeta(3) T_D^3, \tag{8.34}$$

where n_X counts the X and $\overline{X}$ spin degrees of freedom (ζ is the zeta function).

The total entropy density of all other relativistic particles is

$$s_D = g(T_D) \frac{2\pi^2}{45} T_D^3, \tag{8.35}$$

and therefore the ratio of baryons to s is

$$\frac{n_B}{s} \simeq n_{XD} \frac{\Delta B}{s_D} \simeq \frac{45\zeta(3)}{2\pi^4} \frac{N_X}{g(T)} \Delta B \tag{8.36}$$

$$\frac{n_B}{s} \simeq 0.01 \Delta B. \tag{8.37}$$

ΔB is, as before, the baryon number produced per decay of an $(X\overline{X})$ pair. This quantity depends only an the microphysical processes, and is uncertain to the degree to which the magnitude of the CP-violation is uncertain.

The following range of values has been estimated [Kolb and Turner 1983]. If the decays of Higgs bosons are the dominant mechanisms, then

$$\Delta B \sim 10^{-6} \quad \text{to} \quad 10^{-2} \tag{8.38}$$

if gauge-bosons dominate, then

$$(\Delta B) \sim 10^{-10} \quad \text{to} \quad 10^{-6}.$$

[Fry et al. 1980a]. We see that these estimates cover the observationally determined $n_B/s \simeq 10^{-9\pm1}$. Within the minimal $SU(5)$ model, calculations give $\Delta B \simeq 10^{-16}$, however. One remedies this by extending the model beyond the standard three fermion generations, reaching values of $\Delta B \simeq 10^{-8}$ this way, but also making the model depend on more parameters [Nanopoulos and Weinberg 1979; Yanagida and Yoshimura 1980].

The main impact of the result is, perhaps, that it seems possible to calculate a baryon excess within GUT theories that does not largely deviate from the observed value. One can understand why the excess is small and non-zero, but the numerical agreement can hardly be called impressive. Up to now there is no way of deriving a precise value for the CP violation within the theory.

A connection between such calculations and laboratory experiments is established by the hypothetical dipole moment d_n of the neutron. B-violating parameters contribute to d_n, and in specific GUT models, lower limits can be derived such as [Ellis et al. 1981]

$$\left| \frac{d_n}{e} \right| \geq 2.5 \times 10^{-18} \left(\frac{n_B}{n_\gamma} \right) \text{cm}. \tag{8.39}$$

A comparison with the experimental upper limit (Chap. 6) $d_n \le O(10^{-25})$ leads to the bound $n_B/n_\gamma \le O(10^{-7})$. This value itself is not very exciting, but the evident possibility of linking numbers from experiments with processes in the very early universe is there.

Another uncertainty rests with the possibility of a first-order phase transition in the early universe, at the epoch when the GUT symmetry is spontaneously broken (see Chap. 9). This would lead to quite a different thermal history when compared with the picture discussed here.

Let us make a few remarks on the changes induced by such an "inflationary universe". In the inflationary-universe model (see Chap. 9) the thermal history in early stages is different from a standard FL model. In these models, after a stage of exponential expansion the temperature again increases by a transformation of vacuum energy into the energy of radiation fields. This reheating may lead to a temperature at the mass scale of the X-bosons. Then things would remain pretty much as described above. But the reheating temperature may also be smaller than m_X.

The effect of Higgs bosons with small masses, and of a low temperature, smaller than m_X, on the processes of baryogenesis has been investigated within this context [Sato et al. 1985]. The interesting effect emerges that the final value of n_B/s can change drastically, and may even become negative due to the influence of the massive Higgs bosons.

In Fig. 8.10 the result of such a model calculation has been plotted. In the first phase ($T_i \simeq m_X$, when $T \gg m_H$), X-bosons dominate the baryon-number non-conserving processes. Finally, the abundance of X-bosons be-

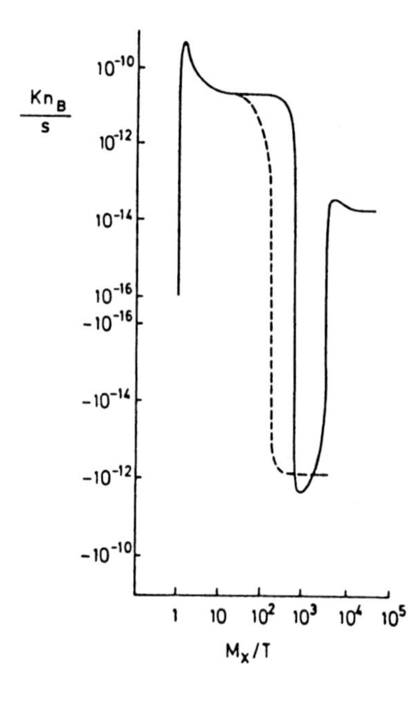

Fig. 8.10. A plot of n_B/s vs. m_X/T: (*solid curve*) $m_H = 10^{12}$ GeV; (*dotted curve*) $m_H = 2.5 \times 10^{13}$ GeV. In both cases the initial temperature was $T_i = m_x$ (after [Sato et al. 1985])

comes negligibly small through decay, and n_B/s approaches a constant value (depending on the CP-violation strength of X-boson decays, as discussed before). In a second stage, when $T \leq m_H$ (the cases $m_H \approx 10^{-3}m_X, m_H = 2.5 \times 10^{-3}m_X$ are shown in Fig. 8.10), the baryon asymmetry develops through reactions mediated by H-bosons. At the beginning of this stage the inverse decays of the H-bosons dominate. These are the processes producing H-bosons from fermions. The asymmetry can even change its sign because of these processes. Higgs boson decays counteract these changes and may eventually lead back to a positive n_B/s. The final outcome depends on the competition of these decay and inverse-decay processes.

The calculations indicate that reasonable results can only be obtained if $T_i > 5 \times 10^{10}$ GeV and $m_H > 3 \times 10^{10}$ GeV. The boundary layer (T interval), between positive and negative values of n_B/s is rather small. Small fluctuations of the reheating temperature may thus produce large fluctuations in n_B/s. This can lead to isothermal density perturbations by a division of the universe into domains of baryons and antibaryons.

These investigations show that the final value of n_B/s may depend in an intricate way an processes involving both X-bosons and Higgs bosons. We may conclude, with some safety, that the calculations do not easily give the correct answer, but that they do explain why n_B/n_γ is a small but non-zero number.

Another interesting result [Fry et al. 1980b] is that any initial baryon asymmetry is strongly damped by the thermal equilibrium phase around $T \simeq m_X$. If the difference between baryon number and lepton number is strictly conserved, $\Delta(B - L) = 0$, as required by certain GUT theories (e.g. the minimal $SU(5)$), then the damping is large enough to restore complete baryon–antibaryon symmetry, diluting any initial asymmetry ($n_B - n_{\overline{B}}$) by a factor of $\simeq 10^{-28}$. We cannot resort to appropriate initial imprints of baryon asymmetry if we take these GUT theories seriously. They not only offer a mechanism for creating a net baryon number, but they also destroy any initial asymmetry. Therefore within these theories one is forced to explain the observed baryon asymmetry by physical processes.

The discrepancies between some of the calculated ratios (n_B/n_γ) and the observed value can perhaps serve to eliminate certain models of GUT. A really fundamental theory of particle interactions would give the ratio (n_B/n_γ) purely in terms of fundamental constants of nature. This ultimate goal lies in the future, but if it can be reached the results for the number (n_B/n_γ) can be used as tests for the first 10^{-35} seconds of an FL universe.

8.5 The Electroweak Phase Transition and Baryon Synthesis

Another open end to this story is the suggestion that anomalous baryon and lepton number violating processes can be very effective around the electroweak phase transition (e.g. [Ringwald 1988; see also Kuzmin et al. 1985]). The non-abelian gauge theory of the weak interactions has $SU(2)$ as a gauge group. The vacuum manifold acquires a non-trivial topology with infinitely many ground states with vanishing energy. The trivial solution has a vanishing gauge field $W_{\rm vac} = 0$, and a Higgs field in the ground state $\phi = \phi_0$. The topologically different vacua are characterized by a winding number n, and different values of ϕ and W. The perturbation calculations correspond to an expansion around the trivial ground state. Transitions to vacua with $n \neq 0$ are not accessible by perturbation theory. A schematic picture is given in Fig. 8.11.

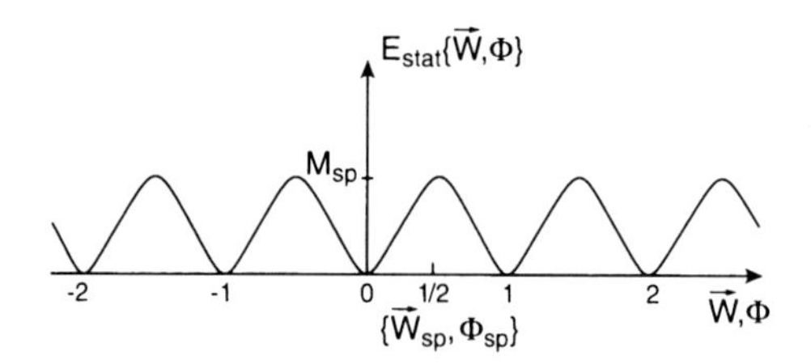

Fig. 8.11. The periodic vacuum structure of the electroweak gauge theory. $\boldsymbol{W}, \phi$ indicates the set of all gauge and Higgs fields. The trivial vacuum has $n = 0$ ($\boldsymbol{W}_{\rm vac} = 0$, $\phi_{\rm vac} = \phi_0$). Energy barriers separate the different vacua ($n = \pm1, \pm2, \dots$). The sphaleron solution ($\boldsymbol{W}_{\rm sp}, \Phi_{\rm sp}$) defines $M_{\rm sp}$, the maximal barrier height. It looks like a maximum of the static energy in this one-dimensional picture, but is in fact a saddle-point, because $E_{\rm stat}$ increases in other directions of the functional space

These non-perturbative transitions are strongly suppressed at energies small compared to the height of the energy barrier between topologically different vacua. This height is given by the energy $M_{\rm sp}$ of an unstable solution of the field equations, a saddle-point configuration (denoted by $\boldsymbol{W}_{\rm sp}, \phi_{\rm sp}$ in Fig. 8.11). The so-called "sphaleron" [Klinkhamer and Manton 1984] can be pictured as a many-particle state of W, Z, and Higgs bosons. Its mass and radius are of the order of

$$
\begin{aligned}
M_{\rm sp} &= \pi \frac{m_W}{\alpha_W} = 7.2\,\mathrm{TeV}, \\
R_{\rm sp} &= m_W^{-1} .
\end{aligned}
\tag{8.40}
$$

($m_W \simeq 80\,\mathrm{GeV}$ is the mass of the W-boson, $\alpha_W \simeq \frac{1}{30}$ the electroweak "fine structure constant"). Transitions between different vacua require energies of this order. Therefore at $\sim 100\,\mathrm{GeV}$ the accelerator experiments do not touch this region, and agree with perturbation calculations.

The non-trivial vacuum manifold is intimately connected with an interesting, non-perturbative violation of the conservation law for (baryon + lepton) number $(B + L)$. The baryon and lepton numbers within the electroweak theory are non-conserved due to quantum mechanical anomalies which initiate transitions between topologically different vacua. Non-zero values of the winding number n produce changes

$$\Delta N_e = \Delta N_\mu = \Delta N_\tau = \frac{1}{9}\Delta N_q = -\Delta n$$

where $N_e(N_\mu, N_\tau)$ are the electron-(μ, τ) lepton number, and N_q is the quark baryon number.

$B - L$ is strictly conserved, even by the anomalous processes. The energy scale of $\sim M_{\text{sp}}$ separating the different vacua leads to an exponential suppression of the $(B + L)$-violating processes. The amplitudes are proportional to the tunnelling probability $\exp(-2\pi/\alpha_W) \simeq 10^{-78}$ [t'Hooft 1976b]. At high temperatures, comparable to M_{sp}, the sphaleron barrier can be broken by classical thermal fluctuations. Instead of the tunnelling factor, the Boltzmann factor $\exp(-M_{\text{sp}}/T)$ now determines the transition probability. For $T \simeq M_{\text{sp}}$ the exponential suppression disappears. Therefore at $T \geq M_{\text{sp}}$ the $(B + L)$-violating processes are very efficient, and any non-zero $B + L$ number imprinted at earlier epochs, e.g. at a GUT phase transition, is damped strongly: $B + L \simeq 0$ after the electroweak phase transition ([Hellmund et al. 1994; Ringwald 1990]).

$B - L$ is conserved, and thus remains zero, if it is zero in the early universe. During the electroweak phase transition, then, $B + L \to 0$, and thus $B = L = 0$ at the end.

A non-zero baryon number can only survive if there is a mechanism in GUT theories allowing $B - L \neq 0$.

Apart from that, the baryon asymmetry must be created during the electroweak phase transition. Since non-equilibrium is an essential ingredient (otherwise $\langle \Delta B \rangle \simeq 0$, as we have shown) the phase transition should be of first order. It can be shown conclusively by lattice and perturbation calculations [Kajantie et al. 1996; Laser et al. 1995; Schmidt 1997] that within the standard model the electroweak phase transition is not of first order. Thus extensions of the standard model are required to give an efficient B-production mechanism. At present nobody seems to have a convincing model.

Exercises

8.1 Assume that initially the universe is baryon–anitbaryon symmetric, $n_B = 0$, and show that from thermal fluctuations

$$\frac{n_B}{n_\gamma} \simeq 10^{-18} \text{ only.}$$

8.2 Compute the number of X-particles within a horizon volume at $t \simeq 10^{-35}$ s in a GUT theory along the basis of equations (8.5) and (8.6). What is the meaning of thermal equilibrium under such conditions?

8.3 Show that the electroweak interactions in the standard model conserve $B - L$.

8.4 Calculate the cosmic epochs when the electroweak interaction is in thermodynamic equilibrium (consult [Ross 1984; Nanopoulos 1981].

8.5 What is the fate of any net baryon asymmetry produced by non-linear effects in the electroweak model, if the phase transition is not of first order, but a gradual change of the expected value of the Higgs field from 0 to ~ 250 GeV? (Consider the behaviour of B-violating processes with temperature.)

8.6 Give an estimate of the proton lifetime if proton decay only happens via classical tunnelling between different electroweak vacua.

9. The Inflationary Universe

"I have read with great interest your talk against Hegelian activities, which represent the Don Quijote element among us theoreticians, or should I say the seducer? A total absence of this vice or evil, however, leaves the game to hopeless Philistines ... "

(Letter from Einstein to Born)

The spontaneous breakdown of a GUT symmetry in the early universe may lead to a phase transition, and subsequently to a thermal history deviating quite strongly from the standard model. This concept of the so-called "inflationary universe" has gained wide publicity – one more reason why it should be discussed thoroughly here, despite its highly speculative nature [Gliner 1965; Sato 1981a,b; Guth 1981; Albrecht and Steinhardt 1982; Linde 1982; Brandenberger 1985, for a review].

Just as the baryon-synthesis processes provide a way of explaining the ratio of $n_B/n_\gamma \simeq 10^{-10}$, instead of setting it up as an initial condition, so the inflationary universe model proposes dealing with other features of the standard model, again connected with the initial conditions of our universe, in a similar way. We shall see in what follows that there exists a certain discrepancy between the claims of the model-builders and the statements that have actually been proven. Thus we can point to one conclusion right at the beginning: the inflationary universe is an appealing concept, but most of its interesting suggestions still remain to be proved.

9.1 Some Puzzles of the Standard Big-Bang Model, or Uneasiness About Certain Initial Conditions

The Friedmann equation,

$$H \equiv \left(\frac{\dot{R}}{R}\right)^2 = \frac{8\pi G}{3}\varrho - \frac{K}{R^2} + \frac{1}{3}\Lambda, \tag{9.1}$$

at the present epoch is

$$H_0^2 + \frac{8\pi G}{3}\varrho_0 - \frac{K}{R_0^2} + \frac{1}{3}\Lambda.$$

We have already seen in Chap. 2 that observations indicate that the terms on the right-hand side are of the same magnitude:

$$\frac{8\pi G}{3H_0^2}\varrho_0 \simeq \frac{|\, K \,|}{R_0^2 H_0^2} \simeq \frac{1}{3}\frac{\Lambda}{H_0^2} \leq O(1).$$

Since $\Omega_0 \equiv 8\pi G\varrho_0/3H_0^2$ is certainly larger than 0.1 and less than 4 (Chap. 2), a similar small interval of values is indicated for $|\, K \,| \,/R_0^2 H_0^2$ and for $\frac{1}{3}\Lambda H_0^{-2}$.

At very early times, however, the terms on the right-hand side of (9.1) are of greatly different magnitude. As

$$\varrho = \frac{\pi^2}{30} g(T) T^4 \tag{9.2}$$

and $R(t)T(t) = \text{const.}$ for isentropic expansion – which is only changed at certain epochs, where annihilation reactions lead to a reheating of the gas of relativistic particles – the density $\varrho \propto R^{-4}$. Thus for $R \to 0$ the term $(8\pi G/3)\varrho$ dominates completely. For later times, however, the cosmological constant Λ will dominate over the ϱ and K/R^2 terms.

One may wonder about this curious fact of a near-balance of three such different terms. There is no doubt that the balance at the present time is very beneficial for our existence and for the existence of the world around us. Indeed, a balance between $G\varrho$ and K/R^2, for a $K = +1$ model, at a time of the order of the Planck time for instance, would be disastrous. In a way similar to the case of the "normal" $K = +1$ universe treated in Chap. 1, we find that the total time of existence before recollapse would be a few Planck times. On the other hand, a balance for $K = -1$ at the Planck time, with $\Omega_{Pl} < 1$, would lead to such rapid expansion that at the present epoch Ω would be catastrophically small – matter condensations could not form in such a universe. It is definitely just as well for us that the curvature and the gravitational term are in near-balance at the present epoch.

This balance of different terms can be reformulated as follows: it can be traced back to the large numerical value of the entropy in a comoving volume $\sim R^3(t)$. In a radiation-dominated early epoch the entropy density s is given as

$$s = \frac{2\pi^2}{45} g(T) T^3, \tag{9.3}$$

and

$$S \equiv sR^3 \tag{9.4}$$

is a constant throughout the whole evolution of the standard big-bang model.

At the present epoch the entropy density is dominated by the background of photons and neutrinos

$$s_0 = 7n_{\gamma 0} = 2.8 \times 10^3\,\text{cm}^{-3}. \tag{9.5}$$

Then the entropy of a comoving volume turns out to be

$$S \simeq 2 \times 10^{87}. \tag{9.6}$$

This number is an initial condition of the standard big-bang model. The fact that there are so many more photons than baryons in our universe is imprinted right at the beginning.

It is instructive to rewrite the Friedmann equation in terms of temperature and entropy:

$$\left(\frac{\dot{T}}{T}^2\right) = \left(\frac{2\pi^2}{45}g\right)^{2/3} \frac{T^4}{M_{Pl}^2} - \frac{K}{S^{2/3}}\frac{2\pi^2}{45}g^{2/3}T^2. \tag{9.7}$$

We see that the curvature term KT^2 is essentially negligible for large temperatures, when S is a large number. The large value of S has another consequence for the early phases of an FL model. Replacing $R(t)$ in (9.1) by the Hubble parameter $H \equiv \dot{R}/R$, and the density parameter $\Omega(t) \equiv (8\pi G/3H^2)\varrho$, we find, for $K \neq 0$:

$$H^2 R^2 \left|1 - \Omega\right| = \left|K\right|. \tag{9.8}$$

Here we have set $\Lambda = 0$, because Λ is completely negligible in early phases. Then

$$\frac{\mid 1-\Omega \mid}{\Omega} = \frac{3}{8\pi G\varrho} \mid K \mid \left(\frac{s}{S}\right)^{2/3}$$

and finally

$$\frac{\mid 1-\Omega \mid}{\Omega} \simeq 0.2 \mid K \mid g^{-1/3} S^{-2/3} \left(\frac{M_{Pl}}{T}\right)^2. \tag{9.9}$$

Inserting numbers, we find

$$\frac{\mid 1-\Omega \mid}{\Omega} \approx 10^{-59} \left(\frac{M_{Pl}}{T}\right)^2. \tag{9.10}$$

Ω must have been incredibly close to 1 for early epochs, because S is so large.

Thus for

$$\begin{aligned} T &= 1\,\mathrm{MeV}, \qquad \frac{\mid \Omega - 1 \mid}{\Omega} \leq 10^{-15}, \\ T &= 10^{14}\,\mathrm{GeV}, \qquad \frac{\mid \Omega - 1 \mid}{\Omega} \leq 10^{-49}. \end{aligned} \tag{9.11}$$

This feature of the standard model, that Ω is close to 1 at all times, has been called the "flatness problem", referring to the large radius of curvature of our universe, or the "entropy problem", referring, of course, to the large value of S [Guth 1981; Linde 1982; Sato 1981a,b]. It does not seem fair to speak of "problems" in the context of the standard big-bang model. In any solution of a differential equation there are certain specific properties of the initial data. If we compute backwards in time we just find the initial data, which are responsible for the state of affairs as we see it now.

But if we look at the present state of the universe as the consequence of certain initial data we might feel a bit uneasy, if the initial data have to be extremely specific – as e.g. in our case requiring $S = 10^{87}$, or $\Omega \approx 1$ at early times. As physicists we would feel more at ease, if we could reach an understanding of such specific conditions in terms of physical processes. Fewer and less specific initial data would be considered as an improved insight into the early universe. Is it possible to start with an S of order 1, and arrive via physical processes at the number $S = 10^{87}$ characteristic of the present epoch? The model of the inflationary universe indicates "yes" as the answer to this question.

Another aspect of the standard model that is somewhat difficult to swallow is the horizon structure of the isotropic and homogeneous space-time. There are particle horizons in these space-times; in other words, there are at any epoch regions that have not had causal contact with each other. The size of the particle horizon at time t and $\chi = 0$ (see Chap. 1) is

$$r_H(t) = R(t) \int_0^t \frac{ds}{R(s)} . \tag{9.12}$$

This is the maximum size of a region in which causal relations can be established at the cosmic time t. In the standard model – during the radiation-dominated phase – $R(t) \sim t^{1/2}$, and the horizon radius is $r_H(t) = 2t$. For $t \to 0$, the horizon r_H shrinks much faster than $R(t)$ (Fig. 9.1). At very early epochs, therefore, most of the regions within the typical dimension $R(t)$ are causally unrelated despite the homogeneity of the model.

This property of the standard model has consequences for the interpretation of the isotropy of the CMB. Until a time t_R radiation and matter were in thermal equilibrium, with an isotropic distribution of photons produced by scattering processes. At $t > t_R$ radiation and matter decoupled, and have evolved independently to the present time. The isotropy of the CMB

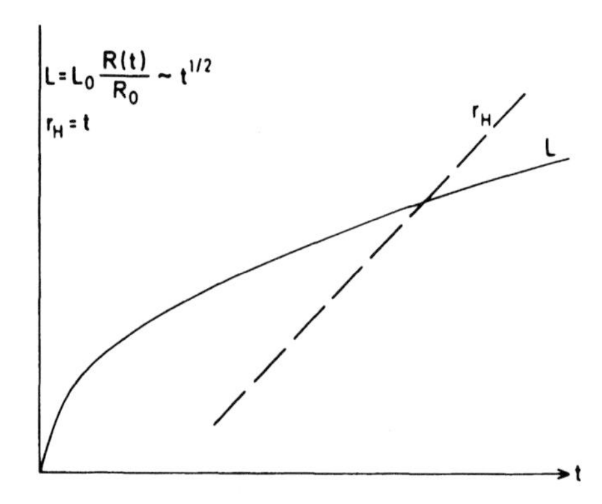

Fig. 9.1. A characteristic length $L = L_0(R/t)/R_0) \propto \sqrt{t}$ and the particle horizon $r_H(t) \propto t$ in the standard big-bang model

implies a similar isotropy at t_R. But we see the same radiation temperature from regions that had not established causal contact with each other at the epoch t_R. The coordinate distance between our epoch t_0 – we place ourselves at $r = 0$ – and the decoupling region at t_R is

$$\chi(t_0, t_R) = \int_{t_R}^{t_0} \frac{ds}{R(s)} \tag{9.13}$$

(the proper distance $r(t) \equiv \chi(t)R(t)$).

By observing in two opposite directions we look at two regions on the "cosmic photosphere" at t_R separated by a coordinate distance (Fig. 9.2)

$$\chi(t_{RA}, t_{RB}) = \chi(t_0, t_R) - \chi(t_R, t_0) \equiv 2\{\chi_H(t_0) - \chi_H(t_R)\}.$$

This distance contains

$$n_{AB} = \int_{t_R}^{t_0} \frac{dt}{R(t)} \left(\int_{t_{PI}}^{t_R} \frac{dt}{R(t)} \right)^{-1} \tag{9.14}$$

horizon lengths (or causality lengths $2r_H$) at time t_R. The factor $\int_{t_R}^{t_0}(dt/R(t))$ is determined by $R(t)$ after the decoupling time, a very well-known expression. The denominator $(\int_{t_{Pl}}^{t_R} dt/R)$ depends on the behaviour of $R(t)$ in the early universe. Changing the standard law $R \sim t^{1/2}$ may lead to drastic changes in the value of the denominator. If we use $R \sim t^{1/2}$ for $t < t_R$, and for $t > t_R$, we have $R \propto t^{2/3}$ (for a $K = 0$ model), $\chi_H \propto t^{1/3} \propto R^{1/2}$, and $n_{AB} = 60$. The fact that $n_{AB} \simeq 60$ is again related to the value of the entropy S: the radius of the observable universe L_0 shrinks to a length $L(t)$ at an earlier epoch t, and $L_0 = (T/T_0)L(t)$. During the radiation-dominated era

$$r_H = 2t \simeq (G\varrho)^{-1/2} \simeq M_{Pl}T^{-2}.$$

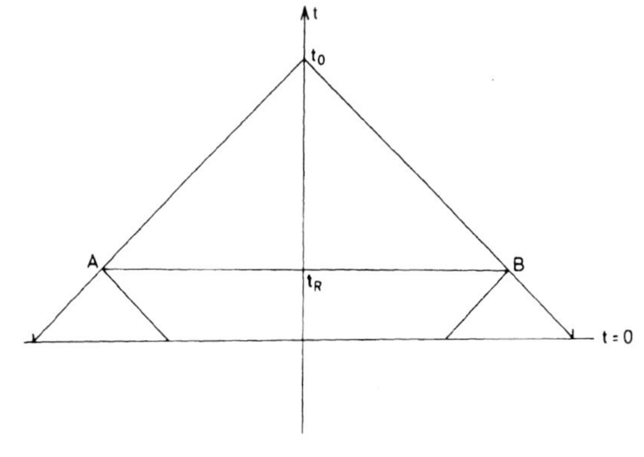

Fig. 9.2. In the standard model two points A and B in opposite directions had no causal contact at the decoupling time t_R (their past light-cones are disjoint)

Therefore the ratio

$$\left(\frac{L(t_R)}{r_H(t_R)}\right)^3 \simeq (L_0 T_0)^3 \left(\frac{T_R}{M_{Pl}}\right)^3 \simeq S \left(\frac{T_R}{M_{Pl}}\right)^3 . \tag{9.15}$$

How is it possible for two regions that were separated by such a large number of causality lengths to acquire an identical temperature at t_R?

The inflationary universe changes the behaviour of $R(t)$ dramatically at early times – the particle horizon increases exponentially, and thus the whole observable universe at t_0 becomes causally connected. n_{AB} in (9.14) becomes much smaller than 1.

The horizon structure of the standard model is called the "horizon problem" – it is, of course, not really a problem of the isotropic and homogeneous Friedmann models. For such models the horizon structure is a trivial consequence of the assumed isotropy and homogeneity. The problem appears, however, if an explanation is sought of the observed isotropy. The solution to this problem might consist in a description starting from a very general inhomogeneous and anisotropic cosmological model. The hope is that one can then show how reasonable physical processes lead to an isotropic and homogeneous Friedmann universe (cf. the discussion in Chap. 4). We shall see below how the inflationary universe attempts to bear out this hope.

Another problem that arises in the classical standard cosmology is the large number of monopoles produced when a GUT model in the early universe is at work. As pointed out in Chap. 7, magnetic monopoles are topological defects in the Higgs field which appear whenever the GUT symmetry is spontaneously broken down to a smaller symmetry group containing a $U(1)$ factor. The correlation length of the Higgs field, ξ, is limited by the causality length in the early universe $\xi \leq r_H(t) \leq 2t$. A lower limit to the density of monopoles can thus be obtained by having one monopole per horizon volume $r_H^3(t)$. Then the ratio of monopole density to entropy density is

$$\frac{n_M}{s} \simeq \left(\frac{T_c}{M_{Pl}}\right)^3 = 10^{-12}, \tag{9.16}$$

where $n_M \propto r_H^{-3}$ has used as well as

$$r_H = 2t = \left(\frac{45}{\pi^3 g_4}\right)^{1/2} \frac{M_{Pl}}{T^2} .$$

The temperature T_c can be expected to be around the typical GUT scale $T_c \simeq 10^{14}$ GeV. Comparing n_M/s to n_B/s, we find that the mass density of monopoles

$$\varrho_M = n_M m_M = 10^{-2} m_p n_B \left(\frac{m_M}{m_p}\right) \approx 10^{14} \varrho_B \qquad (m_M \sim 10^{16} m_p). \tag{9.17}$$

This terrible dominance of monopoles would of course violate all observational limits on the mean density of the universe. This is the monopole problem, the original motivation for the concept of the inflationary universe.

9.2 The Inflationary Universe – A Qualitative Account

The basic idea is very simple. It rests on the assumption that there may be epochs in the early universe where

$$3p + \rho < 0.$$

For this type of matter ("abnormal" matter) the expansion factor can grow faster than just $\propto t$. For example, when $p + \rho = 0$, as for a scalar field with negligible kinetic energy, then $\rho =$ const., and for the $K = 0$ model,

$$R(t) = \exp\left\{\sqrt{\frac{8\pi G}{3}\rho}\, t\right\}. \tag{9.18}$$

The exponential growth of $R(t)$ has been termed "inflation". When should inflation take place? The idea is to have it for a time Δt at some very early epoch near the GUT phase transition $t \simeq 10^{-35}\,\mathrm{s} \approx 10^8 t_{Pl}$. The period of inflation should last until the initial scale factor $R(t_i)$ is blown up by a tremendous factor

$$R(t_f) = R(t_i) \exp\left\{\sqrt{\frac{8\pi G}{3}\rho t}\right\} = 10^{29} \equiv Z. \tag{9.19}$$

If the universe is reheated after t_f to the initial temperature at the start of inflation, then the final entropy

$$S = Z^3 S_0 = 10^{87} S_0. \tag{9.20}$$

Even if the initial entropy S_0 is only $S_0 \simeq 1$, the inflationary phase of the universe will produce the presently observed S. The time interval Δt has to be long enough that $\sqrt{\Lambda}\Delta t \simeq 29 \ln 10$; hence

$$\Delta t \simeq \frac{M_{Pl}}{T_c^2} \times 70 \simeq 7 \times 10^{10} t_{Pl} \simeq 7 \times 10^{-9}\,(\mathrm{GeV})^{-1} \tag{9.21}$$

(for $T_c = 3 \times 10^{14}$ GeV). The time needed for inflation is not an incredibly large value: the temperature $T_c \simeq 10^{14}$ GeV is reached at $t = 10^{-35}\,\mathrm{s} = 10^8 t_{Pl}$, and about 700 times this characteristic expansion time is needed.The value of Δt depends, of course, on T_c^{-2}.

After the inflationary phase the curvature term K/R^2 is reduced by a factor of 10^{58}. This has the consequence that for the inflationary universe the mean density must be extremely close to the critical density $\Omega = 1 \pm 10^{-58}$ at the epoch of the phase transition, and $\Omega = 1 \pm 10^{-6}$ at the present epoch. This is the strongest prediction of this model. As we have seen in Chap. 2, CMB observations indicate that $\Omega_0 = 1.0$ at the present epoch, in agreement with the inflationary model.

The causality problem of the horizon structure of the standard model can also be resolved: a region of the size of the horizon at $t_c \approx 10^8 t_{Pl}$ (about $r_H \approx 10^{-25}$ cm) grows to a size $r_H Z \approx 10^4$ cm through the inflation time Δt. But the radius of the observable universe at the time $t_c + \Delta t$ is only about $c(t + \Delta t) \approx 700\, ct_{\mathrm{GUT}} \approx 10^{-22}$ cm, much smaller than $r_H Z$. Even the whole observable universe at the present epoch is inside just one light-cone at t_{GUT}. All regions that we can observe had - in principle at least – the chance of causal contact (Fig. 9.3).

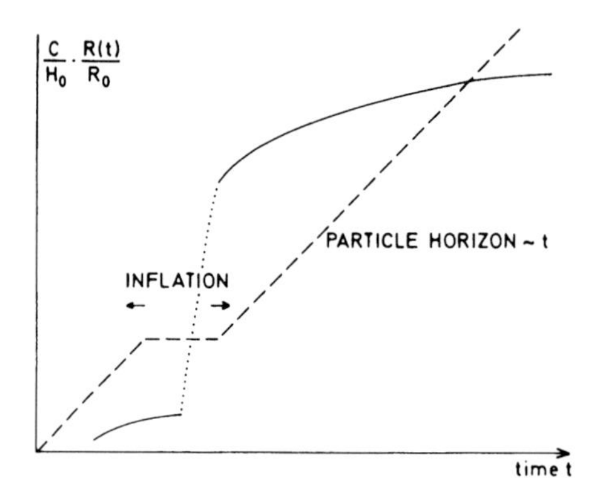

Fig. 9.3. The radius of the observable universe c/H_0 now is reduced to a small fraction of the Hubble horizon length H^{-1} before the inflationary phase

The monopole problem is just inflated away. Because the observable universe now was contained in one light-cone at t_{GUT}, very few monopoles should exist in our universe. The lower limit given in Sect. 9.1 is reduced to just one monopole per causally connected region at t_c. There may be a few more, because the correlation length of the Higgs field could have been smaller than the horizon distance at t_{GUT}. But the problem of having a large overpopulation of monopoles has certainly gone away.

So the inflationary universe concept offers nice rewards for the cosmologist. But the question arises: is it possible? The answer to that question has not yet been given conclusively. To discuss the pros and cons, we will have to look at the inflationary universe in more detail.

9.3 The Old and the New Inflationary Cosmology

One model that was also the basis for the original inflationary model considers the effect of the Higgs fields introduced as elementary scalar fields to enforce the spontaneous symmetry-breaking of a large gauge group to the group of the standard model of the electroweak and strong interactions $SU(3) \times SU(2) \times U(1)$.

The corresponding Higgs potential is temperature-dependent – just as the effective potential in the Ginzburg–Landau theory of superconductivity is temperature-dependent – and has a barrier near $\varphi = 0$, and thus for temperatures below the critical temperature $T_c = 10^{14}$ GeV the universe stays in the symmetric phase and supercools.

Thermal fluctuations or quantum-mechanical tunnelling can then induce the phase transition after a period of cooling below T_c.

The description of spontaneous symmetry-breaking by the introduction of Higgs fields (Chaps. 5 and 6) with an interaction $V(\varphi)$ leads to a specific picture of this phase transition: the Higgs Field φ – look at it simply as a classical scalar field for the moment – is considered as an order parameter, and the potential as an effective thermodynamic potential. This is simply a translation of the concepts of the usual thermodynamic description of a phase transition, as in the case of ferromagnetism, for example – where the order parameter (corresponding to φ here) is the magnetization, and the effective potential is the Legendre transform of the free energy.

During the expansion of the universe the potential V_{eff} is changing with the falling temperature (Fig. 9.4). For $T > T_c$, there is a stable minimum of the potential at $\varphi = 0$ (the symmetric phase); for T$< T_c$ the state $\varphi = 0$ is metastable, and a state $\varphi = \sigma \neq 0$ becomes the new stable minimum. For $T = T_c$ there seems to be the possibility of a phase transition of the first order – where the order-parameter φ changes its value discontinuously from $\varphi = 0$ to $\varphi = \sigma$. Typical values [Guth 1981] are

$$T_c = 10^{14} \quad \text{to} \quad 10^{15}\,\text{GeV}, \tag{9.22}$$

$$\sigma = 4 \times 10^{14}\,\text{GeV},$$

$$V_{\text{eff}}(0) - V_{\text{eff}}(\sigma) = (10^{56} \quad \text{to} \quad 10^{60})\,(\text{GeV})^4 \simeq (T_c)^4. \tag{9.23}$$

Usually one sets $V_{\text{eff}}(\sigma) = 0$, since $\varphi = \sigma$ should correspond to our present state with its very small cosmological constant.

The constant energy density $V_{\text{eff}}(0)$ acts like a cosmological constant in the Friedmann equation

$$\Lambda \equiv \frac{8\pi G}{3} T_c^4 \simeq (10^{15}\,\text{GeV})^4 M_{Pl}^{-2}. \tag{9.24}$$

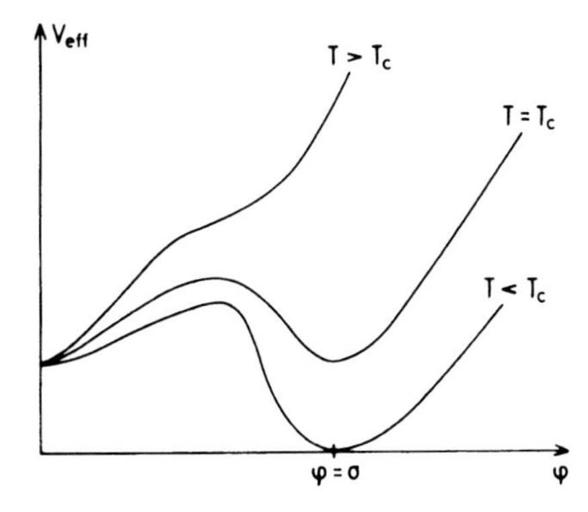

Fig. 9.4. The effective potential V_{eff} as a function of the classical Higgs field φ and the temperature

This GUT cosmological constant is so much larger than the observational limits on a cosmological constant at the present epoch, that it is absolutely necessary to find a way of getting rid of it at later epochs. The claim of the inflationary model is that the "latent heat" ΛM_{Pl}^2 is transformed into radiation when the phase transition is completed.

The Friedmann equation for $K = 0$ can be written as

$$\left(\frac{\dot{R}}{R}\right)^2 \equiv H^2(t) = \frac{8\pi G}{3}\frac{\pi^2}{30}g(T)T^4 + \frac{1}{3}\Lambda, \tag{9.25}$$

with solution

$$R^2(t) = \left(\frac{8\pi G}{\Lambda}\right)^{1/2} C^{1/2}\sinh\left[\sqrt{\Lambda/3}t\right], \tag{9.26}$$

$$C \equiv \frac{\pi^2}{30}g(T)\left(\frac{S}{g(T)}\frac{45}{2\pi^2}\right)^{4/3}.$$

C is a function of the entropy in a comoving volume $S \equiv (k_B T)^3 R^3$, and thus almost constant.

For times $t \geq \Lambda^{-1/2}$, the constant vacuum energy $\sim T_c^4$ dominates the radiation energy density $\propto T^4$, and the solution (9.26) describes an exponentially growing scale factor $R(t)$:

$$R(t) = R(t_c)\exp\left[\sqrt{\Lambda/3}(t - t_c)\right].$$

The basic idea of the inflationary universe lies in the assumption that the universe will "supercool", staying in the phase $\varphi = 0$ for a time interval Δt – and undergoing an exponential growth phase during this time – until the transition $\varphi = 0$ to $\varphi = \sigma$ of the Higgs field induces the phase transition and the subsequent change of the energy density ΛM_{Pl}^2 into radiation. If this scheme works as claimed, then the reheating of the new phase up to T_c can produce entropy.

An inflation by a factor of $Z = 10^{29}$ gives a corresponding decrease of the temperature down to $T \simeq 10^{14} \times 10^{-29}$ GeV $= 10^{-15}$ GeV. If the universe stayed that long in a de Sitter phase, why should it ever come out of it, and change into a Friedmann universe at $\varphi = \sigma$?

The original inflationary model, the "old inflationary universe" [Guth 1981; Sato 1981a,b,c], rested on the assumption that the potential barrier between $\varphi = 0$ and $\varphi = \sigma$ is penetrated by a quantum-mechanical tunnel effect [Coleman 1977]. Thus nuclei of the new stable phase $\varphi = \sigma$ form, and the vacuum energy $V_{\text{eff}}(0)$ is mainly transformed into kinetic energy of the walls of these regions. These bubbles form with zero kinetic energy but then expand, and their expansion velocity soon reaches the velocity of light. Further evolution is then governed by the rate at which bubbles nucleate and collide, and thermalize their kinetic energy.

The nucleation rate must be small to avoid a quick termination of the phase transition – otherwise the inflation would not be sufficient. On the other hand, the bubbles must appear at a certain minimal rate for the phase transition to terminate at all. It has been shown [Sato 1981a,b] that the nucleation rate by quantum tunnelling is too small, that only the possibility remains of driving the transition by thermal fluctuations. These must then have a nucleation rate close to a critical value $\approx 0.3H$. Here

$$H = H_v \equiv \left(\frac{8\pi G}{3}\frac{\pi^2}{30}g(T)T_c^4\right)^{1/2} \equiv \left(\frac{8\pi\varrho_v}{3M_{Pl}^2}\right)^{1/2} \simeq 10^{10}\,\mathrm{GeV}. \qquad (9.27)$$

It has been shown further that the collision frequency of bubbles is too low to efficiently thermalize the energy [Hawking et al. 1982]. The region $\varphi = 0$ outside the bubble is causally decoupled from the interior of the bubbles with $\varphi = \sigma$; outside the bubbles the universe still undergoes a de Sitter expansion. The creation and the growth of bubbles cannot keep up with the exponential expansion of the regions between the bubbles. In comoving coordinates the radius of a bubble

$$\chi(t, t_H) = \int_t^{t_H} \frac{ds}{R(s)}$$

stops increasing, despite its expansion at the speed of light, because the bubble expansion cannot keep pace with the cosmic expansion. Then concentrations of bubbles form, which are each finally dominated by one bubble. This way a very inhomogeneous universe appears, quite in contrast to what we know from observations, and very similar to an Emmental cheese.

A way out of this difficulty has been suggested [Linde 1982; Albrecht and Steinhardt 1982], which keeps the good aspects of the original model. The central idea of this "new inflationary scenario" is a very special way of symmetry-breaking, which has the consequence that the whole observable universe evolves out of a *single* fluctuation region. This is not possible for a generic Higgs potential, because in general the entropy per bubble is rather small.

The model is based on the Coleman–Weinberg potential [Coleman and Weinberg 1973] which is given by

$$\begin{aligned} V_{\mathrm{eff}}(\varphi, T) = {} & \frac{25}{16}a^2\left[\varphi^4 \ln\left(\varphi^2/\sigma^2\right) + \frac{1}{2}(\sigma^4 - \varphi^4)\right] \\ & + \frac{18}{\pi^2}T^4\int_o^\infty dx x^2 \ln\left\{1 - \exp\left[-\left(x^2 + \frac{5}{12}\varphi^2\frac{g^2}{T^2}\right)^{1/2}\right]\right\}. \end{aligned} \qquad (9.28)$$

The Higgs field ϕ has been expressed here as

$$\phi \equiv \varphi\left(\frac{2}{12}\right)^{1/2} \mathrm{diag}\left(1, 1, 1, -\frac{3}{2}, -\frac{3}{2}\right) \qquad (9.29)$$

The old inflationary universe model (many bubbles) and the new one (one bubble)

which corresponds to the symmetry-breaking

$$SU(5) \to SU(3) \times SU(2) \times U(1).$$

The global minimum σ for $T = 0$ has the numerical value

$$\sigma = 1.2 \times 10^{15}\ \mathrm{GeV}.$$

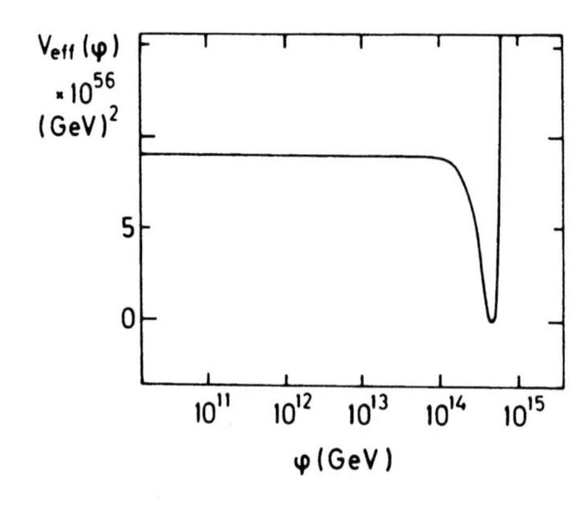

Fig. 9.5. A schematic diagram of the Coleman–Weinberg potential (for $SU(5)$ symmetry-breaking). The potential barrier is very flat and disappears for $T = 0$

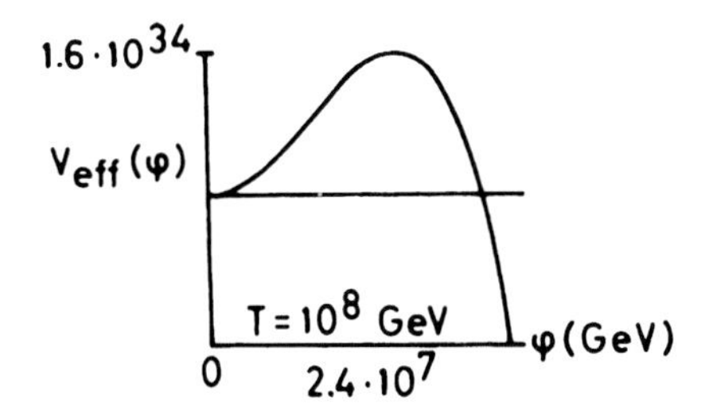

Fig. 9.6. Potential barrier of the $SU(5)$ Coleman-Weinberg potential at $T = 10^8$ GeV

$\alpha = g^2/4 = 1/45$ is the $SU(5)$ gauge coupling constant. For $T = 0$ this potential has a form as shown in Fig. 9.5. For $T > 0$ the value $\varphi = 0$ is a local minimum with curvature

$$\frac{d^2 V_{\text{eff}}}{d\varphi^2}|_{\varphi=0} = \frac{5}{4} g^2 T^2. \tag{9.30}$$

This goes to zero for $T \to 0$. Even for $T > 0$ the potential is very flat, as shown in Fig. 9.5. In this model bubbles nucleate at a rather low temperature $T \ll \sigma$. It is, however, not easy to make precise statements – there are various technical difficulties connected with the tunnelling through the potential barrier (see Sect. 9.8). It is plausible that above $T \approx 10^{10}$ GeV nucleation rates are negligible [Albrecht and Steinhardt 1982]. Below 10^{10} GeV bubbles of the new phase appear at a non-negligible rate, but the vacuum expectation value φ_i of the Higgs field in the bubble is very small too, $\varphi_i \ll \sigma$, because of the small height of the potential barrier $\sim CT^4 \leq \sigma^4$ (C is a constant of order unity). In Fig. 9.6 the small potential barrier is shown for $T = 1 \times 10^8$ GeV. It has been claimed [Albrecht and Steinhardt 1982; Hawking and Moss 1982] that bubbles or domains with a non-vanishing Higgs field expectation $\varphi_i \ll \sigma$ are formed by spinoidal decomposition rather than by nucleation of bubbles.

In Fig. 9.7 we present a "photograph" of the early universe illustrating these different possibilities for the behaviour of bubbles. The picture is, in fact, an electron-microscope photograph of the behaviour of an Fe–Al alloy.

The essential difference between the new model and the old one is the fact that $\varphi_i \ll \sigma$, i.e. that a region with $\varphi = \varphi_i \neq 0$ has almost the same vacuum energy density as the symmetric state $\varphi = 0$. This means that for some time these regions also expand exponentially. This de Sitter expansion continues until the Higgs field has reached the stable minimum $\varphi = \sigma$.

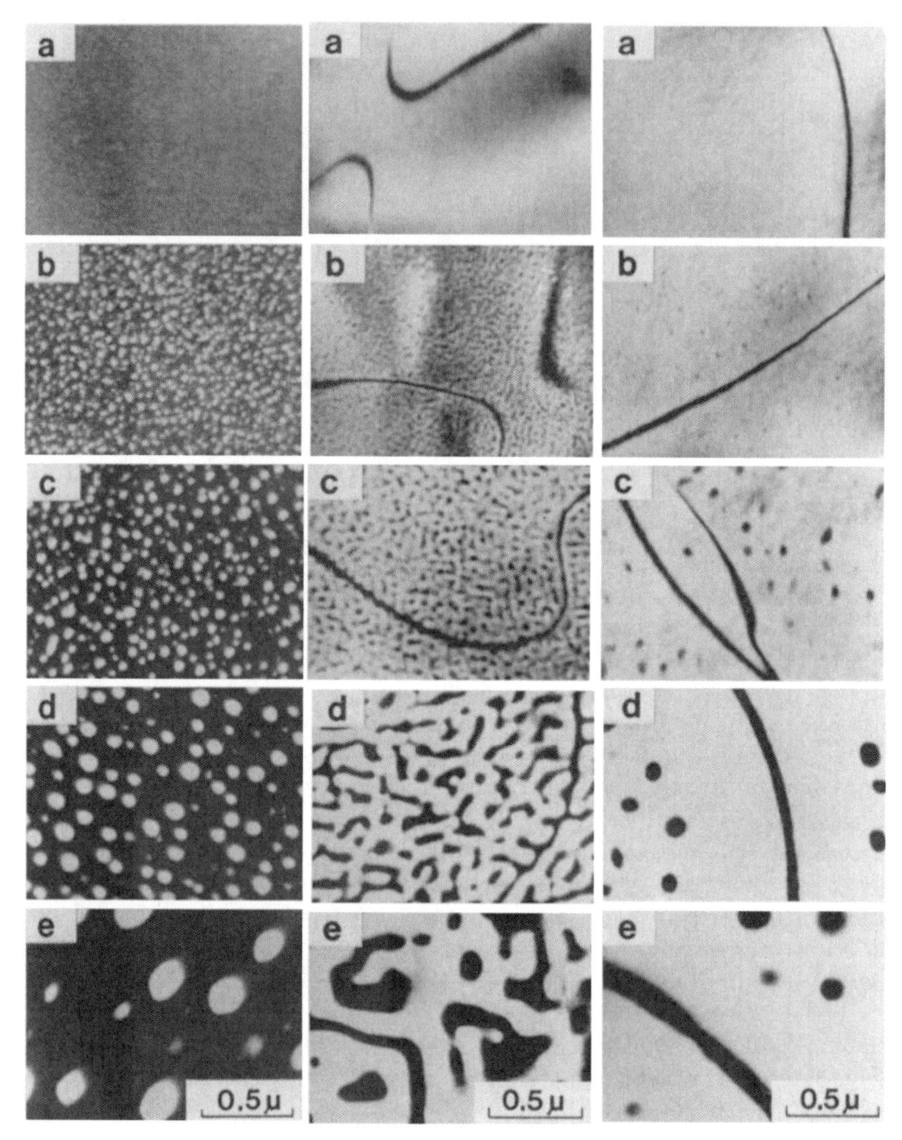

Fig. 9.7. Three conceivable developments of the very early universe are displayed in this photograph. The left- and right-hand sets show the evolution of a metastable phase, i.e. the birth and growth of droplets. Domain walls and spinoidal decomposition can also be seen [Oki et al. 1977]

9.4 Model for the Transition from $\varphi = 0$ to $\varphi = \sigma$ in the Context of the "New" Inflationary Universe

The idea is to have a jump in the expected value of the Higgs field from $\varphi = 0$ to some small initial value $\varphi = \varphi_i \ll \sigma$ by quantum tunnelling or through thermal fluctuations. Starting from φ_i, the Higgs field then follows the classical equation of motion, the Klein–Gordon equation

$$\Box_g \varphi = -\frac{dV}{d\varphi}, \tag{9.31}$$

which for a homogeneous field in a de Sitter space can be written as

$$\ddot{\varphi} + 3H\dot{\varphi} = -\frac{dV}{d\varphi}. \tag{9.32}$$

This should really be viewed as an equation for the expectation values of the quantum field φ,

$$\langle\varphi\rangle^{\cdot\cdot} + 3H\langle\varphi\rangle^{\cdot} = -\langle V'(\varphi)\rangle .$$

The equation usually treated is

$$\langle\varphi\rangle^{\cdot\cdot} + 3H\langle\varphi\rangle^{\cdot} = -V'(\langle\varphi\rangle).$$

What are the conditions for which $\langle V'(\varphi)\rangle = V'(\langle\varphi\rangle)$ holds?

For the Higgs field one takes, as in (9.29),

$$\phi = \varphi\sqrt{\frac{2}{12}}\,\mathrm{diag}\left(1, 1, 1, -\frac{3}{2}, -\frac{3}{2}\right),$$

and for the effective potential the $T = 0$ part is used,

$$V_{\text{eff}} = \frac{25}{16}\alpha^2\left[\varphi^4 \ln\left(\frac{\varphi^2}{\sigma^2}\right) + \frac{1}{2}(\sigma^4 - \varphi^4)\right].$$

For H the constant vacuum scale

$$H_v \equiv \left(\frac{8\pi\varrho_v}{3M_{Pl}^2}\right)^{1/2}$$

provides a reasonable approximation.

Equation (9.32) for the Higgs field can be coupled in a phenomenological way to a radiation density ϱ_r [Albrecht et al. 1982]: a "friction" term δ is added to (9.32) with

$$\delta = a\alpha^2\dot{\varphi}^2\varphi, \tag{9.33}$$

$$\ddot{\varphi} + 3\frac{\dot{R}}{R}\dot{\varphi} = -\frac{dV}{d\varphi} - \delta, \tag{9.34}$$

$$\dot{\varrho}_r + 4\frac{\dot{R}}{R}\varrho_r = \delta, \tag{9.35}$$

$$\left(\frac{\dot{R}}{R}\right)^2 = \frac{8\pi}{3}M_{Pl}^{-2}(\dot{\varphi}^2 + V(\varphi) + \varrho_r). \tag{9.36}$$

The system of coupled equations (9.33) to (9.36) is then solved numerically.

For the classical evolution with initial values $\varphi(0) = \varphi_i = \beta \times 10^8$ GeV – with β a constant of order 1 – and $\dot{\varphi}(0) = 0$, a solution as in Fig. 9.8 is obtained.

The results show that there is a phase $\varphi \sim \varphi(0), \dot{\varphi} \sim 0$, with exponential expansion $R/R_c \sim \exp(t/t_{\text{exp}})$ which lasts until about $t = 190t_{\text{exp}} = 190H_v^{-1}$ and which produces an inflation of the scale factor by a factor of $e^{200} \simeq 10^{87}$, large enough to achieve all the good things in store for the inflation picture.

A bubble of initial size of order $H_v^{-1} \approx 10^{-26}$ cm is thus stretched out to a dimension of $H_v^{-1}\exp(200)$ during the phase transition, i.e. to a dimension 10^{61} cm. The present size of this region is many orders of magnitude larger than the presently observable dimension $\sim H_0^{-1}$. In this new scenario, therefore, we are living in one inflated bubble. All the things we can observe comprise just a tiny part of the universe, having originated from a small fluctuation region within one causal distance. It is clear that the entropy, horizon, and monopole problem are solved as before.

We can see from Fig. 9.8 that φ stays small for most of the time and then increases to the final value $\varphi = \sigma$. The field then executes damped oscillations around the final value, with a characteristic time-scale

$$\tau_{\text{osc}} \simeq 4.8 \times 10^{-4} H_v^{-1}. \tag{9.37}$$

The radiation energy density ϱ_r increases correspondingly to $\sim T_c^4$. Critical damping – a creep solution – is reached for $a \geq 12$.

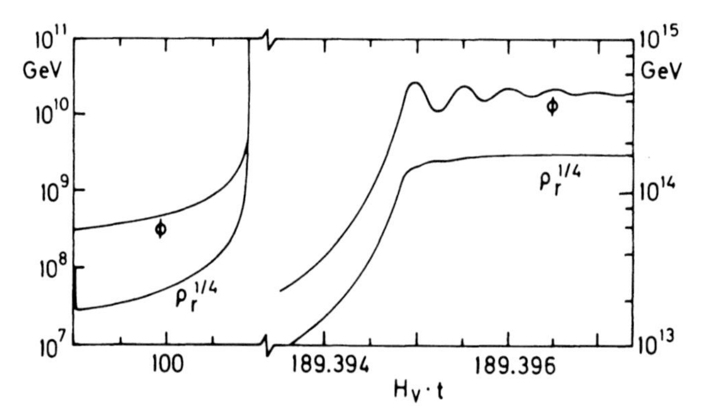

Fig. 9.8. Time evolution of the classical Higgs field coupled to a radiation field. At $t \simeq 190t_{\text{exp}}$ the stable vacuum is reached, $t_{\text{exp}} \equiv H_v^{-1}$

A very critical parameter in these solutions is the starting value of the classical evolution. Only for $\varphi_i \leq 7 \times 10^8$ GeV is sufficient inflation possible.

This has the consequence that a mass of the Higgs field has to satisfy a strong limit: $m^2/\sigma^2 \leq 10^{-10}$. The inflationary model requires m^2 to be close to zero, an extreme fine-tuning with respect to the GUT scale σ. A small, effective mass term, even of the order of $m^2 \approx (10^9\ \text{GeV})^2$, would destroy the inflation. Within the usual GUT theories such mass terms arise naturally, e.g. from the coupling of the Higgs fields to the curvature of the space-time.

9.5 Chaotic Inflation

A suggestion satisfying the necessary constraints without unduly restricting the parameters of the theory relies on a chaotic universe close to the Planck time [Linde 1983, 1985, 1990]. Roughly speaking, the quantum-mechanical uncertainty relation leads to density fluctuations of order $\Delta\varrho \sim M_{Pl}^4$ close to the Planck time $t_{Pl} = M_{Pl}^{-1}$. A simple version of this model is a massive scalar field without self-interaction which is minimally coupled to gravity and has mass $m \ll M_{Pl}$.

The matter Lagrangian is

$$L_m = \frac{1}{2} g^{\mu\nu} \partial_\mu \varphi \partial_\nu \varphi - V(\varphi),$$

with $V(\varphi) = \frac{1}{2} m^2 \varphi^2$.

If there is a sufficiently large region where φ is homogeneous, $\varphi(\boldsymbol{x}, t) = \varphi(t)$, then

$$\varrho = \frac{1}{2}\dot{\varphi}^2 + V(\varphi);$$

$$p = \frac{1}{2}\dot{\varphi}^2 - V(\varphi);$$

for pressure p and density ϱ.

There are three possibilities for the equation of state $p = p(\varrho)$. First,

$$\dot{\varphi}^2 \gg m^2\varphi^2, \quad \text{i.e.} \quad p \sim \varrho.$$

This is an ordinary equation of state $p = (\gamma - 1)\varrho$, with $\gamma = 2$. Second,

$$\dot{\varphi}^2 \ll m^2\varphi^2, \quad \text{i.e.} \quad p \simeq -\varrho.$$

Here $3p + \varrho < 0$, and hence there is inflation. Finally,

$$\frac{1}{2}\dot{\varphi}^2 - \frac{m^2}{2}\varphi^2 = p \simeq 0.$$

This describes dust-like matter, i.e. an ensemble of cold, massive particles. These three possibilities are all realized at different cosmic epochs.

Let us look at the simple case $K = 0$. The Friedmann equation reads

$$H^2 = \frac{8\pi G}{3}\left(\frac{1}{2}\dot{\varphi}^2 + \frac{m^2}{2}\varphi^2\right), \tag{9.38}$$

and the equation of motion for φ

$$\ddot{\varphi} + 3H\dot{\varphi} + m^2\varphi = 0; \tag{9.39}$$

in addition, we have the energy conservation equation

$$\frac{d}{dR}\left(R^3\left(\frac{\dot{\varphi}^2}{2} + \frac{m^2}{2}\varphi^2\right)\right) = 3R^2\left(\frac{\dot{\varphi}^2}{2} - \frac{m^2}{2}\varphi^2\right). \tag{9.40}$$

A simplified analysis is possible when $\dot{\varphi}^2 \ll m^2\varphi^2$, $p = -\varrho$, and $\ddot{\varphi}$ can be neglected. Then

$$\begin{aligned} H^2 &\simeq \frac{4\pi G}{3} m^2\varphi^2, \\ 3H\dot{\varphi} + m^2\varphi &= 0. \end{aligned}$$

Then

$$\varphi(t) = \varphi_0 - \frac{1}{\sqrt{12\pi G}} mt \equiv \frac{1}{\sqrt{12\pi G}} m(t_f - t), \tag{9.41}$$

with the definition $\varphi_0 \equiv \frac{mt_f}{\sqrt{12\pi G}}$.

Inserting this back into the equation for H^2, we have

$$\frac{\dot{R}}{R} = \frac{m^2}{3}(t_f - t),$$

and this leads to a quasi-exponential growth of the expansion factor

$$R(t) = R_f \exp\left\{-\frac{m^2}{6}(t_f - t)^2\right\}.$$

For some initial time t_i, we find

$$R_f/R_i = \exp\left(\frac{m^2}{6}(t_f - t_i)^2\right). \tag{9.42}$$

A moderate change in φ and ϱ produces an enormous change in $R(t)$. This approximation, i.e. to neglect $\ddot{\varphi}$, can be checked by calculating $\dot{\varphi}$ from e.g. (9.41), inserting φ and $\dot{\varphi}$ into (9.39). The condition $\varphi > M_{Pl}$ emerges. Inflation can take place only if the field φ has a value larger than the Planck mass.For arbitrary $V(\varphi)$ we can easily derive a useful expression as follows.

Write $l^2 \equiv \frac{8\pi G}{3}$; then

$$H = l\sqrt{V} \quad \text{and} \quad 3H\dot{\varphi} = V_{,\varphi}$$

are equivalent to

$$\frac{d}{dt}\ln \mathrm{R} = l\sqrt{V} \quad \text{and} \quad \frac{d\varphi}{dt} = -\frac{V_{,\varphi}}{3l\sqrt{V}}.$$

This leads to

$$\frac{d\ln R}{d\varphi} = -3l^2 V/V_{,\varphi}. \tag{9.43}$$

Integrating (9.43) results in

$$R(\varphi) = R_f \exp\left\{-3l^2 \int V/V_{,\varphi} d\varphi\right\}. \tag{9.44}$$

Now, if $V(\varphi) \sim \varphi^n$, such that $V/V_{,\varphi} = \varphi/n$,

$$R(\varphi) = R_f \exp\left\{-\frac{3l^2}{2n}(\varphi^2 - \varphi_f^2)\right\},$$

and

$$R_f/R_i = \exp\left\{\frac{3l^2}{2n}(\varphi_i^2 - \varphi_f^2)\right\}.$$

We see that $R_f/R_i \gg 1$ if $\varphi_i l \gg 1$. In that case, we have inflation.

Complete Classical Analysis

If we insert $H(\varphi)$ from (9.38) into the Klein–Gordon equation (9.39), we obtain

$$\ddot{\varphi} + \frac{3l}{\sqrt{2}}(\dot{\varphi}^2 + m^2\varphi^2)^{1/2}\dot{\varphi} + m^2\varphi = 0. \tag{9.45}$$

The definition of new "phase space" variables

$$x = m\varphi \quad \text{and} \quad y = \dot{\varphi} \tag{9.46}$$

transforms (9.45) into

$$\frac{dy}{dx} = -\frac{1}{my}\left\{\frac{3l}{\sqrt{2}}(y^2 + x^2)^{1/2}y + mx\right\}. \tag{9.47}$$

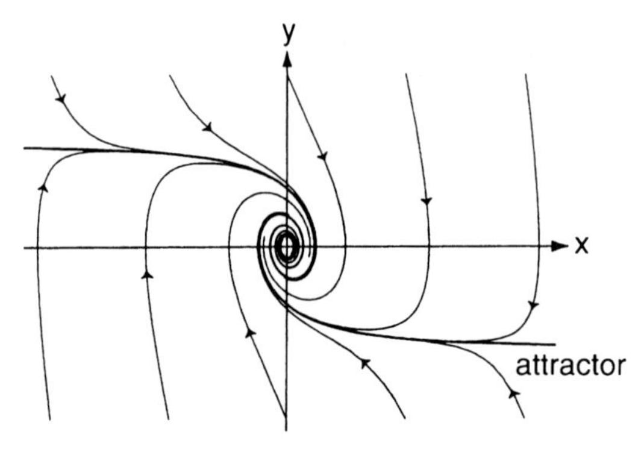

Fig. 9.9. Schematic phase diagram for the motion of a scalar field with potential $V(\varphi) = \frac{1}{2}m^2\varphi^2$. At large energies (large φ) the field approaches rapidly the inflationary phase (attractor) characterized by $\varrho + p \simeq 0$, more or less regardless of the initial condition. Near the origin rapid oscillations set in

Now the phase diagram of Fig. 9.9 can be used to discuss the properties of the solutions:

A) Let $|y| \gg |x|, ly^2 \gg m|x|$.
Then

$$\frac{dy}{dx} = -\frac{3l}{\sqrt{2}}\frac{|y|}{m},$$

with the solution

$$y(x) = y_0 \exp\left\{-\frac{3l}{\sqrt{2}m}x\right\}.$$

In the old variables this corresponds to

$$\dot{\varphi} = y_0 \exp\left\{-\frac{3l}{\sqrt{2}\varphi}\right\},$$

and hence $\varphi = \frac{\sqrt{2}}{3l}\ln t + c$.
From (9.38) for $H(t)$, one finds

$$\ln R = \frac{1}{3}\ln t, \quad \text{i.e.} \quad R \quad \propto \quad t^{1/3}.$$

In this region of phase space, the expansion law for $p = \varrho$ matter holds.

B) Regions with Inflation
Let $\frac{dy}{dx} \approx 0$, and derive from (9.47) that

$$y^2 = -\frac{x^2}{2} \pm \left(\frac{x^4}{4} + \frac{2m^2}{9l^2}x^2\right)^{1/2}.$$

Assuming $m/(lx) \ll 1$,

$$y^2 \simeq \frac{2m^2}{9l^2}.$$

$$y \simeq \begin{cases} -\dfrac{\sqrt{2}}{3l} m \text{ for } x > 0, \\ \dfrac{\sqrt{2}}{3l} m \quad \text{for } x < 0 \end{cases}$$

and $\varphi = -\frac{\sqrt{2}}{3l}\,\mathrm{sign}\,\varphi$.

For $x \gg m/l$, the potential term $\frac{x^2}{2}$ dominates over the kinetic one in this region, and hence there is inflation.

C) $x \ll m/l$; $y \ll m/l$.

New variables $r(t)$ and $\theta(t)$ are defined by

$$y = r\sin\theta, \quad x = r\cos\theta;$$

$$\text{then } \dot{r} = -\frac{3l}{\sqrt{2}} r^2 \sin^2\theta - mr\cos\theta;$$

$$\dot{\theta} = -m - \frac{3l}{\sqrt{2}} r \sin\theta\cos\theta.$$

The average over θ yields

$$r = \frac{2\sqrt{2}}{3l} t^{-1};$$

$$\theta = \theta_0 - mt.$$

Further, we have

$$\dot{\varphi}^2 + m^2\varphi^2 = x^2 + y^2 = r^2,$$

and hence

$$H = \frac{l}{\sqrt{2}} r = \frac{2}{3t}.$$

The solution is equivalent to a dust-dominated solution with $p = 0$. Inside the spirals in Fig. 9.9 the field changes faster than the radius

$$\left\langle \frac{1}{2}\dot{\varphi}^2 - \frac{m^2}{2}\varphi^2 \right\rangle \simeq 0,$$

and the pressure change per period is negligible.

In Fig. 9.9 the different regions are qualitatively drawn in the x, y plane. It is clear from these considerations that there is no reason for the field to take a particular value, i.e. the value $\varphi = 0$, in the whole universe. On the contrary, we may expect all the different values of φ to appear in different regions of space, if they are sufficiently far from each other. In particular, values $\varphi \gg M_{Pl} \simeq 10^{19}$ GeV are quite legitimate.

But we must be aware of the fact that conditions such as $\varrho > M_{Pl}^4$ for the energy density, and $\Delta L < L_{Pl}$ for the size of the domain considered, very likely lead us out of the classical space-time picture into the pre-Planckian era.
The conditions $\varrho < M_{Pl}^4$ and $\Delta L > L_{Pl}$ for a classical space-time, require

$$(\nabla\varphi)^2 < M_{Pl}^4 \quad \text{and} \quad \frac{m^2}{2}\varphi^2 < M_{Pl}^4 .$$

We can see from the phase diagram that the condition $|\nabla\varphi|^2 \leq \frac{1}{2}V(\varphi)$ lets $|\nabla\varphi|$ become exponentially small very quickly. The field moves to the attractor, where $dy/dx \sim 0$, and there is inflation.
This is satisfied when $\varphi > M_{Pl}$.
If one starts with a sufficiently homogeneous region, where $\varphi \sim 10^5 M_{Pl}$, there certainly is inflation.

9.6 Other Models

The basic difference between the old, new, and chaotic inflationary models lies in the choice of the effective potential V_{eff} for the scalar field φ. In Fig. 9.10 I have schematically drawn the different cases. Once you give up the attempt to relate the scalar field to a GUT theory, or a fundamental theory, you are free to choose any kind of φ and V_{eff} to suit your purpose. In the following we list a few examples.

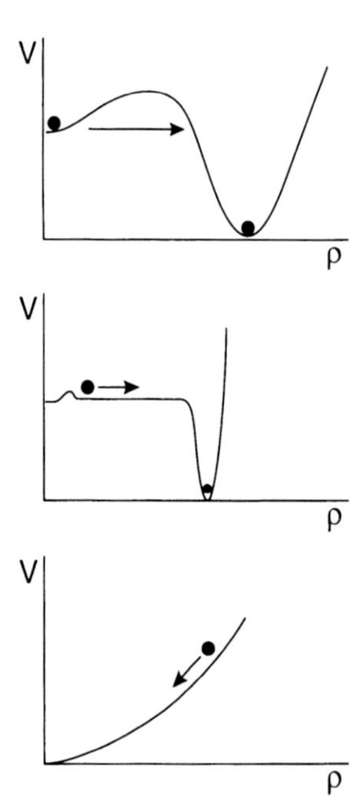

Fig. 9.10. Schematic plot of the essential behaviour of the field φ in old, new, and chaotic inflation

9.6.1 Power Law Inflation

One assumes a potential of the form

$$V(\varphi) \sim \lambda e^{\alpha\varphi}$$

[Lucchin and Matarrese 1985]. This leads to a power-law growth of the expansion factor, $R(t) \sim t^{\alpha'}$, with $\alpha' > 1$. Sufficiently large inflation can be realized only by fine-tuning the parameters. The potential also has no true stable vacuum around which φ can oscillate when inflation ends.

9.6.2 Extended Inflation

This version [La and Steinhardt 1989] adds a Brans–Dicke term to the usual Lagrangian

$$L = \frac{\zeta\sigma^2}{G}R - \frac{1}{2}(\nabla\sigma)^2 + L(\varphi).$$

The first term in L is read as $\sim \tilde{G}^{-1}R$, and $\tilde{G} = G/\zeta\sigma^2(t)$ is interpreted as a time-dependent gravitational constant, $\sigma(t)$ being the Brans–Dicke field. Since

$$H^2 = \frac{8\pi\tilde{G}}{3}\varrho,$$

a decrease in $\tilde{G}$ implies a decrease in H^2, and this leads to power-law inflation. The bubble nucleation rate $\epsilon \propto H^{-4}$ increases and tracks the expansion. In this model bubbles of the true vacuum coalesce if $b \equiv \frac{1}{8\zeta} < 25$. Radar observations limit b to $b > 500$. Thus the model is nicely ruled out.

9.6.3 R^2 Inflation

This is the oldest model, introduced in [Starobinsky 1980]. An R^2 term is added to the Einstein action (R here is the curvature scalar)

$$S = -\frac{1}{8\pi G}\int\sqrt{-g}(R + R^2/6M^2).$$

The model can be reformulated such that it is equivalent to Einstein gravity plus a scalar field with exponential interaction. The effective potential is extremely flat. To get agreement with observations, M must be smaller than M_{Pl} : $M/M_{Pl} \sim 10^{-6}$.

9.6.4 Others

In addition, we have hyperextended inflation, stochastic inflation, eternal inflation, superstring inspired inflation, Kaluza–Klein inflation, multiple inflation, ... About two new models are produced every year.

9.7 The Spectrum of Fluctuations

9.7.1 Basic Features

The inflationary phase affects density perturbations in an interesting way. In fact, the transient inflationary phase is just like an intermediate de Sitter stage in the cosmic evolution. It has been shown in 1981 [Mukhanov and Chibisov 1981] that quantum fluctuations of a polarized vacuum can produce perturbations of the metric of the form

$$Q(k) = 3L_{Pl}M\left(1 + \frac{1}{2}\ln H/k\right), \tag{9.48}$$

where L_{Pl} is the Planck length and M a typical mass of a scalar particle.

The quantity $Q(k)$ is a measure of the amplitude of perturbations with scale dimensions $1/k$ at the time the universe enters the FL expansion stage leaving the de Sitter epoch. Choosing ML_{Pl} appropriately, this result can be consistent with observational requirements.

Where does this special behaviour originate?

In a simple model of a scalar field on de Sitter space the value $\delta\varphi$ of fluctuations around a classical homogeneous solution $\varphi_0(t)$ can be calculated as

$$\delta\varphi = \left\{\frac{H^2}{16\pi^3}\left[1 + \frac{k^2}{H^2}e^{-2Ht}\right]\right\}^{1/2}$$

[Birrel and Davies 1984]. Here $H = H_v$ is the constant expansion rate, k the wave number. The evolution of the physical wavelength $\lambda = (2\pi/k)R(t)$ of a fluctuation can be imagined as shown in Fig. 9.11.

One has to keep in mind that in a de Sitter space $R(t)$ grows proportionally to $\exp(Ht)$. Therefore any physical wavelength $(2\pi/k)R(t)$ grows larger than H^{-1} very quickly. At a time t_1 – during the de Sitter epoch –

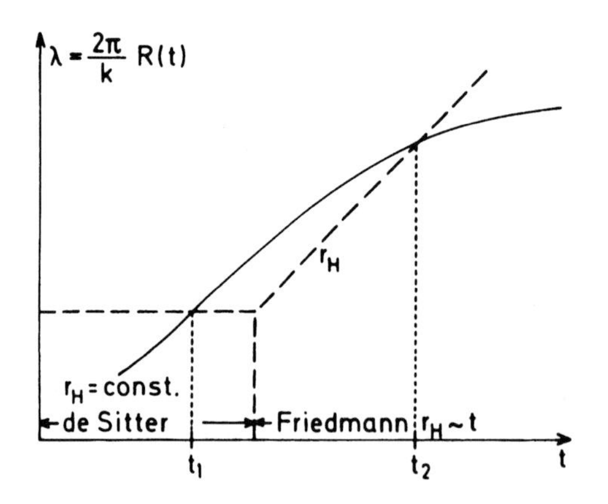

Fig. 9.11. Growth of a physical wavelength λ of a fluctuation compared with the growth of the Hubble radius r_H. For $t_1 < t < t_2$ the wavelength is larger than r_H

the fluctuation exceeds the Hubble radius H^{-1}, and it stays larger up to a time t_2 of the FL epoch. Between t_1 and t_2 the edges of the fluctuation evolve independently of each other (see Chap. 11), i.e. the fluctuation cannot be influenced by microphysical processes during this time interval. During the de Sitter phase

$$\lambda H = 1,$$

when

$$\frac{2\pi}{k} R(t_*) H = 1$$

and

$$t_* \simeq H^{-1} \ln\left(\frac{k}{H}\right).$$

If we insert t_* in the expression for $\delta\varphi$, we find that

$$\delta\varphi(k, t_*) = \text{const.}$$

All perturbations have the same amplitude at $t = t_*$. This fact is essentially responsible for the occurrence of a scale-invariant spectrum – as suggested by Zel'dovich – of the density contrast $\delta_{k/H}$ of the fluctuations at the times $t_2(k)$ when they "cross into the horizon", i.e. when $\lambda H = 1$.

9.7.2 A Detailed Calculation

Fluctuations of the metric can be expressed in terms of gauge-invariant quantities. If we consider only scalar perturbations, then the line element can be written as

$$\begin{aligned} ds^2 = &\ (1 + 2\Phi)dt^2 - (1 - 2\psi)R^2 d\boldsymbol{x}^2 \\ &-2B_{,\alpha}\, dx^\alpha dt + E_{,\alpha\beta}\, dx^\alpha dx^\beta \end{aligned}$$

Choosing a special gauge, we can set $E = B = 0$ (the longitudinal gauge). In the absence of shear $\Phi = -\psi$. Φ, ψ are the Bardeen variables of Chap. 11. The perturbations should be caused by the quantum fluctuations $\delta\varphi$ of the scalar field φ around its classical solution $\varphi_0(t)$:

$$\varphi = \varphi_0(t) + \delta\varphi(\boldsymbol{x}, t).$$

The action for the field φ is

$$S = \int \left[\frac{1}{2}\partial^\mu\varphi\partial_\mu\varphi - V(\varphi)\right] \sqrt{-g}d^4x,$$

and the homogeneous background field $\varphi_o(t)$ gives rise to anenergy density and a pressure

$$\varrho = \frac{1}{2}\dot{\varphi}_0^2 + V(\varphi_0),$$
$$p = \frac{1}{2}\dot{\varphi}_0^2 - V(\varphi_0).$$

We consider the case of a slowly varying $\varphi_0(t)$ such that $\ddot{\varphi}_0 \approx 0$, and

$$3H\dot{\varphi}_0 + V_{,\varphi} = 0.$$

Linearization of the equations of motion, and of the equations of GR in ψ and $\delta\varphi$, results in

$$\delta\ddot{\varphi} + 3H\delta\dot{\varphi} + V_{,\varphi\varphi}\delta\varphi - \frac{\Delta\delta\varphi}{R^2} = -4\dot{\varphi}_0\psi + 2V_{,\varphi}\psi, \tag{9.49}$$
$$\frac{1}{R^2}\Delta\psi - 3H\dot{\psi} - (\dot{H} + 3H^2)\psi = -4\pi G(\dot{\varphi}_0\delta\dot{\varphi} + V_{,\varphi}\delta\varphi). \tag{9.50}$$

The condition $\dot{\varphi}_0 \ll V(\varphi_0)$ or $|\dot{H}| \ll H^2$ leads to inflation. In that case $\alpha \equiv \frac{|\dot{H}|}{H^2}$ is a small parameter, and $\dot{\varphi}_0 \sim \dot{\varphi} \sim 0(\sqrt{\alpha})$. If we take $\dot{H} \sim 0$, i.e. keep only leading terms in (9.49) and (9.50), we arrive at

$$\delta\ddot{\varphi} + 3H\delta\dot{\varphi} + V_{,\varphi\varphi}\delta\varphi - \frac{\Delta\delta\varphi}{R^2} = 2V_{,\varphi}\psi,$$
$$\frac{1}{R^2}\Delta\psi - 3H\psi = -4\pi G V_{,\varphi}\delta\varphi.$$

For long wavelengths, when the physical scale $\lambda_{\text{phys}} > 1/H$, or equivalently $R/k > 1/H$, we find the gauge-invariant relation between the metric fluctuations and $\delta\varphi$:

$$\psi = \frac{1}{2}\frac{V_{,\varphi}}{V}\delta\varphi. \tag{9.51}$$

For $\delta\varphi$ the equation is for $k/R < H$

$$\delta\ddot{\varphi} + 3H\delta\dot{\varphi} + \left(V_{,\varphi\varphi} - \frac{V_{,\varphi}^2}{V}\right)\delta\varphi = 0. \tag{9.52}$$

Inside the horizon, $\lambda_{\text{phys}} < H^{-1}$, we have

$$\frac{\Delta\delta\varphi}{R^2} \gg V_{,\varphi\varphi}\delta\varphi, \quad V_{,\varphi}\psi.$$

Then the equation for $\delta\varphi$ reads

$$\delta\ddot{\varphi} + 3H\delta\dot{\varphi} - \frac{\Delta\delta\varphi}{R^2} = 0. \tag{9.53}$$

The quantization of the field fluctuations makes $\delta\varphi$ into an operator $\delta\hat{\varphi}$, and $\delta\varphi$ in (9.51)–(9.53) is the expectation $\langle\delta\hat{\varphi}^2\rangle^{1/2} = \delta\varphi$.

Now, the scalar field on de Sitter space can be evaluated directly, $\langle\delta\hat{\varphi}^2\rangle = \langle\varphi^2\rangle$ from the basic solutions of the scalar wave equation [Birrel and Davies 1984]:

$$\langle\varphi^2\rangle = \frac{1}{(2\pi)^3}\int\frac{d^3k}{k}\left(\frac{1}{2}+\frac{H^2}{2k^2}\right). \tag{9.54}$$

This expression is infinite, but the "1/2" is removed by the usual renormalization procedure (as in Minkowski space), and the inflationary phase introduces cut-offs in the integral $\int\frac{d^3}{k}\frac{H^2}{2k^2}$.

Wavelengths $\lambda \le \lambda_o = H^{-1}$ are affected initially by the inflation of a region of size $\sim H^{-1}$, and they will undergo inflation for a time t. Thus $\lambda \le H^{-1}e^{Ht}$ or $k \ge He^{-Ht}$. On the other hand, all wavelengths comoving with the de Sitter expansion will have $\lambda > H^{-1}$, thus $k \le H$. Therefore

$$\langle\varphi^2\rangle = \frac{H^2}{(2\pi)^3 2}4\pi\int_{He^{-Ht}}^{H}\frac{dk}{k} \approx \frac{H^3}{4\pi^2}t. \tag{9.55}$$

Let $t \sim H^{-1}$; then $\langle\varphi^2\rangle = H/2\pi$.

We take this value as the initial condition for $\delta\varphi$ on the horizon, and neglect decaying modes, i.e. $\delta\ddot{\varphi} = 0$. Then (9.52) is solved by

$$\delta\varphi = C\frac{V_{,\varphi}}{V} \tag{9.56}$$

and

$$\psi = \frac{1}{2}C\frac{V_{,\varphi}^2}{V^2}. \tag{9.57}$$

The boundary condition

$$\delta\varphi(k = HR) = \frac{H}{2\pi} = C\frac{V_{,\varphi}}{V}$$

determines the constant C.

Finally, the components of ψ in a harmonic decomposition are

$$\psi_{\mathrm{k}} = L_{Pl}\left(\frac{V^{3/2}}{V_{,\varphi}}\right)_{HC}\left(\frac{V_{,\varphi}^2}{V^2}\right)(t). \tag{9.58}$$

The term $L_{Pl}(V^{3/2}/V_{,\varphi})$ evaluated at horizon-crossing, $k = HR$, describes the spectral distribution, while $(V_{,\varphi}/V)^2$ gives the amplitudes.

Equation (9.58) is the general expression for the metric fluctuations in a gauge invariant formalism. For $1/k \sim \lambda > \lambda_{\rm eq}$, the temperature variations measured by COBE are

$$\left|\frac{\Delta T}{T}\right| \simeq 10^{-5} \simeq \frac{2}{3}\psi(t_f). \tag{9.59}$$

t_f marks the end of inflation, when the reheating of the universe leads back to a radiation-dominated FL model. Equation (9.59) is a condition on the amplitude $(V_{,\varphi}/V)^2$. Consider the simplest chaotic inflationary model

$$V = \frac{m^2}{2}\varphi^2.$$

Then $V^{3/2}/V_{,\varphi} = m\varphi^2/\sqrt{8}$. Evaluate $m\varphi^2$ at $k = HR$, i.e.

$$k = HR_f \exp\left(-\frac{3}{4}L_{Pl}^2\varphi^2\right).$$

Then $L_{Pl}^2\varphi^2 = \ln(HR_f/k)$ at horizon-crossing. At the end of inflation $L_{Pl}^2\varphi^2 \sim 1$, and $(V_{,\varphi}/V)^2 \simeq \varphi^{-2} \simeq L_{Pl}^2$.

Finally,

$$\psi_k \simeq L_{Pl} m \ln\{(R_f/k)H\}. \tag{9.60}$$

This is an almost flat spectrum with only a logarithmic dependence on wavelength. Fluctuations on galaxy cluster scales, $l \sim 3\,\text{Mpc} \sim 10^3\,(\text{GeV})^{-1}$, have a comoving wave number $k \simeq R_f 10^{-11}$ GeV,

$$\ln HR_f/k \simeq 70,$$

and

$$\psi_k \simeq mL_{Pl}70 \le 10^{-5}, \quad \text{if} \quad m \le (L_{Pl})^{-1}\frac{10^{-5}}{70} \le 10^{12}\,\text{GeV}.$$

9.8 Summary of a Few Difficulties

The simple picture of the inflationary universe is attractive, but its simplicity is deceptive. There is input into this model from general relativity and from particle physics. Our discussion would be incomplete without talking about some of the unsolved problems.

9.8.1 Tunnelling Probabilities

It is still an open question whether, as required by some models, quantum-mechanical tunnelling can lead from $\varphi = 0$ to $\varphi_i \le 7 \times 10^8$ GeV. A tunnelling probability has been computed from a kind of WKB approximation of the path integral for the Euclidean action

$$\frac{\Gamma}{V} = \frac{S_E[\overline{\varphi}]^2}{4\pi^2\hbar} \exp\left\{-\frac{S_E}{\hbar}\right\} |\det{}'(-\Delta + V''(\overline{\varphi}))\det{}^{-1}(-\Delta + V''(\varphi_0))|^{-1/2} \tag{9.61}$$

[Coleman 1977; Callan and Coleman 1977]. Here $S_E[\overline{\varphi}]$ is the Euclidean action

$$S_E[\overline{\varphi}] \equiv \int d^4x[1/2(\nabla\overline{\varphi}, \nabla\overline{\varphi}) + V(\overline{\varphi})]$$

of the $O(4)$ invariant solution of the classical Euclidean field equation

$$\Box_E\overline{\varphi} \equiv \Delta\overline{\varphi} + V'(\overline{\varphi}),$$

with boundary condition $\lim_{|x|\to\infty} \overline{\varphi}(t, x) = \varphi_0$ (the "bounce" solution), and $\det'$ is the determinant where zero modes of the operator $-\Delta + V''$ have been crossed out. These determinants are infinite, and their regularization is an arbitrary procedure. Equation (9.61) must thus be viewed as ill-defined.

At present it is also not clear how the Euclidean formalism must be generalized to include the effects of gravity. The hope is that in the simple FL space-times a formal transcription of the Euclidean formalism is possible and makes sense (see e.g. [Coleman and de Luccia 1980]).

Even ignoring these cautionary remarks, the computation of the determinant in (9.61) has been achieved only in special cases [Wipf 1985].

A commonly used, rough approximation for the nucleation rate per unit time per unit volume is

$$\frac{\Gamma}{V} \simeq \sigma^4 e^{-S_E}, \tag{9.62}$$

where Γ defines the rate of tunnelling within the volume V. Γ/V has a maximum value for minimum S_E.

Γ/V, or rather S_E, has been evaluated numerically. Some of these results are represented in Fig. 9.12.

It is clearly indicated in Fig. 9.12 how $m^2 = 0$ is a preferred value for the transition. The graph for zero mass looks like a critical curve separating two regions that do not lead to acceptable transition probabilities.

Another suggestion is to tunnel to a specific state $\varphi_{\max}$ [Hawking and Moss 1982], which is a solution of the classical field equation with

$$V''_{\text{eff}}(\varphi)|_{\varphi_{\max}} = 0, \tag{9.63}$$

$$V''_{\text{eff}}(\varphi_{\max}, T) = \max_{\varphi<\sigma} V_{\text{eff}}(\varphi, T).$$

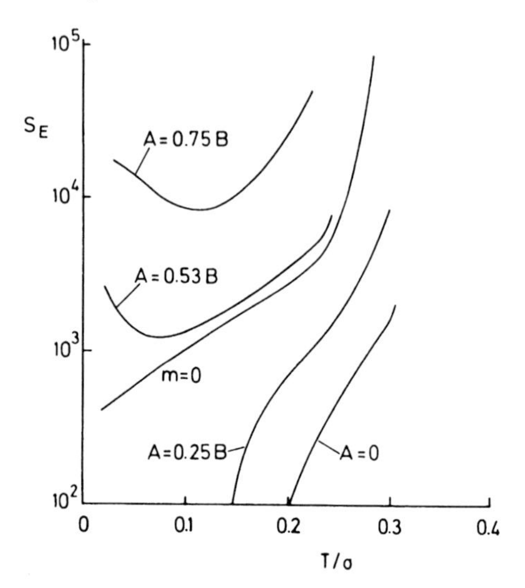

Fig. 9.12. Euclidean action S_E for a Higgs potential $V = (2A - B)\sigma^2\varphi^2 - A\varphi^4 + B\varphi^4 \ln(\varphi^2/\sigma^2) + V_T(\varphi, T)$ (after [Cook and Mahanthappa 1981]). $V_T(\varphi, T)$ is a temperature correction; the "mass term" is $m^2 \equiv (2A - B)\sigma^2$, where $A = (46\,875/2048\pi^2)g^4 - (\lambda_R/4!)$, $B = (5625/1024\pi^2)g^4$, $(g^2/4\pi) = \frac{1}{45}$ and $\lambda_R \equiv$ renormalized φ^4 coupling constant. One must have $A < B$. The plot shows the dimensionless quantity S_E/T as a function of the temperature (in units of σ, $\sigma = 4.5 \times 10^{14}$ GeV). A sufficiently large tunnelling is only possible for $m^2 = 0$

Again, the requirement for a sufficiently long inflationary phase sets a stringent limit

$$\frac{m^2}{\sigma^2} \leq 4 \times 10^{-12}. \tag{9.64}$$

An extreme fine-tuning is necessary.

9.8.2 Inflation in Anisotropic, Inhomogeneous Cosmological Models

The calculations done so far are also quite insufficient with respect to the input from gravitational theory. All the models start with a homogeneous and isotropic FL universe at $t < 10^{-35}$ s. It seems important that the inflationary concept should be applied to a less smooth initial state – an anisotropic and inhomogeneous cosmological model at least. The philosophy of the inflationary universe is indeed that the universe starts in a chaotic phase, and that certain regions, where $\langle\phi\rangle \simeq 0$ is achieved, start to inflate – one such region then evolving to our present universe. This is a rather appealing concept. Not much has been done, however, to support this belief with calculations, equations, or results.

There have been some attempts to investigate inflation in an anisotropic universe. It is found that in homogeneous and anisotropic models (Bianchi type I, for example) the anisotropy is strongly reduced by an inflationary phase (a very clear description can be found in [Rothman and Ellis 1986]).

On the other hand, investigations of homogeneous, anisotropic models also indicate that the initial anisotropy decides the fate of the inflationary mechanisms. If the initial anisotropy is too large, then the re-entry into a thermal stage does not occur – but for reasonably small values the inflationary phase will end with a phase transition leading to a highly isotropic Friedmann universe [Barrow and Turner 1982; Demianski 1984].

These results have to be viewed with caution, since they make use of formulae derived and valid only for Euclidean field theories in flat space. A transfer of the formalism used in these investigations to the more general cases has not yet been achieved. For inhomogeneous models analogous results have been derived: if the inhomogeneities are connected with anisotropies, inflation is stopped for large initial anisotropies.

9.8.3 The Reheating Problem

The coupling between the vacuum energy of the Higgs field and radiation is responsible for the reheating of the universe after the inflationary phase and for the re-entry into an FL phase. It is important to describe this process as precisely as possible. Apart from the phenomenological introduction of a frictional term in the equations that we have discussed (Sect. 9.4) there have been serious attempts to study dissipative effects arising from the presence of quantum fields in a time-dependent classical background [Fujimoto et al. 1985; Morikawa and Sasaki 1984, 1985].

Quantum-theoretical production of particles is in general an entropy-conserving process. Therefore a certain amount of coarse-graining – i.e. collecting physical states into larger classes where the members of each class are considered to be indistinguishable from one another – has to be performed in the relevant density matrix. Entropy production is thus introduced by this coarse-graining, but the resultant dissipation coefficient depends on the way the coarse-graining is done. The dissipation terms obtained have the form $\varphi_c\dot{\varphi}_c$ (where $\varphi_c \equiv \langle\varphi\rangle$ are expectation values of the Higgs field φ); i.e. they are similar to the phenomological model discussed in Sect. 9.4.

Various other ideas of how to convert the vacuum energy into relativistic particles are being discussed. The perturbative decay of the scalar field, now commonly dubbed "inflaton" – [Abbott et al. 1982; Dolgov und Linde 1982; Davidson and Sarkar 2001], can be complicated if the decay products have self-interactions [Khlebnikov and Tkachev 1997; Prokopec und Roos 1997]. The transfer of energy from the inflaton to other fields in a non-perturbative way has been named "preheating" [Kofman et al. 1997]. Preheating, however, is not always an efficient energy-loss mechanism for the inflaton, when its decay products have self-interactions. I do not want here to give a detailed description of these more technical particle physics calculations. Real field-theoretical studies of the energy transfer from the inflaton to other particles have been performed only for toy models.

These investigations have to be considered as preliminary attempts. It is, for instance, not at all clear how to define particles in a time-dependent classical background, but in these calculations a specific particle picture has to be adopted.

Another problem with the appearance of radiation fields in de Sitter space is the choice of a time direction that is spatially homogeneous. Because of the many symmetries of de Sitter space it is possible for the reheated universe to start off in different temporal directions at different spatial locations [Ellis and Rothman 1986]. How is the homogeneity of the re-entry into a hot universe guaranteed?

9.8.4 Convexity and Gauge-Dependence of V_{eff}

The phase structure of the early universe is determined from a one-loop approximation to an effective potential V_{eff} [Guth 1981; Linde 1982]. But it has been realized [Symanzik 1970; Bender and Cooper 1983] that in thermal equilibrium the effective potential has to be convex, a property violated by the naive one-loop expansion schemes. If the exact effective potential is strictly convex, it cannot have the suggested double-well structure even when there is broken symmetry.

The effective potential in field theory is defined in the following way: write the partition sum

$$Z(J) = \left\langle e^{S(\varphi,J)} \right\rangle \equiv e^{W(J)}$$

as a functional of an external, classical field J. Usually the expected value $\varphi_c(J)$ is defined as $\partial W/\partial J \equiv \varphi_c(J)$, but from this the function $J(\varphi_c)$ may not be computable (Fig. 9.13). Therefore one should use

$$\Gamma(\varphi_c) \equiv \inf_J \left(W(J) - \varphi_c J\right),$$

and finally

$$V_{\mathrm{eff}}(\varphi_c) \equiv - \lim_{\Lambda \uparrow R^n} \frac{1}{|\Lambda|} \Gamma(\varphi_\Lambda), \tag{9.65}$$

where $\varphi_\Lambda = \varphi_c$ (constant) in Λ.

One can now easily convince oneself that W, $-\Gamma$ and V_{eff} are all convex. For W,

$$\frac{\partial^2 W}{\partial J^2} = \left\langle \varphi^2 \right\rangle - \left\langle \varphi \right\rangle^2 \geq 0. \tag{9.66}$$

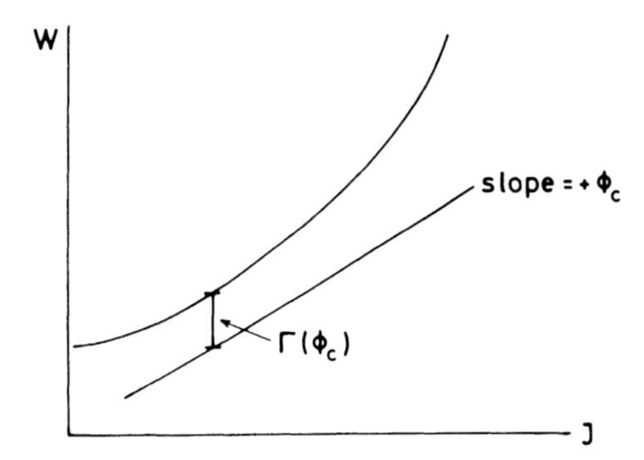

Fig. 9.13. The functional $\Lambda(\varphi_c)$ is defined as the infimum over J of the difference $W - \varphi_c J$

Γ is the infimum over linear functions, hence it must be concave. Therefore V_{eff} must be convex.

A convex V_{eff} must be shaped as in Fig. 9.14, where the dotted line corresponds to a mixture of different phases (Maxwell construction).

In spin systems at low temperatures $W(J)$ has a cusp at $J = 0$, and thus a dotted line, i.e. a phase mixture, exists (Fig. 9.15). It has been shown, however [Elitzur 1975; De Angelis et al. 1978], that in Higgs models (without

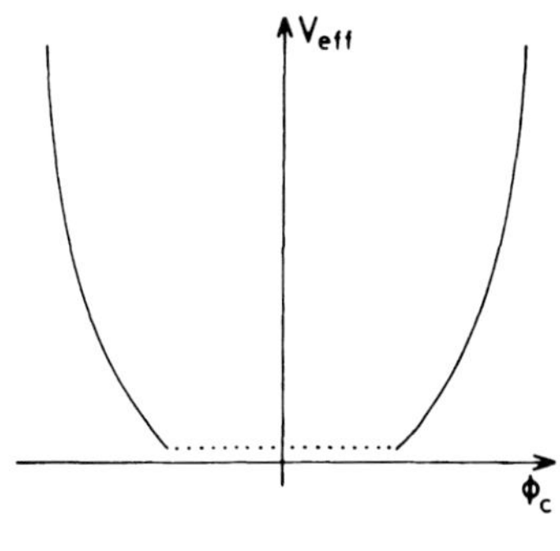

Fig. 9.14. A convex effective potential can describe a phase mixture in equilibrium along the *dotted line*

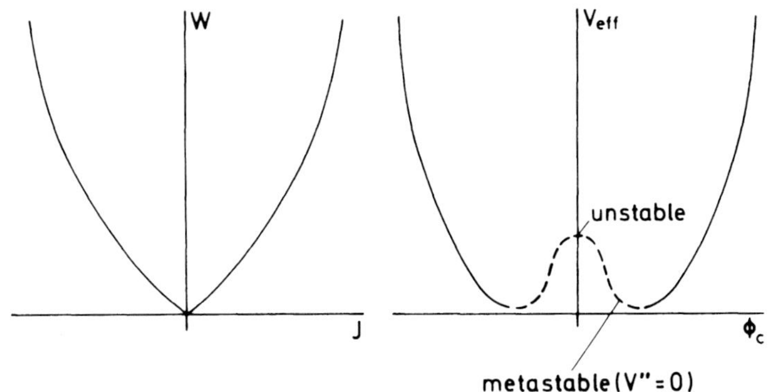

Fig. 9.15. A cusp of $W(J)$ at $J = 0$ gives rise to a region where a metastable phase may be described

gauge fixing) coupled to gauge bosons, W does not have a cusp at $J = 0$. Therefore $V''_{\text{eff}} > 0$, and the dotted line disappears. The loop expansion of the effective potential cannot then be used to describe the phase structure of the thermal states in the early universe. In a fixed gauge, on the other hand, the effective potential $V_{\text{eff}}(\varphi)$ depends on the gauge chosen, and its physical interpretation is unclear.

Eventually the collaboration of many people will bring about a perfect model of inflation

Thus the conclusions drawn for the inflationary phase of the universe do not appear very trustworthy. But perhaps we should not be too strict. At least the success of the Ginzburg–Landau theory for superconductivity indicates that a metastable phase may be described reasonably well by a double-hump potential.

Investigations of gauge theories on a lattice have produced both encouraging and discouraging results. In non-compact, abelian Higgs models there is a non-vanishing expectation value, $\langle\varphi\rangle \neq 0$, in a specific gauge – in the Landau gauge [Kennedy and King 1985; Borgs and Nill 1986a,b]. In such models there is spontaneous symmetry-breaking associated with the Higgs phase. Therefore $\langle\varphi\rangle$ can be interpreted as an order parameter for the phase transition from $\langle\varphi\rangle = 0$ to $\langle\varphi\rangle = \sigma$. In several examples of compact abelian Higgs models, on the other hand, it has been shown that the expectation of the Higgs field vanishes identically [Borgs and Nill 1986a; Kennedy and King 1985].

A derivation of the Coleman–Weinberg potential in terms of gauge-invariant quantities would be highly desirable, since at present it is based completely on a one-loop expansion.

9.9 Concluding Remarks

In the preceding sections I have tried to give a fair account of the present status of the inflationary-universe idea. It may have become clear that the subject is still very ill-defined, under constant need of revision – a steady succession of collapses and resurrections. Although a workable model can be constructed, it seems to require a rather unnatural fine-tuning of microphysical parameters. Thus at present the effort is spent mainly in finding an effective potential with all the desired properties – and not worrying about the relation to a fundamental theory. I want to stress again, however, that an effective potential computed in a one-loop approximation may not be appropriate for the discussion of the phase structure of theories with spontaneously broken local gauge symmetries. Another important problem that remains to be solved in all inflationary models is the thermalization mechanism of the vacuum energy density.

An amusing detail concerning entropy generation in the phase transition after the inflationary epoch should be mentioned here [Straumann 1984a]. For the temperature evolution in the adiabatically expanding universe in the symmetric phase, one has the following equation:

$$\left(\frac{\dot{T}}{T}\right)^2 = \left(\frac{4\pi^3}{45} g(T) \frac{T^2}{M_{Pl}^2} - \frac{K}{S^{2/3}} \frac{2\pi^2}{45} g(T)^{2/3}\right) T^2. \tag{9.67}$$

For $K = 0$, $K = -1$ this equation is well defined. But for $K = +1$, the equation remains well defined only if

$$S > \left(\frac{M_{Pl}}{T}\right)^3 (2\pi)^{-2/3} g^{-1/2} \simeq 10^{-4} \left(\frac{M_{Pl}}{T}\right)^3 . \tag{9.68}$$

For $S \approx 1$, this implies $T > 10^{-4/3} M_{Pl} \approx 10^{17.6}\,\text{GeV}$, i.e. the $K = +1$ universe cannot be used for the inflationary scenario because it collapses before T has dropped to $10^{-4/3} M_{Pl}$. So one has the curious fact that only $K = 0, K = -1$ FL universes can undergo an inflationary phase. A $K = +1$ universe reaches the onset of inflation only if $S > 10^{-4}(M_{Pl}/T)^3 \approx 10^{11}$!

The claim that the concept of the inflationary universe can explain the observed large-scale homogeneity and isotropy of the universe has to be supported by actual computations – one must follow the evolution of anisotropic and inhomogeneous models through an inflationary phase, and see whether one arrives at a smooth Friedmann universe.

The exponential expansion does not lead to a decrease in the vacuum energy density. Therefore the energy of the expanding system increases tremendously. This, at first sight curious, fact has its roots in the non-conservation of the total energy in general relativity. For a Friedmann universe the continuity equation (stemming from $T_{\mu\nu};^{\nu} = 0$) is

$$(\varrho R^3)^{\cdot} = -p(R^3)^{\cdot}.$$

The equation of state for the vacuum is unusual,

$$p_v = -\varrho_v.$$

Positive energy means negative pressure, and vice versa. This equation of state satisfies the continuity equation – for constant ϱ_v – identically. The expansion energy is gained from the negative pressure. Remember that the exponential expansion phase will only occur if the energy–momentum density is dominated by the background vacuum energy density.

It seems that very large anisotropies or inhomogeneities prevent the occurrence of an inflationary phase (cf. Sect. 9.8). In the philosophy of inflationists this may well be accommodated by the picture of a very chaotic global space-time structure - where some tiny part of the cosmos undergoes an inflationary epoch only when its structure is smooth enough. This small part of the universe can then evolve to the smooth FL type structure observed by us. It seems difficult to introduce a quantitative aspect into such a romantic world-view.

While most questions concerning inflationary universe models must be answered by elementary particle theories, the aspects of homogeneity and isotropy have also been investigated along other paths.

The "mixmaster universe" [Misner 1969] proposes smoothing out a chaotic initial state by dissipative effects (e.g. neutrino viscosity, hadron collisions, particle creation). The aim is to produce the observed entropy per baryon of $\simeq 10^{10}$ and the presently observed isotropy. This view suffers from the difficulty that there are specific initial anisotropies which are not smoothed out by the dissipative effects [Stewart 1969].

When looking in somewhat more detail at the thermodynamic system which is presented by the expanding, time-asymmetric universe one may conclude, however, that all these attempts are quite unsatisfactory [Penrose 1979]. Penrose argues against an initial chaos and in favour of a uniform big-bang singularity. The initial state of the expanding universe should have a lower entropy than the final one. Penrose points out that the entropy per baryon of 10^{10} is not a large figure, when gravitational entropy is taken into consideration.

To get an idea of the entropy of gravitation we may consider the entropy of a black hole. There is a formula [Bekenstein 1973; Hawking 1975] which sets the intrinsic entropy S proportional to the surface area A of a black hole's absolute event-horizon

$$S/k_B = A\, c^3/4\hbar G \tag{9.69}$$

($A = 16\pi(G^2M^2/c^4)$ for a Schwarzschild black hole).

Using this formula, we can easily find out that a black hole of $1M_\odot$, formed from baryons, has an entropy per baryon of $\simeq 10^{18}$. A $10^{10}M_\odot$ galaxy, therefore, collapsing mainly to a large central black hole, leads to a value of 10^{28}. Finally, a closed, recollapsing universe containing $\simeq 10^{80}$ baryons that undergoes collapse by a mutual swallowing process of black holes reaches an entropy per baryon of $\simeq 10^{40}$. This large value is mainly produced by the irregularities, the ripples and wrinkles of space-time. The initial entropy per baryon could well have been close to that value had the universe started out in a completely chaotic state. Then the problem is to explain the absurdly low value of 10^{10} observed now. Why is only a fraction $\sim 10^{-30}$ of the available entropy used?

This enormous discrepancy seems to indicate that the explanation for $s/n_B \sim 10^{10}$ cannot just lie in particle physics. Penrose concludes that these values also suggest a complete lack of chaos in the initial geometry. A low-entropy constraint must be imposed on the initial state. But the matter and radiation in the early universe apparently were in thermal equilibrium. So the matter distribution cannot account for the low entropy. Rather a very special initial space-time geometry must be assumed. A precise local condition can be formulated which expresses this idea. But to explain this local condition an appeal must probably be made to a more fundamental theory unifying quantum mechanics and general relativity.

Just how far away we are still from such a fundamental theory is illustrated by the fundamental problem of the cosmological constant. The vacuum does not now support exponential expansion, so the cosmological constant at present must be close to zero.

The observations of the CMB anisotropies in the Boomerang and Maxima (see Sect. 11.6) experiments suggest that $\Omega_{\text{tot}} \approx 1$, and that $\Omega_\Lambda \approx 0.7$. This corresponds to

$$\frac{\Lambda}{8\pi G} \simeq (10^{-12}\,\text{GeV})^4$$

(see e.g. [Weinberg 1989]). This is more than one hundred orders of magnitude below the GUT vacuum energy density. Present-day elementary particle theories have transformed the vacuum into a complicated structure of quantum fluctuations and condensates – because of the many spontaneously broken symmetries. Why is this vacuum transparent to gravity? The changes of the energy densities in the breaking of a GUT, the electroweak gauge symmetry, as well as of the chiral symmetry of QCD are of the order of $(10^{15}\,\text{GeV})^4$, $(10^3\,\text{GeV})^4$, and $(10^{-1}\,\text{GeV})^4$, respectively. In comparison with these numbers the tiny value of the present-day cosmological constant remains a mystery. It seems very courageous to exploit just this mystery to "explain" an early inflationary phase.

Following [Weinberg 2001], we can see that now there are two cosmological constant problems. Why is Λ so small? Why is Ω_Λ just now comparable to the mean mass density? The solution of these problems involves a number of *ad hoc* assumptions which often appear more complicated than the problem they are intended to solve. One attempt is to imagine some scalar field coupled to gravity in such a way that $\frac{\Lambda}{8\pi G} \equiv \varrho_v$ is automatically cancelled or nearly cancelled when the scalar field reaches its equilibrium value. Weinberg has presented arguments to show why this probably does not work [Weinberg 1989]. A second approach is an appeal to an underlying symmetry principle.

Recently an idea, called "quintessence", has become popular [Peebles and Ratra 1988; Wetterich 1988; Zlatev et al. 1999]. The idea of quintessence is that the cosmological constant is small because the universe is old. One considers a uniform scalar field with potential $V(\varphi)$, and postulates a value of φ such that $V_{,\varphi} = 0$. Then φ should approach this value, and close to it change only slowly with time. The matter and energy densities are steadily decreasing, and eventually $V(\varphi)$ dominates, and the universe starts an exponential expansion with $H^2 \simeq \frac{8\pi G}{3} V(\varphi)$.

The problem is, of course, why $V(\varphi)$ should be small or zero at the value φ for which $V'(\varphi) = 0$. You can, however, set it up that way, if you choose your potential in a clever way.

So-called "tracker" solutions have been found, where the cross-over from an early ϱ_m-dominated expansion to a ϱ_φ-dominated expansion can be arranged such that the cross-over occurs just after recombination, and the desired properties arise [Armendariz-Picon et al. 2000].

The anthropic principle can, of course, also be involved. If the observed big bang is just one member of an ensemble of cosmological solutions, and if the vacuum energy density $V(\varphi)$ varies among the different members of the ensemble, then the value observed by us is the way it is because we just happen to be in that particular part of the universe. We are here because this part is suitable for the evolution of intelligent life. This sounds very much like physicists throwing the towel into the ring.

But we should not give up yet – apart from the *ad hoc* constructions mentioned above, to my mind no serious attempt at a solution has been tried so far.

A story told occasionally by W. Pauli [Ch. Enz. A. Thellung: personal communication] fits in this context. Very early in his professional career he had asked himself whether the zero-point energy of the electromagnetic field could be gravitationally effective. Therefore he calculated the curvature radius of an Einstein universe with an energy density given by the zero-point energy of the electromagnetic field with a cut-off at the classical electron radius. He was quite amused to find that this universe "would not even reach out to the Moon".

To conclude, we may say that the idea of the inflationary universe in general is very attractive, explains various cosmological puzzles in a natural way, but we must also stress that most of the detailed models do not inspire much confidence. Although it is not clear at present which features of the concept of the inflationary universe will survive eventually, there is hope that the connection between the laws of particle physics and the properties of the large-scale cosmological structure expressed in this model will initiate interesting and fruitful research activity. Let us hope for models of the inflationary universe that avoid the pitfalls discussed here, but which retain all the good points mentioned.

Exercises

9.1 Consider the mass M contained in a sphere of diameter $\lambda(t)$:

$$M(\lambda) = \frac{4\pi}{3}\varrho(t)\left(\frac{\lambda(t)}{2}\right)^3 .$$

When $\lambda(t) > cH^{-1}(t)$ the scale λ is no longer causally connected. Compute the redshift $z_c(M)$ when this happens. Show that for an object like the Coma cluster $z_c(M)$ is less than z_{eq}, the redshift of the epoch when matter and radiation energy densities are equal.

9.2 Assume that the gravitational potential term $\sim G\varrho$ balances the curvature term K/R^2 at the Planck time in a FL model. Use the formulae of Chap. 1 to compute the lifetime of such a model for $K = +1$. For $K = -1$ calculate the decrease in the initial density.

9.3 Determine the lifetime of a finite $K = +1$, $p = \Lambda = 0$, FL model as a function of the total mass. What happens as the total mass goes to zero? Compare this to the "no matter" limit of $K = 0$, $K = -1$ models. Try to interpret the result for $K = +1$ in terms of a fine-tuning problem.

9.4 Calculate $\dot{\varphi}$ from (9.41), insert it into (9.39), and show that $\ddot{\varphi}$ is small.

9.5 Use the (λ, y) coordinates of exercise 1.5, and write down the wave equation for a scalar field in these coordinates. Find the classical solutions corresponding to plane waves (look up [Birrel and Davies 1984]). Quantize the scalar field in this harmonic composition. Can the modes be interpreted as particles? Compute $\langle\varphi^2\rangle$ for the quantized field, and recover the result of (9.54).

9.6 Consider the idea of quintessence in terms of an FL model with matter and radiation ϱ_m, and a scalar field φ.

Take $\varrho_\varphi = \frac{1}{2}\dot{\varphi}^2 + V(\varphi)$ and $V(\varphi) = M^{4+\alpha}\varphi^{-\alpha}$, where $\alpha > 0$ and M is an adjustable constant. Let the scalar field start with a value $\ll M_{Pl}$, and with $V(\varphi), \dot{\varphi}^2 \ll \varrho_m$.

Take an equation of state $p_m = \frac{1}{3}\varrho_m$. Show that $\varrho_\varphi \propto t^{-2\alpha/(2+\alpha)}$, and $\varrho_m \propto t^{-2}$ using the equation of motion for φ, and the Hubble parameter

$$H^2 = \frac{3}{8\pi G}(\varrho_\varphi + \varrho_m).$$

Eventually ϱ_φ becomes $\simeq \varrho_m$, and then larger. Show that eventually the expansion factor

$$\log R(t) \propto t^{4/(4+\alpha)}.$$

What is the condition on M if $\varrho_\varphi \approx \varrho_m \approx \varrho_{c,o}$ (critical density at the present epoch)?

What happens if we add a constant $\sim M_{Pl}^4$ to $V(\varphi)$?

Part III

Dark Matter and Galaxy Formation

"You will not apply my precept" he said, shaking his head. "How often have I said to you that when you have eliminated the impossible, whatever remains, *however improbable,* must be the truth?"

The Sign of Four, A.C. Doyle

Among the major impacts of particle physics on cosmology have been the new possibilities opened up for solving the problem of galaxy formation. Various weakly interacting non-baryonic particles may contribute significantly to the mass density of the universe, and may even dominate the processes that lead to the clumpy matter distribution observed at present. This fascinating prospect that the structure at very small distances may determine the very large-scale structure will provide the leitmotiv for the three chapters in this part. The cartoon of the snake biting its tail may be an appropriate expression of the basic ideas that will be discussed.

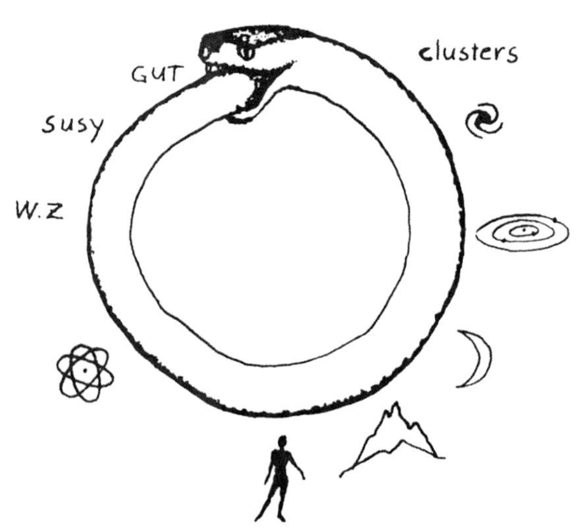

Links between microphysical and cosmic scales are expressed by the symbol of the snake biting its tail. The final goal is the unification involving the scale of the universe and the Planck scale (after [Primack and Blumenthal 1983])

10. Typical Scales – From Observation and Theory

> "Since as the Creation is, so is the Creator also magnified, we may conclude in consequence of an infinity, and an infinite all-active power, that as the visible creation is supposed to be full of siderial systems and planetary worlds, so on, in like similar manner, the endless immensity is an unlimited plenum of creations not unlike the known .... That this in all probability may be the real case, is in some degree made evident by the many cloudy spots, just perceivable by us, as far without our starry Regions, in which tho' visibly luminous spaces, no one star or particular constituent body can possibly be distinguished: those in all likelyhood may be external creation, bordering upon the known one, too remote for even our telescopes to reach."
>
> *Thomas Wright (1750)*

The aim of most modern theories of the structure of the universe is to explain the extreme inhomogeneities observed at present – the size and shapes of galaxies, and the clustering hierarchy of galaxies into groups, Clusters, and superclusters – as a result of the growth by gravitational interaction of initially small perturbations in a smooth background density.

The cosmological model determines the initial increase with time of small density perturbations. The power spectrum of the density fluctuations and the properties of the background cosmological model have a strong influence on the formation of structure. But clearly, there must be some scale, where local physics (radiation-, hydrodynamics) becomes very important in shaping the objects. What is that scale? Is it at the size of clusters, of individual galaxies, or even smaller? Once local physics takes over, there is the chance that some universal properties appear which are independent of the arbitrariness of initial and boundary conditions.

We do not yet know whether the size of individual galaxies is determined by cosmological effects or by effects of the local physics. There are two different general schemes that have been proposed. In some theories, the formation of structures started on small scales and proceeded step by step to the larger structures. Galaxies cluster hierarchically to build groups; groups form clusters, and clusters merge to form superclusters.

This "hierarchical" clustering model [Peebles 1980] has recently found a successful representation in models making use of "cold dark matter" (CDM; see Chaps. 11, 12). In other models, the largest structures collapsed anisotropically, forming thin, dense sheets of matter. This aspect has given to this scenario its popular name, the "pancake" picture [Zel'dovich 1970;

Doroshkevich et al. 1980; Buchert 1989a]. The pancake scenario has its natural representation in "hot dark matter" models (HDM; see Chaps. 11, 12). Recent observations of the abundance of high-redshift objects seem to exclude HDM models.

In this chapter we will try to give an account of the observational status of the large-scale structures. In addition, we shall discuss the arguments for the existence of certain typical scales from basic theoretical considerations, as well as the evidence from observations for such scales.

10.1 The Clustering of Galaxies

10.1.1 Visual Impressions

Only a few neighbouring galaxies of our Milky Way can be seen with the naked eye as diffuse light spots. But large telescopes show us that the sky is literally filled with galaxies. A very impressive observation of this type is shown in Fig. 2.12: the "Hubble Deep Field". The Hubble Space Telescope was pointed for 10 days at a seemingly empty place in the sky, and eventually about 2500 galaxies were registered in that small field of (one arcminute)^2. If this is a typical number, then we may conclude that there are at least 10^{10} galaxies in our observable section of the universe. The galaxies in our neighbourhood are distributed in a very inhomogeneous way. They are apparently arranged in double systems, multiples, groups with up to 100 members, rich clusters with more than 1000 members – with masses of 10^{15} to $10^{16} M_\odot$ and diameters of 5 to 10 Mpc – and superclusters with diameters of 30 to several 100 Mpc. A map of the distribution of the galaxies brighter than the

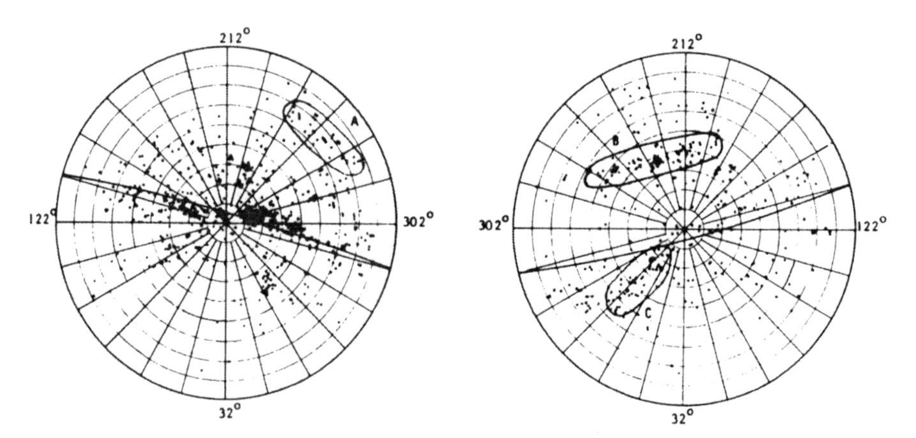

Fig. 10.1. Galaxies brighter than 13^{m}. The left-hand picture shows the northern, the right-hand picture the southern galactic hemisphere. Galactic poles are at the centres, and the spacing of the circles corresponds to 10°. The latitude is shown at the circumference (adapted from [Oort 1983])

13th photographic magnitude, prepared as early as 1932 by Harlow Shapley and Adelaide Ames, showed many of these features (Fig. 10.1).

Soon the astronomers began to collect galaxies systematically. One of the first was the Swiss astronomer Fritz Zwicky, who worked at Caltech in the USA, and who compiled with his collaborators in the fifties a galaxy catalogue which contained all galaxies down to an apparent brightness (in the blue band) of $m_B = 15.7$. This catalogue contains about 30 000 galaxies. Only those glaxies are included in the catalogue whose optical flux is above the threshold given by $m_B = 15.7$. Even very faint near galaxies, but only bright distant ones, are included. This selection effect produces a strong inhomogeneity of the galaxy sample.

A rough estimate shows that reducing the threshold by one magnitude to $m_B = 16.7$ would result in about four times as many galaxies in the catalogue. At present there are sky surveys (with UK Schmidt telescope plates) reaching a limiting magnitude of 21. The Hubble Deep Field is estimated to reach down to magnitude 30 to 32.

Automatic measuring machines are used to obtain large catalogues of exact positions, magnitudes, and information on the type of individual galaxies. The so-called "APM-catalogue" registers galaxies down to $b_j = 20.5$ in the southern hemisphere (declination $\delta < -20°$). This catalogue contains about 4×10^6 galaxies [Maddox et al. 1990a,b].

Fig. 10.2 is derived from two catalogues which list galaxies down to a limiting diameter of one arcminute. This stereographic projection displays the distribution of the bright galaxies on the celestial sphere [Mo 1991]. A small strip around the plane of the Milky Way cannot be observed well, because of absorption by gas and dust inside our galaxy. In galactic coordinates this zone is given by $-20° < b < -2° \, 30'$ (b: galactic latitude; l: galactic longitude). Even in projection the galaxy distribution exhibits distinct patterns.

From small concentrations to large superstructures which extend across the whole sky there is a variety of clustering phenomena. The Virgo cluster in the northern hemisphere (at $l = 290°, b = 70°$) seems to be connected even to our own Local Group galaxies. At $l = 310°$ and $b = 30°$ another concentration of galaxies contains Centaurus, and Hydra to the east ($l = 270°, b = 25°$) and a newly measured system called the "Great Attractor". The well-known Coma cluster is situated close to the northern galactic pole. All these condensations in the galaxy distribution seem to belong to an elongated structure spanning the northern hemisphere along $l = 130°$ and $l = 310°$. This elongated concentration of galaxies determines the supergalactic plane.

An outstanding feature in the southern galactic hemisphere extends around the Perseus-Pisces supercluster ($l \sim 140°, b \sim -30°$) in the shape of a long filament towards Pegasus and Fornax-Eridanus (near $l = 180°$, $b = -50°$). Another remarkable structure begins near Pavo and surrounds the Indos and Capricorn regions in the southwest direction. All these formations appear elongated and rather irregular.

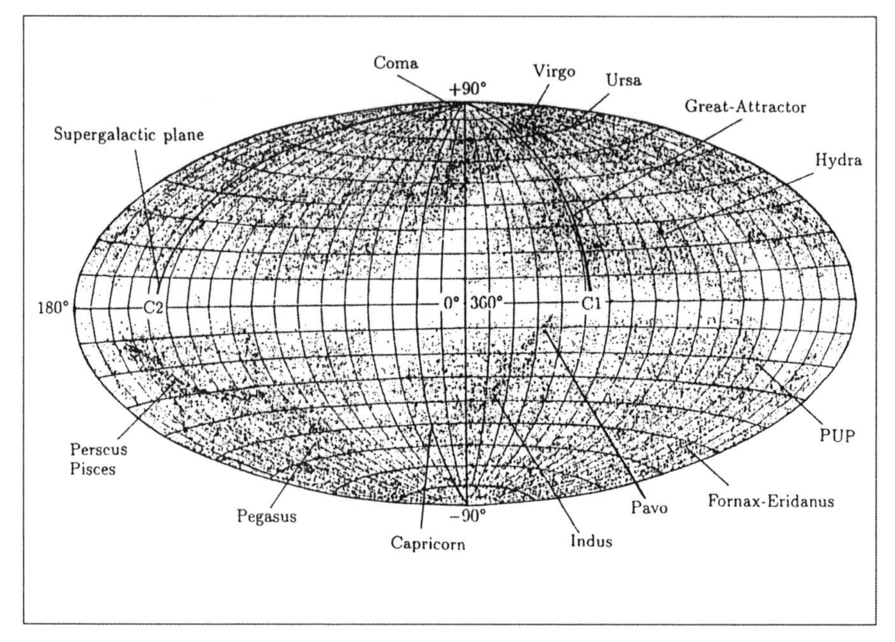

Fig. 10.2. The all-sky distribution of galaxies as derived from a merger of the UGC and ESO catalogues (from [Mo 1991])

In Fig. 10.2 we can also recognize possible connections between these superstructures. Thus the supergalactic plane seems to be linked across point C1 of the galactic plane with the Pavo-Indus-Capricorn condensation. Another example is the possible link of the Perseus-Pisces supercluster with the supergalactic plane at C2. In addition to those regions of high galaxy density there are large, sparsely populated areas. These seem to be less elongated than the high density regions.

The true pattern of the galaxy distribution can only be obtained from the three-dimensional, spatial distribution, since the projection on the sky distorts the real structures. To obtain a three-dimensional image, we must determine the distances to the galaxies. This, however, is time-consuming and imprecise with present-day methods. Therefore one applies the Hubble law

$$dH_o = v_{\text{exp}} = cz$$

to obtain a measure for the distance d from the redshift z. A 3D representation of the galaxies in "redshift space" is derived this way, where the coordinates are the redshift (usually presented as expansion velocity $v_{\text{exp}} = cz$) and the two position coordinates on the celestial sphere (e.g. galactic longitude l and latitude b).

In Fig. 10.3 such a representation is given in pictures corresponding to different redshift intervals. The visual impression is of a kind of cellular

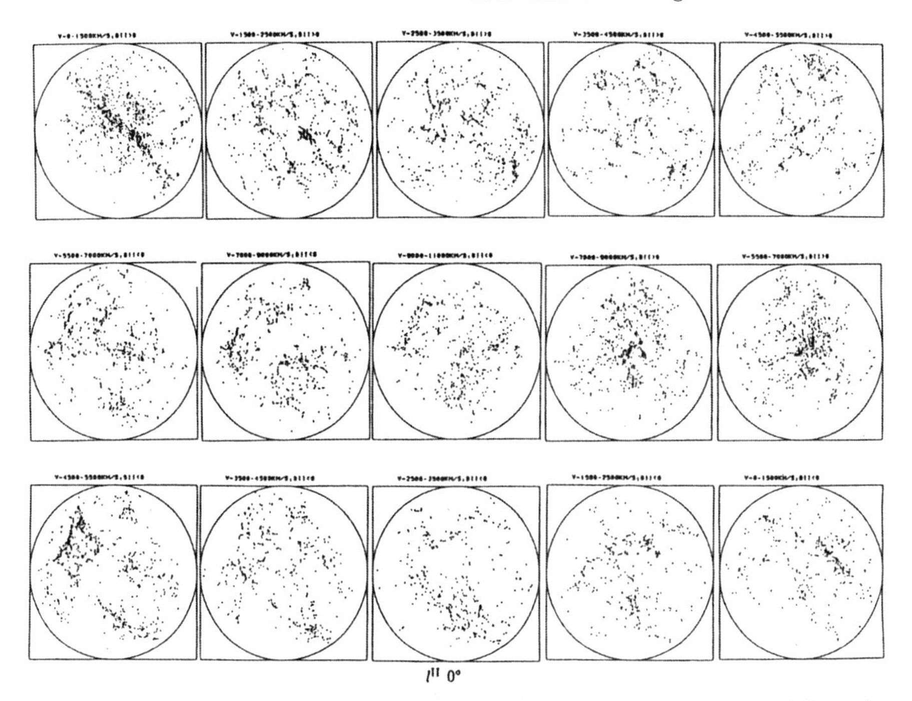

Fig. 10.3. The redshift slices of the galaxy distribution are constructed from the 1988 version of the ZCAT. Although the number of galaxies is quite small compared to present surveys, it serves well to demonstrate the way in which distinct structures appear in redshift space. The picture should be compared to the sky projection shown in Fig. 10.2. Galactic latitude varies from 0° on the big circle to 90° (-90° at the centre) uniformly. Galactic longitude runs clockwise from 0° at the centre of the bottom edge (from [Mo 1991])

structure, where the galaxies are concentrated in thin sheets surrounding large more or less empty regions. For small redshifts the plane of the local supercluster can clearly be seen.

For cz between $3500\,\mathrm{km\,s^{-1}}$ and $5000\,\mathrm{km\,s^{-1}}$ several significant features can be discerned. This range of values corresponds to the redshift of the Perseus-Pisces supercluster in the southern hemisphere, and to the redshift of a region between Coma and Virgo in the northern hemisphere. The galaxies in the Perseus-Pisces supercluster seem to be concentrated in the wall of a big cell with a dimension of $\sim 100h^{-1}$ Mpc. The interior of this cell does not contain any bright galaxies. The galaxy distribution in the northern hemisphere is smoother at those distances; there are several voids with a smaller diameter $\sim 25h^{-1}$ Mpc.

Big redshift surveys have become possible because of the dramatically increased sensitivity of modern detectors. CCD receivers, i.e. detectors based on silicon crystals, register up to 70% of the incoming photons. In contrast, only 1 in every 1000 photons hitting a photographic plate leads to a reaction.

At the time of Hubble a few hours of good visibility were necessary to obtain the spectrum of a bright galaxy with a 2.5 m telescope. Today the same measurement takes just a few minutes on a small 1.5 m telescope. Fibre optics permit simultaneously measurement of many objects in the field of view. In 1956 only about 600 galaxies had a measured redshift, in 1976 around 2700, and by 1990 there were more than 30 000 galaxies with known redshift.

One big survey programme has been carried out at the Center for Astrophysics in Boston [Geller and Huchra 1989]. The first publication from this group displayed a systematic collection of the redshifts of all galaxies of apparent magnitude $m \leq 15.5$, in a region of the sky measuring 6° ($26.5° < \delta < 32.5°$) by 117°, containing about 1100 galaxies [de Lapparent et al. 1986]. The map of this slice, which contains the Coma cluster, is displayed in Fig. 10.4. It gives an impression of strong clustering of galaxies into thin sheets, filaments, or ridges, whilst space is dominated by large, apparently

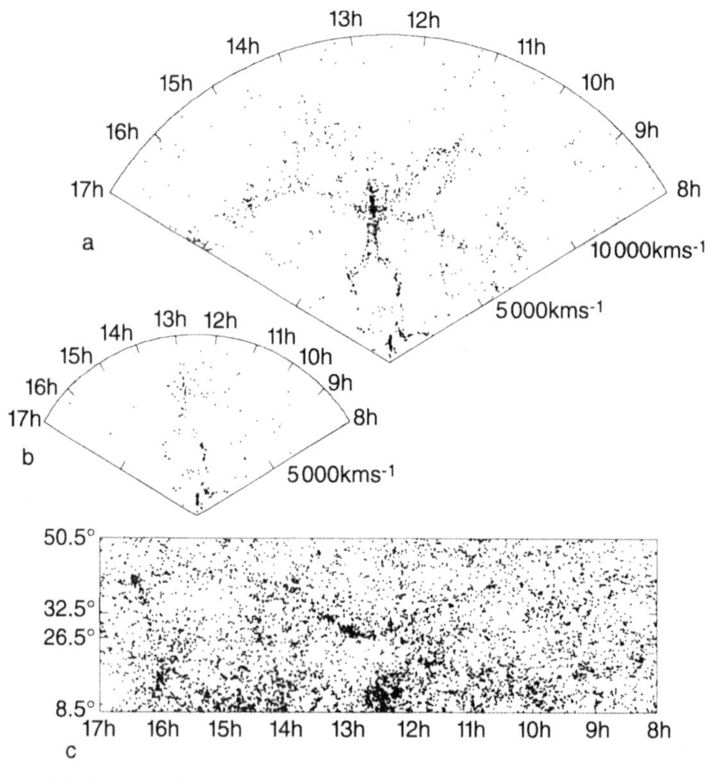

Fig. 10.4. (**a**) A map of the redshifts (expressed as a velocity cz) of 1061 galaxies plotted versus right ascension in a wedge of declination δ, within the interval $26.5° < \delta < 32.5°$. $m_B \geq 15.5$ and $V \geq 15\,000\,\mathrm{kms}^{-1}$. (**b**) The 182 galaxies with $m_B \geq 14.5$, $V \geq 10\,000\,\mathrm{kms}^{-1}$ in the same region. (**c**) The projected map of 7031 objects with $m_B \geq 15.5$, listed by [Zwicky et al. 1958] in the region $8^{\mathrm{h}} \geq \alpha \geq 17^{\mathrm{h}}$; $8.5° < \delta \leq 50.5°$ [de Lapparent et al. 1986]

empty quasi-spherical voids. The universe appears to have a bubbly, foam-like structure. The large Bootes void – a 50 Mpc hole that does not contain any luminous galaxies – which sparked much discussion when it was discovered, is not an exception; structures like it are the rule.

The largest bubbles have diameters of 30 to 50 Mpc. The edges of the layers of galaxies are quite sharply defined. How can such objects arise in a natural way? Various attempts at an explanation will be presented in the following chapters.

A suggestion which can lead very efficiently to voids is "explosive galaxy formation" [Ostriker and Cowie 1981; Ikeuchi 1981]. The blast wave from a large number of supernovae in a newly formed galaxy propagates outward and sweeps up a spherical shell of gas. An almost empty hole is left behind, and the compressed shell fragments into additional galaxies. These again undergo the same process – eventually a collection of bubbles develops which fills most of space, while galaxies sit only on the shell boundaries.

But, while attractive in principle, this theory labours under severe difficulties: the energy output is not sufficient to create the large voids observed. Usually supernovae clear out holes in interstellar space which are tens or sometimes hundreds of parsecs in diameter. Holes on a scale of 10 Mpc presumably need billions of supernovae exploding coherently, i.e. during the time-scale on which sweeping out occurs $\sim 10^8$ years. In addition, the explosive scenario itself needs specific initial conditions that would lead to the formation of the seed galaxies starting the explosion. Instead of this explosive-amplification picture we want to discuss in detail attempts to explain these structures by the work of gravitational instabilities.

Motions relative to the Hubble flow distort the picture of the spatial distribution presented in Fig. 10.4. Such peculiar velocities are especially large in rich clusters of galaxies. For large redshifts ($cz > 5000\,\mathrm{km\,s^{-1}}$) the peculiar velocities no longer strongly distort the pattern, and the true spatial distribution emerges. The statistical analysis of the redshift distortions gives an estimate of the rms peculiar velocity of the galaxies, if the data are good enough (see Chap. 12). Another point of caution is the influence of selection effects in a magnitude-limited survey. Many nearby galaxies are collected in such a survey, and the most distant galaxies sampled are also the most luminous ones. If the 3-dimensional distribution of galaxies is different for different luminosities, a systematic distortion of the derived maps could result. Such uncertainties, however, can never be completely excluded. The interesting result of strong clustering and large voids in the galaxy distribution seems to become more and more firmly established.

Another slice to the north of the Coma cluster [Huchra et al. 1986] intersects the edge of the huge Bootes void which is 124 Mpc in diamter [Kirshner et al. 1981], and the redshift distribution exhibits the same type of sharply defined bubble walls. The CfA redshift survey has been completed for several additional 6° slices in declination adjoining the slice shown in Fig. 10.4. About 15 000 galaxies with measured redshifts are contained in this catalogue which covers the northern sky for latitudes > 40° and a limiting magnitude of $m_B = 15.5$. The features found in Fig. 10.4 turn out to be typical: galaxies lie on sharply defined surfaces surrounding large, almost empty regions. These "voids" are underdense with respect to the mean density by a factor of 5 – 10.

There is no evidence that faint galaxies tend to fill out the voids, instead the test observations made so far indicate that they follow the inhomogeneous structures defined by the bright galaxies.

The most spectacular structure is a broad band of galaxies that extends throughout the 120° in right ascension of the survey between redshifts of 7000 km s^{-1} and 10 000 km s^{-1}. Geller and Huchra have named this structure the "Great Wall". It can be seen in all slices, especially impressive when several slices are superimposed on each other (in Fig. 10.5 this is shown for a declination range from 26.5° to 44.5°). At least $150h^{-1}$ Mpc wide and $10h^{-1}$ Mpc thick, this Great Wall is only limited by the extent of the survey. About half the number of measured galaxies seem to belong to this structure. The visual impression of a "Great Wall" may be somewhat enhanced by selection effects, such as the magnitude limit of the catalogue which leads to a strong reduction of the galaxy density beyond $cz \sim 10\,000$ km s^{-1}. One should also test the statistical significance of this object. What is the probability that it is an accidental arrangement of several smaller "walls"?

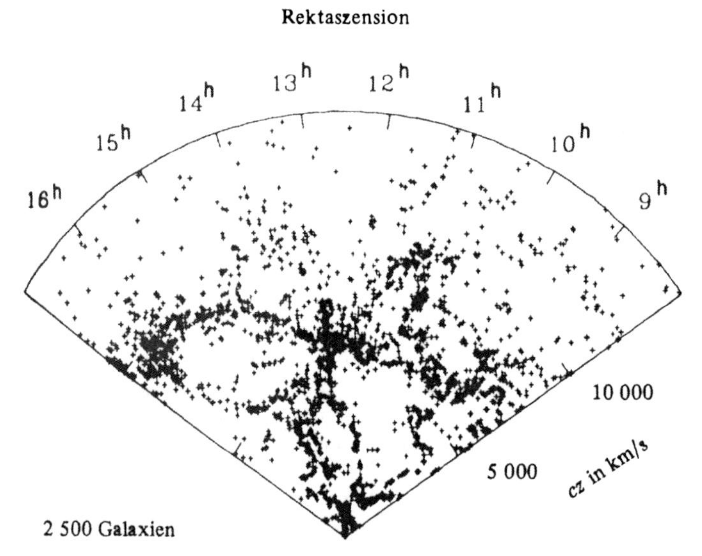

Fig. 10.5. Galaxies from the CfA redshift survey. The declination ranges between 26.5° and 44.5° are projected into one plane. The redshift z is plotted against right ascension. The band of galaxies around $cz \sim 7500$ km s^{-1} is the "Great Wall", a structure at least as large as the dimension of the survey (from [Geller and Huchra 1989])

On the other hand, the result ties in nicely with several other detections of large-scale objects.

A percolation analysis of the angular distribution of Abell clusters [Mo and Börner 1990] showed some evidence for superstructures extending over $350h^{-1}$ Mpc. This analysis was carried out with a smoothed-out density field derived from the point distribution of Abell clusters. The area on the celestial sphere is divided up into cells of a certain size (e.g. 2° by 2°), and for each cell a mean density is calculated. The density profiles of clusters are approximated by a Gaussian profile $\exp(-\theta^2/\lambda^2)/\pi\lambda^2$, where θ is the angular distance, and λ is a smoothing length which has to be chosen in an optimal way. Cells are marked "black" (or "white") when their mean density exceeds (or falls below) a preset threshold density. Black cells belong to a supercluster if they touch each other at a corner or an edge. Then a statistical percolation analysis of these black-cell clusters is compared with the results from a random simulation. There are significant differences up to a mean length of 50° corresponding to $350h^{-1}$ Mpc. The density contrast of these structures is decreasing with increasing scale, such that the superclusters gradually blend into a homogeneous background for dimensions larger than $350h^{-1}$ Mpc.

A distribution of isodensity contours is shown in Fig. 10.6 [Springel 1996] for the galaxies in the PSCz catalogue which contains IRAS point sources [Saunders et al. 2000].

A deep redshift survey of Abell clusters [Huchra et al. 1991] has revealed the existence of a large void in the sample of size $\sim 200h^{-1}$ Mpc. In addition, it was confirmed that the two-point correlation function for clusters is larger than the galaxy–galaxy correlation. The correlation length for Abell clusters is $\sim 20h^{-1}$ Mpc.

In this context the result of a deep survey in a narrow angular range of $\sim 45'$ is remarkable. In such a "pencil beam" survey in the direction of the galactic north and south pole the redshifts of galaxies up to $z = 0.4$ (CfA: $z \leq 0.05$) have been measured. This corresponds to Hubble distances (cz/H_0) of $1200h^{-1}$ Mpc! There are peaks in the distribution of the number of galaxies vs. redshift (Fig. 10.7) according to a pattern which nicely corresponds to the picture of a beam cutting through a series of "Great Walls" one after another. It seems remarkable that even a certain regularity has been discovered corresponding to a spatial distance of $128h^{-1}$ Mpc [Broadhurst et al. 1990]. Several other "pencil beams" in different directions gave similar results [Broadhurst et al. 1991]. It is evident that the firm establishment of such a cellular structure out to large distances (40% of c/H_0, i.e. 20% of the dimension of the observable universe) will force us to modify our concept of a uniform matter distribution.

Hitherto the extent of the largest structures has been determined by the boundaries of the survey, and an indication for a universal upper limit does not yet exist. This leaves us with the sobering thought that the distribution of the luminous matter does not yet reveal the homogeneous structure of the universe – if it really exists. It could be that the part of the universe which

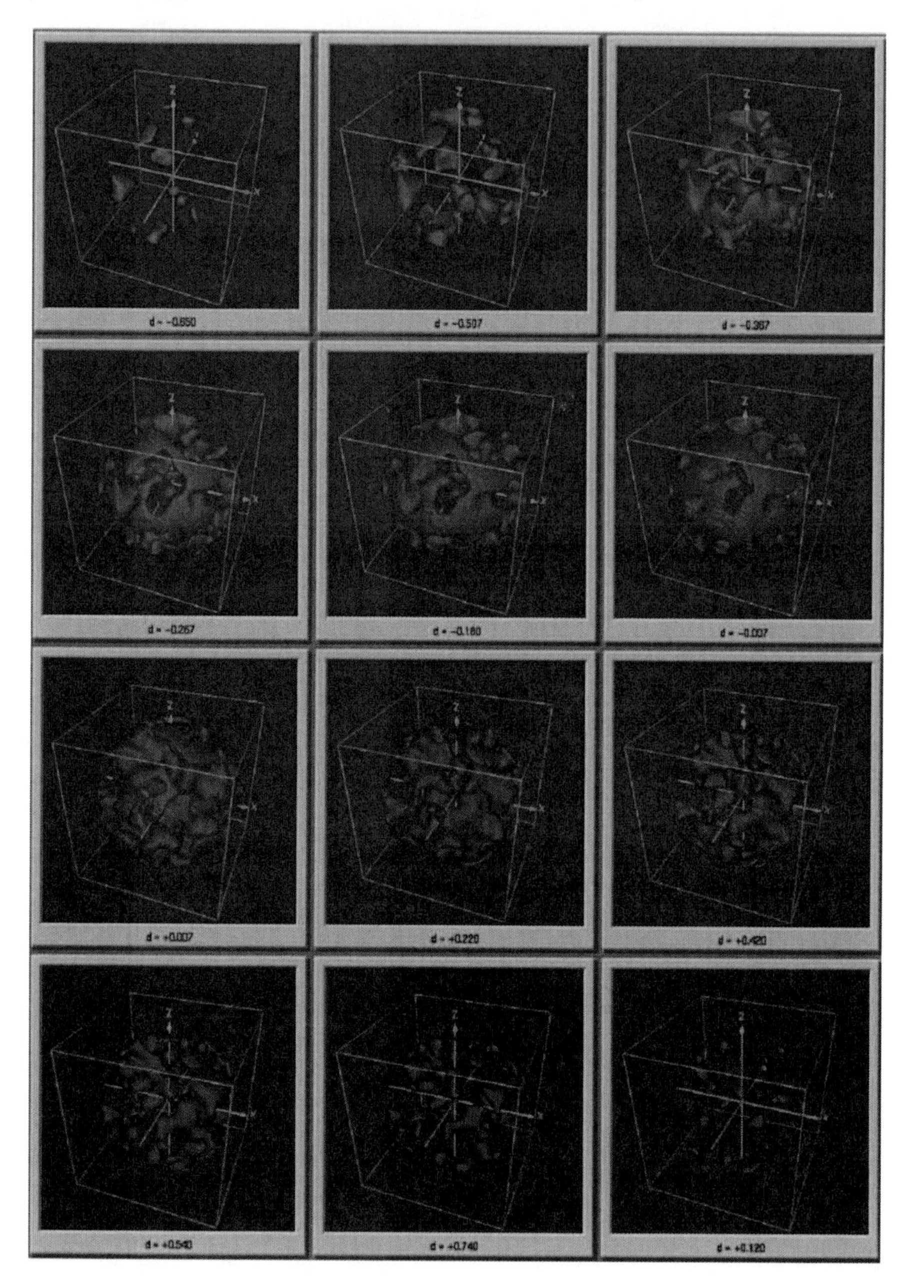

Fig. 10.6. Isodensity contours for the galaxies in the PSCz catalogue (courtesy of V. Springel [Springel 1996])

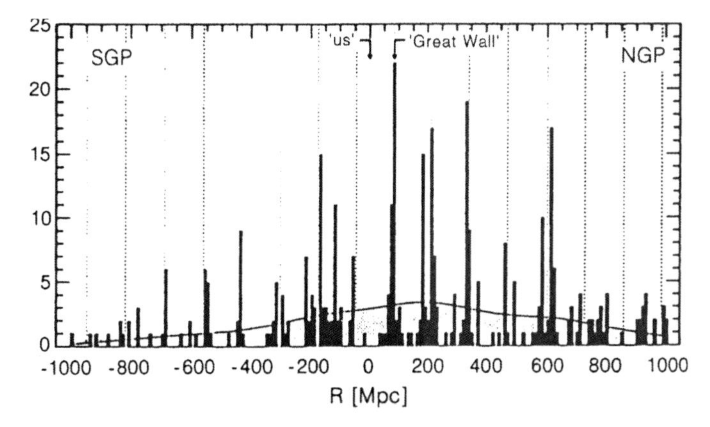

Fig. 10.7. The distribution of galaxies in redshift bins in a small angle beam directed at the galactic north and south pole. This "pencil beam" goes very deep, to redshifts $z = 0.4$. A remarkably regular pattern of peaks and minima seems evident. Those observations, if they are confirmed, really spell trouble for the concept of a homogeneous universe (Broadhurst et al. 1990). The first peak in the north direction is the "Great Wall" of the CfA survey

has been measured is still too small, or has not been investigated with high enough sensitivity to record average characteristics of the galaxy distribution. But it could also be the case that, at least in the population of bright galaxies, these typical properties do not show up.

Another big survey has been carried out by another group at the Center for Astrophysics in Boston [Shectman et al. 1996]. The results have been made publicly available, the Las Campanas Redshift Survey[1]. Over 700 square degrees of the sky were observed in small $1.5^\circ \times 1.5^\circ$ patches. Six long paths spanning about 80° in right ascension were chosen at declinations $\delta = -3^\circ, -6^\circ, -12^\circ$ in the north galactic hemisphere, $\delta = -39^\circ, -42^\circ, -45^\circ$ in the south. With the faint magnitude limit of about 17.5 in the red (a hybrid R-bound of the LCRS [see Shectman et al. 1996]), about 30 000 galaxies were measured in these six bands in the sky. Figure 10.8 is a plot of these galaxies in projection – all the different declination bands are projected into the plane of the graph, and right ascension and redshift z (rather cz) are the remaining coordinates. The catalogue is much more sparsely sampled than the CfA survey, but is much deeper. It reaches a redshift of $z = 0.2$, and thus it really covers a fair fraction of the observable universe. The pattern of the cell-like structure, with voids, sheets, and filaments traced by the galaxies which can be seen in the shallower surveys, clearly repeats itself out to the largest distances. So this seems to be the true basic arrangement of the galaxies.

The statistical analysis of this sample (see below and Chap. 12) has clearly demonstrated that the distribution in the north and south hemisphere is the same within the error limits [Jing et al. 1998]. Thus this portion of the sky

[1] http://manaslu.astro.utoronto.ca/~lin/lcrs.html

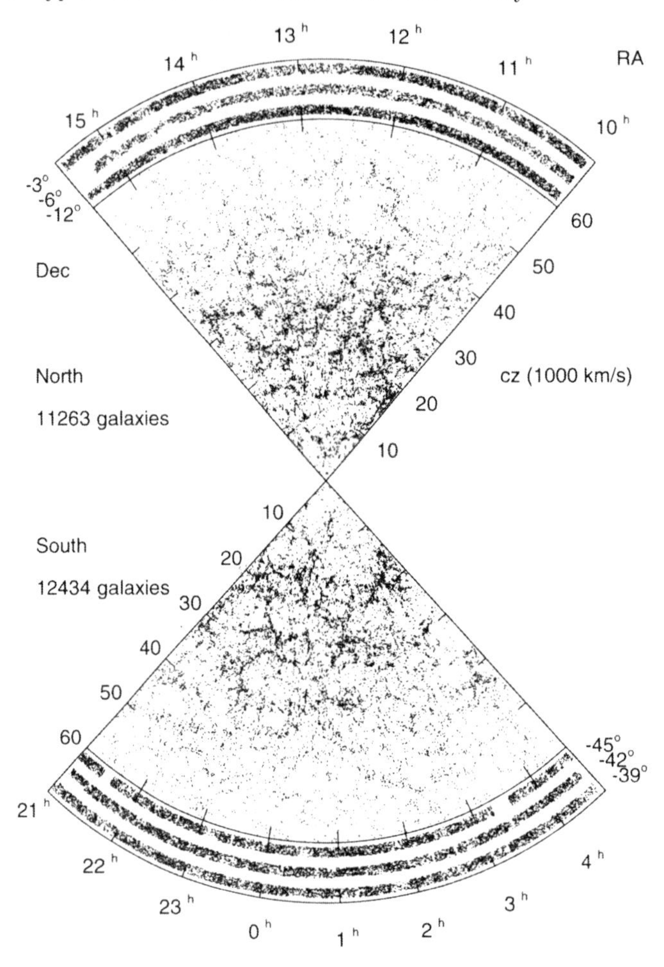

Fig. 10.8. The galaxies in the north and south hemisphere of the LCRS are plotted in redshift and galactic longitude. The observer is at the central point, where the two wedges meet

gives a fair picture of the universe. The quantitative analysis of the LCRS is tricky, because of various observational selection effects. CCD spectroscopy with optical fibres was done only once for each field of $1.5° \times 1.5°$. The observing time for one field was typically 2 hours. In fields with an unusually high density of galaxies there was a "fibre collision effect". There is a mechanical constraint that fibres cannot be too close to each other. This means that objects which are closer to each other than $55''$ cannot both be chosen at random from within the photometric boundaries. About 1100 objects were not observed because of this effect. One can account for this effect by an appropriate treatment of mock catalogues constructed from the numerical simulations.

10.1.2 Correlation Functions

The statistical analysis of galaxy catalogues provides valuable quantitative information in addition to the morphology. Various authors (e.g. [Totsuji and Kihara 1969; Peebles 1980]) have considered the low-order correlation functions as a measure of galaxy clustering in the Zwicky, Lick, and Jagellonian galaxy catalogues.

Complete information on the galaxy distribution consists of the masses and positions of all galaxies in a given volume V. In practice this ideal case never occurs. It is quite difficult to obtain accurate measurements of the angular coordinates and redshifts for the galaxies in a complete sample, whereas the masses are in most cases unknown. Therefore only the point distribution of the galaxy positions is usually analysed.

Suppose that the density distribution $n(\boldsymbol{x})$ can be obtained from the data. One way to extract statistical properties is then to evaluate integrals such as

$$\frac{1}{V}\int_V n(\boldsymbol{x})n(\boldsymbol{x}+\boldsymbol{x}_0)d^3x$$

where $\boldsymbol{x}_0$ is a fixed vector. A dimensionless two-point correlation function $\xi(\boldsymbol{x}_0)$ (Fig. 10.9) is in general use which is constructed from this integral by divsions by the square of the mean density, and by subtraction of 1:

$$\xi(\boldsymbol{x}_0) \equiv \frac{\frac{1}{V}\int_V n(\boldsymbol{x})n(\boldsymbol{x}+\boldsymbol{x}_0)d^3x}{\left(\frac{1}{V}\int_V n(\boldsymbol{x})d^3x\right)^2} - 1. \tag{10.1}$$

It is intuitively clear that this function can be used to describe clustering properties of the point distribution. If the distribution is the realization of a stochastic process without any correlation between neighbouring points, then $\xi = 0$. Strong correlations at a distance $|\boldsymbol{x}_0|$ will show up as large positive values of ξ. Periodic behaviour of ξ would indicate a highly regular structure, like the ordering of points in a crystal lattice.

The integral of $\xi(\boldsymbol{x}_0)$ over the volume V tends to zero for large enough V,

$$\lim_{V\to\infty}\frac{1}{V}\int_V \xi(\boldsymbol{x}_0)d^3x_0 = 0, \tag{10.2}$$

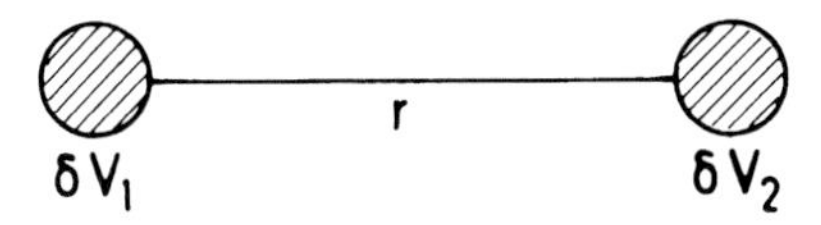

Fig. 10.9. Correlations are studied by looking for deviations from an average distribution. The 2-point correlation function $\xi(r)$ is obtained by searching for pairs of galaxies at a distance r in volumes δV_1 and δV_2 and averaging over the volume occupied by the sample

i.e. ξ describes deviations from the average values. The galaxy catalogues are used to search different volumes of space for correlation properties. The hope is that large enough volumes contain a fair sample of galaxies, such that the same average properties are found in different, well-separated volumes of space. In that sense different parts of the universe can be thought of as independent realizations of the same physical process, and the set of the independent samples approximates a statistical ensemble.

It may, of course, happen that the correlation-function analysis is not the most appropriate tool to extract statistical properties from the distribution. Some criticism has been voiced on the specific form (10.1) of ξ [Pietronero 1987]: the definition of a dimensionless ξ includes normalization by the average density $\langle n \rangle \equiv (1/V) \int n dV$. If, for example, the real distribution is such that the average density changes with the volume V, then (10.1) is not a good definition, because the same set of objects appears with different correlations depending on the volume considered.

Pietronero demonstrates this property with a simple example. A self-similar (fractal) structure is postulated where the number of points within r is

$$N(r) = Br^D$$

(B a constant related to the smallest size system, $B = N_0/r_0^D$; D is the "fractal dimension", also a constant).

First of all an average over all angular directions is formed and a distribution depending only on the distance R is considered. The average density for a sphere of radius R_S is

$$\langle n \rangle = \frac{N(R_S)}{V(R_S)} = \left(\frac{3}{4\pi}\right) BR_S^{-(3-D)}. \tag{10.3}$$

We see that $\langle n \rangle$ is a function of the sample size R_S. The 2-point correlation function is given by

$$\xi(r) = \frac{D}{3}\left(\frac{r}{R_S}\right)^{-(3-D)} - 1.$$

The exponent of the power-law behaviour $\xi \sim r^\gamma$ is an intrinsic property not depending on the sample size. But the normalization of ξ depends explicitly on R_S. This leads to the strange conclusion that the correlation between two points of the same system at distance $r (r < R_S)$ is different according to the size of the volume wherein they are contained.

Such an absurd result can be avoided in this particular example if the average density can be defined independent of position and size of the sample volume. This requires that the length scale R_c, where $\xi(R_c) = 0$, is much smaller than R_S.

There is little or no evidence for a fractal structure of the universe, and the example discussed above may be too extreme. On the other hand, one should expect a similar behaviour in cases where the average density depends on the sample volume. [Pietronero 1987] proposes replacing $\xi(r)$ with the function $\Gamma(r)$ defined as

$$\Gamma(r) = \langle n \rangle \left(\xi(r) + 1\right). \tag{10.4}$$

$\Gamma(r)$ is independent of R_S in the fractal example, and it may or may not be a better measure of correlations than $\xi(r)$ in general. But to determine $\Gamma(r)$ the galaxy catalogues would have to be reanalysed. Therefore we have to be content here to use $\xi(r)$ with the reservation that it may be a good description for the galaxy distribution, but that cluster or supercluster distributions – expecially samples containing large voids with dimensions comparable to the size of the volume considered – probably cannot be analyzed properly by $\xi(r)$.

Earlier work by Peebles and his associates (see [Peebles [1980] for details) concentrated on a two-dimensional analysis of the galaxies in projection. Thus from the catalogues an angular two-point correlation function $w(\theta)$ was derived.

Accidental associations, when galaxies are far apart in distance but close to the same line of sight, can somewhat reduce the reliability of the angular two-point correlation function. The radial two-point correlation function of the projected distribution of galaxies can be obtained from $w(\theta)$ (see [Peebles 1980]).

As a result of the efforts of Peebles and co-workers an approximate power-law behaviour for the two-point correlation function over a certain range of scales has been derived:

$$\begin{aligned} &\xi(r) = \left(\frac{r_0}{r}\right)^{\gamma} \quad \text{for} \quad 0.1 h_0^{-1}\text{Mpc} \le r \le 9 h_0^{-1}\text{Mpc}; \\ &\gamma = 1.77 \pm 0.04; \quad r_0 = (4.3 \pm 0.3) h_0^{-1}\text{Mpc}. \end{aligned} \tag{10.5}$$

The normalization factor r_0 may be an underestimate (it is derived from the redshift catalogue of [Kirshner et al. 1978]).

From an analysis of the Lick catalogue [Groth and Peebles 1977] a sharp break in the power law (10.5) has been found at

$$\xi(r_b) \sim 0.3; \quad r_b \sim 9 h_0^{-1}\text{Mpc}. \tag{10.6}$$

This result indicates that there is negligible clustering on scales greater than r_b. The authors also did not find any evidence for a filamentary structure. This may be due to the fact that the analysis was based on the apparent distribution of galaxies in the sky. The projection tends to reduce the effects of string formation and supercluster structures.

It has been found from an analysis of the local supercluster [Einasto et al. 1984], based on 3-dimensional distributions obtained from redshift catalogues [Huchra et al. 1983], that the 2-point correlation function has a shoulder at $R = 4 - 5$ Mpc. Such a hump has also been found in a correlation function for the Coma-A 1367, Hercules and Perseus supercluster [Gregory et al. 1980]. This indicates that the galaxy distribution shows a definite scale length. Such an effect could be suggestive of a transition from clusters to strings [Einasto et al. 1984].

A study of the CfA redshift catalogue [Davis and Peebles 1983], however, resulted in a 3-dimensional 2-point correlation function with the same power-law behaviour as that found for the projected case

$$\xi(r) \propto r^{-1.8}. \tag{10.7}$$

The discrepancy may indicate that in some cases the system under study may have specific properties – such as a pronounced transiton from clusters to strings in rich clusters – which get lost when average samples are considered. Various selection effects are certainly also important.

There is a suspicion that biases in the galaxy catalogues may prevent a determination of the behaviour of the correlation-function on large scales [Geller et al. 1984]. The disagreement mentioned also indicates that a careful further discussion of these interesting problems is required. The two-point function is, of course, a very limited tool for analysing a 3-dimensional distribution, and the case for the large-scale strings and supercluster networks seems quite convincing. Let us hope that they will not have the same fate as the Martian canals!

Another interesting result has been the investigation of the cluster–cluster (rather than galaxy–galaxy) two-point correlation function. The sample includes more than 100 rich clusters for which redshifts have been obtained from recent surveys. The two-point correlation function has the same slope ($\gamma = 1.8$) as for galaxies, but the correlation is about 20 times larger

$$\xi_{\mathrm{cc}}(r) = 20\xi_{\mathrm{gg}}(r) \tag{10.8}$$

[Bahcall and Soneira 1983; Klypin and Kopylov 1983]. According to Pietronero [Pietronero 1987] the discrepancy can be removed when the dependence of ξ on sample size is taken into account. The simple fractal model relates the sample sizes (typically $R_{\mathrm{c}} = 5R_{\mathrm{g}}$, where R_{c} is cluster sample size and R_{g} galaxy sample size) and the correlation functions

$$\frac{\xi_{\mathrm{cc}}}{\xi_{\mathrm{gg}}} = \left(\frac{R_{\mathrm{c}}}{R_{\mathrm{g}}}\right)^{\gamma}$$

and for $\gamma = 1.8$ the correct factor of 20 is derived. Whereas the galaxy–galaxy correlation becomes noise-dominated at about $r \sim 6\,\mathrm{Mpc}$, coherent structures can be followed to much larger scales $r \sim 25h_0^{-1}$ Mpc. There is, however, a more conventional explanation [Kaiser 1984].

10.1.3 Distribution of Dark Matter

There is convincing evidence that the mean density of luminous matter $\Omega_0 = 0.005$ is only a lower limit, and that the true value of the density parameter is at least 10 times larger (cf. Chap. 2). We have also seen that the mass-to-light ratio is much higher in large clusters than in small systems. But the optical luminosity is not a good tracer of the baryonic content of those structures, since most of the baryons are not in a stellar, but in a gas component. This hot and diffuse intergalactic medium has been detected by its X-ray emissions [Jones and Forman 1983]. The analysis of the emission of X-rays from 14 large clusters gave the surprising result that 5 times as much mass as in the stellar component was present in the form of a 10^8 K hot gas. Thus the total mass visible in stars and galactic gas, plus the hot intergalactic gas, must be taken into account when an attempt is made to trace the mass distribution. The quantity $M_{\rm lum} = M_{\rm Stars} + M_{\rm Gas}$ – which is still less than the mass in baryons, if there are invisible baryons around in the form of Jupiter-like bodies, or if a diffuse ionized intergalactic gas at $t > 10^4$ K exists – is the physically meaningful quantity that should be compared with the total mass M. It turns out [Blumenthal et al. 1984] that $M_{\rm lum}/M$ stays constant within the observational limits from single spiral galaxies up to large clusters. In Table 10.1 and Fig. 10.10 (adopted from [Blumenthal et al. 1984]) the general trend of the observations is presented.

The last line in Table 10.1 refers to preliminary velocity-dispersion data for the dwarf spheroidal galaxies Draco, Carina and Ursa Minor [Aaronson 1983]. Apparently there may be a large amount of dark matter in these systems. the fact that $M/M_{\rm lum}$ is almost constant from galaxies up to large clusters supports the opinion that there is no observational evidence for a large amount of dark matter apart from that present in galaxies and groups of galaxies.

Table 10.1. Mass-to-light ratios and $M/M_{\rm lum}$ on various scales (from [Blumenthal et al. 1984]) (solar units)

Object	$M/M_\odot$	M/L	$M_{\rm Gas}/M_{\rm lum}$	$M/M_{\rm lum}$
Large clusters	10^{15}	316 ± 40	$0.84^{+0.6}_{-0.1}$	$8.4^{+7.0}_{-1.0}$
Small E-dominated groups	5×10^{13}	83^{+80}_{-10}	$0.61^{+0.1}_{-0.1}$	$5.4^{-10.0}_{-2.0}$
Small spiral-dominated groups	2×10^{13}	40^{+50}_{-10}	0	14.2^{+36}_{-6}
Whole Milky Way	10^{12}	50	0	14
Dwarf spheroidals				
stellar M	10^{5-7}	2.5	0	1
Dynamical M	10^{6-8}	30	0	12

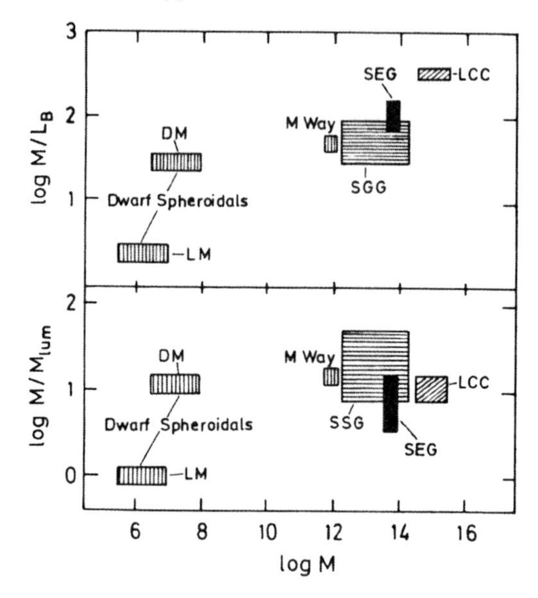

Fig. 10.10. M/L and M/M_{lum} versus $\log M$ (all in units of solar mass ($M_\odot$) and luminosity ($L_\odot$)) (reproduced from Fig. 1 of [Blumenthal et al. 1984]). DM: Dynamical Mass; LM: Luminous Mass; SEG: Small E groups, SSG: Small Spiral Groups; LCC: Large Cluster Cores

Even if we take the limit on the baryon density derived from the deuterium abundances at face value, we cannot exclude that the dark matter may be in the form of baryons. In that case $\Omega_0 \sim 0.1$, about 20 times the density of the luminous matter. The "invisible" baryons must be in a state where they emit very little radiation – a possibility would be to have objects with a mass much smaller than a solar mass (Jupiter-sized objects) so that nuclear-burning reactions do not set in. The main reason why baryonic dark matter fell into disfavour has already been discussed in Chap. 2: the density perturbations have to reach a magnitude $\delta\rho/\rho \sim 1$ before the present epoch – otherwise galaxies and clusters could not form. But in a baryon-dominated universe with adiabatic density perturbations at the decoupling epoch, this implies fluctuations in the 3 K background at the present epoch

$$\frac{\delta T}{T} \geq 10^{-3} \quad \text{on scales } \simeq 3'.$$

Observational limits (Chap. 2) are more than an order of magnitude below this limit $\delta T/T < 5 \times 10^{-5}$ at $3'$! For baryonic dark matter there are two somewhat uncomfortable ways out of this difficulty:

a) A significant reheating of the intergalactic medium could wipe out fluctuations in the 3 K background – but scales $> 10^\circ$ would not be affected.
b) An isothermal promordial fluctuation spectrum has $\delta T = 0$ – but the GUT models of baryon synthesis favour adiabatic fluctuations. Arguments for and against these loopholes will be discussed in Chap. 11.

This problem does not arise for non-baryonic dark matter, since these particles are not coupled to the photons, and the predicted fluctuations in the microwave background can be consistent with the observations. The fluctuations of the baryonic component are small at decoupling, and grow only later to the same size as the fluctuations in the non-baryonic component.

We have already become acquainted with various hypothetical candidates for the non-baryonic component of the dark matter in Chaps. 6 and 7: massive neutrinos with $m_\nu \simeq 5$ to $30\,\mathrm{eV}$; photinos with $m_\gamma > 500\,\mathrm{keV}$; axions with $m_a \simeq 10^{-3}\,\mathrm{eV}$; and gravitinos with $m_G \sim 10^2\,\mathrm{GeV}$. In addition black holes in a certain mass range

$$10^{-16} M_\odot < M_{BH} < 10^6 M_\odot \tag{10.9}$$

may be candidates. The lower limit refers to the decay of smaller black holes by the Hawking radiation, whereas the upper limit is required to avoid the disruption of galactic discs and star clusters [Lacey 1984; Carr 1978].

In Chap. 7 we also mentioned the exotic possibility of the existence of small pieces of quark matter that have not undergone confinement. These different possibilities lead to different scenarios for galaxy formation – as we shall see in the following sections.

10.2 Typical Scales Derived from Theory

10.2.1 The Jeans Mass for an Adiabatic Equation of State

Jeans was the first to consider the behaviour of small density fluctuations on a homogeneous background

$$p_0 = \text{const.}, \quad \varrho_0 = \text{const.}, \quad v = 0.$$

Introducing into the Newtonian equation for the pressure p, the density ϱ, the velocity v, and the gravitational field g, small perturbations p_1, ϱ_1, etc., and linearizing the equations on the homogeneous background, yields

$$\begin{aligned}
&\partial_t \varrho_1 + \varrho_0 \nabla \cdot v_1 = 0, \\
&\partial_t v_1 = (-c_s^2/\varrho_0)\nabla \varrho_1 + \mathbf{g}_1, \\
&\nabla \times \mathbf{g}_1 = 0; \quad \nabla \cdot \mathbf{g}_1 = -4\pi G \varrho_1; \\
&c_s^2 = \frac{p_1}{\varrho_1} = \left(\frac{dp}{d\varrho}\right)_{\text{adiabatic}}.
\end{aligned} \tag{10.10}$$

For such adiabatic perturbations ($p_1 = \varrho_1 c_1^2$) one finally obtains a wave equation for ϱ_1,

$$\partial_{tt} \varrho_1 = c_s^2 \Delta \varrho_1 + 4\pi G \varrho_0 \varrho_1. \tag{10.11}$$

An ansatz of the form $\propto \exp[i(k \cdot x - \omega t)]$ leads to a dispersion equation

$$\omega^2 = k^2 c_s^2 - 4\pi G \varrho_0. \tag{10.12}$$

Wave numbers

$$k \leq k_J = \left(\frac{4\pi G\varrho_0}{c_s^2}\right)^{1/2} \tag{10.13}$$

give rise to exponentially growing modes, with the imaginary part of ω

$$\mathrm{Im}\{\omega\} = c_s(k_J^2 - k^2)^{1/2}. \tag{10.14}$$

k_J is called the Jeans wave number, $\lambda_J \equiv (2\pi/k_J)$ the corresponding Jeans wavelength, i.e., the smallest scale on which gravitational instabilities can be expected. This theory with a static, homogeneous background cannot be used immediately in an expanding universe. First, only with an appropriately chosen cosmological constant can a homogeneous static background be accepted as a sensible configuration. Second, the expansion rate is

$$\frac{\dot{R}}{R} = \left(\frac{8\pi G}{3}\varrho\right)^{1/2} = \left(\frac{2}{3}\right)^{1/2} k_J c_s,$$

i.e. of the same magnitude as $\mathrm{Im}\{\omega\}$.

The use of the Jeans scale can be justified for an expanding universe (cf. Chap. 11). It turns out that perturbations with $k < k_J$ grow as a power t^α of the cosmic time t. Jeans's theory can be used to decide which waves can grow, and which cannot: for large wave numbers, (10.12) reads

$$\omega^2 \approx k^2 c_s^2.$$

This is valid if the following two conditions are satisfied:

a) Gravitational forces are negligible, i.e. the gravitational energy of a sphere of radius k^{-1} is much smaller than its thermal energy

$$\frac{G(\varrho k^{-3})^2}{k^{-1}} \ll \varrho c_s^2 k^{-3}.$$

b) The expansion rate of the universe is much less than the frequency of the wave

$$(G\varrho)^{1/2} \ll |\omega|.$$

Both conditions will be satisfied as long as

$$k \gg k_J. \tag{10.15}$$

Thus we may expect also in an FL cosmological model that there will be a critical wave number, of order k_J, such that disturbances with $k > k_J$ cannot grow, but oscillate like sound waves.

Since in an expanding Friedmann universe $k \propto R^{-1}(t)$, and $nR^3 = \text{const.}$ (for $p \ll \varrho$), the mass within the length $2\pi/k$ is a constant

$$M \equiv \frac{4\pi n m_p}{3}\left(\frac{2\pi}{k}\right)^3 . \tag{10.16}$$

Here n is the number density of nucleons, m_p the proton mass. (The following remarks are taken from [Weinberg 1972].)

Perturbations can grow if $\lambda > \lambda_J$, or if their mass M is greater than the Jeans mass M_J:

$$M_J \equiv \frac{4\pi n m_p}{3}\left(\frac{c_s^2 2\pi}{G\varrho}\right)^{3/2} .$$

From the time of electron–positron annihilation ($T \simeq 10^{10}$ K) to the time of hydrogen combination ($T_R = 4000$ K) we have the non-relativistic baryons and the black-body radiation field in thermal equilibrium at a temperature T. Since the entropy per baryon

$$s_B = \frac{4a\,T^3}{3n_B k_B}$$

is very large ($10^9 - 10^{10}$), the pressure and entropy of the matter can be neglected. The equation of state is

$$\varrho = n m_p + aT^4,$$
$$p = \frac{1}{3} a\,T^4 .$$

The adiabatic sound speed ($s_B = \text{const.}$) is very large, $c_s^2 = 1/3$.

For $T = 10^9$ K, the Jeans mass is

$$M_1 = 6 \times 10^4 M_\odot (\Omega_0 h_0^2).$$

Remember that the specific entropy is

$$s_B = 3.6\frac{n_\gamma}{n_B} = 1.56 \times 10^8 (\Omega_0 h_0^2)^{-1}.$$

The Jeans mass grows as T^{-3} until $(s_B k_B T/m_p)$ is of order 1, i.e., until

$$T \approx \frac{m_p}{k_B s_B} = 6.98 \times 10^4\,\text{K}(\Omega_0 h_0^2).$$

At later times M_J stays at the high value of

$$M_J = 9.06 s_B^2 M_\odot \simeq 2.2 \times 10^{17} (\Omega_0 h_0^2)^{-2} M_\odot . \tag{10.17}$$

The maximum Jeans mass can be written in terms of fundamental quantities

$$M_{J,\max} = 4.8\left(\frac{M_{Pl}^3}{m_p^2}\right)s_B^2.$$

The factor $4.8M_{Pl}^3/m_p^2$ corresponds to the Chandrasekhar mass limit of a body composed of Fermi particles (m_p is the proton mass), while the factor s_B^2 brings in the square of the entropy per baryon, a number characterizing our hot universe.

Immediately after the formation of hydrogen atoms, at t_R, the equations of state change to those of a monatomic ideal gas with $\gamma = 5/3$:

$$\varrho = nm_H + \frac{3}{2}nk_BT;$$
$$p = nk_BT.$$

The sound speed drops from $\sim c/\sqrt{3}$ to

$$c_s^2 = \frac{5}{3}\frac{k_BT}{m_H}.$$

For the Jeans mass this leads to

$$M_J = 4\left(\frac{\pi}{3}\right)^{5/2}\left(\frac{5k_BT}{G}\right)^{3/2} n_B^{-1/2} m_H^{-2}.$$

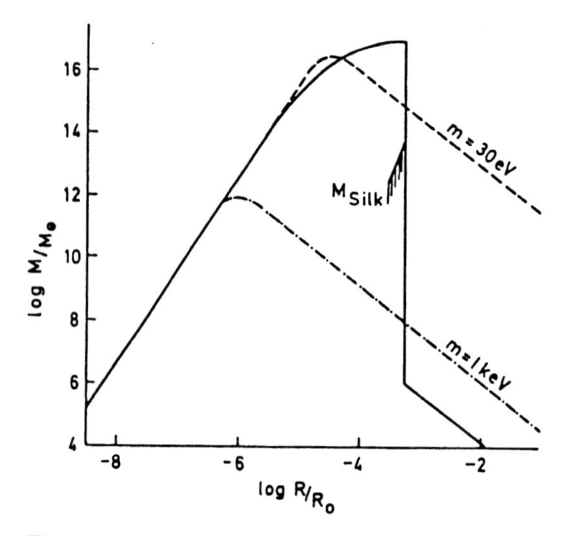

Fig. 10.11. Plot of the Jeans mass versus expansion factor. The *solid line* is for a photon and baryon fluid of ideal gases. The *dotted lines* are for collisionless particles of $m = 30\,\text{eV}$ and $m = 1\,\text{keV}$, respectively. M_{Silk} is the damping scale of Sect. 10.2.3

Immediately after t_R the matter temperature is equal to the temperature of the radiation, and

$$n_B = \frac{4}{3}\frac{a}{k_B}\frac{T^3}{s_B}.$$

From this one obtains the small value

$$M_J = 102 M_\odot s_B^{1/2} = 1.27 \times 10^6 (\Omega_0 h_0^2)^{-1/2} M_\odot. \tag{10.18}$$

For later times $T \propto R^{-2}$, $n \propto R^{-3}$, hence $M_J \propto R^{-3/2}$.

In Fig. 10.11 the variation of Jeans mass with cosmic epoch is shown. Comparing this behaviour with a typical galactic mass scale of $M_G = 10^{11} M_\odot$, we see that before t_R the Jeans mass is much larger $M_J \gg M_G$, whereas after t_R it is much smaller than the mass of a galaxy. The Jeans mass itself does not give any clue to the observed mass distribution on galactic or cluster scales.

10.2.2 The "Jeans Mass" for Collisionless Particles

Non-baryonic dark matter is usually represented by a collisionless system of particles. The interplay between thermal and gravitational energy density which determines the Jeans mass for an adiabatic gas is no longer an adequate description in this case. One can, however, use an effective Jeans mass and Jeans length for the collisionless particles by replacing the sound speed in formulae such as (10.16)–(10.18) by the velocity dispersion owing to the thermal motions of the particles.

Consider a gas of non-relativistic collisionless particles with a phase-space distribution function $f(\boldsymbol{x}, \boldsymbol{p}, t)$. The mean particle density is given by

$$n(\boldsymbol{x}, t) = \int f(\boldsymbol{x}, \boldsymbol{p}, t) d^3 p.$$

For a particular example we can take the particles to be non-relativistic neutrinos. The Vlasov equation describes the time evolution of f:

$$\partial_t f + (\boldsymbol{p}/m) \boldsymbol{\nabla}_{\boldsymbol{x}} f - m \boldsymbol{\nabla} \phi(\boldsymbol{x}, t) \cdot \boldsymbol{\nabla}_p f = 0, \tag{10.19}$$

where $\phi(\boldsymbol{x}, t)$ is the gravitational potential,

$$\Delta \phi = 4\pi G \varrho, \tag{10.20}$$

$$\varrho \equiv mn = \int m f d^3 p. \tag{10.21}$$

Small perturbations on a homogeneous background ($f = f_b + f_1, \phi = \phi_b + \phi_1$) can be described by the linearized system of equations (10.19) to (10.21). Decomposition of the fluctuations into plane waves and integration over $d^3 x$ leads to a dispersion relation

$$0 = 1 + \frac{4\pi G m^2}{\boldsymbol{k}^2} \int d^3 p \frac{k \cdot \boldsymbol{\nabla}_p f_b}{[(\boldsymbol{p}/m)\boldsymbol{k} - \omega]}.$$

This is the exact analogue of the dispersion equation for electrostatic waves in an unmagnetized, collisionless plasma, if e^2 is replaced by $-Gm^2$ and a minus sign is introduced (because of the universally attractive gravitational force). Assuming an isotropic background, one finds

$$0 = 1 + \frac{4\pi G m^4}{k^3} \int dy \frac{df_b(y)}{dy} \frac{1}{y - \omega}. \tag{10.22}$$

The effective Jeans wave number k_J gives $\omega(k_J) = 0$. The integral (10.22) can be evaluated for $\omega = 0$, if the Fermi–Dirac distribution for the background is inserted (e.g. a background of massive neutrinos with temperature T_ν)

$$f_b = (2\pi)^{-3}(e^{p/T_\nu} + 1)^{-1}, \quad \hbar = c = k_B = 1. \tag{10.23}$$

The Jeans wave number k_J is then

$$k_J^2 = \frac{\ln 2}{(2\pi)^3} 16\pi^2 G m^3 T_\nu. \tag{10.24}$$

The mass within the Jeans length $\lambda_J = (\pi/k_J)$ characterizes the fluctuation

$$M_J = \frac{4\pi}{3} \varrho_b \lambda_J^3 = 2^{-5/2} \pi^{7/2} \frac{\zeta(3)\Gamma(3)}{(\ln 2)^{3/2}} \left(\frac{T_\nu}{m}\right)^{3/2} \frac{M_{Pl}^3}{m^2}. \tag{10.25}$$

The maximum of the Jeans mass is reached for $T_\nu = m$, the point where the neutrinos become non-relativistic. (The calculation is, of course, only valid for non-relativistic particles $m \geq T_\nu$; see [Aeppli 1986].)

The critical wave number k_J can be incorporated in a pressure gradient

$$k_J^2 v_s^2 = 4\pi G \varrho_b.$$

Here

$$v_s^2 = \frac{5}{3} (\ln 2)^{-1} \frac{\zeta(5)\Gamma(5)}{(\zeta(3)\Gamma(3))^2} \langle v^2 \rangle$$

describes the velocity dispersion of the neutrinos. Pressure gradient and gravitational force are in approximate balance for $k = k_J$.

We have, approximately,

$$M_J^{\max} \sim \varrho_b \lambda_J^3 \sim M_{Pl}^3 m^{-2}.$$

For a massive particle of $m \sim 30\,\mathrm{eV}$, this corresponds to a value of

$$M_J^{\max} \sim 10^{15} M_\odot,$$

the mass of a cluster of galaxies. We shall see in Chap. 11 how this maximum Jeans mass can define a characteristic scale for collisionless, non-baryonic matter.

10.2.3 The Adiabatic Damping Scale

It is rather disappointing that the Jeans mass does not give us a clue as to why the mass distribution of galaxies is as it is. One such clue to the existence of a specific mass scale has emerged from the consideration of the damping of protogalactic fluctuations during the phase when the wavelengths are less than the Jeans length [Peebles 1980; Weinberg 1971; Silk 1968].

In that phase the fluctuations undergo acoustic oscillations which can be damped by dissipative processes. There can be efficient damping of fluctuations of wavelength λ when the time-scale of damping $t_D(\lambda)$ is less than the expansion scale $t_{\rm exp} = H^{-1}$. Thus a characteristic damping length can be defined by $t_D(\lambda) = t_{\rm exp}$. For all smaller wavelengths the condition for efficient damping $t_D(\lambda) < t_{\rm exp}$ is satisfied.

Dissipative effects arise as deviations from thermal equilibirum – for instance, when the mean free time between collisions of particles is longer than the expansion time. For epochs before decoupling, $t < t_R$, the main interaction mechanism is Thomson scattering by non-relativistic electrons. The photon mean free time between scatterings is

$$\tau_\gamma = (n_e \sigma_T)^{-1}, \tag{10.26}$$

where σ_T is the Thomson cross-section,

$$\sigma_T \equiv \frac{8\pi e^4}{3m_e^2} - 0.6652 \times 10^{-24}\,\mathrm{cm}^2.$$

The mean free time of an electron or proton between Coulomb collisions is [Weinberg 1972] of the order of

$$\tau_c = \left\{ n \left(\frac{k_B T}{m_e} \right)^{1/2} \frac{e^4}{(k_B T)^2} \right\}^{-1}.$$

This time is shorter than t_γ by a factor of order

$$\left(\frac{k_B T}{m_e} \right)^{3/2}.$$

The deviation from perfect thermodynamic equilibrium between matter and radiation is the dominant dissipative effect. The medium of protons, electrons, and photons can be treated in good approximation as an imperfect radiation fluid with coefficients for the shear viscosity η, the bulk viscosity ζ, and the heat conduction χ. Straumann [1976] has improved and corrected earlier results of Weinberg [1971]. When Thomson scattering dominates and the pressure and thermal energy of the matter ar neglected the following expressions are obtained:

$$\eta = \frac{10}{9}\frac{4}{15}\frac{aT^4}{n\sigma_T};$$
$$\zeta = 0;$$
$$\chi = \frac{4}{3}\frac{aT^3}{n\sigma_T}.$$

In general a sound wave in an imperfect fluid will be damped at a rate [Landau and Lifshitz 1959]

$$\Gamma = \frac{k^2}{2(\varrho + p)}\left\{\zeta + \frac{4}{3}\eta + \chi\left(\frac{\partial\varrho}{\partial T}\right)_n^{-1} \times\left[(\varrho + p) - 2T\left(\frac{\partial p}{\partial T}\right)_n + c_s^2 T\left(\frac{\partial\varrho}{\partial T}\right)_n - \frac{n}{c_s^2}\left(\frac{\partial p}{\partial T}\right)_T\right]\right\}. \quad (10.27)$$

Inserting the coefficients, one finds

$$\Gamma = \frac{k^2 aT^4}{6n\sigma_T(nm_P + \frac{2}{3}a\, T^4)}\left\{\frac{160}{135} + \frac{n^2 m_P^2}{aT^4(nm_P + \frac{4}{3}aT^4)}\right\}.$$

Since $k^2 \propto M^{-2/3}$ for the mass M within the length $\pi/|k|$, the sound wave will be damped in amplitude by a factor of the form

$$D \equiv \exp\left\{-\int_{t_A}^{t_0}\Gamma dt\right\} = \exp\left\{-\left(\frac{M_c}{M}\right)^{2/3}\right\}, \quad (10.28)$$

where M_c is a critical mass which is determined from the integral over Γ. For a period before t_R with dominating baryon density – i.e. with a high present density – one has

$$t \approx (6\pi n m_P G)^{-1/2} \quad (10.29)$$

thus

$$\Gamma \simeq \frac{k^2}{6\sigma_T n} \propto t^{2/3}. \quad (10.30)$$

The critical mass M_c from (10.28) is then

$$M_c = \frac{32\pi^4}{3}\left(\frac{m_p}{10\sigma_T}\right)^{3/2}(6\pi G)^{-3/4}(nm_H)^{-3/4}. \quad (10.31)$$

Expressing n through the present density parameter Ω_0 and the Hubble constant $H_0 = h_0\ 100\,\mathrm{km\,s^{-1}\,Mpc^{-1}}$, we find

$$M_c = 3 \times 10^{12}(\Omega_0 h_0^2)^{-5/4} M_\odot.$$

To obtain a relatively low density at the present epoch, the energy-density during the phase of acoustic oscillations had to be dominated by radiation, and

$$t \approx (15.5\pi a T^4 G)^{-1/2},$$

the damping rate

$$\Gamma \simeq \frac{2k^2}{15\sigma_T n} \propto t^{1/2}$$

and

$$M_c \simeq \frac{32\pi^4}{3}\left(\frac{4m_p}{45\sigma_T}\right) \quad (15.5\ a\ T^4 G)^{-3/4}(nm_p)^{-1/2}, \tag{10.32}$$

i.e.

$$M_c \simeq 8 \times 10^{13}(\Omega_0 h_0^2)^{-1/2} M_\odot .$$

Sufficient damping may be expected if the exponent $(M_c/M)^{2/3}$ is about 10. We may conclude that the fluctuations that survive until after the recombination time t_R have to exceed a lower limit, which lies between $1.6 \times 10^{11} M_\odot$, and $6 \times 10^{12} M_\odot$.

From the analysis of adiabatic perturbations another analytic approximation for the damping scale can be derived [Silk and Efstathiou 1983] which agrees quite well with the value quoted above

$$M_{D,S} = 1.3 \times 10^{12}(\Omega_0 h_0^2)^{-3/2} M_\odot .$$

In Fig. 10.12 the results of various numerical efforts to find the damping scale are presented.

Although there is some quantitative disagreement between different authors, the general fact is well established that damping by photon diffusion

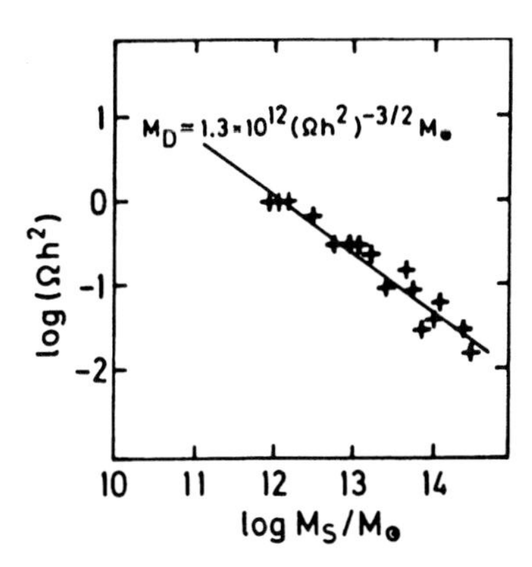

Fig. 10.12. Damping scale as derived by various authors (adapted from [Silk and Efstathiou 1983])

leads to a characteristic mass scale corresponding to groups or clusters of galaxies.

It is easy to derive an order-of-magnitude estimate from the following considerations; the time for the photons to diffuse a length λ is

$$t_{\text{diff}} = \frac{\lambda^2}{c^2}\tau_\gamma^{-1}. \tag{10.33}$$

Thus the sound wave of wavelength λ can be damped by photon diffusion when $t_{\text{diff}} < t_{\text{exp}} \approx t$, i.e. when

$$\lambda < (\tau_\gamma c^2 t)^{1/2} \equiv (l_\gamma l_H)^{1/2} \qquad (l_\gamma \equiv c\tau_\gamma, l_H \equiv ct). \tag{10.34}$$

If for $t < t_R$ there is a baryon-dominated epoch, then $t \propto \varrho^{-1/2}$, $n_e \propto \varrho$, hence $l_\gamma l_H \propto \varrho^{-3/2}$ and the mass scale corresponding to $\lambda_D = (l_\gamma l_H)^{1/2}$ is

$$M_D = 3 \times 10^{12}(\Omega_0 h_0^2)^{-5/4} M_\odot. \tag{10.35}$$

These heuristic arguments are quite successful, as we can see, in determining the damping scale. The conclusion is that after the decoupling of matter and radiation all adiabatic fluctuations with mass less than M_D have been damped out.

10.2.4 The Horizon Scale

The size of the particle horizon,

$$\lambda_H = ct, \tag{10.36}$$

marks a scale beyond which the edges of a perturbation with $\lambda > \lambda_H$ cannot causally affect each other. Again, a characteristic mass scale is connected with λ_H, the total mass within the horizon:

$$M_H = \frac{4}{3}\pi\varrho\left(\frac{\lambda_H}{2}\right)^3 = \frac{\pi}{6}\varrho(ct)^3. \tag{10.37}$$

For a baryon-dominated universe $\varrho \propto t^{-2}$ and $M_H \propto t$. In a radiation-dominated epoch the matter density $\varrho \propto t^{-3/2}$ and $M_H \propto t^{3/2}$.

On scales larger than $\lambda_H = ct$, gravitational instability must be described in a different way. These fluctuations follow their specific FL universe expansion with density $\varrho + \delta\varrho$ until they come within the horizon. On scales less than the Jeans length λ_J perturbations oscillate like acoustic waves, and they are damped out in that phase on all scales $M < M_{D,S}$.

We shall see in the next chapter that density fluctuations of a species of particles grow only slowly, as long as their density is less than that of the radiation. Only after this species of matter dominates can the density perturbations grow without inhibition. Thus the scale of preferred growth is the size of the horizon just at the epoch when matter and radiation have equal energy densities. Since $\varrho_\gamma/\varrho_X = (1+z)$ with $\varrho_\gamma = aT_\gamma^4$ and ϱ_X the density of a species of X particles, we find for the redshift z_{eq} of this epoch

$$(1+z_{eq}) = \frac{\Omega_{X,0}}{\Omega_{\gamma,0}} = 4.2 \times 10^4 (\Omega_{X,0} h_0^2) \vartheta^{-4}, \tag{10.38}$$

where $\vartheta \equiv T_{\gamma 0}/2.7\,\mathrm{K}$. The mass within the horizon scale $\lambda_H = ct_{eq}$ is

$$M_{Heq,X} = 2.2 \times 10^{15} (\Omega_{X,0} h_0^2)^{-2} M_\odot. \tag{10.39}$$

This horizon mass scale is slightly larger than the Jeans mass derived in Sect. 10.2.1. Thus these large scales $M_J \le M \le M_H$ can grow without damping by acoustic oscillations and without inhibition by the radiation energy density. The mass scale can be of the size of large clusters (when $\Omega_X \sim 1$).

10.2.5 Damping by Free Streaming and Directional Dispersion

Collisionless relativistic particles define another damping scale, that scale corresponds to their relativistic motion. Directional dispersion [Peebles 1980; Bond and Szalay 1983] is the effect by which a wave pattern in three dimensions decays. Waves travelling in different directions interfere destructively on average, and consequently the amplitudes decay. This kind of damping is very efficient for extremely relativistic particles. It leads to a damping $\propto t^{-1}$ of fluctuation amplitudes ([Peebles 1980]; see also Chap. 11).

Fluctuations of relativistic particles are also damped by free streaming, i.e. the effect that as long as the particles are relativistic they move freely away from a compression region – the velocity dispersion of the particles is then responsible for the disappearance of the compression region. These processes have the consequence that collisionless relativistic particles damp out all fluctuations on scales less than the horizon.

The first scale that allows growth of fluctuations is thus the horizon size at the epoch when the particles become non-relativistic, $T \approx m_X$. This "free streaming scale" is

$$\lambda_{FS}(T = m_X) = ct(T = m_X) \approx (G\varrho)^{-1/2} \approx M_{Pl} m_X^{-2} g^{1/2}.$$

The corresponding mass is

$$M_{FS} \approx \varrho \lambda_{FS}^3 \approx M_{Pl}^3 m_X^{-2} g^{1/2},$$

where $g(T)$ is the usual statistical factor and $\varrho \equiv (g/2)\varrho_\gamma = (g/2)aT_\gamma^4$.

Exercises

10.1 Calculate the linear evolution of baryonic density fluctuations δ_B in a medium which is a mixture of baryons and CDM (δ_{CDM}). Show that for an $\Omega = 1$ model one has approximately

$$\delta_B(k,t) = \delta_{\mathrm{CDM}}(1 + x_J^2 k^2)^{-1};$$

here x_J is the Jeans length for the sound speed c_s, and $p = c_s^2 \rho$.

10.2 "Silk damping" is the mechanism whereby oscillations of the photon–baryon fluid for mass clumps smaller than the Jeans mass are damped. Consider a multipole expansion of the inhomogeneities, and calculate the multipole indices l for which Silk damping is effective. What does this imply for the CMB anisotropies?

10.3 The relation between the spatial and the angular two-point correlation functions of galaxies can be derived by integrating the 3-dimensional function along the lines of sight. If one takes the relation between the apparent magnitude m of a galaxy at distance r (in Mpc) with absolute magnitude M as

$$m + M = 5 \log r + 25,$$

one can easily relate the angular correlation function $g(m_1, m_2, \theta)$ for two galaxies separated by the angular distance θ on the sky to the spatial function $\Gamma(M_1, M_2, r_{12})$. Derive this integral relation, and show how it simplifies, if the correlations are independent of absolute magnitude

$$\Gamma \equiv \Phi(M_1)\Phi(M_2)\xi(r_{12}).$$

The relation can be further evaluated by taking into account the fact that catalogues usually list galaxies brighter than a limiting magnitude m_0. The angular correlation function $w(\theta)$ is then obtained by integrating the product $gf(m-m_0)$ over magnitudes. f gives the probability of a galaxy of magnitude m being included in the catalogue. Derive the integral equation which gives the relation between $w(\theta)$ and $\xi(r_{12})$ (Limber's equation, see [Peebles 1980]). (Write it in terms of the selection function

$$S(y) = \int dM (\Phi(M)/n) f(M - M_0 + 5 \log y),$$

where $n = \int dM \Phi(M)$.)

10.4 [Pietronero 1987] has given a simple example of a fractal structure, where the number of points within a radius r is $N(r) = Br^D$ (see Sect. 10.1.3).

Compute the mean density and the two-point correlation function for a sphere of radius R_S. How can the result be tested by observations?

10.5 The power spectrum $P(k)$ is the ensemble average of the Fourier components of the density contrast $P(k) = \langle \|\delta(k)\|^2 \rangle$. Show how the power spectrum can be measured from a distribution of objects i:

$$n(\boldsymbol{r}) = \sum_i \delta(\boldsymbol{r} - \boldsymbol{r}_i).$$

(Look it up in [Peebles 1980].)

10.6 Derive the probability that a volume V placed randomly in space is empty of objects. Show that this void probability function is a sum over correlation functions to all orders. Show that, for the hierarchical model in which all correlation functions are given as products of two-point correlation functions, the void probability function depends only on a variable $q = \bar{n}V\bar{\xi}$, where $\bar{n}$ is the mean density,and $\bar{\xi}$ is a double volume integral over the two-point correlation function divided by V^2.

10.7 Let $P(w)$ be the distribution function with mean $\bar{w}$ and variance σ_w^2 of objects $i = 1, 2, \dots, N_g$; $\{w_i\}$ is one observation. The weighted pair count is defined as

$$DD_w(r) = (1/2) \sum w_i w_j U_{ij}(r),$$

where $U_{ij} = 1$ if the separation of objects i and j is within the bin $r \pm dr/2$, and $U_{ij} = 0$ otherwise.
Calculate the mean $E(DD)$ and the variance $\sigma_{DD}^2(w)$ of $DD_w(r)$, and show that

$$\sigma_{DD}^2(w) \simeq \sigma_w^2(\sigma_w^2 + 2\bar{w}^2)E(DD) + (4/N_g^2)\sigma_w^2\bar{w}^2 E(DD^2).$$

($E(DD^2)$ is the mean square of $DD(r)$.)
For the so-called bootstrap resampling, the original data set is used to construct new samples by selecting N_g times one galaxy from the parent sample. Each galaxy in the parent sample is selected with probability $1/N_g$. This set of bootstrap samples is then used to estimate means and variances of pair counts in a data sample. The bootstrap resampling is equivalent to assigning each galaxy (i) in the parent sample a weight w_i equal to the number of times this galaxy is selected. Thus w_i can be $0, 1, \dots, N_g$; $w_i = 0$ means that galaxy i has not been selected. Therefore the weight for each object has a Poisson distribution with mean 1 (selection probability each time is $1/N_g$).
Now, in our example, for bootstrap resampling $\bar{w} = 1$, $\sigma_w^2 = 1$. Compare the bootstrap error to the Poissonian (ensemble) error. Is the bootstrap error a reliable estimate?

11. The Evolution of Small Perturbations

"Worlds on worlds are rolling ever
From creation to decay
Like the bubbles on a river
Sparkling, bursting, borne away."

P.B. Shelley

11.1 Some Remarks on the Case of Spherical Symmetry

11.1.1 Spherical Fluctuations in a Friedmann Universe

Many aspects of a general theory of fluctuations can be guessed by studying a special, simple case – the behaviour of a spherical inhomogeneity in an expanding Friedmann universe filled with dust, i.e. zero-pressure matter.

An inhomogeneity containing a mass M within a sphere of proper radius S leaves the general Friedmann expansion undisturbed if the density within this sphere, obtained by spreading out M over the volume, is equal to the Friedmann background density ϱ_b [Einstein and Strauss 1945]. A model with several such spherical holes in a Friedmann background, with the mass inside the holes concentrated around the centre point, but with a smooth Friedmann universe of density ϱ_b elsewhere, reminded Einstein of "Swiss cheese". Lemaître himself [Lemaître 1931] had shown that a spherical inhomogeneity can be described just like a part of an evolving, homogeneous world-model (see e.g. [Zel'dovich and Novikov 1983]).

Let us consider a sphere of radius r with a density Ω', higher than the background density Ω_b ($\Omega' > \Omega_b$). The consistency of the boundary conditions with Einstein's field equations requires a spherical shell between the radii r and S with a density $\Omega'' < \Omega_b$ (Fig. 11.1).

If $\Omega_b \leq 1$, the two cases $\Omega' \leq 1$ and $\Omega' > 1$ can occur. For $\Omega' \leq 1$, the spherical inhomogeneity expands forever, like the homogeneous FL model with hyperbolic geometry. A bound object does not form in this case. For $\Omega' > 1$, however, the spherical inhomogeneity within r behaves like a closed FL model: it reaches a maximum radius at a certain time t_m, then the expansion ceases, and a contraction towards collapse sets in. Formally the recollapse proceeds to the limit of infinite density within a time interval $2t_m$. In any realistic situation, however, the final state will be some equilibrium state at a finite density. Shock waves will initiate very effective dissipative processes, which turn the energy of collapse into heat. A radial variation of the density $\varrho = \varrho(r)$ is sufficient to lead to the occurrence of shock waves (see [Zel'dovich and Novikov 1983]).

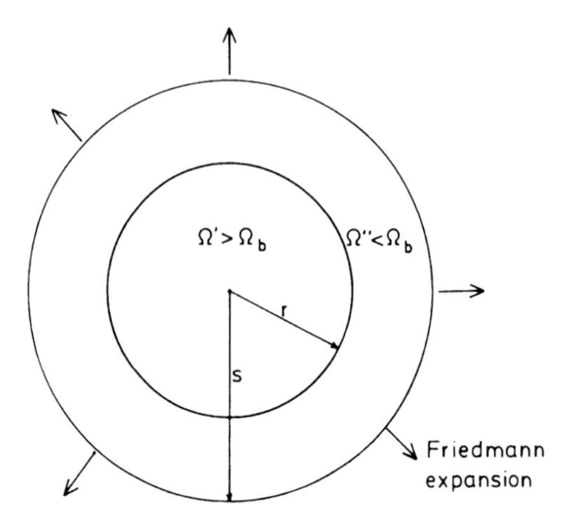

Fig. 11.1. A spherical density inhomogeneity of radius r is surrounded by a void of density $\Omega'' < \Omega_b$. If the mass within S is equal to the background mass within that radius, then the Friedmann expansion proceeds undisturbed even if the mass within r collapses

11.1.2 Linearized Spherical Perturbations

When the density in the perturbed inner sphere is only slighly larger than the background density, $\varrho = \varrho_b + \delta\varrho$, even the simple model sketched above can help us to guess many aspects of a general theory of fluctuations. In this case the expansion factor $a(t)$ of the spherical inhomogeneity is just equal to the background expansion factor $R(t)$ with a small correction

$$a(t) = R(t) + \delta R.$$

Then the Friedmann equations for $a(t)$ and $R(t)$ read

$$\begin{aligned}\left(\frac{\dot{a}}{a}\right)^2 &= \frac{8\pi G}{3}(\varrho_b + \delta\varrho) - \frac{K_S}{a^2},\\ \left(\frac{\dot{R}}{R}\right)^2 &= \frac{8\pi G}{3}\varrho_b - \frac{K_b}{R^2}.\end{aligned}$$

For $p = 0$ the continuity equations $(\varrho_b R^3)^{\cdot} = 0$ and

$$((\varrho_b + \delta\varrho)a^3)^{\cdot} = 0$$

lead to

$$\frac{\varrho - \varrho_b}{\varrho_b} \equiv \delta = -3\frac{\delta R}{R}. \tag{11.1}$$

Differentiation of the Friedmann equations and insertion of the continuity equation lead to an equation for the density contrast δ:

$$\ddot{\delta} + 2\frac{\dot{R}}{R}\dot{\delta} = 4\pi G\varrho_b\delta. \tag{11.2}$$

The time dependence of δ can be read off easily for the case of a $K = 0$ background, where $R \propto t^{2/3}$; $\varrho_b \propto R^{-3}$, and hence $\delta \propto t^{2/3}$. The density contrast of a small spherical perturbation grows in proportion to $t^{2/3}$.

The formulae for the behaviour of δ with time will be rediscovered in the behaviour of general non-spherical fluctuations in a linear approximation.

The early development of a fluctuation when the initial perturbation δ_i is small exhibits the interesting property that only for an initial density contrast larger than a critical value δ_c can bound gravitational objects eventually form [Primack 1984; Peebles 1980]. Consider a spherical fluctuation with initial radius r_i and density contrast δ_i following the expansion of the background at the initial time t_i. The kinetic energy per unit mass at the sphere's boundary is $(H_i \equiv \dot{r}_i/r_i)$

$$K_i = \frac{1}{2} H_i^2 r_i^2,$$

and the potential energy per unit mass at the edge of the fluctuation is

$$W_i = -\frac{1}{2} H_i^2 r_i^2 \Omega_i (1 + \delta_i).$$

The total energy per unit mass is

$$E \equiv K_i + W_i = K_i[1 - \Omega_i(1 + \delta_i)].$$

Only for values $E < 0$ can the expansion of the spherical perturbation stop and turn into a collapse. For $\Omega_0 < 1$, the unperturbed energy $\propto (1 - \Omega_0)$ is positive, and a critical amplitude δ_c exists, with $\delta > \delta_c$ needed for an eventual collapse:

$$\delta_c = \Omega_i^{-1}.$$

For a dust $(p = 0)$ configuration one has

$$\Omega_i = \frac{\Omega_0(1 + z_i)}{1 + z_i \Omega_0};$$

therefore

$$\delta_c = \frac{1 - \Omega_0}{\Omega_0(1 + z_i)}.$$

11.1.3 Non-linear Spherical Fluctuations

As the spherical inhomogeneity reaches its maximum expansion at an epoch t_m, it separates from the continuing background expansion and recollapses. The kinetic energy is zero at the moment of turn-around, and therefore the total energy is equal to the potential energy (per unit mass)

$$E_{\text{tot}} = -\frac{1}{2} \frac{GM}{r_m}.$$

Let r_{eq} be the radius at which a new equilibrium configuration is reached after the collapse. The total energy is conserved, and the virial theorem tells us that the new potential energy is equal to $2E_{\mathrm{tot}}$. Therefore $-\frac{1}{2}\frac{GM}{r_{\mathrm{eq}}} = 2E_{\mathrm{tot}} = -\frac{GM}{r_m}$.

$r_{\mathrm{eq}} = \frac{1}{2}r_m$, and the density $\varrho_{\mathrm{eq}} = 8\varrho_m$. These are only rough estimates, because in a realistic configuration the density will depend on the radius. Deviations from the exact spherical symmetry are also quite probable. In fact, it seems very likely that the collapse will proceed first in one particular direction ("pancake model", see below).

Any dark-matter component is a dissipation-free gas, while dissipative effects can convert the kinetic energy of baryonic matter into radiation. Radiative cooling then allows the baryonic matter to sink to the centre of the collapsed configuration, while the initially well-mixed, dark-matter component separates and forms an extended halo around the central condensation.

If the dissipative collapse of the baryonic matter is halted by angular momentum, a disc is formed. If star formation stops the dissipative collapse of the baryons, a spherical system appears.

The collapse proceeds very rapidly on the free-fall time-scale

$$\tau \simeq (G\varrho)^{-1/2}.$$

The matter condenses into a final state with a central condensation. Figure 11.2 gives an impression of this behaviour from a numerical N-body simulation.

It is quite instructive to compute the density contrast δ_m at the time of maximum expansion t_m. In a background Einstein–de Sitter model ($K = 0$) the mean density $\varrho_b = (6\pi G t^2)^{-1}$, and for a $K = +1$ FL model (cf. Chap. 1)

$$r_m = \frac{2t_m}{\pi}; \quad \frac{2GM}{r_m} = 1.$$

The density in the fluctuation is then

$$\varrho_m = \frac{3\pi}{32 G t_m^2}$$

and

$$\frac{\varrho_m}{\varrho_b} = \frac{9}{16}\pi^2 = 5.6, \tag{11.3}$$

i.e.

$$\delta_m = 4.6. \tag{11.4}$$

This value is independent of initial conditions, and thus it gives a very useful criterion for the turn-around and collapse. If we assume that the collapse proceeds to a virialized state, the sphere shrinks to a radius $r_{\mathrm{vir}} = \frac{1}{2}r_m$,

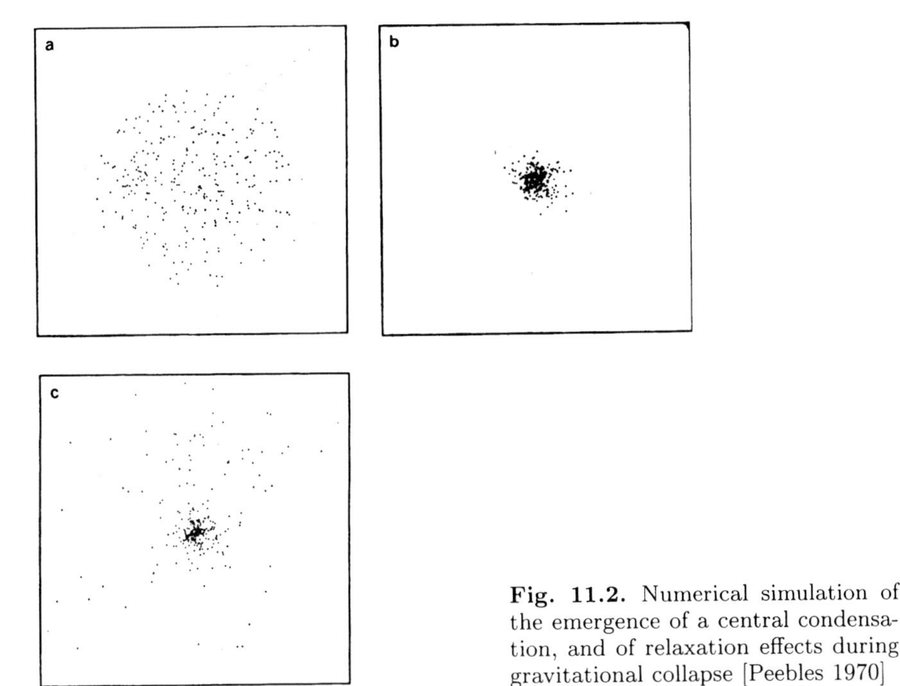

Fig. 11.2. Numerical simulation of the emergence of a central condensation, and of relaxation effects during gravitational collapse [Peebles 1970]

where r_m is the radius at turn-around. The collapse time is $t_c = 2t_m$. Since $\varrho_b(t_m) = 4\varrho_b(t_c)$, the density of the virialized clumps is

$$\varrho_{\text{vir}} = 8\varrho_m = 18\pi^2\varrho_b(t_c), \tag{11.5}$$

about 180 times above the background density. This is again a result which characterizes collapsed virialized clumps of matter independent of initial conditions.

The linear theory gives δ = const. $t^{2/3}$, and $\delta_{\text{lin}} = (\frac{3}{20})(6\pi)^{2/3} = 1.07$ at $t = t_m$. Consequently, small spherical perturbations grow until $\delta_m = 4.6$ (for $\Omega_0 = 1$) or $\delta_{\text{lin}} \approx 1$, then separate from the expansion and collapse. An extrapolation of linear theory gives, for the density contrast of virialized mass concentrations,

$$\delta_{\text{lin,vir}} \simeq 1.69 \equiv \delta_c. \tag{11.6}$$

This is only a formal estimate, useful for defining the collapse time within linear theory. Obviously linear theory can be strictly valid only for $\delta \ll 1$. In the following sections we shall derive results for the case of general fluctuations.

In Chap. 12 we shall briefly discuss how a Gaussian initial distribution of spherical inhomogeneities can reproduce some characteristic features of the galaxy distribution surprisingly well.

11.2 Newtonian Theory of Small Fluctuations

11.2.1 Evolution with Time

The Friedmann equations can be interpreted in a "Newtonian" picture [Weinberg 1972] as

$$\varrho = \varrho_0 \left(\frac{R_0}{R}\right)^3, \tag{11.7}$$

$$\boldsymbol{v} = \boldsymbol{x} \left(\frac{\dot{R}}{R}\right), \quad \boldsymbol{x} = \boldsymbol{x}_0 \left(\frac{R(t)}{R_0}\right) \tag{11.8}$$

$$\boldsymbol{g} = -\boldsymbol{x}\frac{4\pi G\varrho}{3}, \tag{11.9}$$

where $R(t)$ satisfies

$$\dot{R}^2 + K = \frac{8\pi G\varrho R^2}{3}. \tag{11.10}$$

A small perturbation ϱ_1, v_1, g_1 to the background solution satisfies, to first order, the following equations:

$$\partial_t \varrho_1 + \frac{\dot{R}}{R}\varrho_1 + \frac{\dot{R}}{R}(\boldsymbol{x}\cdot\nabla)\varrho_1 + \varrho\nabla\cdot\boldsymbol{v}_1 = 0, \tag{11.11}$$

$$\partial_t \boldsymbol{v}_1 + \frac{\dot{R}}{R}(\boldsymbol{x}\cdot\nabla)\boldsymbol{v}_1 + \frac{\dot{R}}{R}\boldsymbol{v}_1 = -\frac{1}{\varrho}\nabla p_1 + \boldsymbol{g}_1, \tag{11.12}$$

$$\nabla\times\boldsymbol{g}_1 = 0, \quad \nabla\cdot\boldsymbol{g}_1 = -4\pi G\varrho_1.$$

Let us look for adiabatic perturbations

$$p_1 = c_s^2 \varrho_1, \tag{11.13}$$

and plane-wave solutions of the spatially homogeneous equations (11.11), (11.12):

$$(\varrho_1, \boldsymbol{v}_1, \boldsymbol{g}_1) = (\varrho_1(t), \boldsymbol{v}_1(t), \boldsymbol{g}_1(t))\exp(i\boldsymbol{x}\cdot\boldsymbol{q}_1/R(t)). \tag{11.14}$$

Then

$$\partial_t \varrho_1 + 3\frac{\dot{R}}{R}\varrho_1 + i\varrho\frac{\boldsymbol{q}\cdot\boldsymbol{v}_1}{R} = 0, \tag{11.15a}$$

$$\partial_t \boldsymbol{v}_1 + \frac{\dot{R}}{R}\boldsymbol{v}_1 = -\frac{ic_s^2}{\varrho R}\boldsymbol{q}\varrho_1 + \boldsymbol{g}_1, \tag{11.15b}$$

$$\boldsymbol{q}\times\boldsymbol{g}_1 = 0, \quad i\boldsymbol{q}\cdot\boldsymbol{g}_1 = -4\pi GR\varrho_1. \tag{11.15c}$$

Equation (11.15c) can obviously be solved by the ansatz $\boldsymbol{g}_1 = \boldsymbol{g}_1 q$, $iq^2 g_1 = -4\pi GR\varrho_1$; hence

$$g_1 = i\frac{4\pi G\varrho_1 R}{q^2}. \tag{11.16}$$

Decompose $\boldsymbol{v}_1$ into a part $\boldsymbol{v}_{1,\perp}$, perpendicular to $\boldsymbol{q}$, and a part $i\varepsilon$ parallel to $\boldsymbol{q}$:

$$\begin{aligned} \boldsymbol{v}_1 &= \boldsymbol{v}_{1\perp} + i\boldsymbol{q}\varepsilon, \\ \boldsymbol{v}_{1,\perp} \cdot \boldsymbol{q} &= 0, \quad \varepsilon = -i(\boldsymbol{q} \cdot \boldsymbol{v}_1)/q^2. \end{aligned} \tag{11.17}$$

With $\varrho_1(t) \equiv \varrho(t)\delta(t)$, (11.15b) decomposes into

$$\dot{\boldsymbol{v}}_{1\perp} + \frac{\dot{R}}{R}\boldsymbol{v}_{1\perp} = 0 \tag{11.18}$$

and

$$\dot{\varepsilon} + \frac{\dot{R}}{R}\varepsilon = \left(-\frac{c_s^2}{R} + \frac{4\pi G\varrho R}{q^2}\right)\delta. \tag{11.19}$$

Equation (11.15a) gives

$$\dot{\delta} = \varepsilon\frac{q^2}{R}. \tag{11.20}$$

In the linear approximation the rotational modes $v_{1\perp}$ satisfying (11.18) simply decay as $1/R$. We shall discuss below the non-linear behaviour of rotational modes.

The time dependence of δ and ε is more involved: with (11.20), ε can be eliminated from (11.19) and one finally obtains a wave equation for δ:

$$\ddot{\delta} + 2\frac{\dot{R}}{R}\dot{\delta} + \left(\frac{c_s^2 q^2}{R^2} - 4\pi G\rho\right)\delta = 0. \tag{11.21a}$$

From this equation we can recover the Jeans condition (see Sect. 10.2.1). If $R =$ const., set $k := q/R$ (a co-moving wave vector). Then

$$\ddot{\delta} + (c_s^2 k^2 - 4\pi G\varrho)\delta = 0,$$

and

$$\delta = A_k e^{\gamma t + i\boldsymbol{k}\cdot\boldsymbol{x}}$$

implies

$$\gamma^2 + (c_s^2 k^2 - 4\pi G\varrho) = 0;$$

$\gamma = 0$ for $k = k_J = c_s^{-1}\sqrt{4\pi G\varrho}$. For $k \ll k_J$, i.e. $\gamma = \gamma_0 = \sqrt{4\pi G\varrho}$, the exponentially growing mode for the density fluctuation is

$$\delta = e^{\gamma_0 t}\int A_k e^{i\boldsymbol{k}\cdot\boldsymbol{x}} d^3x.$$

$p_1 = 0$ Solutions

If $k \ll k_J$, or equivalently $M_k \gg M_J$ (with $M_k \equiv (4\pi/3)\varrho(2\pi/k)^3$, – satisfied by $M_k > 10^6 M_\odot$ immediately after decoupling), then the pressure term $c_s^2 q^2/R^2$ can be neglected in (11.21a), and we have

$$\ddot{\delta} + 2\frac{\dot{R}}{R}\dot{\delta} - 4\pi G\varrho\delta = 0. \tag{11.21b}$$

Analytic solutions can be found for the FL cosmological models.

A simple method of obtaining these solutions [Peebles 1980] starts from Lemaître's idea that for $p = 0$ and spherical symmetry, each spherical mass shell moves like an FL world-model on its own. The fractional difference of two different homogeneous world-models with slightly different densities ϱ and ϱ_b behaves like the density contrast δ, as we have shown in Sect. 11.1.2.

From $R^3\varrho =$ const., $R \equiv R_b(1-\varepsilon)$ leads to $\delta \equiv (\varrho - \varrho_b)/\varrho_b = 3\varepsilon$. The Friedmann equation $\ddot{R} = (-4\pi/3)G\varrho R$ leads to (11.21b) for δ. If $R(t,K)$ is an FL solution, then $\delta \propto (1/R)(\partial R/\partial K)$ is a solution of (11.21b). The equation $\dot{R}^2 = (8\pi/3)G\varrho R^2 - K \equiv X$ gives solutions $t = \int^R X^{-1/2}dR - C$. K and C should be viewed as arbitrary integration constants. Then

$$\begin{aligned} X^{-1/2}\frac{\partial R}{\partial K} - \frac{1}{2}\int^R dRX^{-3/2} &= 0, \\ X^{-1/2}\frac{\partial R}{\partial C} + 1 &= 0, \end{aligned} \tag{11.22}$$

and the two solutions for $\delta(t)$ are

$$\delta_1 = \frac{X^{1/2}}{R}\int^R X^{-3/2}dR, \quad \delta_2 = \frac{X^{1/2}}{R}. \tag{11.23}$$

Einstein–de Sitter, $K = 0$:
$\delta_1 \propto R$, $\delta_2 \propto R^{-3/2}$.
Direct integration of (11.21b) confirms this result:

$$\begin{aligned} \frac{R}{R_0} &= \left(\frac{H_0 t}{2}\right)^{2/3}, \quad \varrho = (6\pi G t^2)^{-1}, \\ \frac{\dot{R}}{R} &= \frac{2}{3t}; \end{aligned}$$

so (11.21b) reads

$$\ddot{\delta} + \frac{4}{3t}\dot{\delta} - \frac{2}{3t^2}\delta = 0.$$

Solutions are

$$\begin{aligned} \delta_+(t) &= \delta_+(t_R)(t/t_R)^{2/3}, \\ \delta_-(t) &= \delta_-(t_R)(t/t_R)^{-1}. \end{aligned} \tag{11.24}$$

$\delta_+(t_R)$ is the initial amplitude at the time of the decoupling of matter and radiation, t_R. Only for $t > t_R$ is the zero-pressure approximation valid.
$K = +1$ spherical model:
Using (11.22), and remembering that

$$\begin{aligned}\frac{R}{R_0} &= q_0(2q_0 - 1)^{-1}(1 - \cos\theta),\\ H_0 t &= q_0(2q_0 - 1)^{-3/2}(\theta - \sin\theta),\end{aligned}$$

we find the solutions:

$$\begin{aligned}\delta_+(t) \propto D_1 &= \left\{\frac{-3\theta\sin\theta}{(1-\cos\theta)^2} + \frac{5+\cos\theta}{1-\cos\theta}\right\}, \qquad (11.25)\\ \delta_-(t) &\propto \frac{\sin\theta}{(1-\cos\theta)^2}.\end{aligned}$$

$K = -1$, open model:

$$(\cos h\psi - 1)\frac{d^2\delta}{d\psi^2} + \sin h\psi\frac{d\delta}{d\psi} - 3\delta = 0. \qquad (11.26)$$

Solutions:

$$\begin{aligned}\delta_+(t) \propto D_1 &= \left\{\frac{-3\psi\sin h\psi}{(\cos h\psi - 1)^2} + \frac{5+\cos h\psi}{\cos h\psi - 1}\right\},\\ \delta_-(t) &\propto \frac{\sin h\psi}{(\cos h\psi - 1)^2}. \qquad (11.27)\end{aligned}$$

(D_1 is the growth factor, see [Peebles 1980] eqs. (11.9) and (11.16).) For a low-density universe with $\Omega_0 \ll 1$, there is a late time, where $1 + z \ll \Omega_0^{-1}$, and $\sin h \propto t$, $\cos h\psi \propto t$, $\psi \propto \ln t$; thus

$$\delta_+ \sim \delta_+(t_R)\left(-\frac{3t\ln t}{t^2} + \frac{5+t}{t}\right) \sim \text{const.},$$

i.e. the fluctuation does not increase further for large times t. Therefore in a low-density universe the perturbations stop growing at redshifts

$$z_f \simeq \Omega_0^{-1} - 1. \qquad (11.28)$$

In all three cases one finds that for $R(t) \ll R_0$ only δ_+ is a growing solution:

$$\begin{aligned}\delta_+ &\propto t^{2/3},\\ \delta_- &\propto t^{-1}. \qquad (11.29)\end{aligned}$$

Therefore only the δ_+ mode is considered. The present analysis applies for a matter-dominated universe after the time of decoupling, t_R. The density contrast can then grow to the present epoch $t_0 (R_0/R_R \equiv 1 + z_R)$ by a factor [Weinberg 1972]

$$A_0 \equiv \frac{\delta_+(t_0)}{\delta_+(t_R)};$$

A_0 can be computed from (11.23), (11.25), (11.27):

In a $K = +1$ universe:

$$A_0 = \frac{5(1+z_R)}{(1-\cos\theta_0)^3}\{-3\theta_0 \sin\theta_0 + (1-\cos\theta_0)(5+\cos\theta_0)\}. \tag{11.30}$$

$K = 0$ case:

$$A_0 = (1 + z_R). \tag{11.31}$$

$K = -1$ – hyperbolic 3-space

$$A_0 = \frac{5(1+z_R)}{(\cos h\psi - 1)^3}\{-3\psi_0 \sin h\psi_0 + (\cos h\psi_0 - 1)(\cos h\psi_0 + 5)\}. \tag{11.32}$$

For small $q_0 : 0 < q_0 < 1/2$ (i.e. $0 < \Omega_0 < 1$), the perturbations stop growing at $z_f \approx \Omega_0^{-1} - 1$, and

$$A_0 = \frac{5}{2}\Omega_0(1 + z_f). \tag{11.33}$$

For $\Omega_0 = 1$, $A_0 = 1 + z_R$, and for $\Omega_0 \geq 1$ there is an upper limit

$$A_0 = 5(1 + z_R). \tag{11.34}$$

For galaxies to form the non-linear regime must be reached by the linear fluctuations, i.e. they should be able to grow to order unity until the present epoch t_0, $\delta_+(t_0) \sim 1$. Therefore the amplitude at t_R must be larger than $1/A_0$:

$$|\delta_+(t_R)| \geq A_0^{-1}; \tag{11.35}$$

e.g.

$$\delta_+(t_R) \geq (1 + z_R)^{-1} \quad \text{for} \quad \Omega_0 = 1,$$

and

$$\delta_+(t_R) \geq 0.4 \quad \text{for} \quad \Omega_0 = 0.01. \tag{11.36}$$

Of course, the non-linear regime, i.e. the onset of galaxy formations, must have been reached earlier than t_0. The question is, when does δ_+ reach a magnitude of 1? The observations indicate that it must happen at $z \geq 2$.

Adiabatic and Isothermal Fluctuations

So far only adiabatic perturbations have been considered for which the entropy is constant, i.e. $|\delta T/T| = \frac{1}{3}|\delta n/n|$. In the following, isothermal perturbations will occasionally also be discussed. For these perturbations $\delta T = 0$, but the ratio of photon to baryon number in a comoving volume fluctuates. Isothermal fluctuations will be presented in the context of a relativistic description. In general a fluctuation will be a mixture of adiabatic and isothermal fluctuations.

11.2.2 Observational Constraints on Adiabatic Fluctuations

The amplitudes of adiabatic fluctuations at the time of recombination t_R (see (11.35)) have immediate observational consequences for the small-scale angular isotropy of the 3 K background, since $|\delta n/n| = 3|\delta T/T|$. Consider the angular extent θ of an object of linear dimension $D(\theta)$ on a surface with redshift z (Fig. 11.3).

From the derivation following (1.23), we have

$$r_A\theta = D(\theta) \tag{11.37}$$

for small angles θ, and

$$r_A = [H_0q_0^2(1+z)^2]^{-1}[q_0z + (q_0-1)(1+2q_0z)^{1/2} - 1]. \tag{11.38}$$

The mass within $D(\theta)$ is

$$M = \frac{4\pi}{3}D^3(\theta)\varrho, \tag{11.39}$$

and its angular extent is

$$2D(\theta) = \frac{zq_0}{q_0^2(1+z)^2H_0}\theta \quad \text{for} \quad z \gg 1, \tag{11.40}$$

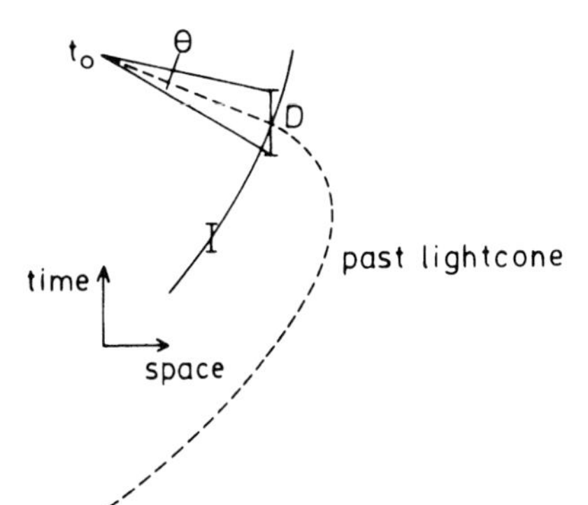

Fig. 11.3. Angular extent of an object at distance r_A.

i.e.

$$\frac{\theta}{2} = q_0 H_0 (1+z) \left(\frac{3M}{4\pi \varrho} \right)^{1/3} = q_0 H_0 \left(\frac{3M}{4\pi \varrho_0} \right)^{1/3} \tag{11.41}$$

since $\varrho \propto (1+z)^{-3}$). Finally

$$\theta = 40''(h_0\Omega_0)(M/10^{11}M_\odot)^{1/3}. \tag{11.42}$$

For $h_0 = 0.7$ ($H_0 = 70$) and $M = 10^{14}M_\odot$ (mass of a cluster),

$$\theta = 4'\Omega_0. \tag{11.43}$$

Measurements of the 3 K microwave background give $|\delta T/T| \simeq 10^{-5}$ on angular scales of 10°, i.e. for adiabatic fluctuations $|\delta n/n| = 3|\delta T/T| \simeq 3 \times 10^{-5}$. This contradicts the limits of (11.36) by an order of magnitude. They require $|\delta n/n| > 10^{-3}$ at $z = z_R$! The model for adiabatic baryonic fluctuations seems to ruled out.

There are three loopholes by which one can hope to escape the fatal conclusions that galaxies cannot be made in an isotropic and homogeneous universe:

1. An epoch of re-ionization at $z \geq 10$ at which the universe became optically thick once more would smooth out the small-scale irregularities. The energy for this re-ionization could be derived from the UV radiation of a generation of high-mass stars. On large angular scales, however, the re-ionization could not affect the radiation anisotropy: at $z = 10$ the horizon size $ct(z=0)$ has an angular extent of $\theta = 36(\Omega_0)^{1/2}$ degrees. A detailed analysis [Silk and Efstathiou 1983; Primack 1986] gives a limit $\theta_c \geq 5(\Omega_0 x)^{1/3}$ degrees, where x is the fractional ionization. Thus at angular scales $\geq 5°$ re-ionization does not erase fluctuations.
2. One can study isothermal fluctuations (see Sect. 11.3), where $\delta T = 0$.
3. Some form of non-baryonic dark matter (massive neutrinos, gravitinos, etc.) may change the picture. This will be discussed below.

11.2.3 Non-linear Newtonian Perturbations

The perturbation variables

$$\delta \equiv \varrho(x,t)/\varrho_b - 1,$$
$$\boldsymbol{v} \equiv R(t)\dot{\boldsymbol{x}}$$

can be inserted into the equations of motion without linearizing the system. Then the evolution equation for δ is

$$\ddot{\delta} + 2\left(\frac{\dot{R}}{R}\right)\dot{\delta} = \frac{\nabla^2 p}{\varrho_b R^2} + \frac{1}{R^2}\boldsymbol{\nabla}\cdot[(1+\delta)\boldsymbol{\nabla}\phi] + \frac{1}{R^2}[(1+\delta)v^i v^j]_{,ij}. \tag{11.44}$$

Here ϕ is the gravitational potential and

$$\nabla^2\phi = 4\pi G\varrho_b R^2\delta.$$

11.3 Relativistic Theory of Small (Linearized) Fluctuations

A relativistic description of fluctuations has to be used at early epochs for several reasons: first, the pressure p/c^2 becomes gravitationally important; second, for a high background density ϱ_b even a small fluctuation δ can have a large effect on the curvature; and third, the horizon scale $ct = 1/H$ varies in proportion to the time. The latter implies any primeval fluctuation with a wavelength $\lambda \propto R \propto \sqrt{t}$, larger than the dimension of the horizon at an epoch close enough to $t = 0$. $t = H^{-1}$, gives in a way the range of effective causal influences in the evolution of linear perturbations. For $\lambda H < 1$ (loosely speaking, "inside the horizon") perturbations of wavelength λ grow, but for $\lambda H > 1$ ("outside the horizon") growth is suppressed.

The usual approach consists in following the evolution of perturbations to an FL background in a linear approximation. The relativistic treatment of such fluctuations [Lifshitz and Khalatnikov 1963; Weinberg 1972; Peebles 1980] has until recently been carried out with a specific choice of coordinate systems, usually the so-called time-orthogonal or synchronous gauge. A perturbation actually transforms the original space-time into a new one, and the two then have to be compared within the same coordinate system. How can one be sure that a real, physical change is described, not just the original space-time in a changed coordinate system? Obviously a description of the perturbation in terms of "gauge-invariant" variables is required, i.e. quantities which do not change under coordinate transformations. Such variables permit a separation of coordinate effects and real physical contributions to the perturbation.

A gauge-invariant formalism has been proposed [Bardeen 1980] which completely specifies the perturbation of an FL cosmological background in terms of gauge-invariant quantities. In the appendix we give a short exposition of these ideas, and in the following we present a few examples. Our main sources of information are the highly recommended reviews of [Kodama and Sasaki 1984; Durrer 1994; Mukhanov et al. 1992].

11.3.1 Gauge-Invariant Formalism

The "3+1" formalism of GR offers a convenient way of describing the basic steps leading to a gauge-invariant formalism (e.g. [Durrer and Straumann 1988; Durrer 1994]. In the Appendix we describe some basic steps necessary in setting up such a formalism. The outcome is that any scalar mode (only scalar modes contribute to density perturbations) can be described by two gauge-invariant metric amplitudes A, B (or equivalently Φ and Ψ, the Bardeen variables) and four gauge-invariant matter variables, Δ (density contrast), Π (anisotropic stress), Γ (entropy production density), and V (shear velocity). Δ is a gauge-invariant variable which is equal to the density contrast in the slicing, where the matter velocity is orthogonal to the family of constant-time hypersurfaces $\{\Sigma_t\}$. V is the amplitude of the shear of the matter motion. $\Gamma = 0$ for adiabatic perturbations, for which

$$\delta p / \delta \varrho = \dot{p} / \dot{\varrho}.$$

It can be shown that Γ is proportional to the divergence of the entropy density $\Gamma \propto s^{\mu};\mu$ in relativistic thermodynamics.

The constraint equations (A.19),(A.20) relate (Δ, V) and (A, B) or (Ψ, ϕ) in an algebraic manner. This fact can be used to rewrite the dynamical equations as coupled differential equations for the matter variables only: (A.21), (A.22). Γ, Π appear as source terms in those dynamical equations for Δ and V. In the following we want to illustrate this formalism with a few simple examples.

11.3.2 Adiabatic Perturbations of a Single Ideal Fluid

In the case of a single perfect fluid with $\Gamma = \Pi = 0$ (adiabatic perturbations) we have $\Psi = -\Phi$. The evolution equations (A.21),(A.22) lead for the simple case $K = 0$ to the following equation:

$$(\varrho R^3 \Delta)'' + (1 + 3c_s^2)\frac{R'}{R}(\varrho R^3 \Delta)' + \left\{ k^2 c_s^2 - \frac{3}{2}(1+w)\left(\frac{R'}{R}\right)^2 \right\} (\varrho R^3 \Delta) = 0. \tag{11.45}$$

(The prime on Δ denotes differentiation $d\Delta/d\eta$, with η a conformal time coordinate $dt = R d\eta$.) Here $w = p/\varrho$, $c_s^2 = p'/\varrho'$; k is the wave number. The term $\frac{8\pi G}{3}\varrho R^2$ from (A.22) has been replaced by $(\frac{R'}{R})^2$. In the special case where w is constant, the background density and $R(\eta)$ follow power laws:

$$R = R_0(\eta/\eta_0)^\beta, \quad \varrho = \frac{3}{8\pi G}\beta^2 R^{-2}\eta^{-2}, \tag{11.46}$$

with $\beta = 2/(3w+1)$.

Let us introduce $x \equiv k\eta$ and

$$f \equiv \varrho R^3 \Delta. \tag{11.47}$$

$x/\beta = kR/R'$ measures the ratio of the size of the Hubble horizon to the wavelength k^{-1}. The function f solves the differential equation

$$\left\{ \frac{d^2}{dx^2} + \frac{2}{x}\frac{d}{dx} + c_s^2 - \frac{\beta(\beta+1)}{x^2} \right\} f = 0. \tag{11.48}$$

Thus the general solution is

$$f_\beta(c_s x) = C_0 j_\beta(c_s x) + D_0 n_\beta(c_s x),$$

where j_β and n_β are spherical Bessel functions of order β, and C_0 and D_0 are arbitrary constants. The gauge-invariant quantities Δ and Ψ are

$$\begin{aligned}\Delta &\simeq \eta^{2-\beta} f_\beta(c_s k\eta) \underbrace{\frac{8\pi G}{3}\beta^{-2}\eta_0^\beta R_0^{-1}}_{A_0}, \\ \Psi &= -\frac{3}{2}k^{-2}\eta^{-\beta} f_\beta(c_s k\eta) A_1, \end{aligned} \tag{11.49}$$

$(A_1 = A_0\beta^2)$.

For $c_s x \ll 1$, i.e. when the wavelength exceeds the sound horizon, the solutions show a power-law behaviour

$$f_\beta \simeq C x^\beta + D x^{-(\beta+1)}.$$

In terms of Δ and Ψ the solution is

$$\begin{aligned}\Delta &= \Delta_0\eta^2 + \Delta_1\eta^{1-2\beta}, \\ \Psi &= \Psi_0 + \Psi_1\eta^{-2\beta-1}, \end{aligned} \tag{11.50}$$

with appropriate constants $\Delta_0, \Delta_1, \Psi_0, \Psi_1$.

The relation between the invariant amplitude Δ and the density perturbation δ depends on the slicing of the space-time. Since $\Delta = \delta$ on velocity-orthogonal hypersurfaces, one identifies $\Delta_0\eta^2$ as a growing mode, and $\Delta_1\eta^{1-2\beta}$ as a decaying mode.

Thus the growing mode behaves as

$$\Delta \propto \begin{cases} R^2 \text{ in the radiation-dominated epoch } (\beta = 1), \\ R \text{ in the matter-dominated era } \qquad (\beta = 2). \end{cases}$$

For $c_s x \gg 1$, i.e. for the wavelength $\lambda = R\frac{2\pi}{k}$ smaller than the sound horizon c_s/H, the solutions show an oscillatory behaviour:

$$f_\beta = B_0 x^{-1}\cos(c_s x - \alpha_\beta) + B_1 x^{-1}\sin(c_s x - \alpha_\beta).$$

Then

$$\begin{aligned}\Delta &= \Delta_0\eta^{1-\beta}\cos(c_s x - \alpha_\beta), \\ \Psi &= \Psi_0\eta^{-(\beta+1)}\cos(c_s x - \alpha_\beta). \end{aligned} \tag{11.51}$$

The "density perturbation" Δ has a growing mode for $\beta < 1$, i.e. for $w > \frac{1}{3}$. All solutions for $c_x x \gg 1$ decay for $\beta > 1$, i.e. $w < \frac{1}{3}$.

For the limiting case $w = \frac{1}{3}$ (radiation or relativistic particles, $p = \frac{1}{3}\varrho$), $\beta = 1$, and $\Delta = \Delta_0\cos(c_s x - \alpha_\beta)$. The mode oscillates. The growing mode is generally identified as a gravitational instability (Jeans criterion). The characteristic Jeans wave number is obtained from the condition that the coefficient of f in (11.48) vanishes, i.e.

$$c_s^2 x_J^2 = \beta(\beta + 1).$$

This in fact gives the usual (relativistic) Jeans criterion,

$$k_J^2 c_s^2 = \frac{3}{2}(1+w)(R'/R)^2,$$

and therefore with $\lambda_J \equiv R\frac{2\pi}{k_J}$, the comoving wavelength,

$$\lambda_J = \left(\frac{\pi c_s^2}{G(\varrho + p)}\right)^{1/2}.$$

We see that the Jeans criterion has a truly invariant meaning. (Also, the present derivation does not suffer from the "Jeans swindle", i.e. the use of a background not obeying the dynamics.)

Another interesting case is matter without pressure $p = 0$, i.e. $c_s^2 = w = 0$. This is called "dust", and it applies to non-interacting, non-relativistic particles.

Since $\varrho R^3 =$ const., the perturbation equation (11.45) reduces to

$$\Delta'' + \frac{R'}{R}\Delta' - \frac{3}{2}\left(\frac{R'}{R}\right)^2 \Delta = 0. \tag{11.52}$$

Since $R \propto \eta^2$, we have

$$\Delta'' + \frac{2}{\eta}\Delta' - \frac{6}{\eta^2}\Delta = 0. \tag{11.53}$$

The general solution $\Delta = A\eta^2 + B\eta^{-3}$ leads to the growing mode solution, as before ($\beta = 2$)

$$\begin{aligned} \Psi &= \Psi_0, \\ \Delta &= \Delta_0 \eta^2. \end{aligned} \tag{11.54}$$

Here, there is no difference for wavelengths of the perturbation inside or outside of the horizon, as would be expected for dust, because these particles move just according to the initial conditions.

Finally, let us look at dust perturbations in an uncoupled radiation field. This is a good model for the physically relevant case of baryons or dark matter particles (ρ_m) embedded in a radiation field (ϱ_r), but without interactions between radiation and matter.

The perturbation equation for Δ is (11.45) as before, with $w = c_s^2 = 0$, but now there are two components, and $\left(\frac{R'}{R}\right)^2$ is given by $\frac{8\pi G}{3}(\varrho_r + \varrho_m)$, where the sum $\varrho_m + \varrho_r$ appears. From (A.21) we see that the last term in (11.45) must be written as

$$-4\pi G \varrho_m R^2.$$

From

$$\Delta'' + \frac{R'}{R}\Delta' - 4\pi G \varrho_m R^2 \Delta = 0, \tag{11.55}$$

we change to a new variable $y \equiv \varrho_m/\varrho_r = R/R_{eq}$, where R_{eq} is the expansion factor at the epoch, when $\varrho_m = \varrho_r$.

Finally, in terms of y,

$$\frac{d^2\Delta}{dy^2} + \frac{2+3y}{(1+y)y}\frac{d\Delta}{dy} - \frac{3}{(1+y)y}\Delta = 0. \tag{11.56}$$

Following [Mészáros 1975] we set $\frac{d^2\Delta}{dy^2} = 0$, and find a solution

$$\Delta = 1 + \frac{3}{2}y. \tag{11.57}$$

For $y \ll 1$, i.e. radiation dominance, $\Delta \simeq$ const. The matter fluctuations stay constant, as long as the uncoupled radiation field dominates ("stagspansion"; "Mészáros effect" [Mészáros 1974, 1975]). The matter distribution is stabilized against fluctuations, when $\varrho_r \gg \varrho_m$, since the time-scale for expansion $(G\varrho_r)^{-\frac{1}{2}}$ is much less than the time-scale for growth $(G\varrho_m)^{-\frac{1}{2}}$ (Jeans criterion). On the other hand, $\varrho_m \gg \varrho_r$, i.e. $y \gg 1$, leads back to the dust case $\Delta \propto \eta^2$.

11.3.3 Perturbation Modes

We do not know the physics of the early universe well enough to derive the amplitudes of the density fluctuations; even the relative amplitudes of perturbations in the different components are not predicted. In these chapters I usually consider the adiabatic mode, which is the fastest growing one. If we consider perturbations of relativistic particles and radiation Δ_R, of baryons Δ_B, and of cold dark matter Δ_{CDM}, then for this mode (which is characterized by $\Gamma = 0$)

$$\Delta_R = \frac{4}{3}\Delta_B = \frac{4}{3}\Delta_{\text{CDM}}. \tag{11.58}$$

The isothermal mode with a constant radiation field, $\Delta_{\text{rad}} = 0$, is not gauge-invariant. The quantity S_{mr} can serve as a gauge-invariant representative, and

$$S_{mr} = \Delta_B = \Delta_{\text{CDM}}. \tag{11.59}$$

The mode with $\Phi = \Psi = 0$ describes entropy perturbations. It is often called isocurvature mode (why?). In models with CDM,

$$\Delta_R = \frac{4}{3}\Delta_B = -\Delta_{\text{CDM}}/A \tag{11.60}$$

(all densities at the initial time $\varrho(i)$). The constant A is determined from the requirement that the total curvature is zero initially.

In models with CDM, neutrinos (ϱ_ν), baryons, and radiation,

$$A = \left(\varrho_\nu + \frac{3}{4}\varrho_B + \varrho_{\rm rad} \right) / \varrho_{\rm CDM} . \tag{11.61}$$

In models with baryons, but no CDM, the isocurvature mode has

$$\Delta_B = -\Delta_{\rm rad} \left(\frac{\varrho_{\rm rad}(i) + \varrho_\nu(i)}{\varrho_B(i)} \right) . \tag{11.62}$$

The adiabatic mode would be expected to be dominant in most scenarios. Only very specific configurations which suppress the growth of the adiabatic mode may lead to isocurvature or other perturbation modes [Efstathiou and Bond 1986].

The baryonic isocurvature mode has been used in models, but it does not have a basis in a well-understood physical process.

11.4 The Power Spectrum of the Density Fluctuations

The basic assumption of observational cosmologists is that the observable part of the universe is typical of the whole, i.e. we can observe a fair sample. Since the galaxies evidently are distributed in a highly structured and clumpy pattern, the basic homogeneity and isotropy of the universe can only be true in a statistical, average sense. The matter distribution in the universe is assumed to be the result of a homogeneous and isotropic random process. Samples taken from well-separated volumes are uncorrelated, and a collection of such samples is like a statistical ensemble generated by many independent applications of the process. Statistics such as correlation functions are averages across the ensemble, and such averages can be replaced by spatial averages if what we see is a fair sample. Whether a particular data set satisfies this assumption must be tested. For the description of the cosmic density field $\varrho(\boldsymbol{x}, t)$ we rely on a fair sample hypothesis, and we consider statistical averages of the density contrast δ as the basic quantities in its description, i.e. we deal with spatial averages

$$\langle \delta \rangle \equiv \frac{1}{V_U} \int \delta(\boldsymbol{x}, t) d^3x \tag{11.63}$$

(V_U is a typical volume). From the defintion (11.63) it is evident that $\langle \delta \rangle$ exhibits the same time dependence as $\delta(\boldsymbol{x}, t)$.

11.4.1 The rms Fluctuation Spectrum

Let us say a few words about the power spectrum of dimensionless density fluctuations $\delta(\boldsymbol{x}, t)$. The definition $\varrho_b(1 + \delta(\boldsymbol{x}, t)) = \varrho(\boldsymbol{x}, t)$ gives an rms fluctuation spectrum

$$\left\langle \frac{\delta\varrho}{\varrho} \right\rangle = (\langle \delta^2 \rangle - \langle \delta \rangle^2)^{1/2}.$$

In linear approximation $\langle \delta \rangle = 0$, and $\langle \delta^2 \rangle \ll 1$. A Fourier decomposition

$$\delta(\boldsymbol{x}, t) = \int d^3k e^{i\boldsymbol{k}\cdot\boldsymbol{x}} \delta_k(t)$$

leads to

$$\int |\delta_k|^2 d^3k \equiv \langle \delta^2 \rangle . \tag{11.64}$$

$P(\boldsymbol{k}) \equiv |\delta_k|^2$ is called the power spectrum of density fluctuations. $P(\boldsymbol{k}) \equiv P(k)$ because of rotational invariance. The correlation function $\xi(x)$ (cf. Chap. 10) can be expressed as

$$\xi(x) = \frac{V_U}{2\pi^2} \int P(k) \frac{\sin kx}{kx} k^2 dk, \tag{11.65}$$

$$P(k) = \frac{4\pi}{V_U} \int \xi(x) \frac{\sin kx}{kx} x^2 dx. \tag{11.66}$$

$\xi(x)$ and $P(k)$ are a Fourier transform pair.

As mentioned in Chap. 10 $\xi(r)$ is the excess probability over a random value of finding objects at separation r. The Fourier components δ_k are complex numbers, where the real amplitude defines a power spectrum; the complex phase factor also plays a role in determining the shape of a fluctuation. $\xi(x)$ and $P(k)$ do not retain the phase information of the fluctuations.

The mass within a sphere of radius r around $\boldsymbol{x}_0$ is [Peebles 1980]

$$\begin{aligned} M(\boldsymbol{x}_0) &= \int d^3x \phi(\boldsymbol{x} - \boldsymbol{x}_0) \varrho(x) \\ &= \int d^3x \phi(\boldsymbol{x} - \boldsymbol{x}_0) \varrho_b (1 + \delta(\boldsymbol{x}, t)) \\ &\equiv \langle M \rangle + \delta M, \end{aligned} \tag{11.67}$$

where

$$\delta M = \int d^3x \phi(\boldsymbol{x} - \boldsymbol{x}_0) \varrho_b \delta.$$

$\phi(x)$ is a selection function defining the shape of the fluctuation that contains M: a spherical step function

$$\phi(s) = \begin{cases} 1, & s \le r, \\ 0, & s > r, \end{cases} \tag{11.68}$$

defines a sharp spherical inhomogeneity; to smooth out the edges one can choose $\phi(s) = \exp(-s^2/r^2)$, a Gaussian selection-function. It follows that

$$\frac{\delta M}{M}(\boldsymbol{x}_0) = \int d^3k e^{i\boldsymbol{k}\cdot\boldsymbol{x}_0} \delta_k W(\boldsymbol{k}), \tag{11.69}$$

where $W(\boldsymbol{k})$ is the "window function" [Peebles 1980]

$$W(\boldsymbol{k}) = \frac{1}{V}\int d^3 s \phi(s) e^{i\boldsymbol{k}\cdot\boldsymbol{s}}. \tag{11.70}$$

For a sharp sphere

$$W(k) = \frac{\sin kr}{kr}. \tag{11.71}$$

For a Gaussian sphere

$$W(k) = e^{-k^2 r^2/2}. \tag{11.72}$$

Averaging over $\boldsymbol{x}_0$ leads to an rms spectrum,

$$\begin{aligned}\left\langle\left(\frac{\delta M}{M}\right)\right\rangle &= \int d^3 k P(k)\left|W(kr)\right|^2 \\ &= 4\pi\int_0^\infty \frac{dk}{k}[k^3 P(k)]W(kr)|^2.\end{aligned} \tag{11.73}$$

$k^3|\delta_k|^2$ is the contribution per wave number interval dk/k. For $W^2 \sim e^{-k^2r^2/2}$ we can approximate the integral by

$$\left\langle\left(\frac{\delta M}{M}\right)^2\right\rangle = 4\pi\int_0^{k_r} dk\; k^2 P(k) \quad \text{with} \quad k_r^{-1} \sim r. \tag{11.74}$$

Generally a power-law spectrum is assumed for the Fourier components,

$$P(k) \propto k^n. \tag{11.75}$$

Then $\langle(\delta M/M)^2\rangle = k_r^{3+n}$.

We can easily refer the power spectrum at a fixed epoch to the spectrum at horizon crossing as follows. The quantity k^{3+n} is characteristic of the rms fluctuation. As long as the fluctuation is outside the horizon it grows in proportion to the square of the expansion factor R^2. The wave number at horizon crossing time is $k_H \propto R^{-1}$.

The relative amplitudes at horizon crossing of two fluctuations k_1^{3+n} and k_2^{3+n} are then simply $(k_2/k_1)^{n+3}(R_2/R_1)^4 = (k_2/k_1)^{n-1}$. Thus we find that a special power-law $n = 1$ [Zel'dovich 1972] gives constant amplitudes at horizon-crossing time. This property can be translated into a statement about the mass contained within a region of size $\sim k^{-1}$:

$$\frac{\delta M}{M} \propto k^{(3+n)/2} \propto M_k^{-\alpha}, \tag{11.76}$$

with $6\alpha - 3 = n$. $a = 2/3$ is the scale-free spectrum.

11.4.2 Change in the Linear $P(k)$ at $t = t_{eq}$

The initial power spectrum $P_i(k)$ must be derived from physical processes in the early universe. Up to now, only the model of the inflationary universe has produced at least an idea of how quantum fluctuations just after the big bang can lead to density perturbations at later epochs. This was discussed in Chap. 9, and some remarks will follow in Sect. 11.5. For the moment we accept an *ad hoc* ansatz of a single power-law form,

$$P_i(k) = A_i k^n. \tag{11.77}$$

An argument which makes this seem not completely *ad hoc* is that (11.77) does not introduce any characteristic scales. This seems reasonable as long as there is no compelling reason from physics to have such scales.

It has previously been pointed out [Harrison 1970] that the index n is constrained because for $n > 1$ there would be an overproduction of small-mass objects, like black holes, while for $n < 1$ the large scales would have too high amplitudes which could destroy the global isotropy and homogeneity of the universe. Harrison concluded that $n = 1$ was a preferred value. As we have seen, $n = 1$ has the consequence that in the matter-dominated epoch fluctuations of different wavelengths $2\pi/k$ have the same amplitude at the time when their scale $k^{-1}R(t)$ is equal to the horizon scale $\lambda_H = H^{-1}(t)$. On each scale the primordial fluctuations larger than the horizon size just follow the evolution equation of an FL model. The spectrum remains unchanged. But once the horizon scale encompasses the fluctuation, the spectrum is changed, because the density perturbations of non-relativistic matter ϱ_m do not grow, as long as the energy density in relativistic particles ϱ_r dominates. This is the Mészáros effect discussed in the previous section (cf. (11.57)). From the time t_{eq}, when $\varrho_m = \varrho_r$, the fluctuations grow. This defines a characteristic scale, namely the size of the horizon at cosmic time t_{eq}:

$$\lambda_{H,eq} \approx \int_0^{t_{eq}} \frac{dt}{R(t)} \propto (\Omega_m h^2)^{-1}.$$

Since perturbations grow $\propto R^2$ for $\frac{2\pi}{k} > \lambda_H$, and stay constant for $\frac{2\pi}{k} < \lambda_H$, as long as $R < R_{eq}$, the initial power spectrum is changed for $k > k_{eq} \equiv \frac{2\pi}{\lambda_{H,eq}}$.

Asymptotically, for $R \gg R_{eq}$, we have

$$P(k) = A_i \begin{cases} k^n & (k < k_{eq}), \\ k^{n-4} & (k > k_{eq}). \end{cases} \tag{11.78}$$

The power spectrum at the present epoch should exhibit a break at k_{eq}. The scale λ_{eq} depends on $(\Omega_m h^2)^{-1}$, and thus the break in the power spectrum measures the energy density of non-relativistic matter.

The components of the cosmic matter also introduce changes in the primordial $P(k)$. We shall comment on these below. In general one writes

$$P_{\rm lin}(k,t) \equiv C(t)P_i(k)T^2_{\rm lin}(k) \tag{11.79}$$

with a linear transfer function describing these changes, and a separate factor $C(t)$ describing the linear growth.

11.4.3 Normalization

The amplitude $C(t)A_i$ must be fixed, and this is usually done by a comparison with astronomical data. Under the assumption that the linear theory is valid until the present epoch, we write $C(t) = [D_1(t)/D_1(t_0)]^2$, where $D_1(t)$ is the linear growth function. To determine A_i, one way is to use the COBE results which apply to very large scales, such that even at present $T_{\rm lin} = 1$.

For flat models

$$A_{\rm COBE} = \left(\frac{96\pi^2}{5}\right)\Omega_0^{-1.54}H_0^{-4}\left(\frac{Q_{\rm rms}}{T_{\gamma 0}}\right)^2,$$

where $Q_{\rm rms}$ is the COBE rms quadrupole fluctuation. Then at large scales, where $T_{\rm lin} = 1$, all the models have the same $P(k)$.

Another popular way to normalize makes use of the amplitude of the mass fluctuation in randomly placed spheres of radius S:

$$\left\langle \left(\frac{\delta M}{M}\right)^2 (S)\right\rangle = \int_0^\infty W^2(kS)P(k)\frac{k^2 dk}{2\pi^2}.$$

W is a filter function. For a top-hat window

$$W_{\rm TH}(x) = \frac{3}{x^3}(\sin x - \cos x).$$

The corresponding mass fluctuations for galaxies can be derived from the two-point correlation function ξ_{gg}. At a scale of $8h^{-1}$Mpc one finds that

$$\sigma_{8,g} \equiv \left(\frac{3}{4\pi S^3}\right)\left(\int_V dV_1 dV_2 \xi_{gg}(r_{12})\right)^{1/2} \simeq 0.96 \tag{11.80}$$

[Peebles 1980].

It has turned out that good fits to the observations can only be achieved in dark matter (DM) models of galaxy formation, if σ_8 from the models is set equal to $b^{-1}\sigma_{8,g}$, where b is called a biasing parameter (e.g. [Davis et al. 1985]).

In a CDM model with $P_i(k) \propto k, \Omega = 1$, a value of $b \simeq 3$ is required to achieve a reasonable fit to ξ_{gg}. The physical basis for the bias is the idea that the galaxies do not trace the DM distribution, but rather that peaks in the DM density field correspond to galaxies. The value of b is introduced *ad hoc*, however, and the uncomfortable feeling remains that here is another quantity which is treated as a free parameter, but which in fact is determined by the physics of galaxy formation.

11.5 Primeval Fluctuation Spectrum

As we have seen in previous sections, the time evolution of a density perturbation, once its scale is less than $1/H$, can be traced in some detail by simple physical arguments. Very little growth is possible for non-baryonic components as long as the radiation energy dominates, and no growth occurs for baryonic fluctuations until the decoupling of matter and radiation. From a redshift z_{eq}, where radiation and matter density are equal, the fluctuations can grow into bound structures until the present epoch. Where does the initial spectrum of density perturbations come from? The general belief is that at some very early epoch some unspecified physical process produced fluctuations of sufficient amplitude.

But at $T_r = 10^{12}$ K, for example, the mass within the horizon is only $M_H = 0.6 M_\odot$. On the other hand, when a galactic mass comes within the horizon the amplitude of any primeval fluctuation is reduced by a factor $N^{-1/2}$, where $N^{1/2} = (M_G/0.6M_\odot)^{1/2} = 10^6$. Even if the amplitude was initially $\delta_H = 1$, the final $\delta \approx 10^{-5}$ is much too small!

In present attempts at describing the galaxy formation process one just *postulates* a spectrum of initial fluctuations. A spectrum of fluctuations as they cross the horizon, i.e. as $\lambda = \int_0^t \frac{dt}{R(t)}$, may be

$$\delta_H = \kappa \left(\frac{M_H}{M_*} \right)^{-\alpha} ,$$

where α and M_* are free parameters. The suggestion by Harrison and Zel'dovich [Harrison 1970; Zel'dovich 1970] is favoured: $\alpha = 0$. The amplitude $\kappa = 10^{-5}$ is set so as to avoid problems with the observational constraints from the CMB. The scale-free, $\alpha = 0$, spectrum seems reasonable, as was mentioned in Sect. 11.4.2. Thus for $\alpha > 0$ a cut-off $\delta \ll 1$ for $M_H \leq M_*$ would be required [Silk and Efstathiou 1983].

There is no compelling physical reason for adopting such a fluctuation spectrum. Recently, however, it has been shown ([Brandenberger and Kahn 1984] and references therein) that the model of the "inflationary universe" predicts a spectrum with $\alpha = 0$. In Sect. 11.5.2 we want to describe briefly the way in which the "inflation" mechanism leads to a definite prediction for the primeval fluctuation spectrum.

11.5.1 Attempts to Derive δ_H

The generation of density perturbations in the early universe can be investigated only in the context of a relativistic formalism, since any specific dimension will be larger than the Hubble radius, $\lambda H(t) > 1$, for early times. Several examples of the generation of density inhomogeneities on superhorizon scales have been investigated [Kodama and Sasaki 1984]. If, by some unspecified mechanism, isotropic stress perturbations occur in a component with density ϱ_1 of the cosmic matter at a temperature T, then the amplitude

$$|\Delta(t_H)| \simeq 1.4 \times 10^{-41}(\varrho_1/\varrho)^{1/2}(T/T_{Pl})^{-3/2}(M_B/10^{12}M_\odot)^{-1/2}, \quad (11.81)$$

where M_B is the mass contained in the fluctuation. Clearly the amplitude is very small for $T \geq 10\,\text{keV} \sim 10^8\,\text{K}$. In a radiation-dust universe one finds that the amplitude of density perturbations when they enter the horizon is at most of the order of the original stress perturbation [Kodama and Sasaki 1984].

The conclusion is that commonplace physical processes are probably not sufficient to generate the fluctuation spectrum. An appeal to exotic phenomena seems necessary. Perhaps the generation of inhomogeneities from quantum phenomena is a viable alternative. The relativistic formalism described above has, however, not yet been adapted to such phenomena.

11.5.2 Perturbations in the Inflationary Universe

The generation of density perturbations within an inflationary model was described in Chap. 9. Here let me just recall a few points. Quite generally, we consider quantum fluctuations $\delta\varphi$ around a classical solution $\varphi_0(t)$ for the scalar field

$$\varphi = \varphi_0(t) + \delta\varphi.$$

The action for the field φ is

$$S = \int \left[\frac{1}{2}\partial^\mu\varphi\partial_\mu\varphi - -V(\varphi)\right]\sqrt{-g}d^4x,$$

and the homogenous background field $\varphi_0(t)$ satisfies

$$\varrho = \frac{1}{2}\dot{\varphi}_0^2 + V(\varphi_0),$$
$$p = \frac{1}{2}\dot{\varphi}_0^2 - V(\varphi_0).$$

Consider scalar fluctuations (no gravitational waves, no vector perturbations) without shear. Then we can write

$$ds^2 = (1 - 2\Psi)dt^2 - (1 - 2\Psi)R^2(t)dl^2, \quad (11.82)$$

where Ψ is the Bardeen potential of (A.20). The equation of motion $\Box_b\varphi = -\frac{\partial V}{\partial \varphi}$ is linearized with respect to Ψ and $\delta\varphi$, and in the long wavelength limit $\lambda > 1/H$ the gauge-invariant relation between metric and $\delta\varphi$ perburbations is

$$\Psi = -\frac{1}{2}\frac{V_{,\varphi}}{V}\delta\varphi. \tag{11.83}$$

For $\delta\varphi$ we insert the expectation value $\delta\varphi_k \sim \langle\hat{\varphi}_k\hat{\varphi}_k\rangle^{1/2}$ of the 2-point function of a quantized scalar field on de Sitter space. Outside the horizon (with $\delta\ddot{\varphi} \simeq \theta$),

$$3H\delta\dot{\varphi} + \left(V_{,\varphi\varphi} - \frac{V_{,\varphi}^2}{V}\right)\delta\varphi = 0,$$

and therefore

$$\delta\varphi = C\frac{V_{,\varphi}}{V},$$
$$\Psi = -\frac{1}{2}C\frac{V_{,\varphi}^2}{V^2}.$$

The boundary condition

$$\delta\varphi(k = HR) = H(k = HR) = C\frac{V_{,\varphi}}{V}$$

determines the constant C. Finally, the components of Ψ in a harmonic decomposition are

$$\Psi_k = L_{Pl}\left(\frac{V^{2/3}}{V_{,\varphi}}\right)\left(\frac{V_{,\varphi}}{V}\right)^2(t), \tag{11.84}$$

where L_{Pl} is the Planck length $\propto G^{-1/2}$. The term $L_{Pl}\left(\frac{V^{2/3}}{V_{,\varphi}}\right)$ evaluated at horizon crossing, $k = HR$, describes the spectral distribution, while $\left(\frac{V_{,\varphi}}{V}\right)^2$ gives the amplitude.

Formula (11.84) is the general expression for the metric fluctuations in a gauge-invariant formalism.

For $\frac{2\pi}{k} \equiv \lambda > \lambda_{eq}$ the temperatures measured by COBE are

$$\left|\frac{\Delta T}{T}\right| \simeq \frac{1}{3}\Psi(t_f) \simeq 10^{-5}. \tag{11.85}$$

T_f marks the end of inflation, when the reheating of the universes brings things back to a radiation-deminated FL model. Expression (11.85) is a condition on the amplitude $\left(\frac{V_{,\varphi}}{V}\right)^2$.

A relevant example is $V(\varphi) = \frac{m^2}{2}\varphi^2$.
$\frac{V^{3/2}}{V_{,\varphi}} \simeq m\varphi^2(\frac{1}{2})^{3/2}$, and $m\varphi^2$ have to be evaluated at $k = HR$, i.e.

$$k = HR_f \exp(-3/4L_{Pl}^2\varphi^2),$$

in the case of Linde's chaotic inflation model. Therefore $L_{Pl}^2\varphi^2 \simeq \ln\left(\frac{HR_f}{k}\right)$ at horizon crossing.
$\left(\frac{V_{,\varphi}}{V}\right)^2 \simeq \varphi^{-2} \simeq L_{Pl}^2$ (at the end of inflation, we have $L_{Pl}^2\varphi^2 \sim 1$). Finally we obtain

$$\Psi_k \simeq L_{Pl} m \ln\left(\frac{R_f H}{k}\right). \tag{11.86}$$

This is an almost flat spectrum with only a logarithmic dependence on wavelength. Fluctuations of galactic scale have a physical wavelength $l \sim 3 \times 10^5$ pc $\sim 10^{38}\,\mathrm{GeV}^{-1}$, i.e. a comoving wave number

$$k \simeq R_f 10^{-11}\,\mathrm{GeV}.$$

Since $H \sim 10^{12}$ to 10^{19} GeV, we have at most $H\frac{R_f}{k} \sim 10^{30}$, and $\ln H\frac{R_f}{k} \simeq 70$. Thus $\Psi_k \simeq mL_{Pl}70$ is less than 10^{-5} if

$$m \lesssim 10^{12}\,\mathrm{GeV}.$$

11.6 CMB Anisotropies

The CMB photons propagating in the real universe will not be quite isotropic, because the structures forming from small initial fluctuations will leave an imprint on their distribution. The properties of the background cosmological model, the spectrum of the density fluctuations, and the nature of the dark matter can all be found in the CMB anisotropies; we just have to learn how to read this information. The MAP and PLANCK satellite missions will observe the CMB with high angular resolution, and will provide a detailed map of $\frac{\Delta T}{T}(\Theta)$ over the sky. It is therefore necessary to take a close look at the theoretical expectations.

There exist complete numerical solutions of the radiation transport equations (e.g. [Bond and Efstathiou 1987]), but the physical content can be explained in a less demanding formalism. Following [Hu 1995; Hu et al. 1997], I want to discuss the relevant physical processes leading to CMB anisotropies, as they are displayed in Fig. 11.4, in a qualitative, but – perhaps – more intuitive way. If you really want to calculate yourself, you should look up more technical treatments ([Hu and Sugiyama 1995a,b] for example). The basic physics has already been discussed in an earlier paper by [Sunyaev and

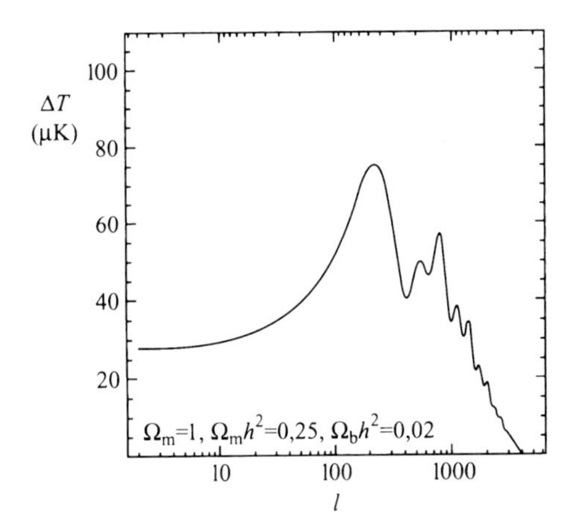

Fig. 11.4. The mean square temperature fluctuations of the CMB are expected to depend on angle or equivalently on multipole index in a characteristic way, showing a sequence of peaks at distinct angular scales

Zel'dovich 1970], and I will also freely quote from this publication in the following.

Before recombination, photons and baryons are tightly coupled, because the photons scatter off the electrons (Compton scattering) and the electrons and protons are attracted towards each other by the electrical force (Coulomb interaction).

This component of the cosmic matter is essentially a photon–baryon fluid, and perturbations in that fluid behave as described by the solutions (11.50): Inside the sound horizon $c_s x \gg 1$ ($x = k\frac{R}{R'}$) they show an oscillating behaviour. These acoustic oscillations occur because the pressure of the photons resists the compression of the fluid by its own gravity. The maxima and minima in Fig. 11.4 look like traces of such acoustic oscillations, and this is in fact so. It is, however, not quite obvious how such features can be seen in the CMB sky today. This will be explained in the following subsections. Clearly, the picture leading to the solutions (11.50) is much too simplified. The sound speed c_s is, for example, assumed to be constant, but it will definitely change during the recombination era. The sound oscillations must somehow be in phase, because otherwise they will interfere with each other and cannot be preserved. This is achieved by the synchronization due to the big bang – all sound waves of the same wavelength have had the same travel time since the big bang. Thus, at least, the large ones of the order of the horizon size are in phase at the recombination epoch. Also, the cosmic matter contains a dark matter component, and therefore the perturbations of the baryon–photon fluid will experience gravitational interactions. Most important is the change of conditions during recombination: during this epoch hydrogen atoms are formed, and the CMB photons acquire a large mean free path. They experience a last scattering, and then they cross the observable universe (of typical scale $\frac{2c}{H_0}$) essentially freely. Again this ideal picture is not quite correct, since recombination does not happen instantaneously, and there is some time interval during which photons and baryons gradually separate.

In addition, there may be a re-ionization of the atoms at epochs somewhere between $z = 100$ at $z = 10$.

Regions of compression and rarefaction at the surface of last scattering represent hot and cold spots in the CMB sky today. Photons are also red-shifted when they climb in and out of potential wells on the last scattering surface.

Within this picture, we want to understand how the anisotropies shown in Fig. 11.4 are formed. First, we have to look more closely at the epoch of recombination.

11.6.1 Recombination

For redshifts $z \approx 1500$ (i.e. $T_\gamma = T_{\gamma 0} z \approx 4000\,\mathrm{K}$) the number densities of protons, electrons, and hydrogen atoms are equal (in agreement with the Saha equation)

$$n_e = n_p = n_H .$$

Further expansion, and cooling of the CMB below 4000 K, changes the equilibrium picture described by the Saha equation. Free electrons and radiation are only found in equilibrium with the excited levels of atomic hydrogen. The recombination rate is determined by two-quantum decays of the 2s level. This is described by the equation

$$\begin{aligned} \frac{dn_p}{dt} &= -3n_H n_p - wH_{23}, \\ H_{23} &= \mathrm{const.} n_p^2 e^{-B/z}, \end{aligned} \tag{11.87}$$

where $B = \frac{I}{4k_B T_{\gamma 0}} = 1.458 \times 10^4$; $I = 13.6\,\mathrm{eV}$, the ionization potential of atomic hydrogen and $w = 8\,\mathrm{s}^{-1}$ the probability of a two-quantum transition from the 2s state of an hydrogen atom to the ground state

$$\mathrm{H(2s)} \to \mathrm{H(1s)} + \gamma_1 + \gamma_2 .$$

The solution of this equation is

$$x(z) = \frac{n_p}{n_p + n_H} = \frac{n_e}{n_e + n_H} = \frac{A}{z\sqrt{\Omega}} e^{-B/z}, \tag{11.88}$$

where

$$A = \frac{(2\pi m_e k_B T_{\gamma 0})^{3/2} H_0 I}{4wn_c k_B T_0 h^3} \approx 6 \times 10^6 ;$$

n_c is the critical density, $n_c = 10^{-5} h^2\,\mathrm{cm}^{-3}$. For redshifts less than 900, thermodynamic equilibrium between hydrogen states and free protons and electrons cannot be maintained, the ionization time from the 2s levels by collisions with quanta and electrons becomes greater than w^{-1}, and therefore the solution (11.88) is correct only in the interval $900 < z < 1500$. The optical depth of the universe due to Thomson scattering equals

$$\tau(z) = \Omega^{1/2}\sigma_T n_c H_0^{-1} \int_0^z \eta^{1/2} x(\eta) d\eta = \tau_0 + a z^{3/2} e^{-B/z}, \quad (11.89)$$

where $a = \Omega^{1/2}\sigma_T n_c H_0^{-1}(A/B) = 27.3$. The constant

$$\tau_0 = \Omega^{1/2}\sigma_T n_c H_0^{-1} \int_0^{900} x(z) z^{1/2} dz = 0.4,$$

[Sunyaev and Zel'dovich 1970]. We can use this to estimate the thickness of the last scattering surface. The scattering of photons on moving electrons leads to a frequency shift due to the Doppler effect. To a first approximation this effect is proportional to the optical depth of the moving matter and decreases as the radiation propagates from some condensing proto-object to the observer, i.e.

$$\frac{\Delta T}{T} = \int_0^{\infty} \frac{\boldsymbol{u}\cdot\boldsymbol{r}}{r} \times c e^{-\tau(z)} \frac{d\tau}{dz} dz; \quad (11.90)$$

$\boldsymbol{u}\cdot\boldsymbol{r}$ is the projection of the velocity along the line of sight, and the integral is carried out along the radial ray $r = r(z), t = t(z)$.

The function $e^{-\tau}\frac{d\tau}{dz}$ can be computed from (11.89). It has a sharp maximum for $z_{\max} = 1055$ (with a value of 3.32×10^{-3}), and decreases exponentially towards smaller and larger z-values, reaching half its maximum value at $z = 960$ and 1135. This function can be approximated conveniently by a Gaussian function with dispersion $\sigma_z = 75$ and integral = 1. Thus $\Delta z = 75$ is a good measure of the thickness of the surface of last scattering. Slightly condensed objects become transparent at $z = 1055$, which is a sharp maximum, so that we actually see the structure of this surface in the CMB anisotropies – unless there is some later epoch of re-ionization.

There is a smearing out of fluctuations due to the finite thickness ($\sigma_z = 75$). This corresponds to an angular scale of about $\Theta \sim 0.1°$. So below that angle cosmic fluctuations cannot be observed.

11.6.2 Anisotropy Formation

For a photon–baryon system the interaction by Compton scattering can be approximated by an expansion of the Boltzmann equation and the Euler equation in terms of the Compton scattering time. To zeroth order,

$$\begin{aligned}\Delta'_\gamma &= \frac{4}{3}\Delta'_B \quad \text{(or)}\ \Theta'_0 = \frac{1}{3}\delta'_B;\\ \Theta_1 \equiv V_\gamma &= V_B;\\ \Theta_l &= 0 \qquad \text{for } l \geq 2.\end{aligned}$$

Here the radiation is isotropic in the baryon rest-frame, and the radiation fluctuations grow adiabatically with the baryons. The iterative first-order solution depends on the sound speed $c_s^2 = \frac{c^2}{3}\frac{1}{1+R_s}$, with

$$R_s = \frac{3\rho_B}{4\rho_\gamma} = \frac{450}{(1+z)}\left(\frac{h^2\Omega_B}{0.015}\right)$$

proportional to the ratio of the baryon to photon density.

The first-order equations are

$$\begin{aligned}\Theta'_0 &= -\frac{k}{3}V_\gamma - \Phi',\\ V'_\gamma &= -\frac{R'_s}{1+R_s}V_\gamma + \frac{1}{1+R_s}k\Theta_0 + k\Psi.\end{aligned}$$

We can rewrite these equations as a single second-order equation

$$\Theta''_0 + \frac{R'_s}{1+R_s}\Theta'_0 + k^2c_s^2\Theta_0 = F, \tag{11.91}$$

with

$$F = -\Phi'' - \frac{R'_s}{1+R_s}H\Phi' - \frac{k^2}{3}\Psi.$$

Ψ is the peculiar gravitational potential, Φ is the perturbation to the spatial curvature. In linear perturbation theory $\Phi = -\Psi$. The homogeneous equation $F = 0$ yields the two fundamental solutions

$$\begin{aligned}\Theta_a &= (1+R_s)^{-1/4}\cos kr_s,\\ \Theta_b &= (1+R_s)^{-1/4}\sin kr_s.\end{aligned} \tag{11.92}$$

The sound horizon r_s is given by

$$r_s \equiv \int_0^\eta c_s d\eta'.$$

The particular solution can be constructed by Green's method as

$$\hat{\Theta}_0(\eta) = C_1\Theta_a(\eta) + \Theta_b(\eta) + \int_0^\eta \frac{\Theta_a(\eta')\Theta_b(\eta) - \Theta_a(\eta)\Theta_b(\eta')}{\Theta_a(\eta')\Theta'_b(\eta') - \Theta'_a(\eta')\Theta_b(\eta')}F(\eta')d\eta'.$$

Substituting (11.92) and replacing C_1, C_2 by the initial conditions $\Theta_0(0)$ and $\Theta_0'(0)$, this reads

$$
\begin{aligned}
(1+R_s(\eta))^{1/4}\hat{\Theta}_0(\eta) &= \Theta_0(0)\cos\ k_s + \frac{\sqrt{3}}{k}\left[\Theta_0'(0) + \frac{1}{4}R_s'(0)\Theta(0)\right]\sin\ kr_s \\
&+ \frac{\sqrt{3}}{k}\int_0^\eta d\eta'(1+R_s)^{3/4}\sin\{kr_s(\eta) - kr_s(\eta')F(\eta')\}, \qquad (11.93) \\
k\Theta_1 &= -3(\Theta_0' + \Phi').
\end{aligned}
$$

Some basic features of acoustic oscillations can be seen most easily in a simplified case. Let us consider a static potential: $\Psi' = 0 = \Phi'$. This corresponds to a model which was always matter-dominated. Assume also that R_s is constant. This is not true for the real universe, but the generalization to more realistic cases will be pointed out later. The solution (with $F = -(k^2/3)\Psi$) in this case is

$$
\begin{aligned}
\hat{\Theta}_0(\eta) &= [\Theta_0(0) + (1+R_s)\Psi]\cos(kr_s) \\
&+ \frac{1}{kc_s}\Theta_0'(0)\sin(kr_s) - (1+R_s)\Psi. \qquad (11.94)
\end{aligned}
$$

The sound horizon here is $r_s = c_s\eta$.

The zero-point of the oscillation

$$
\Theta_0 = -(1+R_s)\Psi \qquad (11.95)
$$

is shifted with the baryon content ($R_s \equiv \frac{3}{4}\rho_B/\rho_\gamma$), and the amplitude of the oscillation increases with R_s. Adiabatic initial conditions, $\Theta_0(0) = -\frac{2}{3}\Psi$, $\Theta_0' = 0$, stimulate the cosine harmonic, while isocurvature initial conditions, $\Theta_0'(0) \neq 0$, stimulate the sine harmonic.

If we neglect the baryons ($R_s = 0$), then $\Theta_0 + \Psi$ is the quantity that oscillates around zero. When the baryons are included, the fluid acquires gravitational and inertial mass. For constant R and Φ the oscillator equation for Θ_0 can be written as

$$
(1+R_s)\Theta_0'' + \frac{k^2}{3}\Theta_0 = -(1+R_s)\frac{k^2}{3}\Psi. \qquad (11.96)
$$

The effective mass of the oscillator is

$$
m_{\text{eff}} = 1 + R_s.
$$

The baryons drag the photons along, and thus inside the potential wells there is greater compression, shifting the zero point of the effective temperature $\Theta_0 + \Psi$ to $-R_s\Psi$. All compressional phases will be enhanced over rarefaction phases. The oscillations increase in amplitude because of the zero point shift $-R_s\psi$, since the initial zero point displacement becomes larger.

At the time of last scattering η_* the photons decouple from the baryons and stream out of the potential wells. If we look at the oscillations not as fluctuations in time with a fixed wave number k, but rather as functions of k at a fixed time η_*, then we see a series of peaks with

$$k_m = m\pi/c_s\eta_*$$

for the mth peak.

Odd peaks represent the compression phase (temperature crests), and even peaks the rarefaction phase (temperature troughs) inside the potential wells. For isocurvature conditions the compression phase is reached for even m peaks, because isocurvature conditions resist the gravitational attraction.

The bulk velocity of the fluid along the line of sight $V_\gamma/\sqrt{3}$ causes a Doppler shift. Since the turning points of the fluid's oscillation are at the extrema of $\Theta_0 + \Psi$, the velocity is shifted in phase by 90° from the density. We have

$$V_\gamma(\eta)/\sqrt{3} = \frac{\sqrt{3}}{k}\Theta_0' = \sqrt{3}\{\Theta_0(0) + (1 + R_s)\Psi\}c_s \sin(kr_s) + \frac{\sqrt{3}}{k}\Theta_0'(0)\cos(kr_s). \tag{11.97}$$

Comparing with (11.94), we see that, indeed, the velocity is phase-shifted from the temperature by $\pi/2$. The zero point of the oscillation is not displaced. The amplitude of the Doppler shift $V_\gamma/\sqrt{3}$ is reduced by a factor of $\sqrt{3}c_s = (1 + R_s)^{-\frac{1}{2}}$ compared with the temperature. If the sound velocity is exactly $\frac{1}{\sqrt{3}}$ (if $R_s = 0$), however, then on average the Doppler shift peaks cancel the acoustic oscillations, and the peaks are gone. It is unfortunate that the name "Doppler peaks" for the oscillations seems to have stuck, whereas "acoustic peaks" would be more appropriate.

The addition of baryons significantly changes the relative velocity contributions. The velocity decreases by a factor $(1+R_s)^{-\frac{1}{2}}$, and the compressional peaks of the temperature fluctuations are increased by a factor of $1+6R_s$ over the $R_s = 0$ case. The velocity peaks are reduced by a factor of $\frac{1+3R_s}{\sqrt{1+R_s}(1+6R_s)}$ compared to the Θ_0-peaks, and the acoustic peaks reappear.

Although this toy model illustrates well the fact that acoustic peaks appear, it is not an accurate solution. But it has been found [Hu and Sugiyama 1995a,b] that a WKB approximation can be applied to the oscillator equation

$$\Theta'' + \left(\frac{R_S'}{1 + R_S}\right)\Theta' + (k^2c_s^2)\Theta = R,$$

because R_s is changing on a time-scale much longer than the time-scale on which Θ oscillates. Thus the solution (11.94) with constant R_s describes the situation quite accurately, if the driving force F is modelled with sufficient accuracy.

The steep decrease of the power spectrum of $\Theta_0\left(\frac{\Delta T}{T}\right)$ to the right in Fig. 11.4 is caused by photon diffusion during recombination, and the finite thickness of the surface of last scattering.

11.6.3 Dependence on Model Parameters

The 3D modes of Θ_0 are observed as 2D projections on the celestial sphere. The angular size of a given physical scale depends on the curvature of space-time, i.e. on Ω. Then the angular extent of the first acoustic peak is determined by the sound horizon kc_s, and by Ω. Lower Ω means that a smaller angle corresponds to the same scale, i.e. the peak will shift to the right. The overall scaling is roughly $l \propto \Omega^{-\frac{1}{2}}$.

If $R_s = 0$ there are no peaks in the power spectrum, as we have mentioned already. Since $R_s \propto h^2\Omega_B$ the compression peaks (number $1, 3, \ldots$) will be higher than the rarefaction peaks (number $2, 4, \ldots$) if $\Omega_B > 0$. The amplitude difference will tell us about Ω_B.

The gravitational potential is constant only if matter dominates the dynamics completely. Realistic models have time-varying potentials leading to a dependence of the peak heights on the total density $\Omega_0 h^2$, on Ω_Λ, and on the curvature $1 - \Omega_m - \Omega_\Lambda$. A decay of the potential $|\Psi|$ decreases the gravitational redshift, leading to an effective blueshift in the well. The implied decay of the curvature perturbation $|\Psi|$ leads to a time dilatation blueshifting the photons and boosting the oscillation amplitude to $\simeq \frac{3}{2}\Psi(0)$.

Only perturbations of short wavelengths which cross the sound horizon during the radiation-dominated epoch experience this enhancement. For example, for the models with $\Omega_m h^2 \simeq 0.25$, the sound horizon at equality is several times smaller than at last scattering η_* (for $z_{eq} = 2.4\times 10^4\Omega_m h^2$ which is the value for three flavours of massless neutrinos, $\eta_{eq} < \eta_*$). Increasing η_{eq}, by lowering $\Omega_m h^2$ or increasing the number of relativistic species, makes the effect important for the first few peaks too, and this boosts their amplitudes.

In the free streaming limit after the last scattering, the gravitational redshift and time-dilatation effects combine to form the "integrated Sachs–Wolfe" (ISW) effect. The contributions at small scales more or less cancel each other when the photon traverses many wavelengths during the decay of the potential. This decay does not occur for modes that cross the sound horizon between last scattering and the epoch of full matter domination. In summary, both the epoch when the cosmological model changes from radiation-dominated to matter-dominated and the epoch when Ω_Λ becomes dominant leave their imprint on the CMB.

These temperature fluctuations in space are not what is actually observed: the temperature anisotropies on the sky correspond to spatial fluctuations on a distant surface. Physical scales, like the sound horizon $c_s\eta_*$, subtend a certain angle in the sky. The relation between the measuring rod and the angle is defined by the cosmological model used. Thus the angular scale of the peaks in the power spectrum tells us about the parameters of the cosmological model: $H_0, \Omega_0, \Omega_B, \Omega_\Lambda$.

The angular scale of a feature in the power spectrum is determined by an expansion in spherical harmonics:

$$Y_{lm}(\Theta, \varphi) = \left(\frac{(2l+1)(l-m)!}{4\pi(l+m)!} \right)^{\frac{1}{2}} P_l^m(\cos\Theta) e^{im\varphi};$$

P_l^m are the associated Legende functions; l and m are integers, and $l \geq 0$, $|m| \leq l$.

The spherical harmonics form a complete orthonormal set on the unit sphere.

The temperature difference $\Delta \boldsymbol{n}$, often denoted by $\Delta T/T$,

$$\Delta(\boldsymbol{n}) = (T(\boldsymbol{n}) / \langle T(\boldsymbol{n}) \rangle) - 1,$$

where $\boldsymbol{n}$ is the unit vector $(\sin\theta\cos\varphi, \sin\theta\sin\varphi, \cos\theta)$, can be expanded in terms of spherical harmonics:

$$\Delta(\theta, \varphi) \equiv \sum_{l=0}^{\infty} \sum_{m=-l}^{+l} a_{lm} Y_{lm}(\theta, \varphi).$$

The angular correlation function

$$C(\theta) = \langle \Delta(\boldsymbol{n}\Delta(\boldsymbol{n}')) \rangle ,$$

where $\boldsymbol{n} \cdot \boldsymbol{n}' = \cos\theta$. Using the addition theorem, one obtains

$$C(\theta) = \sum_{l=0}^{n} (2l+1) C_l P_l(\cos\theta).$$

l corresponds roughly to $\frac{100}{\theta}$, with θ in degrees. $C(\theta)$ is plotted in Figs. 2.44 and 11.7 showing CMB anisotropies versus l.

Recently a substantial part of the sky has been covered by balloon experiments (Boomerang and Maxima) and by interferometric measurements [Jaffe et al. 2001; Pryke et al. 2002]. The angular resolution (~ 0.5 degrees) is sufficient to detect the CMB power spectrum up to a multipole index $\ell \simeq 800$. The launch of the Boomerang experiment at the South Pole is shown in Fig. 11.5. The patchy structure of hot and cold spots detected by these experiments is displayed for a small patch of the sky in Fig. 11.6.

Finally, the plot of $C(\theta)$ is shown in Fig. 11.7. There is a pronounced maximum at $\ell \simeq 200$. This corresponds to the largest acoustic oscillation of size $c_s \eta_*$. The fact that this physically defined length appears under an angle of half a degree, implies that the universe has zero curvature, i.e. that $\Omega_{\text{tot}} = 1$. The exact analysis results in a value

$$\Omega_{\text{tot}} = 1.02 \pm 0.06.$$

Fig. 11.5. The launch of the Boomerang ballon at the South Pole with Mt. Erebus in the background

Observations at the South Pole get more and more popular

A small positive or negative curvature is still permitted by the observational uncertainties. The compression is followed by a rarefaction in the acoustic oscillation which appears as a second maximum in Fig. 11.7 (it is positive, since the square of $\Delta T/T$ is plotted.) The compression amplitude relative to the rarefaction amplitude tells us about the relative magnitude of the baryon density. The fits to the curve in Fig. 11.7 give constraints

$$\Omega_B h^2 = 0.022 {}^{+0.004}_{-0.003}$$

Fig. 11.6. A small piece of the sky, about the size of one of the COBE patches, shows a number of spots of different CMB temperature

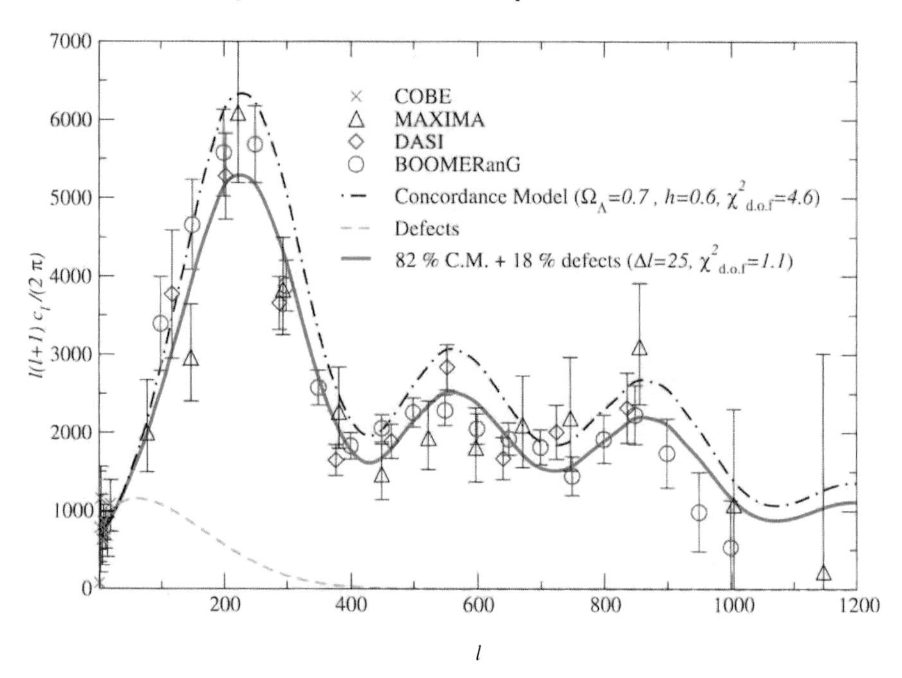

Fig. 11.7. Similarly to Fig. 2.44, the mean square of the CMB temperature anisotropies is shown for various balloon and interferometric experiments with model-fitting curves (from [Bouchet et al. 2002])

for the baryonic, and

$$\Omega_{\mathrm{CDM}} h^2 = 0.13 \pm 0.005$$

for the dark non-relativistic matter ($h = H_0/100$ is the Hubble constant in units of 100; typical values are $h = 0.65 \pm 0.07$).

Baryonic matter and cold dark matter together fall short of the critical value $\Omega_{\mathrm{tot}} = 1$. The deficit must be accounted for by another component of the cosmic matter which is distributed quite uniformly without any clumping on the scale of galaxy clusters. A constant or nearly constant energy density Ω_Λ must exist, with

$$\Omega_\Lambda = 0.71 \pm 0.13.$$

In Chap. 9 there is further discussion of this mysterious dark energy. A best fit for all these cosmic parameters is

$$\begin{aligned} \Omega_{\mathrm{tot}} &= 1, \\ \Omega_\Lambda &= 0.7, \\ \Omega_{\mathrm{CDM}} &= 0.28, \\ \Omega_B &= 0.02, \\ H_0 &= 68. \end{aligned}$$

The DASI interferometric instrument located at the Amundsen–Scott South Pole research station has detected polarization in the CMB [Kovac et al. 2002], at a level as expected from current theories. CMB polarization is thought to arise from scattering by electrons of a radiation field with a local quadrupole moment. This is caused mainly by the Doppler shifts induced by the local velocities of plasma blobs around the epoch of decoupling. Thus future precision measurements of the polarized components of the CMB will directly probe the dynamics of the density fluctuations at the time of decoupling of matter and radiation.

11.7 Non-baryonic "Hot" and "Warm" Dark Matter

The arguments against an origin of the galaxy formation process in adiabatic or isothermal small perturbations of a purely baryonic background are quite strong, as we have seen. It seems reasonable therefore to investigate the influence of a non-baryonic background.

A useful terminology for dark-matter candidates is to classify them according to their random velocity at the time of their decoupling [Bond et al. 1983; Primack and Blumenthal 1983]. "Hot DM" refers to light particles ($m \le 100\,\mathrm{eV}$) which are still relativistic when the mass within the horizon corresponds to a galactic mass $M_H \simeq 10^{12} M_\odot$. These particles have a number density comparable to the density of the photons. A typical example is a massive neutrino with a mass $m_\nu \simeq 30\,\mathrm{eV}$.

"Warm DM" consists of particles which have decoupled at higher temperatures and which just become non-relativistic when a galactic mass is within the horizon. As a consequence, their number density is an order of magnitude lower than the number density of hot DM particles. A natural candidate, a massive gravitino, experienced a readjustment of its mass in recent theories of SUSY breaking, and is now (with $m_G \geq 2\,\mathrm{GeV}$) a candidate for "cold DM".

"Cold" DM particles decouple very early. They are either very heavy particles (such as photinos, gravitinos, neutralinos or right-handed neutrinos) which are non-relativistic very early ($M_H \ll 10^{12} M_\odot$), or they are created with essentially zero random velocity (a Bose condensate of axions, quark lumps).

The essential difference between these types of particles in the context of galaxy formation is the scale on which density fluctuations are damped by "free streaming" of the relativistic particles (see Chap. 10): neutrinos, i.e. hot DM particles, destroy all fluctuations on scales less than $\sim 10^{15} M_\odot$ (typical supercluster mass), warm DM particles would have a "free-streaming" scale of $10^{12} M_\odot$, and for cold DM the "free-streaming" scale is of no importance.

The attractive feature of a dominant non-baryonic DM particle lies in the rather stringent predictions that can be derived if just one or two kinds of non-interacting particles have to be taken into account. The fact that only simple scenarios have been investigated in detail does not mean, however, that more complicated mechanisms – such as a baryonic DM for small systems, and massive neutrinos to form the large-scale structures, or interacting DM – can be excluded.

The experiments on neutrinos discussed in Chap. 7 have led to evidence for ν-oscillations, i.e. for $m_\nu \neq 0$. Solar neutrino experiments, and the observations of atmospheric neutrinos by the Japanese Super-Kamiokande experiment, give $(\Delta m)^2 \leq 10^{-2}\,\mathrm{eV}^2$, and this leads to the sad fact that the masses $m_{\nu_e}, m_{\nu_\mu}, m_{\nu_\tau}$ are all less than 0.1 eV. This makes the neutrinos cosmologically irrelevant if the masses are hierarchically ordered: $m_{\nu_e} \ll m_{\nu_\mu} \ll m_{\nu_\tau}$. If they are all approximately the same, $m_{\nu_e} \simeq m_{\nu_\mu} \simeq m_{\nu_\tau}$, then a neutrino mass of a few eV is still a possibility. So, very likely, the neutrinos cannot be the cosmic DM, but it is nevertheless instructive to consider the workings of such particles in the course of cosmic evolution. This we shall do in the following few pages. And, after all, there may be a sterile neutrino with a significant contribution to the cosmic matter density.

11.7.1 Neutrino Stars

Massive neutrinos can cool and form Fermi condensates, which may be the dark halos of galaxies or larger systems. The neutrinos must satisfy the Pauli

Table 11.1. The mass–radius relation for neutrino spheres

Object	$M/M_\odot$	R	m_ν
Galactic halo	10^{12}	100 kpc	30 eV
Dwarf halo	10^{8}	10 kpc	75 eV
Group of galaxies	5×10^{13}	1 Mpc	10 eV
Supercluster of galaxies	10^{15}	10 Mpc	2.5 eV

principle in their distribution, and therefore their phase-space configuration is fixed. The simplest case is to assume that the neutrinos form a "white-dwarf-like" structure. For a stable configuration, i.e. for a total mass less than the critical mass, the following mass radius relation can be derived:

$$MR^3 = \frac{91.9\hbar^6}{G^3 m_\nu^8};$$

just replace the electron mass m_e by the neutrino mass m_ν in the usual formulae (e.g. [Weinberg 1972]). We can use this relation to derive limits on the neutrino mass, if for M and R we insert typical dimensions of observed structures. In Table 11.1 some of these limits are listed.

The smallest objects for which the neutrinos could cluster are galaxies: $M = 10^{12} M_\odot$ and $R = 100\,\text{kpc}$ gives a neutrino mass of $m_\nu \approx 30\,\text{eV}$. Clustering on smaller scales requires masses in excess of 30 eV, in conflict with cosmological limits.

We see that the experimental limits on ν-masses already exclude their clustering on supercluster scales. Therefore another DM component is needed if clustered objects are to form.

11.7.2 Typical Scales for Hot DM

Even for collisionless particles the concept of the Jeans mass is a valid one. The ordinary pressure (sound speed) of a collisional fluid is replaced by the velocity dispersion $\langle v^2 \rangle$ of the collisionless particles ($c_s^2 = \frac{1}{3}\langle v^2 \rangle$). We have seen in Sect. 10.2 how the Jeans mass for neutrinos is obtained from a solution of the Vlasov equation

$$M_{J,\nu} = 2^{-5/2}\pi^{7/2}\frac{\zeta(3)\Gamma(3)}{(\ln 2)^{3/2}}\frac{M_{Pl}^3}{m_\nu^2}\left(\frac{T_\nu}{m_\nu}\right)^{3/2},$$

where ζ is the zeta function, Γ the gamma function, and m_ν the neutrino mass.

The maximum occurs for $T_\nu = m_\nu$ (the calculation holds for non-relativistic neutrinos $m_\nu \leq T_\nu$):

$$M_{J,\nu}^{\max} = 1.8\frac{M_{Pl}^3}{m_\nu^2} = 3.2 \times 10^{15} M_\odot \left(\frac{m_\nu}{30\text{eV}}\right)^{-2}.$$

For scales below the maximum Jeans mass there are very efficient damping mechanisms for collisionless relativistic particles: damping by free streaming and by directional dispersion (e.g. [Bond and Szalay 1983; Peebles 1980]). Directional dispersion is very effective for extremely relativistic particles. In one dimension a wave pattern on the neutrino background would be preserved, but in three dimensions peaks and troughs of waves travelling in different directions cross, and consequently the amplitude of the wave decays. A simple illustration is provided by the solutions of the Vlasov equation (neglecting gravity) with a velocity-dependent term:

$$\partial_t f + \boldsymbol{v} \cdot \boldsymbol{\nabla} f = 0.$$

For the relativistic plane wave $f = f_k(t)e^{i\boldsymbol{k}\cdot\boldsymbol{x}}$, the equation is

$$\dot{f}_{\boldsymbol{k}} + ik \cdot \boldsymbol{v} f_k = 0,$$

$$f_k(t) = f_k(0)e^{ikvt\mu} \quad (\mu \equiv k \cdot v/kv).$$

A disturbance is a superposition of such waves

$$\delta(x) = e^{i\boldsymbol{k}\cdot\boldsymbol{x}} f_k(0) \int d^3v\delta(|\boldsymbol{v}| - c)e^{ikvt\mu}$$

$$= e^{ik\cdot x} f_k(0) \frac{\sin(kct)}{kct}.$$

There is strong damping $|\delta| \sim t^{-1}$.

Relativistic particles also experience damping by free streaming. As long as the particles are relativistic they move freely from a compression region – faster particles move a longer distance than slow ones, and therefore the density enhancement is damped out at a rate which depends on the velocity dispersion (this is exactly the phenomenon of Landau damping [Landau 1946]). The neutrinos are relativistic until the temperature has dropped below m_ν. At that time the horizon size is of an order of magnitude

$$\lambda_H \equiv d_\nu = ct(T = m_\nu) = (G\varrho)^{-1/2} \simeq M_{Pl} m_\nu^{-2}$$

($\varrho = am_\nu^4$ at $T = m_\nu$).

All fluctuations of a size less than $d_\nu \sim M_{Pl} m_\nu^{-2}$ are damped by the "free streaming" of the relativistic collisionless neutrinos. the mass-scale corresponding to this length is

$$M_{\mathrm{FS},\nu} \equiv d_\nu^3 m_\nu n_\nu (T = m_\nu) \simeq d_\nu^3 m_\nu^4 \equiv M_{Pl}^3 m_\nu^{-2}.$$

A more detailed calculation gives [Bond 1980]

$$M_{\mathrm{FS},\nu} = 1.77 M_{Pl}^3 m_\nu^{-2} = 2.88 \times 10^{18} M_\odot \left(\frac{m_\nu}{1\,\mathrm{eV}}\right)^{-2} \approx M_{J,\nu}^{\max}. \qquad (11.98)$$

This corresponds to the mass scale of large clusters only for $m_\nu \leq 30\,\mathrm{eV}$.

For the comoving damping-length scale the precise formula is

$$d_\nu = 1.23 \left(\frac{m_\nu}{1\,\text{eV}}\right)^{-1} (1+z)^{-1}\, 10^3\,\text{Mpc}.$$

A "neutrino star" has a mass of $M_{Pl}^3 m_\nu^{-2}$, and a radius of $(M_{Pl}/m_\nu)m_\nu^{-1}$.

The free-streaming scale of the neutrinos is equal to the maximum Jeans mass. Therefore fluctuations on all scales below the maximum Jeans mass are very efficiently damped out.

The observed clusters and superclusters cannot be formed by the clumping of neutrinos if the experimental mass limits give $m_\nu \ll 1\,\text{eV}$.

As we have seen earlier, light or massless left-handed neutrinos remain in thermal equilibrium until the temperature drops below 1 MeV. For each ν-species, we have today (taking into account the heating of the photon gas from e^+e^- annihilation) the relation

$$n_{\nu 0} = \frac{3}{4}\frac{4}{11} n_{\gamma 0} = 109 \left(\frac{T_{\gamma 0}}{2.7\,\text{K}}\right)^3 \quad \text{cm}^{-3}.$$

The contribution to the cosmological density is – if we consider one species with mass m_ν –

$$\varrho_{\nu 0} = m_\nu n_{\nu 0} = 109\ \text{eV cm}^{-3} \left(\frac{m_\nu}{1\,\text{eV}}\right) \vartheta^3 \left(\vartheta \equiv \frac{T_{\gamma 0}}{2.7\,\text{K}}\right).$$

The critical density is

$$\varrho_c = 10\,500 h_0^2\,\text{eV cm}^{-3}, \quad \text{so} \quad \Omega_\nu \equiv \frac{\varrho_\nu}{\varrho_c} = 0.01 h_0^{-2} \vartheta^3 \left(\frac{m_\nu}{1\,\text{eV}}\right). \tag{11.99}$$

The scale of the damping by free streaming is

$$d_\nu = 13(\Omega_\nu h_0^2)^{-1} \vartheta^{-2} (1+z)^{-1}\,\text{Mpc}. \tag{11.100}$$

11.7.3 The Hot-DM Fluctuation Spectrum

For one type of massive neutrino we have $h_0^2 \Omega_\nu = 0.01(m_\nu/1)\,\text{eV}$, and therefore the mass within the horizon, when the energy density in relativistic particles equals that in non-relativistic particles, is

$$M_{H,eq} = 0.7 \times 10^{15} \left(\frac{m_\nu}{1\,\text{eV}}\right)^{-2} M_\odot \simeq 7 \times M_{J,\nu}^{\max}.$$

As the two scales are so close we may conclude that the neutrino fluctuations grow essentially on one scale $M = M_{J,\nu}^{\max} \approx M_{\text{FS},\nu} \sim M_{H,eq}$. The "hot" DM structures are, to a good approximation, independent of the initial fluctuation spectrum; their evolution is determined just by the scale of the maximum Jeans mass.

The fluctuations of wavelength smaller than the horizon size are prevented from growing any further as long as the energy density in the non-relativistic DM is smaller than the energy density in relativistic particles. Since the energy density in relativistic particles relative to non-relativistic matter falls as $1+z$, there is always an epoch t_{eq}, where the two energy densities are equal. For $t > t_{eq}$ the non-relativistic DM dominates and fluctuations grow.

For neutrinos the situation depends somewhat on the different types of massive neutrinos: one massive species with mass m_ν, and two massless neutrino types, plus the photon, yield the relation (for $T \leq 0.5\,\mathrm{MeV}$)

$$\varrho_\nu = \left(\frac{7}{8} g_\nu \left(\frac{T_\nu}{T_\gamma}\right)^4\right) \frac{\pi^2}{30} \varrho_\gamma$$

and

$$\varrho_{\mathrm{rel}} = \left(2 + 2\frac{7}{8} g_\nu \left(\frac{T_\nu}{T_\gamma}\right)^4\right) \frac{\pi^2}{30} \varrho_\gamma$$

at the time when the neutrino becomes non-relativistic, $T_\nu = m_\nu$. Then $\varrho_\nu/\varrho_{\mathrm{rel}} = 0.09$ ($(T_\nu/T_\gamma) = (4/11)^{1/3}$ if $m_\nu < 1\,\mathrm{MeV}$).

The two energy densities become equal at z_{eq}, and $(1+z_{\mathrm{nr}})/(1+z_{eq}) = 1/0.09 \simeq 11$. Since $t \propto (1+z)^{-2}$, the cross-over time $t_{eq} \sim 100 t_{\mathrm{nr}}$. If all three types of neutrinos are massive with the same mass m_ν, then at $T = m_\nu$, $\varrho_\nu/\varrho_{\mathrm{rel}} = \frac{1.36}{2} = 0.68$. In that case

$$(1+z_{\mathrm{nr}}) = 1.4(1+z_{eq}), \quad \text{and} \quad t_{eq} \approx 2 t_{\mathrm{nr}}.$$

Since

$$(1+z_{\mathrm{nr}}) \simeq 10^4 \left(\frac{m_\nu}{1\,\mathrm{eV}}\right), \tag{11.101}$$

we can safely assume that for neutrinos, the epoch when they become non-relativistic and the epoch when they start to dominate more or less coincide.

Therefore, in a universe dominated by hot DM the primeval fluctuation spectrum is not very important, because as the fluctuations enter the horizon they grow as $(1+z)^{-1}$. Therefore scales of order $M_{J,\nu}^{\max}$ go non-linear first, while masses $M > M_{J,\nu}^{\max}$ enter the horizon later, and thus have less time to grow. Thus the spectrum is always strongly peaked at $M_{J,\nu}^{\max}$. Qualitatively, then, the spectrum has the form shown in Fig. 11.8, where we have plotted $k^{3/2}|\delta_k|$, i.e. the amplitude $\delta\varrho/\varrho$ as a function of the mass within the wavelength $2\pi/k$ ($M = (4\pi^4/3)\varrho_0/k^3$).

The neutrino fluctuations δ_ν grow from the time at which $T = m_\nu$ and they are (on scales larger than d_ν) not subject to acoustic oscillations or

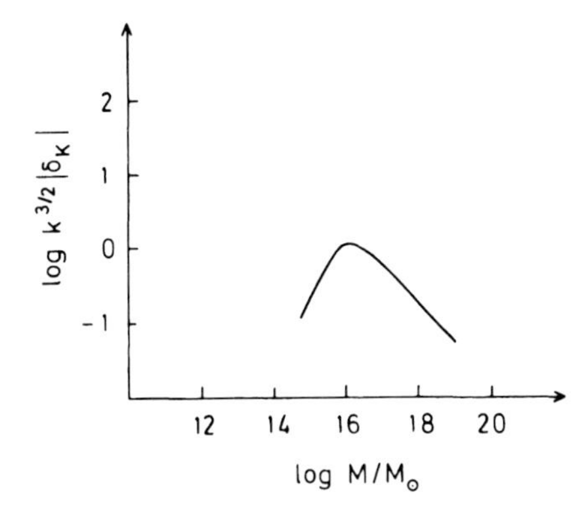

Fig. 11.8. Schematic density fluctuation spectrum (DFS) for hot dark matter at decoupling time $t = t_R$

damping as are the baryons. $T = m_\nu$ corresponds to approximately $z \approx 10^4$. If a fluctuation of total mass M_J comes within the horizon at that time, then the density contrast δ_ν continues to grow linearly with $(1+z)^{-1}$ (in a $K = 0$, $\Omega = 1$ model) until the decoupling time t_R at $z_R \sim 10^3$. The density contrast in the mixture of radiation and baryons δ_{RB}, on the other hand, oscillates and stops growing during that period, because the fluctuation is below the baryonic Jeans mass.

Thus if initially at horizon crossing $\delta_{RB} = \delta_\nu$, then at decoupling $\delta_\nu \simeq 10\delta_{RB}$. The neutrino fluctuations have time to grow to non-linear amplitudes, while the small δ_{RB} corresponds to small-angle fluctuations $\delta T/T \leq 10^{-5}$ – just within the limits of observation.

11.7.4 The "Pancake Model" for Galaxy Formation

In a neutrino-dominated universe structures of mass $\sim M_{J,\nu}^{\max}$ (at $T = m_\nu$), i.e. $M_{J,\nu}^{\max} \sim 10^{15} M_\odot$, are the first to start growing, the first to reach the non-linear regime, the first to collapse. Subsequent fragmentation of such supercluster structures is then the cause of galaxy formation (see the schematic in Fig. 11.9).

It was shown some time ago that in general this collapse will proceed asymmetrically. An analytic method has been developed to incorporate some non-linear effects [Zel'dovich 1970].

The theory was originally proposed for baryonic adiabatic fluctuations with a large scale, below which all substructure is suppressed. Thus it applies also to neutrino structures. Zel'dovich starts out with a description of particle trajectories. If ζ is the comoving coordinate of a particle, the position in Eulerian coordinates is expressed as

$$r = R(t)\left(\zeta - \frac{b(t)}{R(t)}\nabla\psi(\zeta)\right). \qquad (11.102)$$

The first term describes the Hubble expansion for a uniform distribution of matter, and the second term the effect of an inhomogeneity on the motion.

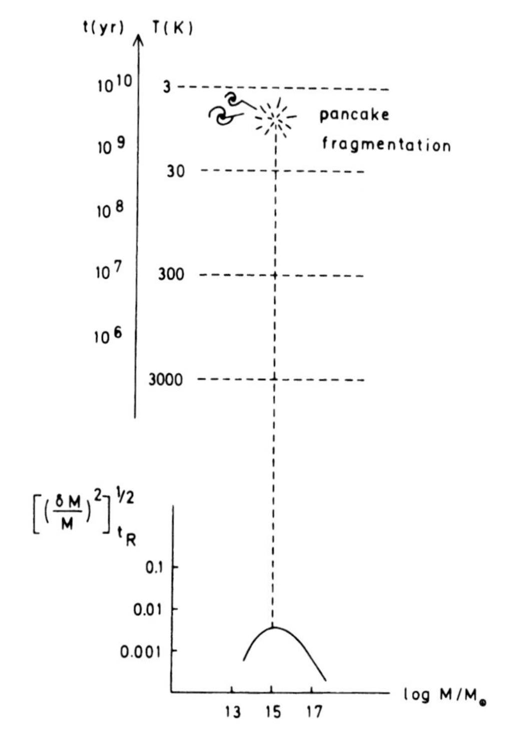

Fig. 11.9. DFS at t_R and subsequent collapse of large-scale structures in a schematic diagram (after [Rees 1984])

$\psi(\zeta)$ is a function of the comoving coordinates which gives the spatial dependence of a gravitationally unstable perturbation,

$$\delta\varrho = \varrho_b e^{\gamma_0 t}\psi(\zeta), \tag{11.103}$$

where $\gamma_0 \equiv \sqrt{4\pi G\varrho_b}$ (Jeans criterion).

The function $b(t)$ grows more rapidly than $R(t)$ (e.g. for a $K = 0$ FL background, $R \sim t^{2/3}, b \sim t^{4/3}$). For a given law of motion $r(t,\zeta)$ the evolution of the density can be easily evaluated. It follows from the relation

$$dm = \varrho dxdydz = \varrho_b d\zeta_1 d\zeta_2 d\zeta_3 R^3$$

that the density $\varrho(\zeta, t)$ can be written as

$$\varrho(\zeta,t) = \varrho_b(t)\left\{\det\left|\delta_{ij} - \frac{b}{R}\frac{\partial^2\psi}{\partial\zeta_i\partial\zeta_j}\right|\right\}^{-1}. \tag{11.104}$$

If the eigenvalues of the matrix are $\alpha > \beta > \gamma$, then the evolution of ϱ reduces in the linear limit ($\alpha(b/R) \ll 1$) to $\delta \propto (\alpha+\beta+\gamma)R \propto t^{2/3}$. But as $\alpha(b/R)$ approaches 1, the density ϱ becomes infinite, $\delta\varrho/\varrho \propto [1-(b/R)\alpha]^{-1}$.

The collapse of an inhomogeneity thus occurs first in one particular direction, leading to transient sheet-like or "pancake" structures.

"Cooking pancakes"

The validity of these approximations in the non-linear regime must be justified. In one dimension (11.104) is an exact solution of the Euler–Poisson equations. One can find solutions similar to (11.104) in three space dimensions if certain constraints are accepted [Buchert and Götz 1987; Buchert 1989a,b]. As the collapse proceeds towards the sheet-like structure, there is, of course, a build-up of pressure, and a shock will propagate outwards from the mid-plane. There will be radiative cooling of the baryons behind the shock. Numerical studies show that the cooling layer, with thickness less than 0.1%, contains 30% or more of the shocked gas. The cooled layer will probably fragment by gravitational instability. Hopefully this would lead eventually to the formation of galaxies. In Fig. 11.10 a schematic picture of the "pancake" collapse has been drawn. Figure 11.11 gives a colourful impression of the density fields in such a model. The observed large-scale filamentary structure

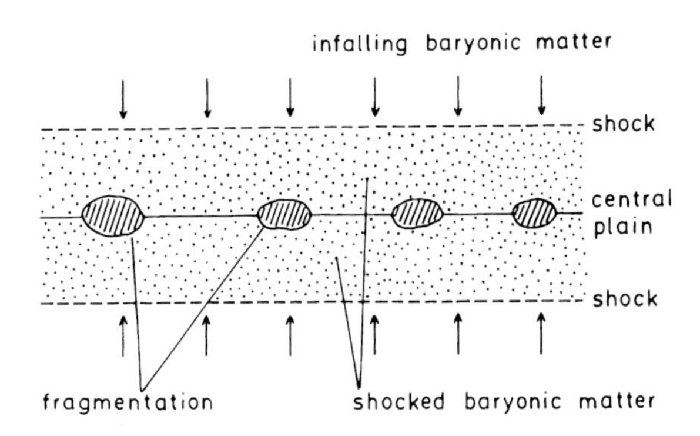

Fig. 11.10. Schematic picture of a neutrino pancake

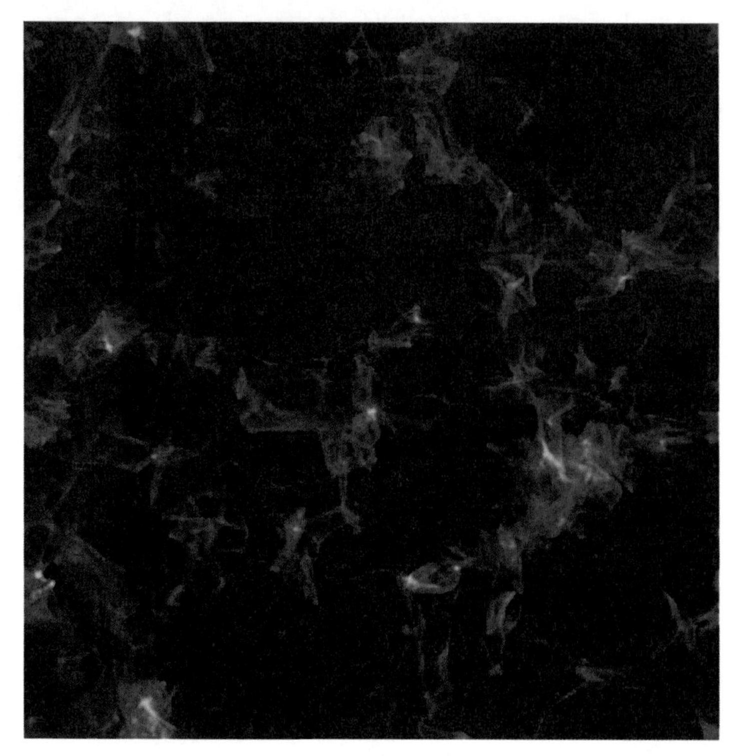

Fig. 11.11. The density field of the Zel'dovich approximation in a 250 Mpc box (courtesy of J. Schmalzing [Schmalzing 1999])

in regions of size 50 Mpc seems to be reproduced quite well by the collapse of the large scales in this picture (see Chap. 12).

11.7.5 Problems with Massive Neutrinos

In numerical studies (cf. Chap. 12) a number of difficulties have shown up:

1. The fragmentation process after the supercluster collapse is apparently not easy to conceive. Numerical simulations (see Chap. 12) give $z < 2$ for the epoch of galaxy formation.
2. About 85% of the baryons do not condense into galaxies, but remain as a hot intergalactic gas.
3. The neutrinos cannot be the dark matter observed in galaxies or in dwarf spheroidals, because they cannot cluster on these scales.
4. The baryons in the supercluster potential wells produce X-rays to such an extent that the limits on the X-ray background are violated.

Some of these difficulties are discussed in more detail in Chap. 12. They have, in particular, more or less eliminated hot DM as a successful candidate. The hope now lies with warm DM and cold DM.

11.7.6 Candidates for Warm DM

Since there are, at present, no good candidates for this species – the "non-hot" particles predicted by supersymmetry such as the gravitino or the photino must be attributed to the cold DM components – we can just assume that there is an X-particle of a hypothetical mass $\sim 1\,\text{keV}$ which interacts very weakly with ordinary matter. It should interact more weakly than the neutrino, such that its decoupling temperature T_{Xd} is much larger than the neutrino decopling temperature.

The usual entropy conservation argument yields, for the density of the X-particle at the present epoch,

$$n_{X0} = \frac{2g_x}{g} n_{\gamma 0}\,[\text{cm}^{-3}]. \tag{11.105}$$

As always, g is the effective number of helicity states of interacting bosons and fermions at $T \leq T_{Xd}$; g_x is the number of helicity states of the X-particle.

The conventional standard model of QCD and electroweak interactions, as well as the GUT theories, tell us that for $1\,\text{GeV} \leq T_{Xd} < T_{\text{GUT}}, g \sim 100$, and hence (with $n_{\gamma 0} = 400\vartheta^3\,\text{cm}^{-3}$)

$$n_{X0} \simeq 8g_X\,[\text{cm}^{-3}]\vartheta^3. \tag{11.106}$$

Expressed in terms of the density parameter,

$$h_0^2 \Omega_{X,0} = 1.1 \left(\frac{g_X}{1.5}\right)\left(\frac{100}{g}\right)\frac{m_X}{1\,\text{keV}}\vartheta^3. \tag{11.107}$$

$\Omega_X \approx 1$ for $m_X \approx 1\,\text{keV}$, and hence the X-particles should have a mass in that range. With $m_X \simeq 1\,\text{keV}$ the X-particles become non-relativistic at a temperature $T \sim m_X > m_\nu$. Therefore their free-streaming mass limit is smaller than the corresponding neutrino mass limit

$$M_{\text{FS},X} \sim M_{Pl}^3 m_X^{-2} = 10^{12}\left(\frac{m_X}{1\,\text{keV}}\right)^{-2} M_\odot. \tag{11.108}$$

The free streaming scale corresponds to the scale of a galactic mass.

11.7.7 Fluctuation Spectrum for Warm DM

$T \sim M_X$ occurs long before the X-particles become the dominant species. With Ω_X from (11.106) they start to dominate over the radiation energy density (taking into account only the photons) $\Omega_\gamma = 2.4 \times 10^{-5} h_0^{-2}$ after a time t_{eq} (redshift z_{eq}) for which

$$1 + z_{eq} = \frac{\Omega_X}{\Omega_\gamma} = 4.2 \times 10^4 (\Omega_X h_0^2)\vartheta^{-4}. \tag{11.109}$$

The horizon scale at this time (see Chap. 10) is

$$ct_{eq} = 3.17 \times 10^{19} c (1 + z_{eq})^{-2}, \tag{11.110}$$

and, referred to a comoving wavelength and to the present time,

$$\lambda_H = (1 + z_{eq}) ct_{eq} = 9 (\Omega_X h_0^2)^{-1} \vartheta \,\text{Mpc}. \tag{11.111}$$

This is the scale of the horizon when the X-particles start to dominate. The mass within λ_H is

$$M_{X,eq} = 2.2 \times 10^{15} (\Omega_x h_0^2)^{-2} M_\odot. \tag{11.112}$$

The main difference from the neutrino case is the fact that here a range of mass scales $M_{\mathrm{FS},X} < M < M_{X,eq}$ is available for the growth of fluctuations.

The second difference is the "stagnation" phenomenon (Mészáros effect) for scales $\lambda_{\mathrm{FS}} \leq \lambda \leq \lambda_{eq}$. When these scales are crossing the horizon, the X-particles have less energy density than the radiation, $z \gg z_{eq}$, and therefore (according to Sect. 11.3) the perturbations stay constant.

During this phase of constant δ the power spectrum of the fluctuations is changed [Peebles 1980, p. 358 ff]. Write

$$\delta_k(f) = A(k) \delta_k(i), \tag{11.113}$$

where $\delta_k(i)$ is the amplitude at high z and $\delta_k(f)$ the amplitude at $z_f = z_d \simeq 10^3$. $\delta_k(t)$ grows as $(1 + z)^{-2}$ for $z \gg z_{eq}$, until $\lambda \leq \lambda_H$, then it stays almost constant until $z = z_{eq}$. The horizon is crossed at a redshift $(1 + z_k) \propto R^{-1} \propto k$, since $\lambda = ct \propto R^2$ occurs for a comoving wavelength $\lambda \equiv 2\pi R(t)/k$ at $\lambda \sim R^2$, i.e. $k^{-1} \simeq R$.

Thus the initial power spectrum $|\delta_k(i)|^2 \propto k^n$ is tilted, for any part that has survived as a fluctuation within the horizon ($\lambda > \lambda_{\mathrm{FS},X}$ or $k < k_{\mathrm{FS};X}$), to a form

$$|\delta_k(f)|^2 \propto k^{n-4}, \tag{11.114}$$

[Peebles 1980, (92.57)].

The resulting density-fluctuation spectrum is schematically depicted in Fig. 11.12. A similar effect occurs when the different scales enter the horizon with the same amplitude, but the fluctuation at larger wavelengths comes in at a later time.

With the power law $|\delta_k|^2$ for the Fourier components of the density distribution, the density contrast on a scale k_i is

$$\left(\frac{\delta\varrho}{\varrho}\right)^2_{k_i} \propto \int_0^{k_i} k^2 \, |\delta_k|^2 dk$$

or

$$\left(\frac{\delta\varrho}{\varrho}\right)_{k_i} \propto k_i^{(n+3)/2}.$$

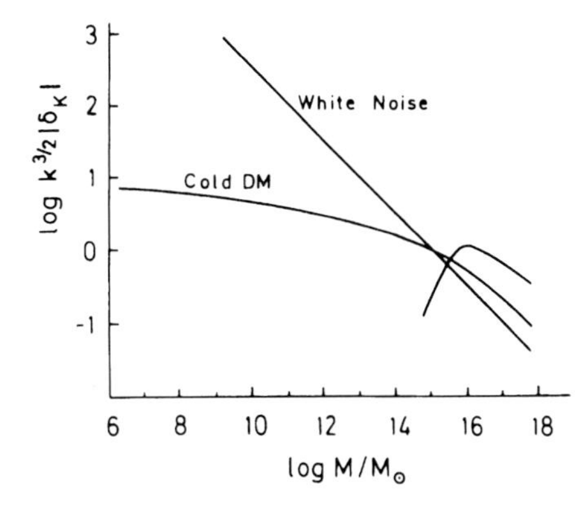

Fig. 11.12. A schematic diagram of the spectra at decoupling times of white noise, hot, and cold DM fluctuations (after [Primack 1986])

For all scales less than $k_H = 2\pi/\lambda_H$, the growth time until z_R is reduced by a factor $(1+z_k)^{-2} \propto k^{-2}$, and hence for $k_i < k_H$ the spectrum is tilted to $k_i^{(n-1)/2}$.

Detailed calculations (e.g. [Bond and Szalay 1983]) show that there is still some small amount of growth for fluctuations with M between $M_{\rm FS}$ and M_{eq}, leading to a DFS at the time of decoupling, which has a fairly broad peak at $M \sim M_{{\rm FS},X}$.

Thus in the warm DM picture objects of $M \geq M_{{\rm FS},X} \geq 10^{12} M_\odot$ – galaxies and small groups – are the first to form, and larger structures collapse later, as $\delta_{\rm DM}$ grows to exceed unity on successively larger mass-scales.

The fluctuation spectrum on a scale $M \equiv R^3(4\pi mn/3)(\pi/k)^3$ is, as we have seen, $\delta(M) = k^{3/2}|\delta_k| \propto M^{-(n/6-1/2)}$. Thus, if $n \geq 0$, one finds qualitatively a spectrum which rises in amplitude up to $M_{H,eq}$, then falls off. The fall-off is $\propto M^{-2/3}$ for $n = 1$.

11.7.8 Problems with Warm DM

There may be various problems with warm DM which have to be discussed thoroughly when the numerical simulations are scrutinized. A similar problem to the case of hot DM may be the fact that smaller systems (like dwarf galaxies) have to form via fragmentation following the collapse of structures of $M \geq M_{\rm FS}$. Since the typical velocity of X-particles is expected to be $\simeq 100\,{\rm km\,s^{-1}}$ (the rotation velocity of spirals, if X-DM clusters on that scale) the dwarf spheroidals could bind only a small fraction (escape velocity $\simeq 10\,{\rm km\,s^{-1}}$).

Therefore the ratio $M/M_{\rm lum}$ would be expected to be much smaller for dwarf galaxies than for large spirals – contrary to observations.

The main problem with warm DM is, of course, that there does not seem to be a good – even if hypothetical – candidate for this type of matter. As one unknown cosmologist put it: "An elementary particle that does not exist in particle theories should also not exist in cosmology."

11.8 Non-baryonic Cold Dark Matter

The lack of experimental evidence for particles which can form the cold DM makes this scenario somewhat speculative too. On the other hand, there is a wealth of candidates, like black holes, quark nuggets, photinos, neutralinos, and axions – and the hope is that some of the planned experiments may produce evidence for these or other new kinds of matter (for the axions see [Sikivie 1983]).

From a fundamental point of view the situation is rather deplorable: all these hypothetical candidates are introduced *ad hoc* without any fundamental explanation or prediction of quantities such as the ratio of the density of dark matter to the density of luminous matter.

11.8.1 The Growth of Fluctuations

The characteristic property of cold DM (CDM) is the unimportance of the free-streaming mass limit; this limit is $M_{\rm FS} \leq 10^8\ M_\odot$ for all the different possible CDM candidates. Therefore it is negligible for the problem of galaxy formation. The DFS is then determined by the primordial spectrum on scales larger than the horizon, and by the "stagspansion" period (Mészáros effect [Mészáros 1975]).

Again only adiabatic fluctuations are considered here. In the standard formalism fluctuations δ grow as $\delta \propto R^2$ on scales larger than the horizon. When a fluctuation becomes smaller than the horizon scale, it cannot continue to grow, since until $z \leq z_{eq}$ the radiation energy dominates, preventing the unimpeded growth of fluctuations.

With the inclusion of N_ν different ($m_\nu = 0$) ν-species, equality of the matter and radiation energy density occurs at

$$1 + z_{eq} = 4.2 \times 10^5 \Omega_0 h_0^2 (1 + 0.68 N_\nu)^{-1}.$$

Thus the spectrum will be distorted after the epoch of equality. The exact form of the distortion can be calculated from the linearized Boltzmann equations for photon, baryon, neutrino, and dark matter distribution functions. The power spectrum in the linear regime is given by

$$P(k,t) = C(t) P_i(k) T_{\rm lin}(k, \Omega_0 h^2) \quad \text{for } t > t_{eq}.$$

$C(t)$ is the fluctuation amplitude at cosmic time t, $P_i(k)$ is the initial spectrum.

A useful fitting formula for the linear transfer function $T_{\rm lin}$ in CDM models is [Bardeen et al. 1986]

$$\begin{aligned} T_{\rm lin}(k, \Omega_0 h^2) = & \left[\frac{\ln(1 + 2.34q)}{2.34q}\right]^2 \\ & \times (1 + 3.89q + (16.1q)^2 + (5.46q)^3 + (6.71q)^4)^{-2}, \end{aligned} \tag{11.115}$$

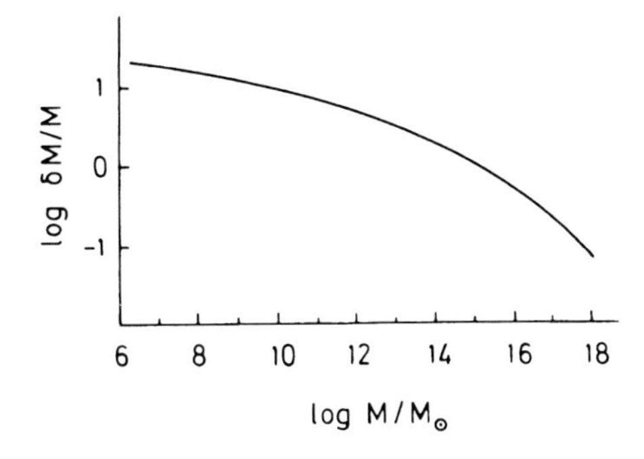

Fig. 11.13. A plot of $\delta M/M$ for $n = 1$. The curve is relatively flat for $M < 10^9 M_\odot$, and steepens to $\alpha M^{-2/3}$ at $M \gg M_{eq}$ (after [Primack 1986])

where $q = k/(\Omega_0 h^2)[h\ \text{Mpc}^{-1}]$. This form is now used in many numerical computations, because the CDM models have become widely used, and successful in comparison with the observations. We shall discuss them more in Chap. 12.

In Fig. 11.12 the hot, warm and cold DM fluctuation spectra are sketched. Figure 11.13 contains a plot of $\delta M/M$ $(n = 1)$. $\delta M/M$ is relatively flat for $M < 10^9\ M_\odot$ and steepens to $\propto M^{-2/3}$ at large values of $M \gg M_{eq}$.

11.8.2 Galaxy and Cluster Formation

When $\delta M/M$ grows to ≈ 1, non-linear gravitational effects become important; the object separates from the Hubble flow and begins to contract. The collapse converts enough gravitational energy into the energy of random motions for the virial theorem to be satisfied: $\langle V \rangle = -2\langle T \rangle$. (The average potential energy equals twice the average kinetic energy.) After virialization the mean density within a fluctuation is about 200 times the density at the maximum radius of expansion; see Sect. 11.1.

As $\delta M/M$ of CDM has the largest amplitude for small masses (see Fig. 11.13), these masses will first undergo collapse. It is evident that a spectrum as in Fig. 11.12 will lead to a hierarchical clustering picture, where the systems forming first have a mass just above the Jeans mass of the baryons at decoupling ($M_{JB} \sim 10^5 M_\odot$), with larger and larger objects following the collapse – a process that may in principle go on until the present epoch. Real galaxies can, however, form only when the baryons can cool to a sufficient degree – such that further collapse can take place. Gas of primordial composition cannot cool significantly unless it is first heated to $\gg 10^4$ K, above its ionization temperature.

With a primordial Zel'dovich spectrum normalized to $\delta M/M = 1$ at the present epoch at a radius [Peebles 1980] of $r \approx 8 h_0^{-1}$ Mpc the masses $M \geq 10^9 M_\odot$ can heat the gas sufficiently (above 10^4 K), so that it can radiate rapidly and contract. An upper bound of $M \leq 10^{12} M_\odot$ can be derived by the requirement that the cooling time should be shorter than the collapse time [Rees and Ostriker 1977].

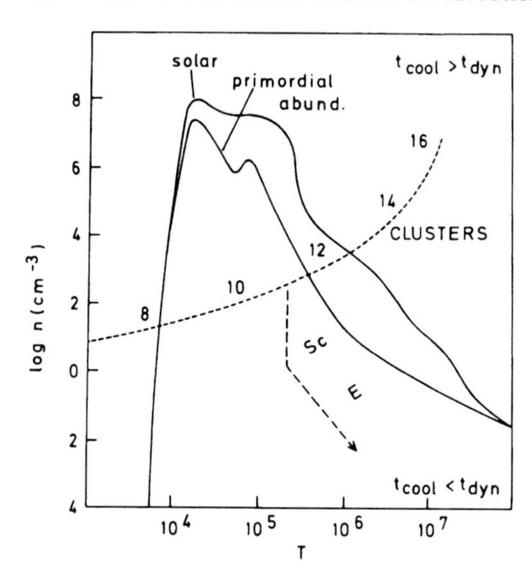

Fig. 11.14. Baryon density versus virial temperature of fluctuations with $\langle\delta\rangle = 1$ ($n = 1$, $\Omega h_0 = 1$, $\varrho_B/\varrho_{\rm tot} = 0.07$). Below the cooling curve (*solid lines*) are the regions where the baryons can cool within a dynamical time-scale. *Dashed line*: possible evolutionary path for the dissipation of baryon energy. The number along the virial temperature curve (*dotted*) is $\log M/M_\odot$ [after Primack 1986]

In Fig. 11.14 these limits are displayed in a plot of the density of baryonic matter versus internal kinetic energy of fluctuations just after virialization ($\delta M/M$ is calculated for $\Omega = h_0 = 1$). Further contraction of the baryon components is possible in protogalaxies for which the cooling time is short compared to the dynamical time. Thus a separation of non-baryonic DM and the baryons occurs because the DM components stay at the radius determined by the virial theorem. We will have more to say about galaxy formation in Chap. 12.

11.8.3 Problems with Cold DM

Again, the real problem is the question of the existence of any DM particle. Another problem might be the existence of the large voids such as that in Bootes, where the number density of galaxies is at least a factor of 4 lower than the average density. In the case of high $\Omega_0(=1)$, some means of suppressing galaxy formation in large regions must be found to hide most of the mass. In a low-Ω_0 universe a similar mechanism must be at work, because linear fluctuation stop growing for $z \leq \Omega_0^{-1}$, and large voids (regions of lower than average density) would stop expanding in comoving coordinates. This implies that the luminous galaxies do not accurately mirror the distribution of the cold DM on large scales. The suggestion that galaxies form preferably in regions of higher density ("biased" galaxy formation) is a first attempt in dealing with these questions. We shall discuss these problems and successes of the DM scenario in more detail in the context of an N-body simulation (Chap. 12) where the non-linear effects can also be studied.

11.9 What's Wrong with Baryonic Galaxy Formation?

"One argument is sufficient, if it were convincing"

R. Feynman

11.9.1 Excluding Baryons from Galactic Halos

The DM in the halos of galaxies cannot be in gaseous form – it would have to be hot to be pressure-supported, and hence produce too much X-ray radiation. It cannot be dust grains – this would contradict the formation of low-metallicity Population III stars. It cannot be in the form of "Jupiter-like bodies" – it seems difficult to make many stars not burning nuclear fuel, without forming a few larger ones that do burn it. (This seems to be a rather weak argument.) It cannot be collapsed stars – too many high-charge nuclei would be ejected during collapse.

11.9.2 Deuterium Limit

The deuterium abundance – if of cosmic origin – limits the baryon density to $\Omega_B h^2 = 0.02 \pm 0.01$ (cf. Chap. 3).

11.9.3 The Growth of Fluctuations

As we have shown, purely baryonic matter with $\Omega_B \leq 0.05$ has adiabatic fluctuations ($\delta_r = \frac{4}{3}\delta_B$) growing only between $z_r \sim 10^3$ and $z = \Omega^{-1} \sim 20$. Therefore we can only expect growth by a factor of 100. The amplitude at z_r must be $\delta\varrho/\varrho \geq 10^{-2}$ for masses above the Silk mass M_{DS}. This mass at t_R corresponds to an angular scale of $1'$ today, and to temperature fluctuations $\delta T/T = \frac{1}{3}\delta\varrho/\varrho \geq 3 \times 10^{-3}$ on a scale of arcmin – this is two orders of magnitude above the observational upper limits! Therefore a purely baryonic universe made from adiabatic fluctuations seems impossible.

Loopholes

1) *Isentropic perturbations* ($\Gamma = 0$) would not be constrained by limits on $\delta T/T$. But they are not easy to understand from a physical point of view. In addition, they initiate adiabatic perturbations of similar amplitude – another possible conflict with $\delta T/T$ limits (see Sect. 11.3).
2) *Re-ionization* of the matter at some redshift $z_i \geq 10$ by UV photons (perhaps from an early generation of stars) would smear out original fluctuations by rescattering if the optical depth of the re-ionized matter was of order unity or greater. This loophole seems to close gradually through the realization that in most models of galaxy formation the bulk motion of the scattering electrons

will be the cause of additional fluctuations of the background radiation field [Vishniac 1987].

The original argument is very simple. The assumption of complete re-ionization at z_i would give an optical depth (per Hubble radius)

$$\begin{aligned}\tau(z_i) &= (ct_H)^{-1} \int_0^{z_i} \sigma_T n_e dl \\ &= \frac{H}{c} \int_0^{z_i} \sigma_T n_{B,0}(1+z)^3(1+z)^{-2}(1+\Omega z)^{-1/2} dz \qquad (11.116) \\ &= 4.7 \times 10^{-3} \frac{\Omega_{B,0}}{0.1} h_0 z_i^{3/2} .\end{aligned}$$

Here σ_T is the Thomson cross-section.

Set $\tau(z_i) = 1$, and obtain

$$z_i = 36 \left[\frac{\Omega_{B,0} h_0}{0.1} \right]^{-2/3} . \qquad (11.117)$$

Re-ionization must have occurred before z_i, to have the required effect of washing out $\delta T/T$ fluctuations. For $h_0 = 1/2$, $\Omega_{B,0} = 0.2$ we have $z_i \simeq 36$, and for $h_0 = 1/2$, $\Omega_{B,0} = 0.02$ we have $z_i \simeq 120$. It is difficult, however, to imagine physically plausible scenarios where efficient sources of the ionizing UV radiation exist at such early times.

This scenario has the interesting possibility that the scattering processes give rise to additional temperature anisotropies which are often larger than the original fluctuations. There are basically two different mechanisms that can produce such effects. Both were first mentioned by [Zel'dovich and Sunyaev 1969].

a) The Comptonization of the background photons by a hot electron gas reduces the black-body temperature at long wavelengths. This "Sunyaev-Zel'dovich" effect (see Chap. 2) depends on the distribution of the hot gas, and significant temperature anisotropies occur only in particular galaxy formation models.

b) The bulk motion of clouds of ionized gas causes a temperature distortion due to the scattering of the photons

$$\frac{\Delta T}{T} = - \int_0^{t_0} (\boldsymbol{n} \cdot \boldsymbol{v}) n_e \sigma_T e^{-\tau} dt ,$$

where n_e is the electron density, and n is the unit vector along the line of sight; $\boldsymbol{v}$ is the velocity; σ_T is the Thomson scattering cross-section. The scattering processes over the past history of a given line of sight are evaluated in the integral up to the present epoch t_0. Multiple scattering effects are approximated by an optical depth factor $e^{-\tau}$. [Vishniac 1987] demonstrates that this effect can produce significant temperature fluctuations on arcminute scales.

11.10 Topological Defects and Galaxy Formation

Topological defects (see Chaps. 6 and 7) – if they existed in reality – would be seeds for the accumulation of matter. If they were produced in the early universe they would create a system of centres of gravitational attraction. Even in an initially uniform configuration density perturbations would be introduced. Their distribution is determined by their physical properties and the history of their formation and evolution. Thus they might open the path to a well-defined theory of structure formation. This is all very well, but a precisely defined theory has the drawback that each new measurement rattles the foundations, if the theory is not the right one. Actually, the few data available now already suffice to rule out the simple topological defect models. Thus they have to be revived by adding new properties. So, for instance, in the case of cosmic strings we have seen scenarios with strings, with loops, with moving strings, with superconducting strings, with strings with neutrons, etc. (see [Shellard and Vilenkin 1994; Hindmarsh and Kibble 1995] for reviews).

Fortunately we do not have to learn the details of all these models – although anybody who wants is free to do so – since the measurement of the first acoustic peak in the CMB by the Boomerang and Maxima experiments has put them out of business. Figure 11.15 illustrates the situation (after [Bouchet et al. 2002]).

The data clearly show an acoustic peak at a multipole index $l \simeq 200$ (cf. Sect. 11.6). A typical cosmic string scenario just gives a single, broad hump at $l \simeq 100$ with an amplitude about a factor of 4 lower (indicated by the dashed line in Fig. 11.15). Global defect models (e.g. textures) generally predict a peak around $l \simeq 350$ with a height about 1/4 of the measured one.

Cosmic strings from the symmetry-breaking of a local gauge group give rise to a single wide hump, or an almost flat spectrum with a rapidly decreasing $\Delta T/T$ for high l values.

The reason for this difference from adiabatic perturbations is that in topological defect models the individual fluctuation modes are coupled due to the non-linear structure of the stress–energy tensor. This smoothes the spectrum and removes the acoustic peaks.

So, at the moment, these models are no longer acceptable. It is, however, still worthwhile to mention some of their basic properties. No cosmological structure formation model is completely dead until such time as we have found the one correct model.

Topological defect models can be included in the perturbation scheme simply by adding an inhomogeneous term. Then these false vacuum configu-

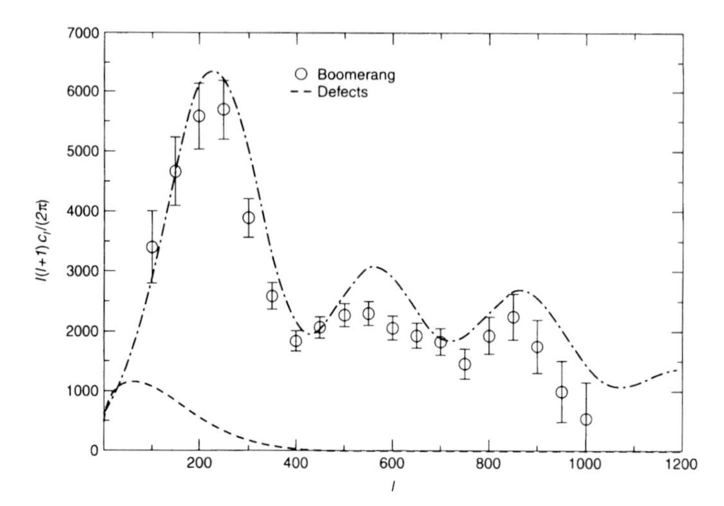

Fig. 11.15. The dashed line shows the CMB anisotropy expected from a standard cosmic string scenario. The discrepancy with the observations is obvious

rations act as seeds for density fluctuations even in an initially unperturbed space-time. This simple description is valid if the mean density of the seeds is much smaller than the background density of the cosmological model, and if the seeds interact only by gravitation with the other matter components.

The energy–momentum tensor of the seeds $T^{\mu\nu}_{(s)}$ has no contributions from the homogeneous background, and is therefore by itself gauge-invariant. It is determined by the unperturbed equations of motion of the seeds. The geometrical perturbations can be split into a part caused by the seeds (Ψ_s, Φ_s), and a part caused by the perturbations of the remaining matter components:

$$\Psi = \Psi_s + \Psi_m, \quad \Phi = \Phi_s + \Phi_m, \quad \text{etc.}$$

The formalism is presented in detail in [Durrer 1994]. The simple case of a single fluid, only scalar perturbations, $\Pi = \Gamma = 0$ (adiabatic, isotropic), and a flat model ($K = 0$) may serve as an illustration. These conditions lead to the following equations for the density contrast D:

$$\begin{aligned} D'' &+ k^2 c_s^2 D + (1 + 3c_s^2 - 6w)(R'/R)D' \\ &- 3[w(R''/R) - 3(R'/R)^2(c_s^2 - w) + (1+w)\frac{4\pi}{3}G\varrho R^2]D = S, \end{aligned} \tag{11.118}$$

where the source term

$$S = (1 + w)4\pi G(f_\varrho + 3f_p).$$

f_ϱ and f_p are the scalar contributions to the $T^{\mu\nu}_{(s)}$, corresponding to the density and pressure of the seeds. The perturbation induced by S can be computed from the homogeneous solutions D_1 and D_2 of (11.118)

$$D = c_1 D_1 + c_2 D_2,$$
$$c_1 = -\int (SD_2/W)dt, \quad c_2 = -\int (SD_1/W)dt;$$

$W = D_1 D_2' - D_2' D_2$ is the Wronskian determinant of the homogeneous solutions. The behaviour is determined by the typical frequencies of the source. If the time dependence of D_1, D_2 and S can be approximated by power laws, then D is proportional to S as long as $S \neq 0$. If D_1 and D_2 are waves with frequency ω and approximately constant amplitude, then D can be approximated by a wave with amplitude proportional to

$$\omega^{-1} \int e^{i\omega t} S \, dt.$$

Let us look at some properties of cosmic strings and textures.

11.10.1 Strings and Galaxy Formation

In Chap. 6 some of the properties of cosmic strings were discussed. The crucial dimensionless number $G\mu$, which is approximately $\sim 10^{-6}$ in GUT theories, determines most of their cosmological significance. μ is the mass per unit length of the string.

There may be some difficulty in forming strings after a period of inflation. Strings that existed before an inflationary epoch would be diluted, just like monopoles. The reheating after inflation may reach temperatures that are high enough to allow another phase transition which produces strings. A potential $V_{\text{eff}} = \lambda(\phi^2 - v^2)^2$ gives $\mu \sim \pi v^2$, while the temperature of the phase transition is $T_c \sim \lambda^{1/2} v$. With $v = 10^{16}$ GeV, $\lambda = 10^{-5}$ one can satisfy the constraint that T_c is less than the reheating temperature $\sim 10^{14}$ GeV, and $G\mu \sim 10^{-6}$. This is, however, not a natural model – some fine-tuning of the parameters is necessary.

Strings come into being with infinite length or as closed loops, but generally in a complicated configuration of a tangled network of loops and lines. The correlation length is limited by the horizon scale λ_H at their time of formation. The interaction and self-interaction of strings leads to the production of loops with radius $r \sim \lambda_H$, and a subsequent decay into smaller loops. The loops keep almost fixed, comoving locations, and their oscillations lead to the emission of gravitational waves (cf. Chap. 6). Some properties of the string network have been found in numerical simulations [Turok 1984, 1986; Vachaspati and Vilenkin 1984]. The computations are quite complex, however, and up to now the predictions from cosmic string models show some dependence on the simulation procedure, and the group of scientists involved in it ([Bouchet et al. 2002] and references therein).

The matter inhomogeneities in a universe filled with strings develop by accretion onto the loops. According to the size of the loops, a hierarchy of objects is generated, smaller ones inside larger ones, and these can be identified eventually with galaxies and clusters. The initial fluctuation spectrum is unimportant; the distribution of the loops is determining the structures.

In the simplest models the 2-point correlation function $\xi(r)$ for loops of mean separation d is a dimensionless function of r/d. The reason is that all loops are produced in the same way, and the scale determining the formation process is the horizon scale at that epoch. Therefore at the time of formation all loops show identical correlations. The cosmic expansion subsequently stretches all separations in proportion to the expansion factor $R(t)$, so $\xi(r/d)$ does not change. This is interesting, but it seems to be different from observational results (compare Chap. 12).

The numerical simulations have been done for simple cases such as the model dominated by cold dark matter with $\Omega = 1$. Perturbations start to grow in that case at t_{eq}, when radiation and matter density are equal. The mean separation of loops of size r is $\sim r$ when they are formed. The requirement that loops be massive enough to accrete objects with the mass of an Abell cluster up to the present epoch leads to a value

$$G\mu \simeq 2 \times 10^{-6}.$$

There is a chance to test this value experimentally. The gravitational waves radiated by oscillating loops and strings contribute to a background of gravitational radiation. Such a "noisy" background would lead to a "timing noise" in pulsars which could in principle be detected. J.H. Taylor and co-workers have found a good candidate for such a high-precision measurement in the extremely stable period of the millisecond pulsar PSR 1937+21 [Rawley et al. 1987]. This object has since 1988 kept its period of ~ 1.6 ms very precisely and stably. The analysis of timing residuals allows a safe upper limit to be put on the density parameter Ω_g of a noise background of gravity waves

$$\Omega_g < 4 \times 10^{-7} h_0^{-2} \qquad (11.119)$$

at frequencies of 0.23 cycles per year [Rawley et al. 1987].

This is already close to the background expected from cosmic strings which may roughly be estimated as follows [see Straumann 1988]. Loops surviving until the epoch t in the radiation era have a typical size $L \sim G\mu t$, and produce gravitational waves of frequency $\sim L^{-1}$ and energy density

$$\varrho_g^s \simeq (\dot{M}_g t) n(L) \simeq \mu L n(L)$$

(compare (6.130)), where $n(L)$ is the density of loops larger than L; $n \sim t^{-3/2} L^{-3/2}$ in the radiation-dominated era, and thus

$$\varrho_g^s \simeq (G\mu)^{1/2} (gt^2)^{-1}.$$

Comparing this expression to the temporal behaviour of the photon energy density $\varrho_\gamma \sim (30Gt^2)^{-1}$, we find

$$\Omega_g^s \simeq 30(G\mu)^{1/2}\Omega_\gamma \simeq 6 \times 10^{-7} h_0^{-2} \mu_6^{1/2} \qquad (\mu_6 \equiv G\mu/10^{-6}). \qquad (11.120)$$

A more exact calculation [Brandenberger et al. 1986] results in a somewhat lower estimate,

$$\Omega_g^s = 4 \times 10^{-8} h_0^{-2} \Omega_0^{-1} \mu_6^{1/2}. \qquad (11.121)$$

There is also some uncertainty in this result due to several uncertain numerical factors occurring in string formation scenarios. (A product $\beta\nu\gamma$ appears on the right-hand side of (11.120), where ν characterizes the number-density distribution, β the mass – $M = \beta\mu L$ – and ν the efficiency of gravitational radiation emission. Favoured values of $\nu = 0.01$, $\beta = 10$, and $\gamma = 5$ have been used to obtain the result of (11.121).)

Cosmic string models usually require some additional DM component. They do not give a consistent structure formation model just by themselves. It seems to me more economical to investigate models with just DM, such models may be sufficient.

Textures and Structure Formation

The behaviour of cosmic dust ($p = 0$ matter) in the field of a spherically symmetric texture (see Chap. 6 and [Durrer 1994]) can be described with a source term

$$S = 16\pi\eta^2(1 + y^2)^{-2}, \qquad (11.122)$$

where $y = (t - t_c)/r$. t_c is the epoch of collapse of the texture ($4\pi G\rho t_c \ll 1$). The density perturbation of the dust

$$D = 16\pi\eta^2 t/r[\arctan(t/r) + \pi/2] + 1. \qquad (11.123)$$

Asymptotically $D(t \to -\infty) = 0$, and $\lim_{r\to\infty} D(t,r) = 16\pi\eta^2$. For $t/r \ll 1$, the solution is $D \simeq 16\pi\eta^2 t/r$. Near the epoch of collapse, $t/r \gg 1$ and D is of the order of $16\pi\eta^2$. At a given time t_* after the collapse D has the following profile:

$$D \simeq 16\pi\eta^2 \quad \text{at} \quad t = t_* \quad D \simeq 16\pi^2\eta^2 t_*/r, \quad \text{and}$$
$$D \to \infty \quad \text{for} \quad r \to 0.$$

This divergence leads to the early formation of non-linear structures on small scales.

The distortions induced in the CMB are typically

$$\frac{\Delta T}{T}(\text{at } 2^\circ) = (4 \pm 0.8) \times 10^{-5}$$

i.e. higher than the COBE limits. As I have already mentioned, the acoustic peak measured by the CMB experiments cannot be reproduced. Additional DM components are needed in texture models. The texture scenario just sets the initial perturbation distribution.

Exercises

11.1 There are solutions of the Einstein equations which describe a spherical mass inside an expanding cosmological model. A simple case is the Schwarzschild solution (black hole) surrounded by an expanding FL model. One can show that the Schwarzschild solution inside a radius r_0 can be joined smoothly to the metric of an FL model (with $\Lambda = 0$), if it has the form

$$\begin{aligned} ds^2 = &\, c^2 dt^2 (1 - 2m/r)(1 - r\rho_0^2)(1 - 2m/(\rho_0 R))^{-2} \\ &- dr^2 (1 - 2m/r)^{-1} - r^2 d\Omega^2 . \end{aligned}$$

Here $m = GM/c^2$, where M is the Schwarzschild mass; ρ_0 is a constant

$$\rho_0 = (6m/(8\pi G \rho_m R^3)^{1/3}),$$

where $\rho_m(t)$ is the mean mass density of the FL model; $R(t)$ the expansion factor. The radius, where the Schwarzschild sphere joins the FL solution, is $r_0 = (6m/(8\pi G \rho_m))^{1/3}$.
Derive these relations. Effects of the expansion are felt inside r_0 only through the correction term containing $R(t)$. How large are these effects for a typical cluster of galaxies, for a supercluster? What is the size of r_0 for typical astronomical masses M?

11.2 What happens if r_0 is less (greater) than the value given in Exercise 11.1?
The mass M of the solution in Exercise 11.1 could be concentrated in a thin shell at a radius $r \leq r_0$. How does such a mass shell evolve? (Compare the solution given in [Sato and Maeda 1983].)

11.3 Calculate the free-streaming length for warm dark matter.

11.4 What is the CMB anisotropy expected from one domain wall? The signal from a distribution of such objects can also be estimated. Is there a sequence of peaks as in the CDM models?

11.5 Compute the CMB anisotropy signal expected from textures.

11.6 Compute (numerically, if necessary) the curves for the growth of linear density perturbations in the adiabatic case, and for cosmological models with

$\Omega_m = 1.0$, with $\Omega_\Lambda = 0.7, \Omega_m = 0.3$, and with $\Omega_\Lambda = 0, \Omega_m = 0.1$. What is the difference for structure formation, if the different models are all normalized to the same value at the present time?

11.7 In simple models of quintessence the equation of state is $p/\rho = w$. Assume w to be constant and compute $H_0 t_0$ for different values of w. Can the measurements already constrain w?

11.8 Can a positive cosmological constant influence the gravitational equilibrium of large masses? Consider a spherical configuration to a post-Newtonian approximation, and derive the changes in the gravitational force due to the cosmological constant.
Is the Coma cluster still a bound object? How is the result modified in an expanding cosmological model?

12. Non-linear Structure Formation

"There are difficulties; there are certainly difficulties ...
Have you any alternative theory which will meet the facts?"

The Sign of Four - A.C.Doyle

It is an evolution from the simple to the complex, from
unordered chaos to highly differential entities, from
the unorganized to the organized.

V. Weisskopf: Knowledge and Wonder (1977)

12.1 General Remarks

The discussion in Chap. 10 was concerned with the evidence for structure in the universe. The galaxies seem to be concentrated on the surface of spherical bubbles. The interior of these bubbles apparently does not contain any luminous matter, while the walls of the cells often contain sheet-like structures of galaxies. The walls are, at some locations, strengthened by strings of clusters which converge on very rich knots. The centre of a knot is usually a prominent cluster of galaxies (Virgo, Coma, Perseus, ...). These superstructures contain typically $10^{15}-10^{16}M_{\odot}$, between the superstructures are large regions, seemingly empty of luminous matter, the so-called "voids". The key question is to explain this type of structure in an expanding universe.

Two different scenarios have been suggested:

a) The hierarchical model starts with the formation of galaxies from small fluctuations ($z \geq 10$), which become non-linear on very small scales first.

Schematic model of the cell-structure of the universe: "It's like Swiss cheese, only more so."

The large-scale structure is then expected to develop on an expanding background from these first units.

b) The "pancake model" attributes the cell structure to large-scale adiabatic density fluctuations (on a scale of $100h_0^{-1}$ Mpc at the present epoch). The problem then is to understand how galaxies form within these large structures.

Observations and further theoretical developments show that the true stiuation probably can be described neither by well-defined clumps in equilibrium merging together, nor by a fragmentation process. Rather it seems to be a process continuing on all scales, where smaller clumps which are not equilibrated yet merge, while on larger scales the structures are forming already.

Clearly, such complex processes can no longer be described analytically. Therefore many attempts have been made to run computer simulations of the evolution of an expanding universe, and then to extract the observable statistical properties from the numerical results.

A very helpful fact in such computations is the similarity of gravitating systems on an expanding background to non-magnetic plasmas. The essential difference between Newtonian gravity and the Coulomb interaction is the all-pervading, attractive nature of the gravitational force. A system of gravitating particles necessarily has a positive mass, while a plasma is treated as electrically neutral. Therefore stationary gravitating systems are necessarily inhomogeneous or balanced by an external force. The situation is different in the case of a non-stationary, quasi-homogeneous system on an expanding background: the gravitational attraction is compensated by the expansion, and the treatment of inhomogeneities on the background becomes very similar to the plasma-physics problem – the density fluctuations compared to the mean density can be positive or negative, and the total mass fluctuation is zero ("neutrality"). This similarity makes detailed simulations possible. In Table 12.1 we list the basic features of these N-body simulations.

Table 12.1. Simulations without coherence length on an expanding background. N is the number of particles

Ref.	N
Peebles [1971]	100
Aarseth et al. [1979]	$1000, 4000\,(P\;P)$
Efstathiou and Eastwood [1981]	$20\,000\,(P^3M)$
Miller [1984]	$10^5, 64^3\,(PM)$
Davis et al. [1985]	$20\,000\,(P^3M)$
Jing [1998]	$256^3\,(P^3M)$
Jenkins et al. (Virgo Collab.) [2001]	$256^3\,(P^3M)$
Jing [2001]	$512^3\,(P^3M)$

The ideal situation would be to include the radiation hydrodynamics of baryonic matter on galactic scales in addition to the purely gravitational interaction of the dark matter. This is not possible, however, since the observations show structures from 100 kpc to 100 Mpc, suggesting a dynamical range of at least 10^3. Therefore, to resolve the structures on all scales, about 10^9 particles should be simulated just to cover the purely gravitational interaction adequately. In addition, we might want $\sim$ 100 particles per galaxy to arrive at a rough picture of the formation process. Altogether a straightforward model would need more than 10^{11} particles, whereas sophisticated codes on presently available computers describe less than 10^9 particles.

It is clear that we are a long way from the ideal situation. There must be approximations to the physics in each type of numerical simulation.

In the following I will exhibit the salient features of the N-body simulations with the example of the computations by Yipeng Jing [Jing 1998, 2001; Jing and Suto 1998] which are at the moment the best, and most easily accessible to me.

12.2 N-Particle Simulations

The aim of the simulations is to find out whether the clustering of the galaxies does really result from the growth via gravitational instability of small initial fluctuations. In particular, the quantitative statistical analysis of the correlation functions, and their dependence on the spectrum of the fluctuations, are to be tested.

12.2.1 Equations of Motion

Consider the motion of N gravitating particles in a finite volume of an expanding Friedmann–Lemaître background model. After decoupling ($z \leq z_d$) the expansion factor satisfies the Friedmann equation for $p = 0$ matter:

$$\dot{R}^2 - \frac{8\pi G}{3}\bar{\varrho}(t)R^2 = -K \tag{12.1}$$

($\bar{\varrho}$ is the average density, $\bar{\varrho}R^3$ is constant).

With the help of

$$H(t) \equiv \frac{\dot{R}}{R}, \quad \varrho_c = \frac{3H^2}{8\pi G}, \quad \Omega(t) \equiv \frac{\bar{\varrho}}{\varrho_c} = \frac{8\pi G}{3H^2}\bar{\varrho},$$

and

$$t' \equiv tH(0),$$

(12.1) can be written in dimensionless form

$$\frac{dR(t')}{dt'} = \left(\frac{\Omega(0)}{R(t')} + 1 - \Omega(0) \right)^{1/2} . \tag{12.2}$$

Here $R(0) = 1$, and $t = 0$ marks a starting point close to the time of decoupling t_d. The time $t' = tH(0)$ is measured in units of the initial Hubble time $H^{-1}(0)$. $\Omega(0)$ and $R(0) = 1$ are the initial conditions that determine the evolution via (12.2). Note that for $K = 0, \Omega(t) = 1$ always, whereas $\Omega(t)$ is a decreasing function of t for $K = -1$.

The motion of the particles is described in comoving coordinates x_i':

$$\begin{aligned} x_i' &\equiv \frac{x_i(t)}{R(t)}, \\ x_i'(0) &\equiv x_i(0). \end{aligned} \tag{12.3}$$

Without gravitation between the particles in comoving coordinates, they would keep their position $x_i'(0)$ and $\bar{\varrho}$ would remain constant. The equations of motion are (suppressing subscripts i):

$$\dot{\boldsymbol{v}} = -\nabla\phi, \tag{12.4}$$

$$\dot{\boldsymbol{x}} = v, \tag{12.5}$$

$$\Delta\phi = 4\pi G\varrho(t, \boldsymbol{x}). \tag{12.6}$$

The continuity equation is

$$\varrho R^3 = \text{const.}$$

Transformed to comoving coordinates ($\boldsymbol{v}' = \dot{\boldsymbol{x}}'$), (12.4) reads

$$\dot{\boldsymbol{v}}' + 2H(t)\boldsymbol{v}' = -\frac{\nabla\phi}{R(t)} - \frac{\ddot{R}}{R}\boldsymbol{x}'. \tag{12.7}$$

Note that the effect of the expansion $2H\boldsymbol{v}'$ can be transformed away by setting $dt' \equiv dt/R^2(t), \hat{\boldsymbol{v}}' = d\boldsymbol{x}'/dt' = R^2\boldsymbol{v}'$. The term $\propto (\ddot{R}/R)\boldsymbol{x}'$ represents the tidal forces ($\approx \ddot{R}/R$) which always appear in such a transformation to comoving coordinates.

Since $\nabla\phi = R^{-1}\nabla'\phi, \nabla'(\frac{1}{2}(x')^2) = \boldsymbol{x}'$, one can re-express the equation as

$$\dot{\boldsymbol{v}}' + 2H(t)\boldsymbol{v}' = -R^{-3}\nabla'\phi', \quad \text{with} \quad \phi' \equiv R(t)\phi + \frac{R^2\ddot{R}}{2}x'^2. \tag{12.8}$$

Therefore the Poisson equation in comoving coordinates is

$$\Delta'\phi' = 4\pi G\varrho R^3 + 3R^2\ddot{R}, \qquad \varrho' \equiv \varrho R^3, \tag{12.9}$$

where ϱ' is the density in the comoving system. With $p = 0$,

$$\ddot{R} = -\frac{4}{3}\pi G\bar{\varrho}(t)R(t),$$

and hence

$$\Delta'\phi' = 4\pi G(\varrho'(t, \boldsymbol{x}') - \bar{\varrho}(0)). \tag{12.10}$$

A constant, negative mass density $\bar{\varrho}(0)$ has appeared in the Poisson equation because of the transformation to comoving coordinates. It is, of course, eminently reasonable that in comoving coordinates only the difference $\varrho' - \bar{\varrho}(0)$ acts as a source for the gravitational potential ϕ'. This has, in addition, the consequence that the total mass fluctuation is zero in a sufficiently large system. The N galaxies of the simulation are usually placed in a cubic box with 3-fold periodic boundary conditions – for such a configuration the total mass must be zero as a necessary condition for a solution of the Poisson equation. Clusters have a density excess, $(\varrho' - \bar{\varrho}) > 0$, voids have $(\varrho' - \bar{\varrho}) < 0$, hence clusters and voids repel each other. This saturation of the comoving gravitational force is the reason for the "clustering phenomenon" in the expanding universe.

12.2.2 The Model of the Simulation

The following set of equations has to be integrated numerically:

$$\boldsymbol{v}'_i = \boldsymbol{x}'_i \qquad (i = 1, \dots, N);$$

$$\dot{\boldsymbol{v}}'_i + 2H(t)\boldsymbol{v}'_i = \boldsymbol{F}'_i; \tag{12.11}$$

$$\boldsymbol{F}'_i = -R\ ^3\nabla'\phi'_{|x'_i}; \tag{12.12}$$

$$\Delta\phi' = 4\pi G(\varrho'(t, \boldsymbol{x}'_i) - \bar{\varrho}(0)); \tag{12.13}$$

$$\frac{dR}{dt} = \left(\frac{\Omega(0)}{R(t)} + (1 - \Omega(0))\right)^{1/2}, \quad R(0) = 1; \tag{12.14}$$

$$R^3\bar{\varrho} = \varrho(0). \tag{12.15}$$

The simulations are always concerned with a cubic volume as a part of the universe with periodic boundary conditions in all three space directions. Within this volume N galaxies move under the influence of their relative gravitational force (measured relative to the constant mean comoving density).

1st Step: Equations (12.11) to (12.15) are solved by a difference method, a type of leap-frog scheme (x' evaluated at $t \pm \Delta t$ intervals, v' at $t \pm \frac{1}{2}\Delta t$). After n iterations of this procedure (denoting $f(t + n\Delta t)$ by f^n for any function f):

$$\boldsymbol{v}'^{n+1/2} = \boldsymbol{v}'^{n-1/2}\left[\frac{1 - H(t)\Delta t}{1 + H(t)\Delta t}\right] + \frac{\mathbf{F}'^n\Delta t}{1 + H(t)\Delta t}; \tag{12.16}$$

$$\boldsymbol{x}'^{n+1} = \boldsymbol{x}'^n + \boldsymbol{v}'^{n+1/2}\Delta t. \tag{12.17}$$

2nd Step: The Friedmann equation is integrated in a Taylor expansion

$$R^{n+1} = R^n + \dot{R}^n \Delta t + \frac{1}{2}\ddot{R}^n (\Delta t)^2 + \dots , \tag{12.18}$$

$$\dot{R}^{n+1} = \dot{R}^n + \ddot{R}^n \Delta t + \dots . \tag{12.19}$$

The higher derivatives are determined by the Friedmann equation (time unit $H^{-1}(0)$):

$$\ddot{R} = -\Omega(0)/R^2, \qquad \dddot{R} = 2\Omega(0)\dot{R}/R^3; \quad \text{etc.} \tag{12.20}$$

The precision of the integration procedures is tested by comparison with (12.2). The same time step as in the integration of the equations of motion is used.

3rd Step: To compute the force, essentially two different schemes are in use.

a) PP – the "Particle–Particle" Method (Aarseth Integrator)

The force $\boldsymbol{F}'_i$ is obtained from a summation over all pair interactions

$$\boldsymbol{F}'_i = -\sum_{j \neq i} G m_j \frac{\boldsymbol{x}'_i - \boldsymbol{x}'_j}{|\boldsymbol{x}'_i - \boldsymbol{x}'_j|^3} .$$

This method is very time-consuming and therefore has a limited range of application – the number of particles $N \leq 1000$. Since there are $N(N-1)/2$ pairs of particles the number of computing operations is also proportional to $N(N-1)/2$. The nominal CPU time per time-step can be estimated as $\approx 10N^2 t_{1\mathrm{op}}$, where $t_{1\mathrm{op}}$ is the nominal time for 1 operation [Aarseth 1984]. Now $t_{1\mathrm{op}}$ is about 10^{-9} s for a processor. Thus to follow 10^5 particles for about 1000 time-steps takes about one day. The limitations of this method are evident. Improvements can be made, e.g. by introducing a time-step for each particle. Particles with slowly varying forces are given big time-steps. This reduces the time requirements somewhat, but it still does not allow large-scale simulations. ($N = 1000$ simulations were first carried out by [Aarseth et al. 1979].)

b) P^{3}M: "Particle–Particle–Particle–Mesh Method"

The pair force $F(j \leftrightarrow i)$ is divided into two parts,

$$\boldsymbol{F}(j-i) = \boldsymbol{F}^{sr}(j-i) + \boldsymbol{F}^{m}(j-i) \tag{12.21}$$

[Hockney and Eastwood 1981]. $\boldsymbol{F}^{sr}(j-i)$ is representative of a rapidly varying short-range component to which only the nearest neighbours contribute. A more slowly varying part $\boldsymbol{F}^{m}$ ("mesh-force") is approximated by a lattice force (PM method – cf. Sect. 12.2.2c). The number of operations per time-step is no longer N^2, it increases instead as

$$N_{Op} = \alpha N + \beta(N_m) + \gamma N_n N$$

where N_n is the number of nearest neighbours, N_m the number of lattice points. Typical values are $\alpha = 20, \beta = 5N_m^3$ for an $N_m \times N_m$ lattice. With $N_m = 32, N = 10^5 : \alpha N + \beta(N_m) = 4\frac{1}{2}$ s for $t_{1\mathrm{op}} = 1$ µs.

A recent version of direct summation [Appel 1985; cf. also Monaghan 1985] uses hierarchical clusters of particles and reduces the number of operations to $N \log N$, more efficient than direct summation for $N \geq 10^3$.

c) "PM" Model (Particle–Mesh)

The name PM stands for the direct solution of Poisson's equation on a grid [Hockney and Eastwood 1981]. It has been common practice to use an algorithm with moderate accuracy on a regular grid. Very fine grids may often be needed, because the accuracy of these algorithms is often very poor. It may prove to be an advantage to spend more computing time on a more accurate version of a Poisson solving algorithm, e.g. a multi-grid code (see [Monaghan 1985] for an outline of the principles). For the moment we are content to describe the procedure in its basic form, where the calculation of the mesh-force proceeds as follows [Hockney and Eastwood 1981].

A finite volume in the universe is divided up into a lattice $M \times M \times M$ with lattice constant H, and total length $L = MH$. The grid points are designated by $X_p = pH$, $p = (p_1, p_2, p_3)$; while $x_i(t)$ describes the particle positions in the lattice. In the following $(x' \to x, t' \to t)$. Then the calculation of the force $F(x_i)$ proceeds in four steps:

Step 1: A mass is ascribed to each lattice point X_p for a given distribution of the N particles x_i:

$$\varrho(\boldsymbol{p}) = \frac{M}{V_c} \sum_{i=1}^{N} W(\boldsymbol{x}_i - \mathbf{X}_p), \tag{12.22}$$

where $V_c = H^3$ and $W(x - X_p)$ is the mass distribution function. Equation (12.22) is effectively an interpolation formula from disordered points ("particles") to the grid.

Step 2: The Poisson equation is solved on the lattice by Fourier transform

$$\phi(\boldsymbol{p}) = 4\pi G V_c \sum_{p'} G(\boldsymbol{p} - \boldsymbol{p}')\varrho(\boldsymbol{p}'), \tag{12.23}$$

where $G(\boldsymbol{p})$ is Green's function on the lattice.

Step 3: The gravitational force at the lattice points $\mathbf{X}_p$ is computed from

$$E(\boldsymbol{p}) = -\mathrm{D}\phi(\boldsymbol{p}), \tag{12.24}$$

where D is the derivative on the lattice.
Step 4: The force is interpolated from the intermediate positions x_i:

$$\boldsymbol{F}(\boldsymbol{x}_i) = \sum_p W(\boldsymbol{x}_i - \mathbf{X}_p)\boldsymbol{E}(\boldsymbol{p}). \tag{12.25}$$

Steps 2 and 3 are done with the fast Fourier transform. In solving the Poisson equation it is important to control the errors. At each step one tries to correct for the error by changing $G(\mathbf{p} - \mathbf{p}')$ in the sense of minimizing the deviation of the calculated force from some reference force. This procedure is essential for the PM scheme.

The short-range behaviour is smoothed by assuming finite-size particles with a shape function proportional to $(1 - r/R)$, i.e. the point particle is replaced by a small spherical ball, whose density falls linearly from the centre $(r = 0)$ to the boundary $(r = R)$. The radius R is chosen differently for PM $(R \sim 2$ to $4H)$ and PP $(R \sim 9.2H)$ calculations.

The form of $W(x)$ essentially determines the numerical error. A reasonable choice for $W(x)$ should satisfy a few criteria:

(W1) For particle spacings $\gg H$ the fluctuations should become smaller ("spatial localization of errors").
(W2) $\phi(\boldsymbol{p})$ and $\mathbf{F}(\boldsymbol{x}_i)$ for a particle in a cell should change slowly when the particle is moving within the cell ("fluctuations of spatially localized errors should also stay small").
(W3) The total momentum should be conserved.

Condition (W3) is satisfied if the same function W is used for the mass asignment function and for the force interpolation function – as has been done in Step 4. It should be noted that energy and angular momentum are not conserved in the equations of motion resulting from this set-up. The violation of the conservation laws is, however, usually quite small in simulations, with a relative error of $\sim 1\%$. This is sufficiently accurate for these calculations. The CiC scheme ("cloud in cell") distributes the mass only onto the nearest neighbours: In one dimension,

$$W(x) = \begin{cases} 1 - \frac{|x|}{H} & |x| < H, \\ 0 & |x| \geq H; \end{cases} \tag{12.26}$$

and in three dimensions,

$$W(\boldsymbol{x}) = W(x)W(y)W(z). \tag{12.27}$$

A fundamental set of interpolating functions has been derived [Schoenberg 1973] in the form

$$\bar{f}(x, H) = \sum_j f(X_j) M_n(x - X_j, H), \tag{12.28}$$

where $M_n(u, H)$ are basic splines, which read in the simplest cases [Monaghan 1985] as follows:

$$M_1(u, H) = \begin{cases} 1 & 0 \le \frac{u}{H} \le \frac{1}{2}, \\ 0 & \text{otherwise.} \end{cases} \tag{12.29}$$

M_1 is the nearest grid-point interpolation formula.

$$M_2(u, H) = \begin{cases} 1 - \frac{|u|}{H} & 0 \le \frac{u}{H} \le 1, \\ 0 & \text{otherwise.} \end{cases} \tag{12.30}$$

This is the linear interpolation formula used in (12.26).

$$M_3(u, H) = \begin{cases} \frac{3}{4} - \frac{u^2}{H^2} & 0 \le \frac{u}{H} \le \frac{1}{2}, \\ \frac{1}{2}\left(\frac{3}{2} - \frac{|u|}{H}\right)^2 & \frac{1}{2} \le \frac{u}{H} \le \frac{3}{2}, \\ 0 & \text{otherwise.} \end{cases} \tag{12.31}$$

This is the first basic spline with a continuous first derivative. M_3 was derived in [Hockney and Eastwood 1981], where the big improvement in accuracy over simpler schemes was also noted.

d) The Short-Range Force

Even if only the nearest neighbours are taken into account, it is very time-consuming to calculate the short-range force exactly for each pair. Therefore [Efstathiou et al. 1985], in many cases the ratio $|F_s|/r$ is tabulated at many points spaced linearly over the radius, and the short-range force at any specific location is found by linear interpolation. At some radius r_s the short-range force is set to zero. This leads to a small discontinuity (a jump of a few percent if $r_s \propto M^{-1}$) in the total force at r_s. In Fig. 12.1 (from [Efstathiou et al. 1985]) an illustration is given of the force law used in the P^3M calculation. The mesh-force apparently is close to the correct force only for scales larger than the size of one mesh cell. The fits to the short-range force allow the reproduction of the correct force on scales much smaller than one mesh cell.

As stated above, a more accurate numerical solution of the Poisson equation may actually be worthwhile, although it may use up somewhat more computing time.

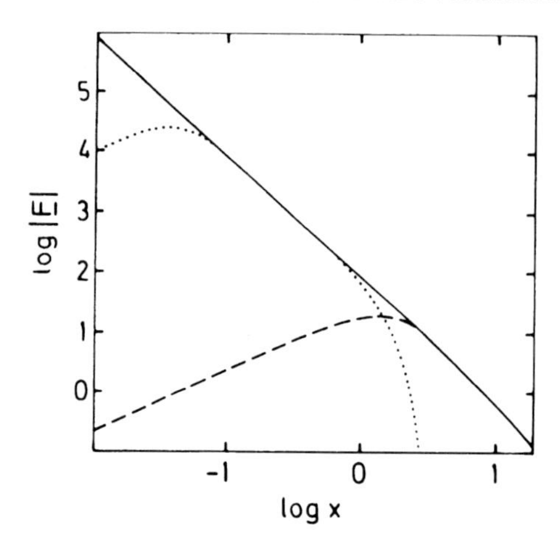

Fig. 12.1. An illstration of the force law used in the $\mathrm{P}^3\mathrm{M}$ code. *Dashed line*: PM contribution (cf. Sect. 12.3); *Dotted line*: short-range contribution with a softening introduced at very small distances (after [Efstathiou et al. 1985]); $x \equiv r/H$

12.3 Simulations with Cold Dark Matter

Here, we shall discuss the simple picture of structure formation by purely gravitational interactions of collisionless particles starting from an initially Gaussian distribution of density perturbations.

12.3.1 Initial Conditions

A given primordial spectrum, denoted by the power spectrum $P_i(k)$ for example, changes with time even during the linear evolution phase (cf. Chaps. 10, 11). If the dark matter consists of weakly interacting massive particles (neutralinos, axions) the fluctuations show no damping at the physically relevant scales. The relativistic free-streaming length is much smaller than, say, the size of a galaxy. As a consequence the fluctuation spectrum has large amplitudes on all relevant scales.

After the equality of the energy densities of matter and radiation the precise evolution of the spectrum is determined by the linearized Boltzmann equations for the distribution function of photons and dark matter particles (baryons must also be included, but we will not do so here).

The CDM power spectrum computed in this way is given by

$$P(k,R) = C(R)P_i(k)T^2_{\text{lin}}(k,\Omega_o h^2). \tag{12.32}$$

Here $C(R)$ is the fluctuation amplitude as a function of the expansion factor $R(t)$. We define $P_i(k)$, as well as the 2-point correlation function $\xi(x)$, through the Fourier amplitudes of the density contrast $|\delta_k|$. But $\delta_k \equiv |\delta_k| \exp(i\phi_k)$ is a complex field, and it might be interesting to recover information on the phase ϕ_k. This question has not been dealt with in great detail so far.

To set up acceptable initial conditions the statistical distribution of the density fluctuations must be prescribed. A Gaussian random field is generally assumed, and this results in a probability function for the Fourier components δ_k:

$$\Pi(|\delta_k|, \phi_k) d|\delta_k| d\phi_k = \frac{2|\delta_k|}{P_i(k)} \exp\left(-\frac{|\delta_k|^2}{P_i(k)}\right) d|\delta_k| \frac{d\phi_k}{2\pi}. \tag{12.33}$$

The distribution function is completely determined by $P_i(k)$ as expected. The ansatz for $\delta(\boldsymbol{x})$ as a Gaussian random field becomes invalid when the fluctuations become non-linear. $\delta(x) \geq -1$ by definition, but for a random Gaussian field one must formally have

$$-\infty < \delta < +\infty.$$

In the linear regime $|\delta| \ll 1$ there is no problem, but for non-linear fluctuations the distribution function will loose its Gaussian nature.

A useful fitting formula for T_{lin} is [Bardeen et al. 1986]

$$\begin{aligned} T_{\text{lin}} = & \ln(1 + 2.34q)/2.34q[1 + 3.89q + (16.1q)^2 \\ & + (5.46q)^3 + (6.71q)^4]^{-1/4}, \end{aligned} \tag{12.34}$$

where $q = k/\Gamma h$ $[M_{pc}^{-1}]$, and the "shape parameter" Γ has been introduced here in concordance with the way these spectra are generally presented in the literature [Efstathiou et al. 1992; Bardeen et al. 1986]. $\Gamma = \Omega h$ in CDM models. This scaling for $k(k/\Omega h^2)$ occurs because the characteristic value during the linear evolution $k_{H,eq}$ is proportional to $\Omega_o h^2$.

For small q, this form of T_{lin} reproduces the asymptotic behaviour of $P(k)$ for times after the equality of matter and radiation energy density,

$$P(k) = \begin{cases} k^n & \text{for} \quad k \ll k_{H,eq}, \\ k^{n-4} & \text{for} \quad k \gg k_{H,eq}, \end{cases} \tag{12.35}$$

if the initial power spectrum is

$$P_i(k) \propto k^n. \tag{12.36}$$

To fix the amplitude $C(R)$ one can use the mass fluctuations in randomly placed spheres of radius S (see also Sects. 11.4 and 11.8). The fluctuations $\langle (\frac{\delta M}{M})^2 (S) \rangle$ are expressed as

$$\left\langle \left(\frac{\delta M}{M}\right)^2 (S) \right\rangle = \int_0^\infty W^2(kS) P(k) \frac{k^2 dk}{2\pi^2}$$

in terms of the power spectrum $P(k)$, and of the filter function W, often used as a top-hat window:

$$W_{\text{th}}(x) = \frac{3}{x^3}(\sin x - x\cos x).$$

This quantity is then compared to an observational quantity derived from the counts of galaxies:

$$\sigma_{8,g} \equiv \left(\left\langle\left(\frac{\delta N}{N}\right)^2_{gg}(S = 8h^{-1}Mpc)\right\rangle\right)^{1/2} \simeq 0.96. \tag{12.37}$$

$\sigma_{8,g}$ is the root of the galaxy number fluctuations within randomly placed spheres of radius S.

Although not quite correct, we may assume that (12.32) holds until the present time t_o and determine $C(R_o)$ such that the mass fluctuation $\sigma_{8,m}$ at $S = 8h^{-1}$ Mpc, given by (12.36), is equal to $\sigma_{8,g}/b$. The factor b is called a bias factor. It measures the difference in clustering between galaxies and mass fluctuations. With $\sigma_{8,g} \sim 1$, we have $b^2 = 1/\sigma^2_{8,m}$, i.e. the fluctuation in galaxies is b times bigger than that in total mass.

Another way to determine the amplitude $C(R_o)$ is derived from the CMB measurements. If at the present epoch the power spectrum $P(k, R_o)$ is approximated by Ak for small values of k, then (as we have remarked in Chap. 11) the CMB quadrupole $Q_{\rm rms}$ is related to A by

$$A = \left(\frac{96\pi^2}{5}\right)\Omega_0^{-1.54}H_0^{-4}\left(\frac{Q_{\rm rms}}{T_{\gamma,0}}\right)^2 \tag{12.38}$$

for $K = 0$ cosmological models. For other models the relation is a little bit messier.

The linear evolution then leads to a power spectrum $P(k, z_i)$ which is taken as the starting point for the numerical computations.

The implementation of a particle distribution representing the given $P(k, z_i)$ requires some care. The method generally in use now is the Zel'dovich approximation (compare Sect. 11.7). It consists basically in a mapping from the Lagrangian (unperturbed) position $\boldsymbol{q}$, i.e. the coordinate moving with the particle, to the Eulerian position $\boldsymbol{x}$:

$$\boldsymbol{x} = \boldsymbol{q} + b(t)\mathbf{S}(\boldsymbol{q}). \tag{12.39}$$

$b(t)$ is a solution of the evolution equation for the linear density contrast

$$\ddot{b} + 2\frac{\dot{R}}{R}\dot{b}(t) - 4\pi G\varrho_b b(t) = 0,$$

and $\mathbf{S}(\boldsymbol{q})$ is an arbitrary function of $\boldsymbol{q}$ (see also Sect. 12.5).

The growing mode solution $b(t) \propto D_1(t)$ (see Sect. 11.2.1) gives rise to an approximate solution

$$\begin{aligned}\boldsymbol{x} &= \boldsymbol{q} + D_1(t)\mathbf{S}(\boldsymbol{q}),\\ \boldsymbol{q} &= \dot{D}_1(t)\mathbf{S}(\boldsymbol{q}),\\ \delta(\boldsymbol{x}) &= -D_1(t)\,\mathrm{div}\,\mathbf{S}.\end{aligned} \tag{12.40}$$

The function $\mathbf{S}$ describes the perturbations in Lagrangian space. It must be chosen such that $\delta(\boldsymbol{x})$ reproduces via its Fourier components the given power spectrum $P(k, z_i)$:

$$\mathbf{S}(\boldsymbol{q}) = \sum_{j=1}^{N_k} \sin(\boldsymbol{k}_j \cdot \boldsymbol{q} + \phi(\boldsymbol{k}_j)) \frac{\boldsymbol{k}_j}{|\boldsymbol{k}_j^2|} \left(|\delta(\boldsymbol{k}_j)|^2 / w(k_j) \right)^{1/2} . \tag{12.41}$$

The orientation of $\boldsymbol{k}_i$ is random, and the mode density between k and $k + dk$ is $w(k)$. N_k is the total number of waves. The resulting density fluctuations have a random Gaussian distribution with the desired spectrum $P(k, z_i)$, as long as $|\boldsymbol{q}| \gg |D_1(t)\mathbf{S}\boldsymbol{q}|$ everywhere, and $|\delta(\boldsymbol{k})|$ and $\phi(\boldsymbol{k})$ follow the probability distribution function (12.33). Each numerical simulation is the realization of a random point process, whose distribution functions are given by $P(z, k_i)$ and $\mathbf{S}(\boldsymbol{q})$. Thus the statistical quantities and the deterministic equations of motion are connected.

12.3.2 The Simulations

The simulations I present here are those of Yipeng Jing [Jing 1998; Jing and Suto 1998; Jing and Börner 2001]. We shall discuss his results for three typical CDM models. The once proudly named "standard CDM" model (SCDM) with $\Omega_m = 1$ now has difficulties when we try to describe the real world with it. The open model with a mass density $\Omega_m = 0.3$, less than the critical, and a vanishing cosmological constant $\Omega_\Lambda = 0$ (OCDM) seems now to be in conflict with the cosmic parameters derived from the CMB anisotropies. Despite the fact that these two models seem to rule themselves out already, we will present results for them here and there. It is quite instructive to compare them to the model that is considered the "standard" at present (June 2002): this is a flat CDM model with $\Omega_m = 0.3$ and $\Omega_\Lambda = 0.7$ (LCDM). As we shall see below, even this model cannot fit all the astronomical data that are now available. The model parameters Ω_m, Ω_Λ, as well as the shape parameter Γ and the normalization σ_8 are shown in Table 12.2.

The initial density power spectrum, together with the cosmological model, completely fixes the clustering of the dark matter particles in these CDM models. For these simulations the linear CDM power spectrum of [Bardeen et al. 1986] is used as the initial spectrum. Each model was calculated with

Table 12.2. Model parameters

Model	Ω_m	Ω_Λ	Γ	σ_8
SCDM	1.0	0.0	0.5	0.6
OCDM	0.3	0.0	0.25	1.0
LCDM	0.3	0.7	0.21	1.0

Jing's vectorized P^3M code on the Fujitsu VPP 300/16 R supercomputer at the National Astronomical Observatory of Japan. Each simulation was performed with 256^3($\sim$ 17 million) particles and with a force resolution of $\eta \sim \frac{1}{2000} L$ (where L is the size of the simulation box). Two different box sizes, $100\,h^{-1}$Mpc and $300\,h^{-1}$Mpc, have been employed for each CDM model. Then the mass of a simulation particle is

$$m_p = 1.7 \times 10^{10} \Omega_m \ h^{-1} M_\odot \qquad \text{for the smaller box of } 100\,h^{-1}\text{Mpc},$$

and

$$m_p = 4.6 \times 10^{11} \Omega_m \ h^{-1} M_\odot \qquad \text{for the } 300\,h^{-1} \text{ Mpc box.}$$

In both cases the mass resolution is sufficient to identify the DM halo of a single galaxy, of say $10^{12} M_\odot$. In the simulations the clustering of the dark matter particles can be calculated accurately. These clumps of DM particles are, however, not galaxies. The formation of real galaxies is not calculated because only CDM particles are considered, and not the dissipative baryon component. To compare the simulations with observations, one has to identify condensations of DM particles as a kind of "pseudo-galaxy", and then compare the distribution of such objects with the real galaxy distribution.

Jing has used quite a standard method to find such condensations [Jing 1998]. DM halos in the simulations are identified with a friends-of-friends algorithm with a linking parameter 0.2 times the mean particle separation. In such a scheme the separation of each pair of particles is examined, and if the separation is smaller than the linking parameter, the pair is linked. Thus one obtains a set of distinct groups which consist of particles linked together. In each group the particles whose total energy is positive are removed. When particle groups contain only gravitationally bound particles, they are considered as possible "pseudo-galaxies". Finally, the DM simulation box contains a set of DM halos which can be identified as "pseudo-galaxies" if their mass is less than a certain limit.

In Figs. 12.2 to 12.5 the DM density field for a LCDM model is shown in its time evolution from redshift $z = 9.00$ to $z = 0.0$. (The fact that $z = 2.86$, $z = 1.04$ are chosen as cosmic epochs is a technical subtlety concerning the read-out times in the numerical simulation.) The model produces the well-known pattern of filaments, walls and voids, indicative of a roughly cell-like structure. In these four figures we can also see very nicely how the initially small density contours (at $z = 9$) become more and more pronounced, until the bright red spots appear which are indicative of the formation of a gravitationally bound system. The smallest red dots visible in the figures correspond to galaxies, while the larger red ones are halos of much larger mass corresponding to rich clusters, perhaps.

In Figs. 12.2 to 12.5 the full simulation box of $(100\,h^{-1}\text{Mpc})^3$ is projected into the plane of the picture. The underdense regions can clearly be discerned even though the "voids" do not appear completely empty, because of objects

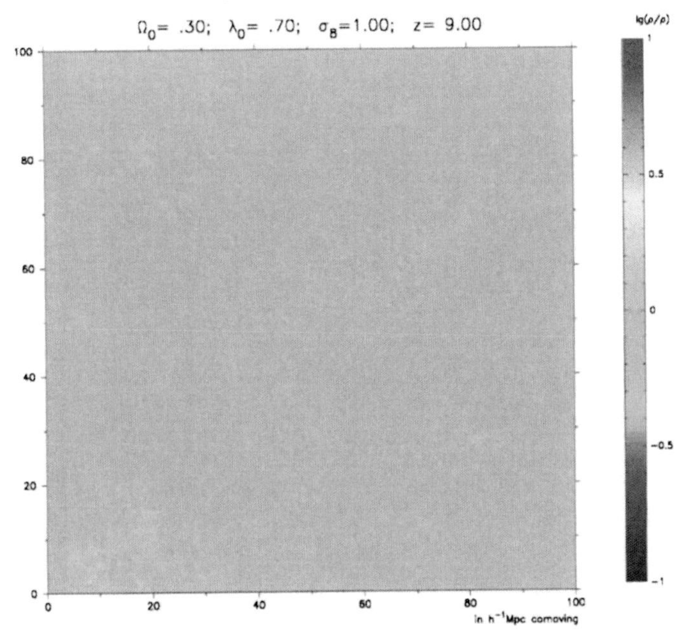

Fig. 12.2. The density field of an LCDM model is shown. The box-size is $100h^{-1}$ Mpc, and 16 777 216 particles used. The redshift is $z = 9$ (Courtesy of Y.P. Jing, Shanghai)

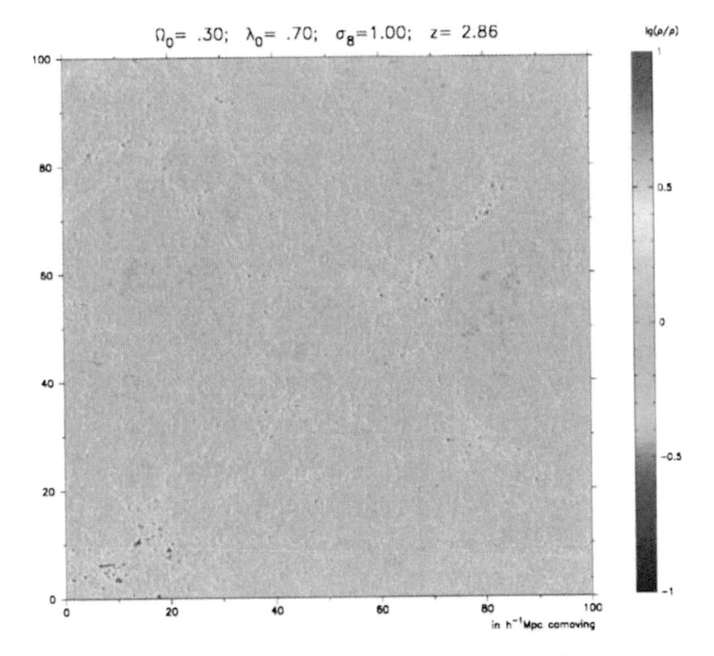

Fig. 12.3. As Fig. 12.2, but at $z = 2.86$

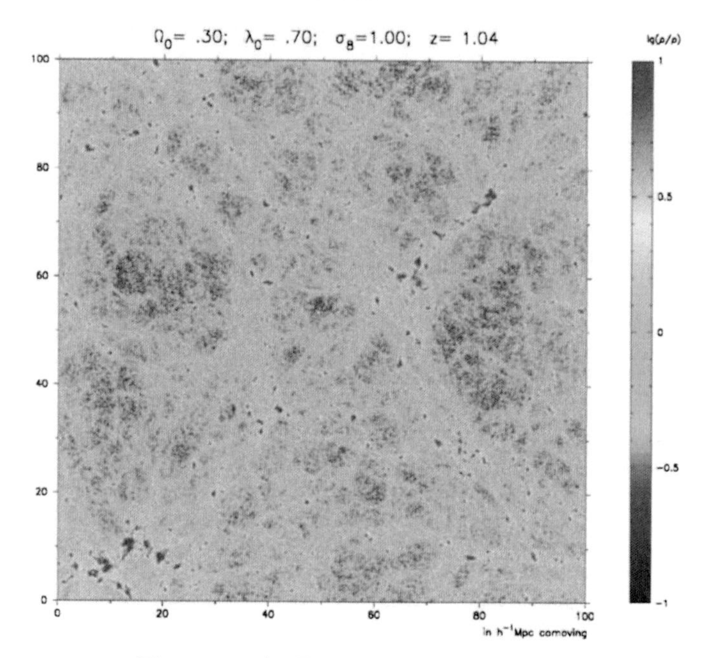

Fig. 12.4. As Fig. 12.2, but at $z = 1.04$

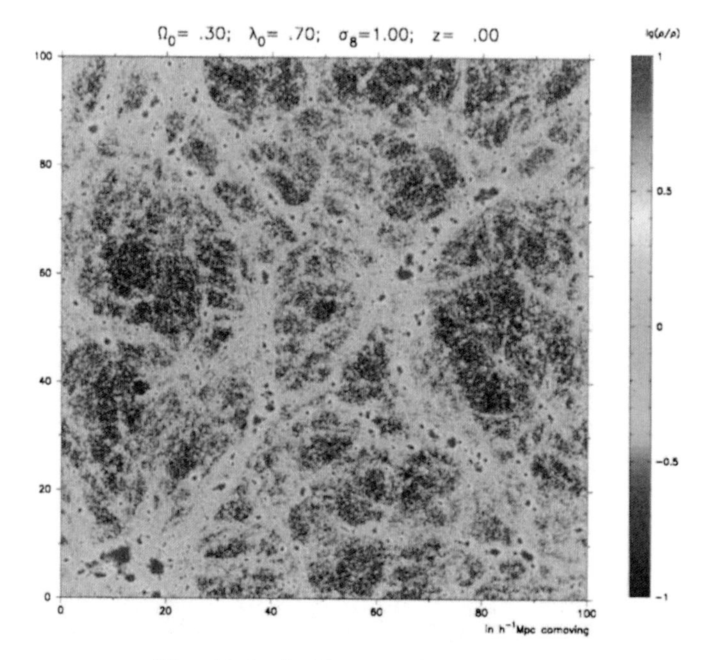

Fig. 12.5. As Fig. 12.2, but at $z = 0$

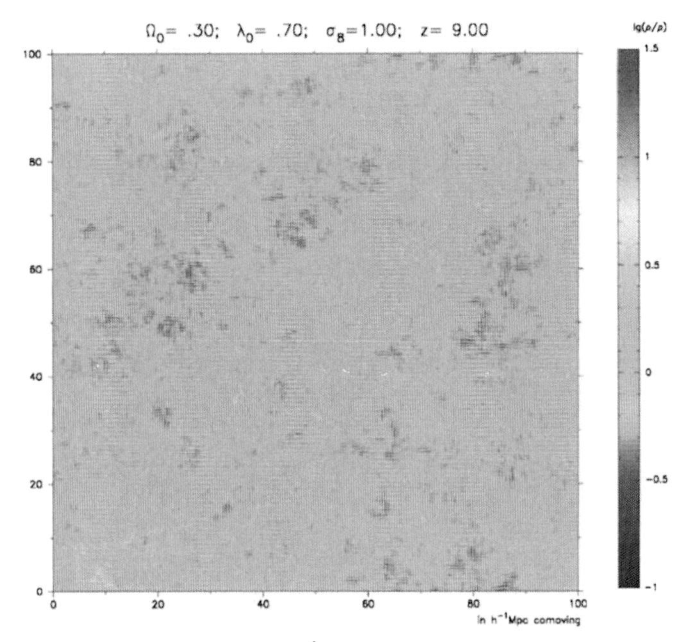

Fig. 12.6. A slice of thickness $10h^{-1}$ Mpc is shown of the LCDM simulation of Jing. The redshift is $z = 9$ (Courtesy of Y.P. Jing, Shanghai)

Fig. 12.7. As Fig. 12.6, but at $z = 1.04$

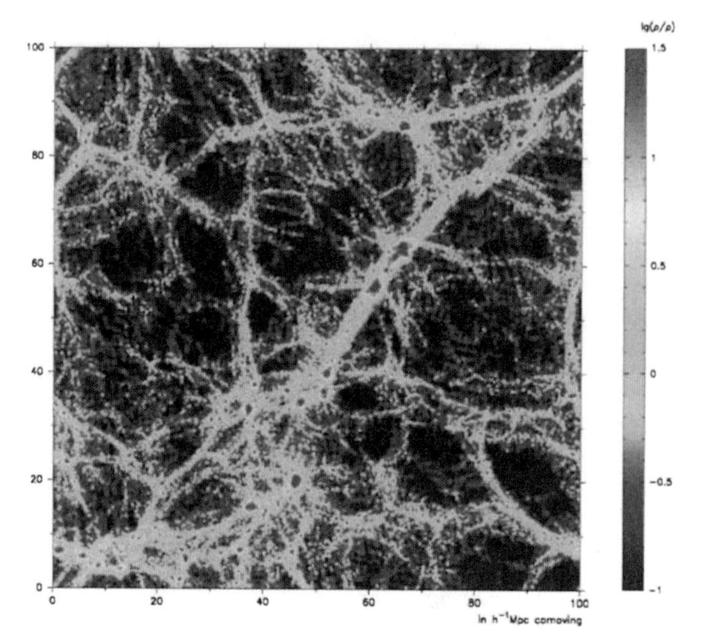

Fig. 12.8. As Fig. 12.6, but at $z = 0$

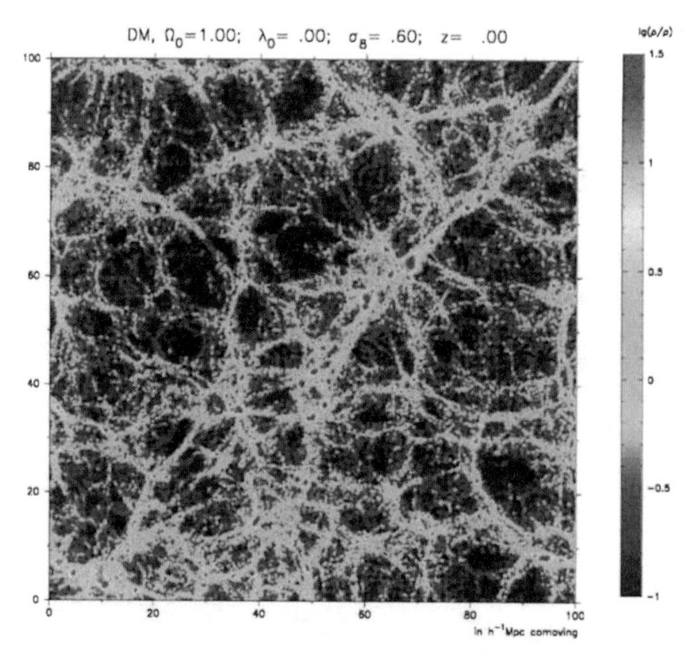

Fig. 12.9. A slice through an SCDM simulation is shown at $z = 0$. This should be compared to Fig. 12.7 (Courtesy of Y.P. Jing, Shanghai)

projected onto them. The cell and filament structure can be brought out very clearly if we project out a slice through the simulation box of thickness $10\,h^{-1}$ Mpc (Figs. 12.6 to 12.8). Another remarkable feature are the very extended filamentary, or sheet-like structures formed by galaxies and clusters which span the whole extent of the simulation box. In the lower left-hand corner an extremely massive structure has formed.

Different realizations of the same numerical model lead to slightly different pictures. Qualitatively they look very similar, but in the detail there are some modifications. These changes are actually larger than the differences between different models in the set considered here. For instance, the SCDM simulation at $z = 0.0$ looks almost identical to the LCDM result at $z = 1.04$ (Fig. 12.9). This seems to be a general characteristic of such CDM modelling: contrary to statements found here and there in publications, there is no eye-catching distinction between models, if you are aware of the fact that their time evolution is different, and so the patterns must be compared at the appropriate cosmic epoch for each model. Quantitatively the models are quite different, as we shall see in the next section.

12.4 Comparison with Observations

Quantitative comparisons between different data sets (model data or real astronomical data) can reveal properties of the clustering pattern which are not apparent when we just look at the pictures (such as Figs. 12.2 to 12.9). Various statistics can be useful for this purpose. We shall in the following discuss properties of the two-point correlation function, the pairwise peculiar velocity dispersion, and the three-point correlation function for the CDM models described in Sect. 12.3, and for two galaxy catalogues, the Las Campanas Redshift Survey (see also Chap. 10), and the PSCz (the IRAS point source catalogue). In addition some results from the Sloan Digital Sky Survey (SDSS) will be presented.

The analysis of the distribution of the DM particles in a simulation is straightforward, because the positions and velocities of all the particles are known. A galaxy catalogue requires a careful discussion of selection effects before meaningful information can be extracted. For example, galaxy catalogues are magnitude-limited in general, i.e. there is a minimal luminosity

$$L_{\text{min}} = 4\pi r^2 l_{\text{min}}$$

which is sufficient for detection at a distance r. l_{min} is the limiting flux that can be detected ($l \geq l_{\text{min}}$ corresponds to a magnitude limit $m_{\text{lim}} - m \leq m_{\text{lim}}$). To define a selection function $S(r)$ which denotes the probability that a galaxy at distance r will be included in the catalogue we can determine the luminosity function $\phi(L)$ which gives the number density of galaxies in the catalogue with luminosities between L and $L + dL$. $S(r)$ is just the ratio of the number of galaxies with luminosities above $L_{\min}(r)$ to the total number of galaxies in the survey:

$$S(r) = \frac{\int_{L_{\min}(r)}^{\infty} \phi(L)dL}{\int_{L_{\min}(r_o)}^{\infty} \phi(L)dL}.$$

r_o denotes the distance within which every galaxy will be included. At distance r the number $S(r)$, less than 1, gives the probability of a galaxy being included in the catalogue. (Obviously we have chosen a coordinate system where the observer is at the centre.)

The function $S(r)$ can now be used to construct a "mock catalogue". To generate a mock catalogue, a random position is selected within the simulation box. This is taken to be the position of the observer. Then a catalogue is generated according to the angular boundaries. For each DM particle j in the simulation a random number p_j between 0 and 1 is produced on the computer. If $p_j \geq S(r_j)$ the DM particle will be included as a "pseudo-galaxy" in the mock catalogue. This procedure can be repeated for different parts of the simulation. It produces a number of mock catalogues which provide reliable estimates of the sampling effects and enable the study of the effects of incompleteness in apparent magnitude and surface-brightness. These mock samples are also very important in studying other observational selection effects and their influence on the statistical properties of the data.

12.4.1 Correlation Functions and the Las Campanas Redshift Survey

As an example we discuss some results obtained for the Las Campanas Redshift Survey [Shectman et al. 1996]. This survey (see Chap. 10) is the largest redshift survey of optically selected galaxies which is now publicly available. The PSCz survey of similar size consists of infrared selected galaxies, and will be discussed in the next section.

The LCRS consists of about 30 000 galaxies, observed in $\sim 1.5° \times 1.5°$ fields in the sky, in three slices each across the north and south hemisphere. The analysis of [Jing et al. 1998] which we report here, was done for a subsample of the LCRS which consists of all galaxies with recession velocities between 10 000 and 45 000 km s^{-1}, and with absolute magnitudes (in the LCRS hybrid R band) between -18.0 and -23.0. There are 19 558 galaxies in this sample, of which 9480 are in the three north slices, and the rest in the three south slices.

The LCRS is a well-calibrated sample of galaxies, ideally suited for statistical studies of the large-scale structure. All known systematic effects are well quantified and documented, and thus can be corrected easily in a statistical analysis.

One interesting exception is the 'fibre collision' limitation: the observers used 112 fibre detectors or 50 fibre detectors for their scans, but they did not come back to observe very dense regions on the sky again. Thus even in a very crowded $1.5° \times 1.5°$ field a maximum of only 112 redshifts of galaxies was recorded. In other words, if two galaxies are closer than $55''$ in the sky, only one was included in the survey. Obviously such a procedure can also be implemented for the mock symples, and its influence on statistical quantities can be studied this way [Jing et al. 1998].

For the statistical analysis the LCRS presents a point distribution in a certain volume of space with the radial distance given by the redshift. Now the redshift follows in general not the perfect Hubble law

$$cz = H_0 d,$$

but has contributions from the peculiar velocities of the galaxies in clusters, or other bound systems. The Hubble relation is therefore modified:

$$cz = H_0 d + v_p,$$

where v_p is the radial component of the peculiar velocity. As a consequence the 2-point correlation function in redshift space ξ_z is not just a function of the distance

$$s^2 = r_p^2 + \pi^2,$$

but depends anisotropically on the separations of a galaxy pair perpedicular to (r_p) and along (π) the line of sight, $\xi_z(r_p, \pi)$. Jing et al. estimate the function $\xi_z(v_p, \pi)$ by the expression

$$\xi_z(r_p, \pi) = 4\ RR(r_p, \pi) \cdot DD(r_p, \pi)[(DR(r_p, \pi)]^{-2} - 1. \tag{12.42}$$

Here $DD(r_p, \pi)$ is the count of distinct galaxy–galaxy pairs with perpendicular separations in the bin $r_p \pm 0.5\Delta r_p$, and with radial separations in the bin $\pi \pm 0.5\Delta\pi$. In symbols:

$$DD(r_p, \pi) = \sum_{i,j} u_{ij},$$

where $u_{ij} = 1$ if

$$r_p - \Delta r_p \leq r_p(ij) \leq r_p + \Delta r_p$$

and

$$\pi - \Delta\pi \leq \pi_{ij} \leq \pi + \Delta\pi;$$

$r_p(ij)$ and π_{ij} are the separations of galaxies i and j perpendicular to and along the line of sight, respectively.

In computing the pair counts one gives each galaxy (i) a weight $S^{-1}(r_i)$ to account for selection effects in the catalogue. Therefore one really uses

$$DD(r_p, \pi) = \sum_{i,j} S^{-1}(r_i) S^{-1}(r_j) u_{ij}. \tag{12.43}$$

$RR(r_p, \pi)$ and $DR(r_p, \pi)$ are similar counts of pairs formed by two points from a random distribution, and by one galaxy and one random point, respectively [Hamilton 1993].

The random sample, which contains 100 000 points, is generated in the same way as the mock samples, except that the points are originally randomly distributed in space.

The two-point correlation function in redshift space is shown in Fig. 12.10. It is general practice to integrate $\xi_z(r_p, \pi)$ over the variable π. A function which depends only on r_p results, the so-called "projected" two-point correlation function $w(r_p)$:

$$w(r_p) \equiv \int_0^\infty \xi_z(r_p, \pi) d\pi = \sum_i \xi_z(r_p, \pi_i) \Delta\pi_i. \tag{12.44}$$

The projected two-point correlation function (2PCF) for the LCRS survey is presented in Fig. 12.11.

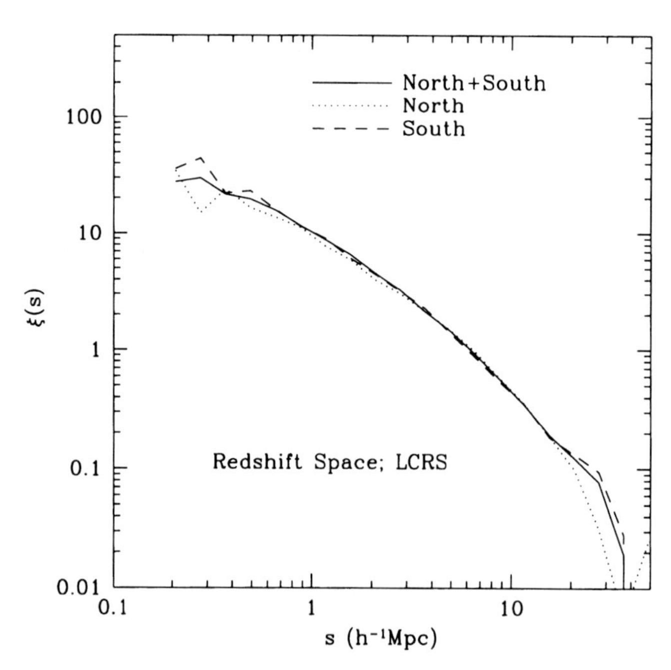

Fig. 12.10. The two-point correlation function in redshift space for the galaxies of the LCRS is plotted as a function of the redshift distance s

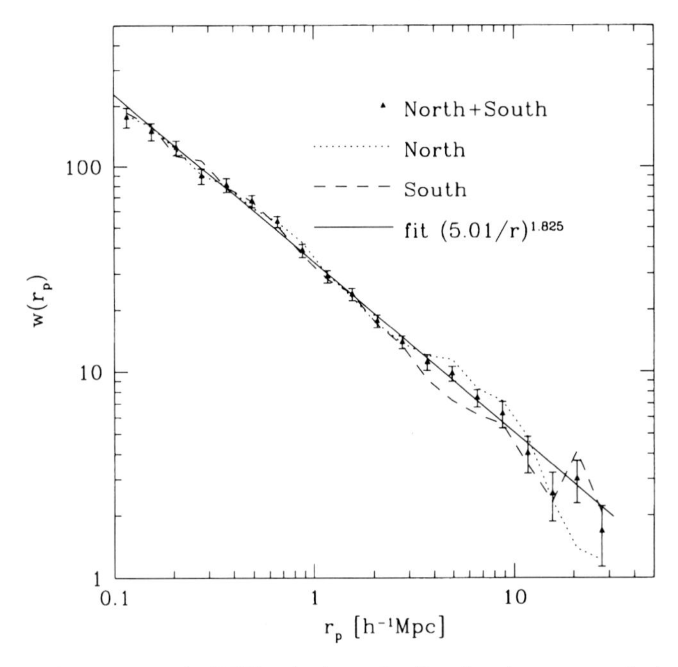

Fig. 12.11. The projected 2PCF $w(r_p)$ can be fitted quite accurately by a power law

Triangles show the result for the whole sample. Error bars have been estimated in [Jing et al. 1998] by a bootstrap resampling technique. One hundred bootstrap samples have been generated, and the error bars are the scatter of $w(r_p)$ among these bootstrap samples. Jing et al. have also shown that the standard errors among different mock samples are comparable (to within a factor of 2) to the bootstrap errors, and so it does not matter much which error estimated is used in the discussion. $w(r_p)$ is related to the "real space" correlation function $\xi(r)$ by

$$w(r_p) = 2 \int_0^\infty dy \xi[(r_p^2 + y^2)^{1/2}] \tag{12.45}$$

(The proof of this relation is left as an exercise).

If $w(r_p)$ in Fig. 12.11 is fitted by a power law (for a certain range of r_p)

$$w(r_p) = A\, r_p^{1-\gamma},$$

then $\xi(r) = (r_o/r)^\gamma$, and

$$r_o^\gamma = A \frac{\Gamma(\gamma/2)}{\Gamma(1/2)\Gamma[(\gamma-1)/2]},$$

where Γ is the gamma function.

$w(r_p)$ can be fitted to a power law for $r_p < 28h^{-1}$Mpc. Then a fit to $\xi(r) = (r_o/r)^\gamma$ yields

$$r_o = 5.01 \pm 0.05h^{-1}\text{Mpc},$$
$$\gamma = 1.825 \pm 0.018. \tag{12.46}$$

These values are similar to the fits obtained in earlier work [see Peebles 1980].

The fibre collision effect of the LCRS reduces the correlation function by a small amount on small scales. Such a suppression can be corrected with the help of mock samples.

These corrections, and the increase of the error bars by a factor of about 2 from the bootstrap values, lead to a set of values

$$r_o = 5.05 \pm 0.12h^{-1}\ \text{Mpc}, \quad \gamma = 1.862 \pm 0.034.$$

It can also be seen from Figs. 12.10 and 12.11 that the north and the south samples of the LCRS agree with each other, and with the whole sample, reasonably well, especially for $r_p < 5h^{-1}$Mpc. The error bars for the subsamples are about 1.5 times as large as those of the whole sample, which emphasizes the consistency of the results. The LCRS is the first observational sample large enough for different subvolumes to have the same 2PCF. The results can be compared to the APM survey correlation function. This is an angular correlation function, and a real-space 2PCF can be derived from the measurements [cf. Peebles 1980]. The 2PCF of the LCRS is significantly steeper than the APM 2PCF. At the moment it is not clear whether some unknown systematic feature in the two surveys, or the fact that galaxies are selected differently (in a bluer band in the APM), is the reason for this.

When we compare Figs. 12.10 and 12.11, we see clearly that the redshift-space correlation function ξ_z has a much lower slope than $w(r_p)$. This is an illustration of the fact that the peculiar motions of the galaxies at small separation tend to weaken the apparent clustering. This peculiar motion can also be measured in a redshift survey. The 3-dimensional velocity difference of a pair of galaxies at points $\boldsymbol{x}$ and $\boldsymbol{x}+\boldsymbol{r}$, i.e. at separation $\boldsymbol{r}$, is

$$\boldsymbol{v}_{12}(\boldsymbol{r}) = \boldsymbol{v}(\boldsymbol{x}) - \boldsymbol{v}(\boldsymbol{x}+\boldsymbol{r}).$$

This quantity cannot be measured directly. Its root mean square variation in the radial direction is called the "pairwise velocity dispersion" (PVD) $\sigma_{12}(r)$.

Write $\boldsymbol{v}_{12}\cdot\frac{\boldsymbol{r}}{r} \equiv v_{12}$. Then $v_{12}(\boldsymbol{r}) = v_{12}(r)$ generally, because of the general isotropy. Then

$$\langle(v_{12}(r) - \langle v_{12}(r)\rangle)^2\rangle^{1/2} \equiv \sigma_{12}(r) = \langle(\boldsymbol{v}_{12}(r) - \langle\boldsymbol{v}_{12}(r)\rangle)^2/3\rangle^{1/2}.$$

$\sigma_{12}(r)$ can be determined from the anisotropy of the redshift space $\xi_z(r_p,\pi)$. The basic idea is that these anisotropies are due to the random motions of the galaxies. If these motions can be described by a distribution function $f(v_{12})$

of the relative velocity of galaxy pairs along the line of sight, then $\xi_z(r_p, \pi)$ can be written as the convolution of the real-space 2PCF with $f(v_{12})$:

$$\xi_z(r_p, \pi) = \int f(v_{12}) \xi((r_p^2 + (\pi - v_{12})^2)^{1/2}) dv_{12}. \qquad (12.47)$$

$f(v_{12})$ is unknown, and to make any progress we must make an assumption about the functional form of $f(v_{12})$. Based on observational and theoretical considerations an exponential form is adopted for $f(v_{12})$:

$$f(v_{12}) \equiv \frac{1}{\sqrt{\sigma_{12}}} \exp\left(-\frac{\sqrt{2}}{\sigma_{12}} |v_{12} - \langle v_{12} \rangle|\right), \qquad (12.48)$$

where $\langle v_{12} \rangle$ is the mean and σ_{12} the dispersion of the one-dimensional pairwise peculiar velocities.

Assuming a model for $\langle v_{12} \rangle$, and also modelling $\xi(r)$ from the projected 2PCF $w(r_p)$, one can estimate the PVD σ_{12} by comparing the observed $\xi_z(r_p, \pi)$ with the modelled one, given by the right-hand side of (12.47). σ_{12} is generally a function of r_p.

We should realize that the PVD measured in this way is not identical to the PVD given directly by the peculiar velocities of galaxies. In the reconstruction of σ_{12} from (12.47) and (12.48) the "infall" velocity $\langle v_{12}(r) \rangle$ is not known for the observed galaxies.

For $f(v_{12})$ and $\xi(r)$ one has approximations which may be very accurate, but we do not know. Furthermore, the PVD estimated from the redshift distortion is a kind of average of the true PVD along the line of sight, and since the true PVD depends on the separations of galaxies in real space, the two quantities are different by definition.

Figure 12.12 demonstrates this effect: the PVD reconstructed from 20 mock samples (for the model with $\Omega_m = 0.2$) according to the above procedure is compared to the true PVD

$$\langle [(\boldsymbol{v}_{12}(r) - \langle \boldsymbol{v}_{12}(r) \rangle)^2 / 3] \rangle^{1/2}$$

of the full simulation. Even though the two functions agree qualitatively, there are significant differences. These arise from the fact that the true PVD depends significantly on the separation of pairs in real space and that the distribution of the pairwise velocities is not perfectly exponential. To obtain a meaningful comparison between models and observations, these systematics must be treated very carefully. Unfortunately at the moment we do not have a convincing method for taking care of these problems. The most objective approach seems to be to construct a large set of mock samples from the simulation models, and to analyse them in the same way as the real samples. [Jing et al. 1998] have assumed an infall model based on the following self-similar ansatz:

$$\langle v_{12}(r) \rangle = -\frac{y}{1 + (r/r_*)^2},$$

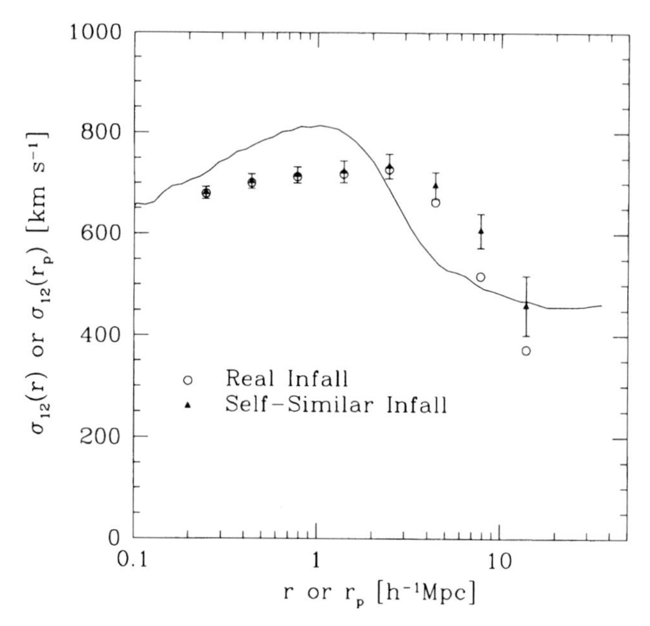

Fig. 12.12. The PVD from mock samples is compared to the PVD from the full simulation

where $r_* = 5\ h^{-1}$Mpc and y is the radial separation in real space. This ansatz is a good approximation to the infall pattern seen in CDM models.

In Fig. 12.13 the result for the PVD of the LCRS is plotted. The PVD of the LCRS galaxies is about

$$(500 \pm 50)\ \mathrm{km\ s}^{-1} \quad \text{at} \quad r_p = 1\ h^{-1}\ \mathrm{Mpc}.$$

Considering the uncertainties of the bootstrap errors, and of the fibre collision effect, a best estimate is

$$\sigma_{12}(r_p = 1\ h^{-1}\mathrm{Mpc}) = (570 \pm 80)\ \mathrm{km\ s}^{-1}. \tag{12.49}$$

This result is quite robust, as can be seen from the values obtained for the north and south samples. Recently this result has been confirmed by the Sloan Digital Sky Survey.

It is instructive to see the comparison of these LCRS results to models. The same statistical procedure used for the real data is applied to mock samples derived from the N-body simulations [Jing et al. 1998]. The projected 2PCF and the PVD of dark matter particles are estimated for each mock sample. The averages of these two quantities, and the 1σ scatter among the mock samples, are plotted in Fig. 12.14 for the three theoretical models. The LCRS results are included to facilitate comparison. The $w(r_p)$ function predicted by the $\Omega_m = 0.3$ and the $\Omega_m = 0.2$ model is in good agreement with the observed one on scales larger than $\sim 5h^{-1}$ Mpc. On smaller scales, however, the predictions of both models lie above the observational result. The

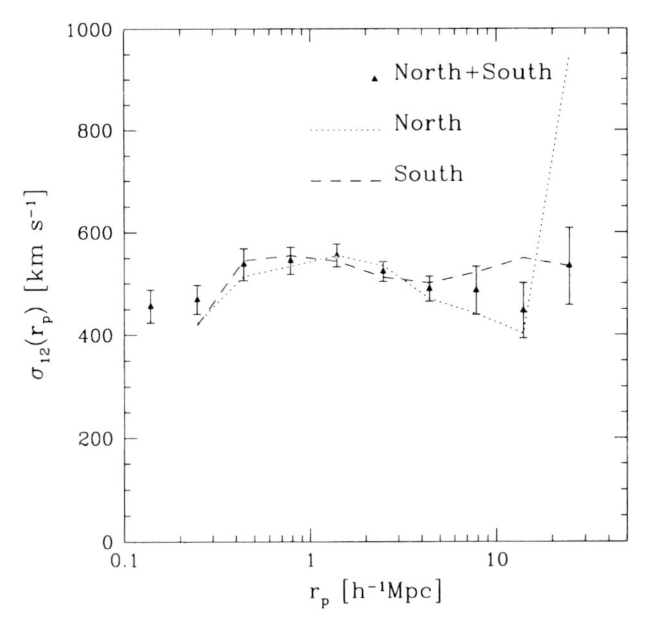

Fig. 12.13. The PVD for the LCRS

projected 2PCF predicted by the SCDM model is lower than that obtained from the LCRS data. The reason is the lower value of the normalization, $\sigma_8 = 0.62$, in this model. If the model prediction is shifted upwards by a factor of $1/(0.62)^2 \simeq 2.6$, as implied by a linear biasing factor $b = 1/\sigma_8 = 1.6$, the SCDM model fits the observed $w(r_p)$ on intermediate scales. The discrepancy on scales $r_p \geq 5\,h^{-1}$Mpc is due to the fact that this model does not have enough power on large scales. This fact is well known from several other observations, notably the APM survey [Maddox et al. 1990b; Efstathiou et al. 1990]. It is also apparent from the figure that this model predicts too steep a $w(r_p)$ around $v_p = 1h^{-1}$ Mpc. For all three models the discrepancy betwen theoretical predictions and observations is significant. Thus a scale-dependent bias is required by all three models to make them compatible with the observed real-space correlation function given by the LCRS.

The PVDs of the dark matter particles are higher than the observed value in all three models for $r_p < 5\,h^{-1}$ Mpc (Fig. 12.14). On larger separations the fluctuations become very large, and the result is very sensitive to the infall model. This is also a well-known problem for numerical simulations. The value from the LCRS of $\sigma_{12} = 570 \pm 80$ kms^{-1} at $1\,h^{-1}$Mpc eases the difficulty somewhat compared to the earlier value of σ_{12} ($r_p = 1\,h^{-1}$Mpc) $= 340 \pm 40$ kms^{-1} [Davis and Peebles 1983b] which for 10 years was really a thorn in the flesh of DM simulations. Then [Mo et al. 1993] corrected this early work, and identified various selection effects which had badly distorted previous results. It has been shown that the absence or presence of rich clusters in

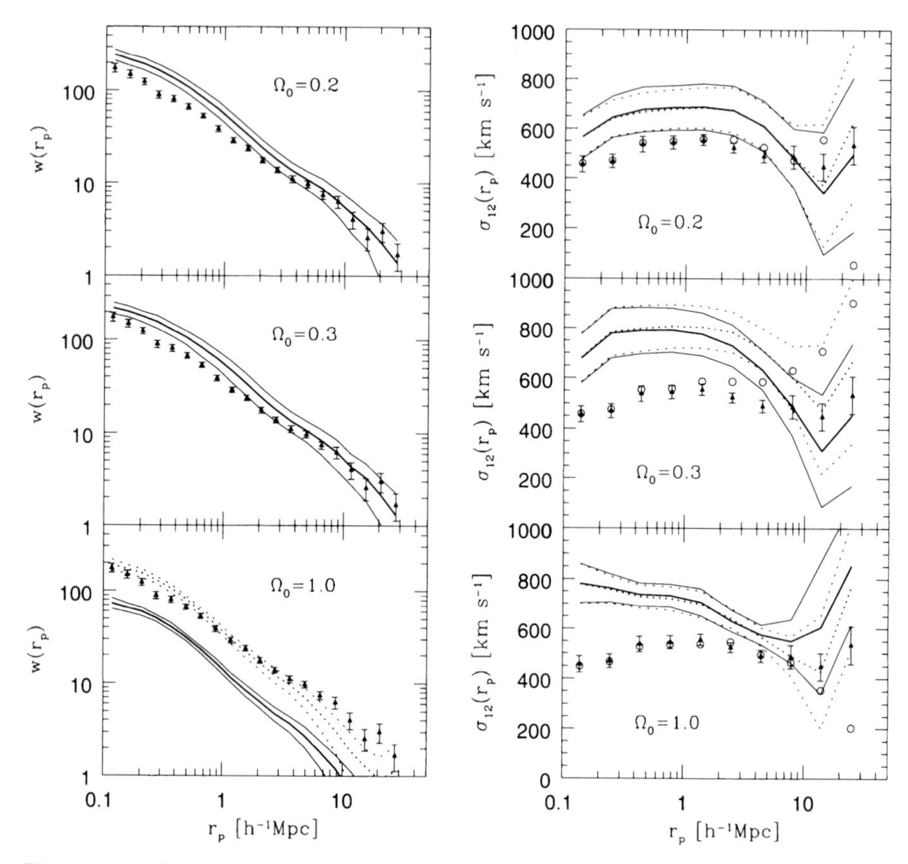

Fig. 12.14. Comparison of the predictions of CDM models with the LCRS results. The left panels show the projected 2PCF. Triangles show the observational results. The mean value and the $1-\sigma$ limits of the CDM models are shown by thick and thin lines, respectively. The left panels show the PVD as obtained from the model of self-similar infall (*triangles, solid lines*), and from the infall pattern of the simulation (*circles, dashed lines*)

galaxy samples is especially important. The LCRS contains about 30 rich clusters – a sufficient number to permit an accurate measurment of the PVD. The numerical simulations still produce too high a PVD, though.

The PVD for the mass on small scales is proportional to $\sigma_8\Omega_m^{0.5}$ [Mo et al. 1997]. From the observed abundance of galaxy clusters it has been argued that $\sigma_8\Omega_m^{0.6}$ is about 0.5 [White et al. 1993; Kitayama and Suto 1997]. But lower values, such as $\sigma_8\Omega_m^{0.6} \simeq 0.4$, may still be consistent with observations. The values which apply to the models of Jing discussed here are $\sigma_8\Omega_m^{0.6} \simeq 0.38$ for the $\Omega_m = 0.2$ model and $\sigma_8\Omega_m^{0.6} \simeq 0.5$ for the $\Omega_m = 0.3$ model.

For SCDM the value $\sigma_8\Omega_m^{0.6} = 0.6$ is a little bit higher than the observational result. Therefore, the higher PVD values also argue for a bias scheme. Another question is the dependence of the PVD on $\langle v_{12}\rangle$ models. In Fig. 12.14

the self-similar infall model and the infall pattern derived directly from the CDM simulations are both shown (solid and dotted lines). It seems that the results are not sensitive to the specific model used.

12.4.2 A Phenomenological Bias Model

From the discussion above it seems obvious that the CDM models currently favoured in the theory of structure formation can only be viable if the galaxy distribution is biased relative to the underlying clustering of the mass.

The "bias" is, of course, completely determined by the physical processes leading to the formation of a galaxy. Since models describing the physics adequately do not yet exist, one can try to investigate the consequences of a simple bias prescription. An educated guess that also improves the fit to the data may well be not too far off the true situation.

In [Jing et al. 1998] it has been suggested that the number of pseudo-galaxies per unit mass (N/M) may be lower in more massive halos. Specifically, it was assumed that

$$N/M = (M/M_*)^{-\alpha}, \tag{12.50}$$

where M is the halo mass, M_* a reference mass $\sim 10M_G$, N the number of 'galaxies' in the halo, and α the parameter to be determined by a fitting procedure.

Since the predicted 2PCF is steeper, and the PVD is higher than the observed value on a scale $\sim 1h^{-1}$Mpc, the parameter α should be positive.

Jing et al. have incorporated this bias model in the correlation analysis by giving each "mock galaxy" a weight which is proportional to $M^{-\alpha}$, where M is the mass of the cluster in which the "galaxy" resides. In Fig. 12.15 we can see how the agreement between models and data can be improved by this bias scheme. Here the "cluster-under-weight" bias was applied only for $M > M_*$ with M_* equal to about 10 times the mass of a typical galaxy. The parameter value $\alpha = 0.08$ turned out to give the best fit. The 2PCF now fits very well for the $\Omega_m = 0.2$ and $\Omega_m = 0.3$ models. The agreement between the predictions of the models and the LCRS data is also improved substantially for the PVD. For both low-Ω models the shape of the predicted $\sigma_{12}(r_p)$ is similar to the observed one. The amplitude of the PVD predicted by the $\Omega_m = 0.2$ model is consistent with the observed value, while the prediction of the $\Omega_m = 0.3$ model is about 30% too high. This corresponds to the difference in $\sigma_8\Omega_m^{0.6}$ for these two models. It seems that $\Omega_m = 0.2$ is the preferred CDM model.

We may ask whether such a bias is plausible. The cluster-underweight (CLW) bias scheme is actually opposite to the bias proposed for the SCDM model [White et al. 1987], since it has the consequence that the number of galaxies formed per unit mass is reduced rather than enhanced in high-density regions. One possible interpretation would be that galaxies are systematically more massive, and so the number of galaxies formed per unit mass is lower

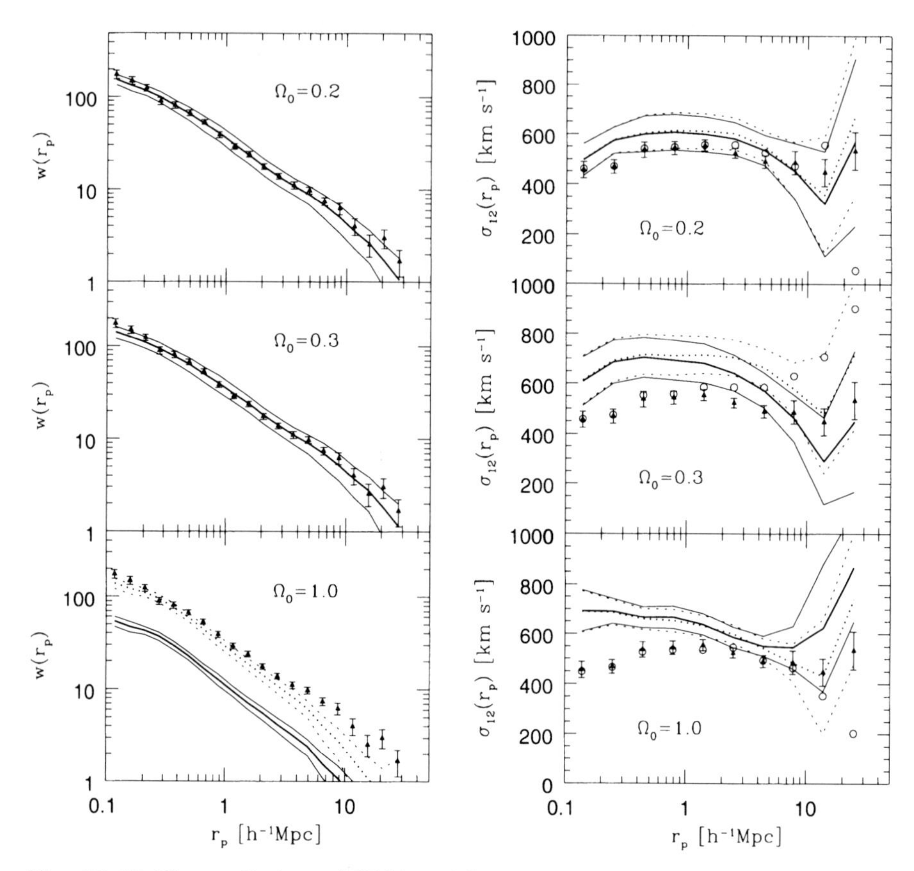

Fig. 12.15. The predictions of CDM models are compared to observations, a simple bias model as in (12.50) is employed. Lines and symbols have the same meaning as in Fig. 12.14

(although the luminosity per unit mass might be higher) in rich clusters than in poor clusters and groups. There is also some support from observations. Based on detailed photometric and spectroscopic observations of rich clusters in the CNOC survey, [Carlberg et al. 1996] found that the number of galaxies (brighter than −18.5 mag in the K band) per unit mass is systematically lower in clusters with higher velocity dispersions. The data show that the number of galaxies per unit mass in clusters with velocity dispersions ≥ 1000 kms^{-1} is lower by about a factor of 1.5 as compared to poorer clusters.

12.4.3 The PSCz Survey

The PSCz surveyconsists of galaxies selected by the IRAS satellite in the infrared part of the spectrum at a wavelength of 60 μm with flux larger than 0.6 Jy [Saunders et al. 2000].

The analysis of this data set shows very clearly that a strong bias is required to fit the 2PCF [Jing et al. 2002]. In Fig. 12.16 a fit to the projected $w(r_p)$ is shown, for the $\Omega_m = 0.3$ model and a CLW bias with $\alpha = 0.25$ [Jing et al. 2002]. Such a strong bias is required because the method of selecting the PSCz galaxies tends to avoid cluster regions, and to pick out preferably spiral galaxies in the field.

The Sloan Digital Sky Survey is the largest galaxy survey so far. Observations for this survey are now being carried out, and a large set of data has been analysed. In Fig. 12.17 we show the PVD σ_{12} for SDSS galaxies, and for PSCz galaxies. The steep decline of the PVD at small separations is

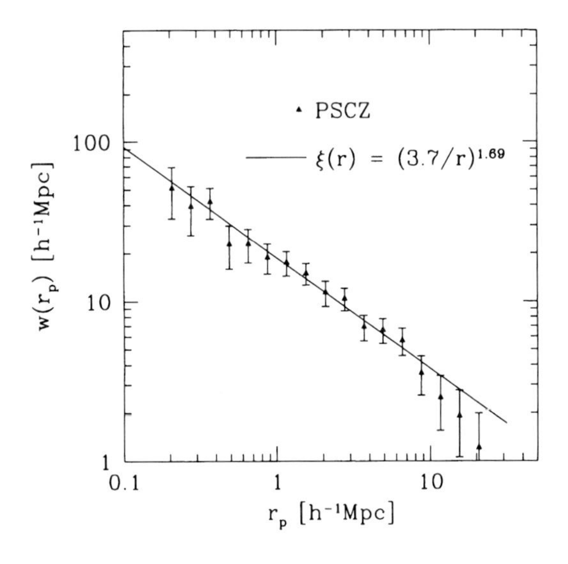

Fig. 12.16. The projected 2PCF measured from the PSCz catalogue (*filled triangles*) is smaller and less steep than the LCRS function. The error bars are 1-σ deviations from a bootstrap resampling. The *solid line* is the best power-law fit

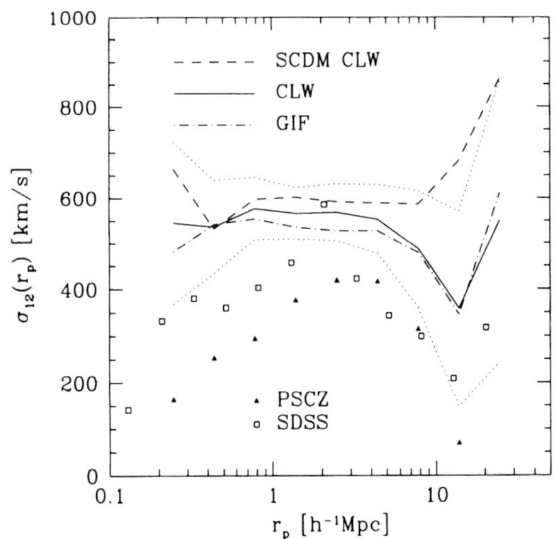

Fig. 12.17. The PVD from the PSCz is compared to the same quantity obtained for the SDSS blue and red galaxies

also found for the blue galaxies in the SDSS analysis, while the red galaxies agree more or less with the results of the LCRS [Zehavi et al. 2002]. Since the blue SDSS galaxies are mainly in groups, it may be that the PVD at small separations is mainly determined by the way the dark matter halo of the group is populated with galaxies. This question must be investigated by a combination of N-body simulations and hydrodynamic calculations.

12.4.4 Three-Point Correlation Functions

The detailed analysis of the LCRS sample has been carried one step further, i.e. the three-point correlation function (3PCF) has also been determined [Jing and Börner 1998]. The 3PCF $\zeta(r_{12}, r_{23}, r_{31})$ is defined by the joint probability dP_{123} of finding one galaxy simultaneously in each of the three volume elements dV_1, dV_2, and dV_3 at positions $\boldsymbol{r}_1, \boldsymbol{r}_2, \boldsymbol{r}_3$ respectively [Peebles 1980].

$$\begin{aligned} dP_{123} = \bar{n}(\boldsymbol{r}_1)\bar{n}(\boldsymbol{r}_2)\bar{n}(\boldsymbol{r}_3)[1 + \xi(r_{12}) + \xi(r_{23}) + \xi(r_{31}) \\ + \zeta(r_{12}, r_{23}, r_{31})]dV_1 dV_2 dV_3; \end{aligned} \tag{12.51}$$

where $r_{ij} = |\boldsymbol{r}_i - \boldsymbol{r}_j|$, and $\bar{n}(r_i)$ is the mean density of galaxies at $\boldsymbol{r}_i$.

The determination of the 3PCF was pioneered by Peebles and his co-workers in the seventies [Peebles 1980]. Based on their careful analysis of the Lick and Zwicky angular catalogues of galaxies, they proposed a so-called "hierarchical" form

$$\zeta(r_{12}, r_{23}, r_{31}) = Q[\xi(r_{12})\xi(r_{23}) + \xi(r_{23})\xi(r_{31}) + \xi(r_{31})\xi(r_{12})], \tag{12.52}$$

and found the constant $Q \simeq 1.29 \pm 0.2$. For small redshift samples of galaxies the Q value was found to be around 0.6, much smaller than the value advocated by Peebles and his co-workers. This difference may be due to the redshift distortion effect which reduces the number of close neighbours at small separations in a galaxy sample, and thus also reduces the Q value.

The LCRS is the first sample which is large enough to allow an examination of the validity of the hierarchical form. The relation (12.52) can be expressed as an equation for a normalized function $Q_{\mathrm{red}}(s_{12}, s_{23}, s_{31})$:

$$\begin{aligned} Q_{\mathrm{red}}(s_{12}, s_{23}, s_{31}) = \zeta(s_{12}, s_{23}, s_{31})\{\xi(s_{12})\xi(s_{23}) + \xi(s_{23})\xi(s_{31}) \\ + \xi(s_{12})\xi(s_{31})\}^{-1}; \end{aligned}$$

s signifies redshift-space coordinates as before.

ζ can be measured from the counts of triplets of galaxies. It proves convenient (following [Peebles 1980]) to introduce variables which desribe the shape of the triangles formed by the triplets of galaxies. For a triangle with the three sides $s_{12} \leq s_{23} \leq s_{31}$, one defines variables s, u, v as

$$s = s_{12}, \quad u = s_{23}/s_{12}, \quad v = (s_{31} - s_{23})/s_{12}.$$

The 3PCF in redshift space $Q_{\rm red}(s,u,v)$ depends both on the real space distribution of galaxies, and on their peculiar motions. Although the information contained in $Q_{\rm red}$ is useful for the study of large-scale structures, it seems quite clear that $Q_{\rm red}$ is different from the real space $Q(r,u,v)$. By analogy with the analysis for the 2PCF, Jing and Börner have integrated over the line of sight peculiar velocities $\pi_{12}, \pi_{23}, \pi_{31}$ to obtain a projected function $Q_{\rm proj}$ from the measured $Q_{\rm red}$.

The redshift space 3PCF $\zeta(r_{p12}, r_{p23}, r_{p31}, \pi_{12}, \pi_{13})$ is defined by

$$\begin{aligned} dP^z_{123} &= \bar{n}(\boldsymbol{s}_1)\bar{n}(\boldsymbol{s}_2)\bar{n}(\boldsymbol{s}_3)[1+\xi_z(r_{p12},\pi_{12})+\xi_z(r_{p23},\pi_{23})+\xi_z(r_{p31},\pi_{31}) \\ &\quad +\zeta_z(r_{p12},r_{p23},r_{p31},\pi_{12},\pi_{13})]d\boldsymbol{s}_1 d\boldsymbol{s}_2 d\boldsymbol{s}_3; \end{aligned} \tag{12.53}$$

where dP^z_{123} is the joint probability of finding one object simultaneously in each of the three volume elements $d\boldsymbol{s}_1, d\boldsymbol{s}_2$, and $d\boldsymbol{s}_3$ at positions $\boldsymbol{s}_1, \boldsymbol{s}_2$, and $\boldsymbol{s}_3$; $\xi_z(r_p,\pi)$ is the redshift space 2PCF, and r_{pij} and π_{ij} are the separations of objects i and j perpendicular and along the line of sight, respectively. The projected 3PCF $\Pi(r_{p12}, r_{p23}, r_{p31})$ is then defined as

$$\Pi(r_{p12},r_{p23},r_{p31}) = \int \xi_z(r_{p12},r_{p23},r_{p31},\pi_{12},\pi_{31})d\pi_{12}d\pi_{23}.$$

r_p, u, v coordinates are introduced to quantify a triangle with $r_{p12} \le r_{p23} \le r_{p31}$ on the projected plane, analogously to s, u, v in redshift space. Then

$$\begin{aligned} Q_{\rm proj}(r_p,u,v) = \Pi(r_p,u,v)\{&w(r_{p12})w(r_{p23})+w(r_{p23})w(r_{p31}) \\ &+w(r_{p31})w(r_{p12})\}; \end{aligned} \tag{12.54}$$

where r_p is the projected 2PCF.

An interesting property of the projected 3PCF is that the normalized $Q_{\rm proj}(r_p,u,v)$ is not only a constant, but also equal to Q. Therefore the measurement of $Q_{\rm proj}$ can be used to test the hierarchical form for the 3PCF.

The results from the LCRS indicate that the 3PCFs, both in redshift space and in real space, exhibit small but significant deviations from the hierarchical form. The 3PCF in redshift space can be fitted by

$$Q_{\rm red}(s,u,v) = 0.5 \cdot 10^{[0.2+0.1(s/s+1)^2]v^2} \tag{12.55}$$

for $0.8 < s_{12} < 8h^{-1}$ Mpc and $s_{31} < 16\ h^{-1}$ Mpc. The projected 3PCF can be fitted by

$$Q_{\rm proj}(r_p,u,v) = 0.7r_p^{-0.3} \tag{12.56}$$

for $0.2 < r_{p12} < 3\ h^{-1}$ Mpc and $r_{p31} < 6\ h^{-1}$ Mpc (s and r_p are also in units of h^{-1} Mpc).

Since the real space 3PCF $\zeta(r, u, v)$ may depend on r, u and v in a complicated way, it is not possible to invert the equation

$$\Pi(r_{p12}, r_{p23}, r_{p31}) = \int \zeta\left(\sqrt{r_{p1}^2 + y_{12}^2}, \sqrt{r_{p23}^2 + y_{23}^2}, \sqrt{r_{p31}^2 + (y_{12} + y_{23})^2}\right) dy_{12} dy_{23},$$

to obtain ζ from Π. It has been shown [Jing and Börner 1998], however, that the projected $Q_{\rm proj}(r_p, u, v)$ can be modelled very well by half the value of $Q_{\rm proj}^{\rm CDM}(r_p, u, v)$, the value obtained for the mock catalogues for the $\Omega_m = 0.2$ simulation.

It follows that

$$Q(r, u, v) = 0.5 Q^{\rm CDM}(r, u, v), \tag{12.57}$$

where $Q^{\rm CDM}(r, u, v)$ is the real-space 3PCF of the CDM model with $\Omega_m = 0.2$.

[Jing and Börner 1998] have computed the 3PCF with and without the CLW bias for the $\Omega_m = 0.2$ model, and found that the value of $Q(r, u, v)$ does not change much. The result (12.57) may indicate that the gravitational interaction alone is not sufficient to describe the clustering of galaxies, and physical processes of gas and radiation hydrodynamics connected with galaxy formation must be taken into account.

At any rate, more sophisticated bias models or a more sophisticated combination of model parameters must be considered.

12.4.5 A Few More Observations

A) Bulk Motion. The measurements of a bulk motion for a sample of 200 elliptical galaxies [Dressler et al. 1987] and the subsequent (undeserved?) publicity for the "seven samurai" and their "Great Attractor" have motivated several groups to use various distance indicators to get more accurate peculiar velocities. The real uncertainties in the use of semi-empirical relations such as the infrared Tully–Fisher relation for spirals (luminosity linewidth) or the diameter–velocity dispersion (D_n–σ) relation for ellipticals are difficult to assess (see also Chap. 2). The prediction of the "Great Attractor" [Lynden-Bell et al. 1988] depends crucially on the calibration of the D_n–σ relation $\log D_n = x \log \sigma + f(r)$ (where $r = zc/H_0$, D_n is the photometrically defined angular size, σ the velocity dispersion). A higher accuracy in the distance determination can be achieved only in a statistical sense, by measuring the distance to groups of galaxies, and then reducing the error by dividing by the square root of the number of galaxies in the group. Systematic errors in this procedure, like the dependence of D_n on a third parameter, may really bring trouble. The algorithm defining the different groups of galaxies is of particularly crucial importance. It has been shown that a slightly different calibration [Kates and Weigelt 1990] makes the model of a "Great Attractor" statistically less favourable (in a χ^2 test) than even simple models of a nearly constant velocity field varying linearly with distance.

Other attempts to improve the modelling involve a method of reconstructing the velocity field with a specific smoothing length [Bertschinger and Dekel 1989]. The authors claim that by this reconstruction one can achieve from the few data points a higher accuracy than the actual observations. This claim seems doubtful, since the smoothing procedure must have a great influence on the result. A comparison of the mass density field derived in this way, and of the IRAS field, gives $\Omega_0^{0.6}/b = 1.0 \pm 0.3$, higher than, but consistent with the estimate from the IRAS dipole. The application of this procedure to a sample of spirals (Tully–Fisher distances) and ellipticals (D_n–σ distances) roughly confirmed the Great Attractor model [Bertschinger et al. 1990].

B) Observations of Large Scales. Several observations point to the existence of interesting large-scale features in the galaxy distribution. Whether or not such features can constrain structure formation models is not clear at the moment. They may be singular features which could evolve from a special initial condition. Here we mention a few such cases.

A percolation analysis of the angular distribution of Abell clusters [Mo and Börner 1989] showed some evidence for superstructures extending over $350\,h^{-1}$Mpc. This analysis was presented in Chap. 10. "Pencil beam" surveys were also mentioned in Chap. 10. The galaxies seem to cluster according to a pattern which nicely corresponds to the picture of a beam cutting through a series of "Great Walls" one after another. It seems remarkable that a certain regularity has been discovered corresponding to a spatial distance of $128\,h^{-1}$Mpc [Broadhurst et al. 1990].

The LCRS survey is not suitable for definitely checking this, because the band structure in the declination of this survey limits all reliable results to scales which are smaller than the width of the declination bands.

Up to now the extent of the largest structures is determined by the boundaries of the survey; an indication for a universal upper limit does not yet exist. This leaves us with the sobering thought that the distribution of the luminous matter may not yet reveal the homogeneous structure of the universe – if it really exists. It could be that the part of the universe which has been measured is still too small, or has not been investigated with high enough sensitivity to record average characteristics of the galaxy distribution.

C) Faint Galaxies. A number of CCD observations [Tyson 1988; Cowie 1989] have extended galaxy counts for faint magnitudes $B \simeq 27$, and have shown the existence of a large population of faint blue objects which probably are galaxies in their early stages of formation. The total number of such faint blue objects is estimated at $\sim 10^{10}$.

The APM survey [Maddox et al. 1990a] registers galaxies brighter than magnitude 20.5 in the pass-band used (called "b_J"; see [Maddox et al. 1990a]).

The number of galaxies per magnitude bin and per square degree $N(m)$ is compared with the predictions of a cosmological model without evolution and with the conservation law $n_G R^3$ = const. If the counts are normalized to the no-evolution model at $b_J = 17$, one finds about twice the number predicted at $b_J = 20.5$. This implies large evolution effects within the galaxy population already at the rather moderate redshifts of $z = 0.1$ to 0.2 [Maddox et al. 1990b]. Dramatic evolution effects of this type and so close to the present epoch were unexpected and spell trouble for all kinds of cosmological considerations. Galaxies can apparently no longer be treated as a kind of standard marker of cosmic structure; galaxy formation must be a fairly recent phenomenon. But our own galaxy at a time in the past corresponding to $z =$ 0.1 should not really look much different from its present state. Interestingly enough, there is more and more evidence accumulating for evolutionary effects in the galaxy population at low redshifts. The IRAS faint source catalogue shows a number count excess at $0.3J_y$ (about $z \sim 0.1$; [Lonsdale et al. 1990]).

Deep redshift surveys [Broadhurst et al. 1992] support the idea of strong evolution, while at the same time excluding several possible explanations. The changes in $N(m)$ cannot be due to a large number of local low-luminosity objects (dwarf galaxies). The evolution of the luminosity of individual galaxies also cannot account for this effect. One possible way out (suggested by [Broadhurst et al. 1992]) would be to accept epochs of strongly increased star formation (starbursts), perhaps caused by several mergers with other galaxies, of each galaxy during its lifetime. Every galaxy thus has experienced about five mergers in the time interval between $z = 1$ and $z = 0$. The number of galaxies is not conserved in this model; there were many more (and smaller) galaxies at earlier times. This scheme allows an excellent fit to the data, not only for the B-band (*blue*) number counts, but also for $N(m)$ in the K-band (*red*). B-band number counts are increased by this mechanism, but the small stars mainly contributing to the red K-band build up rather slowly, and are not strongly affected by starbursts and mergers in their evolution.

12.4.6 High-Redshift Objects

A) Counts of Damped Lyman-alpha (DL) Systems. Damped Lyman-alpha systems [Wolfe 1993] seem to be a very good probe to study galaxy formation and evolution at high redshifts. DL systems are detected as absorption line systems along the line of sight to a high-redshift quasar. They have high HI column densities of 10^{19-22} cm^{-2}. The column density distribution $f(N_{\mathrm{HI}}, z)dN_{\mathrm{HI}}$ at a certain redshift can be used to estimate the mass density of neutral hydrogen in DL systems

$$\Omega_D = H_0 \frac{\mu m_{\mathrm{H}}}{\varrho_c} \int N_{\mathrm{HI}} f(N_{\mathrm{HI}}, z) dN_{\mathrm{HI}}, \tag{12.58}$$

where ϱ_c is the critical density, $\mu = 1.33$.

These data [Lanzetta et al. 1993] are compared to model values derived from an estimate of $n(M, z)$ from the Press-Schechter scheme [Mo and Miralda-Escudé 1994]. All the models have some problems in accounting for the high-redshift data, because there are not enough collapsed halos of DM at high z.

Considering that the neutral hydrogen may form disc-like structures [Mo and Miralda-Escudé 1994] there are indications that for simple models the discrepancy gets larger.

B) Lyman Break Galaxies. The Lyman break at 912 Å in the rest system is a distinctive feature in the UV spectrum of a galaxy, where its radiation becomes absorbed by the hydrogen atoms in the galaxy itself. A sharp break in the continuum appears. At redshifts 3 to 4 this feature is in the optical part of the spectrum. Such galaxies can be found by skilfully arranging narrow-band filters. Galaxies which disappear at some particular band immediately tell us their redshift. Such a method has led to the detection of a large number of so-called "Lyman break" galaxies [Steidel et al. 1998]. The suggestion has been advanced that these galaxies are the progenitors of present-day massive galaxies. A series of high-resolution CDM simulations has been carried out with 17 million particles in a comoving box of $(100\,h^{-1}\mathrm{Mpc}^3$. The observed number density of Lyman break galaxies at redshifts ~ 3 corresponds to the density of dark matter halos of mass $\simeq 10^{12} M_\odot$ in the simulations. The clustering found for those objects in one 8.74×17.64 arcmin sky area [Steidel et al. 1998] cannot be reproduced easily by CDM models – one similar concentration in every six fields of the given area is found in an $(\Omega_m = 0.3, \Omega_\Lambda = 0.7)$ model [Jing and Suto 1998].

12.4.7 Conclusion

Many more observations have been accumulated. Here I have just discussed the comparison between theory and observations using a few outstanding examples. This is a rapidly developing field. As new surveys and larger simulations become available, it will be interesting to see which of the results presented above will remain.

The general conclusion may be that CDM models can explain the observed correlations and abundances of galaxies, except for the 3PCF, if a flat, low-density cosmological model is employed, together with the CLW bias scheme.

12.5 Neutrinos and Large-Scale Structure Formation

12.5.1 The Coherence Length

The original suggestion by Zel'dovich and co-workers [Zel'dovich 1970] that the observed large structure could be explained as the result of the gravitational instability of adiabatic density fluctuations has been revived in recent years, because of the possibility that the universe has a mass density that is dominated by massive neutrinos.

A general feature of this scenario is the existence of a coherence length corresponding to the "Silk damping mass" in the baryon-dominated models, and to the "free-streaming" damping length for massive collision-free particles (neutrinos, X-particles). In Chap. 11 we presented the formula

$$\lambda_d = 13(\Omega_X h_0^2)^{-1}\vartheta^{-2}(1+z)^{-1} \text{ Mpc}$$

for the free-streaming length of collisionless particles.

Fluctuations with wavelengths less than λ_d are strongly damped [Bond and Szalay 1983]:

$$|\delta_k|^2 \propto k^n 10^{-2(k/k_d)^{1.5}}, \qquad k_d \equiv \frac{2\pi}{\lambda_d}. \tag{12.59}$$

We refer all wavelengths to the comoving system of coordinates, and to the present epoch $t = t_0$. For neutrinos one obtains

$$\lambda_\nu = 14 \text{ Mpc } m_{\nu,10}^{-1}\vartheta^{-1},$$
$$m_{\nu,10} \equiv \frac{m_\nu}{10\text{eV}}, \qquad \vartheta \equiv T_{\gamma 0}/2.7\,\text{K}. \tag{12.60}$$

A neutrino mass of a few eV gives damping lengths of $\sim$ 100 Mpc. This shows that the pure neutrino model is not viable for structure formation. There is no power on cluster scales, let alone galaxy scales. Also dissipative processes seem to be unable to give reasonable small scale structure via fragmentation processes. The neutrino is discussed here because it is instructive to study the effect of a coherence length. Also it may be that a small admixture of neutrinos to CDM is possible – leading to mixed dark matter (MDM) models.

In a ν-dominated universe density perturbations with $\lambda = \lambda_\nu$ reach the non-linear regime ($\delta\varrho \approx \varrho$) long before galaxies are formed. The collision-dominated baryonic matter initially has no influence at all on the collapse of such fluctuations. Therefore tiny initial anisotropies in the fluctuation spectrum can be amplified strongly, since there are no counteracting pressure terms. The gravitational collapse occurs preferentially in one dimension (the so-called "pancake" instability). For a strictly pressureless medium, caustics ($\varrho \to \infty$) would form along the central plane of a pancake (see Chap. 11).

The baryonic component, however, resolves the caustic into a shock. The infalling baryonic matter runs through this shock, is decelerated, heated (to temperatures $T \approx 10^7$ K to 10^8 K), and its energy is dissipated. The highest density is in the central plane. This matter can cool most rapidly, fragmenting into clouds which evolve into protogalaxies.

12.5.2 The Vlasov Equation for Massive Neutrinos

The universe consisting of massive ν's (or X's) is described as a system with a large number of identical gravitating point particles. The system is then completely characterized by the N-particle distribution function

$$\mu = \mu(t; \boldsymbol{x}_i, \boldsymbol{v}_i), \quad i = 1, \dots, N.$$

At redshifts $z \leq 1500$ the neutrinos are non-relativistic, and a Newtonian description is possible. The time evolution of the system follows from the Liouville equation – which simply expresses the conservation of probability in the $6N$-dimensional phase space:

$$\begin{aligned} \frac{d\mu}{dt} &= \frac{\partial \mu}{\partial t} + \sum_{i=1}^{N} \nabla_{\boldsymbol{x}_i} \left(\mu \frac{d\boldsymbol{x}_i}{dt} \right) + \sum_{i=1}^{N} \nabla_{\boldsymbol{v}_i} \left(\mu \frac{d\boldsymbol{v}_i}{dt} \right) \\ &= \frac{\partial \mu}{\partial t} + \sum_{i=1}^{N} \{ \boldsymbol{v}_i \cdot \nabla_{\boldsymbol{x}_i} \mu + \mathbf{F}_i \cdot \nabla_{\boldsymbol{v}_i} \mu \} = 0. \end{aligned} \tag{12.61}$$

$\boldsymbol{F}_i$ is the total gravitational force on particle i, and

$$\boldsymbol{F} = \sum_{j \neq i} \boldsymbol{F}(j - i) = \sum_{j \neq i} -Gm \frac{\boldsymbol{x}_i - \boldsymbol{x}_j}{|\boldsymbol{x}_i - \boldsymbol{x}_j|^3}.$$

The background universe is thought to be homogeneous, and by a transformation to comoving coordinates one can take care of the overall expansion.

As before,

$$\begin{aligned} \boldsymbol{x}_i' &= \frac{R_0}{R(t)} \boldsymbol{x}_i, \\ \boldsymbol{v}_i' &= \frac{R_0}{R(t)} (\boldsymbol{v}_i - H\boldsymbol{x}_i), \\ dt' &= (R_0^2 / R^2) dt. \end{aligned}$$

Since $\mu'(t'; x_i', v_i) = \mu(t; x_i, v_i)$ (conservation of the normalization) we find

$$\frac{\partial \mu'}{\partial t'} + \sum_{i=1}^{N} \{ \boldsymbol{v}_i' \nabla_{\boldsymbol{x}_i}' \mu' + \boldsymbol{K}_i' \nabla_{\boldsymbol{v}_i}' \mu' \} = 0, \tag{12.62}$$

where

$$\begin{aligned} \boldsymbol{K_i} &= \frac{R^3(t)}{R_0^3} \left[\sum_{i \neq i} \boldsymbol{F}(j - i) - (\dot{H} + H^2) \boldsymbol{x}_i' \right] \\ &= \frac{R(t)}{R_0} \left[\sum_{j \neq i} \boldsymbol{F}(j - i) - \frac{R^2}{R_0^2} Y(t) \boldsymbol{x}_i' \right]; \end{aligned} \tag{12.63}$$

$Y(t) \equiv \dot{H} + H^2$ is the tidal force in the comoving system. One can also write

$$\boldsymbol{K}'_i = \sum_{j \neq i} \boldsymbol{K}'(j-i)$$

with

$$\boldsymbol{K}'(j-i) = \frac{R(t)}{R_0}\left[\boldsymbol{F}'(j-i) - \frac{1}{N-1} Y \boldsymbol{x}'_i \frac{R^2}{R_0^2}\right]. \tag{12.64}$$

The first term in (12.63) represents the pair force, while the second term represents the average contribution of the total "force-like" term resulting from the expansion.

Then (12.61) is a normal Liouville equation for N particles which interact with a time-dependent force. This identification allows the use of the formalism of statistical mechanics:

i) Define the particle distribution function as follows:

$f_1(t'; \boldsymbol{x}'_i, \boldsymbol{v}'_i) \equiv \prod_{j \neq i} d\boldsymbol{\sigma}_j \mu'(1, \ldots, N)$ 1-particle distribution function;

$f_2(t'; \boldsymbol{x}'_i, \boldsymbol{x}'_j, \boldsymbol{v}'_j) \equiv \int \prod_{k \neq i,j} d\boldsymbol{\sigma}_k \mu'(1, \ldots, N)$ 2-particle distribution function;

$$d\boldsymbol{\sigma}_j \equiv d^3x'_j d^3v'_j. \tag{12.65}$$

If μ' is symmetric with respect to the interchange of pairs, then the functions f_1, f_2 do not depend on the specific particles chosen; $f_1 = f_1(t'; x', v')$, and Nf_1 is the density in the 6-dimensional phase space.

ii) The Liouville equation implies dynamic equations for the s-particle distribution functions, e.g. via the BBGKY hierarchy. For strongly correlated systems this hierarchy is not applicable. But for the 1-particle distribution function f_1 an exact, closed, integro-differential equation can be derived (e.g. [Balescu 1975]):

$$\begin{aligned} &\frac{\partial}{\partial t'} f_1(t';1) + \boldsymbol{v}'_1 \nabla'_{\boldsymbol{x}_1} f_1(t';1) + \langle \boldsymbol{K}'_1 \rangle \nabla'_{\boldsymbol{v}_i} f_1(t';1) = S_1[f_1], \\ &\langle \boldsymbol{K}'_1 \rangle \equiv \int \boldsymbol{K}'_1 \prod_{j \neq i} f_1(t';j) d\sigma_j. \end{aligned} \tag{12.66}$$

Initial conditions must supplement this equation.

If $S_1[f_1]$ is set equal to zero, then a Vlasov equation in comoving coordinates is obtained:

$$\begin{aligned} &\partial_{t'} f + \boldsymbol{v}' \nabla_{\boldsymbol{x}'} f + \boldsymbol{F}' \nabla_{\boldsymbol{v}'} f = 0, \\ &\boldsymbol{F}' = \frac{R^3(t)}{R_0^3(t)} (\boldsymbol{F} - \boldsymbol{x}' Y), \end{aligned} \tag{12.67}$$

where $\nabla \cdot \boldsymbol{F} = -4\pi G \varrho = -4\pi G \int f d^3v$.

Since, for $p = 0$,

$$Y = -\frac{4\pi}{3}\bar{\varrho},$$

where $\bar{\varrho}$ is the mean density, we can rewrite $\boldsymbol{F}'$ as

$$\boldsymbol{F}' = \frac{R^3(t)}{R_0^3}\left(\boldsymbol{F} + \frac{4\pi G}{3}\bar{\varrho}\boldsymbol{x}'\right),$$

and

$$\nabla' \cdot \boldsymbol{F}' = -4\pi G\frac{R(t)}{R_0}\left[\varrho' - \frac{R^3(t)}{R_0^2}\bar{\varrho}\right] = -4\pi G'[\int f' d^3v' - \bar{\varrho}_0], \tag{12.68}$$

where use has been made of $R^3(t)\bar{\varrho} = \text{const.}$ and $G' \equiv GR(t)/R_0$ is like an effective coupling constant increasing with time.

Equations (12.68) and (12.67) determine the time evolution of the 1-particle distribution function. "Mass neutrality" is only guaranteed on average (because of mass conservation), but since ϱ' is completely inhomogeneous, there are positive and negative effective mass densities in (12.68). Only quantities with the same sign attract each other; with opposite sign they repel each other;

12.5.3 Cooking Pancakes

In a neutrino-dominated universe structures of mass $M_{Pl}^2 m_\nu^{-2} \sim 10^{15} M_\odot$ are the first to start growing, the first to reach the non-linear regime, the first to collapse. Subsequent fragmentation of such supercluster stuctures should then be the cause of galaxy formation. Although the fragmentation process is not well understood, it seems that the simple adiabatic collapse cannot produce a sufficient number of condensed objects on smaller scales. Furthermore, the observations point to a hierarchical model, where galaxies form first, and then build up the large structures.

Thus the "pancake" model cannot by itself serve as a model for galaxy formation. But some features seem worth preserving, especially the excellent description it gives of the large-scale structure observed – the arrangement of galaxies in sheets and filaments in surrounding large empty regions.

Lin, Mestel, and Shu discussed the "pancake-instability" for pressure-loss collapse in 1965. In a collision-free medium there is no pressure to counteract slight initial anisotropies of the collapse, and thus 1-dimensional structures, the "pancakes", form [Lin et al. 1965].

Zel'dovich in 1970 stressed the importance of the mechanism for the collapse of primordial fluctuations [Zel'dovich 1970]. It has been shown in 3D simulations that the formation of pancakes is a generic feature of the growth of fluctuations with a cut-off [Centrella and Melott 1983]. For $\delta > 1$ a cell-like structure is formed containing pancakes and filaments. At the same time large voids of approximate spherical symmetry form ("Swiss cheese" structure). The Zel'dovich approximation was discussed in Chap. 11. Interesting attempts to justify the extrapolation of this approximation into the non-linear regime (the 1D Zel'dovich approximation is an exact solution of the Euler–Poisson equations) have been carried out [Buchert 1989]. The best argument in favour of this analytic scheme is the excellent agreement with N-body simulations [Buchert et al. 1995].

In Fig. 11.11 a high-resolution Zel'dovich approximation is shown [Schmalzing 1999]. The free-streaming length d_ν was implemented as a cut-off scale in a flat initial spectrum. A cellular pattern of the large-scale structure clearly emerges, where pancakes as filamentary or sheet-like objects surround large areas of very low density.

This gives an eye impression similar to the patterns seen in galaxy surveys. Since the neutrino-dominated models cannot overcome the problems with forming galaxies, one might want to preserve some aspects of this model by looking at a mixture of hot and cold DM, the "mixed dark matter" models.

12.5.4 Mixed DM Models

The basic idea of MDM models is to change the CDM power spectrum somewhat by a small admixture of neutrinos to the CDM particles. This could lead

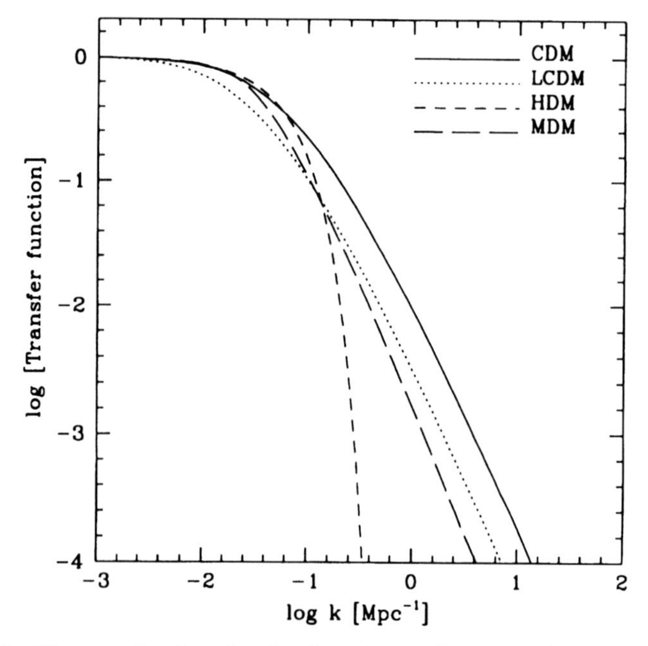

Fig. 12.18. The transfer function is plotted as a function of k for various dark matter models

to a reduction of the small-scale power of the CDM model, because of the cut-off in the HDM component, and to an enhancement of the large-scale power, where e.g. the SCDM model has not enough to explain the APM correlations. The properties of simulations of MDM models have been explored to some extent [Jing et al. 1993]. About 20% of neutrinos can be admixed, and still a good fit to collapsed objects at high redshifts can be obtained. In Fig. 12.18 the transfer function for some linearly evolved power spectra is shown. The steep cut-off of the HDM model is changed to a moderate fall-off at smaller wavelength for the MDM model.

12.6 Analytic Approaches

The Zel'dovich approximation, as well as the evolution of spherical inhomogeneities, are analytic approximations to structure formation in the universe. These two approaches were introduced in Chap. 11 and in Sect. 12.5.3. Here we want to describe briefly the analytic treatment of the evolution of statistical quantities, and the Press–Schechter (PS) theory.

12.6.1 Dynamics and Statistics

This approach, pioneered by [Peebles 1980], consists in finding a way to follow the galaxy distribution from reasonable initial conditions through the normal dynamics to a final state that matches the observations reasonably well.

The dynamics used by Peebles is the BBGKY hierarchy adapted from plasma physics. The first equation in the hierarchy is a relation between the one- and two-point correlation functions; it shows how the peculiar velocity of a randomly chosen particle is affected by the perturbations of neighbours. The second equation describes how the relative numbers and the motions of pairs of particles depend on the perturbations of the neighbouring particles; this is an equation between the two- and three-point correlation functions. The sequence continues in an infinite set of equations containing all the n-point correlation functions. For practical purposes the BBGKY hierarchy must be terminated at some level by posing some evident constraint, or a reasonable model for correlation functions of higher order. Here, I want to mention a few interesting models that have been set up for the low-order statistics which concern the first two equations of the BBGKY hierarchy – up to the 3PCF.

First, we introduce the concepts and repeat some definitions and arguments from [Peebles 1980]. Take a small patch Γ around the spatial coordinate $\boldsymbol{x}$, and the momentum $\boldsymbol{p}$ in the six-dimensional single particle phase space. Consider the actual particle system as one realization of a random process. Γ should be so small that it either contains one or no particle of the realization. A large number of realizations gives a statistical ensemble of particle systems. The matter in Γ for realization number i is described by a single particle distribution function

$$f_i \equiv \delta(\boldsymbol{x} - \boldsymbol{x}_i(t))\delta(\boldsymbol{p} - \boldsymbol{p}_i(t))$$

if Γ contains a particle, $f_i = 0$ otherwise. The Liouville equation describes the conservation of the phase space density:

$$\left(\frac{\partial}{\partial t} + \frac{dx^m}{dt}\frac{\partial}{\partial x^m} + \frac{dp^m}{dt}\frac{\partial}{\partial p^m}\right) f_i = 0. \tag{12.69}$$

Averaging f_i over the ensemble gives the one-point function

$$\langle f_i(\boldsymbol{x}, \boldsymbol{p}, t)\rangle \equiv b(p, t),$$

which is just the probability of finding a particle in the patch Γ. If we write $\Gamma = d^3x d^3p$, and take into account that the point distribution is derived from a homogeneous and isotropic random process, then

$$dP = b(p, t)d^3x d^3p$$

is the probability, and b depends only on t and $p = |\boldsymbol{p}|$. The normalization is given by integrating over momenta

$$\int b d^3p = nR^3 = \text{const.},$$

where n is the mean number density, $R(t)$ the cosmic expansion factor.

The time derivatives in (12.69) are given by

$$\begin{aligned}\frac{dx^m}{dt} &= -\frac{Tp^m}{MR^3},\\ \frac{dp^m}{dt} &= -\frac{GM^2}{R}\sum_\ell (x_l^m - x^m)/|\boldsymbol{x}_l - \boldsymbol{x}|^3.\end{aligned}$$

Therefore the ensemble average of (12.69) is given by

$$\frac{\partial}{\partial t}b(p_1, t) + \frac{GM^2}{R}\frac{\partial}{\partial p_1^m}\int d^3x_2 d^3p_2 c(1,2)(x_2^m - x_1^m)/r_{12}^3 = 0. \tag{12.70}$$

This is the first BBGKY equation. It has been derived via the introduction of the 2PCF as the joint probability

$$dP = \varrho_2(1,2)d^3x_1 d^3x_2 d^3p_1 d^3p_2$$

of finding a particle in $d^3x_1 d^3p_1$ with momentum $\boldsymbol{p}_1$, and simultaneously a particle in $d^3x_2 d^3p_2$ with momentum $\boldsymbol{p}_2$. One usually defines a reduced 2PCF $c(1,2)$ (1 stands for $\boldsymbol{x}_1, \boldsymbol{p}_1$ etc.):

$$\varrho_2(1,2) \equiv b(1)b(2) + c(1,2).$$

The normalization is

$$\int c(1,2)d^6p = n^2R^6\xi(r_{12};t).$$

Similarly, the 3PCF is

$$\varrho_3(1,2,3) = [b(1)b(2)b(3) + b(1)c(2,3) + b(2)c(3,1) + b(3)c(1,2) + d(1,2,3)]d^9xd^9p,$$

and

$$\int dd^9p = n^3R^9\zeta(1,2,3). \tag{12.71}$$

Now averaging $\langle f_i d\boldsymbol{p}_1/dt\rangle$ gives

$$\frac{GM^2}{R}\left\langle f_i\sum_\ell(\boldsymbol{x}_l-\boldsymbol{x}_1)/|\boldsymbol{x}_l-\boldsymbol{x}_1|^3\right\rangle = \frac{GM^2}{R}\int d^3x_2d^3p_2\varrho_2(1,2)(\boldsymbol{x}_2-\boldsymbol{x}_1)/|\boldsymbol{x}_2-\boldsymbol{x}_1|^3,$$

and hence one arrives at (12.71).

For the next equation two small separate patches Γ_1 and Γ_2 in phase space are considered, with the two-particle distribution function $f_i(1,2)$. Averaging the Liouville equation leads eventually to the second BBGKY equation [Peebles 1980, p. 261]

$$\frac{\partial}{\partial t}\varrho_2(1,2) + \frac{p_1^m}{MR^2}\frac{\partial}{\partial x_1^m}\varrho_2(1,2) + \frac{GM^2}{R}\frac{(x_2-x_1)^m}{r_{12}^3}\frac{\partial}{\partial p_1^m}\varrho_2(1,2) + \frac{GM^2}{R}\frac{\partial}{\partial p_1^m}\int d^3x_3d^3p_3\varrho_3(1,2,3)\frac{(x_2-x_1)^m}{r_{31}^3} + (1\leftrightarrow 2) = 0 \tag{12.72}$$

($(1 \leftrightarrow 2)$ indicates that the terms where particles 1 and 2 are exchanged should be added).

Integrating equations (12.70) over the momenta p_1 and p_2 eliminates the momenta derivatives, and with the symmetry properties of $c(1,2)$ eventually leads to

$$\frac{\partial}{\partial t}\xi + \frac{1}{r^2}\frac{1}{R}\frac{\partial}{\partial r}[r^2(1+\xi(r,t))\langle v_{12}(r,t)\rangle] = 0. \tag{12.73}$$

This equation expresses the conservation of particle pairs with separation r which can be seen by rewriting it as

$$\frac{\partial}{\partial t}\left\{\bar{n}R^3\int_0^r 4\pi r^2dr[1+\xi(r,t)]\right\} + 4\pi R^2r^2\bar{n}[1+\xi(r,t)]\langle v_{12}(r,t)\rangle = 0; \tag{12.74}$$

$\langle v_{12}(r,t)\rangle$ is the mean pairwise peculiar velocity. The first term in (12.74) expresses the change in the number of pairs within the sphere of radius r, and the second term describes the average "flux" across this sphere.

In the strongly non-linear regime ($\xi \gg 1$) the clustering is expected to be statistically "stable" in the sense that the average proper separation of pairs remains constant. This is equivalent to saying that

$$\langle v_{12}(r,t)\rangle + \dot{R}r = 0. \tag{12.75}$$

Equation (12.73) can be rewritten

$$\frac{\partial}{\partial t}(1+\xi) = \frac{\dot{R}}{R}\frac{1}{r^2}\frac{\partial}{\partial r}[r^2(1+\xi)];$$

the solution of this equation is

$$1+\xi(r,R) = R^3 g(Rr),$$

where g is an arbitrary function. Since $\xi \gg 1$ was assumed in the beginning of this derivation, the solution is

$$\xi = R^3 g(Rr). \tag{12.76}$$

In the special case where the BBGKY hierarchy admits similarity solutions, we can find out more about the behaviour of ξ. Similarity solutions exist in an Einstein–de Sitter cosmology, i.e. when $R \propto t^{2/3}$, and when the fluid limit applies. Then the one-particle distribution function admits solutions of the form

$$f(\boldsymbol{x},\boldsymbol{p},t) = t^{-3\beta}\hat{f}(\boldsymbol{x}/t^{\beta-\frac{1}{3}},\boldsymbol{p}/t^{\beta}).$$

We obtain

$$\xi(x,t) = \hat{\xi}(x/t^{\beta-\frac{1}{3}}). \tag{12.77}$$

These equations describe a situation in which the matter distributions at different times are similar with characteristic lengths scaled as $t^{\beta-\frac{1}{3}}$, and characteristic momenta scaled as t^{β} [Peebles 1980]. It only works for $R \propto t^{2/3}$, and for scale-free initial fluctuation spectra

$$P(k) \propto k^n.$$

From linear theory (i.e. for $\xi \ll 1$)

$$\xi(x,t) \propto x^{-(n+3)}t^{4/3}, \tag{12.78}$$

and, using (12.79),

$$\beta = \frac{4}{3(n+3)} + \frac{1}{3}.$$

Now (12.76), together with the scaling behaviour, leads for $\xi \gg 1$ to the form $\xi(x,t) \propto x^{-\gamma} t^{\delta}$, with

$$\gamma = 3(n+3)/(n+5),$$
$$\delta = 4/(n+5).$$

For $n = 1$, for example, $\xi \propto x^{-2} t^{2/3}$.

This example clearly shows that even in the strongly non-linear regime the two-point correlation function keeps track of the initial conditions (by its dependence on n). The functional form of ξ is not a universal characteristic of the clustering process, but depends sensitively on the initial fluctuation spectra (see also [Suto 1993], where this point is elaborated further).

12.6.2 The Cosmic Virial Theorem

The first moment of the second BBGKY equation integrated over momenta can be written as [Peebles 1980]

$$\begin{aligned} &\frac{\partial}{\partial t}(1+\xi)v^{\alpha} + \frac{\dot{R}}{R(1+\xi)v^{\alpha}} + \frac{1}{R}\frac{\partial}{\partial x^{\beta}}(1+\xi)\langle v_{12}^{\alpha} v_{12}^{\beta}\rangle \\ &+\frac{2Gm}{R^2}\frac{x^{\alpha}}{x^3}(1+\xi) + 2G\bar{\varrho}R\frac{x^{\alpha}}{x^3}\int_0^x d^3x\,\xi(x) \\ &+2G\bar{\varrho}R\int d^3x_3\,\zeta(1,2,3)\frac{x_{13}^{\alpha}}{x_{13}^3} = 0. \end{aligned} \tag{12.79}$$

Consider a particle pair located at $\boldsymbol{x}_1$ and $\boldsymbol{x}_2$. If the separation $r_{12} \equiv |R(\boldsymbol{x}_1 - \boldsymbol{x}_2)|$ is small, the effective pressure term due to the relative peculiar velocity dispersion is balanced in equilibrium by the gravitational force. The dominant terms in this "statistical equilibrium" condition are the third and the last in (12.79) [cf. Peebles 1980, Suto 1993]. The simplified equation for statistical equilibrium is then

$$\frac{1}{R}\frac{\partial}{\partial x^{\beta}}[\xi\langle v_{12}^{\alpha} v_{12}^{\beta}\rangle] + 2G\bar{\varrho}R\int d^3x_3\,\zeta(1,2,3)\frac{x_{13}^{\alpha}}{x_{13}^3} = 0. \tag{12.80}$$

$\langle v_{12}^{\alpha}\rangle \equiv v^{\alpha}$ in the above equations is the α-component of the mean peculiar velocity $\langle(\boldsymbol{v}(\boldsymbol{x}) - \boldsymbol{v}(\boldsymbol{x}+\boldsymbol{r}))^{\alpha}\rangle$; ζ is the 3PCF.

If the velocity dispersions are isotropic,

$$\langle v_{12}^{\alpha} v_{12}^{\beta}\rangle \equiv \frac{1}{3}\langle v_{12}^2\rangle\delta^{\alpha\beta},$$

then we can integrate (12.80) once to obtain

$$\langle v_{12}^2(r)\rangle = \frac{6G\bar{\varrho}}{\xi(r)}\int_r^{\infty}\frac{dr'}{r'}\int d^3z\,\frac{\boldsymbol{r}'\cdot\boldsymbol{z}}{z^3}\zeta(r', z, |\boldsymbol{r}' - \boldsymbol{x}|). \tag{12.81}$$

This result is called the cosmic virial theorem (CVT). The equation can be further evaluated if we take a power law for $\xi : \xi = \left(\frac{r_0}{r}\right)^\gamma$, and the hierarchical form

$$\zeta = Q[\xi(1)\xi(2) + \xi(2)\xi(3) + \xi(3)\xi(1)]$$

for the 3PCF.

Finally, one obtains [Suto 1993] for the 1-dimensional PVD

$$\sigma^2(r) \equiv \frac{\langle v_{12}^2(r)\rangle}{3} = \frac{2\pi G\bar{\varrho}Qr_0^\gamma r^{2-\gamma}J(\gamma)}{(\gamma-1)(2-\gamma)(4-\gamma)};$$

here J is an integral with $J(1.8) = 3.70$.

The analysis of observational data and of numerical simulations does not lead to a satisfactory agreement with this approximate solution (cf. [Jing and Börner 1998]).

12.6.3 Analytic Solutions

Making use of the relations of the BBGKY hierarchy, semi-analytic models can be constructed for all low-order moments of the distribution function [Mo et al. 1997]. These models allow such moments to be calculated directly from the initial density perturbation power spectrum for a given cosmological model. Although this is an incomplete description it provides a clear picture of how these moments are related to various parameters of the cosmological models.

It all starts with a fitting formula for the evolved 2PCF [Hamilton 1992; Peacock and Dodds 1996]. We define a quantity

$$\Delta^2(k) \equiv \frac{1}{2\pi^2}k^3P(k),$$

the "power variance". The evolved 2PCF $\xi(r)$ is related to the evolved power variance $\Delta_E(k)$ by

$$\xi(r) = \int_0^\infty \frac{dk}{k}\Delta_E^2(k)\frac{\sin kr}{kr}. \tag{12.82}$$

It has been possible to obtain a fitting formula – using arguments first given in [Hamilton 1992] – which relates Δ_E to the initial density spectrum for a given cosmological model. The evolved power variance is assumed to be a function of its linear counterpart Δ:

$$\Delta_E^2(k_E) = f\{\Delta^2(k_L)\},$$

where

$$k_L = (1 + \Delta_E^2(k_E))^{-1/3}k_E.$$

The functional form of $f(x)$ is guessed from the linear, and highly non-linear approximations, and from a detailed comparison with N-body simulations. $f(x)$ depends on the cosmological parameters $R(t), \Omega_m(t)$, and $\Omega_\Lambda(t)$. It is clear that Δ_E is determined by Δ for a given cosmology, and (12.82) then can be used to obtain $\xi(r)$.

The 2PCF $\xi(r)$ can be used to derive further quantities.

From the pair conservation equation the average of the pairwise peculiar velocity

$$\langle v_{12}(r)\rangle \equiv \left\langle [\boldsymbol{v}(\boldsymbol{x}) - \boldsymbol{v}(\boldsymbol{x}+\boldsymbol{r})] \cdot \frac{\boldsymbol{r}}{r} \right\rangle$$

can be written as

$$\frac{\langle v_{12}(r)\rangle}{Hr} = -\frac{1}{3}[1+\xi(y,R)]\frac{\partial\bar{\xi}(y,R)}{\partial \ln R}, \tag{12.83}$$

where r is the proper and y the comoving separation between the pairs;

$$H = H_0\left[\Omega_{\Lambda,0}(1-R^{-2}) + \Omega_{m,0}(R^{-3}-R^{-2}) + R^{-2}\right]^{1/2}$$

is the value of the Hubble parameter for an expansion factor $R(t)$; and

$$\bar{\xi}(y,R) = \frac{3}{y^3}\int_0^y y^2 dy \xi(y,R).$$

Thus in order to obtain $\langle v_{12}(r)\rangle$ the derivative $\partial\Delta_E(k,R)/\partial R$ must be worked out.

The mean square peculiar velocity of mass particles $\langle v_1^2\rangle$ is related to the 2PCF by the cosmic energy equation

$$\frac{d}{dR}R^2\langle v_1^2\rangle = 4\pi G\bar{\varrho}R^3\frac{\partial I_2(R)}{\partial \ln R}, \tag{12.84}$$

where $\bar{\varrho}$ is the mean cosmic density and

$$I_2(R) \equiv \int_0^\infty y dy \xi(y,R).$$

Even for the pairwise PVD a reasonably accurate fitting formula can be derived. The PVD projected along the separations of particle pairs ($\langle v_{12}^2(r)\rangle^{1/2}$) from N-body simulations is compared to a fitting formula

$$\langle v_{12}^2(r)\rangle^{1/2} = \Omega_m^{0.5} H r_c \phi(r/r_c); \tag{12.85}$$

ϕ is a universal function – appropriately chosen – and r_c is a non-linear scale (the virial radius of M^* haloes). The main features of $\langle v_{12}^2(r)\rangle^{1/2}$ are

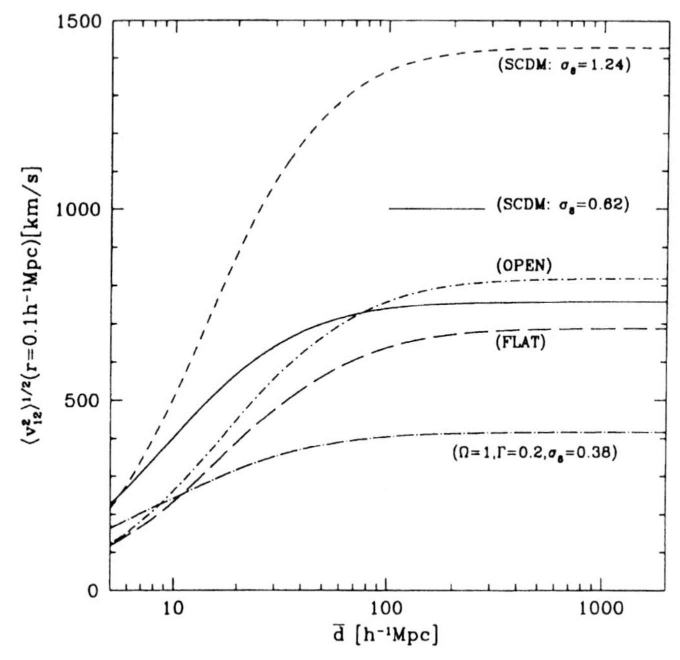

Fig. 12.19. The pairwise velocity dispersion for a separation of $0.1h^{-1}$Mpc is shown vs. the mean separation of haloes removed from the sample (see text for more details)

a) monotonic rise at small r;
b) saturation at large r.
c) a maximum at medium r.

For more details consult [Mo et al. 1997]. In Fig. 12.19 an interesting aspect of this model-building is shown. One can remove all haloes above a certain mass from a sample, and calculate $\langle v_{12}^2 \rangle$ for the smaller sample. The mean separation of haloes with mass exceeding M is $\bar{d}(M) \equiv [N(>M)]^{-1/3}$. Thus the dependence of $\langle v_{12}^2(r) \rangle$ on the separation $\bar{d}(M)$ of haloes removed from the sample is computed. The result for $r = 0.1h^{-1}$ Mpc is displayed in Fig. 12.19. We see that $\langle v_{12}^2(r) \rangle^{1/2}$ does not reach the asymptotic value even for $\bar{d} \approx 50h^{-1}$ Mpc, a typical value for the mean separation of rich clusters. Thus to have the small-scale PVD fairly sampled, one needs a sample containing many rich clusters. The first redshift catalogue satisfying this criterion is the LCRS.

12.6.4 The Press–Schechter Mass Function

An important application of the spherical collapse calculation is an estimate of the number density of objects of mass M which have turned around and collapsed at redshift z. A simplifying assumption, that spherical random

Gaussian density fluctuations collapse and form bound objects, leads to the differential mass function per unit comoving volume $n(M,z)dM$ at redshift z. This occurs when the density contrast is smoothed over the mass scale M, and computed in linear theory, reaching δ_c.

$$M = -\left(\frac{2}{\pi}\right)^{1/2} \frac{\varrho_0}{M} \frac{\delta_c}{\Delta^2(M,z)} \frac{d\Delta(M,z)}{dM} \times \exp\left\{-\frac{\delta_c^2}{2\Delta^2(M,z)}\right\} dM. \tag{12.86}$$

$n(M,z)$ is the comoving number density at redshift z of objects with mass between M and $M+dM$; ϱ_0 is the present mass density; and $\Delta(M,z)$ is the density fluctuation computed from linear theory,

$$\Delta^2(M,z) = \frac{D^2(z)}{2\pi^2} \int_0^\infty P(k,t_0) W^2(kr_0) k^2 dk \tag{12.87}$$

where $D(z)$ is the linear growth factor $D(z) = \frac{D_1(R(z))}{D_1(R_o)}$, and $W(kr_0)$ is the top-hat window function, with $\frac{4\pi}{3}\varrho_0 r_0^3 \equiv M$.

This expression was originally derived in the Einstein–de Sitter universe, but it is also valid in $\Omega_\Lambda < 1$ and $\Omega_m \neq 0$ models [Lilje 1992]. Studies with N-body simulations [Lacey and Cole 1994] show that the PS estimate is quite a good approximation provided δ_c is close to the critical value 1.69. It seems that this model can be used quite successfully to build an evolution model of galaxy formation, where smaller objects aggregate to form successively larger structures as time progresses. Such a hierarchical model for CDM has been worked out in detail, and combined with some approximate modelling of star formation, radiation processes, supernovae, etc. to get an idea of how realistic galaxy formation might work [Kauffmann et al. 1999].

The formula for $n(M,z)$ can be used to estimate the mass density of objects $\Omega(M,z)dM$:

$$\Omega(M,z)dM = 3.60\ h^4 \frac{M}{10^{12} M_\odot} n(M,z)dM\ (\mathrm{Mpc})^{-3}. \tag{12.88}$$

The mass function $n(M,z)$ can also be written in terms of the circular velocity of bound objects

$$V_c = \sqrt{\frac{GM}{S}}.$$

If we take S to be the virial radius $r_{\mathrm{vir}} = (1/2)\ r_m(M,z)$ of dissipationless spherical collapse, it may be related to the comoving radius $r_0 = r_0(M)$.

With the defintion $\delta_{\max}(z) \equiv \frac{r_0(M)}{r_m(M,z)}$, we find

$$V_c = (GM/2r_m)^{1/2} = H_0 \Omega_m^{1/2} \delta_{\max}^{1/6} r_0(M) \tag{12.89}$$

and finally

$$n(V_c, z) dV_c = \frac{3}{(2\pi)^{2/3}} \delta_c H_0^2 \Omega_m \delta_{\max}^{1/3} \Delta^{-2}(M, z) \frac{d\Delta}{dr_0} \times \exp\left\{-\frac{\delta_c^2}{2\Delta^2}\right\} dV_c. \tag{12.90}$$

For arbitrary Ω_m and Ω_Λ the collapse factor $\delta_{\max}(z)$ has to be found numerically on the basis of the spherical non-linear model (see e.g. [Suto 1993]).

12.7 How Did Galaxies Form?

Galaxy formation is still a problem, both observationally and theoretically, despite intensive work and much progress in recent years. Proper treatment of this topic would require a whole book to itself. In this section I will just touch upon a few interesting aspects. I will make use of the very readable group report published in [Börner and Gottlöber, 1997]. There the discussions were organized around five questions, as follows.

12.7.1 What Does Galaxy Formation Mean?

This is actually a non-trivial question. So far we have considered the clustering of dark matter particles, but galaxies also consist of stars, gas and dust. The description of their formation cannot be complete without a model of star formation, itself one of the most challenging problems in astrophysics. Furthermore, the well-defined shapes of these composite objects must have been assembled in the course of time. Which feature marks the appearance of a galaxy? An astronomer looking for high-redshift galaxies would look for objects in which the first 1% of stars and the first significant amount of metals have formed. An observer trying to understand the Hubble sequence would find it most important to investigate objects where the first 50% of the stars have formed. Each would give quite a different definition of galaxy formation. It seems reasonable not to restrict ourselves to a given definite model.

Let us accept the idea that the properties of galaxies should be traceable back to initial density fluctuations present after recombination. The action of gravity is quite well understood, but there remains the important question how the visible appearance of galaxies is affected by the non-linear physical processes of radiation and hydrodynamics.

Observations of galaxies at $z > 0.5$ with the Keck and Hubble telescopes indicate that many large galaxies have acquired a morphology similar to present-day galaxies by $z \geq 1$. On the other hand, numerous faint blue galaxies may have a rather different morphology. The observation of the Lyman break galaxies at $z \geq 4$ may have a good deal more to tell us about galaxy formation. Perhaps a rapid burst of star formation marks the onset of galaxy formation [Partridge and Peebles 1967], or rather – as seems more likely according to recent observations – a series of modest starbursts.

An alternative approach is not to ask about individual galaxies, but rather to consider the entire population of diffuse, optically luminous objects as it changes with redshift. We might identify the phenomenon of galaxy formation through the evolution of the luminosity function, or the integrated metal content in the universe. Vigorous star formation, for example, may lead to a peak in the luminosity function.

The initial question should perhaps be replaced by more observationally driven questions. Did stars form before or after galaxies acquired their gross morphology? When, and in what order, were the major components of galaxies assembled?

12.7.2 What Sets the Observed Distributions of Galaxy Properties?

Is there something like the Hertzsprung–Russell diagram for galaxies? The tight correlation between luminosity and colour for stars played a strong role in the understanding of stellar structure. Similar regularities among the properties of galaxies have not yet been found. Most analysis over the last two decades has concentrated on the statistics of simple properties such as absolute magnitude, colour, metallicity, velocity dispersion (for bulges) or neutral hydrogen line widths (for gaseous discs), and sizes. We have mentioned in Chap. 2 the "fundamental plane" of ellipticals, and the Tully–Fisher relation for disc galaxies.

One well-known relation that has a tentative explanation in the context of hierarchical clustering models of galaxy formation is the optical luminosity distribution of galaxies. This distribution, the "luminosity function" [Schechter 1976] is fitted well by the 3-parameter function

$$\phi(L)dL = \phi^*(L/L_*)^\alpha \exp(-L/L^*)dL/L^*,$$

where $\phi(L)$ is the number of galaxies with luminosities between L and $L+dL$; L^* is a characteristic luminosity; α determines the relative numbers of faint galaxies ($L \ll L^*$); and ϕ^* sets the overall normalization. These parameters are measured in several different surveys (see Chap. 2). What is the physics which determines these parameters? Hierarchical clustering alone cannot account for it: high-resolution N-body simulations of CDM (as well as analytical models) show that hierarchical clustering leads to a mass distribution of dark haloes which follows the Schechter form. However, the characteristic mass M^* is of the order of $10^{13} M_\odot$, the mass of a galaxy group. Thus, characteristic galaxy masses cannot be explained by gravity alone. Furthermore, considering the baryons, we can easily see that gravity alone is not sufficient: the entropy of the baryonic gas

$$S = (T/\mathrm{keV})(n/\mathrm{cm}^{-3})^{-2/3}$$

cannot be changed by gravity. But the primordial (post-recombination) value is $S \simeq 10^{-6}$. The intergalactic medium in clusters has $S \simeq 10$, the warm interstellar medium $S \simeq 10^{-3}$, and for a star like the Sun the central entropy is $S \simeq 10^{-13}$.

The primordial baryonic gas must be shock-heated to raise its entropy, and then it must cool efficiently to settle into galaxies and stars. A detailed description of the complex physical processes involved is not possible at the moment. Some indication of the necessity of detailed feedback mechanisms is given by semi-analytic and numerical model calculations. In many cases the cooling processes are so efficient that most of the baryonic gas condenses out at early epochs. On the other hand, the upper mass limits for hot gas clumps that can cool within the Hubble time $1/H_0$ are too large, $\simeq 10^{13} M_{\odot}$, to explain the characteristic mass scale. The feedback mechanism which must operate to regulate galaxy formation will include photo-ionization of the intergalactic medium by stars and active galactic nuclei, supernova heating, stellar winds, etc. Perhaps observations can eventually help to identify important feedback processes.

Another indication of the need for feedback processes is the faint-end slope α of the galaxy luminosity function. Plausible hierarchical clustering models all seem to have too many faint galaxies left over from earlier times, i.e. too negative a value of α. When models are normalized to fit the abundances of clusters and massive galaxies, then both the Press–Schechter analytic scheme and N-body simulations predict $\simeq 10$ times too many galaxies today with $v_c \simeq 100$ kms^{-1}.

Another problem in this context is the explanation of the morphology–density relation. Why are ellipticals more abundant in a dense environment?

It seems that answers to these and many other questions require the identification of a feedback mechanism on galactic scales.

12.7.3 What Does Galaxy Evolution Tell Us About Galaxy Formation?

Information on this topic is provided by the observations of high-z galaxies with the HST and large, ground-based telescopes like the Keck and the VLT. There have been remarkable findings of an excess of morphologically peculiar faint blue galaxies. The red early-type galaxies show much less evolution in morphology and number. The spectacular discovery of several hundred Lyman break galaxies has revealed that at $z \simeq 3.5$ galaxy formation was well under way.

Where are the remnants of the faint blue galaxies today? Does the population of "normal" galaxies seen in deep surveys contain all or only a fraction of the galaxies seen nearby?

12.7.4 What Is Galaxy Bias?

The luminous parts of galaxies represent only a small fraction of the total mass in the universe. Do galaxies provide a representative sample of the mass? The possibility that galaxies may be distributed differently from the mass must be faced. We speak of "biased galaxy formation" in this context. Biases are statistical differences arising when different populations of objects are compared. Thus biases involve two factors: the selection of the objects and the measurement to be performed. Here we would be interested in comparing the average distribution of the mass with the clustering and the motions of galaxies. A nice illustration is contained in a recent paper ([Jing et al. 2002]), where the statistics of galaxies selected in the infrared with the IRAS satellite are compared to optically selected galaxies from the LCRS. It turns out that the infrared galaxies have a bias quite different from the optically selected ones. The bias model used here was derived from the consideration of giving less weight to dense clusters when the two-point correlation function is calculated. In the CDM simulation the dark matter particles from massive haloes are removed randomly such that a reduction of particle number N with halo mass M of the type

$$N \propto M^{-\alpha}$$

occurs. The parameter α is different for IRAS and LCRS galaxies. This so-called cluster-underweight bias is a promising scheme, and it works better in many cases than the so-called high peaks model [Bardeen et al. 1986]. This supposes that bright galaxies form at the maxima of the initial density fluctuation field smoothed on some scale (usually a few Mpc in order to obtain the correct number density of peaks as galaxies). The two-point correlation function of such peaks has a higher amplitude than the underlying density field, if regions above some threshold for the overdensity are considered [Kaiser 1984]. If only peaks above a high threshold are selected, the result for a Gaussian random field is

$$\xi_{gg}(r) = \exp\left[b^2 \xi_{mm}(r)\right] - 1,$$

where ξ_{gg} is the two-point correlation function for the galaxies, ξ_{mm} for the mass. On large scales, where $|\xi_{gg}(r)| \ll 1$, the result is

$$\xi_{gg}(r) \approx b^2 \xi_{mm}(r).$$

The constant of proportionality is the "bias factor". This peaks bias model was first introduced by Kaiser to account for the stronger clustering of Abell clusters compared with galaxies.

Another bias model supposes that the relative fluctuations of galaxies and mass are proportional:

$$\frac{\delta n_g(\boldsymbol{x})}{n_g(\boldsymbol{x})} = b(m)\frac{\delta m}{m}.$$

Here the galaxy number density is represented as a continuous field $n_g(\boldsymbol{x})$ (e.g. by smoothing the point distribution with some window function). If b is a constant this is called the "linear bias model", if b is a function of m this is called the "local bias model". Such statistical models of biased galaxy formation can only be simplified approaches to the true situation. Galaxies are not just markers on a fixed Gaussian background field. They form through non-linear gravitational and gas dynamic processes starting from small fluctuations. This non-linear evolution itself generates a bias of the dense clumps relative to the mass. An interesting question which may soon be answered is whether there are physically realistic models for galaxy formation that predict strong bias.

12.7.5 How Are Protogalaxies Related to the Intergalactic Medium at High Redshift?

Absorption lines seen against background quasars provide a powerful probe of gas at high redshift. The Lyman α forest, Lyman limit systems, and damped Lyman α systems produce a growing amount of detailed information about galaxy and star formation at high redshift. It seems that the Lyman α forest can be formed with approximately the correct column density distribution (for which photo-ionization is especially important) by the web of structure in a hierarchical clustering model. This model was originally suggested by [Bi 1993], and rediscovered several years later.

The metallicity of the Lyman α forest "clouds" is about 1% of the solar one. This pollution of the hydrogen gas with metals may have come from first stars which formed in clumps much smaller than a galaxy in the cosmic web. This points to an onset of star formation even before galaxies assembled. From here on, we go into the wide topic of galaxy formation which requires more than one book to adequately describe its many facets. It is a fascinating field, but a textbook on cosmology should stop before it becomes too deeply entangled in it.

Exercises

12.1 Calculate the entropy of the baryonic gas for the primordial (post-recombination) epoch, for the intergalactic medium in clusters, for the warm interstellar medium, and for the centre of a star like the Sun.

12.2 Show that the Zel'dovich approximation is an exact solution in one spatial dimension.

12.3 Prove that $w(r_p)$, as it is defined in equation (12.44), is related to the real-space correlation function by (12.45).

12.4 The two- and three-point correlation function, as well as the pairwise velocity dispersion, can be measured for the LCRS (Sect. 12.4). Do they satisfy the cosmic virial theorem? Try to do the calculations using the values given in Sect. 12.4, and discuss the result.

12.5 The cluster-under-weight bias model presented in Sect. 12.4.2 can be compared to the peaks bias model of Sect. 12.7.4, and to the bias model of [Mo and White 1996]. Find a way to present the different schemes graphically, and discuss the differences.

12.6 Show that from linear perturbation theory, and with a linear galaxy bias, the redshift power spectrum $P_l^S(k,\mu)$ is related to the linear power spectrum in real space $P_l^R(k)$ by

$$P_l^S(k,\mu) = P_l^R(k)[1+\beta\mu^2]^2;$$

μ is the cosine of the angle between the wave vector k and the line of sight; $\beta = \Omega_m^{0.6}/b_l$, where b_l is the linear bias parameter, Ω_m the cosmic density parameter.

A. The Gauge-Invariant Theory of Perturbations

The discussion here follows [Kodama and Sasaki 1984; Durrer and Straumann 1988; Durrer 1994].

A.1 The "3+1" Formalism

Consider a space-time (M, g) that allows the choice of a time coordinate t, and therefore the choice of a family of constant-time hypersurfaces Σ_t. In other words, the space-time admits a one-parameter family of embeddings i_t: $\sigma \to M$, such that the slices $\Sigma_t \equiv i_t(\Sigma)$ are space-like, and the curves $i_t(x)$, $x \in \Sigma$ (fixed) are time-like (Fig. A.1).

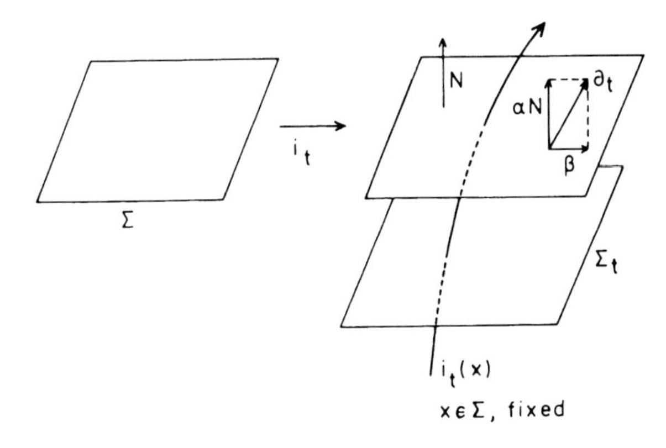

Fig. A.1. The choice of a family of hypersurfaces Σ_t involves slicing the space-time such that Σ_t are space-like and $i_t(x)$ are time-like curves

The time-like vector field ∂_t, defined by these curves, can be decomposed into normal and parallel components relative to the hypersurfaces Σ_t:

$$\partial_t = \alpha N_t + \beta.$$

Here N_t is the unit normal of Σ_t. α is the "lapse function" which represents the ratio of the proper time interval to the coordinate-time interval Δt between two hypersurfaces Σ_t and $\Sigma_{t+\Delta t}$. β is the "shift vector" which measures the rate of deviation of a line with a constant space coordinate from a line normal to Σ_t hypersurfaces.

The metric γ_t induced on Σ_t and its second fundamental form K_t are related by

$$\partial_t \gamma_t = -2\alpha K_t + L_\beta \gamma_t,$$

where L_β is the Lie derivative with vector β.

The Einstein and Ricci tensors of (M, g) can be split into components parallel and normal to N_t. The equations of Gauss and Codazzi-Mainardi give (suppressing the suffix t):

$$\perp G_{\mu\nu} \perp = \frac{1}{2}[R(\gamma) + (Tr\{K\})^2 - Tr\{K^2\}],$$
$$\| G_{\mu\nu} \perp = \nabla(Tr\{K\}) - \nabla \cdot K.$$

With Einstein's field equations $G_{\mu\nu} = 8\pi G T_{\mu\nu}$, these give rise to the so-called *constraints.* They do not contain second-order t-derivatives and they are independent of the lapse function α and the shift-vector field β. The *dynamical* equations are obtained from the components

$$|R_{\mu\nu}(g)| = R_{\mu\nu}(\gamma) + (Tr\{K\})K - K \cdot K$$
$$-\frac{1}{\alpha}(\partial_t K - L_\beta K) - \frac{1}{\alpha}\mathrm{Hess}(\alpha).$$

A.1.1 Unperturbed Solutions

For FL background models one has, in standard coordinates $\alpha = 1, \beta = 0$, and $\gamma = R^2(t)h$, with the scale factor $R(t)$ and metric h_{ij} such that (Σ, h) is a 3-dimensional Riemannian manifold with constant curvature $K = 0, \pm 1$. The energy-momentum tensor must be of the perfect fluid type

$$T_{\mu\nu} = (\varrho + p)u_\mu u_\nu + p g_{\mu\nu}.$$

In addition, there are matter variables for the various components such as baryons, neutrinos, axions, and strings, which contribute to the density ϱ and the pressure p.

For $\alpha = 1, \beta = 0$, and the perfect-fluid energy–momentum tensor, the equations above give rise to

$$\frac{6K}{R^2} + 6\frac{\dot{R}^2}{R^2} = 16\pi G\varrho$$

and

$$\tilde{R} = -\frac{4\pi}{3}(\varrho + 3p)R.$$

A.1.2 Small Perturbations

Linearize all the basic equations in the usual manner, $g = g^{(0)} + \delta g$ etc. In the linearization scheme the infinitesimal change in δg corresponding to the infinitesimal diffeomorphism (coordinate transformation) $x \to x + \xi$ is

$$\delta g \to \delta g + L_\xi g^{(0)}.$$

For the matter variables, for example,

$$\delta \varrho \to \delta \varrho + L_\xi \varrho^{(0)} \equiv \delta \varrho + \xi \cdot \nabla_\xi \varrho^{(0)}.$$

These transformations lead to corresponding changes in the perturbed quantities $\delta\alpha, \delta\beta, \delta\gamma, \delta K$, etc. The aim of a gauge-invariant formalism is the construction of quantities which are unchanged by these gauge transformations of the background. The gauge-invariant variables are zero for such coordinate transformations of the background.

Since the background spaces (Σ, h) are homogeneous and isotropic one performs a harmonic analysis of the perturbations on Σ. The decomposition into "scalar", "vector", and "tensor" contributions in (Σ, h) is in general not unique. Only for finite Σ is the harmonic analysis straightforward. For infinite Σ (hyperbolic spaces) there are mathematical intricacies which we cannot discuss here.

Any vector V on Σ can be decomposed as

$$V = V^v + V^s$$

where V^s is a (smooth) gradient and V^v has vanishing divergence. Similarly, any symmetric, second-rank tensor t can be decomposed as

$$t = t^t + t^v + t^s,$$

with

$$\begin{aligned} t^s_{ij} &= Tr\{t\}h_{ij} + (\nabla_i \nabla_j - \frac{1}{3}\delta_{ij}\Delta)f, \\ t^v_{ij} &= \nabla_i X_j + \nabla_j X_i, \quad \nabla \cdot X = 0, \\ Tr\{t^t\} &= 0, \qquad \nabla \cdot t^t = 0. \end{aligned}$$

t^s, V^s, and scalars are called "scalar-type" constitutents; t^v, V^v "vector-type", and t^t "tensor-type".

In a FL geometry these decompositions lead to a decoupling of covariant, linear, (at most) second-order differential equations, such that equations result that contain only components of one type. Only the scalar modes are interesting for the problems of galaxy formation, because vector and tensor modes do not contribute to density perturbations.

A.2 Perturbations of $g_{\mu\nu}$ and $T_{\mu\nu}$

The background space-time is described by the Robertson–Walker metric (cf. Chap. 1)

$$ds^2 = -dt^2 + R^2(t)d\sigma^2,$$

where $d\sigma^2 = h_{ij}dx^i dx^j$.

Scalar quantities are expanded in terms of scalar harmonic functions $Y(x)$ which satisfy the equation

$$(\Delta + k^2)Y(x) = 0. \tag{A.1}$$

k^2 represents an eigenvalue of the operator Δ on Σ. Scalar-type components of vectors on Σ are expanded in terms of

$$Y_i = -k^{-1}\nabla_i Y. \tag{A.2}$$

Scalar-type components of tensors can be written as

$$Y_{ij} = k^{-2}\nabla_i\nabla_j + \frac{1}{3}h_{ij}Y. \tag{A.3}$$

The homogeneity and isotropy of Σ guarantee that there is no coupling between the expansion coefficients of different eigenfunctions in linear perturbation coefficients. Hence we can discuss the equations for one representative mode with eigenvalue k. The functions $Y_k(x)$ do not appear in their explicit complicated shape in the following. k corresponds to the wave number, i.e. to $2\pi/\lambda$, of the perturbation.

For the scalar perturbation the perturbed metric tensor $\tilde{g}_{\mu\nu}$ can generally be expressed in terms of four independent functions a, b, H_L and H_T:

$$\begin{aligned}\tilde{g}_{00} &= -R^2(1 + 2aY);\\ \tilde{g}_{0j} &= -R^2 bY_j;\\ \tilde{g}_{ij} &= R^2[h_{ij} + 2H_L Y h_{ij} + 2H_T Y_{ij}].\end{aligned} \tag{A.4}$$

a is the perturbation of the lapse function, b the amplitude of the shift vector perturbation. H_L corresponds to the amplitude of the perturbation of a unit spatial volume, H_T describes anisotropic distortions of each constant time hypersurface.

The perturbed energy-momentum tensor $\tilde{T}^{\mu\nu}$ can be characterized by its algebraically independent components. The starting point is the definition of the perturbed $(3+1)$ velocity $\tilde{u}^\mu$ of matter as the time-like eigenvector with unit norm of $\tilde{T}^{\mu\nu}$, and of the perturbed proper density $\tilde{\varrho}$ as the corresponding eigenvalue:

$$\begin{aligned}\tilde{T}^\mu_\nu \tilde{u}^\nu &= -\tilde{\varrho}\tilde{u}^\mu,\\ \tilde{u}_\mu\tilde{u}^\mu &= -1.\end{aligned} \tag{A.5}$$

The remaining freedom can be described by the spatial stress tensor $\tilde{\tau}_{\mu\nu}$, which is orthogonal to $\tilde{u}^{\mu}$:

$$\tilde{\tau}_{\mu\nu} \equiv (\delta^{\alpha}_{\mu} + \tilde{u}_{\mu}\tilde{u}^{\alpha})(\delta^{\beta}_{\nu} + \tilde{u}_{\nu}\tilde{u}^{\beta})\tilde{T}_{\alpha\beta}, \qquad \tilde{u}^{\mu}\tilde{\tau}_{\mu\nu} = 0. \tag{A.6}$$

Since $\tilde{\varrho}$ is a scalar it can be written as

$$\tilde{\varrho} = \varrho(1 + \delta Y); \tag{A.7}$$

δ is the amplitude of the density perturbation.

Further,

$$\frac{\tilde{u}^i}{\tilde{u}^0} = vY^i, \tag{A.8}$$

$$\begin{aligned} \tilde{\tau}^0_0 &= 0, \\ \tilde{\tau}^0_j &= p(v - b)Y_j, \\ \tilde{\tau}^j_0 &= -pvY_j, \\ \tilde{\tau}^i_j &= \tilde{T}^i_j, \quad i = 1, 2, 3. \end{aligned} \tag{A.9}$$

Since $\tilde{T}^i_j$ is a second-rank symmetric tensor with respect to spatial coordinate transformations, it can be expressed as

$$\tilde{T}^i_j = p(\delta_j + \Pi_L Y \delta^i_j + \Pi_T Y^i_j). \tag{A.10}$$

Π_L and Π_T obviously have the properties of an isotropic and an anisotropic pressure perturbation, respectively.

Finally, one obtains the expressions

$$\begin{aligned} \tilde{T}^0_0 &= -\varrho(1 + \delta Y), \\ \tilde{T}^0_j &= (\varrho + p)(v - b)Y_j, \\ \tilde{T}^j_0 &= -(\varrho + p)vY^j. \end{aligned} \tag{A.11}$$

A.3 Gauge-Invariant Variables

Infinitesimal coordinate transformations of scalar type

$$\begin{aligned} \bar{\eta} &= \eta + TY, \\ \bar{x}^j &= x^j + LY^j \end{aligned}$$

(where T, L are arbitrary functions of time, and η is the conformal time coordinate $d\eta \equiv dt/R(t)$) produce corresponding gauge transformations in the perturbed variables. To first order, one obtains ($T', L' \equiv dT/d\eta, dL/d\eta$)

$$\begin{aligned}
\bar{a} &= a - T' - \left(\frac{R'}{R}\right) T, \\
\bar{b} &= b + L' + kT, \\
\bar{H}_L &= H_L - \left(\frac{k}{3}\right) L - \left(\frac{R'}{R}\right) T, \\
\bar{H}_T &= H_T + kL.
\end{aligned} \tag{A.12}$$

$\sigma_g \equiv k^{-1} H_T' - b$ is the amplitude of the shear perturbation.

The gauge transformations contain two arbitrary functions of time, and therefore two independent gauge invariants can be constructed from a, b, H_L, and H_T. One possible choice of two such invariants is

$$\begin{aligned}
A &\equiv a - R^{-1} \left[\left(\frac{R^2}{R'}\right) \left(H_L + \frac{1}{3} H_T\right) \right]', \\
B &= b + \left(\frac{R'}{R}\right)^{-1} k \left(H_L + \frac{1}{3} H_T\right) - k^{-1} H_T'.
\end{aligned} \tag{A.13}$$

Any linear combination of A and B and their time derivatives with arbitrary functions of time as coefficients is also gauge-invariant, of course. An example is given by Bardeen's invariants [Bardeen 1980] Φ and Ψ defined as

$$\begin{aligned}
\Phi &\equiv k^{-1} \left(\frac{R'}{R}\right) B, \\
\Psi &\equiv A + (kR)^{-1} (RB)'.
\end{aligned} \tag{A.14}$$

The geometrical interpretation of Φ and Ψ is especially simple: in a slicing where the shear σ_g of the unit-normal vector field $\tilde{N}_\mu$ is zero ($\sigma_g = 0$), Φ represents the amplitude of the perturbation in the intrinsic curvature of a constant time hypersurface $\tilde{\Sigma}$, Ψ the perturbation amplitude in the gravitational potential (i.e. of a quantity which becomes the gravitational potential in the Newtonian limit). The matter variables have the following gauge-transformation properties

$$\begin{aligned}
\bar{v} &= v + L', \\
\bar{\delta} &= \delta + (1+w) 3 \frac{R'}{R} T, \quad \text{where } w = p/\varrho, \\
\bar{\Pi}_L &= \Pi_L + 3 \frac{c_s^2}{w} (1+w) \frac{R'}{R} T, \quad \text{where } c_s^2 = \frac{p'}{\varrho'}. \\
\bar{\Pi}_T &= \Pi_T.
\end{aligned} \tag{A.15}$$

Obviously two gauge invariants can be constructed from the matter variables alone:

$$
\begin{aligned}
\Pi &\equiv \Pi_T, \\
\Gamma &\equiv \Pi_L - \frac{c_s^2}{w}\delta, \\
\Pi_L &\equiv \delta p, \quad \frac{\delta\varrho}{\varrho} \equiv \delta.
\end{aligned}
\tag{A.16}
$$

For adiabatic perturbations, $\delta p/\delta\varrho = c_s^2, \Gamma = 0$, i.e. Γ describes an entropy perturbation. We see that entropy perturbations and adiabatic perturbations are gauge-invariant concepts.

Gauge-invariant quantities corresponding to v and δ consist of a combination of metric and matter variables. It turns out that a very convenient choice is to use

$$
V \equiv v - k^{-1}H_T'. \tag{A.17}
$$

V is the amplitude of the shear of the matter motion. In the slicing where $\sigma_g = 0$ one has $V = v - b$, i.e. V represents the velocity of matter relative to observers moving along the normal lines.

There does not exist a unique natural definition of a gauge-invariant density perturbation. Various combinations are possible which represent the density contrast for different choices of slicing. A convenient choice leading to a simple form of the Einstein equation is

$$
\Delta \equiv \delta + 3(1+w)\frac{R'}{R}k^{-1}(v - b). \tag{A.18}
$$

Δ represents the density contrast in the slicing which is chosen such that the matter (3+1) velocity is orthogonal to the hypersurfaces of constant time. For wavelengths much smaller than the horizon size, $R'/(Rk) \ll 1$, the ambiguity in the definition of Δ disappears, and it is simply equal to the density contrast δ.

From the perturbed Einstein equation the Bardeen variables Φ and Ψ are expressed algebraically in terms of matter variables:

$$
2(k^2 - 3K)\Phi = 8\pi G R^2 \varrho\Delta, \tag{A.19}
$$

$$
k^2(\Phi + \Psi) = -8\pi G R^2 p\Pi. \tag{A.20}
$$

When $\Pi = 0$, as for ideal fluids, we find

$$
\Psi = -\Phi;
$$

and for $K = 0$,

$$
-k^2\Psi = 4\pi G R^2 \varrho\Delta.
$$

This is the equation for the Newtonian gravitational potential of the perturbation.

A.4 Linearized Einstein Equations for Gauge-Invariant Variables

The constraint equations yield two algebraic relations between the metric and the matter variables:

$$8\pi G \varrho \Delta = 2\frac{R'}{R}\frac{k}{R^2}\left(1 - \frac{3K}{k^2}\right) B,$$

$$8\pi G(\varrho + p)V = 2\frac{R'}{R}\frac{k}{R^2}A + \left\{8\pi G(\varrho + p) - 2\frac{K}{R^2}\right\} B. \qquad \text{(A.21)}$$

These equations can be used to eliminate A and B from the dynamical equations, resulting in a set of equations for the matter variables alone:

$$\Delta' - 3w\frac{R'}{R}\Delta = -\left(1 - \frac{3K}{k^2}\right)(1 + w)kV - 2\left(1 - \frac{3K}{k^2}\right)\frac{R'}{R}w\Pi,$$

$$V' + \frac{R'}{R}V = -k\left[\frac{1}{2}8\pi G\frac{R^2\varrho}{k^2 - 3K} - \frac{c_s^2}{1 + w}\right] + k\frac{w}{1 + w}\Gamma \qquad \text{(A.22)}$$

$$-k\left[\frac{2}{3}\left(1 - \frac{3K}{k^2}\right)\frac{1}{1 + w} + 8\pi G\frac{R^2\varrho}{k^2}\right] w\Pi.$$

Equations (A.22) can be combined to give a second-order differential equation for Δ:

A little bit of mathematics is unavoidable

$$\Delta'' - \{3(2w - c_s^2) - 1\}\frac{R'}{R}\Delta' + 3\left[\left\{\frac{3}{2}w^2 - 4w - \frac{1}{2} + 3c_s^2\right\}\left(\frac{R'}{R}\right)^2 + \frac{3w^2 - 1}{2}K + \frac{k^2 - 3K}{3}c_s^2\right]\Delta = S, \tag{A.23}$$

where

$$S \equiv -(k^2 - 3K)w\Gamma - 2\left(1 - \frac{3K}{k^2}\right)\frac{R'}{R}w\Pi' + \left[\{3(w^2 + c_s^2) - 2w\}\left(\frac{R'}{R}\right)^2 + w(3w + 2)K + \frac{k^2 - 3K}{3}c_s^2\right] \times \left(1 - \frac{3K}{k^2}\right)2\Pi. \tag{A.24}$$

From (A.23) and (A.24) we see that entropy (Γ) and anisotropic stress (Π) perturbations act as sources for density perturbations Δ. They must be specified as functions of time or in terms of Δ and V. Often Γ and Π appear as a consequence of the intrinsic structure of matter. In the case of, for example, a multi-component structure, Γ and Π are given by density and velocity perturbations of the components.

A.5 The General Solution

The general solution of the perturbation equations for the total matter density Δ has been discussed in many papers. A very clear exposition is given by [Hu 1995]. We just quote a few of the results. The growing solution can be written as a combination

$$\Delta = \Phi(0)U_\Lambda + S(0)U_I; \tag{A.25}$$

U_Λ and U_I are functions of $R(t)$ specifying the time evolution of the solutions of (A.23) and (A.24) for the case of adiabatic ($\Gamma = 0$) and entropy perturbations ($\Phi(0) = 0$), also called isocurvature perturbations. Two quantities, the initial curvature perturbation $\Phi(0)$ and the initial entropy fluctuation, are sufficient to specify the growing solution. Any mixture of adiabatic and isocurvature modes is, of course, also possible.

For a mixture of photons, baryons, and cold dark matter, the entropy becomes

$$S = \Delta_m - \frac{3}{4}\Delta_r = \left(1 - \frac{\Omega_B}{\Omega_0}\right)\Delta_{\mathrm{CDM}} + \frac{\Omega_B}{\Omega_C DM}\Delta_B - \frac{3}{4}\Delta_\gamma. \tag{A.26}$$

Here $\Delta_{\rm CDM}, \Delta_B, \Delta_\gamma$ are the perturbed energy densities of CDM, baryons, and photons, respectively. The implications for the adiabatic mode are as follows. During the radiation (RD), and matter (MD) dominated epochs the total density fluctuation takes the form

$$\Delta/\Phi(0) \begin{cases} \frac{4}{3}\left(\frac{k}{k_{eq}}\right)^2\left(1-\frac{3K}{k^2}\right)R^2 & \text{(RD)}, \\ \frac{6}{5}\left(\frac{k}{k_{eq}}\right)^2\left(1-\frac{3K}{k^2}\right)R & \text{(MD)}. \end{cases} \tag{A.27}$$

Moreover, since $S=0$, the components evolve together:

$$\Delta_{\rm CDM} = \Delta_B = \frac{3}{4}\Delta_\gamma.$$

The velocity and potential Ψ are given by

$$V/\Phi(0) = \begin{cases} -\frac{\sqrt{2}}{2}\left(\frac{k}{k_{eq}}\right)R & \text{(RD)}, \\ -\frac{3\sqrt{2}}{5}\left(\frac{k}{k_{eq}}\right)\sqrt{R} & \text{(MD)}; \end{cases} \tag{A.28}$$

$$\Psi/\Phi(0) = -\Phi/\Phi(0) = \begin{cases} -1 & \text{(RD)}, \\ -\frac{9}{10} & \text{(MD)}. \end{cases}$$

The temperature perturbation in the Newtonian gauge can be derived from the definition

$$\Theta_0 = \frac{\Delta_\gamma}{4} - \frac{\dot{R}}{R}\frac{V}{k}: \tag{A.29}$$

$$\Theta_0/\Phi(0) = \begin{cases} \frac{1}{2} & \text{(RD)}, \\ \frac{3}{5} & \text{(MD)}. \end{cases} \tag{A.30}$$

For the isocurvature perturbations in a purely baryonic case,

$$\Delta/S(0) = \begin{cases} \frac{1}{6}(\frac{k}{k_{eq}})^2(1-\frac{3K}{k^2})R^3 & \text{(RD)}, \\ \frac{4}{15}(\frac{k}{k_{eq}})^2(1-\frac{3K}{k^2})R & \text{(MD)}; \end{cases} \tag{A.31}$$

$$\Delta_B = \frac{1}{4+3R}(4S+3(1+R)\Delta);$$

$$\Delta_\gamma = \frac{4}{3}(\Delta_B - S).$$

Therefore,

$$\Delta_B/S(0) = \begin{cases} 1-\frac{3}{4}R & \text{(RD)}, \\ \frac{4}{3}\left[R^{-1}+\frac{1}{5}\left(\frac{k}{k_{eq}}\right)^2\left(1-\frac{3K}{k^2}\right)R\right] & \text{(MD)}; \end{cases} \tag{A.32}$$

$$\Delta_\gamma/S(0) = \begin{cases} -R & \text{(RD)}, \\ \frac{4}{3}\left[-1+\frac{4}{15}\left(\frac{k}{k_{eq}}\right)^2\left(1-\frac{3K}{k^2}\right)R\right] & \text{(MD)}; \end{cases} \tag{A.33}$$

$$V/S(0) = \begin{cases} -\frac{\sqrt{2}}{8}\left(\frac{k}{k_{eq}}\right)R^2 & \text{(RD)}, \\ -\frac{2\sqrt{2}}{15}\left(\frac{k}{k_{eq}}\right)\sqrt{R} & \text{(MD)}; \end{cases} \tag{A.34}$$

$$\Psi/S(0) = -\Phi/S(0) = \begin{cases} -\frac{1}{8}R & \text{(RD)}, \\ -\frac{1}{5} & \text{(MD)}; \end{cases} \tag{A.35}$$

$$\Theta_0/S(0) = \begin{cases} -\frac{1}{8}R & \text{(RD)}, \\ -\frac{1}{5} & \text{(MD)}. \end{cases} \tag{A.36}$$

B. Recent Developments First Year Results from WMAP

B.1 Temperature Anisotropies

In 2003 the results obtained during the first year of operation of the MAP satellite were published in a special issue of the Supplements of the Astrophysical Journal (Volume 148). The MAP mission was renamed WMAP in honor of David T. Wilkinson, one of the pioneers of CMB research, who died just before completion of the satellite experiment. Figure B.1 is a representative sky map, in which maps in the five frequency bands of WMAP (23, 33, 41, 61, and 94 GHz) are combined, and the foregrounds have been eliminated as far as possible. So this figure gives an impression of the CMB sky structure. The multipole analysis of the maps results in an angular power spectrum, as it is described in Chap. 11.

The top panel 1 Fig. B.2 displays this quantity. The clear detection of acoustic peaks (and of temperature-polarization correlations, see lower panel) supports the concept of adiabatic primordial fluctuations. The WMAP data alone can already accurately determine many of the cosmological parameters [Spergel et al. 2003]. Considering a flat universe with radiation, baryons, cold dark matter and cosmological constant, and a power-law spectrum of adiabatic primordial fluctuations, the WMAP data alone lead to values for the baryon density $\Omega_{\mathrm{B}} h^2 = (0.024 \pm 0.001)$, the matter density $\Omega_{\mathrm{CDM}} h^2 = (0.14 \pm 0.02)$, Hubble constant $h = (0.72 \pm 0.05)$, and spectral index $n = (0.99 \pm 0.04)$. A combination of the WMAP results with the COBE determination of the CMB temperature, the CMB measurements of the CBI and ACBAR experiments, and the 2dF survey's galaxy power spectrum lead to a "best fit cosmological model" [Bennett et al. 2003, Spergel et al. 2003]. For this fit the total density is

$$\Omega_{\mathrm{tot}} = 1.02 \pm 0.002. \tag{B.1}$$

The baryon density

$$\Omega_{\mathrm{B}} h^2 = 0.0224 \pm 0.0009; \tag{B.2}$$

the dark matter density

$$\Omega_{\mathrm{CDM}} h^2 = 0.135 \pm 0.009; \tag{B.3}$$

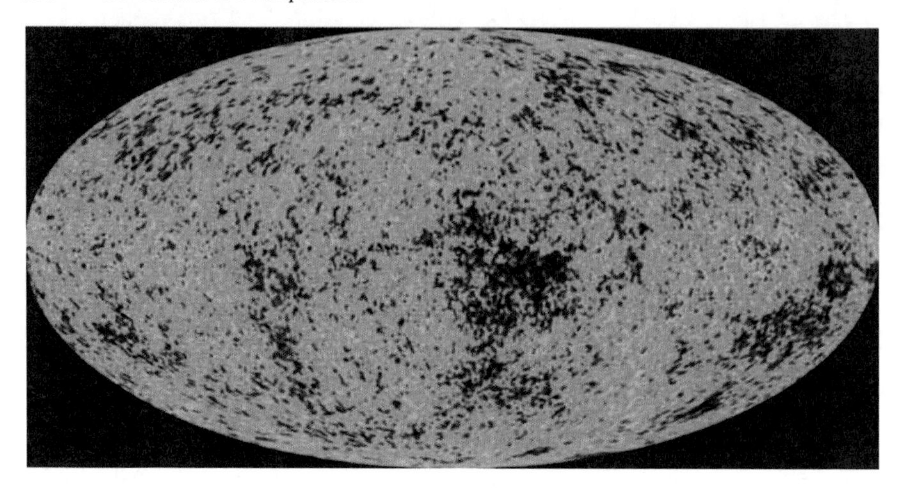

Fig. B.1. A sky map showing the cosmic CMB temperature fluctuations as measured by WMAP. This so-called "internal linear combination" map combines maps in the five frequency bands in such a way as to maintain unity response to the CMB while minimizing foreground contamination. The angular resolution is about 12′ [Bennett et al. 2003; courtesy of the WMAP team]

then the dark energy density

$$\Omega_\Lambda = 0.73 \pm 0.04; \tag{B.4}$$

here the Hubble constant h has been used

$$h = 0.71 \pm 0.04. \tag{B.5}$$

Other quantities that can be determined are the fluctuation amplitude in spheres of $8h^{-1}$Mpc radius

$$\sigma_8 = 0.84 \pm 0.04, \tag{B.6}$$

and the spectral index of the power spectrum

$$n = 0.93 \pm 0.03. \tag{B.7}$$

The redshift at decoupling is determined as

$$z_{\mathrm{dec}} = 1089 \pm 1, \tag{B.8}$$

the decoupling time

$$t_{\mathrm{dec}} = (379 \pm 8) \times 10^3 \,\text{years}, \tag{B.9}$$

and the spread of the decoupling epoch is

$$\Delta z = 195 \pm 2. \tag{B.10}$$

B.2 Polarization Signal

The isotropic CMB radiation field does not produce any polarization, but the temperature anisotropies (especially the quadrupole term) can lead to a detectable signal from Thomson scattering on matter inhomogeneities. The instruments aboard WMAP have measured the Stokes parameters I, Q, and V across the sky at frequencies of 23, 33, 41, 61, and 94 GHz. The incident radiation is split into two orthogonal polarization modes by orthomode

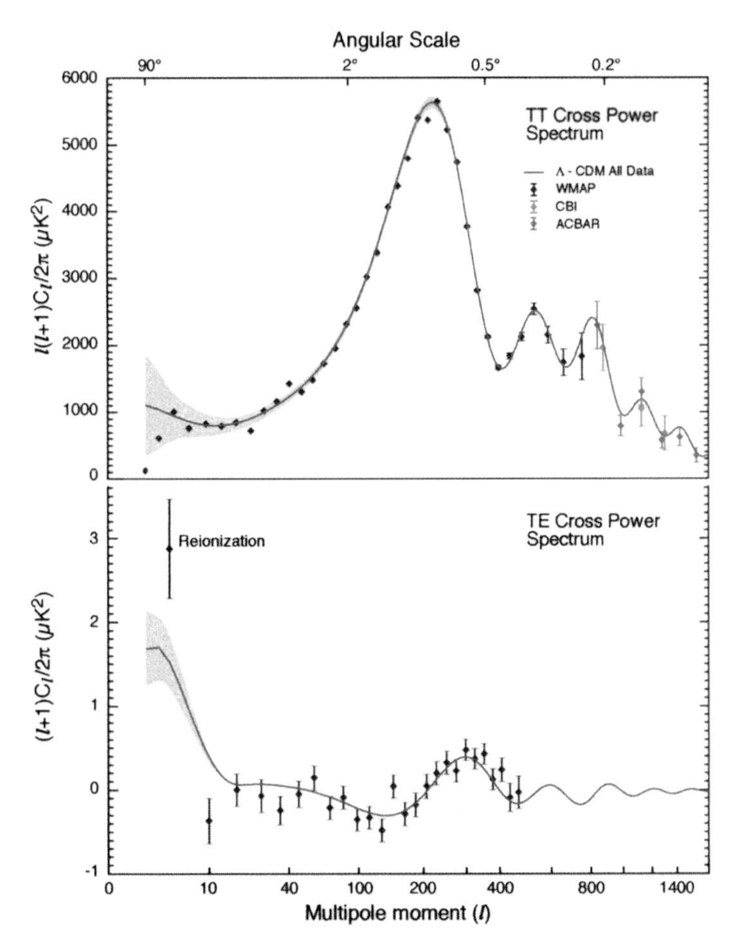

Fig. B.2. The WMAP angular power spectrum for temperature variations is shown in the top panel. The grey band represents the cosmic variance expected for the best-fit cosmological model (*solid line*). The quadrupole has a surprisingly low amplitude. Excursions from a smooth spectrum (e.g. at $\ell \simeq 40, \ell \simeq 210$) are only slightly larger than expected statistically. The bottom panel shows the temperature–polarization (TE) cross-power spectrum $(\ell + 1)c_e/2\pi$. The peak near $\ell \sim 300$ is out of phase with the temperature difference power-spectrum, as expected for adiabatic perturbations [Bennett et al. 2003; courtesy of the WMAP team]

transducers. Each orthogonal mode is registered by one of two independent radiometers ("±") giving a decomposition of the sky signal from direction $\underline{n}$ as

$$T_{\pm}(\underline{n}) = I(\underline{n}) \pm Q(\underline{n}) \cos 2\gamma \pm U(\underline{n}) \sin 2\gamma; \tag{B.11}$$

with a suitably defined angle γ (cf. Kogut et al. 2003). Each of the two radiometers measures the difference of the signal between two points in the sky separated by $\sim 140°$. The difference of the output of the two radiometers is then proportional to the polarization. Thus a sky map of the Q and U values is obtained. For the analysis of the first-year data of WMAP the two-point angular correlation function between temperature and polarization was evaluated

$$C^{IQ}(G) = \frac{\sum\limits_{ij} I_i Q'_j w_i w_j}{\sum\limits_{ij} w_i w_j}. \tag{B.12}$$

Unit weight ($w_i = 1$) was assigned to the CMB temperature anisotropy I_i. Q' and U' are coordinate independent quantities calculated for each pair of pixels by a rotation of the Q and U values. Q' is thus a linear combination of Q and U.

The lower panel in Fig. B.2 shows this temperature–polarization cross-correlation as a function of the multipole index ℓ.

There is excess power on large angular scales compared to the predictions based on the ΔT power spectrum alone. This excess power probably is due to an epoch of reionization in the universe at redshifts between 11 and 30, depending on the reionizing scenarios considered. A reasonably safe conclusion is that at these epochs the first stars must have formed. The optical depth for the CMB photons is estimated as $\tau = 0.17 \pm 0.04$, larger than expected given the detection of a Gunn–Peterson trough in the absorption spectra of distant quasars. This implies that the universe has a complex ionization history.

References

Aaronson, M., 1983, Astrophys. J., **266**, L11
Aarseth, S.J., 1984, in *Methods of Computational Physics*, ed. by J.U. Brackhill, B.J. Cohen (Academic, New York), p. 1
Aarseth, S.J., Gott, J.R., Turner, E.L., 1979, Astrophys. J., **236**, 43
Abbott, L.F., Farhi, E., Wise, M.B., 1982, Phys. Lett., **B117**, 29
Abe, F. et al. (CDF collab.), 1994, Phys. Rev. **D50**, 2966
Abraham, R.G. et al., 1996, Mon. Not. R. Astron. Soc. **279**, L47
Abrikosov, A.A., 1961, Sov. Phys.-JETP, **13**, 451
Aeppli, B., 1986, Diploma Thesis, Univ. Zürich
Alard, C., 1995, in *Astrophys. Applications of Gravitational Lensing*, Proc. IAU Symp. 173 (Boston, Kluwer; ed. by C.S. Kochanek, J.N. Hewitt)
Albrecht, A., Steinhardt, P., 1982, Phys. Rev. Lett., **48**, 1220
Albrecht, A., Steinhardt, P., Turner, M.S., Wilczek, F., 1982, Phys. Rev. Lett., **48**, 1437
Alcock, C. et al., 1993, Nature **365**, 621
Alcock, C. et al., 1997, Astrophys. J. **479**, 119
Amaldi, U., de Boer, W., Furstenau, H., 1991, Phys. Lett. **B260**, 447
Anders, E., Ebihara, M., 1982, Geochim. Cosmochim. Acta, **46**, 2363
Apollonio, M. et al. (Chooz Collab.), 1999, Phys. Lett. **B466**, 415
Appel, A.W., 1985, *An Investigation of Galaxy Clustering Using an Asymptotically Fast N-body Algorithm*, Princeton University Thesis
Armendariz-Picon, C., Mukhanov, V., Steinhardt, P.J., 2000, Phys. Rev. Lett. **85**, 4438
Arnett, W.D., Branch, D., Wheeler, J.C., 1985, Nature, **314**, 337
Ashtekar, A., 1995, Les Houches 1992, *Gravitation and Quantisation*, Elsevier, ed. by B. Julia, J. Zinn-Justin, p. 181
Aubourg, E. et al., 1993, Nature **365**, 623
Audouze, J., 1984, in Proc. First ESO-CERN Symp., ed. by G. Setti, L. van Hove, p. 293
Bach, J., Fröhlich, J., Signal, I.M., 2000, J. Math. Phys. Sci. **41**, 3985B
Bahcall, N.A., Soneira, R.M., 1983, Astrophys. J., **270**, 20
Bahcall, J.N., Pinsonneault, M.H., 1996, Astrophys. J. **467**, 475 (see also http://www.sns.ias.edu/~jnb/)
Bahcall, J.N., Pinsonneault, M.H., Basu, S., 2001, Astrophys. J. **555**, 990B
Balescu, R., 1975, *Equilibrium and Nonequilibrium Statistical Mechanics* (Wiley, New York)
Balser, D.S. et al., 1994, Astrophys. J. **430**, 667
Baluni, V., 1979, Phys. Rev., **D19**, 2227
Bardeen, J.M., 1980, Phys. Rev., **D22**, 1882
Bardeen, J., Cooper, L.N., Schrieffer, J.R., 1957, Phys. Rev., **108**, 1125
Bardeen, J.M., Bond, J.R., Kaiser, S., Szalay, A.S., 1986, Astrophys. J. **304**, 15
Barrow, J.D., 1980, Mon. Not. R. Astron. Soc., **192**, 427

Barrow, J.D., 1983, Fundam. Cosmic Phys., **8**, 83
Barrow, J.D., Turner, M.S., 1982, Nature, **292**, 35
Barrow, J.D., Juszkiewicz, R., Sonoda, D.H., 1983, Nature, **305**, 397
Bartelmann, M., Narayan, R., 1995, Astrophys. J. **451**, 60
Bartelmann, M., Narayan, R., 1999, in *Proc. of 1995 Jerusalem Winter School*, ed. by A. Dekel, J.P. Ostriker (Cambridge Univ. Press), p. 360
Battye, R.A., Shellard, F.D.S., 1994, Nucl. Phys. **B423**, 260
Baum, W.A., 1962, in *Astronomical Techniques* ed. by W. Hüttner (University of Chicago Press, Chicago)
Bean, A.J. et al., 1983, MNRAS, **205**, 605
Bekenstein, J.D., 1973, Phys. Rev., **D7**, 2333
Bekenstein, J.D., Milgrom, M., 1984, Astrophys. J. **286**, 7
Bender, C.M., Cooper, F., 1983, Nucl. Phys., **B224**, 403
Bender, R., 1988, Astron. Astrophys. **193**, L7
Bender, R., 1990, in *Dynamics and Interactions of Galaxies*, ed. by R. Wielen (Springer, Heidelberg)
Bender, R., Burstein, D., Faber, S.M., 1992, Astrophys. J. **399**, 462
Bender, R. et al. 1989, Astron. Astrophys. **217**, L35
Bennett, C.L., et al., 2003, Astrophys. J. Suppl. **148**, 1
Bernstein, J., Brown, L.M., Feinberg, G., 1989, Rev. Mod. Phys. **61**, 25
Bertschinger, E., Dekel, A., 1989, Astrophys. J., **336**, L5
Bertschinger, E. et al., 1990, Astrophys. J., **364**, 370
Bi, H.G., 1993, Astrophys. J. **405**, 479
Bi, H.G., Börner, G., Chu, Y., 1992, Astron. Astrophys. **206**, 1
Binney, J., Tremaine, S., 1987, *Galactic Dynamics* (Princeton Univ. Press)
Birkinshaw, M., Hughes, J.P., 1994, Astrophys. J. **420**, 33
Birrel, N., Davies, P.C.W., 1984, *Quantum Fields in Curved Space* (Cambridge Univ. Press)
Bjorken, J.D., Drell, S., 1965, *Relativistic Quantum Fields* (McGraw-Hill, New York)
Blumenthal, G.R. et al., 1984, Nature, **311**, 517
Boesgaard, A.M., Steigman, G., 1985, Ann. Rev. Astron. Astrophys., **23**, 319
Böhringer, H., 1995, Ann. N.Y. Acad. Sci. **759**, 67
Bond, H.E., 1980, Astrophys. J. Suppl., **44**, 517
Bond, J.R., Efstathiou, G., Silk, J., 1980, Phys. Rev. Lett. **45**, 1980
Bond, J.R., Efstathiou, G., 1987, Mon. Not. R. Astron. Soc., **226**, 655
Bond, J.R., Szalay, A.S., 1983, Astrophys. J., **274**, 443
Bond, J.R. et al., 1983, Proc. 1983 Moriond Conf. (ed. by Audouze, J., Tran Thanh Van; Reidel), 87
Bonifacio, P., Molaro, P., 1997, Mon. Not. Roy. Astron. Soc. **285**, 847
Bonnell, J.T., Nemiroff, R.J., Goldstein, J.J., 1996, Pub. Astron. Soc. Pac. **108**, 1065 ff.
Borgs, C., Nill, F., 1986a, Commun. Math. Phys., **104**, 349
Borgs, C., Nill, F., 1986b, Phys. Lett., **171B**, 289
Boris, S. et al., 1985, Phys. Lett., **159B**, 217
Börner, G., Ehlers, J., 1988, Astron. Astrophys. **204**, 1
Börner, G., Mo, H.J., 1990, Astron. Astrophys., **227**, 324
Bosch, F. et al., 1996, Phys. Rev. Lett. **77**, 5190
Börner, G., Gottlöber, S., ed., *The evolution of the universe*, J. Wiley 1997
Bottema, R., 1993, Astron. Astrophys. **275**, 16
Bottema, R., 1997, Astron. Astrophys. **328**, 517
Bouchet, F.R. et al., 2002, Phys. Rev. **D65**, 021301
Braginsky, V.B., 1974, in *Experimental Gravitation*, Academic Press, New York, p. 235

Brandenberger, R.H., 1985, Rev. Mod. Phys. **57**, 1
Brandenberger, R., Kahn, R., 1984, Phys. Lett. **141B**, 317
Brandenberger, R., Albrecht, A., Turok, N., 1986, Nucl. Phys. **B277**, 605
Broadhurst, T.J., Ellis, R.S., Koo, D.C., Szalay, A.S., 1990, Nature **343**, 726
Broadhurst, T.J., Ellis, R.S., Glazebrook, K., 1992, Nature **355**, 55
Broadhurst, T.J. et al., 1991, Proc. Texas-ESO-CERN Symp., Brighton, UK, Dec. 1990
Buchert, T., 1989a, Astron. Astrophys. **223**, 9
Buchert, T., 1989b, Rev. Mod. Astron. **2**, 267
Buchert, T., Götz, G., 1987, J. Math. Phys. **28**, 271
Buchert, T., Melott, A.L., Weiss, A.G., 1995, Astron. Astrophys. **5**, 84
Buras, A.J., Gaillard, M.K., Nanapoulos, D.V., 1978, Nucl. Phys. **B135**, 66
Burbidge, E.M., Burbidge, G.R., Fowler, W.A., Hoyle, F., 1957, Rev. Mod. Phys. **29**, 547
Burles, S., Tytler, D., 1998, Astrophys. J. **499**, 699;
Burles, S., Tytler, D., 1998, Astrophys. J. **507**, 732
Burstein, D., 1979, J. Astrophys. **234**, 435
Cabrera, B., 1982, Phys. Rev. Lett., **48**, 1378
Callan, C., Coleman, S., 1977, Phys. Rev. D **16**, 1762
Cameron, A.G.W., 1982, *Essays in Nuclear Astrophysics* ed. by C.A. Barnes, D.N. Clayton, D.N. Schramm (Cambridge University Press) p. 23
Candelas, P., Horowitz, G.T., Strominger, A., Witten, E., 1985, Nucl. Phys. **B258**, 46
Carignan, C., Freeman, K.C., 1985, Astrophys. J., **294**, 494
Carlberg, R.G. et al., 1996, Astrophys. J. **462**, 32
Carroll, S.M., Press, W.H., Turner, E.L., 1992, Annu. Rev. Astron. Astrophys. **30**, 499
Carr, B.J., 1978, Comments on Astrophys. Space Phys. **7**, 161
Centrella, J., Melott, A.L., 1983, Nature **305**, 196
Chaboyer, B., 1995, Astrophys. J. **444**, 9
Chaboyer, B., Demarque, P., Sarajedini, A., 1996, Astrophys. J., **459**, 558
Chaichian, M., Nelipa, N.F., 1984, *Introduction to Gauge Field Theory*, Tets and Monographs in Physics (Springer-Verlag Berlin, Heidelberg)
Chandrasekhar, S., 1968, *Ellipsoidal Figures of Equilibrium* (Yale University Press, New Haven)
Chiu, H.Y., 1965, Phys. Rev. Lett. **17**, 712
Clementini, G. et al., 1995, Mon. Not. R. Astron. Soc. **280**, 309
Cleveland, B. et al., 1998, Astrophys. J. **496**, 505
Coleman, S., 1977, Phys. Rev. **D15**, 2929
Coleman, S., de Luccia, F., 1980, Phys. Rev. **D21**, 3305
Coleman, S., Weinberg, E., 1973, Phys. Rev. **D7**, 1888
Combes, F. et al., 1995, *Galaxies and Cosmology* (Springer Verlag)
Cook, G.P., Mahanthappa, K.T., 1981, Phys. Rev. **D23**, 1321
Copi, C.J., Schramm, D.N., Turner, M.S., 1995, Science **267**, 192
Courteau, S., 1997, Astron. J. **114**, 2402
Cowan, J.J., Cameron, A.G.W., Truran, J.W., 1983, Astrophys. J., **265**, 429
Cowan, J.J., Thielemann, F.-K., Truran, J.W., 1991, Ann. Rev. Astron. Astrophys. **29**, 447
Cowie, L.L., 1989, in *The Epoch of Galaxy Formation*, ed. by C.S. Frenk et al. Kluwer (Dordrecht)
Darriulat, P., 1984, First ESO-CERN Symp. ed. by G. Setti, L. van Hove, p. 17
David, L.P. et al., 1994, Astrophys. J. **428**, 544
Davidson, S., Sarkar, S., 2000, JHEP 0011, 012
Davies, R.L., Sadler, E., Peletier, R., 1993, Mon. Not. R. Astron. Soc. **262**, 650

Davies, R.L. et al., 1983, Mon. Not. R. Astron. Soc. **262**, 650
Davies, R.L. et al., 1987, Astrophys. J. **266**, 41
Davis, M., Peebles, P.J.E., 1983a, Annu. Rev. Astron. Astrophys. 21, 109
Davis, M., Peebles, P.J.E., 1983b, Astrophys. J. **267**, 465
Davis, M., Efstathiou, G., Frenk, C.S., White, S.D.M., 1985, Astrophys. J. **292**, 371
Davis, Jr., R., 1980, in Proc. Neutrino Mass Miniconf. ed. by V. Barger, D. Cline (Telemark, Wisconsin) p. 38
De Angelis, G.F., de Falco, D., Guerra, F., 1978, Phys. Rev. **D17**, 1624
De Lapparent, V., Geller, M.J., Huchra, J.P., 1986, Astrophys. J. **302**, L1
De Rujula, A., 1985, Nucl. Phys. **A434**, 605
De Vaucouleurs, G., 1958, Astron. J. **63**, 253
De Vaucouleurs, G., 1959, in *Astrophysics IV: Stellar Systems*, Handbuch der Physik, Vol. 53 ed. by S. Flügge (Springer-Verlag, Berlin, Heidelberg), p. 311
De Vaucouleurs, G., 1981, Proc. 10th Texas Symp. in Rel. Astrophysics (New York Acad. Sci.), p. 90
De Vaucouleurs, G., 1982, Nature **299**, 303
Dekel, A., 1994, Ann. Rev. Astron. Astrophys. **32**, 371
Demianski, M., 1984, Nature **307**, 140
Deser, S., Zumino, B., 1977, Phys. Rev. Lett. **35**, 827
Dicke, R.H., 1964, in *Experimental Relativity, Relativity, Groups, and Topology*, ed. by C. De Witt, B. De Witt (Gordon & Breach, New York), p. 165
Dicus, D.A., Kolb, E.W., Gleeson, A.M., Sudarshan, E.C.G., Teplitz, V.I. et al., 1982, Phys. Rev. **D26**, 2694
Dine, M., Fischler, W., Srednicki, M., 1981, Phys. Lett. **104B**, 199
Disney, M., 1976, Nature **236**, 573
Djorgovski, S., Davis, M., 1987, Astrophys. J. **313**, 59
Dolgov, A.D., 1980, Zh. Eksp. Teor. Fiz. **79**, 337
Dolgov, A.D., Linde, A., 1982, Phys. Lett. **B116**, 329
Doroshkevich, A.G., Khlopov, M.Yu., Sunyaev, R.A., Szalay, A.S., Zel'dovich, Ya.B., 1980, Ann. N.Y. Acad. Sci. **375**, 32
Doroshkevich, A.G., Kotok, E.V., Novikov, I.D., Polyudov, A.N., Shandarin, S.F., Sigor, Yu.S., 1980, Mon. Not. R. Astron. Soc. **192**, 321
Dress, W.B., Miller, P.D., Pendlebury, J.M., Perrin, P., Ramsey, N.F., 1977, Phys. Rev. **D18**, 9
Dressler, A., 1980, Astrophys. J. **236**, 531
Dressler, A., 1984, Astrophys. J. **281**, 512
Dressler, A., Faber, S.M., Burstein, D., Davies, R.L., Lynden-Bell, D., Terlevich, R.J., Wegner, G., 1987, Astrophys. J. **313**, L37
Durrel, P.R., Harris, W.E., 1993, Astron. J. **105**, 1420
Durrer, R., 1994, Fundamentals of Cosmic Physics **15**, 209
Durrer, R., Straumann, N., 1988, Helvet. Phys. Acta **61**, 1027
Eckart, A., Genzel, R., 1996, Nature **383**, 415
Eckart, A., Genzel, R., 1997, Mon. Not. R. Astron. Soc. **284**, 576
Efstathiou, G., Bond, J.R., 1986, Mon. Not. R. Astron. Soc. **218**, 103
Efstathiou, G., Eastwood, J.W., 1981, Mon. Not. R. Astron. Soc. **194**, 503
Efstathiou, G., Silk, J., 1983, Fund. Cosmic Phys., **9**, 1
Efstathiou, G., Davis, M., Frenk, C.S., White, M.D., 1985, Astrophys. J. Suppl. **57**, 241
Efstathiou, G., Ellis, R.S., Peterson, B.A., 1988, Mon. Not. R. Astron. Soc. **232**, 431
Efstathiou, G., Sutherland, W.J., Maddox, S.J., 1990, Nature **438**, 705
Efstathiou, G., Bond, J.R., White, S.D.M., 1992, Mon. Not. R. Astron. Soc. **258**, 1p

Ehlers, J., 1971, Proc. Enrico Fermi Summer School, Varenna (1969), ed. by R.K. Sachs (Academic Press, New York), p. 1
Ehlers, J., 1976, AG Mitt. **38**, 41
Ehlers, J., Geren, P., Sachs, R.K., 1968, J. Math. Phys. **8**, 1344
Ehlers, J., Rindler, W., 1987, Astron. Astrophys. **174**, 1
Einasto, J., Klypin, A.A., Saar, E., Shandarin, S.F., 1984, Mon. Not. R. Astron. Soc. **206**, 529
Einstein, A., Strauss, E.C., 1945, Rev. Mod. Phys. **17**, 120
Elitzur, S., 1975, Phys. Rev. **D12**, 3979
Ellis, G.F.R., 1971, Proc. Enrico Fermi Summer School, Varenna 1969, ed. by R.K. Sachs (Academic Press, New York), p. 104
Ellis, G.F.R., 1979, Gen. Rel. Grav. **11**, 281
Ellis, G.F.R., 1980, Proc. IX Texas Symp. Ann. N.Y. Acad. Sci. **336**, 130
Ellis, G.F.R., 1984, in *General Relativity and Gravitation*, ed. by B. Bertotti et al. (Reidel, Dordrecht)
Ellis, G.F.R., 1986, Univ. Cape Town, preprint 86/15
Ellis, G.F.R., Perry, J.J., 1979, Mon. Not. R. Astron. Soc. **187**, 357
Ellis, G.F.R., Rothmann, T., 1986, Univ. of Capetown, preprint 86/16
Ellis, G.F.R., Schmidt, B.G., 1977, Gen. Rel. Grav. **8**, 915
Ellis, G.F.R., Schreiber, G., 1986, Phys. Lett. **A115**, 97
Ellis, G.F.R., Maartens, R., Nel, S.D., 1978, Mon. Not. R. Astron. Soc. **184**, 439
Ellis, G.F.R., Nel, S.D., Maartens, R., Stoeger, W.R., Whitman, A.P., 1985, Phys. Rep. **124**, 315
Ellis, J., 1984, in Proc. First ESO-CERN Symp. ed. by G. Setti, L. van Hove, p. 435
Ellis, J., 1986, Nature **323**, 595
Ellis, J., Steigman, G., 1979, Phys. Lett. **B84**, 1986
Ellis, J., Gaillard, M.K., Nanopoulos, D.V., 1979, Phys. Lett. **B80**, 360
Ellis, J., Gaillard, M.K., Zumino, B., 1980, Phys. Lett. **94B**, 343
Ellis, J., Gaillard, M.K., Nanopoulos, D.V., 1981, Nature **293**, 42
ESA, *The Hipparcos Catalogue*, 1997, ESA SP-1200
Esmailzadeh, R., Starkman, G.D., Dimopoulos, S., 1991, Astrophys. J. **378**, 504
Etherington, I.M.H., 1933, Phil. Mag. **15**, 761
Ezawa, H., Swieca, J.A., 1967, Commun. Math. Phys. **5**, 330
Faber, S., Gallagher, J., 1979, Annu. Rev. Astron. Astrophys. **17**, 135
Faber, S.M., Jackson, R.E., 1976, Astrophys. J. **204**, 668
Faber, S.M. et al., 1997, Astron. J. **114**, 1771
Fabian, A.C., 1981, Ann. N.Y. Acad. Sci. **375**, 235
Fahlmann, G. et al., 1994, Astrophys. J. **437**, 56
Farhi, E., Jaffe, R.L., 1984, Phys. Rev. **D30**, 2379
Fayet, P., 1975, Phys. Lett. **58B**, 67
Fayet, P., 1977, Phys. Lett. **69B**, 489
Fayet, P., 1979, Phys. Lett. **86B**, 272
Fayet, P., 1984, First ESO-CERN Symp., ed. by G. Setti, L. van Hove, p. 35
Fayet, P., Ferrara, S., 1977, Phys. Rep. **32C**, 250
Feast, M.W., 1984, Mon. Not. R. Astron. Soc. **211**, 51p
Feast, M.W., Catchpole, R.M., 1997, Mon. Not. R. Astron. Soc. **286**, L1
Ferrarese, L. et al., 1996, Astrophys. J. **468**, L95
Ferrarese, L. et al., 1996, Astrophys. J. **464**, 568
Field, G.B., Perrenod, S.C., 1977, Astrophys. J. **215**, 717
Fiorini, E., 1983, Acta Phys. Austriaca, Suppl. **25**, 143
Fiorini, E., 1984, First ESO-CERN Symp., ed. by G. Setti, L. van Hove, p. 81
Fischbach, E., Sudarsky, D., Szafer, A., Talmadge, C., Aronson, S.H., 1986, Phys. Rev. Lett. **56**, 3
Fisher, K.B. et al., 1994, Mon. Not. R. Astron. Soc. **267**, 927

Fixsen, D. et al., 1996, Astrophys. J. **437**, 576
Florentin-Nielsen, R., 1984, Astron. Astrophys. **138**, L19
Fowler, W.A., 1972, in *Cosmology, Fusion, and Other Matters*, ed. by F. Reines (Boulder, Colorado), p. 67
Fowler, W.A., Hoyle, F., 1960, Ann. Phys. **10**, 280
Fowler, W.A., Meisl, C.C., 1985, preprint OAP-660 Symp. on Cosmogonical Processes (Boulder, Colorado)
Freeman, K.C., 1970, Astrophys. J. **160**, 811
Freedman, W.L. et al., 1994a, Nature **371**, 757
Freedman, W.L. et al., 1994b, Astrophys. J. **427**, 628
Freese, K., Adams, F., 1989, Phys. Rev. D.
Friedmann, A., 1922, Z. f. Physik **10**, 377
Fritschi, M., Holzschuh, E., Kündig, W., Petersen, J.W., Pixley, R.E., Stüssi, H., 1986, Phys. Lett. **173B**, 485
Fröhlich, J., Morchio, G., Strocchi, F., 1981, Nucl. Phys. **B190**, 553
Fry, J.N., Olive, K.A., Turner, M.S., 1980a, Phys. Rev. **D22**, 2977
Fry, J.N., Olive, K.A., Turner, M.S., 1980b, Phys. Rev. Lett. **45**, 2074
Fukuda, S. et al., 2000, Phys. Rev. Lett. **85**, 4003
Fukuda, S. et al. (Super-Kamiokande coll.), 2001, Phys. Rev. Lett. **86**, 5651
Gale, N.H. et al., 1975, Earth Planet Sci. Lett. **26**, 195
Garnavich, P. et al., 1998, Astrophys. J. **493**, 53
Geiss, J., 1993, *Origin and Evolution of the Elements*, p. 89, ed. by N. Prantzos, E. Vangioni-Flam, M. Cassé (Cambridge Univ. Press)
Geller, M.J., Huchra, J., 1989, Science **246**, 897
Geller, M.J., de Lapparent, V., Kurtz, M.J., 1984, Astrophys. J. **287**, L55
Gell-Mann, M., Ne'eman, Y., 1964, *The Eightfold Way*, (Benjamin, New York)
Gendreau, K.C. et al., 1995, PASJ **47**, L5
Georgi, H., Quinn, H.R., Weinberg, S., 1974, Phys. Rev. Lett. **33**, 451
Geroch, R.P., 1968, Ann. Phys. N.Y. **48**, 526
Giacomelli, G., 1984, Prod. First ESO-CERN Symposium, ed. by. G. Setti, L. van Hove, p. 111
Gilli, R., Salvati, M., Hasinger, G., 2001, Astron. Astrophys. **366**, 407
Ginsparg, P., Glashow, S., 1986, Physics Today (May), 7
Ginzburg, V., Landau, L.D., 1950, Sov. Phys. JETP **20**, 1064
Giovanelli, R. et al., 1997, Astrophys. J. **477**, L1
Glashow, S.L., 1958, Ph.D. Thesis (Harvard Univ.)
Glashow, S.L., de Rujula, A., 1980, Phys. Rev. Lett. **45**, 942
Gliner, E.B., 1965, Sov. Phys. JETP **22**, 378
GNO Collab., 2000, Phys. Lett. **B490**, 16
Goldberg, H., 1983, Phys. Rev. Lett. **50**, 1419
Goldstone, J., 1961, Nuovo Cim. **19**, 154
Gould, A, Bahcall, J.N., Flynn, C., 1996, Astrophys. J. **465**, 759
Graaf, D., Freese, K., 1996, Astrophys. J. **456**, L49
Green, M.B., 1986, in 2nd ESO-CERN Symp. ed. by G. Setti, L. van Hove, p. 121
Green, M.B., Schwarz, J.H., 1984, Phys. Lett. **149B**, 117
Green, M.B., Schwarz, J.H., 1985, Phys. Lett. **151B**, 21; Nucl. Phys. **B255**, 93
Gregory, A., Thompson, L.A., Tifft, W.A., 1980, Astrophys. J. **243**, 411
Groenewegen, M.A.T., de Jong, T., 1998, Astron. Astrophys. **337**, 797
Groom, D.E. et al., 2000, Particle Data Group, Eur. Phys. J. **C15**, 1
Gross, D., Wilczek, F., 1973, Phys. Rev. Lett. **30**, 1343
Groth, E.J., Peebles, P.J.E., 1977, Astrophys. J. **217**, 385
Gruber, D.E. et al., 1999, Astrophys. J. **520**, 124
Gunn, J.E., Lee, B.W., Lerche, I., Schramm, D.N., Steigman, G., 1978, Astrophys. J. **223**, 1015

Guth, A.H., 1981, Phys. Rev. **D23**, 347
Haber, H., Kane, G.L., 1985, Phys. Rep. **117**
Haehnelt, M.G., Steinmetz, M., Rauch, M., 1996, Astrophys. J. **465**, L95
Hagman, C.A. et al., 1998, Phys. Rev. Lett. **80**, 2043
Hamilton, A.J.S., 1992, Astrophys. J. **385**, 5
Hamilton, A.J.S., 1993, Astrophys. Lett. **406**, L47
Han, M.Y., Nambu, Y., 1965, Phys. Rev. **139**, 1006
Harrison, E.R., 1970, Phys. Rev. **D1**, 2726
Harrison, E.R., 1973, Ann. Rev. Astron. Astrophys. **11**, 155
Hasenfratz, P., 1986, Swiss Math. Society Meeting on Math. Phys. (Zürich, Dec. 1986)
Hasinger, G., 2002, Proc. Symp. *New Visions of the X-ray Universe in the XMM-Newton and Chandra Era* (ESTEC); astro-ph/0202430v2
Hausmann, M.A., Ostriker, J.P., 1978, Astrophys. J. **224**, 320
Hawking, S.W., 1974, Nature **248**, 30
Hawking, S.W., 1975, Comm. Math. Phys. **43**, 199
Hawking, S.W., Ellis, G.F.R., 1973, *The Large Scale Structure of Spacetime* (Cambridge Univ. Press)
Hawking, S.W., Penrose, R., 1970, Proc. R. Soc. London **A314**, 529
Hawking, S.W., Moss, I.G., 1982, Phys. Lett. **110B**, 35
Hawking, S.W., Moss, I.G., Stewart, J.M., 1982, Phys. Rev. **D26**, 2681
Hayakawa, S., 1984, Adv. Space Res. **3**, 449
Hayashi, C., 1950, Prog. Theor. Phys. **5**, 224
Heidmann, J., 1980, *Relativistic Cosmology* (Springer, Berlin, Heidelberg)
Hellings, R.W., Adams, P.J., Anderson, J.D., Keesey, M.S., Law, E.L., Standish, E.M., Canuto, V.M., Goldman, I., 1983, Phys. Rev. Lett. **51**, 1609
Hellings, R.W., Adams, P.J., Canuto, V.M., Goldman, I., 1983, Phys. Rev. **D28**, 1822
Hellmund, M., Kripfganz, J., Schmidt, M.G., 1994, Phys. Rev. **D50**, 7650
Henriksen, M.J., Mamon, G.A., 1994, Astrophys. J. Lett. **421**, L63
Hernanz, M. et al., 1994, Astrophys. J. **434**, 652
Higgs, P.W.B., 1964, Phys. Lett. **12**, 132
Higgs, P.W.B., 1966, Phys. Rev. **145**, 1156
Hillebrandt, W., 1978, Space Sci. Rev. **21**, 639
Hillebrandt, W., Höflich, P., Truran, J.W., Weiss, A., 1987, Nature **327**, 597
Hillebrandt, W., Niemeyer, J.C., 2000, Ann. Rev. Astron. Astrophys. **38**, 191
Hindmarsh, M.B., Kibble, T.W.B., 1995, Rep. Prog. Phys. **58**, 477
Hirata, K. et al., 1987, Phys. Rev. Lett. **58**, 1490
Hockney, R.W., Eastwood, J.W., 1981, *Computer Simulation Using Particles* (McGraw-Hill, New York)
Höflich, P. et al., 1996, Astrophys. J. **472**, L81
Hofmann, K.H. et al., 1997, in 'Science with the VLT Interferometer' (ESO Workshop), (Springer), p. 367
Hogan, C.J., Rees, M.J., 1984, Nature **311**, 109
Hoyle, F., Burbidge, G., 1998, Astrophys. J. **509**, L1
Hu, W., 1995, Ph.D. thesis (Princeton Univ.)
Hu, W., Sugiyama, N., 1995a, Ap. J. **436**, 456
Hu, W., Sugiyama, N., 1995b, Phys. Rev. **D 51**, 2599
Hu, W., Sugiyama, N., Silk, J., 1997, Nature **386**, 37
Hubble, E., 1929, Proc. NAS **15**, 168
Hubble, E., 1958, Mon. Not. R. Astron. Soc. **113**, 666
Huchra, J.P., Davis, M., Lathan, D.W., Tonry, L.J., 1983, Astrophys. J. Suppl. **53**, 89
Huchra, J.P., Postman, M., Geller, M.J., 1986, Astron. J. **92**, 1239

Huchra, J. et al., 1991, Astrophys. J. **365**, 66
Hudson, B., Kennedy, B.M., Podesek, F.A., Hohenberg, C.M., 1984, Geochim. Cosmochim. Acta
Hut, P., Olive, K., 1979, in *Physical Cosmology*, ed. by R. Balian, J. Audouze, D. Schramm (North-Holland, New York)
Ikeuchi, S., 1981, Publ. Astron. Soc. Japan **33**, 211
Itzykson, C., Nauenberg, M., 1966, Rev. Mod. Phys. **38**, 95
Izotov, Y.I., Thuan T.X., Lipovetsky, V.A., 1994, Astrophys. J. **435**, 647
Izotov, Y.I., Thuan T.X., Lipovetsky, V.A., 1997, Astrophys. J. Suppl. **108**, 1
Jacoby, G.H. et al., 1992, Publ. Astron. Soc. Pac. **104**, 599
Jaffe, A., Taubes, C., 1980, *Vortices and Monopoles* (Birkhäuser, Boston, Mass.)
Jaffe, A.H. et al., 2001, Phys. Rev. Lett. **86**, 3475
Jedamzik, K., 1998, astro-ph/9805156, Proc. *Neutrino Astrophysics* (Ringberg 1997)
Jedamzik, K., Fuller, G.M., 1995, Astrophys. J. **452**, 33
Jerjen, H., Tammann, G.A., 1993, Astron. Astrophys. **276**, 1
Jenkins, A. et al. (VIRGO collab.), 2001, Mon. Not. R. Astron. Soc. **321**, 272
Jing, Y.P., 1998, Astrophys. J. **503**, L9
Jing, Y.P., 2001, Astrophys. J. Lett. **550**, L125
Jing, Y.P., Börner, G., 1998, Astrophys. J. **503**, 37
Jing, Y.P., Börner, G., 2001, Mon. Not. R. Astron. Soc. **325**, 1389
Jing, Y.P., Suto, Y., 1998, Astrophys. J. **494**, L5
Jing, Y.P., Mo, H.J., Börner, G., Fang, L.Z., 1993, Astrophys. J. **411**, 450
Jing, Y.P., Mo, H.J., Börner, G., 1998, Astrophys. J. **494**, 1
Jing, Y.P., Börner, G., Suto, Y., 2002, Astrophys. J. **564**, 15
Johnson, H., Morgan, W., 1953, Astrophys. J. **117**, 313
Jones, B.T., 1976, Rev. Mod. Phys. **48**, 107
Jones, C., Forman, W., 1983, Astrophys. J. **272**, 439
Jüttner, F., 1928, Z. f. Physik **47**, 542
Kafka, P., 1968, Nature **213**, 346
Kaiser, N., 1984, Astrophys. J. **284**, L9
Kaiser, N., Squires, G., 1993, Astrophys. J. **404**, 441
Kaiser, N., Stebbins, A., 1984, Nature **310**, 391
Kajantie, K., Laine, M., Rumukainen, K., Shaposhnikov, M., 1996, PRL **77**, 2887
Kaluza, T., 1921, Sitzungsber. Preuss. Akad. Wiss. Phys. Math. Kl. **966**
Käppeler, F., Beer, H., Wisshak, K., Clayton, D., Macklin, R.L., Ward, R.D., 1982, Astrophys. J. **257**, 821
Kates, R.E., Weigelt, U.V., 1990, Astron. Astrophys. **240**, 1
Kauffmann, G. et al., 1999, Mon. Not. R. Astron. Soc. **307**, 529
Kawano, L., 1992, Fermilab-Pub-92/04-A
Keeton, C.R., Kochanek, C.S., 1997, Astrophys. J. **487**, 42
Kennedy, T., King, C., 1985, Phys. Rev. Lett. **55**, 776
Khlebnikov, S.Yu., Tkachev, I.I., 1997, Phys. Rev. Lett. **79**, 1607
Kiefer, C., 1997, *The Evolution of the Universe*, p. 97 (Dahlem workshop 1995; J. Wiley, ed. by G. Börner, S. Gottlöber)
Kirshner, R.P., Oemler, A., Schechter, P.L., 1978, Astrophys. J. **83**, 1549
Kirshner, R.P., Oemler, A. Jr., Shectman, S.A., 1981, Astrophys. J. **248**, L57
Kirsten, T., 1978, *The Origin of the Solar System*, ed. by S.F. Dermott (Wiley & Sons, New York), p. 267
Kitayama, T., Suto, Y., 1997, Astrophys. J. **490**, 557
Kittel, *Quantum Theory of Solids*
Kittel, C., 1976, *Introduction to Solid State Physics* (Wiley, New York)
Klapdor-Kleingrothaus, H.V., Päs, H., Smirnov, A.Y., 2001, Phys. Rev. **D63**, 073005

Klein, O., 1926, Z. f. Physik **37**, 895
Klinkhamer, F.R., Manton, N.S., 1984, Phys. Rev. **D30**, 2214
Klypin, A.A., Kopylov, A.I., 1983, Sov. Astron. Lett. **9**, 75
Kobayashi, S., Sasaki, S., Suto, Y., 1996, Publ. Astron. Soc. Japan **48**, L107
Kochanek, C.S., 1996, Astrophys. J. **466**, 638
Kodama, H., Sasaki, M., 1984, Prog. Theor. Phys. Suppl. **78**
Kofman, L., Linde, A., Starobinsky, A.A., 1997, Phys. Rev. **D56**, 3258
Kogut, A. et al., 2003, Astrophys. J. Suppl. **148**, 161
Kolb, E., 1986, Nucl. Phys. **B252**, 321
Kolb, E.W., Turner, S.M., 1983, Annu. Rev. Nucl. Part. Sci. **33**, 645
Kolb, E.W., Turner, S.M., 1990, *The Early Universe* (Addison-Wesley)
Kolb, E.W., Wolfram, S., 1980, Nucl. Phys. **B172**, 224
Kolb, E.W., Seckel, D., Turner, M.S., 1985, Nature **314**, 415
Koo, D.C., Kron, R.G., 1992, Ann. Rev. Astron. Ap. **30**, 613
Kormendy, J., 1982, in *Morphology and Dynamics of Galaxies*, 12th Course Swiss Astron. Soc., ed. by L. Marinet, M. Mayor
Kovac, J. et al., 2002, Nature **420**, 772
Kramer, D., Stephani, H., Heret, E., MacCallum, M., 1980, *Exact Solutions of Einstein's Field Equations*, ed. by E. Schmutzer (Cambridge Univ. Press and VEB Dt. Verl. d. Wiss. Berlin, DDR)
Kristian, J., Sachs, R.K., 1966, Astrophys. J. **143**, 379
Krumlinde, J., Möller, P., Wene, C.O., Howard, W.M., 1981, Proc. 4th Int. Conf. on Nuclei far from Stability, CERN 81-09, p. 26
Kundić, T. et al., 1997, Astrophys. J. **482**, 75
Kuzmin, V.A., Rubakov, V.A., Shaposhnikov, M.E., 1985, Phys. Lett. **B155**, 36
La, D., Steinhardt, P.J., 1989, Phys. Rev. Lett. **62**, 376
Lacey, C.H., 1984, in *Formation and Evolution of Galaxies*, ed. by J. Audouze, J. Tran. Thanh Van (Reidel, Boston, Mass.), p. 351
Lacey, C., Cole, S., 1994, Mon. Not. R. Astron. Soc. **271**, 676
Lahav, O., 1987, Mon. Not. R. Astron. Soc. **237**, 129
Landau, L.D., 1946, J. Phys. (USSR) **10**, 25
Landau, L.D., Lifshitz, E.M., 1959, *Fluid Mechanics* (Pergamon, Oxford)
Landau, L.D., Lifshitz, E.M., 1979, *Classical Theory of Fields*, 4th ed. (Pergamon, London)
Langacker, P., 1981, Phys. Rep. **72C**, 185
Lanzetta, K.M., Turnshek, D.A., Wolfe, A.M., 1993, Astrophys. J. Suppl. **84**, 1
Lanzetta, K.M., Wolfe, A.M., Turnshek, D.A., 1995, Astrophys. J. **440**, 435
Laser, A., Kripfganz, J., Schmidt, M.G., 1995, Nucl. Phys. **B433**, 467
Lauer, T.R. et al., 1995, Astron. J. **110**, 2622
Learned J., 2001, *Current Aspects of Neutrino Physics*, ed. by D.O. Caldwell (Physics and Astronomy on-line library, Springer), p. 89
Lederer, C., Shirley, V.S., 1978, Table of Isotopes, Seventh Edition (J. Wiley & Sons, New York)
Lee, B.W., Zinn-Justin, J., 1972, Phys. Rev. **D5**, 3121
Lee, T.D., 1979, preprint, CU-TP-170 (Columbia University)
Lee, W.B., Weinberg, S., 1977, Phys. Rev. Lett. **39**, 165
Leibundgut, B., 2001, Ann. Rev. Astron. Astrophys. **39**, 67
Lemaître, G., 1927, Ann. Soc. Sci. Bruxelles **47A**, 49
Lemaître, G., 1931, Mon. Not. R. Astron. Soc. **91**, 490
Levshakov, S.A., Kegel, W.H., Takahara, F., 1998, Astron. Astrophys. **336**, L29
Lifshitz, E.M., Khalatnikov, I., 1963, Adv. Phys. **12**, 185
Lilje, P.B., 1992, Astrophys. J. **386**, 33
Lilly, S.J. et al., 1995, Astrophys. J. **455**, 108
Lin, C.C., Mestel, L., Shu, F.U., 1965, Astrophys. J. **142**, 1431

Lin, H. et al., 1996, Astrophys. J. **464**, 60
Linde, A.D., 1982, Phys. Lett. **108B**, 389
Linde, A.D., 1983, JETP Lett. **38**, 176
Linde, A.D., 1985, Comments Astrophys. Space Phys. **16**, 229
Linde, A.D., 1990, *Particle Physics and Inflationary Cosmology* (Harvard Acad. Publ.)
Lindner, M., 1986, Z. f. Physik, **C31**, 295
Linsky, J.L. et al., 1995, Astrophys. J. **451**, 335
Lobashev, V.M. et al., 1999, Phys. Lett. **B460**, 227
Lonsdale, C.J., Hacking, P.B., Conrow, T.P., Rowan-Robinson, M., 1990, Astrophys. J. **358**
Loveday, J., 1998, astro-ph/9805255; ASP Conf. 130, p. 68 (1999)
Loveday, J. et al., 1992, Astrophys. J. **390**, 338
Lu, N.Y., Salpeter, E.E., Hoffmann, G.L., 1994, Astrophys. J. **426**, 473
Lucchini, F., Matarrese, S., 1985, Phys. Rev. **D32**, 1316
Lutz, T.E., Kelker, D.H., 1973, PASP **85**, 573
Lynden-Bell, D., Lahav, O., Burstein, D., 1989, Mon. Not. R. Astron. Soc. **241**, 325
Lynden-Bell, D. et al., 1971, Mon. Not. R. Astron. Soc. **15**, 95
Lynden-Bell, D. et al., 1988, Astrophys. J. **326**, 19
Lyubimov, V.A. et al., 1980, Phys. Lett. **94B**, 266
L3 Collab., 1989, Phys. Lett. **B231**, 509
Maoz, D., Rix, H.W., 1993, Astrophys. J. **416**, 425
Maddox, S.J., Sutherland, W.J., Efstathiou, G., Loveday, J., 1990a, Mon. Not. R. Astron. Soc. **243**, 692
Maddox, S.J. et al., 1990b, Mon. Not. R. Astron. Soc. **242**, 43
Mahoney, W.A., Ling, J.C., Jacobson, A.S., Lingenfelter, R.E., 1982, Astrophys. J. **262** 742
Mahoney, W.A., Ling, J.C., Wheaton, W.A., Jacobson, A.S., 1984, Astrophys. J. **286**, 578
Malmquist, K.G., 1920, Medd. Lund. Obs. Ser. **2**, no. 22
Mampe, W., Ageron, P., Bates, J., Pendeburg, J., Stegerl, A., 1989, Phys. Rev. Lett. **63**, 593
Marciano, W.J., 1998, *Fundamental Particles and Interactions* (Woodbury, NY), p. 176; (AIP Conference Proceedings **423**, ed. by R.S. Panvini, T.J. Weiler)
Materne, J., Hopp, U., 1984, Mitt. Astron. Ges. **60**
Mather, J.C. et al., 1990, Astrophys. J. **354**, L37
Mathewson, D.S., Ford, V.L., 1996, Proc. IAU Symp. 168, p. 175 (Kluwer, ed. by M.C. Kafatos, Y. Kondo)
Mattig, W., 1958, Astr. Nachr. **284**, 109
McMillan, R., Ciardullo, R., Jacoby, G.H., 1993, Astrophys. J. **416**, 62
Mendéz, R. et al., 1993, Astron. Astrophys. **275**, 534
Mermin, N.D., Wagner, H., 1966, Phys. Rev. Lett. **17**, 1133
Mermin, N.D., 1967, J. Math. Phys. **8**, 1061
Mészáros, P., 1974, Astron. Astrophys. **37**, 225
Mészáros, P., 1975, Astron. Astrophys. **38**, 5
Meyer, H., 1986, pers. commun.
Milgrom, M., 1983, Astrophys. J. **270**, 365
Mikheyev, S.P., Smirnov, A., 1986, Il Nuovo Cimento **C9**, 17
Miller, R.H., 1984, Astron. Astrophys. **138**, 121
Misner, C.W., 1968, Astrophys. J. **151**, 431
Misner, C.W., 1969, Phys. Rev. Lett. **22**, 1071
Misner, C.W., Thorne, K.S., Wheeler, J.A., 1973, *Gravitation* (W.H. Freeman & Co., San Francisco)

Miyaji, T., Hasinger, G., Schmidt, M., 2001, Astron. Astrophys. **369**, 49
Mo, H.J., 1991, Ph.D. thesis, Munich University
Mo, H.J., Börner, G., 1989, Astron. Astrophys. **224**, 1
Mo, H.J., Miralda-Escudé, J., 1994, Astrophys. J. **429**, L63
Mo, H.J., Jing, Y.P., Börner, G., 1993, Mon. Not. R. Astron. Soc. **264**, 825
Mo, H.J., Jing, Y.P., Börner, G., 1997, Mon. Not. R. Astron. Soc. **286**, 979
Mo, H.J., Mao, S.D., White, S.D.M., 1998, Mon. Not. R. Astron. Soc. **295**, 319
Mo, H.J., Mao, S.D., White, S.D.M., 1999, Mon. Not. R. Astron. Soc. **304**, 175
Monaghan, J.J., 1985, Computer Phys. Rep. **3**, 71
Morikawa, M., Sasaki, M., 1984, Prog. Theor. Phys. **72**, 782
Morikawa, M., Sasaki, M., 1985, Phys. Lett. **165B**, 59
Mössbauer, R.L., 1984, Proc. First ESO-CERN Symposium, ed. by G. Setti, L. van Hove, p. 273
Mould, J. et al., 1995, Astrophys. J. **449**, 413
Mukhanov, V.F., Chibisov, G.V., 1981, Pis'ma Zh. Eksp. Teor. Fiz. **33**, 544
Mukhanov, V.F., Feldman, F.A., Brandenberger, R.H., 1992, Phys. Rep. **215**, 203
Muller, R.A., 1978, Sci. Am. **238**, 64
Muller, R.A., 1980, Proc. IX Texas-Symp., Ann. N.Y. Acad. Sci. **336**, 116
Nanopoulos, D.V., 1981, Prog. Part. Nucl. Phys. **6**, 23
Nanopoulos, D.V., Weinberg, S., 1979, Phys. Rev. **D20**, 2454
Narayan, R., Bartelmann, M., 1998, Proc. 1995 Jerusalem Winter School (in press; MPA preprint 961 (1996))
Navarro, J.F., Frenk, C., White, S.D.M., 1997, Astrophys. J. **490**, 493
Nicolai, H., 1983, Acta Phys. Austriaca Suppl. **25**, 71
Nicolai, H., Niedermaier, M., 1989, Phys. Bl. **45**, 459
Nielsen, H.B., Olesen, P., 1973, Nucl. Phys. **B61**, 45
Nielsen, N.K., Olesen, P., 1978, Nucl. Phys. **B144**, 376
Niemeyer, J.C., Hillebrandt, W., Woosley, S.E., 1996, Astrophys. J.
Ogawa, I., Matsuki, S., Yamamoto, K., 1996, Phys. Rev. **D53**, R1740
Okada, A. et al., 2000, astro-ph 00070003, Proc. 30th Internat. Conf. on High Energy Physics
Oki, K., Sagane, H., Eguchi, T., 1977, J. de Physique Colloque C7, 414
Olive, K.A., 1998, Proc. 12th Lake Louise Winter Institute, ed. by A. Astbury, B.A. Campbell, F.C. Khanna, J.L. Pinfold (World Sci. Press), p. 60
Olive, K.A., Skillman, E., Steigman, G., 1997, Astrophys. J. **483**
Oort, J.H., 1983, Annu. Rev. Astron. Astrophys. **21**, 373
Oort, J.H., 1984, Proc. First ESO-CERN Symp., ed. by G. Setti, L. van Hove, p. 209
Öpik, E., 1922, Astrophys. J. **55**, 406
O'Raifeartaigh, L., 1975, Nucl. Phys. **B96**, 331
Osmer, P.S. 1982, Astrophys. J. **253**, 28
Ostriker, J.P., Cowie, L.L., 1981, Astrophys. J. Lett. **243**, L127
Oudmaijer, R.D., Groenewegen, A.T., Schrijver, H., 1998, Mon. Not. R. Astron. Soc.
Paczyński, B., 1986, Nature **319**, 567
Padmanabhan, T., 1988, *Highlights in Gravitation and Cosmology*, p. 156 (Goa, India, Knudsen)
Pagel, B.E.J. et al., 1992, Mon. Not. Roy. Astron. Soc. **255**, 325
Parker, E.N., 1970, Astrophys. J. **160**, 383
Parker, E.N., Turner, M.S., Bogdan, T.J., 1982, Phys. Rev. **D26**, 1296
Particle Data Group, 1996, Phys. Rev. **D53**, 6065
Partridge, R.B., 1995, *3K: The Cosmic Microwave Background Radiation* (Cambridge Univ. Press, Cambridge, UK)

Partridge, R.B., Peebles, P.J.E., 1967, Astrophys. J. **147**, 868
Peacock, J.A., 1999, *Cosmological Physics* (Cambridge Univ. Press)
Peacock, J.A., Dodds, S.J., 1996, Mon. Not. R. Astron. Soc. **280**, L19
Peacock, J.A., Gull, S.F., 1981, Mon. Not. R. Astron. Soc. **196**, 611
Peccei, R.D., Quinn, H., 1977, Phys. Rev. Lett. **38**, 1440
Peebles, P.J.E., 1966, Astron. J. **146**, 542
Peebles, P.J.E., 1970, Astron. J. **75**, 13
Peebles, P.J.E., 1971, *Physical Cosmology* (Princeton University Press, Princeton, NJ)
Peebles, P.J.E., 1980, *The Large Scale Structure of the Universe* (Princeton University Press, Princeton, NJ)
Peebles, P.J.E., 1993, *Principles of Physical Cosmology* (Princeton Univ. Press)
Peebles, P.J.E., Groth, E.J., 1975, Astrophys. J. **196**, 1
Peebles, P.J.E., Seldner, M., 1978, Astrophys. J. **225**, 7
Peebles, P.J.E., Ratra, B., 1988, Astrophys. J. **325**, L17
Penrose, R., 1966, in *Perspectives in Geometry and Relativity*, ed. by B. Hoffmann (Indian University Press, Indiana) p. 271
Penrose, R., 1979, in *General Relativity, an Einstein Centenary Survey* ed. by S.W. Hawking, W. Israel (Cambridge Univ. Press) p. 581
Penzias, A.A., Wilson, R.W., 1965, Astrophys. J. **142**, 419
Perkins, D.H., 1984, Annu. Rev. Nucl. Part. Sci. **34**, 1
Perlmutter, S. et al., 1998, Nature **391**, 51
Peterman, A., 1979, Phys. Rep. **53**, 157
Pettini, M., Smith, L.J., Hunstead, R.W., King, D.L., 1994, Astrophys. J. **426**, 79
Pierce, M.J. et al., 1994, Nature **371**, 385
Pietronero, L., 1987, Physica **A144**, 257
Planck, M., 1899, Mitt. Thermodyn. Folg. 5
Plohr, B., 1981, J. Math. Phys. **22**, 2184
Politzer, H.D., 1973, Phys. Rev. Lett. **30**, 1346
Polyakov, A.M., 1974, JETP Lett. **20**, 194
Preskill, J., Wise, M.B., Wilczek, F., 1983, Phys. Lett. **120B**, 127
Press, W.H., Schechter, P., 1974, Astrophys. J. **187**, 425
Primack, J.H., 1986, in Proc. of Enrico Fermi School, Varenna 1984
Primack, J.R., 1984, Varenna Intern. School of Astrophysics, STiN-8527786 P
Primack, J.R., Blumenthal, G.R., 1983, in Proc. 1983 Moriond Conf., 163 ed. by Audouze, J., Tran Thanh Van (Reidel, Boston, Mass.)
Prokopec, T., Roos, T.G., 1997, Phys. Rev. **D55**, 3768
Pryke C. et al., 2002, Astrophys. J. **568**, 46
Raffelt, G.G., 1986, Phys. Lett. **166B**, 402
Raffelt, G.G., 1986, Ph.D. Thesis (Munich University)
Raffelt, G.G., 1990, Phys. Reports **198**, 1
Raffelt, G.G., 1996, *Stars as Laboratories for Fundamental Physics* (Univ. Chicago Press)
Raffelt, G.G., 2000, in [Groom et al. 2000]
Rawley, L.A., Taylor, J.W., Davis, M.M., Allan, D.W., 1987, Science **238**, 761
Rees, M., 1984, First ESO-CERN Symp., ed. by G. Setti, L. van Hove,
Rees, M.J., Ostriker, J.P., 1977, Mon. Not. R. Astron. Soc. **179**, 541
Refsdal, S., 1964, Mon. Not. R. Astron. Soc. **128**, 307
Rephaeli, Y., 1995, Astron. Astrophys. **33**, 541
Riess, A.G., Press, W.H., Kirshner, R.P., 1995, Astrophys. J. **438**, L17
Riess, A.G. et al., 1998, Astron. J. **116**, 1009
Riess, A.G. et al., 1998, Astrophys. J. **504**, 935
Ringwald, A., 1988, Phys. Lett. **B201**, 510
Ringwald, A., 1990, Nucl. Phys. **B330**, 1

Ritson, D.M., 1982, SLAC Pub. 2977
Rood, R.T., Bania, T.M., Wilson, T.L., 1992, Nature **355**, 618
Ross, G.G., 1983, Acta Phys. Austriaca Suppl. XXV p. 145
Ross, G.G., 1984, *Grand Unified Theories* (Benjamin)
Rothman, T., Ellis, G.F.R., 1986, Univ. of Capetown preprint
Rowan-Robinson, M., 1985, *The Cosmological Distance Ladder* (Freeman, New York)
Rowan-Robinson, M. et al., 1990, Mon. Not. R. Astron. Soc. **242**, 318
Rubakov, V., 1981, JETP Lett. **33**, 644
Rubin, V.C., 1984, in First ESO-CERN Symp. ed. by G. Setti, L. van Hove, p. 204
Rubin, V.C. et al., 1985, Astrophys. J. **289**, 81
Ryle, M., 1968, Annu. Rev. Astron. Astrophys. **6**, 249
SAGE Collab., 1999, Phys. Rev. Lett. **83**, 4686
Saha, A. et al., 1997, Astrophys. J. **486**, 1
Sakharov, A.D., 1967, Zh. Eksp. Teor. Fiz. Pisma Red. **5**, 32
Salam, A., 1968, Proc. 8th Nobel Symp. on El. Particle Theory ed. by Svartholm (Almquist, Stockholm)
Salam, A., Strathdee, J., 1982, Ann. Phys. **141**, 316
Salam, A., Weinberg, S., 1962, Phys. Rev. **127**, 965
Salaris, M., Degl'Innocenti, S., Weiss, A., 1997, Astrophys. J. **479**, 665
Salpeter, E.E., 1984, First ESO-CERN Symp. ed. by G. Setti, L. van Hove, p. 180
Sandage, A.R., 1972, Astrophys. J. **178**, 1
Sandage, A.R., Kristian, J., Westphal, J.A., 1976, Astrophys. J. **205**, 688
Sandage, A.R., Tammann, G.A., 1974, Astrophys. J. **190**, 525
Sandage, A.R., Tammann, G.A., 1976, Astrophys. J. **207**, L1
Sandage, A.R., Tammann, G.A., 1982, in Astrophysical Cosmology, Scripta Varia **48**, 23
Sandage, A.R., Tammann, G.A., 1982, Astrophys. J. **256**, 339
Sandage, A.R., Tammann, G.A., 1984a, Proc. First ESO-CERN Symposium ed. by G. Setti, L. van Hove, p. 127
Sandage, A.R., Tammann, G.A., 1984b, Nature **307**, 326
Sandage, A.R. et al., 1996, Astrophys. J. Lett. **460**, L15
Sanders, R.H., 1986, Mon. Not. R. Astron. Soc. **223**, 539
Sanders, R.H., 1989, Mon. Not. R. Astron. Soc. **241**, 135
Sarkar, S., 1996, Reports on Progress in Physics **59**, 1493 S
Sato, H., Maeda, K., 1983, Prog. Theor. Phys. **70**, 119
Sato, K., 1981a, Mon. Not. R. Astron. Soc. **195**, 487
Sato, K., 1981b, Mon. Not. R. Astron. Soc. **99b**, 66
Sato, K., 1981c, Mon. Not. R. Astron. Soc. **195**, 198
Sato, K., Kobayashi, H., 1978, Prog. Theor. Phys. **58**, 1775
Sato, K. et al., 1982, Phys. Rev. Lett. **1088**, 103
Sato, N., Kodama, H., Sato, K., 1985, Prog. Theor. Phys. **74**, 405
Satz, H., 1985, Annu. Rev. Nucl. Part. Sci. **35**, 245
Saunders, W. et al., 2000, Mon. Not. R. Astron. Soc. **317**, 55
Schade, D. et al., 1995, Astrophys. J. **451**, L1
Schechter, P.L., 1976, Astrophys. J. **203**, 297
Schindler, S., 1996, Astron. Astrophys. **305**, 756
Schlattl, H., 2001, Phys. Rev. **D64**, 013009
Schlattl, H., Weiss, A., 1999, www.mpa-garching.mpg.de/
Schlattl, H., Weiss, A., Ludwig, H.G., 1997, Astron. Astrophys. **322**, 646
Schmalzing, J., 1999, *On Statistics and Dynamics of Cosmic Structure*, (Dissertation, Univ. of München)
Schmidt, B.G., 1971, Gen. Rel. Grav. **1**, 269
Schmidt, M., 1968, Astrophys. J. **151**, 393

Schmidt, M.G., 1997, in *Beyond the Desert*, 756 (IOP Publishing, H.V. Klein-grotheus, H. Päs (ed))
Schneider, P., Ehlers, J., Falco, E.E., 1992, *Gravitational Lenses* (Springer)
Schoenberg, I.J., 1973, *Cardinal Spline Interpolation* (SIAM, Philadelphia)
Schramm, D.N., Wagoner, R.V., 1977, Annu. Rev. Nucl. Sci. **27**, 37
Schreckenbach, K., Mampe, W., 1992, J. Phys. **G18**, 1
Schweizer, F., Whitmore, B.C., Rubin, V.C., 1983, Astron. J. **88**, 909
Schwinger, J., 1956, Ann. Phys. **2**, 407
Scully, S.T. et al., 1996, Astrophys. J. **462**, 960
Segal, I.E., Nicoll, J.F., 1980, Astron. Astrophys. **82**, L3
Segal, I.E., Nicoll, J.F., 1996, Astrophys. J. **459**, 496
Seitz, S., Schneider, P., 1996, Astron. Astrophys. **305**, 383
Seitz, S. et al., 1996, Astron. Astrophys. **314**, 707
Shectman, S.A. et al., 1996, Astrophys. J. **470**, 172
Shellard, E.P.S., Vilenkin, A., 1994, *Cosmic Strings and Other Topological Defects* (Cambridge Univ. Press)
Sigad Y. et al., 1998, Astrophys. J. **495**, 516
Sikivie, P., 1983, Phys. Rev. Lett. **51**, 1415
Sikivie, P., 1983, Erratum, Phys. Rev. Lett. **52**, 695
Silk, J., 1968, Astrophys. J. **151**, 459
Silk, J., 1995, Astrophys. J. Lett. **438**, L41
Skillman, E. et al., 1994, Astrophys. J. **431**, 172
Smail, I. et al., 1995, Mon. Not. R. Astron. Soc. **237**, 277
Smith, P.F., 1986, Proc. 2nd ESO-CERN Symposium ed. by G. Setti, L. van Hove, p. 237
Smoot, G.F. et al., 1992, Astrophys. J. Lett. **396**, L1
Smoot, G., Gorenstein, F., Muller, R.A., 1977, Phys. Rev. Lett. **39**, 898
SNO collaboration, 2001, Phys. Rev. Lett. **87**, 07130
SNO collaboration, 2002, preprint, April 20 (2002)
Spergel, D., Bahcall, J.N., 1988, Phys. Lett. **B200**, 366
Spergel, D.N., et al., 2003, Astrophys. J. Suppl. **148**, 175
Spite, F., Spite, M., 1982, Astron. Astrophys. **115**, 357
Spite, F., Spite, M., 1982, Nature **297**, 483
Springel, V., 1996, Diplomarbeit, Univ. of Munich
Squires, G. et al., 1996, Astrophys. J. **461**, 572
Starobinsky, A.A., 1980, Phys. Lett. **B91**, 90
Stecker, F., 1980, Phys. Rev. Lett. **45**, 1460
Steidel, C.C., Dickinson, M., Persson, S.E., 1994, Astrophys. J. **437**, L75
Steidel, C. et al., 1998, Astrophys. J. **492**, 428
Steigman, G., 1976, Annu. Rev. Astron. Astrophys. **14**, 339
Steigman, G., 1979, Annu. Rev. Nucl. Part. Sci. **29**, 313
Stewart, J.M., 1969, Mon. Not. R. Astron. Soc. **145**, 347
Straumann, N., 1976, Helv. Phys. Acta **49**, 269
Straumann, N., 1982, Lecture notes of SIN Spring School (Zuoz, 1982)
Straumann, N., 1984a, Univ. of Zürich, Seminar on Cosmology
Straumann, N., 1984b, Lecture Notes (Univ. Zürich) Introduction to GUTs
Straumann, N., 1984c, *General Relativity and Relativistic Astrophysics* (Springer, Berlin, Heidelberg)
Straumann, N., 1988, in Proc. 2nd Workshop MPG-CAS on High Energy Astrophysics ed. by G. Börner (Springer, Berlin, Heidelberg)
Strauss, M.A., 1989, Ph.D. thesis, Univ. of Calif., Berkeley
Sunyaev, R.A., 1978, in *Large Scale Structure of the Universe* ed. by Y. Einasto, M.S. Longair (Reidel, Boston, Mass.)
Sunyaev, R.A., Zel'dovich, Ya.B., 1970, Astrophys. Space Sci. **7**, 3

Sunyaev, R.A., Zel'dovich, Ya.B., 1980, Annu. Rev. Astron. Astrophys. **18**, 537
Suto, Y., 1993, Prog. Theor. Phys. **90**, 1173
Syer, D., Mao, S.D., Mo, H.J., 1999, Mon. Not. R. Astron. Soc. **305**, 357
Symanzik, K., 1970, Commun. Math. Phys. **16**, 48
Symbalisty, E.M.D., Schramm, D.N., 1981, Rep. Prog. Phys. **44**, 293
Takahashi, K., 1998, Proc. Tours Nucl. Symp. III, AiP (in press)
Takahashi, K., Janka, H.-Th., 1997, Proc. of Atami conf., p. 213 (Origin of Matter and Evolution of Galaxies in the Universe) (World Sci., ed. by T. Kajino, Y. Yoshii, S. Kubono)
Takeda, M. et al., 1999, Astrophys. J. **522**, 225
Tammann, G.A., Sandage, A., 1985, Astrophys. J. **294**, 81
Tammann, G.A. et al., 1997, in *Thermonuclear Supernovae*, ed. by P. Ruiz-Lapuente, R. Canal, J. Isern (Kluwer) p. 735
Tatsumoto, M. et al., 1976, Geochim. Cosmochim. Acta **40**, 617
Taylor, J.H., 1986, in Proc. of GRG 11 (Stockholm 1986)
Thielemann, F.-K., 1984, in Stellar Nucleosynthesis Proc. Erice Workshop 1983, ed. by C. Chiosi, A. Renzini
Thielemann, F.-K., Metzinger, J., 1983, Astron. Astrophys. **123**, 162
Thielemann, F.-K., Truran, J.W., 1985, Proc. 5th Moriond Astrophysics Meeting Nucleosynthesis and its Implications on Nuclear and Particle Physics, ed. by J. Audouze (Reidel, Boston, Mass.)
Thielemann, F.-K., Wiescher, M., 1990, in Proc. *Primordial Nucleosynthesis Symp.* (Chapel Hill, NC, Oct. 89)
Thielemann, F.-K., Arnould, M., Hillebrandt, W., 1979, Astron. Astrophys. **74**, 175
Thielemann, F.-K., Metzinger, J., Klapdor, H.V., 1983, Z. Phys. **A309**, 301
t'Hooft, G., 1971, Nucl. Phys. **B35**, 167
t'Hooft, G., 1974, Nucl. Phys. **B79**, 276
t'Hooft, G., 1976a, Nucl. Phys. **B105**, 538
t'Hooft, G., 1976b, Phys. Rev. Lett. **37**, 8
Thurston, P.W., Weekes, J.R., 1984, Scientific American (July) 94
Tinsley, B.M., 1977a, Astrophys. J. **211**, 621
Tinsley, B.M., 1977b, Astrophys. J. **216**, 548
Tinsley, B.M., 1980, Fund. Cosm. Phys. **5**, 287
Tipler, F.J., 1976, Ph.D. thesis (Univ. of Maryland)
Tipler, F.J., 1977, Annals of Phys. **108**, 1
Toomre, A., 1981, in *The Structure and Evolution of Normal Galaxies*, ed. by S.M. Fall, D. Lynden-Bell (Cambridge Univ. Press)
Totsuji, H., Kihara, T., 1969, Publ. Astron. Soc. Japan **21**, 221
Treimann, S.B., Wilczek, F., 1980, Phys. Lett. **B95**, 222
Trowers, W.P., 1983, Acta Phys. Austr. Suppl. **25**, 101
Truran, J.W., Cowan, J.J., Cameron, A.G.W., 1978, Astrophys. J. **222**, L63
Tully, R.B., Fisher, J.R., 1977, Astron. Astrophys. J. **54**, 661
Turok, N., 1984, Nucl. Phys. **B242**, 520
Turok, N., 1986, Proc. 2nd ESO-CERN Symposium ed. by G. Setti, L. van Hove, p. 175
Turok, N., 1989, Phys. Rev. Lett. **63**, 2625
Tyson, J.A., 1988, Astron. J. **96**, 1
Udalski, A. et al., 1992, Acta Astron. **42**, 253
Ullmann, O., 1997, Ph.D. thesis, Munich University
Unsöld, A., 1991, *Der neue Kosmos*, 5th ed. (Springer, Berlin, Heidelberg)
Vachaspati, T., Vilenkin, A., 1984, Phys. Rev. **D30**, 2036
Vandenberg, D.A., Bolte, M., Stetson, P.B., 1990, Astron. J. **100**, 445
van den Bergh, S., 1983, Comments on Astrophys. **10**, 27
van den Bergh, S., 1996, Astrophys. J. **472**, 431

Van Waerbeke, L. et al., 2002, Astron. Astrophys. **393**, 369
Vilenkin, A., 1984, Astrophys. J. **282**, L51
Vilenkin, A., 1985, Phys. Rep. **121**, 263
Vishniac, E.T., 1987, Astrophys. J. **322**, 597
Wagoner, R.V., Fowler, W.A., Hoyle, F., 1967, Science **155**, 1369
Wald, R.M., 1984, *General Relativity* (Univ. Chicago Press, Chicago)
Wall, J.V., Pearson, T.J., Longair, M.S., 1980, Mon. Not. R. Astron. Soc. **193**, 683
Wall, J.V., Pearson, T.J., Longair, M.S., 1981, Mon. Not. R. Astron. Soc. **196**, 597
Walker, T.P. et al., 1991, Astrophys. J. **376**, 51
Walsh, D., Carswell, R.F., Wegmann, R.J., 1979, Nature **279**, 381
Warren, S.J., Hewett, P.C., Irwin, M.J., McMahon, R.G., Bridgelund, M.T., Bunclark, P.S., Kibblewhite, E.J., 1987, Nature **325**, 131
Webb, J.K. et al., 1997, Nature **388**, 250
Weinberg, E.J., 1964, in *Lectures on Particles & Fields* ed. by S. Deser, K. Ford (Prentice Hall, Englewood Cliffs, NJ)
Weinberg, S., 1967, Phys. Rev. Lett. **19**, 1264
Weinberg, S., 1971, Astrophys. J. **168**, 175
Weinberg, S., 1972, *Gravitation and Cosmology* (Wiley, New York)
Weinberg, S., 1977, *The First Three Minutes*, Basic Books
Weinberg, S., 1982, Phys. Rev. Lett. **48**, 1303
Weinberg, S., 1989, Rev. Mod. Phys. **61**, 1
Weinberg, S., 2001, Proc. 4th Symposium *Dark Matter*, ed. by D.B. Cline (Springer Verlag), p. 18
Weinberg, S., Salam, A., Glashow, S., 1980, Rev. Mod. Phys. **52**, 515
Weinheimer, C. et al., 1999, Phys. Lett. **B460**, 219
Weiss, R., 1980, Annu. Rev. Astron. Astrophys. **18**, 489
Weisskopf, V., 1981a, Phys. Today **34**, 69
Weisskopf, V., 1981b, Contemp. Phys. **22**, 375
Wess, J., Zumino, B., 1974, Nucl. Phys. **B70**, 39
Wetterich, C., 1985, Nucl. Phys. **B252**, 309
Wetterich, C., 1988, Nucl. Phys. **302**, 668
Weyl, H., 1919, Ann. Phys. **59**, 101
Weyl, H., 1923, *Raum, Zeit, Materie*, 5th ed. (Springer, Berlin, Heidelberg)
White, M., Scott, D., Silk, J., 1994, Ann. Rev. Astron. Astrophys. **32**, 329
White, S.D.M. et al., 1987, Nature **330**, 451
White, S.D.M. et al., 1993, Nature **366**, 429
Whitmore, B.C. et al., 1990, Astron. J. **100**, 1489
Wickramasinghe, N.C., Edmunds, M.G., Chitre, S.M., Narlikar, J.V., Ramadurai, S., 1975, Astrophys. Space Sci. **35**, L9
Will, C.M., 1981, *Theory and Experiment in Gravitational Physics* (Cambridge Univ. Press)
Willick, J.A. et al., 1997, Astrophys. J. Suppl. **109**, 333
Winter, K., 1986, in Proc. of 2nd ESO-CERN Symposium, ed. by G. Setti, L. van Hove (ESO, München, p. 3)
Wipf, A, 1985, Helvet. Phys. Acta **58**, 531
Wise, M.B., Georgi, H., Glashow, S.L., 1981, Phys. Rev. Lett. **47**, 402
Witten, E., 1980, Phys. Lett. **91B**, 81
Witten, E., 1981, Phys. Lett. **105B**, 267
Witten, E., 1984, Phys. Rev. **D30**, 272
Witten, E., 1985, Phys. Lett. **153B**, 243
Wolf, J.A., 1967, *Spaces of Constant Curvature* (McGraw-Hill, New York)
Wolfe, A.M., 1993, Ann. N.Y. Acad. Sci. **688**, 281
Wolfe, A.M., Lanzetta, K.M., Folte, C.B., Chaffee, F.H., 1995, Astrophys. J. **454**, 698

Wolfram, S., 1979, Phys. Lett. **B82**, 65
Woodroofe, M., 1985, Annals Statist. **13**, 163
Woody, D.P., Richards, P.L., 1979, Phys. Rev. Lett. **42**, 925
Yahil, A., Tammann, G.A., Sandage, A., 1977, Astrophys. J. **217**, 903
Yanagida, T., Yoshimura, M., 1980, Nucl. Phys. **168**, 534
Yang, C.N., Mills, R.L., 1954, Phys. Rev. **96**, 191
Yokoi, K., Takahashi, K., Arnould, M., 1983, Astron. Astrophys. **117**, 65
Yoshikawa, K., Itoh, M., Suto, Y., 1998, Pub. Astron. Soc. Japan **50**, 203
Zehavi, I. et al., 2002, Astrophys. J. **571**, 172
Zel'dovich, Ya.B., 1970, Astron. Astrophys. **5**, 84
Zel'dovich, Ya.B., 1972, Mon. Not. R. Astron. Soc. **160**, 1p
Zel'dovich, Ya.B., 1980, Mon. Not. R. Astron. Soc. **192**, 663
Zel'dovich, Ya.B., Novikov, I.D., 1983, *The Structure and Evolution of the Universe* (Univ. Chicago Press, Chicago, Ill.)
Zel'dovich, Ya.B., Okun, L.B., Pikel'ner, S.B., 1965, Sov. Phys. Usp. **84**, 113
Zel'dovich, Ya.B., Sunyaev, R.A., 1969, Astrophys. Space Sci. **4**, 301
Zlatev, I., Wang, L., Steinhardt, P.J., 1999, Phys. Rev. Lett. **82**, 896
Zucca, E. et al., 1997, Astron. Astrophys. **326**, 477
Zwicky, F., Herzog, E., Wild, P., Karpowicz, M., Kowal, C.T., 1961–1968, Catalogue of Galaxies and of Clusters of Galaxies, Vols. 1 to 6 (Caltech, Pasadena)

Index

ASTRONOMY AND ASTROPHYSICS LIBRARY

The Stars By E. L. Schatzman and F. Praderie

Modern Astrometry 2nd Edition
By J. Kovalevsky

The Physics and Dynamics of Planetary Nebulae By G. A. Gurzadyan

Galaxies and Cosmology By F. Combes, P. Boissé, A. Mazure and A. Blanchard

Observational Astrophysics 2nd Edition
By P. Léna, F. Lebrun and F. Mignard

Physics of Planetary Rings Celestial Mechanics of Continuous Media
By A. M. Fridman and N. N. Gorkavyi

Tools of Radio Astronomy 4th Edition
By K. Rohlfs and T. L. Wilson

Astrophysical Formulae 3rd Edition (2 volumes)
Volume I: Radiation, Gas Processes and High Energy Astrophysics
Volume II: Space, Time, Matter and Cosmology
By K. R. Lang

Tools of Radio Astronomy Problems and Solutions By T. L. Wilson and S. Hüttemeister

Galaxy Formation By M. S. Longair

Astrophysical Concepts 2nd Edition
By M. Harwit

Astrometry of Fundamental Catalogues The Evolution from Optical to Radio Reference Frames
By H. G. Walter and O. J. Sovers

Compact Stars. Nuclear Physics, Particle Physics and General Relativity 2nd Edition
By N. K. Glendenning

The Sun from Space By K. R. Lang

Stellar Physics (2 volumes)
Volume 1: Fundamental Concepts and Stellar Equilibrium
By G. S. Bisnovatyi-Kogan

Stellar Physics (2 volumes)
Volume 2: Stellar Evolution and Stability
By G. S. Bisnovatyi-Kogan

Theory of Orbits (2 volumes)
Volume 1: Integrable Systems and Non-perturbative Methods
Volume 2: Perturbative and Geometrical Methods
By D. Boccaletti and G. Pucacco

Black Hole Gravitohydromagnetics
By B. Punsly

Stellar Structure and Evolution
By R. Kippenhahn and A. Weigert

Gravitational Lenses By P. Schneider, J. Ehlers and E. E. Falco

Reflecting Telescope Optics (2 volumes)
Volume I: Basic Design Theory and its Historical Development. 2nd Edition
Volume II: Manufacture, Testing, Alignment, Modern Techniques
By R. N. Wilson

Interplanetary Dust
By E. Grün, B. Å. S. Gustafson, S. Dermott and H. Fechtig (Eds.)

The Universe in Gamma Rays
By V. Schönfelder

Astrophysics. A Primer By W. Kundt

Cosmic Ray Astrophysics
By R. Schlickeiser

Astrophysics of the Diffuse Universe
By M. A. Dopita and R. S. Sutherland

The Sun An Introduction. 2nd Edition
By M. Stix

Order and Chaos in Dynamical Astronomy
By G. J. Contopoulos

Astronomical Image and Data Analysis
By J.-L. Starck and F. Murtagh

ASTRONOMY AND ASTROPHYSICS LIBRARY

The Early Universe Facts and Fiction
4th Edition By G. Börner

The Design and Construction of Large Optical Telescopes By P. Y. Bely

The Solar System 4th Edition
By T. Encrenaz, J.-P. Bibring, M. Blanc, M. A. Barucci, F. Roques, Ph. Zarka

General Relativity, Astrophysics, and Cosmology By A. K. Raychaudhuri, S. Banerji, and A. Banerjee

Stellar Interiors Physical Principles, Structure, and Evolution 2nd Edition
By C. J. Hansen, S. D. Kawaler, and V. Trimble

Asymptotic Giant Branch Stars
By H. J. Habing and H. Olofsson

The Interstellar Medium
By J. Lequeux

Printed by Publishers' Graphics LLC
DBT140121.15.18.246